SAAB 900

16 VALVE

Official Service Manual 1985, 1986, 1987, 1988, 1989, 1990, 1991, 1992, 1993

BentleyPublishers.com

Selected Books and Repair Information From Bentley Publishers

Motorsports

Alex Zanardi: My Sweetest Victory
Alex Zanardi and Gianluca Gasparini
ISBN 978-0-8376-1249-2

The Unfair Advantage *Mark Donohue and Paul van Valkenburgh*
ISBN 978-0-8376-0069-7

Equations of Motion - Adventure, Risk and Innovation
William F. Milliken
ISBN 978-0-8376-1570-7

Engineering

Physics for Gearheads
Randy Beikmann
ISBN 978-0-8376-1615-5

Bosch Automotive Handbook
Robert Bosch GmbH
ISBN 978-0-8376-1732-9

Bosch Fuel Injection and Engine Management *Charles O. Probst, SAE*
ISBN 978-0-8376-0300-1

Maximum Boost: Designing, Testing, and Installing Turbocharger Systems
Corky Bell ISBN 978-0-8376-0160-1

Supercharged! Design, Testing and Installation of Supercharger Systems
Corky Bell ISBN 978-0-8376-0168-7

Scientific Design of Exhaust and Intake Systems *Phillip H. Smith & John C. Morrison* ISBN 978-0-8376-0309-4

Audi

Audi A4 Service Manual: 2002-2008, 1.8L Turbo, 2.0L Turbo, 3.0L, 3.2L
Bentley Publishers
ISBN 978-0-8376-1574-5

Audi A4 Service Manual: 1996-2001, 1.8L Turbo, 2.8L *Bentley Publishers*
ISBN 978-0-8376-1675-9

Audi TT Service Manual: 2000-2006, 1.8L turbo, 3.2 L *Bentley Publishers*
ISBN 978-0-8376-1625-4

Audi A6 (C5 platform) Service Manual: 1998-2004 *Bentley Publishers*
ISBN 978-0-8376-1670-4

BMW

Memoirs of a Hack Mechanic
Rob Siegel ISBN 978-0-8376-1720-6

BMW 5 Series (E60, E61) Service Manual: 2004-2010 *Bentley Publishers*
ISBN 978-0-8376-1689-6

BMW X3 (E83) Service Manual: 2004-2010 *Bentley Publishers*
ISBN 978-0-8376-1731-2

BMW X5 (53) Service Manual: 2000-2006 *Bentley Publishers*
ISBN 978-0-8376-1643-8

BMW 3 Series (E90, E91, E92, E93) Service Manual: 2006-2011
Bentley Publishers ISBN 978-0-8376-1723-7

BMW 3 Series (E46) Service Manual: 1999-2005 *Bentley Publishers*
ISBN 978-0-8376-1657-5

BMW 3 Series (E36) Service Manual: 1992-1998 *Bentley Publishers*
ISBN 978-0-8376-1709-1

BMW 3 Series (E30) Service Manual: 1984-1990 *Bentley Publishers*
ISBN 978-0-8376-1647-6

BMW Z3 (E36/7) Service Manual: 1996-2002 *Bentley Publishers*
ISBN 978-0-8376-1617-9

BMW 5 Series (E60, E61) Service Manual: 2004-2010 *Bentley Publishers*
ISBN 978-0-8376-1689-6

BMW 5 Series (E39) Service Manual: 1997-2003 *Bentley Publishers*
ISBN 978-0-8376-1672-8

Corvette

Corvette: America's Star-Spangled Sports Car *Karl Ludvigsen*
ISBN 978-0-8376-1659-9

Corvette by the Numbers: The Essential Corvette Parts Reference 1955-1982 *Alan Covin*
ISBN 978-0-8376-0228-2

Zora Arkus-Duntov: The Legend Behind Corvette *Jerry Burton*
ISBN 978-08376-0858-7

Porsche

Porsche 911 (996) Service Manual: 1999-2005 *Bentley Publishers*
ISBN 978-0-8376-1710-7

Porsche 911 (993) Service Manual: 1995-1998 *Bentley Publishers*
ISBN 978-0-8376-1719-0

Porsche Boxster Service Manual: 1997-2004 *Bentley Publishers*
ISBN 978-0-8376-1645-2

Porsche 911 Carrera Service Manual: 1984-1989 *Bentley Publishers*
ISBN 978-0-8376-1696-4

Porsche 911 SC Service Manual: 1978-1983 *Bentley Publishers*
ISBN 978-0-8376-1705-3

Porsche: Excellence Was Expected
Karl Ludvigsen ISBN 978-0-8376-0235-6

Porsche — Origin of the Species
Karl Ludvigsen ISBN 978-0-8376-1331-4

Volkswagen

Volkswagen Rabbit, GTI Service Manual: 2006-2009 *Bentley Publishers*
ISBN 978-0-8376-1664-3

Volkswagen Jetta Service Manual: 2005-2010 *Bentley Publishers*
ISBN 978-0-8376-1616-2

Volkswagen Jetta, Golf, GTI Service Manual: 1999-2005 *Bentley Publishers*
ISBN 978-0-8376-1678-0

Volkswagen Jetta, Golf, GTI: 1993-1999, Cabrio: 1995-2002 Service Manual *Bentley Publishers*
ISBN 978-0-8376-1660-5

Volkswagen GTI, Golf, Jetta Service Manual: 1985-1992 *Bentley Publishers*
ISBN 978-0-8376-1637-7

Volkswagen Corrado Repair Manual: 1990-1994 *Bentley Publishers*
ISBN 978-0-8376-1699-5

Volkswagen Passat, Passat Wagon Service Manual: 1998-2005
Bentley Publishers
ISBN 978-0-8376-1669-8

MINI Repair Manuals

MINI Cooper Service Manual: 2007-2013 *Bentley Publishers*
ISBN 978-0-8376-1730-5

MINI Cooper Service Manual: 2002-2006 *Bentley Publishers*
ISBN 978-0-8376-1639-1

Mercedes-Benz

Mercedes-Benz C-Class (W202) Service Manual 1994-2000
Bentley Publishers
ISBN 978-0-8376-1692-6

Mercedes Benz E-Class (W124) Owner's Bible: 1986-1995
Bentley Publishers
ISBN 978-0-8376-0230-1

Mercedes-Benz Technical Companion
Staff of The Star and members of Mercedes-Benz Club of America
ISBN 978-0-8376-1033-7

BentleyPublishers®
Automotive Reference

Bentley Publishers has published service manuals and automobile books since 1950. For more information, please contact Bentley Publishers at 1734 Massachusetts Avenue, Cambridge, MA 02138 USA, or visit our web site at **BentleyPublishers.com**

SAAB 900

16 VALVE

Official Service Manual
1985, 1986, 1987, 1988, 1989, 1990, 1991, 1992, 1993

[B] BentleyPublishers®
.com

BENTLEY PUBLISHERS® | Automotive Reference™

Bentley Publishers, a division of Robert Bentley, Inc.
1734 Massachusetts Avenue
Cambridge, MA 02138 USA
BentleyPublishers®.com

Information that makes the difference®

Technical contact information

We welcome your feedback. Please submit corrections and additions to our technical discussion forum at:

`http://www.BentleyPublishers.com`

Updates and corrections

We will evaluate submissions and post appropriate editorial changes online as updates or tech discussion. Appropriate updates and corrections will be added to the book in future printings. Check for updates and corrections for this book before beginning work on your vehicle. See the following web address for additional information:

`http://www.BentleyPublishers.com/updates/`

WARNING—Important Safety Notice

Do not use this manual for repairs unless you are familiar with basic automotive repair procedures and safe workshop practices. This manual illustrates the workshop procedures for some maintenance and service work. It is not a substitute for full and up-to-date information from the vehicle manufacturer or for proper training as an automotive technician. Note that it is not possible for us to anticipate all of the ways or conditions under which vehicles may be serviced or to provide cautions as to all of the possible hazards that may result.

We have endeavored to ensure the accuracy of the information in this manual. Please note, however, that considering the vast quantity and the complexity of the service information involved, we cannot warrant the accuracy or completeness of the information contained in this manual.

FOR THESE REASONS, NEITHER THE PUBLISHER NOR THE AUTHOR MAKES ANY WARRANTIES, EXPRESS OR IMPLIED, THAT THE INFORMATION IN THIS MANUAL IS FREE OF ERRORS OR OMISSIONS, AND WE EXPRESSLY DISCLAIM THE IMPLIED WARRANTIES OF MERCHANTABILITY AND OF FITNESS FOR A PARTICULAR PURPOSE, EVEN IF THE PUBLISHER OR AUTHOR HAVE BEEN ADVISED OF A PARTICULAR PURPOSE, AND EVEN IF A PARTICULAR PURPOSE IS INDICATED IN THE MANUAL. THE PUBLISHER AND AUTHOR ALSO DISCLAIM ALL LIABILITY FOR DIRECT, INDIRECT, INCIDENTAL OR CONSEQUENTIAL DAMAGES THAT RESULT FROM ANY USE OF THE EXAMPLES, INSTRUCTIONS OR OTHER INFORMATION IN THIS MANUAL. IN NO EVENT SHALL OUR LIABILITY WHETHER IN TORT, CONTRACT OR OTHERWISE EXCEED THE COST OF THIS MANUAL.

Before attempting any work on your Saab, read **ix Warnings and Cautions** and any **WARNING** or **CAUTION** that accompanies a procedure in the manual. Review the **WARNINGS** and **CAUTIONS** each time you prepare to work on your Saab.

Your common sense and good judgment are crucial to safe and successful service work. Read procedures through before starting them. Think about whether the condition of your car, your level of mechanical skill, or your level of reading comprehension might result in or contribute in some way to an occurrence which might cause you injury, damage your car, or result in an unsafe repair. If you have doubts for these or other reasons about your ability to perform safe repair work on your car, have the work done at an authorized Saab dealer or other qualified shop.

Part numbers listed in this manual are for identification purposes only, not for ordering. Always check with your authorized Saab dealer to verify part numbers and availability before beginning service work that may require new parts.

Special tools required to perform certain service operations are identified in the manual and are recommended for use. Use of improper tools may be detrimental to the car's safe operation as well as the safety of the person servicing the car.

The vehicle manufacturer will continue to issue service information updates and parts retrofits after the editorial closing of this manual. Some of these updates and retrofits will apply to procedures and specifications in this manual. We regret that we cannot supply updates to purchasers of this manual.

This manual is prepared, published and distributed by Bentley Publishers, 1734 Massachusetts Avenue, Cambridge, Massachusetts 02138. All information contained in this manual is based on the information available to the publisher at the time of editorial closing. Saab has not reviewed and does not vouch for the accuracy or completeness of the technical specifications and work procedures described and given in this manual.

Library of Congress Catalog Card No. for 1993 edition 93-70727
Saab P/N 02 16 861

ISBN 978-0-8376-1693-3
Bentley Stock No. S993
Editorial closing 06 / 1993
Mfg. code: S993-16

Manufactured in the United States of America.

Foreword

This Saab 900 16 Valve Service Manual covers all1985 through 1993 900 models with 16 valve engines and is specifically designed to cover these models built for sale in the United States only.

All Saab 900 16 valve models:

1985-1993	Turbo, including Turbo SPG (all)
1987-1993	Convertible (all)
1986-1993	900S (all)
1989-1992	900 (all)

This manual has been organized using the Saab Official numbered indexing system. This numbered system will be primarily helpful to Saab factory-trained technicians experienced with the multi-volume Official Saab 900 Service Manual binder set, but we hope it will also facilitate information retrieval for all other users of the manual.

Saab uses an internal code for the various engines used in the 900 models. All 16 valve engines used in the 900 models are grouped under the B202 or B212 heading. The letter "B" designates the fuel type, or gasoline (benzine). The first two numbers designate the engine size, e.g. 2.0 or 2.1 liter. The last number designates the number of camshafts. All 16 valve engines are of the twin overhead camshaft design. The single overhead camshaft Saab 2.0 liter 8 valve engine has an engine code of B201. This information is primarily used internally by Saab, but may be helpful if you have access to Saab technical information.

For the Saab owner with automobile repair skills and training, this manual gives detailed maintenance and repair information. In addition, the Saab owner who has no intention of working on his or her own car will find that owning and reading this manual will make it possible to be better informed and to discuss repairs more intelligently with a professional technician. This manual has been prepared from the repair information that Saab provides to its factory-trained technicians and has been developed primarily with the experienced do-it-yourself Saab owner in mind. The aim throughout has been clarity and understanding with practical functional descriptions, step-by-step procedures, and accurate specifications.

The Saab owner intending to do maintenance and repair should have a set of tools including a set of metric wrenches and sockets, screwdrivers, a torque wrench, and feeler gauges, since these basic tools will be used to do the majority of the maintenance and repair procedures described in this manual. This manual includes detailed information on these basic tools and other tips for the beginner in the first section of the manual, entitled 100 Fundamentals for the Do-it-yourself Owner. For some of the repairs described in this manual, Saab technicians use special tools. The text will note when a repair requires these special tools and, where possible, will recommend practical alternatives.

We have endeavored to ensure the accuracy of the information in this manual . When the vast array of data presented in the manual is taken into account, however, no claim to infallibility can be made. The publisher therefore cannot be responsible for the result of any errors that may have crept into the text. The publisher encourages comments from the readers of this manual in regard to any errors and, also, suggestions for improvement in the presentation of technical material. These communications have been and will be carefully considered in the preparation of future printings of this and other manuals. Please write to Robert Bentley, Inc., Cambridge, Massachusetts 02138.

Saab offers extensive warranties, especially on components of the fuel delivery and emissions control systems. Therefore, before deciding to repair a Saab that may still be covered wholly or in part by any warranties issued by Saab Cars USA Inc., consult your authorized Saab dealer. You may find that he can make the repair either free or at minimum cost.

Regardless of its age and whether or not it is still protected by warranty, your Saab is both an easy car to service and an easy car to get serviced. So, if at any time a repair is needed that you feel is too difficult or dangerous to do yourself, a trained Saab technician is ready to do the job for you. Each authorized Saab dealer service department has made a significant investment in service and diagnostic test equipment, specials tools, and Saab original parts. He is also the best source of the most up-to-date repair and service techniques, which includes factory training and technical literature. Servicing your Saab through an authorized Saab dealer will insure that your investment will be protected while maintaining the highest degree of service standards.

Robert Bentley, Inc.

Please read these warnings and cautions before proceeding with maintenance and repair work.

WARNING—

- Some repairs may be beyond your capability. If you lack the skills, tools and equipment, or a suitable workplace for any procedure described in this manual, we suggest you leave such repairs to an authorized Saab dealer service department, or other qualified shop.

- Saab is constantly improving its cars. Sometimes these changes, both in parts and specifications, are made applicable to earlier models. Therefore, before starting any major jobs or repairs to components on which passenger safety may depend, consult your authorized Saab dealer about Technical Bulletins that may have been issued since the editorial closing of this manual.

- Do not re-use any fasteners that are worn or deformed in normal use. Many fasteners are designed to be used only once and become unreliable and may fail when used a second time. This includes, but is not limited to, nuts, bolts, washers, self-locking nuts or bolts, circlips and cotter pins. Always replace these fasteners with new parts.

- Never work under a lifted car unless it is solidly supported on stands designed for the purpose. Do not support a car on cinder blocks, hollow tiles or other props that may crumble under continuous load. Never work under a car that is supported solely by a jack. Never work under the car while the engine is running.

- If you are going to work under a car on the ground, make sure that the ground is level. Block the wheels to keep the car from rolling. Disconnect the battery negative (–) terminal (Ground strap) to prevent others from starting the car while you are under it.

- Never run the engine unless the work area is well ventilated. Carbon monoxide kills.

- Finger rings, bracelets and other jewelry should be removed so that they cannot cause electrical shorts, get caught in running machinery, or be crushed by heavy parts.

- Tie long hair behind your head. Do not wear a necktie, a scarf, loose clothing, or a necklace when you work near machine tools or running engines. If your hair, clothing, or jewelry were to get caught in the machinery, severe injury could result.

- Do not attempt to work on your car if you do not feel well. You increase the danger of injury to yourself and others if you are tired, upset or have taken medication or any other substance that may keep you from being fully alert.

- Illuminate your work area adequately but safely. Use a portable safety light for working inside or under the car. Make sure the bulb is enclosed by a wire cage. The hot filament of an accidentally broken bulb can ignite spilled fuel or oil.

- Catch draining fuel, oil, or brake fluid in suitable containers. Do not use food or beverage containers that might mislead someone into drinking from them. Store flammable fluids away from fire hazards. Wipe up spills at once, but do not store the oily rags, which can ignite and burn spontaneously.

- Always observe good workshop practices. Wear goggles when you operate machine tools or work with battery acid. Gloves or other protective clothing should be worn whenever the job requires working with harmful substances.

- Friction materials such as brake or clutch discs may contain asbestos fibers. Do not create dust by grinding, sanding, or by cleaning with compressed air. Avoid breathing asbestos fibers and asbestos dust. Breathing asbestos can cause serious diseases such as asbestosis or cancer, and may result in death.

- Disconnect the battery negative (–) terminal (Ground strap) whenever you work on the fuel system or the electrical system. Do not smoke or work near heaters or other fire hazards. Keep an approved fire extinguisher handy.

- Batteries give off explosive hydrogen gas during charging. Keep sparks, lighted matches and open flame away from the top of the battery. If hydrogen gas escaping from the cap vents is ignited, it will ignite gas trapped in the cells and cause the battery to explode.

- Connect and disconnect battery cables, jumper cables or a battery charger only with the ignition switched off, to prevent sparks. Do not disconnect the battery while the engine is running.

- Do not quick-charge the battery (for boost starting) for longer than one minute. Wait at least one minute before boosting the battery a second time.

- Do not allow battery charging voltage to exceed 15.0 volts. If the battery begins producing gas or boiling violently, reduce the charging rate. Boosting a sulfated battery at a high charging rate can cause an explosion.

- The air-conditioning system is filled with chemical refrigerant, which is hazardous. The A/C system should be serviced only by trained technicians using approved refrigerant recovery/recycling equipment, trained in related safety precautions, and familiar with regulations governing the discharging and disposal of automotive chemical refrigerants.

- Do not expose any part of the A/C system to high temperatures such as open flame. Excessive heat will increase system pressure and may cause the system to burst.

- Some aerosol tire inflators are highly flammable. Be extremely cautious when repairing a tire that may have been inflated using an aerosol tire inflator. Keep sparks, open flame or other sources of ignition away from the tire repair area. Inflate and deflate the tire at least four times before breaking the bead from the rim. Completely remove the tire from the rim before attempting any repair.

- Some cars covered by this manual are equipped with a supplemental restraint system (SRS), that automatically deploys an airbag in the event of a frontal impact. The airbag is inflated by an explosive device. Handled improperly or without adequate safeguards, it can be accidently activated and cause serious injury.

- To prevent personal injury or airbag system failure, **only trained Saab dealer service technicians** should test, disassemble or service the airbag system.

- The airbag unit should not be removed, nor should any other work be performed on the airbag system, until at least 20 minutes after the battery is disconnected.

continued on next page

Please read these warnings and cautions before proceeding with maintenance and repair work.

WARNING (continued) —

- On airbag-equipped cars, never apply stickers or any other type of covering on the steering wheel. Do not let chemical cleaners, oil or grease come into contact with the vinyl covering of the airbag unit.

- Never open or otherwise attempt to repair airbag system parts. Always use new parts. Never leave airbag parts or the partially disassembled airbag system unattended.

- Do not expose the airbag unit to temperature above 194°F (90°C), even for brief periods. Keep clear of heat sources such as hot plates, soldering irons, heat lamps and welding equipment.

- When driving or riding in an airbag-equipped vehicle never hold test equipment in your hands or lap while the vehicle is in motion. Objects between you and the airbag can increase the risk of injury in an accident.

- On cars equipped with anti-lock brakes (ABS), the ABS modular assembly is capable of "self-pressuring" up to 210 bar (3045 psi). Serious injury may result if the unit is not properly depressurized before servicing the system.

- Greases, lubricants and other automotive chemicals contain toxic substances, many of which are absorbed directly through the skin. Read manufacturer's instructions and warnings carefully. Use hand and eye protection. Avoid direct skin contact.

CAUTION—

- If you lack the skills, tools and equipment, or a suitable workshop for any procedure described in this manual, we suggest you leave such repairs to an authorized Saab dealer or other qualified shop.

- Saab offers extensive warranties, especially on components of fuel delivery and emission control systems. Therefore, before deciding to repair a Saab that may still be covered wholly or in part by any warranties issued by Saab Cars USA, Inc., consult your authorized Saab dealer. You may find that he can make the repair for free, or at minimal cost.

- Saab part numbers listed in this manual are for identification purposes only, not for ordering. Always check with your authorized Saab dealer to verify part numbers and availability before beginning service work that may require new parts.

- Before starting a job, make certain that you have all the necessary tools and parts on hand. Read all the instructions thoroughly, do not attempt shortcuts. Use tools appropriate to the work and use only replacement parts meeting Saab specifications. Makeshift tools, parts and procedures will not make good repairs.

- Use pneumatic and electric tools only to loosen threaded parts and fasteners. Never use these tools to tighten fasteners, especially on light alloy parts. Always use a torque wrench to tighten fasteners to the tightening torque specification listed.

- Be mindful of the environment and ecology. Before you drain the crankcase, find out the proper way to dispose of the oil. Do not pour oil onto the ground, down a drain, or into a stream, pond or lake. Consult local ordinances that govern the disposal of wastes.

- On cars equipped with the anti-lock brakes (ABS), the ABS control unit cannot withstand temperatures from a paint-drying booth or a heat lamp in excess of 203°F (95°C) and should not be subjected to temperatures in excess of 185°F (85°C) for more than two hours.

- Before doing any electrical welding on cars equipped with ABS, disconnect the battery negative (–) terminal (Ground strap) and the ABS control unit connector.

- On cars equipped with anti-theft radios, make sure you know the correct radio activation code before disconnecting the battery or removing the radio. If the wrong code is entered into the radio when power is restored, that radio may lock up and be rendered inoperable, even if the correct code is then entered.

- Connect and disconnect a battery charger only with the battery charger switched off.

- Do not quick-charge the battery (for boost starting) for longer than one minute. Wait at least one minute before boosting the battery a second time.

- Sealed or "maintenance free" batteries should be slow-charged only, at an amperage rate that is approximately 10% of the battery's ampere-hour (Ah) rating.

- Do not allow battery charging voltage to exceed 15.0 volts. If the battery begins producing gas or boiling violently, reduce the charging rate. Boosting a sulfated battery at a high charging rate can cause an explosion.

010 Technical Data—General

TECHNICAL DATA—GENERAL

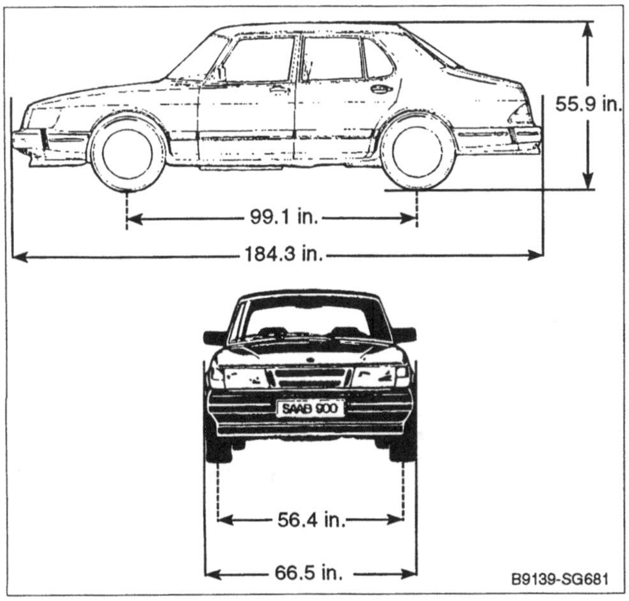

General	mm	in.
Overall length (sedans, convertibles)	4680	184.3
Overall length (hatchback)	4687	184.5
Overall width	1690	66.5
Overall height (at curb weight)	1419	55.9
Overall height with convertible at highest point	2230	87.8
Road clearance (full load)		
SPG models	120	4.7
all other models	137	5.4
Track (front)		
alloy wheels	1430	56.4
steel wheels	1437	56.6

	mm	in.
Track (rear)	1440	56.7
Wheelbase	2517	99.1
Turning radius	5600	220.5
	kg.	**lb.**
Curb weight (depending on equipment)	1252-1411	2755-3105
Gross vehicle weight (depending on equipment)	1677-1768	3690-3890
Weight distribution		
at curb weight	59-62% front	
at gross weight	52-55% front	
Vehicle capacity weight—ex. conv. (five people, 180 lb. luggage)	422	930
Maximum roof rack load—ex. conv.	100	220
Maximum trailer weight:		
trailer with brakes	900	2000
trailer without brakes	450	1000
Maximum trailer tongue weight	75	165
Trunk volume (SAE)		
hatchback		14.9 cu.ft.
hatchback, parcel shelf removed		19.1 cu.ft.
sedan		14.2 cu.ft.
convertible		10.7 cu.ft.

Engine Specifications

Model	Year	Displacement cc (cu. in.)	Compression ratio	Horsepower @ rpm SAE net	Torque lb-ft @ rpm SAE net	Fuel system
900/900S/Convertible	1986-1987	1985 (121)	10.1:1	125 @ 5500	128 @ 3000	Bosch LH 2.2
	1988	1985 (121)	10.1:1	128 @ 6000	128 @ 3000	Bosch LH 2.4
	1989-1990	1985 (121)	10.1:1	128 @ 6000	128 @ 3000	Bosch LH 2.4
	1991-1993	2118 (129)	10.1:1	140 @ 6000	133 @ 2900	Bosch LH 2.4.2
900 Turbo/Turbo Convertible	1985-1988	1985 (121)	9.0:1	160 @ 5500	188 @ 3000	Bosch LH 2.2
	1989-1993	1985 (121)	9.0:1	160 @ 5500	188 @ 3000	Bosch LH 2.4
900 SPG	1985-1986	1985 (121)	9.0:1	160 @ 5500	188 @ 3000	Bosch LH 2.2
	1987-1988	1985 (121)	9:0:1	165 @ 5500	195 @ 3000	Bosch LH 2.2
	1989	1985 (121)	9.0:1	165 @ 5500	195 @ 3000	Bosch LH 2.4
	1990-1991	1985 (121)	9.0:1	175 @ 5500	195 @ 3000	Bosch LH 2.4

LUBRICATION AND MAINTENANCE—QUICK DATA

Engine Oil Change	
Saab recommended engine oil	SAE 10W-30, API Service Rating SG or SF/CD
Alternate for extremely hot climates	15W-40, API Service Rating SG or SF/CD
Engine oil drain plug tightening torque (13 mm or 19mm wrench size)	29-39 Nm (21-29 ft-lb)
Engine Oil Capacity (including filter)	4.2 quarts (4.0 liters)

Ignition System Applications	
1985-1992 turbo models	basic Hall ignition
1986-1992 non-turbo models	EZK ignition —Hall-effect with knock sensor

Spark Plug Applications	
Model	**Spark Plug**
1986-1988 Normally aspirated	NGK BCP 6ES
alternate	NGK BCP 6EV
1989-1993 Normally-aspirated	NGK BCP 5ES

Spark Plug Applications	
Model	**Spark Plug**
1985-1993 Turbo	
normal driving	NGK BCP 7EV
alternate	NGK BCP 7ES
city driving	NGK BCP 6EV
alternate	NGK BCP 6ES
Spark plug electrode gap	0.024-0.028 in. (0.6-0.7 mm)
Tightening torque—sparkplugs	25 to 29 Nm (18 to 21 ft-lb)

Ignition Timing	
Distributor vacuum hose disconnected and plugged, where applicable	
Turbo models	16°BTDC @ 850 rpm
Normally aspirated models	14°BTDC @ 850 rpm

Cooling System	
Capacity (50/50 mixture anti-freeze and water)	10.5 quarts (10 liters)

Engine Drive Belt Tensions (measure using special belt tensioning gauge)		
Drive belt tension	**N**	**(lb)**
Alternator		
checking (minimum)		
one belt	200	(45)
two belts	420	(95)
adjusting, used belts		
one belt	310±20	(70±5)
two belts	645±20	(145±5)
adjusting, new belts		
one belt	535±45	(120±10)
A/C compressor		
checking (minimum)	245	(55)
adjusting, used belt	355±20	(80±5)
adjusting, new belt	535±45	(120±10)
Power steering		
checking (minimum)	220	(50)
adjusting, used belt	310±20	(70±5)
adjusting, new belt	445±45	(100±10)

Manual Transmission	
Grade of oil	SAE 10W-30 SF/CC, SF/CD, SG
Alternate grade of oil	SAE EP 75 API-GL-4 or API-GL-5
Oil capacity	3.1 quarts (3.0 liters)

Automatic Transmission	
ATF	Ford Specification M2C-33F (alternate: Ford Specification G)
ATF capacity	8.5 quarts (8.0 liters)
A/T final drive grade of oil	SAE EP 80 or 75 API-GL-4 or API-GL-5
Alternate grade of oil	10W-30 engine oil
A/T final drive capacity	1.3 quarts (1.25 liters)
ATF drain plug tightening torque	5-8 Nm (48-72 in-lb)
A/T final drive drain or filler plug tightening torque	39-59 Nm (28-44 ft-lb)

Brake System	
Brake Fluid Grade	SAE Dot 4
Brake Pad Wear Limit	
1985-1987 models	1.0 mm (0.04 in.)
1988 and later models	4.0 mm (0.16 in.)

WARNING
Do not use SAE DOT 5 brake fluid. Brake system failure may result.

Steering and Wheel Alignment	
Power Steering Fluid	GM Power Steering Fluid (GM 9985010), Texaco TL4634 or equivalent

022 Engine—Technical Data

ENGINE TIGHTENING TORQUES

B9081

Tightening Torques	Nm	ft-lb (in-lb)
Camshaft sprocket to camshaft	63	46
Camshaft bearing caps to cylinder head	15	11
Camshaft timing chain tensioner to timing case	63	46
Clutch slave cylinder to transmission case	6–14	(53–124 in-lb)
Coolant drain plug to engine block	15	11
Coolant temperature sensor (LH sensor)	20	15
Connecting rod cap to connecting rod	55	41
Crankshaft pulley to crankshaft		
1985-1990	190	140
1991 and later	175	129

continued

Tightening Torques	Nm	ft-lb (in-lb)
Cylinder block end plate to engine (flywheel end)	20	15
Cylinder head to cylinder block		
stage 1	60	44
stage 2	80 plus and additional 90° (1/4 turn)	59 plus and additional 90° (1/4 turn)
Engine to transmission		
automatic transmission	33–39	24–29
manual transmission	25±3	18±2.2
Engine oil cooler line to oil filter housing	7–10	(62–89 in-lb)
Engine oil drain plug to case	29–39	21–29
EGR valve to cylinder head	8	(71 in-lb)
Exhaust manifold to cylinder head		
turbo models	25	18
normally aspirated models	20	15
Flywheel to crankshaft (17 mm bolt head)	60	44
Flywheel to crankshaft (19 mm bolt head)	85	63
Fuel pump to fuel tank (threaded collar)—1990 and later	55	41
Ignition distributor to cylinder head	20	15
Intake manifold to cylinder head	22	16

continued on next page

Tightening Torques	Nm	ft-lb (in-lb)
Main bearing caps to cylinder block	110	81
Oil filter to oil filter housing	10	7.4
Oil pressure switch to oil filter housing		
small (1/4"-18 NPTF)	10	(89 in-lb)
large (M14X1.5)	30	22
Oil pump to timing chain cover	8	(71 in-lb)
Oxygen sensor	40	30
Spark plugs to cylinder head	28	21
Thermostat housing to cylinder head	18	13
Timing chain cover to engine block and cylinder head	20	15
Timing chain tensioner to cylinder head	63	46
Valve cover to cylinder head	15	11
All other fasteners		
M5 bolt	5	(44 in-lb)
M6 bolt	10	(89 in-lb)
M8 bolt	20	15
M10 bolt	40	30

CONNECTING RODS AND PISTON PINS

B9082

Connecting rods	mm	in.
Diameter of big end (bearing shells removed)	56.000–56.019	2.2047–2.2055
Diameter of small end (bushing installed)	24.005–24.010	0.9451–0.9453
Big-end clearance	0.026–0.062	0.0010–0.0024
Maximum permissible weight variation per set	9 g (0.32 oz.)	

Piston pins	mm	in.
Piston pin diameter	23.996–24.000	0.9447–0.9449
Piston pin to connecting rod clearance		
B202 engine	0.005–0.014	0.0002–0.0006
B212 engine	0.002–0.011	0.00008–0.0004

PISTON RINGS

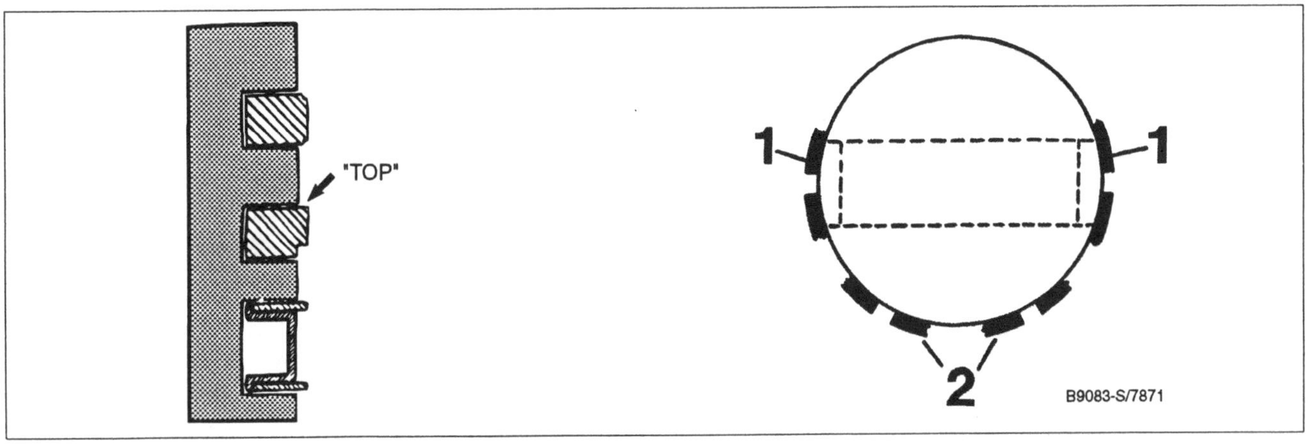

Piston Ring	Top compression ring	Second compression ring	Oil scraper ring
End gap	0.35–0.55 mm 0.0138–0.0217 in.	0.30–0.45 mm 0.0118–0.0177 in.	0.38–1.40 mm 0.0150–0.0551 in.*
Width (thickness)	1.73–1.75 mm 0.0681–0.0689 in.	1.98–1.99 mm 0.0780–0.0783 in.	2.63–2.73 mm 0.1035–0.1075 in.**
Side clearance in piston groove	0.05–0.09 mm 0.0020–0.0035 in.	0.04–0.07 mm 0.0016–0.0028 in.	—

*Applies to segment part of ring
**Oil scraper ring segment width (thickness) 0.58–0.64 mm 0.0028–0.02552 in.

PISTONS

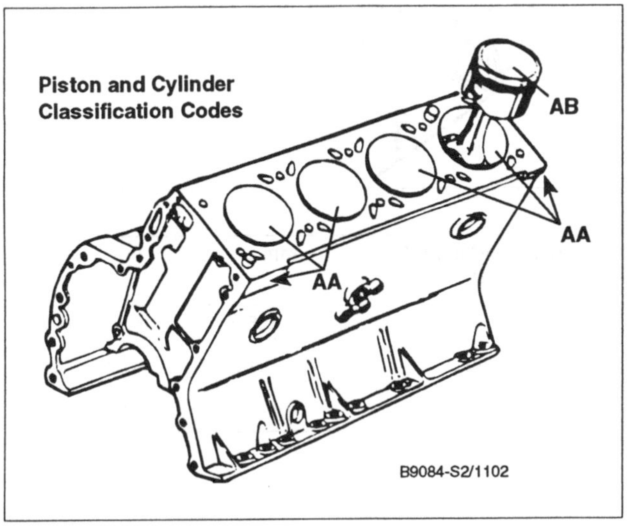

Piston diameter	mm	in.
Turbo engine with Mahle pistons		
standard A	89.960–89.970	3.5417–3.5421
standard AB	89.970–89.978	3.5421–3.5424
standard B	89.978–89.986	3.5424–3.5427
standard C	89.986–90.002	3.5427–3.5434

continued

Piston diameter	mm	in.
Turbo engine with Mahle pistons (cont'd)		
first oversize (0.5 mm)	90.460–90.475	3.5614–3.5620
second oversize (1.0 mm)	90.960–89.975	3.5811–3.5817
Normally aspirated engine with Mahle or KS pistons (ex. B212 engine)		
standard A	89.978–89.988	3.5424–3.5428
standard AB	89.988–89.996	3.5428–3.5431
standard B	89.996–90.004	3.5431–3.5435
standard C	90.004–90.020	3.5435–3.5441
first oversize (0.5 mm)	90.482–90.497	3.5623–3.5629
second oversize (1.0 mm)	90.982–89.997	3.5820–3.5826
Normally aspirated engine with Hepolite pistons (ex. B212 engine)		
standard A	89.977–89.985	3.5424–3.5427
standard AB	89.985–89.991	3.5427–3.5430
standard B	89.991–89.999	3.5430–3.5433
standard C	89.999–90.015	3.5433–3.5439
Normally aspirated engine with Hepolite pistons (ex. B212 engine)		
first oversize (0.5 mm)	90.477–90.492	3.5621–3.5627
second oversize (1.0 mm)	90.977–89.992	3.5818–3.5824

continued on next page

Piston diameter	mm	in.
B212 engine (2.1 liter engine)		
standard A	92.982–92.992	3.6607–3.6611
standard B	92.993–93.002	3.6611–3.6615
standard B	93.003–93.012	3.6615–3.6619
first oversize (0.5 mm)	93.482–93.492	3.6804–3.6808
second oversize (1.0 mm)	93.982–93.992	3.7001–3.7005
nominal piston clearance	0.009–0.035	0.00035–0.00137

Piston-to-Cylinder Clearance (mm)			
piston/ cylinder classification	non-turbo (KS)	turbo	non-turbo (Hepolite)
B202 engine (2.0 L)			
A/A	0.012–0.032	0.030–0.050	0.015–0.033
AB/A	0.004–0.022	0.022–0.040	0.009–0.025
AB/B	0.014–0.032	0.032–0.050	0.019–0.035
B/A	—	0.014–0.032	0.001–0.019
B/B	0.006–0.024	0.024–0.042	0.011–0.029
C/B	—	0.008–0.034	—
B212 engine (2.1 L)			
A/A	—	—	0.008–0.028
B/B	—	—	0.009–0.027
B /B	—	—	0.009–0.027

CYLINDER HEAD AND VALVE MECHANISM

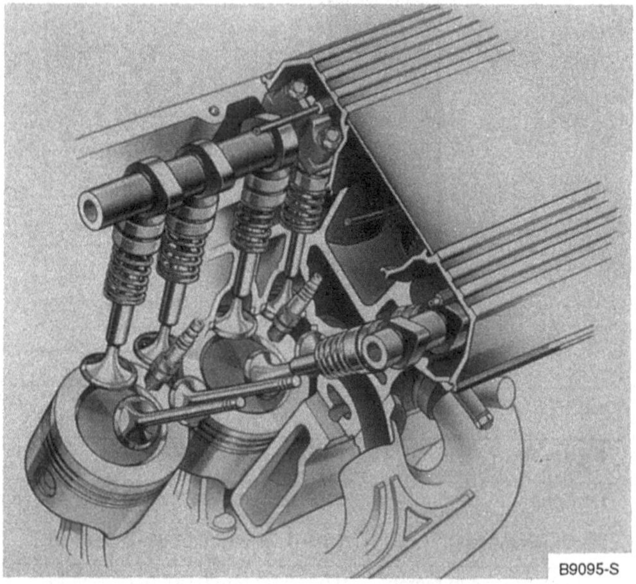

B9095-S

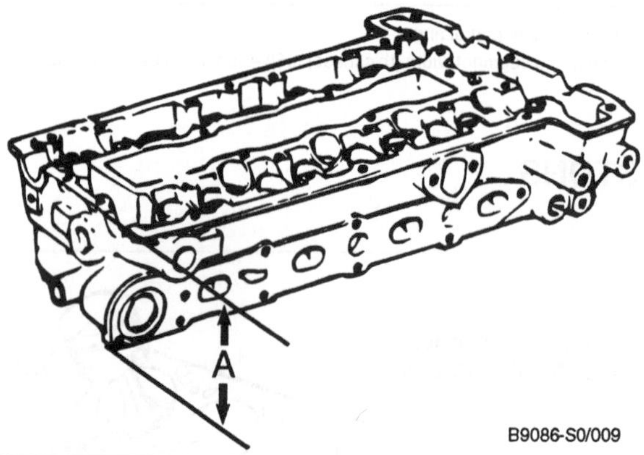

B9086-S0/009

Cylinder head height (dimension A)	mm	in.
New	140.5±0.1	5.533±0.004
After machining	140.1±0.05	5.516±0.004
Machining limit	0.4	0.0016

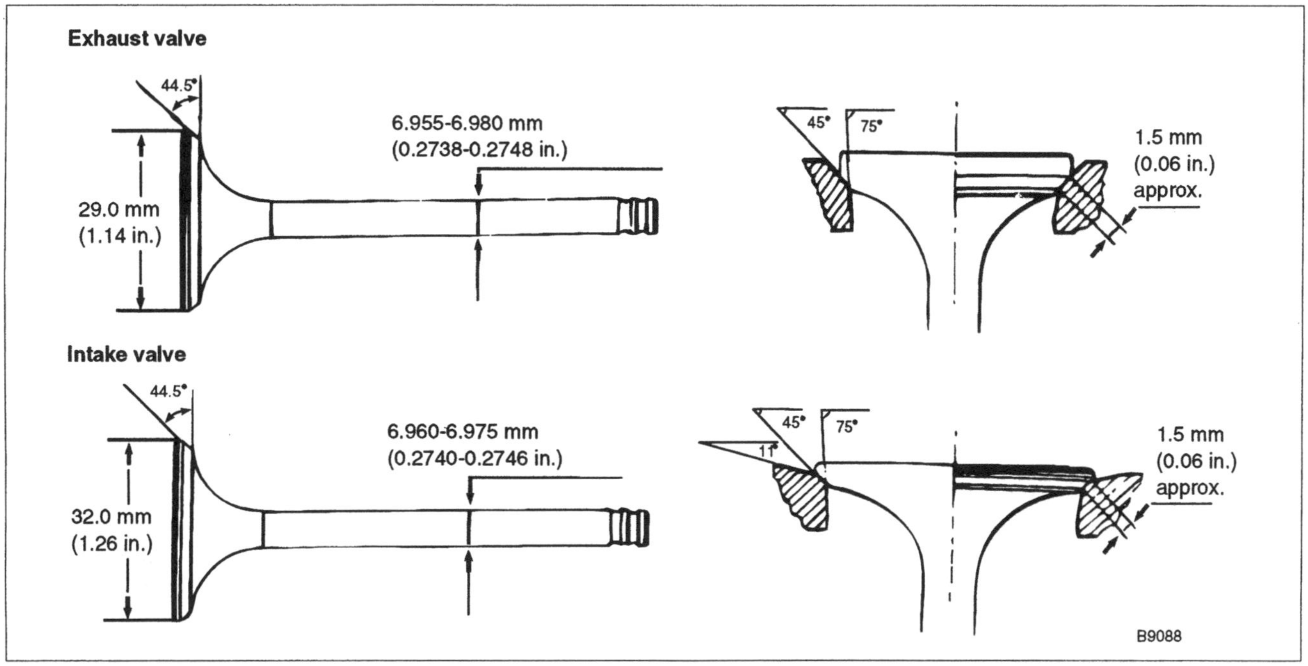

Valves	mm	in.
Valve clearance	non-adjustable	
Valve guides		
valve guide wear (clearance between valve guide and valve stem)		
maximum permissible	0.5	0.02
length	49.00	1.929
outside diameter	12.039–12.050	0.4740–0.4744
bore for valve guide in cylinder head	12.000–12.018	0.4724–0.4731
Valve Springs		
out-of-square (maximum)	1.75	0.07
free height	45.0±1.5	1.77±0.06
installed height	37.0	1.46
height when compressed with pressure of 131–141 lbs.	28.4	1.12

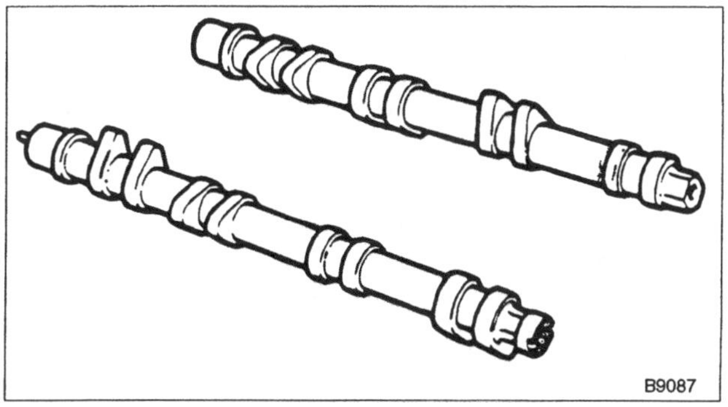

Camshaft	mm	in.
Bearing diameter	28.922–28.935	1.1387–1.1392
Axial play (maximum permissible)	0.08–0.35	0.0031–0.0138

Cam followers	mm	in.
Diameter	32.96–32.98	1.2976–1.2984
Height	26	1.02
Bore for cam followers in camshaft bridge	33.000–33.016	1.2992–1.2998

CRANKSHAFT

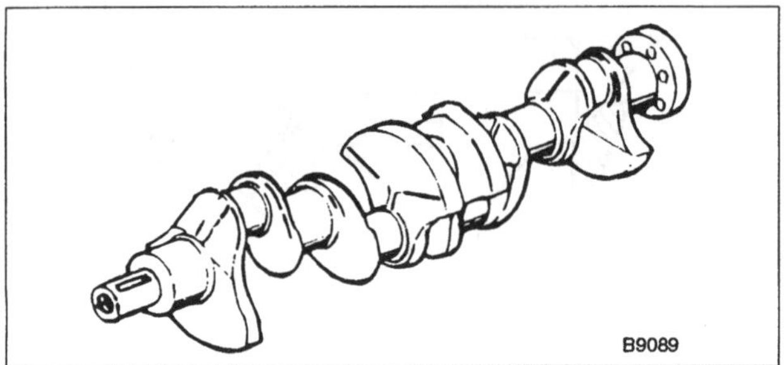

B9089

Crankshaft	mm	in.
Maximum variation in straightness (runout)	0.10	0.004
Axial play (end float)	0.08–0.28	0.003–0.011
Maximum out-of-round of journals	0.05	0.002
Radius of main journal fillet	2.2–2.5	0.09–0.10
Main bearing oil clearance	0.020–0.062	0.0008–0.0024
Big-end bearing oil clearance	0.026–0.062	0.0010–0.0024
Crankpin diameter		
standard	51.981–52.000	2.0465–2.0472
first undersize	51.731–51.750	2.0367–2.0374
second undersize	51.481–51.500	2.0268–2.0276
third undersize	51.237–51.250	2.0172–2.0177
fourth undersize	50.987–51.000	2.0074–2.0079
Main journal diameter		
standard	57.981–58.000	2.2827–2.2835
first undersize	57.731–57.750	2.2729–2.2736
second undersize	57.481–57.500	2.2630–2.2638
third undersize	57.237–57.250	2.2534–2.2539
fourth undersize	56.987–57.000	2.2436–2.2441

Color markings, main and big-end bearing shells		
standard	first oversize	second oversize
thin-red	thin-yellow	thin-white
thick-blue	thick-green	thick-brown

LUBRICATION SYSTEM

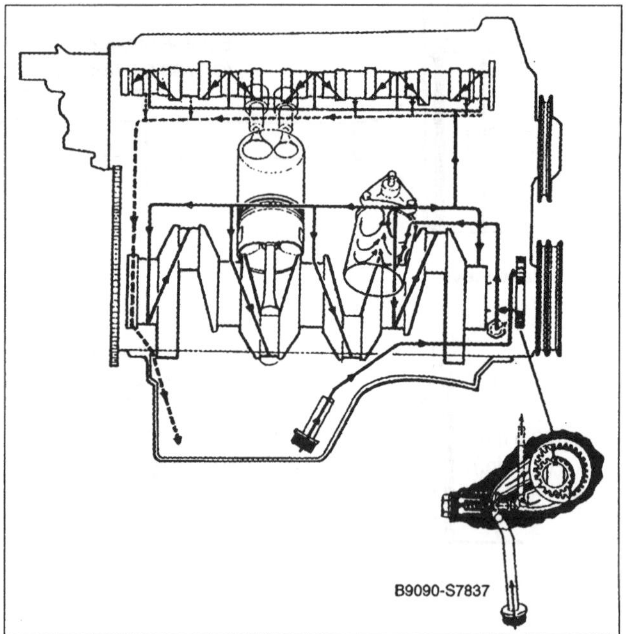

B9090-S7837

Lubrication System		
Oil pressure at 2000 rpm, engine temp. 80°C (176°F)	at least 2.7 bar	(39 psi)
Oil pressure relief valve opening pressure	3.6–5.3 bar	(52–77 psi)
Warning light on pressure	below 0.3–0.5 bar	(4.4–7.3 psi)
Oil pump end float	0.03–0.08 mm	(0.0012–0.0031 in.)

LH FUEL INJECTION

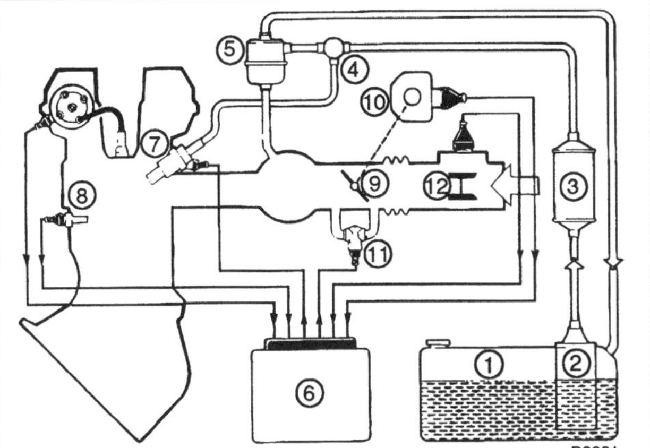

B9091

1. Fuel tank
2. Fuel pump
3. Fuel filter
4. Fuel distribution
5. Pressure regulator
6. LH control unit
7. Fuel injector
8. Coolant temperature sensor
9. Throttle plate
10. Throttle switch
11. AIC valve
12. Air mass meter

System	Year/Model	Identification
LH 2.2	1985–1988 turbo 1986–1987 normally aspirated	Three wire AIC idle valve, metal air mass meter with sealed mixture adjustment screw
LH 2.4	1989 and later turbo 1988–1990 normally aspirated	Two wire AIC idle valve, plastic air mass meter without mixture adjustment
LH 2.4.2	1991 and later normally aspirated	Three wire AIC idle valve, similar plastic air mass meter without mixture adjustment

Fuel Pump	
Fuel Pump Delivery Volume (minimum)	900 ml (30 oz.) in 30 seconds
Fuel system pressure — fuel pump running, engine off	
normally aspirated models	3.0 bar (43.5 psi)
turbo models	2.5±0.05 bar (36.3±0.7 psi)
Fuel system pressure — engine idling	
normally aspirated models	2.4 bar (34.8 psi)
turbo models	1.9 bar (27.6 psi)
Residual fuel pressure	
all models	0.1–0.2 bar (1.5–3.0 psi) below system pressure, minimum after 10 minutes

B9092

Fuel Injection	
LH Coolant Temperature Sensor Resistance	
Temperature	**Resistance (Ohms, ±10%)**
–4°F (–20°C)	14,000
14°F (–10°C)	9,000
32°F (0°C)	5,800
50°F (10°C)	3,800
58°F (15°C)	3,000
68°F (20°C)	2,600
76°F (25°C)	2,000
86°F (30°C)	1,700
176°F (80°C)	320

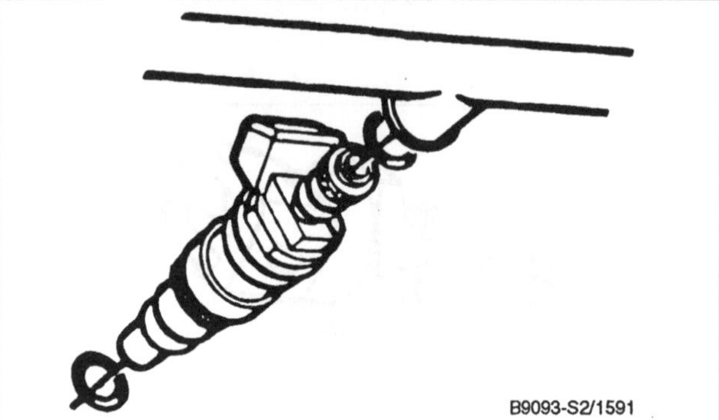

B9093-S2/1591

Fuel Injection		
Fuel injector resistance	16 Ohms at 68°F (20°C)	
Oxygen Sensor Specifications		
Oxygen sensor voltage at idle (engine warm)	0.4–1 volt	
Oxygen sensor preheater resistance	approx. 4 Ohms	
AIC Valve Resistance		
Fuel System	Test terminals	Resistance (Ohms)
LH 2.2	1 and 2	20±2
	2 and 3	20±2
	1 and 3	40±4
LH 2.4	1 and 2	8±2
LH 2.4.2	1 and 2	10–15
	2 and 3	10–15
Idle speed (adjustable only on LH 2.2 injection)		850±50 rpm
Idle %CO (adjustable only on LH 2.2 injection)		
normally aspirated	0.5–1.5%	
turbo	0.9–1.6%	
Dashpot Adjustment Specifications (LH 2.2)	speed at which dashpot rod touches throttle lever	
normally aspirated	2500±100 rpm	
turbo	2600±100 rpm	

COOLING SYSTEM

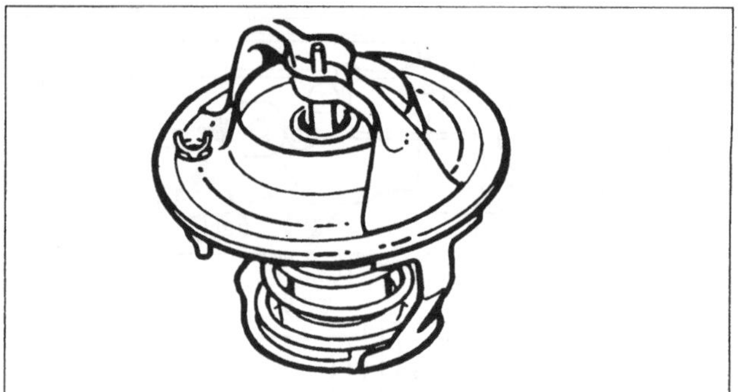

B9094-S5207

Cooling System	
Cooling system capacity	10 liters (10.5 qts.)
Permissible coolant leakage	1 qt. over 60,000 miles (1 liter/100,000 km)
Coolant thermoswitch closing temperature	194–203°F (90–95°C)
Coolant thermoswitch opening temperature	185–194°F (85–90°C)
Coolant type	Phosphate-free 50% mixture anti-freeze 50% water
Coolant temperature sender resistance	51.2±4.3 Ohms at 90°C (194°F)
Cooling system test pressure (maximum)	1.2 bar (17.4 psi.)

TURBOCHARGER

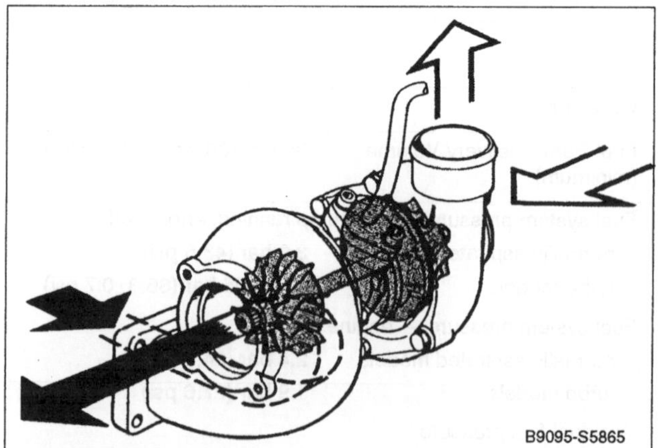

B9095-S5865

Turbocharger basic boost pressure	bar	psi
All except 1987 and later SPG models	0.35±0.03	5.0±0.4
1987 and later SPG models	0.40±0.03	5.8±0.4

023 Electrical System—Technical Data

ELECTRICAL SYSTEM

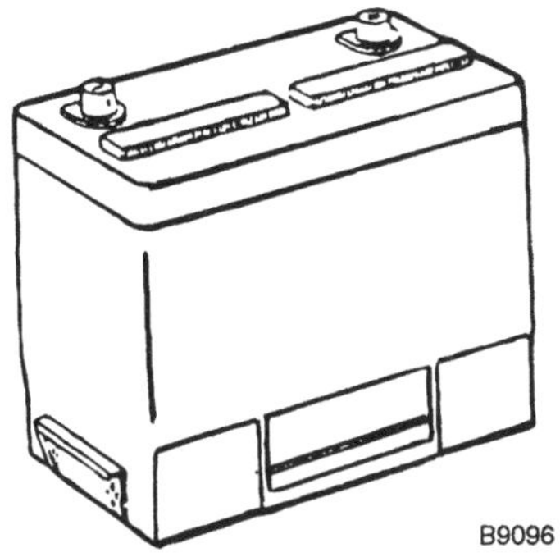

B9096

Tightening Torques	Nm	ft-lb (in-lb)
Ignition distributor hold-down bolt	15	11
Knock sensor to engine block	20	15
Solenoid to starter	4.5–5.5	(40–49 in-lb)
Starter field winding strap and battery cable to solenoid	7–9	(62–80 in-lb)
all other fasteners		
M5 bolt	5	(44 in-lb)
M6 bolt	10	(89 in-lb)
M8 bolt	20	15
M10 bolt	40	30

Charging System	
Alternator	
rated output	14 volts

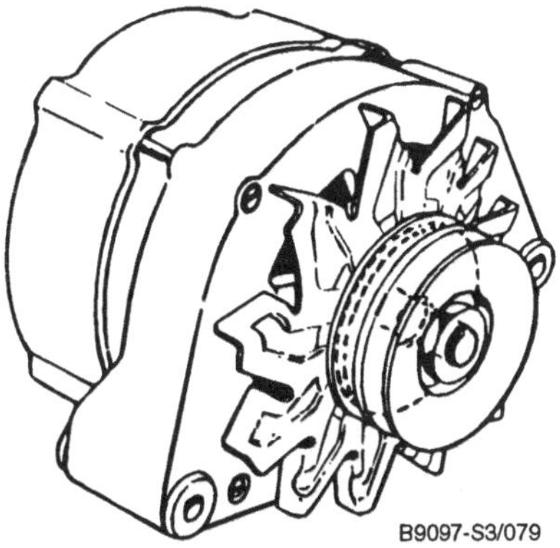

B9097-S3/079

Charging System	
Alternator	
rated speed	1900 rpm
minimum brush length	5 mm (0.2 in.)
Output test values	
36 amps	@ 1500 rpm
54 amps	@ 1900 rpm
80 amps	@ 6000 rpm

Ignition System	
Ignition system type	

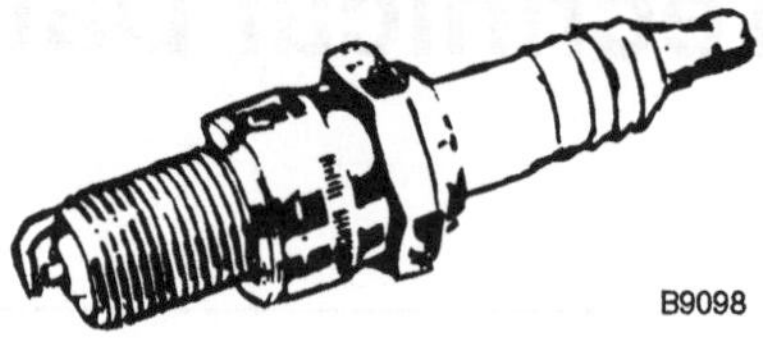
B9098

Ignition System	
1985 and later, turbo	basic Hall ignition
1986 and later, non-turbo	EZK ignition-Hall-effect with knock sensor
Ignition firing order	1-3-4-2 (no. 1 cylinder next to firewall)
Ignition timing (vacuum hose disconnected)	
Turbo	16°BTDC @ 850 rpm
Non-turbo	14°BTDC @ 850 rpm
Ignition coil	
primary winding resistance*	0.5–0.9 Ohms
secondary winding resistance*	7200–8200 Ohms
Spark plug wires and rotor	
ignition rotor resistance	1000 Ohms
coil high tension lead resistance (complete with end connectors)	500-1500 Ohms
spark plug lead resistance (complete with end connectors)	2000-4000 Ohms

*Measure with all wires disconnected from coil. See **340 Ignition System** for testing information

024 Transmission—Technical Data

TRANSMISSION

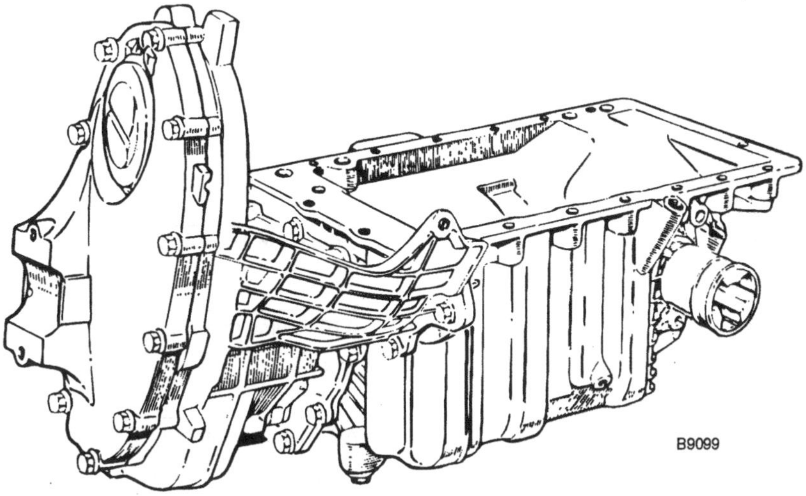

Tightening Torques	Nm	ft-lb (in-lb)
ATF pressure tap (automatic transmission)	5–7	(44–62 in-lb)
ATF sump pan bolts (automatic transmission)	8–12	(71–106 in-lb)
ATF sump pan drain plug (automatic transmission)	5–8	(44–71 in-lb)
ATF cooler connection nut (automatic transmission)	13–16	10–12
Clutch slave cylinder mounting screws	6–14	(53–124 in-lb)
Gearbox drain plug (manual transmission)	39–59	29–44
Selector cable setscrew tightening torque (automatic transmission)	2.5	(22 in-lb)

Tightening Torques	Nm	ft-lb (in-lb)
Shift rod clamp bolt (manual transmission)	30–35	22–26
All other fasteners		
M5 bolt	5	(44 in-lb)
M6 bolt	10	(89 in-lb)
M8 bolt	20	15
M10 bolt	40	30

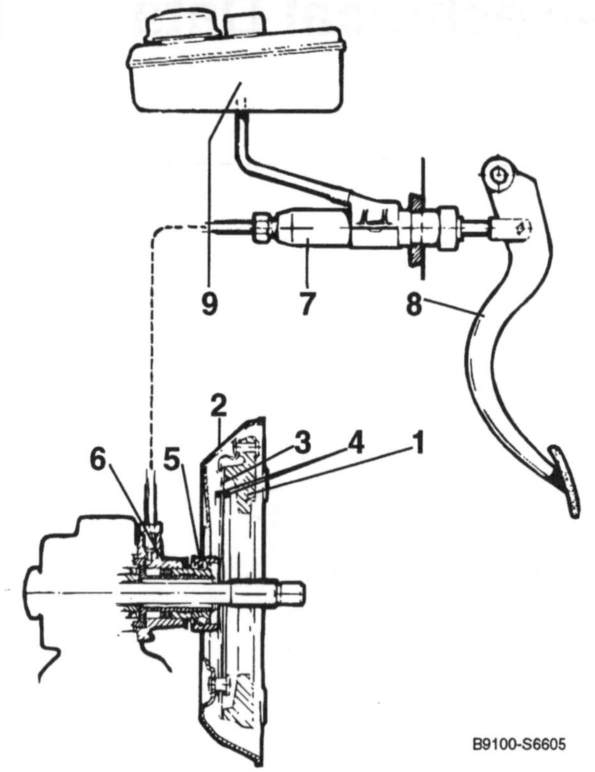

1. Pressure plate
2. Housing
3. Diaphragm spring
4. Pivot rings
5. Release bearing
6. Slave cylinder
7. Master cylinder
8. Clutch pedal
9. Clutch fluid reservoir

Manual Transmission	
Clutch type	Fichtel & Sachs
Clutch diameter non-turbo turbo	 204 mm (8 in.) 216 mm (8.5 in.)
Clutch adjustment	automatic (non-adjustable)
Manual transmission oil capacity	3.0 l (3.5 U.S. qts.)
Grade of oil	SAE 10W-30 SF/CC, SF/CD, SG
Alternate grade of oil	SAE EP 75 API-GL-4 or API GL-5
Clutch/brake master cylinder fluid	DOT 4 brake fluid

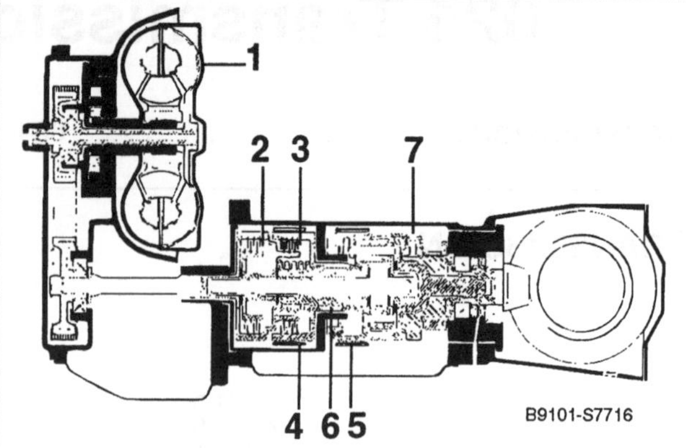

1. Torque convertor
2. Front clutch
3. Rear clutch
4. Front band
5. Rear band
6. One way clutch
7. Governor

Automatic Transmission	
Type	Borg Warner 37
ATF volume (transmission)	8.0 l (8.5 U.S. qts.)
ATF	Ford specification M2C-33F (alternate Ford specification G)
A/T final drive capacity	1.25 l (1.3 U.S. qts.)
A/T final drive grade of oil	SAE EP80 or 75 API-GL-4 or AP I-GL5
Alternate grade	SAE 10W-30 engine oil
General stall speed (APC solenoid unplugged)	2250–2700 rpm
Line pressure at idle (gear selector in D)	4.2–4.9 bar (61–71 psi)

025 Brakes—Technical Data

BRAKES

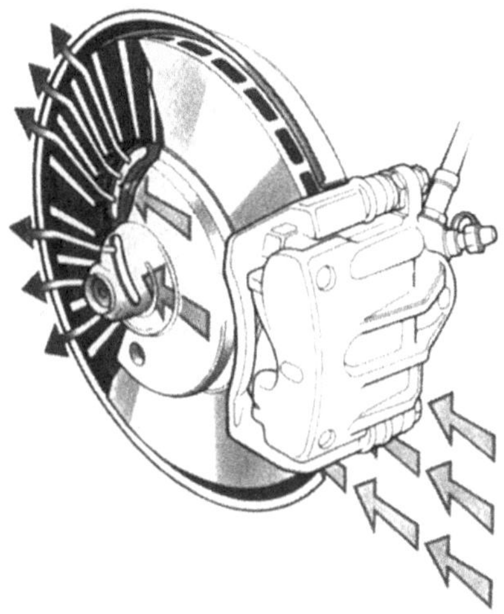

B9102

Tightening Torques	Nm	ft-lb (in-lb)
ABS hydraulic unit to bulkhead	26±4	19±3
Front caliper mounting bolt to guide pin (1988 models and later)	30–35	22–26
Front caliper to steering member (1985 to 1987)	110–130	81–96
Front pad carrier to steering member (1988 models and later)	70–110	52–81
Rear caliper to axle (1985 to 1987)	70–90	52–66
Rear caliper to pad carrier (pins) (1988 models and later)	25–30	18–22
Rear pad carrier to axle (1988 models and later)	40–54	30–40
All other fasteners		
M5 bolt	5	(44 in-lb)
M6 bolt	10	(89 in-lb)
M8 bolt	20	15
M10 bolt	40	30

Front Brake Disc Applications	
Model year and Model	**Disc type**
1985 through 1987 900	Solid
1985 and 1986 900s	Solid
1985 Turbo	Solid
1987 900s	Ventilated
1986 through 1987 Turbo, including convertible	Ventilated
1988 and later	All Ventilated

Front Brake Disc	Solid disc	Ventilated disc (1986-1987)	Ventilated disc (1988 and later)
Thickness (new)	12.7 mm (0.50 in.)	20 mm (0.787 in.)	23.5 mm (0.93 in.)
Thickness (wear limit)	11.2 mm (0.441 in.)	18 mm (0.709 in.)	21.5 mm (0.85 in.)
Brake surface lateral runout (maximum)	0.1 mm (0.004 in.)	0.1 mm (0.004 in.)	0.08 mm (0.003 in.)

Rear Brake Disc	
Thickness (new)	9.0 mm (0.354 in.)
Thickness (wear limit)	7.0 mm (0.276 in.)
Brake surface lateral runout (maximum)	0.1 mm (0.004 in.)

ABS Specification	
ABS wheel speed sensor air gap	0.65 mm (0.026 in.)

026 Steering and Wheel Alignment—Technical Data

STEERING AND WHEEL ALIGNMENT

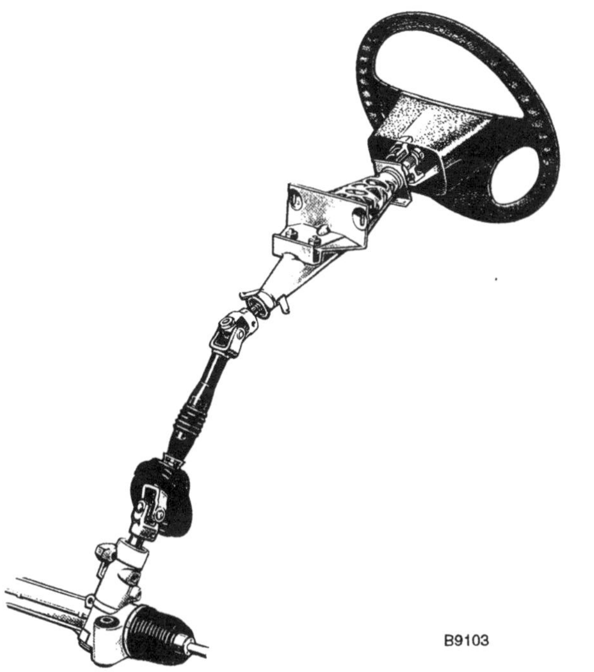

B9103

Tightening Torques	Nm	ft-lb (in-lb)
Ball joint to control arm (upper or lower)	40–55	29–41
Ball joint to control arm bolts (self-locking nuts)	40–55	29–41
Ball joint to steering swivel member (self-locking nut)	35–50	26–37
Bearing bracket to control arm (locknut and lock washer)		
lower control arm bracket to control arm	75–90	55–66
upper control arm bracket to control arm	55–70	41–52
Coil spring seat to control arm	55–110	41–81

Tightening Torques	Nm	ft-lb (in-lb)
Lower control arm bearing bracket (front or rear) to body		
M12 dry joint	70–95	52–70
M12 lubricated joint (waxed nut)	45–55	33–41
Lower control arm bracket to body	25–35	18–26
Power steering rack to body	60–80	44–59
Road wheel to wheel hub		
lug nuts	90–110	66–81
lug bolts	105–125	77–92
Steering column to steering rack (clamping bolt, self-locking nut)	20–27	15–20
Steering column to body (bolts, use with Loctite)	20–27	15–20
Steering wheel to steering column (self-locking nut)	25–28	18–21
Universal joint to pinion shaft (clamping bolt and self-locking nut)	25–34	18–25
Power steering hose fittings	20–34	15–25
Tie rod end to tie rod	60–80	44–59
Upper control arm bearing bracket (front or rear) to body	40–55	30–41
All other fasteners		
M5 bolt	5	(44 in-lb)
M6 bolt	10	(89 in-lb)
M8 bolt	20	15
M10 bolt	40	30

Steering and Wheel Alignment Specifications	
Chassis Ride Height (measured between top of wheel rim and fender/hood gap)	
Except sport chassis	243 mm (9.56 in.)
Sport chassis	
with 15 in. wheel	230 mm (9.06 in.)
with 16 in. wheel	217 mm (8.54 in.)

Wheel Alignment Specifications		
Measurement	**Model**	
	except sport chassis	sport chassis
Front		
caster	2°±0.5	2°±0.25
camber	0.5°±0.5	0.25°±0.25
toe-in	2±1 mm (0.08±0.04 in.)	1.5±0.5 mm (0.06±0.02 in)
Swivel pin inclination	11.5°±1	11.5°±1
Steering angle—outer wheel	20°	20°
Steering angle—inner wheel	20.75°±0.50	20.75°±0.50
Rear wheels (non-adjustable)		
toe-in	4±1 mm (0.16±0.04 in.)	4±1 mm (0.16±0.04 in.)
camber	-0.5°±0.25	-0.5°±0.25
Ball joint wear limit		
maximum radial play	1 mm	0.040 in.
maximum axial play	2 mm	0.078 in.
Power Steering Fluid		
GM power steering fluid (GM 9985010), Texaco TL4634 or equivalent	75 cl	0.8 qt

027 Suspension—Technical Data

SUSPENSION

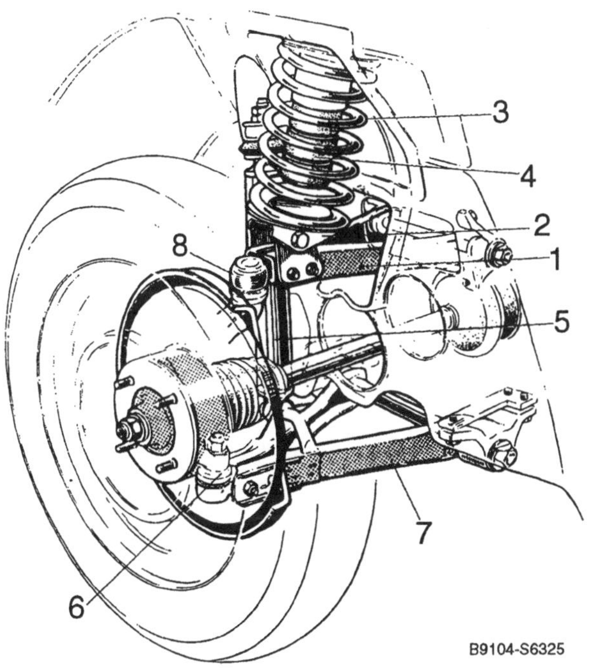

1. Upper control arm
2. Spring seat
3. Coil spring
4. Bump stop
5. Shock absorber
6. Lower ball joint
7. Lower control arm
8. Upper ball joint

Tightening Torques	Nm	ft-lb (in-lb)
Front shock absorber to lower control arm	90–100	66–74
Front suspension coil spring seat to control arm	55–110	41–81
Anti-roll bar bracket to lower control arm	40–55	30–41
Power steering rack to body	60–80	44–59
Spring link front mount to body	70–90	52–66

Tightening Torques	Nm	ft-lb (in-lb)
Panhard rod to body or rear axle	40–70	30–52
Torque arm to body or rear axle		
lubricated bushings	21–35	15–26
dry bushings	40–70	30–52
Wheel hubs to drive axle (locking nut)	290–310	214–229
Wheel lug bolts	105–125	77–92
Wheel lug nuts	90–110	66–81
All other fasteners		
M5 bolt	5	(44 in-lb)
M6 bolt	10	(89 in-lb)
M8 bolt	20	15
M10 bolt	40	30

Front Coil Spring Specifications		
Spring color codes	**Spring free length**	
Standard chassis		
green, light green, black, or white	373 mm	14.7 in.
yellow or red	380 mm	15.0 in.
pink or brown	388 mm	15.3 in.
blue or light blue	372 mm	14.6 in.
Sport chassis		
silver or bronze	301 mm	11.8 in.

Rear Coil Spring Specifications		
Spring color codes	**Spring free length**	
Standard chassis		
black or white (left spring)	311 mm	12.24 in.
green or light green (right spring)	308 mm	12.12 in.
Sport chassis		
silver or bronze	293 mm	11.5 in.

028 Body—Technical Data

BODY

B9105

Tightening Torques	Nm	ft-lb (in-lb)
A/C system tightening torques		
A/C compressor-to-evaporator hose		
connection at compressor (pad-type)	22–27	16–20
hose fitting at evaporator	28–39	21–29
A/C compressor-to-condensor hose		
connection at compressor (pad-type)	22–27	16–20
hose fitting at condensor	21–28	15–21
A/C condensor-to-receiver/dryer hose		
hose fitting at condensor	14–20	10–15
hose fitting at receiver/dryer	14–20	10–15
A/C evaporator-to-receiver/dryer hose		
hose fitting at evaporator	14–20	10–15
hose fitting at receiver/dryer	14–20	10–15
A/C expansion valve to evaporator	21–27	15–20
A/C expansion valve pressure equalization fitting to evaporator	7–10	(62–89 in-lb)
Oil fill plug to A/C compressor	8–12	(71–106 in-lb)
Convertible top hydraulic hose fittings	5.1–6.2	(45–55 in-lb)
Seatbelt guide bolt to body (convertible models)	24–40	18–30
All other seat belt mounting bolts	45±10	33±7
All other fasteners		
M5 bolt	5	(44 in-lb)
M6 bolt	10	(89 in-lb)

continued

Tightening Torques	Nm	ft-lb (in-lb)
All other fasteners		
M8 bolt	20	15
M10 bolt	40	30

Body Specifications		
Convertible pump hydraulic fluid	Aeroshell Fluid 4/Shell Code 60 421 (Saab part no. 30 18 694	
Convertible top hydraulic pump minimum working pressure	25 bar	365 psi

A/C System Specifications		
Refrigerant oil capacity (compressor replacement)	1.75 dl	4.9 oz.
Refrigerant oil capacity per component (due to sudden leakage)		
evaporator	0.5 dl	1.4 oz.
receiver/dryer	0.2 dl	0.57 oz.
condenser	0.2 dl	0.57 oz.
hose	0.2 dl	0.57 oz.
A/C refrigerant pressure switch (low gas pressure)		
opening pressure	2.9 bar	41 psi
A/C Coolant temperature switch		
opening temperature	115°C	239°F
A/C Anti-frost switch		
switch closed	4.0–6.2°C	39.2–43.1°F
switch open	0.4–2.6°C	33–36.6°F

100 Fundamentals for the Do-it-yourself Owner

TABLES

GENERAL

Although the Saab 900 is a sophisticated and complex machine, nearly all basic maintenance and most repairs can be accomplished by any interested owner with basic mechanical skills and the right information. While some of the repairs covered in this manual are complicated and require special knowledge and equipment, most of the care that is required in the lifetime of the average Saab is well within the capabilities of the do-it-yourselfer.

CAUTION

Do not use this manual unless you are familiar with basic automotive repair procedures and safe workshop practices. This manual illustrates the workshop procedures required for most service work; it is not a substitute for full and up-to-date information from the vehicle manufacturer or for proper training as an automotive technician. Note that it is not possible for us to anticipate all of the ways or conditions under which vehicles may be serviced or to provide cautions as to all of the possible hazards that may result.

CAUTION
Your common sense and good judgment are crucial to safe and successful service work. Read procedures through before starting them. Think about whether the condition of your car, your level of mechanical skill, or your level of reading comprehension might result in or contribute in some way to an occurrence which might cause you injury, damage your car, or result in an unsafe repair. If you have doubts for these or other reasons about your ability to perform safe repair work on your car, have the work done at an authorized Saab dealer or other qualified shop.

This section of the manual is intended to help the beginner get started smartly and safely with Saab maintenance and repair. The section begins with **Form and Function,** a general description of the car and its individual systems, followed by a discussion on **How To Use This Manual.**

Tips on mechanic's skills and workshop practices that can help the beginner do a faster, complete, and more thorough job can be found under **Getting Started. Tools** describes the basic tools needed to do most of the procedures in this manual.

The section ends with a quick reference guide to emergencies—what to do when the car won't start or when a warning light comes on, including basic troubleshooting and information on how to gauge the seriousness of a problem.

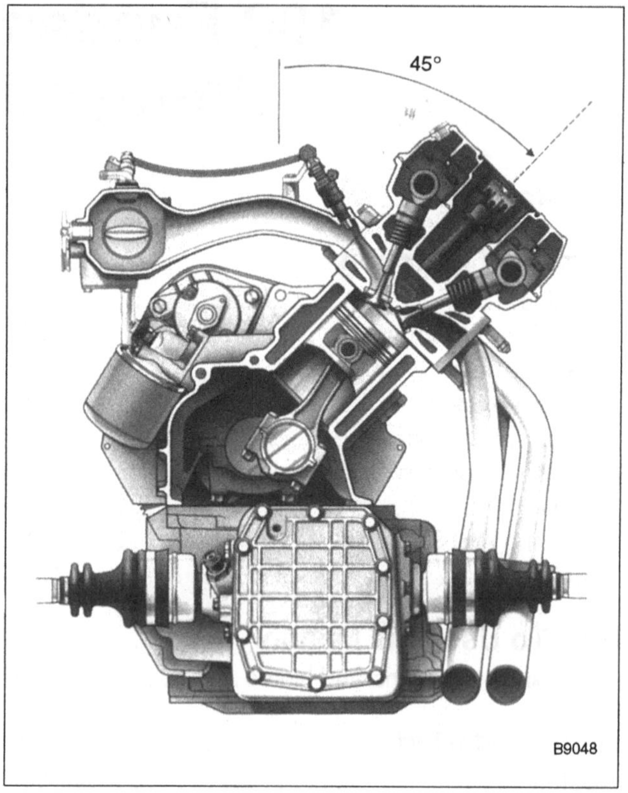

Fig. 1. Cutaway of Saab 900 16 valve overhead camshaft engine. The engine is sloped at 45% to lower the car's center of gravity and to allow for a low hood line. The transmission is bolted to the bottom of the engine.

FORM AND FUNCTION

While the complexity of the Saab 900 may at first seem overwhelming to the do-it-yourself owner, it can be simplified and more easily understood by viewing the car as an assembly of simpler systems, each performing its own independent functions.

Engine

The Saab 900 models covered by this manual are powered by a liquid-cooled, four-cylinder, 16-valve, in-line engine. The power train is integrated with the clutch, transmission, and differential in a compact, lightweight power unit that occupies very little space. See Fig. 1.

The 16 valve cylinder head uses self-adjusting hydraulic cam followers to actuate the valves. Each valve has a large flow area to allow the engine to breath efficiently. Because the cylinder head has four valves per cylinder instead of two, individual valves are subject to far less demanding working conditions. See Fig. 2.

Fig. 2. Cutaway view of valve train.

Saab Turbo with APC

In addition to the naturally aspirated engine family, the 900 is available with a turbocharged engine. See Fig. 3. The turbo engine delivers extra power at low engine speeds when demanded, but during normal driving it is as economical as the naturally aspirated engine. The turbocharged engine offers the same acceleration as many six-cylinder and eight-cylinder engines, but without the well-known disadvantages of excessive weight, unwieldy bulk with too many moving parts, and high fuel consumption.

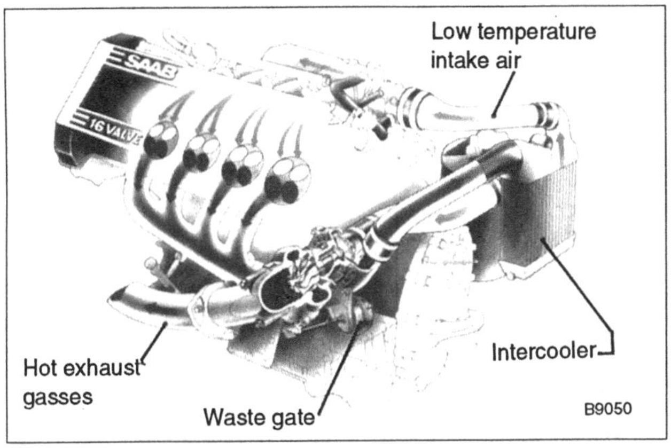

Fig. 3. Cutaway of turbocharger system. The intake air is compressed by the exhaust-driven turbocharger. It is then cooled by as much as 110°F in the intercooler before it enters the engine. The intercooler increases power and reduces the thermal stresses in the engine.

Safety margins must be built into the design of an internal combustion engine to allow for manufacturing tolerances, changes in the condition of the engine, and varying climatic conditions and fuel quality. As a result, the energy content of the fuel cannot be fully utilized. To answer this problem, the Saab APC system was designed to ensure maximum utilization of the energy in the fuel at all times. The boost pressure delivered by the turbocharger is continually adjusted to suit the knocking limit of the air/fuel mixture in the cylinders. Therefore, the engine adjusts itself automatically to the fuel being used. The higher the octane, the higher the maximum engine power. See Fig. 4.

NOTE

On Saab Turbo SPG models, premium unleaded fuel is required at all times.

The APC system consists of simple electronics. A knock sensor on the engine block senses the onset and degree of engine knock and transmits an electrical signal to the control unit. The control unit also receives signals from the intake manifold pressure sensor and the ignition system (engine rpm). The data is processed by the control unit, which transmits a signal to the APC solenoid valve to regulate the boost pressure (via the wastegate) of the turbocharger. Therefore, maximum operating boost pressure under full engine load is electronically governed. In the event of an APC system malfunction, the wastegate is mechanically adjusted to provide a safe, low boost pressure limit.

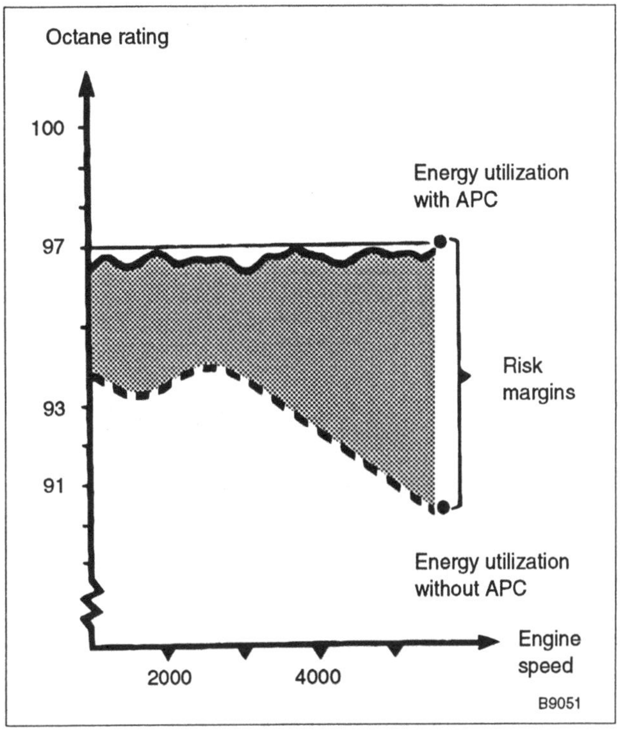

Fig. 4. A conventional engine must incorporate safety margins for engine tolerances, various climatic conditions, and variations in fuel grades available. The Saab APC turbo engine adjusts itself automatically to suit the operating conditions.

Fuel Injection

All Saab 900 16-valve models use an electronically controlled Bosch LH-Jetronic fuel injection system. See Fig. 5. LH is a German acronym for hot wire, or "Luft masse messen mit hitzdraht". Hot wire refers to the operation of the air mass meter, which is used to measure the air entering the engine.

The intake air flows across an electrically heated platinum wire in the tubular air meter. See Fig. 6. The control unit then measures the electrical energy necessary to maintain the wire at a constant temperature as a measure of the air mass. At the same time the control unit senses the engine speed and the engine temperature, which enables it to meter the correct quantity of fuel to suit the requirements of the engine at all times.

Fuel is delivered to the engine through the fuel injectors. The intake manifold is equipped with one fuel injector per cylinder. Each injector is fitted upstream of the corresponding intake valves and actuated electro-magnetically by signals from the electronic control unit. The control unit calculates the opening time for the injectors so that the quantity of the fuel injected will be correct in relation to the quantity of air flowing to the engine. See Fig. 7.

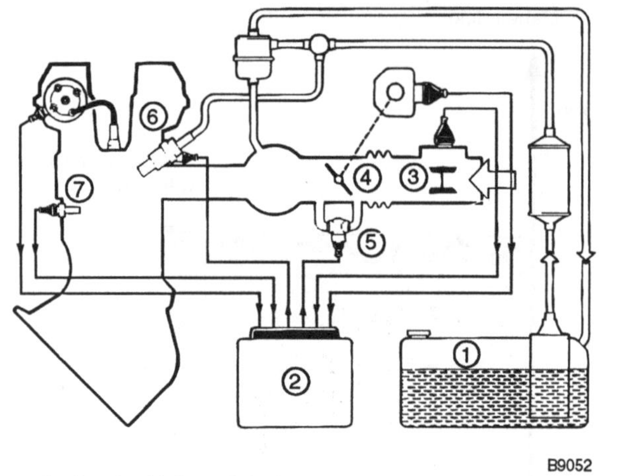

1. Fuel tank with in-tank fuel pump
2. Electronic control unit
3. Air mass meter
4. Throttle valve
5. Automatic Idle Control (AIC)
6. Fuel injector
7. Temperature sensor

Fig. 5. Schematic of LH-Jetronic fuel injection system used on all Saab 16 valve engines.

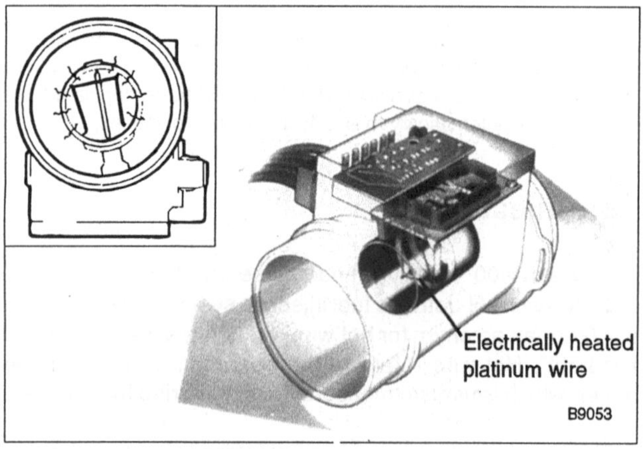

Fig. 6. View of LH (hot wire) air mass meter. The platinum heated wire is maintained at a constant temperature and measures the mass of the air flow going into the engine. It uses solid-state circuitry and has no moving parts to wear out. The inset shows a front view of air mass meter.

Ignition System

The electronic ignition system creates the high-voltage spark necessary to ignite the combustible air/fuel mixture in the cylinders. The ignition coil boosts the voltage so that the spark will be hot enough to ignite the air/fuel mixture. The timing of the spark is controlled by the ignition control unit.

The ignition distributor, synchronized to the rotation of the engine, delivers the spark to the right cylinder at precisely the

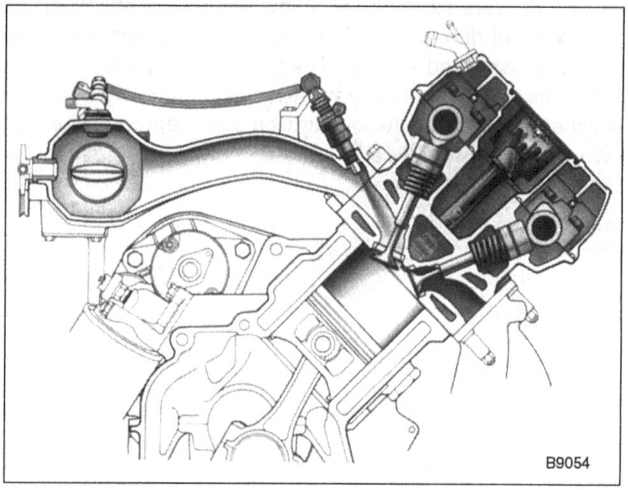

Fig. 7. Cutaway view of engine showing operation of the electromagnetic fuel injector. The injectors are mounted in the intake manifold, one for each cylinder.

right time. Since each cylinder has to have a spark once for every two revolutions of the crankshaft, the distributor turns at one-half crankshaft speed. The basic system is shown schematically in Fig. 8.

Electrical System

Many components, and all electrical accessories, are powered by the car's electrical system. The electrical system uses a battery to store energy, an engine-driven alternator to generate electricity and recharge the battery, and various wiring harnesses and other circuits to distribute electric power to the rest of the car. The electrical system is represented in Fig. 9.

The flow of electricity depends upon a closed-loop path—a complete circuit. Electrical current flows through wires to the consumer, a light bulb for example, and back to the battery in a complete circuit. The electrical route back to the source, which completes the circuit, is called a path to ground. Every consumer of electrical power in the car must have a source of power and a path to ground in order to operate.

Commonly, the electrically conductive metal structure of the automobile is used as a ground path. The negative (–) terminal of the battery connects to the car body, and all of the electrical consumers in the car make a ground connection to the car body, thus eliminating the need for many feet of additional wire.

Electrical components near the engine are often grounded directly to the engine, which is then grounded to the body. Some components are grounded through their housings which are bolted to a ground. Electrically, the effect is the same.

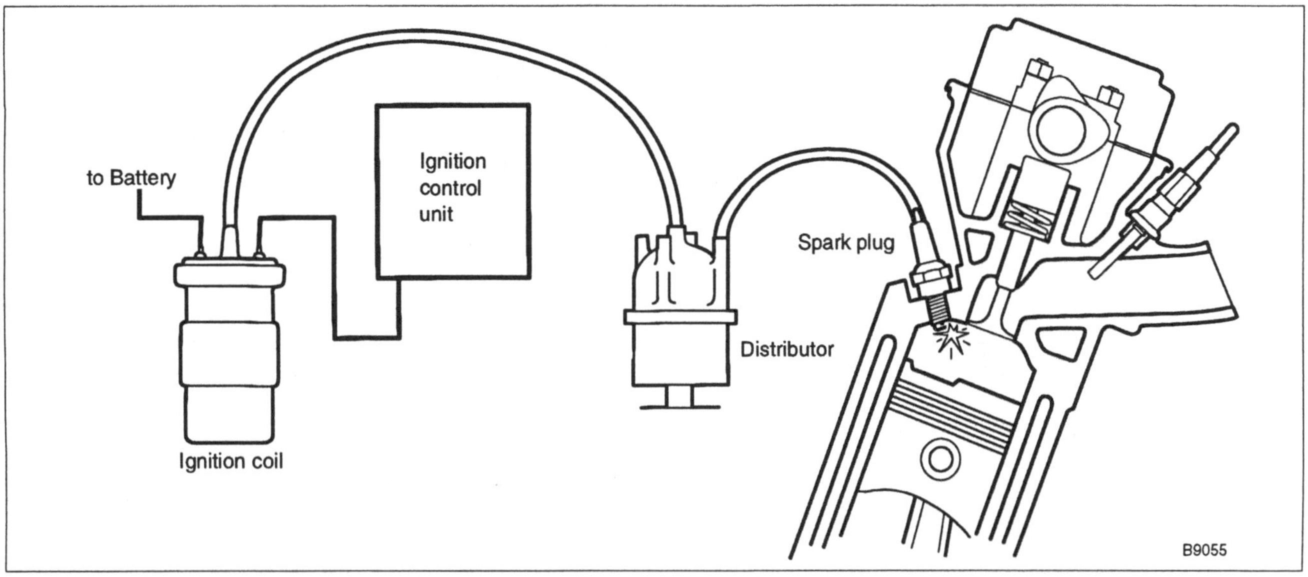

Fig. 8. Schematic of typical ignition system.

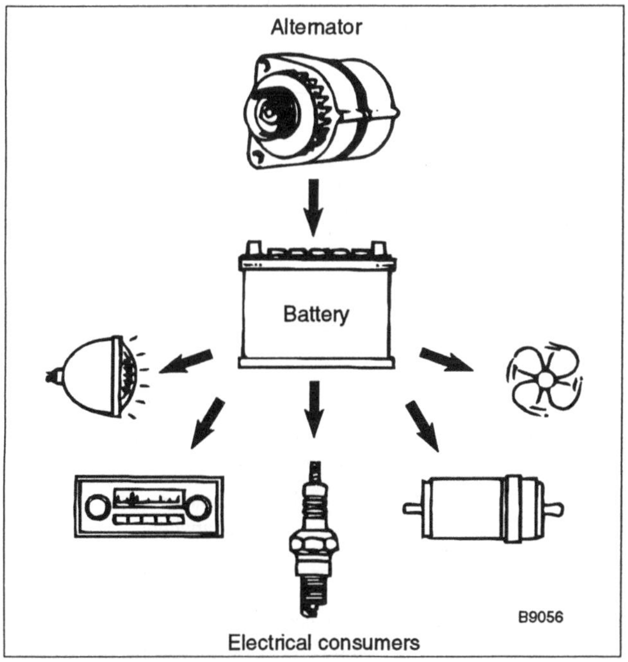

Fig. 9. The alternator generates electricity to recharge battery and power other electrical consumers.

Transmission and Drive Train

The transmission is mounted beneath the engine and is arranged in a housing together with the final drive. See Fig. 10. The transmission is chain driven off the engine via the clutch.

Although the Saab engine develops a substantial amount of power, it does so best at relatively high revolutions per minute (rpm). To handle all driving conditions, it is necessary to use gearing to change the ratio of engine rpm to vehicle speed. See Fig. 11.

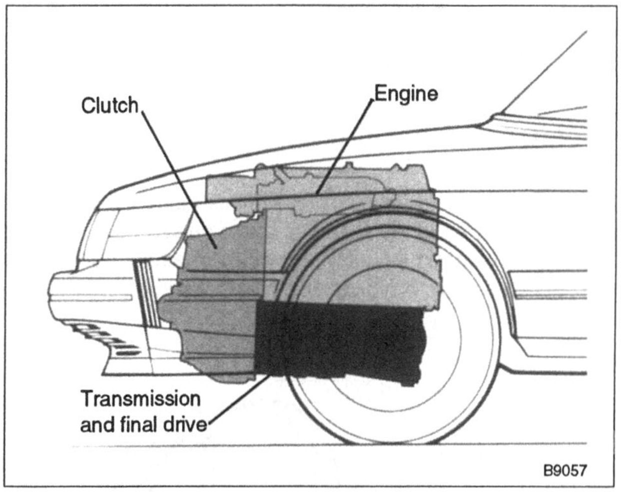

Fig. 10. The transmission and final drive are mounted beneath engine.

A manual transmission arranges several sets of gears in a common housing. A set of two gears determines a gear ratio, each suited to a particular range of driving speeds. A shifting mechanism allows the driver to change from one gear ratio to the next to match vehicle speed.

In an automatic transmission, hydraulic fluid under pressure in a complex network of passages, valves and control mechanisms engage and disengage constantly meshed planetary gear sets. Hydraulic controls responding to vehicle speed, engine load, throttle position and gear shift position select the appropriate gear ratio.

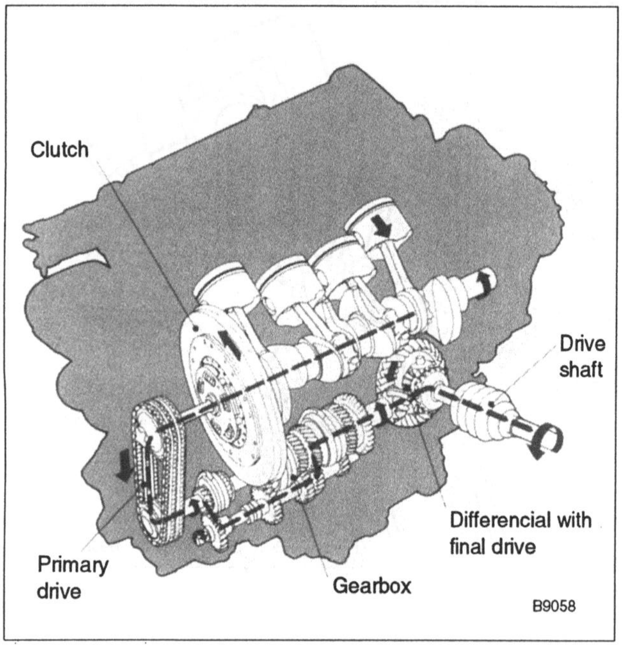

Fig. 11. The gear reduction between the engine and the driven wheels takes place in three stages: the primary chain drive, the transmission gears, and the final drive.

Brakes

All of the Saab models covered in this manual feature disc brakes at all four wheels. A disc brake squeezes pads lined with friction material against both sides of a flat, round brake disc, called a rotor. A typical disc brake assembly is shown in Fig. 12.

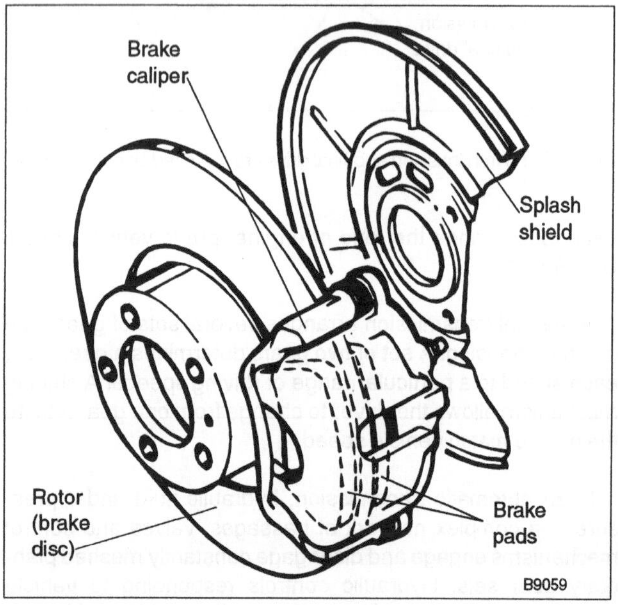

Fig. 12. Disc brake assembly showing disc, caliper, and splash shield. Caliper assembly holds pads with friction material.

The brakes act to slow or stop the car by causing friction. Since cars are relatively heavy, the friction required to stop safely and effectively is quite high, and generating this friction requires considerable force. The cars covered by this manual use either an engine vacuum boost system or hydraulic boost system (cars with ABS) to multiply the force applied to the brake pedal and to distribute it uniformly to the wheels.

The brake pedal is connected by a mechanical linkage to the master cylinder, mounted on the firewall at the back of the engine compartment. A piston in the master cylinder creates hydraulic pressure in the brake lines going to the wheels.

At each wheel, the hydraulic pressure acts on the brake caliper to cause friction and slow the wheel. The sizes of the hydraulic components are such that the driver's force applied to the brake pedal is multiplied many times by the time it reaches the wheels.

1990 and later 900 models are equipped with an anti-lock braking system (ABS). As the name implies, the purpose of this system is to prevent the wheels from locking during hard braking. Speed sensors at each wheel sense when the wheel is about to lock, and an electronic system modulates the braking force to that wheel. See Fig. 13.

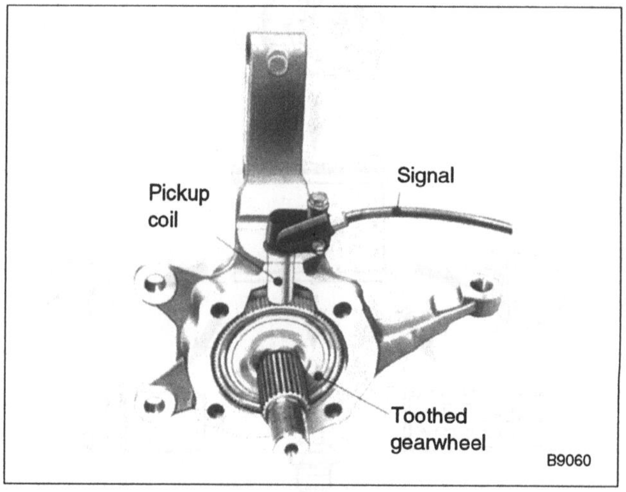

Fig. 13. A moisture-proof sensor senses the rolling speed of each wheel. The sensor is inductive in that an electric current is induced in the pickup coil every time a tooth on the gearwheel passes the sensor.

Steering and Suspension

The suspension and steering systems are what allow the wheels to move and turn for a smooth ride, stability and directional control. See Fig. 14. The suspension system is the combination of springs, shock absorbers, and other stabilizing devices that support the weight of the car and cushion the effects of bumps. On some models, stabilizer bars aid stability by transferring some of the cornering force acting on the suspension.

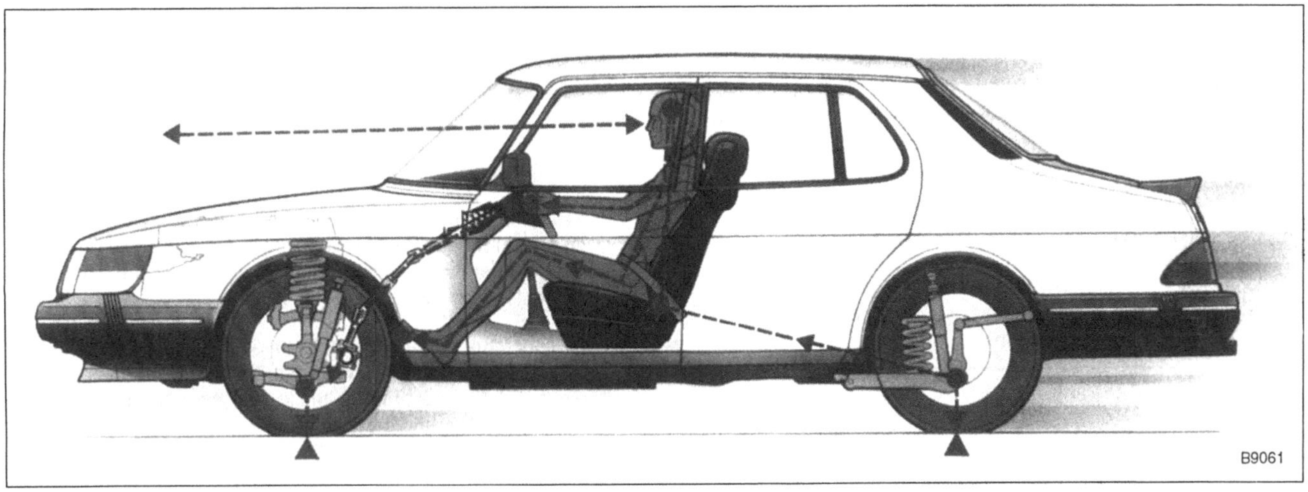

Fig. 14. The suspension and steering system gives the driver control over the car by supplying constant feedback on the car's reactions.

The steering system consists of the steering rack mechanisms, linkages, and a belt-driven hydraulic pump. Power-assisted steering uses hydraulic fluid under pressure to do some of the work normally done by the driver turning the steering wheel.

Body

The body is the basic building block. All of the Saab models covered in this manual feature unitized body construction, meaning that they do not have a separate frame. A complex body shell is the main structural platform to which all the other systems are attached. Subassemblies attach engine, drivetrain, suspension, and steering systems to the basic body structure.

The doors, the instrument panel, the seats, and other interior trim pieces are also added to the body shell. Other parts of the body shell function as mounting points for the other major and minor subsystems.

How To Use This Manual

The manual is divided into 9 sections:

0 TECHNICAL DATA
1 LUBRICATION AND MAINTENANCE
2 ENGINE
3 ELECTRICAL SYSTEM
4 TRANSMISSION
5 BRAKES
6 STEERING AND WHEEL ALIGNMENT
7 SUSPENSION
8 BODY AND STEERING

0 TECHNICAL DATA lists all of the specifications used throughout the manual and is intended to be used as a quick reference guide for the more experienced technician familiar with Saab 900 cars. **1 LUBRICATION AND MAINTENANCE** covers the maintenance schedules and service procedures needed to do all of the Saab recommended scheduled maintenance work.

The remaining seven sections (2 through 8) are repair oriented and are divided into multiple repair groups. For clarity and ease of use, each major section begins with a **General** repair group, e.g. **200 Engine—General.** These "00" (double zero) groups are mostly descriptive in nature, covering topics such as theory of operation and troubleshooting. The remainder of the repair groups contain the more involved and more detailed system repair information.

A master listing of the 9 sections and the corresponding 69 individual repair groups can be found at the beginning of each section. Thumb tabs on the first page of each repair group page help locate the groups quickly.

Each repair group has its own Table of Contents listing the major subject headings within the group, and the pages on which they begin. Page numbers throughout the manual are organized according to the repair group system. For example, you can expect to find information on the turbocharger (Repair Group 291) beginning on page 291-1. A comprehensive index is found at the back of the manual.

Warnings, Cautions and Notes

Throughout this manual are many passages with the headings **WARNING, CAUTION,** or **NOTE.** These very important headings have different meanings.

WARNING

A warning is the most serious of the three. It warns of unsafe practices that are very likely to cause injury, either by direct threat to the person(s) doing the work or by increased risk of accident or mechanical failure while driving.

CAUTION

A caution calls attention to important precautions to be observed during the repair work that will help prevent accidentally damaging the car or its parts.

NOTE

A note contains helpful information, tips that will help in doing a better job and completing it more easily.

Please read every **WARNING, CAUTION**, and **NOTE** at the front of the manual and as they appear in repair procedures. They are very important. Read them before you begin any maintenance or repair job.

Some **WARNING**s and **CAUTION**s are repeated wherever they apply. Read them all. Do not skip any. These messages are important, even to the owner who never intends to work on the car.

GETTING STARTED

Most of the necessary maintenance and minor repair that a Saab will need can be done with ordinary tools, even by owners with little or no experience in car repair. Below is some important information on how to work safely, a discussion of what tools will be needed and how to use them, and a series of mechanic's tips on methods and workmanship.

Safety

Although an automobile presents many hazards, common sense and good equipment can ensure safety. Accidents happen because of carelessness. Pay attention and stick to these few important safety rules.

WARNING

- Never run the engine in the work area unless it is well-ventilated. The exhaust should be vented to the outside. Carbon Monoxide (CO) in the exhaust kills.

- Remove all neckties, scarfs, loose clothing, or jewelry when working near running engines or power tools. Tuck in shirts. Tie long hair and secure it under a cap. Severe injury can result from these things being caught in rotating parts.

WARNING

- Remove rings, watches, and bracelets. Aside from the dangers of moving parts, metallic jewelry conducts electricity and may cause shorts, sparks, burns, or damage to the electrical system when accidentally contacting the battery or other electrical terminals.

- Disconnect the battery negative (–) cable whenever working on or near the fuel system or anything that is electrically powered. Accidental electrical contact may damage the electrical system or cause fire.

- Never work under a lifted car unless it is solidly supported on jack stands that are intended for that purpose. Do not support a car on cinder blocks, bricks, or other objects that may shift or crumble under continuous load. Never work under a car that is supported only by the lifting jack.

- The fuel system is designed to retain pressure even when the ignition is off. When working with the fuel system, loosen the fuel lines very slowly to allow the residual pressure to dissipate gradually. Avoid spraying fuel.

- Fuel is highly flammable. When working around fuel, do not smoke or work near heaters or other fire hazards. Keep an approved fire extinguisher handy.

- Illuminate the work area adequately and safely. Use a portable safety light for working inside or under the car. A fluorescent type light is best because it gives off less heat. If using a light with a normal incandescent bulb, use rough service bulbs to avoid breakage. The hot filament of an accidentally broken bulb can ignite spilled fuel or oil.

- Keep sparks, lighted matches, and open flame away from the top of the battery. Hydrogen gas emitted by the battery is highly flammable. Any nearby source of ignition may cause the battery to explode.

- Never lay tools or parts in the engine compartment or on top of the battery. They may fall into confined spaces and be difficult to retrieve, become caught in belts or other rotating parts when the engine is started, or cause electrical shorts and damage to the electrical system.

- Some of the cars covered by this manual are equipped with a Supplemental Restraint System (SRS) that automatically deploys an airbag. The airbag unit uses an explosive device to electrically ignite a powerful gas. On cars so equipped, any work involving the steering wheel should only be performed by an authorized Saab dealer. Performing repairs without disarming the SRS may cause serious personal injury.

Lifting the Car

For those repairs that require raising the car, the proper jacking points should be used to raise the car safely and avoid damage. There are six jacking points from which the car can be safely raised. The jack supplied with the car by Saab can only be used at the four side points—just behind the front wheel or just in front of the rear wheel. In addition to the four side points, there are front and rear center jacking points that can be used to lift one end. See Fig. 15.

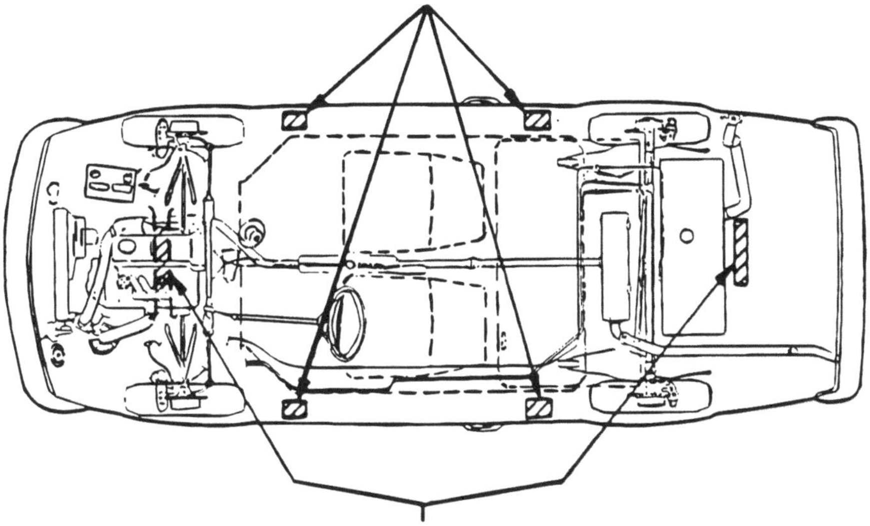

Fig. 15. Saab jacking points.

CAUTION
When raising the car using a floor jack or a hydraulic lift, carefully position the jack pad so that it does not damage the body. A suitable liner (wood, rubber, etc.) should be placed between the jack and the car so that the underbody will not be damaged.

To raise car safely

1. Park the car on a flat, level surface.

2. Place the jack in position. See Fig. 16. Make sure the jack is resting on flat, solid ground. Use a board or other support to provide a firm surface for the jack, if necessary.

Fig. 16. Saab-supplied jack correctly installed at front left jacking point.

3. Raise the car slowly.

WARNING
Watch the jack closely. Make it stays stable and does not shift or tilt. As the car is raised, the car will want to roll slightly and the jack will want to shift.

---

4. Once the car is raised, block the wheel that is opposite and farthest from the jack to prevent the car from unexpectedly rolling.

WARNING
Do not rely on the transmission or the emergency brake to keep the car from rolling. While they will help, they are not a substitute for positively blocking the opposite wheel.

Never work under a car that is supported only by a jack. Use jack stands that are properly designed to support the car. See **Tools**.

To work safely under a car

1. Disconnect the battery negative (–) cable so that no one else can start the car. Let others know what you will be doing.

2. Raise the car slowly as described above.

3. Use at least two jack stands to support the car. A jack is a temporary lifting device and should not be used alone to support the car while you are under it. Use positively locking jack stands that are designed for the purpose of supporting a car. For more information on jack stands, see **Tools** below.

WARNING

Do not use wood, concrete blocks, or bricks to support a car. Wood may split. Blocks or bricks, while strong, are not designed for that kind of load, and may break or collapse.

---

4. Place jack stands on a firm, solid surface, just like the jack. If necessary, use a flat board or similar solid object to provide a firm footing.

5. After placing the jack stands, lower the car slowly until its weight is fully supported by the jack stands. Watch to make sure that the jack stands do not tip or lean as the car settles on them, and that they are placed solidly and will not move.

6. Observe all jacking precautions again when raising the car to remove the jack stands.

GENERAL ADVICE FOR THE BEGINNER

The tips in the paragraphs that follow are general advice to help any do-it-yourself Saab owner perform repairs and maintenance tasks more easily and more professionally.

Planning Ahead

Most of the repairs and maintenance tasks described in this manual can be successfully completed by anyone with basic tools and abilities. Some cannot. To prevent getting in too deep, know what the whole job requires before starting. Read the procedure thoroughly, from beginning to end, in order to know just what to expect and what parts will have to be replaced.

Cleanliness

Keeping things organized, neat, and clean is essential to doing a good job, and a more satisfying way to work. When working under the hood, fender covers will protect the finish from scratches and other damage. Make sure the car is relatively clean so that dirt under the cover does not scratch.

Avoid getting tools or clothing near the battery. Battery electrolyte is a corrosive acid. Be careful with brake fluid, as it can cause permanent damage to the car's paint. Finally, keep rubber parts such as hoses and belts free from oil or gasoline, as they will cause the material to soften and fail prematurely.

Non-reusable Fasteners

Many fasteners used on the cars covered by this manual must be replaced with new ones when they are removed. These include but are not limited to: bolts, nuts (self-locking, nylock etc.), cotter pins, studs, brake fittings, roll pins, pins, clips and washers. Genuine Saab parts should only be used for this purpose.

Some bolts, for example, are designed to stretch during assembly and are permanently altered rendering them unusable again. Always replace fasteners where instructed to do so. See an authorized Saab dealer for applications and ordering information.

Tightening Fasteners

When tightening the bolts or nuts that attach a component, it is always good practice to tighten the bolts gradually and evenly to avoid misalignment or over stressing any one portion of the component. For components sealed with gaskets, this method helps to ensure that the gasket will seal properly and completely.

Where there are several fasteners, tighten them in a sequence alternating between opposite sides of the component. Fig. 17 shows such a sequence for tightening six bolts attaching a typical component. Repeat the sequence until all the bolts are evenly tightened to the proper specification.

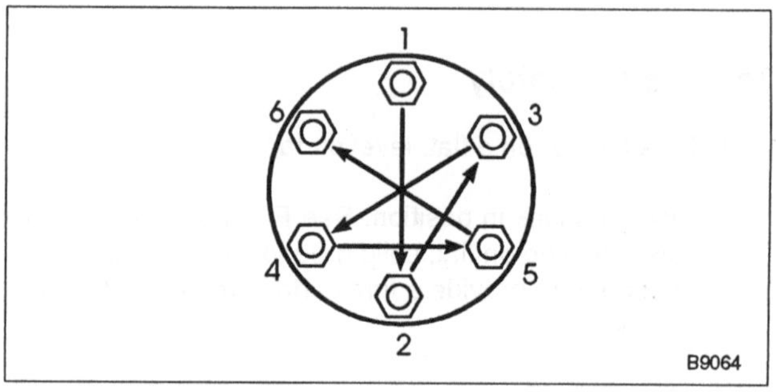

Fig. 17. Sequence for alternately tightening multiple fasteners.

For some repairs a specific tightening sequence is necessary, or a particular order of assembly is required. Such special conditions are noted in the text, and the necessary sequence is described or illustrated.

Bolt Torque

Tightening fasteners to a specified torque value using a torque wrench is a good way to ensure that bolts are correctly tightened. If a torque wrench is not used there is a danger of going too far and damaging the fastener or the threads in the mating part.

Too little torque on a fastener can also cause problems. Vibration of assembled parts can subject fasteners to stress alternating in opposite directions that will eventually cause them to loosen. To counter this loosening, fasteners are tightened more and actually stretched a small amount.

When special tightening torques are required, they are listed in the text where the fastener is being installed. If there is no torque listed for a specific fastener, use **Table a** as a general guide. The sizes listed are for the bolt thread diameter, not the size of the wrench. **Table b** lists the most common wrench sizes for the bolts used on the cars covered by this manual.

Table a. General Tightening Torques (unless noted otherwise in text)

Bolt diameter	Nm	ft-lb
M5	5	3.5 (44 in-lb)
M6	10	7.5 (89 in-lb)
M8	20	15
M10	40	30

Table b. Bolt Diameter and Wrench Size

Bolt diameter	Most Common Wrench Size
M5	8 mm
M6	10 mm
M8	12mm or 13mm
M10	17mm
M12	19mm
M14	22mm

Gaskets

The smoothest metal mating surfaces still have imperfections that can allow leakage. To prevent leakage at critical joints, gaskets of soft, form-fitting material are used to fill in the imperfections.

To be most effective, gaskets are designed to crush and become thinner as the mating parts are bolted together. Once a gasket has been used and crushed, it is no longer capable of making as good a seal as when new, and is much more likely to leak. For this reason, gaskets should not be reused. Always plan to use new gaskets for any reassembly. Some gaskets — such as headgaskets are directional. Make sure that these are being installed correctly. This same logic applies to any part used for sealing, including rubber O-rings and copper sealing washers.

Seals

In places where a shaft must pass through a housing, flexible lip seals are used to keep the lubricating oil or grease from leaking out past the rotating shaft.

Seals are designed to be installed in the housing only once and should never be reused. As long as they are not removed from the housing and not leaking, they need not be replaced. Seals, however, do age and deteriorate, and there is no easier time to replace them than when the car is already apart for some other repair.

When doing repairs that require removing a seal, be very careful not to scratch or otherwise damage the metal surfaces. Even minor damage to sealing surfaces can cause seal damage and leakage.

The key to seal installation is to get the seal in straight without damaging it. Use an object that is the same diameter as the seal housing to gently and evenly drive it into place. If a proper size seal driver is not available, a socket of the right size will do.

Coat the entire seal with oil to help it go in more easily. Seals are directional. Make sure that it is being installed with the lip facing the correct way. Normally the lip faces the inside. Notice the installation direction of the old seal before removing it.

Cleaning

Any repair job will be less troublesome if the parts are clean. For cleaning old parts, there are any number of solvents and parts cleaners available commercially.

For cleaning parts prior to assembly, commercially available aerosol cans of carburetor cleaner or brake cleaner are handy to use, and the cleaner will evaporate completely, leaving no residue.

WARNING
Virtually all solvents used for cleaning parts are highly flammable, especially in aerosol form. Use with extreme care. Do not smoke. Do not use these products near any source of sparks or flame.

Let any solvent or cleaning product dry completely. Low-pressure, dry compressed air is helpful if available. Also, use only lint-free rags for cleaning and drying.

Electrical Testing

A great many electrical problems can be understood and solved with only a little fundamental knowledge of how electrical circuits function.

Electric current only flows in a complete circuit. To operate, every electrical device in the car requires a complete circuit including a voltage source and a path to ground. The positive (+) side of the battery is the original voltage source, and ground is any return path to the negative (–) side of the battery, whether through the wiring harness or the car body. Except for portions of the charging system, all electrical current in the car is direct current (DC) and flows from positive (+) to negative (–).

Switches are used to turn components on or off by completing or interrupting the circuit. A switch is "open" when the circuit is interrupted, and "closed" when the circuit is completed. Fig. 18 shows a complete circuit schematically. See **3 ELECTRICAL SYSTEM** for electrical troubleshooting.

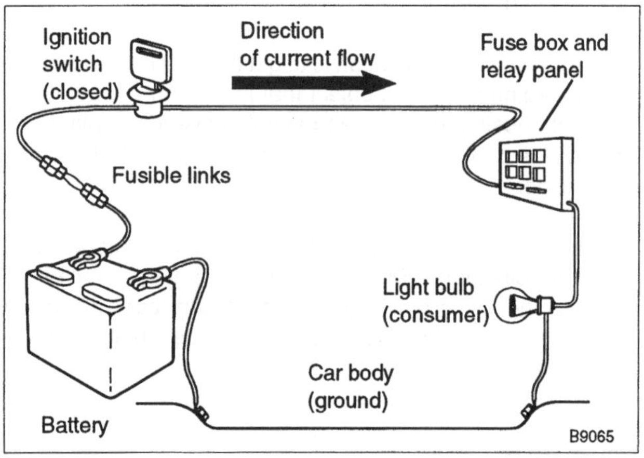

Fig. 18. Schematic representation of simple circuit for light bulb. Switch is shown closed, making circuit complete.

Wire Repairs

Repairs to a wiring harness to reconnect broken wires or correct shorts to ground deserve special care to make the repair permanent. The wire ends must be clean. If frayed or otherwise damaged, cut off the end. If the wire is too short, splice in a new piece of wire of the same size and make two connections.

Use connectors that are designed for the purpose. Crimped-on or soldered-on connectors are best. Crimp connectors and special crimping pliers are widely available. If soldering, use a needlenose pliers to hold the wire near the solder joint and create a "heat dam". This keeps the heat and the solder from traveling up the wire. Always use a solder made specifically for electrical work.

NOTE

Twisting wires together is really only a temporary repair, since corrosion and vibration will eventually spoil the connection.

Insulate the finished connection. Electronics stores can supply heat-shrinkable insulating tubing that can be placed onto the wire before connecting, slid over the finished joint, and shrunk to a tight fit with a heat gun or hair dryer. The next best alternative is electrical tape. Make sure the wire is clean and free of solder flux or other contamination. Wrap the joint tightly and completely to seal out moisture.

BUYING PARTS

Many of the maintenance and repair tasks in this manual call for the installation of new parts, or the use of new gaskets and other materials when reinstalling parts. Most often, the parts that will be needed should be on hand before beginning the job. Read the introductory text and the complete procedure to determine which parts will be needed.

NOTE

For some bigger jobs, partial disassembly and inspection are required to determine a complete parts list. Read the procedure carefully and, if necessary, make other arrangements to get the necessary parts while your car is disassembled.

Information You Need To Know

Fig. 19 shows the locations of the important information that may be necessary to have on hand when buying parts or having work done on the car.

Genuine Saab Parts

Genuine Saab replacement parts from an authorized Saab dealer are designed and manufactured to the same high standards as the original parts. They will be the correct material, manufactured to the same specifications, and guaranteed to fit and work as intended by the engineers who designed the car. Some genuine Saab parts have a limited warranty.

Many independent repair shops make a point of using genuine Saab parts, even though they may at times be more expensive. They know the value of doing the job right with the right parts. Parts from other sources can be as good, particularly if manufactured by one of Saabs original equipment suppliers, but it is often difficult to know.

Saab is constantly updating and improving their cars, often making improvements during a given model year. Saab may recommend a newer, improved part as a replacement, and your authorized dealer's parts department will know about it and provide it. The Saab parts organization is best equipped to deal with any Saab parts needs.

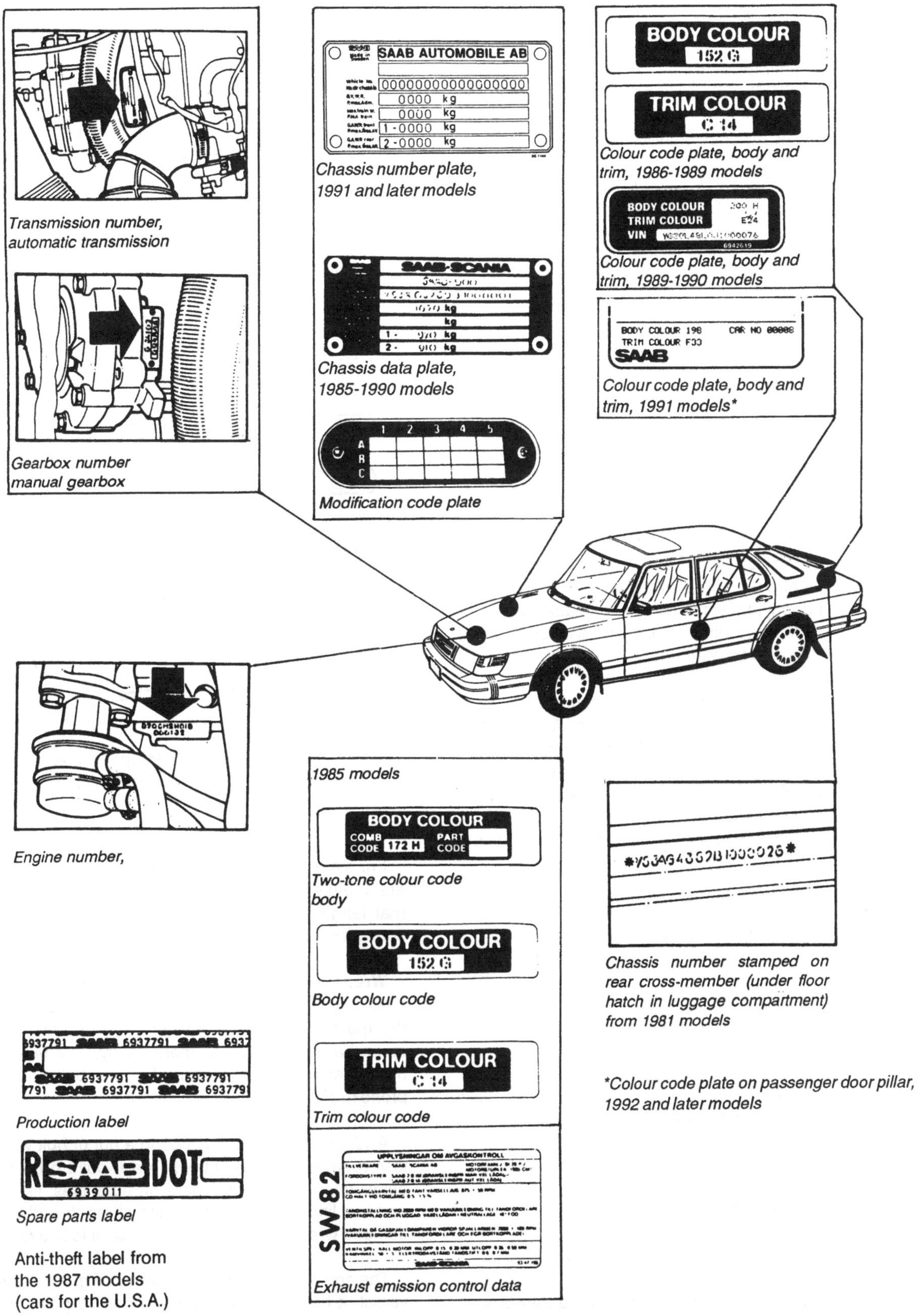

Fig. 19. Locations of important information that may be required when buying parts.

Non-returnable Parts

Some parts cannot be returned for credit, even if they are the wrong parts for the car. The best example is electrical parts, which are almost universally considered non-returnable because they are so easily damaged internally.

Buy electrical parts carefully, and be as sure as possible that a replacement is needed, especially for expensive parts such as control units. It may be wise to let an authorized Saab dealer or other qualified shop confirm your diagnosis before replacing an expensive part that cannot be returned.

TOOLS

Most maintenance can be accomplished with a small selection of the right tools. Tools range in quality from inexpensive junk, which may break at first use, to very expensive and well-made tools which, to the professional, are worth every bit of their high cost. The best tools for most do-it-yourself Saab owners lie somewhere in between.

Cheap tools are not a bargain. They often do not hold up to even casual use, and they present a greater risk of personal injury. If they fit poorly, they can actually damage the fasteners they are intended to remove, making it that much harder to use a good tool the next time around.

Many reputable tool manufacturers offer good quality, moderately priced tools with a lifetime guarantee. A broken tool can be exchanged for a new one, for the life of the tool. These are your best buy. They cost a little more, but they are good quality tools that will do what is expected of them. Sears' Craftsman® line is one such source of good quality, reasonably priced, and guaranteed tools. Other sources of special tools are:

Schley Products Inc.
5350 E. Hunter Ave., Anaheim Hills, CA 92807
(714) 693-7666

Baum Tools Unltd. Inc.
P.O. Box 87, Longboat Key, FL 34228
(800) 848-6657

Assenmacher Specialty Tools
6440 Odell Place, Boulder, CO 80301
(303) 530-2424

Basic Tool Requirements

NOTE
Saabs are delivered with a tool kit mounted to the underside of the trunk lid. The kit contains a basic selection of tools that may fulfill some of the requirements listed in this section.

The basic hand tools described below can be used to accomplish most of the simple maintenance and repair tasks.

Screwdrivers. Three types, the common flat-blade type, the Phillips type will handle almost all screws used on Saabs. Two or three different sizes of each type will be best, since a screwdriver of the wrong size will damage the screw head. On late models Saabs, Torx® screw are quite common. A T25 size Torx driver will handle most of these screws.

A complete set of screwdrivers can often be purchased for about the same money as the four or six individual ones that are really necessary. See Fig. 20.

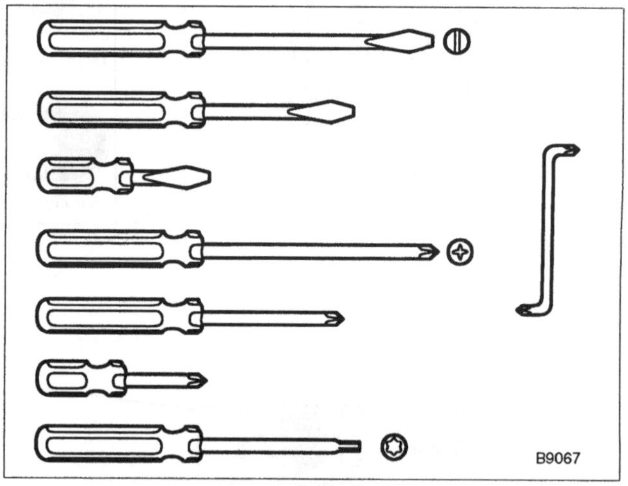

Fig. 20. Common flat-blade (top), Phillips (middle) and Torx® (bottom) screwdrivers. Offset screwdriver (right) is used for screws with limited access.

For a more complete tool box, include "stubby" screwdrivers or offset screwdrivers for use in tight spots where a normal length screwdriver will not easily fit and Torx®head screwdrivers.

Wrenches. Wrenches come in different styles for different uses. Fig. 21 shows several. The basic open-end wrench is the most widely used, but grips on only two sides. It can spread apart and slip off more easily. The box-end wrench has better grip, on all six sides of a nut or bolt, and is much less prone to slip.

A 12-point box-end can loosen a nut or bolt where there is less room for movement, while a 6-point box-end provides better grip. For hex fasteners on fluid lines, like brake lines and fuel lines, a flare-nut wrench offers the advantages of a box-end wrench with a slot that allows it to fit over the line.

The combination wrench, shown in Fig. 22, is the most universal. It has one open-end and one 12-point box-end. For Saabs, 10mm and 13mm wrenches are the most common sizes needed. On most models, a 19mm wrench is needed to loosen and tighten the engine oil drain plug. A complete set should also include 6mm, 7mm, 8mm, 9mm, 11mm, 12mm, 14mm, and 15mm.

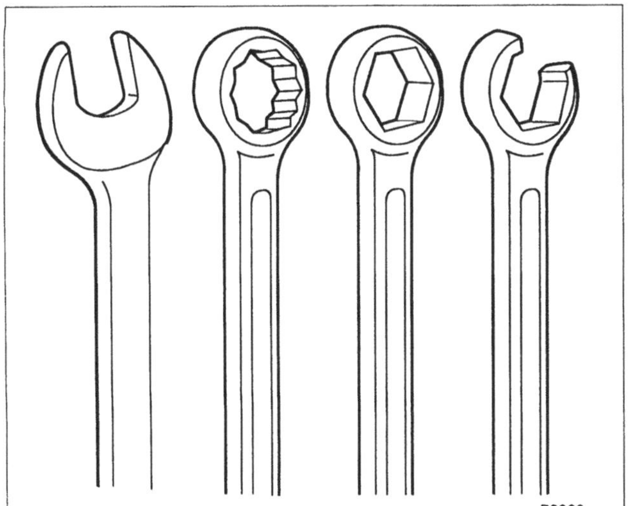

Fig. 21. Types of wrench heads. From left, open-end, 12-point box-end, 6-point box-end, flare nut.

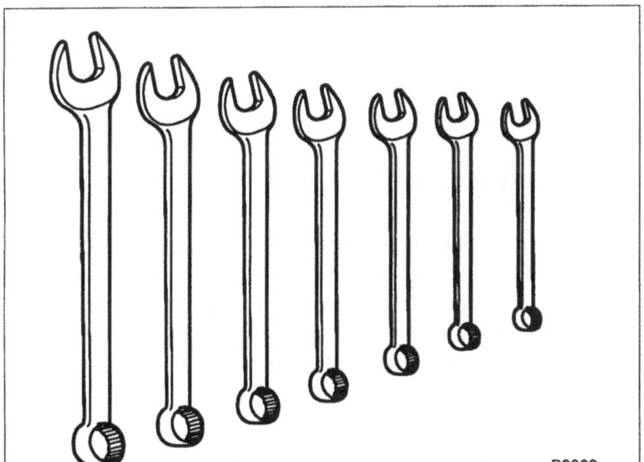

Fig. 22. Combination wrenches with one open-end and one 12-point box-end.

Sockets. Sockets perform the same job as box-end wrenches, but offer greater flexibility. They are normally used with a ratchet handle for speed and convenience, and can be combined with extensions to reach fasteners more easily.

Standard sockets come in 6-point and 12-point styles. For use with a ratchet the 6-point offers a better grip on tight nuts and bolts. As with wrenches, 6mm to 15mm, 17mm, and 19mm are the most needed sizes. See Fig. 23.

Sockets come with different size connections to drive handles or extensions, called the drive size. The most common drive sizes are 1/4 in., 3/8 in., and 1/2 in.

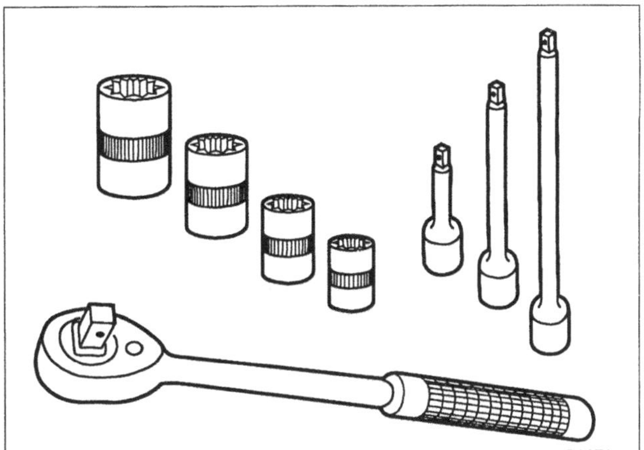

Fig. 23. Sockets, extensions, and a ratchet handle.

As a start, 6-point sockets with a 3/8 in. square drive, two or three 3/8 in. extensions of different lengths, and a 3/8 in. drive ratchet handle will be suitable for most jobs.

For a more complete tool box, add deep sockets and a greater variety of handles and extensions. A universal joint extension can allow access from an angle where a straight extension will not quite fit.

Spark Plug Socket. A special socket for spark plugs is the correct size, is deep enough to accommodate a spark plug's length, and includes a rubber insert to both protect the spark plug from damage and grip it for easier removal. The spark plugs used in Saab engines require a 5/8 in. socket. See Fig. 24.

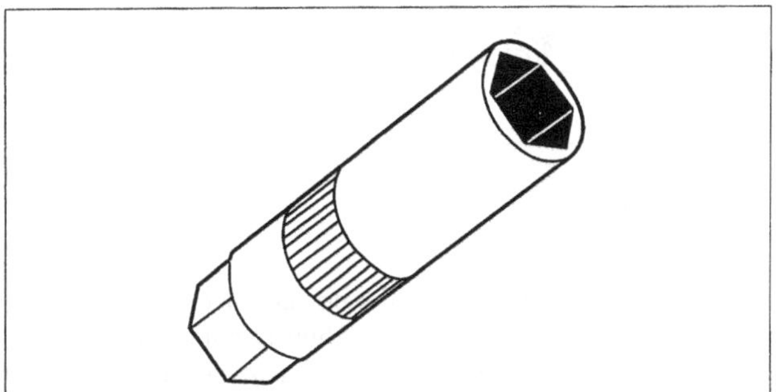

Fig. 24. Spark plug socket (5/8 in.).

Pliers. A few of the many types of pliers are shown in Fig. 25. Most are used for holding irregular objects, bending, or crimping. Some have special applications.

A needlenose plier is used for gripping small and poorly accessible objects, and is useful for wiring and other electrical work. A locking plier such as the well-known Vise-Grip® is useful because of its tight grip.

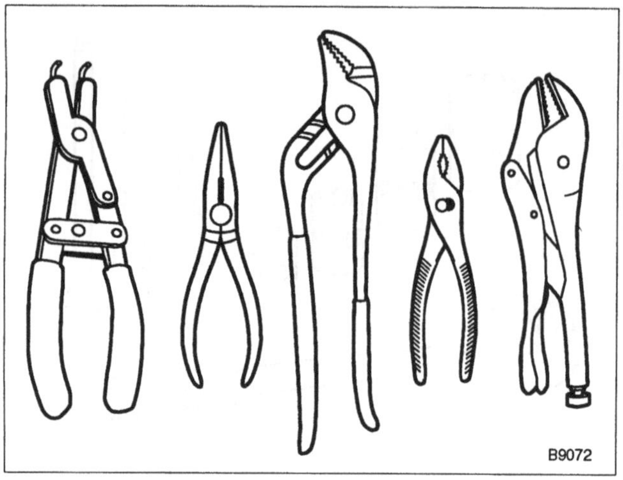

Fig. 25. Pliers. From left, snap-ring, needlenose, Channel-lock®, common, locking.

Snap-ring and circlip pliers with special tipped jaws are used to remove and install snap-rings or circlips. A Channel-lock® or water pump plier has adjustable jaws that can be quickly changed to match the size of the object being held to give greater leverage.

Adjustable wrench can be a useful addition to a small tool kit. See Fig. 26. It can substitute in a pinch, if two wrenches of the same size are needed to remove a nut and bolt. Use extra care with adjustable wrenches, as they especially tend to loosen, slip, and damage fasteners.

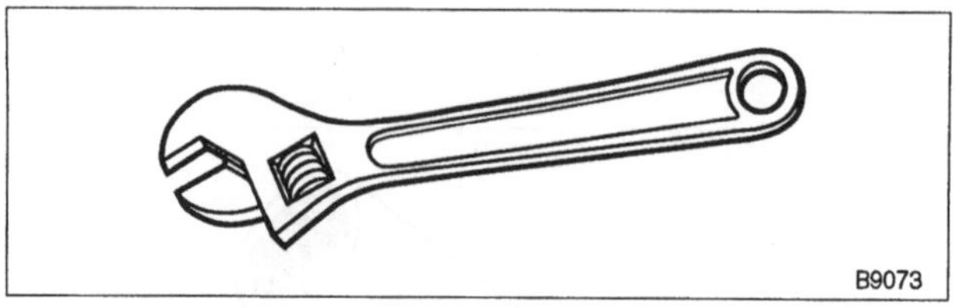

Fig. 26. Adjustable wrench.

Compared to a wrench of the correct size, an adjustable wrench is always second best. They should only be used when the correct size wrench is not available. Choose one of average size range, about 6 to 8 inches in length.

Jack Stands

Strong jack stands are extremely important for any work that is done under the car. Jacks are designed only for short term use and are not solid enough to support the car for a long period. A jack should never be used alone to support the car while working underneath.

Use only jack stands that are designed for the purpose. Blocks of wood, concrete, bricks, etc. are not safe or suitable substitutes.

Jack stands are available in several styles. A typical jack stand is shown in Fig. 27. The best ones are made of heavy material for strength, have a wide base for stability, and are equipped to positively lock in their raised positions. Get the best ones available.

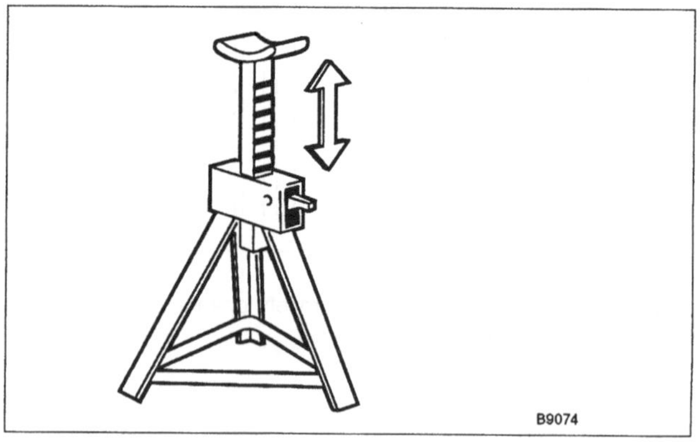

Fig. 27. Jack stand for safely supporting car to work underneath.

Oil Change Equipment

Changing engine oil requires a box-end wrench or socket to loosen and tighten the drain plug 19mm (13mm on 1991 and later cars), a drain pan (at least 7 qt. capacity), and an oil filter wrench. These items are shown in Fig. 28. A wide, low drain pan will fit more easily under the car. Use a funnel to pour the new oil into the engine.

An oil filter wrench is needed for some models to remove the oil filter. Be sure to get a filter wrench that will grip the Saab oil filter tightly.

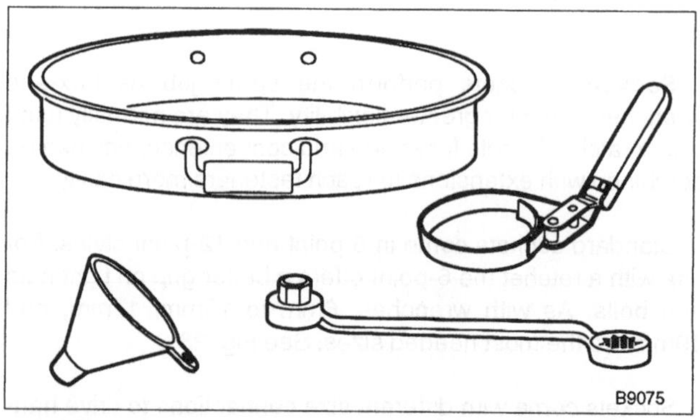

Fig. 28. Oil change equipment includes drain plug wrench (19mm or 13mm), 7 qt. drain pan, oil filter wrench, and funnel.

Torque Wrench

A torque wrench is used to precisely tighten threaded fasteners to a predetermined value. Nearly all of the repair procedures in this manual include Saab-specified torque values in Newton-meters (Nm) and the equivalent values in foot-pounds (ft-lb).

Several types of torque wrenches are widely available. They all do the same job, but offer different convenience features at different prices. Two typical torque wrenches are shown in Fig. 29. The most convenient ones have a built-in ratchet, and can be preset to indicate when a specific torque value has been reached. Follow the wrench manufacturer's directions for use to achieve the greatest accuracy.

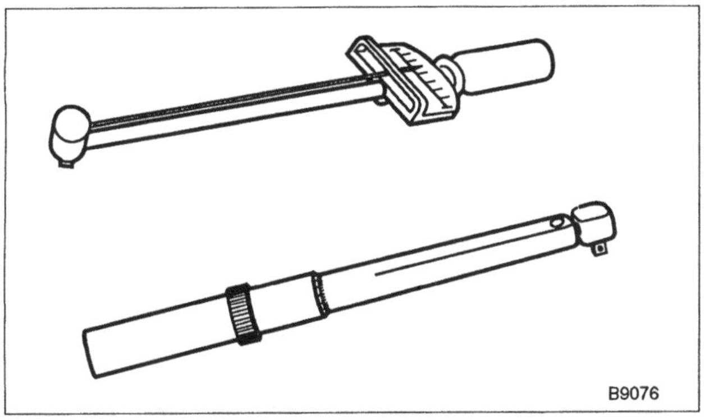

Fig. 29. Torque wrenches. Inexpensive beam-type (top) is adequate but must be read visually. Ratchet-type (bottom) can be preset to indicate when torque value has been reached.

A torque wrench with a range up to about 250 Nm (185 ft-lb) has adequate capacity for most of the repairs covered in this manual. For recommended torque values of 10 Nm or below, the English system equivalent is given in inch-pounds (in-lb). These small values may be most easily reached using a torque wrench calibrated in inch-pounds. To convert foot-pounds to inch-pounds, multiply by 12. To convert inch-pounds to foot-pounds, divide by 12.

Feeler Gauges

Feeler gauges are thin metal strips of precise thickness, used to measure small clearances. They are normally available as a set, covering a range of sizes. For Saabs, metric feeler gauges (in millimeters) are the best choice. Fig. 30 shows a set of feeler gauges.

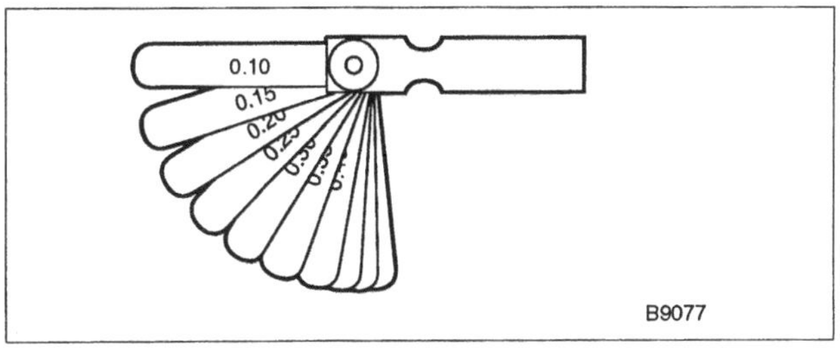

Fig. 30. Feeler gauge set, used for precise measurement of clearances between parts.

Volt-Ohm Meter (VOM) or Multimeter

Many of the electrical tests in this manual call for the measurement of resistance (ohms) or voltage values. For safe and accurate tests of ignition, fuel injection, and emission control systems, the multimeter, shown in Fig. 31, should be digital, with high (at least 10,000 ohms) input impedance. Some meters have automotive functions such as dwell and pulse width that are useful for troubleshooting ignition and fuel injection problems.

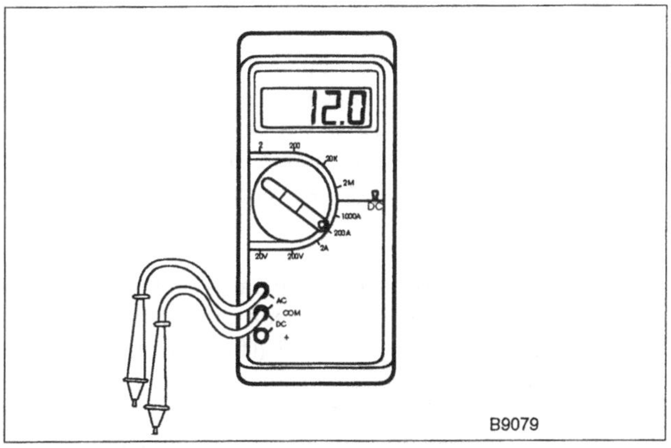

Fig. 31. Multimeter with test probes.

CAUTION

Ignition, fuel injection, emission controls and other electronic systems may be damaged by the high current draw of a test light with a normal incandescent bulb. As a general rule, use a high impedance digital multimeter or an LED test light for all electrical testing.

Saab Special Tools

Some of the more challenging repairs covered in this manual call for the use of Saab special tools. This, however, does not automatically mean that the job is too complicated or out of reach of the novice.

Many of the Saab special tools mentioned in this manual are inexpensive and are simply the best thing to use to do the job correctly. In these cases, the tool is identified with a Saab part number. See your authorized Saab dealer parts department for information on how to order special tools.

There are some jobs for which expensive special tools are essential, and not a cost-effective purchase for one-time repair by the do-it-yourself owner. This manual includes such repairs for the benefit of those with the necessary experience and access to tools. For the do-it-yourselfer, the need for special tools is noted in the text, and whether or not Saab dealer service is recommended.

EMERGENCIES

Changing a Tire

If a tire goes flat while driving, pull well off the road. Changing a tire on a busy street or highway is very dangerous. If necessary, drive a short distance on the flat tire to get to a safe place. It is much better to ruin a tire or rim than to risk being hit.

Stop the car on as flat a surface as possible, in a place where you can be easily seen by other drivers. Avoid stopping just over the crest of a hill. Turn on the emergency flashers, and set out flares or emergency markers well behind the car. Passengers should get out of the car and stand well away from the road. Take the jack, tools, and spare wheel from the trunk. Chock the wheel diagonally opposite to the one being changed.

Loosen the wheel bolts while the car is on the ground, but leave them a little snug. Place the jack under in the lifting point nearest the wheel being changed (lifting points are shown above in Fig. 15). Use a board to provide a firm footing for the jack if the ground is soft. Raise the car only far enough so that the wheel is off the ground, and then remove the wheel bolts and the wheel. Remove the lug bolts or nuts and remove the tire.

Install the spare wheel. Install the wheel bolts or nuts and tighten them by hand, then lower the car. With all wheels on the ground, fully tighten the bolts in a cross-wise pattern. Torque the wheel bolts when installing the wheel.

Tightening torques

- wheel lug bolts . . . . . . 105 to 125 Nm (77 to 92 ft-lb)
- wheel lug nuts . . . . 90 to 110 Nm (66 to 81 ft-lb)

If torquing the wheel fasteners is not possible, tighten them as much as possible, then loosen and retorque the bolts to the proper specification at the earliest opportunity. Check the inflation pressure of the spare tire.

Car Will Not Start

If the engine turns over slowly or not at all, especially on cold mornings, the battery may not be sufficiently charged. Jump-starting the battery from another car may help. Jump-starting is described below.

NOTE
Be sure to read the cautions under ***Jump Starting*** prior to jump starting a low battery. Failure to follow the cautions may result in damage to the electronic components in the car.

Push starting (or tow starting) a car with an insufficiently charged battery is another option. To push start the car, turn on the ignition, put the car in second or third gear and push in the clutch pedal. Push or tow the car. When the car is moving at a fair speed (10 to 15 mph), release the clutch pedal. After the engine has started, push the clutch pedal back in and allow the engine to idle. Rev the engine for a few minutes to help restore the charge to the battery.

WARNING
Use extreme caution when push starting a car. Be aware of other traffic. Use the emergency flashers.

NOTE
On cars with automatic transmissions, the design of the transmission makes it impossible to start the engine by pushing the car.

If the engine is turning over at normal speed with the starter motor, the battery and starter are fine. Check to make sure that there is fuel in the tank. Don't rely on the fuel gauge, it may be faulty. Instead, remove the gas filler cap and rock the car. If there is gas in the tank, you should hear a sloshing sound at the filler neck. If so, turn the ignition on and listen for the sound of the fuel pump. It should run for a few seconds, then stop. If it doesn't, fuel may not be reaching the engine.

The engine also may have difficulty starting because it has too much fuel, because the fuel system is vapor-locked on a hot day, or because the ignition system is wet on a very damp day. There will probably be a strong smell of gas if the engine has too much fuel (referred to as "flooded").

Try holding the accelerator pedal fully to the floor while starting. This electrically turns off the fuel injectors. If the engine stills does not start, wait for a few minutes and then try starting the engine again. If you suspect vapor-lock, raise the hood, let the engine cool, and then try to start the engine.

On damp days, check the distributor cap and spark plug wires for condensation. If they are wet, remove and replace the wires one at a time and dry them off with a clean dry cloth, then remove the distributor cap and wipe it dry inside and out.

Jump-Starting

Cars with partially discharged or completely dead batteries can be jump-started using the good battery from another car.

When jump-starting the engine, always heed the following warnings and cautions.

WARNING

• Battery acid (electrolyte) can cause severe burns, and will damage the car and clothing. If electrolyte is spilled, wash the surface with large quantities of water. If it gets into eyes, flush them with water for several minutes and call a doctor.

• Batteries produce explosive and noxious gasses. Keep sparks and flames away. Do not smoke near batteries.

• Do not jump-start the engine if you suspect that the battery is frozen. Trapped gas may explode. Allow the battery to thaw first.

• Do not quick-charge the battery (for boost starting) for longer than one minute, and do not exceed 15 volts at the battery with the boosting cables attached. Wait at least one minute before boosting the battery a second time.

To jump-start the engine, place the cars close together, but do not allow them to touch. Turn off the engine of the car with the good battery. Connect the jumper cables as shown in Fig. 32.

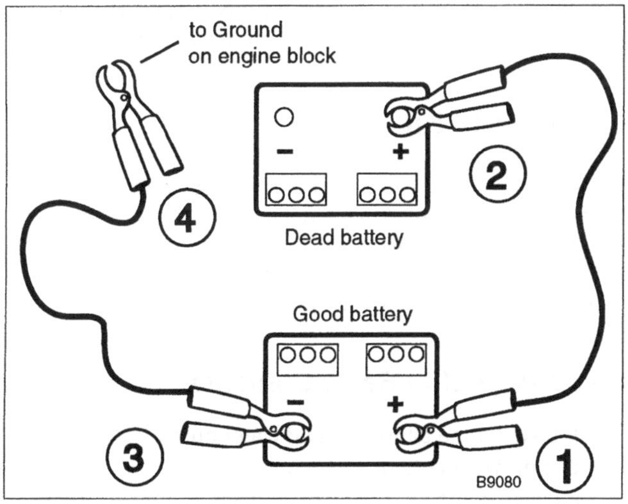

Fig. 32. Battery jumper cables connections. Numbers indicate correct sequence for cable attachment.

The battery is mounted in the front right corner of the engine compartment. Connect the end of one cable to the positive post of the good battery, and the other end of the same cable to the positive post of the dead battery. The positive post is marked with a plus (+) sign.

Connect one end of the other cable to the negative (–) post of the good battery, and connect the other end of the same cable to the engine block of the car with the dead battery. Make the connection as far away from the battery as possible, as there may be sparks.

Have a helper start the car with the good battery and race the engine slightly, then start the car with the dead battery. Leave the cars running and disconnect the cables in the reverse order in which they were installed. The car with the dead battery will need to run for about a 1/2 hour to recharge the battery.

Overheating

If the coolant temperature is too high, find a safe place to stop and turn the engine off. Open the hood and allow the engine to cool until the temperature gauge needle is at the lower third of the scale. Continuing to drive an overheated car can cause expensive engine damage.

WARNING

Do not remove the coolant reservoir or radiator cap with the engine hot. Undoing either could spray hot coolant, and cause burns, or damage the engine.

NOTE

If the engine cannot be safely turned off, make sure the air conditioner is off and turn the heater to high. This will help cool the engine until a safe stopping place can be reached.

Overheating may be caused by low coolant level or a damaged V-belt. Visually check the coolant level and V-belts as described in **110 Maintenance Program**. If coolant is lost, check the filler cap, hoses, clamps and radiator for signs of leakage. Check for leaks at the water pump on the rear of the engine.

If no leaks are found, add coolant after the engine has cooled. The car can be driven, but have the cooling system thoroughly checked as soon as possible. If replacement coolant is not available, then plain water can be used, but the coolant should later be drained and refilled with the proper mixture of anti-freeze and water.

CAUTION

Do not add cold water or coolant to a hot engine. Severe engine damage could result from the sudden temperature change.

If steam is coming from the engine compartment then there is most likely a burst coolant hose or a large leak in the cooling system. To find the leak, look for signs of coolant leakage on hoses, at hose connections, or on the radiator. Let the engine cool thoroughly, then add coolant or water to fill the system and start the engine. If a great deal of water or coolant flows out of the hole, then the car should not be driven until repairs are made. If there is a slight seepage, then it may be possible to drive a short distance, adding coolant as needed.

Oil Pressure Warning Light

If the oil pressure warning light does not go out immediately after the engine starts, or if it comes on while driving the car, stop the engine immediately to prevent severe engine damage.

Check the oil level as described in **110 Maintenance Program**. If the level is low, add oil to the correct level and start the engine. If the light is still on, do not run the car at all. Have it towed.

Brake Fluid Level Warning Light

The red brake fluid level warning light is an indicator of brake fluid loss. Problems with the brake system should be checked and repaired immediately. See **5 BRAKES** for more information.

Anti-Lock Brake System Warning Indicator

If the anti-lock brake system warning indicator comes on at normal driving speeds, the anti-lock braking system is out of service. Under normal conditions, there will be no change in the effectiveness of the brakes. In an emergency situation, however, the normal anti-lock function is lost and the brakes could lock. Check the system as described in **5 BRAKES**.

Dim Lights

Headlights that are dim or gradually getting dimmer generally indicate a problem with the battery or charging system. The battery charge indicator light may come on as the lights are dimming. In either case, the engine and accessories are running off of the battery alone, and will soon discharge it altogether.

If possible, do not stop the engine unless you have the capability to jump start it. There may not be enough power in the starting system to restart the engine. Instead, turn off as many electrical consumers as possible. This will reduce the current drain and will allow the car to be driven farther before you lose all battery power.

With the engine and ignition off, check to see if the battery cables are firmly attached, or if there are any loose wires leading to the battery or to the alternator. Look for heavily corroded (covered by fluffy white deposits) wires and connectors.

Disconnecting, cleaning, and reinstalling corroded wires and connectors may solve the problem. Also check V-belt tension as described in **110 Maintenance Program.**

Towing

The cars covered by this manual should be towed with a tow truck using wheel lift or flat bed equipment. Do not tow the car on all four wheels except for very short distances to move it to a safe place. If flat-towing the car, use the towing eyes at the front of the car under the bumper. Set the transmission in neutral.

When towing the car from the rear with the front wheels on the ground, a maximum distance of 30 miles is acceptable. If the car needs to be towed further, have the front wheels placed on dollies.

NOTE

Do not tow with sling-type equipment. The front spoilers and the bumpers may sustain damage.

Towing a Saab with an automatic transmission with the front wheels on the ground can result in transmission damage due to lack of lubrication. Always tow the car with the transmission lever in "N" (neutral) and the key in position "1". Saab recommends that cars with automatic transmission be towed with the rear wheels on the ground for no more than 30 miles (50 km), at no more than 25 mph (40 km/h). Be sure the transmission fluid has been topped off before starting the tow.

Spare Parts Kit

Carrying a basic set of spare parts can prevent a minor breakdown from turning into a major annoyance. Many of the following items won't allow you to do major repair work on the car, but they will help in the event of the failure of something that can disable the car or compromise its safety.

Spare Parts Kit - Basic Contents:

- V-belts for the alternator and water pump
- one or two quarts of engine oil
- a gallon container of engine coolant (premixed 50/50 anti-freeze and water)
- spare fuel pump relay, also spare main relay
- a new, unopened bottle of brake fluid (DOT 4 specifications)
- 10 amp, 15 amp, and 20 amp fuses
- upper and lower radiator hoses

Spare Parts Kit - Additional Contents

- replacement headlight (sealed beam or bulb)
- brake light, turn signal light, and tail light bulbs
- wiper blades
- distributor cap and rotor

110 Maintenance Program

TABLES

GENERAL

The useful life of any car depends on the kind of maintenance it receives. The procedures described in this section of the manual include all of the routine checks and maintenance steps that are both required by Saab under the terms of their warranty protection and recommended by Saab to ensure long and reliable operation of your car. Also included are some instructions and recommendations for more basic car care.

Some maintenance procedures, such as oil change service, require no special tools and can be carried out by almost any interested Saab owner, regardless of his or her previous mechanical experience. Certain other maintenance tasks require special tools and equipment. Idle speed and ignition timing adjustments are some examples. If you lack the tools or a suitable workplace for doing any of the maintenance described in this section, we suggest you leave this work to an authorized Saab dealer or other qualified shop. We especially urge you to consult an authorized Saab dealer before beginning any repairs on a car still covered by the manufacturer's warranty.

All of the maintenance work described here is important and should be carried out at the correct time or mileage interval. Your Saab should not be thought of as a maintenance-free machine. Correct care will protect your investment and help you to get many years of driving reliability and enjoyment from your Saab.

Saab is constantly updating their recommended maintenance procedures and requirements. The information contained here is as accurate as possible at the time of publication. If there is any doubt about what procedures apply to a specific model or model year, or what intervals should be followed, remember that an authorized Saab dealer always has the latest information on factory-recommended maintenance.

SAAB RECOMMENDED MAINTENANCE PROGRAM

Table a and **Table b** list the maintenance tasks specified by Saab that should be done at specified mileage intervals to ensure the proper function, safety and durability of the Saab automobile under normal use. The schedules are divided into two parts; **Oil Change/Safety Inspection** and **Major Service.**

NOTE
Aside from keeping your car in the best possible condition, proper maintenance plays a role in maintaining full protection under Saab's new-car warranty coverage. If in doubt about the terms and conditions of your car's warranty, an authorized Saab dealer should be able to explain them.

The **Oil Change/Safety Inspection** covers the most basic level of routine maintenance. Saab's required Oil Change/-Safety Inspection specifies changing the engine oil and oil filter, rotating the tires, and checking other systems and their fluids. This service should be made at 7,500 mile intervals on all models except 1992 nad later, models. On 1992 and later models, the service should be made at 10,000 mile intervals, beginning at 15,000 miles.

NOTE
On 1992 and later, Saab specifies a one-time 5,000 mile "First Service", which should be done by an authorized dealer. There is no charge for this service. For more information on the Saab maintenance system, see your glove box information or an authorized Saab dealer.

The **Major Service** interval signals the need for more comprehensive maintenance and inspection. This service should be performed every 30,000 miles.

NOTE
On 1992 and later models, the first Major Service should be done at 35,000 miles, then at 30,000 mile intervals thereafter.

Except where noted, the maintenance items listed apply to all models covered by this manual. The columns on the right side of each table give quick-reference information about the job—whether tools are needed, whether the procedure requires new parts, whether the car should be warmed-up to normal operating temperature and, in some cases, a recommendation that the job be turned over to an authorized Saab dealer because of the need for special equipment or expertise.

Following the Saab recommended maintenance tables are detailed descriptions of all of the tasks, in the order in which they appear in the tables.

Table a. Oil Service/Safety Inspection
1985-1991: every 7,500 miles
1992 and later: every 10,000 miles

Maintenance item	Tools required	New parts required	Warm engine required	Dealer service recommended
Engine maintenance				
Change oil and oil filter	*	*	*	
Check engine coolant level and add as required. Check coolant freezing point. Inspect coolant hoses and hose clamps	*	*		
Check engine drive belt (V-belt) condition and tension	*			
Inspect exhaust system				
Inspect fuel system components, fuel lines, and vacuum hoses				
Electrical System Maintenance				
Check battery terminal connections and electrolyte level. Clean terminals if necessary				
Check for proper function of exterior and interior lights				

continued on next page

Table a. Oil Service/Safety Inspection (continued)
1985-1991: every 7,500 miles
1992 and later: every 10,000 miles

Maintenance item	Tools required	New parts required	Warm engine required	Dealer service recommended
Transmission Maintenance				
Check manual transmission oil level and add as required		*		
Check automatic transmission fluid level and add as required		*		
Check automatic transmission final drive oil level and add as required	*	*		
Inspect rubber boots for inner and outer drive axle joints				
Chassis Maintenance				
Inspect brake system for damaged hoses and lines, leaks or damage				
Check brake pad wear (road wheels removed)	*			
Check brake (and clutch) fluid level and add as required		*		
Check parking brake operation				
Check power steering fluid level and add as required		*		
Check windshield washer fluid level and add as required				
Check shock absorber bushings and shock function				
Inspect rubber boots for ball joints, tie-rod ends, and steering rack				
Check tire pressure, including spare	*			
Rotate tires. Inspect tires for uneven wear	*			
Check front wheel toe-in alignment (1985-1991 models only)	*		*	*
Body Maintenance				
Check operation of doors. Lubricate door hinges, stops, and locks				
Road Test				
Check engine, transmission, clutch, steering, and brake performance. Check cruise control control and ventilation system. Check for any noises (brake squeal, rattle, etc.)				

Table b. Major Service
(every 30,000 miles*)

Maintenance item	Tools required	New parts required	Warm engine required	Dealer service recommended
Engine Maintenance				
Change oil and oil filter	*	*	*	
Flush and replace engine coolant (maximum 2-year intervals)	*	*		
Inspect coolant hoses and hose clamps	*			
Check engine drive belt (V-belt) condition and tension*	*			
Replace distributor cap and rotor**	*	*		
Replace and gap spark plugs	*	*		
Inspect and check resistance of spark plug wires***	*			
Check and adjust ignition timing** (all except models with crankshaft mounted EZK Ignition Hall Sender. See **340 Ignition System** for application information)	*		*	*
Check and adjust idle speed (cars with LH 2.2 Fuel Injection only, see **240 Fuel Injection** for application information)	*		*	*
Check and adjust deceleration system** (cars with LH 2.2 Fuel Injection only, see **240 Fuel Injection** for application information)	*		*	*
Check turbo overpressure safety switch (1985-1988 Turbo models only)	*	*	*	*
Replace oxygen sensor** (cars with LH 2.2 Fuel Injection only, see **240 Fuel Injection** for application information)	*	*		
Inspect crankcase ventilation system***				
Inspect Evaporative Loss Control Device (ELCD) system***				
Replace charcoal canister (1985-1990 models)**	*	*		
Replace air filter element		*		
Inspect fuel system components, fuel lines, and vacuum hoses				
Lubricate accelerator and throttle linkage				
Replace fuel filter	*	*		
Inspect exhaust system				
Electrical System Maintenance				
Check battery terminal connections and electrolyte level. Clean terminals if necessary				
Check for proper function of exterior and interior lights				
Check headlight aim	*			*
Transmission Maintenance				
Check manual transmission oil level and add as required	*	*		
Drain and refill automatic transmission fluid. Remove front pan and clean transmission filter****	*	*	*	*
Check automatic transmission final drive oil level and add as required	*	*		
Inspect rubber boots for inner and outer drive axle joints				

continued on next page

Table b. Major Service (continued)
(every 30,000 miles*)

Maintenance item	Tools required	New parts required	Warm engine required	Dealer service recommended
Chassis Maintenance				
Inspect brake system for damaged hoses and lines, leaks or damage				
Check brake pad wear (road wheels removed)	*			
Replace brake fluid and flush and bleed brake system	*	*		*
Check parking brake operation				
Check power steering fluid level				
Check windshield washer fluid level				
Check shock absorber bushings and shock function				
Inspect rubber boots for ball joints, tie-rod ends, and steering rack				
Check tire pressure, including spare	*			
Rotate tires. Inspect tires for uneven wear	*			
Check front wheel alignment (camber, caster, toe-in)	*	*		*
Body Maintenance				
Check operation of doors. Lubricate door hinges, stops, and locks				
Check and lubricate convertible top latches				
Road Test				
Check engine, transmission, clutch, steering, and brake performance. Check function of instruments, controls, horn, windshield wipers, cruise control control and ventilation system. Check for any noises (brake squeal, rattle, etc.)				

* Every 12 months or 30,000 miles, which ever comes first only
** Every 60,000 miles
*** At the first 60,000 miles, then every 12 months thereafter
**** An automatic transmission service is recommended every 15,000 miles if the vehicle is under the severe service conditions of trailer towing, extensive city driving or driving in hot climates.

ENGINE MAINTENANCE

Engine Oil and Oil Filter

The engine oil level is checked with a dipstick located on the in the engine block on the side of the intake manifold. See Fig. 1. Make sure the car is on level ground. If the engine has been run, allow it to cool for about 3 minutes for the most accurate check.

Check the level by pulling out the dipstick and wiping it clean. Reinsert it all the way and withdraw it again. The oil level is correct if it is between the maximum and minimum marks on the end of the stick. The distance between the marks corresponds to approximately one quart of oil.

Top up the oil through the dipstick tube. Add only the amount needed to bring the level to the maximum mark on the dipstick, using an oil of the correct viscosity and grade as listed below. Too much oil can be just as harmful as too little. Make sure the dipstick is screwed down properly after use.

CAUTION
The use of fluids or lubricants that do not meet Saab's specifications may impair performance and reliability, and may void warranty coverage.

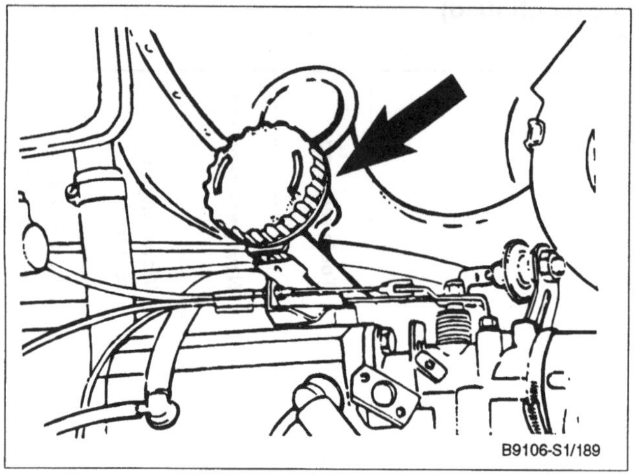

Fig. 1. Location of oil filler cap with dipstick (arrow). Remove and install by turning.

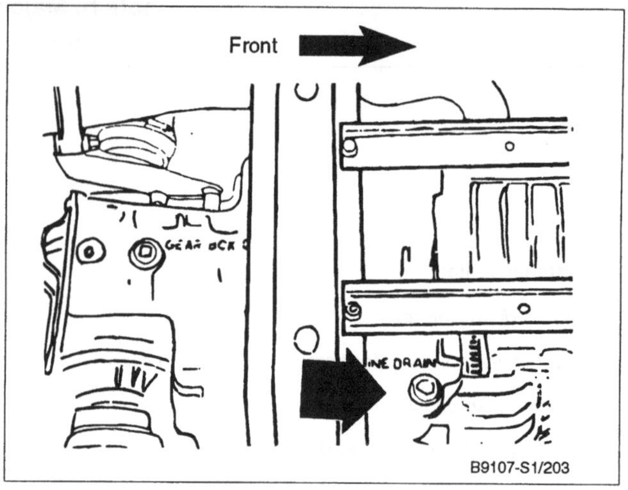

Fig. 2. Engine oil drain plug (arrow) in oil pan underneath engine.

Saab Recommended Engine Oil

- SAE 10W-30, API Service Rating SG or SF/CD
- Alternate for extremely hot climates:
  15W-40, API Service Rating SG or SF/CD

NOTE

The amount of oil that needs to be added between oil changes varies from one engine to another. Generally, a new engine or an engine operated routinely at high speeds will consume more oil. It is helpful to become familiar with the rate at which a particular engine requires oil. A sudden increase may be an early warning of engine mechanical problems.

If the car is used primarily for short trips in slow moving traffic, or routinely operated aggressively, the oil should be changed at more frequent intervals (twice as often) to promote longer engine life.

A complete oil change requires new oil, a new oil filter, and a new drain plug sealing washer. The tools needed are a drain plug socket or box wrench (19 or 13 mm), a drain pan of at least 6 US qt. (5.6 L) capacity, and an oil filter wrench.

To change oil and filter

1. Run the car for a few minutes to slightly warm the engine and the oil, then shut the engine off.

2. With the car on level ground, place a drain pan under the oil drain plug shown in Fig. 2.

NOTE

The car will not need to be raised if a shallow drain pan is used.

---

3. Using a socket or box wrench, loosen the drain plug. By hand, remove the plug and let the oil drain into the pan.

WARNING

Pull the loose plug away from the hole quickly to avoid being burned by the hot oil. If possible, use gloves to protect your hands.

---

4. When the oil flow has diminished to an occasional drip, reinstall the drain plug with a new sealing washer and torque the plug.

CAUTION

If using a "fast-lube" service facility for oil changes, make sure the technician hand-starts and torques the engine oil drain plug using only hand-tools. Using power tools can strip the threads of the plug and the oil pan.

Tightening torque

- engine oil drain plug (13 mm
  or 19mm wrench size) . . . . . . 29-39 Nm (21-29 ft-lb)

NOTE

On mid-1991 and later cars with manual transmission, the engine oil drain plug was changed from a 19mm wrench size to a 13mm wrench size. The plugs are identical except for the bolt head size. This change was made to reduce the possibility of overtightening the larger head drain plug and cracking the transmission case. If replacing the larger drain plug with the smaller plug, be sure to use a new steel washer with the rubber bonded to the perimeter. Do not use the old design copper drain plug gasket with the larger head bolt.

---

5. Position the drain pan directly under the oil filter. Using an oil filter wrench, loosen it in a counterclockwise direction, then remove it by hand. See Fig. 3.

Fig. 3. Engine oil filter on left-hand (driver's) side of engine.

6. After the oil stops dripping, wipe clean the oil filter gasket surface on the filter mounting flange. Lubricate the rubber gasket of the new oil filter with a light coating of clean engine oil.

7. Install the filter by hand until the gasket contacts the mounting flange, then turn the filter another 1/2 turn to tighten it.

CAUTION
Overtightening the oil filter will make the next change much more difficult.

---

8. Refill the crankcase with oil. Use the dipstick to check for the correct oil level.

Engine Oil Capacity (including filter)
- 4.2 quarts (4.0 liters)

Saab Recommended Engine Oil
- SAE 10W-30, API Service Rating SG or SF/CD
- Alternate for extremely hot climates:
  15W-40, API Service Rating SG or SF/CD

---

9. Start the engine and check that the oil pressure warning light immediately goes out. Allow the engine to run for a few minutes to circulate the new oil, then check for leaks at the drain plug and around the oil filter. Stop the engine and recheck the oil level.

NOTE
Dispose of the used oil properly. Use tight-sealing containers and mark them clearly. Check with the place of purchase about disposal.

Coolant (Antifreeze) and Cooling System

Cooling system maintenance consists of maintaining the coolant level and its freezing point as well inspecting the hoses. Because the coolant's anti-corrosion and anti-freeze additives gradually lose their effectiveness, replacement of the coolant is recommended every 30,000 miles or 2 years, whichever comes first. Drain and refill the coolant as described in **261 Radiator and Cooling System.**

Cooling System Capacity
- 10.5 quarts (10 liters)

CAUTION
Use only Saab original anti-freeze when filling the cooling system. Use of any other anti-freeze may be harmful to the cooling system. If Saab original anti-freeze is not used, Saab recommends changing the coolant every 12 months.

A translucent expansion tank, or overflow reservoir, provides easy monitoring of coolant level without opening the system. The coolant level should always be checked when the engine is cold. The coolant level should be between the maximum and the minimum mark on the expansion tank, as shown in Fig. 4.

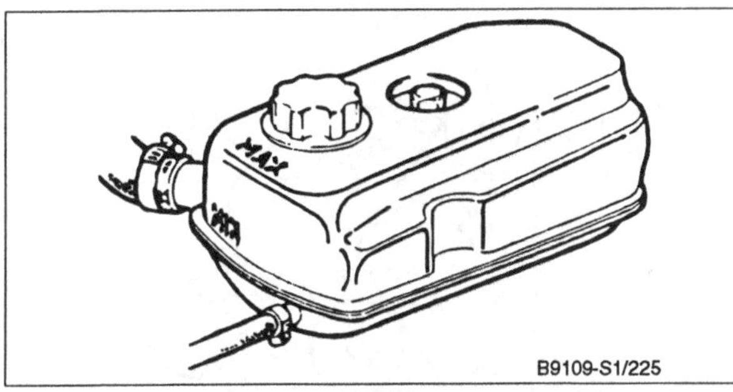

Fig. 4. Fill mark on coolant expansion tank. Coolant level should be between MAX and MIN marks.

Inspect the hose by first checking that all connections are tight and dry. Coolant seepage indicates either that the hose clamp is loose, that the hose is damaged, or that the connection is dirty or corroded. Dried coolant has a chalky appearance. Check the hose condition by pinching them. Hoses should be firm and springy. Replace any hose that is cracked, that has become soft and limp, or has been contaminated by oil. See Fig. 5.

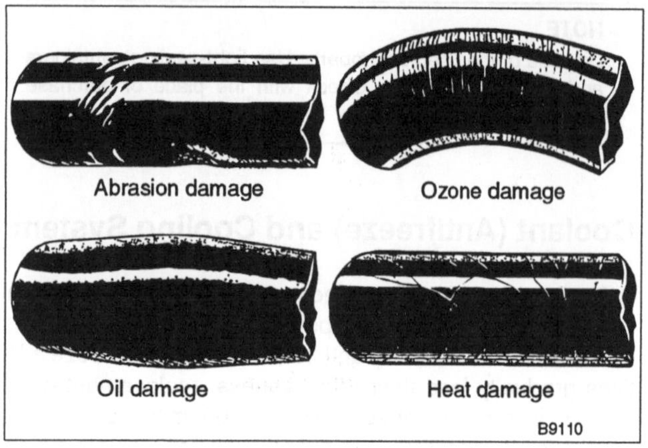

Fig. 5. Examples of damaged coolant hoses. Any of conditions shown is cause for replacement. Courtesy of Gates Rubber Inc.

To check the freezing point of the coolant, use a hydrometer. With the engine cold, remove the cap from the expansion tank and draw up some coolant with the hydrometer. Saab recommends using a 50/50 mixture of anti-freeze and water, which has a freezing point of -35°F (-38°C). If the freezing point is too high, drain a small amount of coolant from the system and add new anti-freeze until the proper protection level is obtained.

Drive Belts

Drive belts (or V-belts) and pulleys transfer power from the engine crankshaft to various accessories. Cars covered by this manual have four drive belts. The alternator and coolant pump are driven by two belts, side by side. See Fig. 6.

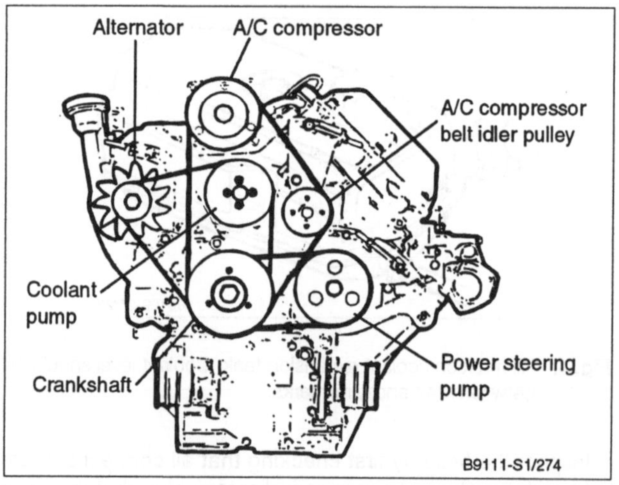

Fig. 6. Drive belt configuration.

Incorrect drive belt tension can decrease the life of the belt and the component it drives. Inspect belts with the engine off. Twist the belt to inspect its sidewalls and bottom. Belt structural damage, glazed or shiny sidewalls caused by a loose belt, or separation caused by oil contamination are all reasons to replace a belt. Some of these faults are illustrated in Fig. 7.

NOTE

Always replace both alternator belts as a matched set.

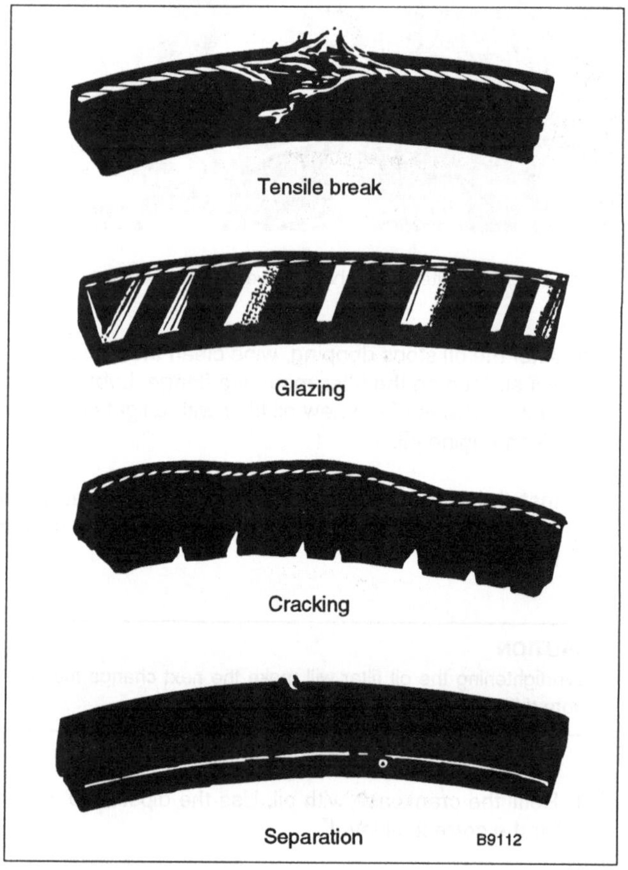

Fig. 7. Examples of belt failure. Courtesy of Gates Rubber Inc.

Drive belt squealing is normally caused by incorrect belt tension (too loose) or by a worn belt. Extremely loud squealing may only be corrected by replacing the belt. To accurately tension the drive belts, a special drive belt tensioning tool should be used. Correct belt tensions are listed in **Table c.**

Table c. Drive Belt Tensions (measure using special belt tensioning gauge)

Drive belt	Tension N (lb)
Alternator	
checking (minimum)	
one belt	200 (45)
two belts	420 (95)
adjusting, used belts	
one belt	310±20 (70±5)
two belts	645±20 (145±5)
adjusting, new belts	
one belt	535±45 (120±10)
A/C compressor	
checking (minimum)	245 (55)
adjusting, used belt	355±20 (80±5)
adjusting, new belt	535±45 (120±10)
Power steering	
checking (minimum)	220 (50)
adjusting, used belt	310±20 (70±5)
adjusting, new belt	445±45 (100±10)

Adjust the alternator belt by first disconnecting the negative (–) battery cable. Loosen the lower (pivot) mounting bolt and then adjust the belt tension through the adjusting nut. See Fig. 8. When the belt is correctly tensioned, there should be approximately 0.2 in (5 mm) of deflection in either direction at the midpoint of the belt. For the most accurate check, use a belt tensioning gauge.

Fig. 8. Alternator drive belt tension being adjusted.

The A/C compressor drive belt is adjusted through the belt's idler pulley. See Fig. 7 above. To adjust the A/C compressor belt, simply loosen the idler pulley mounting bolts, adjust the position of the pulley until the belt tension is correct, and tighten the mounting bolts. Check the belt tension using a drive belt tensioning gauge.

To adjust the power steering drive belt, first loosen the pump's mounting bolts. Then adjust the belt tension by turning the adjusting nut. See Fig. 9. Check the belt tension using a drive belt tensioning gauge.

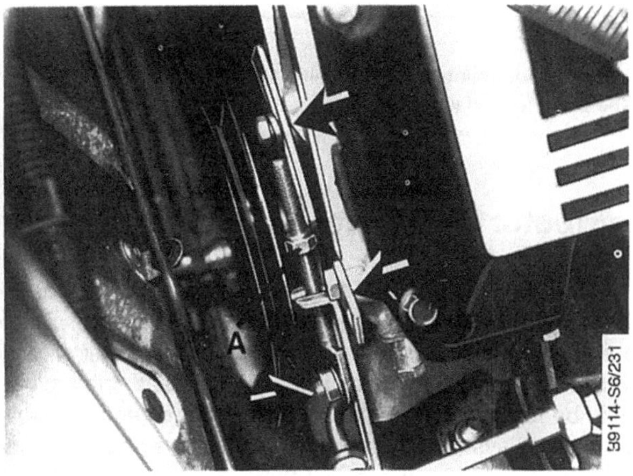

Fig. 9. To adjust power steering drive belt, loosen pump mounting bolts (arrows), and then turn adjusting nut (**A**).

To reduce the chance of drive belt failure while driving, replacement of the belts every four years is recommended. Loosen the mounting and belt adjusting bolts and adjust until the belt tension is as loose as possible, then remove the belt by slipping it over the pulleys. In some cases it may be necessary to remove one drive belt to get to another. Cross section and length determine belt size. Use the old belt for comparison, or make sure that the new belt fits into the pulley groove as shown in Fig. 10.

NOTE

When belts are replaced with new ones, keep the old set in the luggage compartment by the spare tire for emergency use.

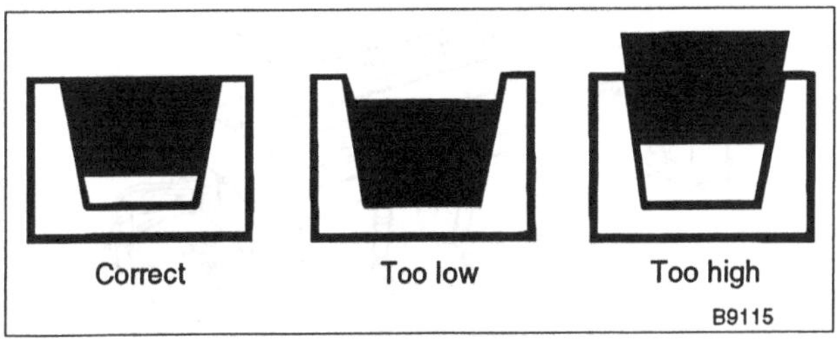

Fig. 10. Cross-section of correct drive belt position in pulley. Courtesy of Gates Rubber Inc.

With the belt off, clean the pulleys using a suitable solvent. Inspect the pulleys for wear or damage that may cause early failure of the new belt. This is also a good opportunity to inspect the belt-driven accessory, checking for bearing wear and excess play, for example. When installing the new belt, gently pry it over the pulleys. Too much force may damage the belt or the accessory. Tension the belt(s), run the engine for a few minutes (at least 1500 rpm), then recheck the belt tension.

CAUTION
Do not overtighten the drive belts. Overtightening will cause the bearings to fail prematurely.

Distributor Cap, Rotor, and Spark Plugs

The distributor cap, the rotor, and the spark plug wires deliver a high-voltage spark to the spark plugs. See Fig. 11. They are subject to insulation breakdown, corrosion fouling, and electrode wear and damage. The cap, rotor and spark plugs should be replaced at the specified intervals to ensure maximum ignition system efficiency. There is no replacement intervals for spark plug wires, but they should be periodically tested and inspected. Guidelines for visual inspection and testing, and instructions for distributor cap and rotor replacement are found in **340 Ignition System.**

WARNING
Lethal voltages are present at the distributor, ignition coil, spark plug wires, and spark plugs when the key is turned on or the engine is running.

NOTE
When removing the spark plug wires, label their positions so that they can be reinstalled in the proper places. If the wires get mixed up, see **340 Ignition System** for more information on the firing order.

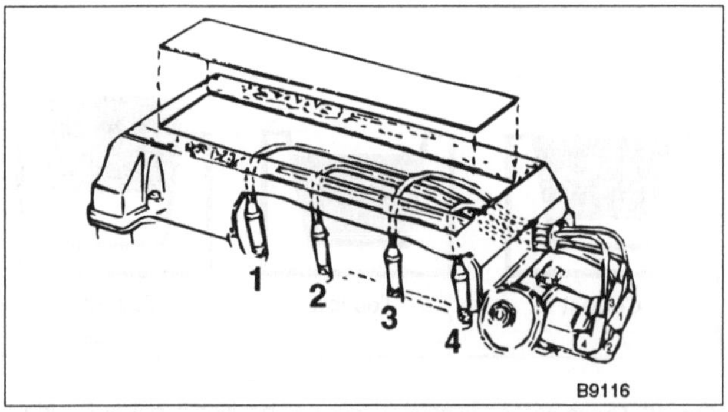

Fig. 11. Ignition system.

To replace the spark plugs, first remove the center cover above the plugs. Gently remove the spark plug wire by pulling on the protective boot, and blow or brush away any dirt from around the base of the plug to prevent it from entering the engine when the plug is removed.

Use a 5/8 in. spark plug socket to remove spark plugs. See Fig. 12. The correct spark plugs for the different engines covered by this manual are listed in **Table d.** Use a spark plug gap gauge to check the gap. If necessary, bend the outer electrode slightly to adjust the gap to meet the specification. Do not bend or file the center electrode. See Fig. 13.

Fig. 12. Spark plug being removed.

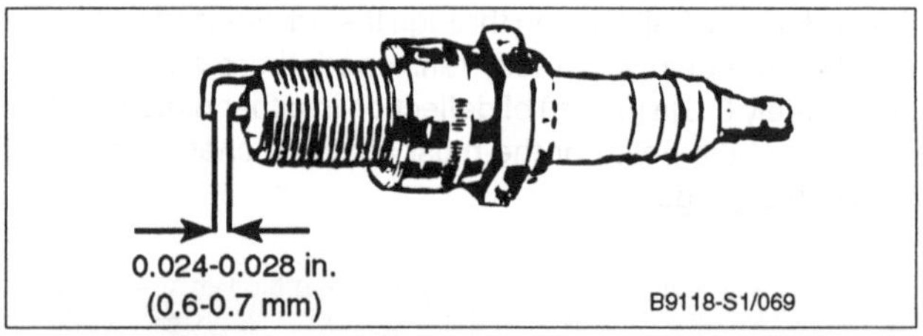

Fig. 13. Spark plug electrode gap.

Spark Plug Electrode Gap
• All . . . . 0.024-0.028 in. (0.6-0.7 mm)

Table d. Spark Plug Applications

Model	Spark Plug
1986 and later, Non-turbo	NGK BCP 6ES
alternate	NGK BCP 6EV
1989 and later, Non-turbo	NGK BCP 5ES
1985 and later, Turbo	
normal driving	NGK BCP 7EV
alternate	NGK BCP 7ES
city driving*	NGK BCP 6EV
alternate	NGK BCP 6ES

* On 1985 through 1987 turbo models, the Saab recommended spark plug replacement interval for city driving is 15,000 miles. On 1988 and later turbo, this interval is 30,000 miles.

Lightly lubricate the new spark plug threads with a little oil. Thread the plugs into the cylinder head by hand to prevent cross-threading. Torque the spark plugs.

Tightening torque
- sparkplugs . . . . .25 to 29 Nm (18 to 22 ft-lb)

Ignition Timing, Idle Speed, Deceleration System

Ignition timing should be checked every 60,000 miles on all cars with distributor-mounted Hall sensors. See **340 Ignition System** for Hall sensor applications and procedures for checking and adjusting ignition timing. It is not necessary to check/adjust ignition timing on cars with a crankshaft-mounted Hall sensor.

Ignition Timing (distributor vacuum hose disconnected and plug, where applicable)
- Turbo models. . . . . 16°BTDC @ 850 rpm
- Normally aspirated models . . . 14°BTDC @ 850 rpm

On all cars covered by this manual, the idle speed is electronically controlled by the Automatic Idle Control (AIC) System. On cars with LH 2.2 Fuel Injection, the idle speed should be checked and adjusted through the AIC system at the specified mileage interval as described in **240 Fuel Injection.**

NOTE
It is not necessary to check the idle speed on cars with LH 2.4 or LH 2.4.2 Fuel Injection, as the idle speed is fully adaptive (self-correcting) through the LH control unit.

On cars with LH 2.2 Fuel injection and automatic transmission, a mechanical deceleration dashpot is mounted to the throttle housing to hold the throttle open for a short period when the throttle is quickly closed. The dashpot plunger should be adjusted to touch the throttle lever at a specified engine speed. See **240 Fuel Injection** for dashpot adjustment procedures and specifications.

Turbo Overpressure Safety Switch (1985-1998 Turbo models)

The turbo overpressure safety switch turns the fuel pump off when the manifold pressure reaches a higher than normal level. This is a safety feature in the event the turbocharger wastegate malfunctions. Checking the switch requires a hand-held pressure/vacuum pump and an accurate pressure gauge. This procedure is described in detail in **234 Fuel Pump and Fuel Tank.**

Oxygen Sensor

The oxygen sensor monitors engine combustion efficiency by measuring the oxygen content of the exhaust gasses. That information in turn is used to control the fuel injection system and reduce exhaust emissions. Any problems with the oxygen sensor will directly affect exhaust emissions and the way the engine runs.

Replacement of the oxygen sensor at the specified interval ensures that the engine and emission control system will continue to operate as designed. The sensor is mounted in the front exhaust pipe, ahead of the catalytic convertor.

NOTE
On 1991 and later models, there is no recommended replacement interval for the oxygen sensor. When the sensor is no longer operating correctly the Check Engine light will come on signaling the need for the sensor's replacement.

The sensor is threaded into place and has wires extending from the tip. Trace the sensor wiring back from the sensor and disconnect the electrical connector. When installing a new sensor, apply a light coat of anti-seize compound to the sensor threads. Torque the sensor and reconnect the wiring.

Tightening torque
- oxygen sensor to exhaust pipe . . . . 40 Nm (30 ft-lb)

CAUTION
Do not get any anti-seize compound on the sensor tip or in the sensor slits. The anti-seize compound will quickly foul the sensor element and render the sensor inoperative.

NOTE
Special sockets for replacing the oxygen sensor are available from most automotive parts stores to allow the sensor to be installed without damaging the wiring.

Crankcase Ventilation and Vacuum Hoses

The rubber vacuum hoses for the crankcase ventilation system should be carefully inspected for cracks or deterioration. Gently bend the hoses at all connecting points to check for cracks. Fig. 14 shows a typical hose routing diagram.

Evaporative Loss Control Device (ELCD)

The ELCD system collect fuel vapors from the fuel tank. Check all of the rubber vapor hoses for cracks or deteriora-

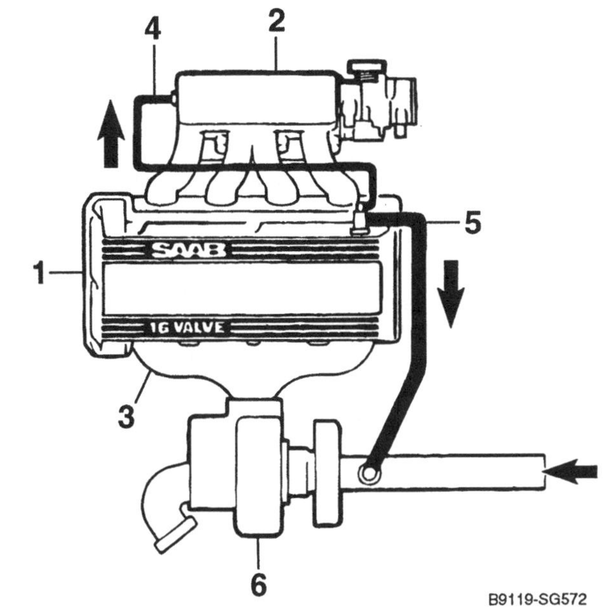

1. Crankcase
2. Intake manifold
3. Exhaust manifold
4. Hose connected to intake manifold
5. Hose
6. Turbocharger

Fig. 14. Hose routing for typical crankcase ventilation system.

tion. On all models up to 1990, the charcoal canister should be replaced at 60,000 mile intervals. The canister is mounted in the left-front corner of the engine compartment and is easily removed. Fig. 15 shows the canister. For more information on the ELCD system, see **254 Exhaust Emission Control.**

NOTE
On 1991 and later models, there is no recommended replacement interval for the charcoal canister.

Fuel System

A fuel system safety inspection should be made at the specified mileage intervals listed in the maintenance program tables given earlier. Inspect the fuel tank, fuel lines and hoses, and fuel system for damage or leaks. If fuel odors are detected in the passenger compartment, the fuel pump seal (beneath the luggage compartment) may be faulty. Also check the ELCD vapor hoses that run overhead through the headliner. See **234 Fuel Tank and Fuel Pump** for replacement of the fuel pump seal. Inspect all electrical connections for corrosion or damage or loose connections.

Air Filter Element

The specified replacement interval (30,000 miles) for the air filter element is based on normal use. If the car is operated primarily in dusty conditions, the air filter should be serviced more frequently. A dirty air filter starves the engine for air, reducing power output and increasing fuel consumption. Fig. 16 shows the typical air filter housing for the engines covered by this manual.

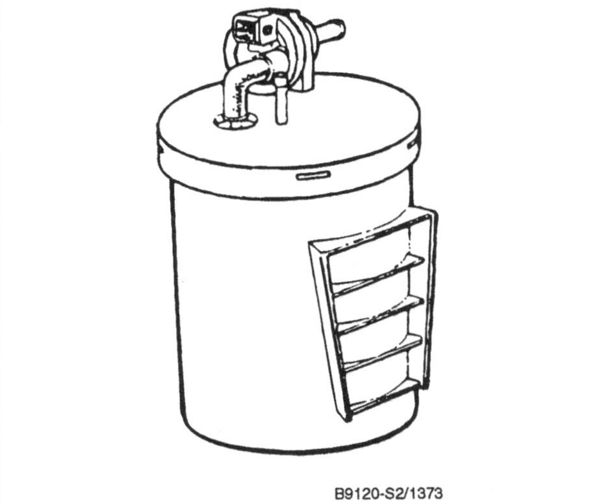

Fig. 15. Charcoal canister for ELCD system.

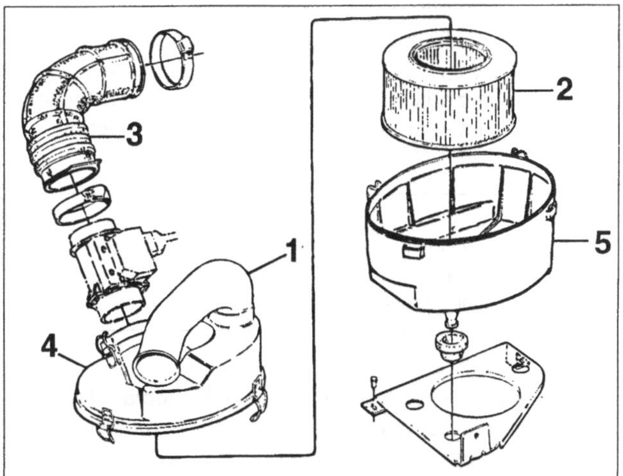

1. Air intake
2. Air filter
3. Air intake bellows
4. Air filter housing, upper
5. Air filter housing, lower

Fig. 16. Typical air filter housing in front left corner of engine compartment.

The upper and lower parts of the air filter housing are fastened together with spring clips around the outside edge. To replace the air filter element, unfasten the clips and separate the upper air filter housing from the lower housing just enough to remove the filter element. Take note of the filter's installed position. Wipe the inside of the air filter housing using a lint-free cloth and install the new filter and housing.

Fuel Filter

Because of varying quality of gasoline, the fuel filter may become contaminated or clogged enough to restrict fuel flow. To prevent any such problems, and to guarantee continued good performance, the filter should be replaced every 30,000 miles. The fuel filter is located beneath the right rear of the car, just in front of the gas tank.

When replacing the fuel filter, disconnect the battery negative (–) cable and clamp the filter inlet and outlet hoses to lessen fuel spillage. Loosen the center mounting bracket and the two banjo fittings on either end of the filter. See Fig. 17. Note the arrow or markings indicating direction of flow on the new filter. Install the filter and using new sealing washers at the banjo fitting.

WARNING

Fuel will be expelled when the filter is removed. Do not smoke or work near heaters or other fire hazards. Keep a fire extinguisher handy.

CAUTION

Clean thoroughly around the filter connections before removing them, and make sure that no dirt gets into the fuel lines.

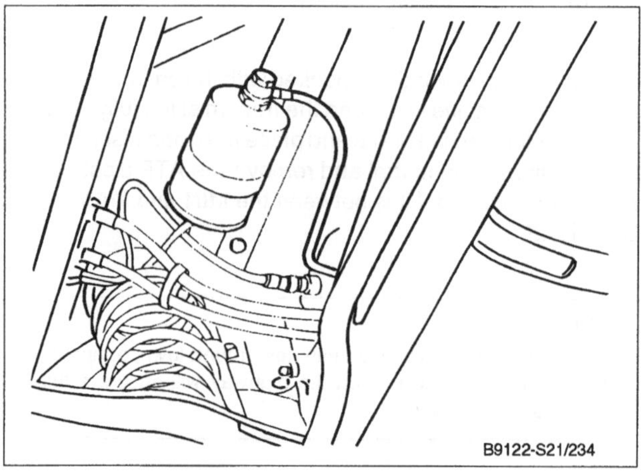

Fig. 17. Fuel filter beneath car near right rear wheel. Direction of flow is indicated by arrow or markings on filter housing.

Accelerator and Throttle Linkage

The accelerator and throttle linkage should be lubricated at the intervals specified in the tables given earlier. Use a general purpose oil on the joints and bearings of the linkage. Use a multipurpose grease on the bearing points of the throttle plate.

Exhaust System

Scheduled maintenance of the exhaust system is limited to inspection. Check for restrictions due to dents or kinks. Check for weakness or perforation due to rust. Check to see that all the hangers are in place and properly supporting the system and that the system does not strike the body. Alignment of the system and the location of the hangers are described in **252 Exhaust System**.

Exhaust system life varies widely according to driving habits and environmental conditions. If short-distance driving predominates, the moisture and condensation in the system will not fully dry out. This will lead to early corrosion damage and more frequent replacement.

ELECTRICAL SYSTEM AND BATTERY SERVICE

Electrical System Functional Check

Electrical system service includes checking that all electrical components, including interior and exterior lighting, are working correctly. Replace any faulty bulbs found. If more involved electrical work is required, refer to **3 ELECTRICAL SYSTEM** for system troubleshooting and wiring diagrams.

Battery Service

Battery maintenance includes keeping the battery and terminals clean and keeping the connections tight. See Fig. 18. Even a thin layer of dust containing conductive acid salts can cause the battery to slowly discharge. Inspect the battery cables for chafing and deterioration of the insulation caused by high heat. For a more detailed discussion of the battery, including testing and charging, see **311 Battery**.

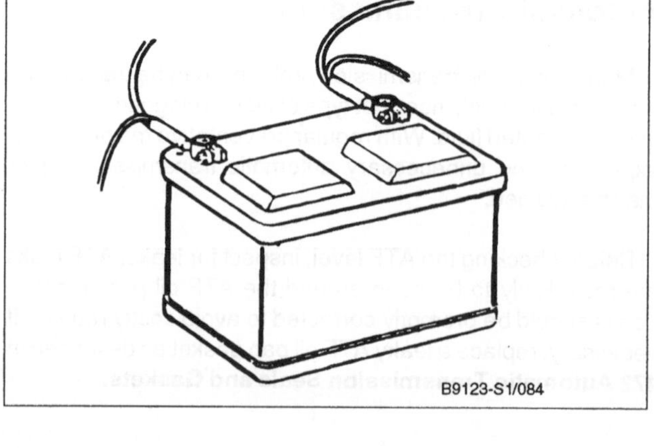

Fig. 18. Battery.

To remove battery corrosion, begin by disconnecting the cables. Disconnect the negative (–) cable first. Clean the terminal posts and the cable clamps with a wire brush. Clean the main chassis ground terminal on the opposite end of the negative (–) cable.

Corrosion can be washed away with a baking soda and water solution that will neutralize the acid. Apply the solution carefully, though, since it will also neutralize the acid inside the battery. Avoid getting the solution into the battery cells through vent holes. Reconnect the cable clamps, positive (+) cable first. Lightly coat the outsides of the terminals, hold down screw, and clamps with petroleum jelly, grease, or a commercial battery terminal corrosion inhibitor.

WARNING
Battery acid is extremely dangerous. Take care to keep it from contacting eyes, skin, or clothing. Wear eye protection. Extinguish all smoking materials and do not work near any open flames.

CAUTION
Disconnecting the battery cables with the engine running, or reconnecting the cables to the incorrect posts will damage the electrical system.

Battery electrolyte should be maintained at the correct level just above the battery plates and their separators. The correct level is approximately 5 mm (1/4 in.) above the top of battery plates or to the top of the indicator marks (if applicable). The battery plates and the indicator marks can be seen once the filler caps are removed. If the electrolyte level is low, replenish it by adding distilled water only.

TRANSMISSION SERVICE

Automatic Transmission

Many automatic transmission problems can be traced to an incorrect fluid level, incorrect type of fluid, a clogged ATF filter, or contaminated fluid. With regular preventative maintenance, expensive and unnecessary automatic transmission repair may be avoided.

Before checking the ATF level, inspect for leaks. ATF leaks are most likely to be seen around the ATF oil pan gaskets. Leaks should be promptly corrected to avoid costly repairs. If necessary, replace a leaky ATF oil pan gasket as described in **472 Automatic Transmission Seals and Gaskets.**

CAUTION
Extreme cleanliness is important when working on the automatic transmission. Use lint-free rags to check the level, and use a clean funnel when adding fluid.

The location of the dipstick for checking the ATF is shown in Fig. 19. The area between the **MIN** mark and the **MAX** mark on the dipstick represents approximately 1 pint (1/2 liter) of ATF.

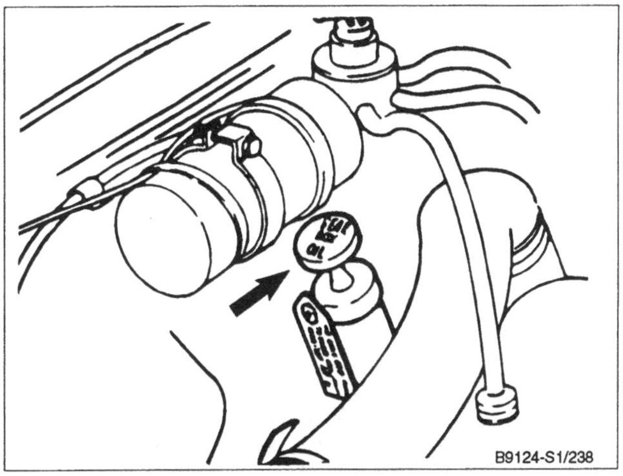

Fig. 19. ATF dipstick location (arrow).

The ATF level should be checked with the car on a level surface with the engine idling and the transmission fully warmed. Firmly set the parking brake and place the transmission selector lever in park or neutral and remove the ATF dipstick. The ATF level is correct if it is between the **MIN** and **MAX** marks. See Fig. 20.

NOTE
Driving the car for five to ten minutes around town, or approximately 12 miles (20 km) on the highway will ensure a fully warmed transmission.

If the level is too low, use a clean funnel to add ATF through the dipstick/filler tube until the fluid level is correct.

Automatic Transmission Fluid (ATF)
- Ford Specification M2C-33F (alternate: Ford Specification G)

Check the condition of the ATF by rubbing some between your fingers and sniffing it. The ATF should not be foamy, gritty, or have a burnt odor. Contaminated ATF should be drained

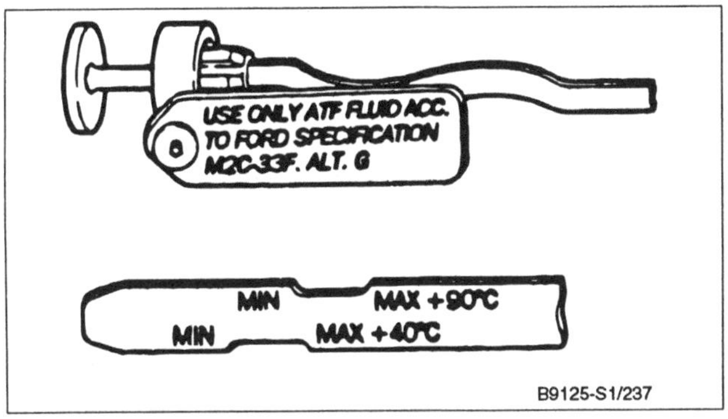

Fig. 20. ATF dipstick and fluid level marks.

and replaced to prevent further damage, but doing so will not repair any internal transmission damage that has already occurred.

NOTE
Because ATF is a red/brown color that discolors to black/-brown during normal use, ATF color may not be a good indicator of its condition.

To replace ATF and clean ATF filter

The ATF should be changed and the filter cleaned every 30,000 miles under normal conditions. If the car is operated under severe service conditions of trailer towing, extensive city driving or driving in hot climates Saab recommends that the fluid be changed every 15,000 miles.

CAUTION
Towing the car or running the engine without ATF in the transmission will severely damage the transmission. See **100 Fundamentals for the Do-it-yourself Owner** for towing instructions.

---

1. Raise the car and support it on jackstands.

2. Remove the support member or protective plate from beneath the transmission.

3. Place a drain pan of at least 10 quarts (10 liters) capacity under the transmission and remove the transmission drain plug. See Fig. 21.

4. Clean the front transmission pan (sump) and surrounding areas. Remove the front pan mounting bolts and remove the pan from the transmission.

5. Remove the filter mounting bolts and remove the filter. See Fig. 22. Clean the filter using a suitable solvent or fresh ATF.

Fig. 21. Automatic transmission fluid drain plug (**1**). Final drive drain plug is shown at **2**.

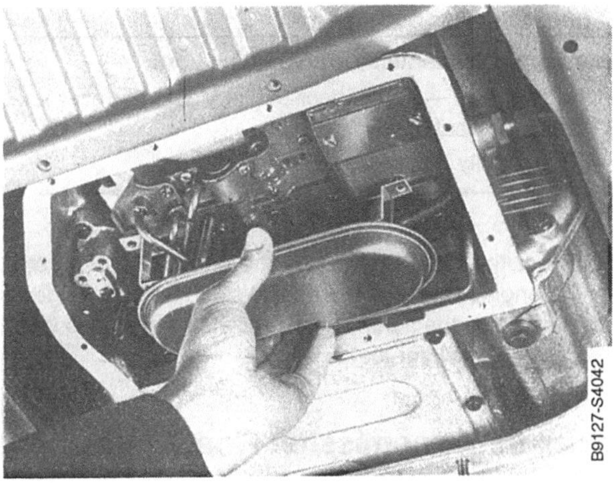

Fig. 22. ATF filter being removed.

6. Remove and clean the magnet in the oil pan. Note the position of the magnet.

7. Clean the oil pan and install the pan magnet. See Fig. 23. Make sure the pan is completely dry. Remount the filter using a new O-ring and tighten the mounting bolts.

8. Using a new pan gasket without any sealer, install the front pan. Install the ATF drain plug. Tighten the bolts and the plug to the torque listed below.

Tightening torques
- ATF pan mounting bolts. . . . . 8-12 Nm (71-106 in-lb)
- ATF drain plug . . . . 5-8 Nm (44-71 in-lb)

---

9. Install the support member or protective plate.

10. Refill the transmission with fluid according to the type and amount specified below.

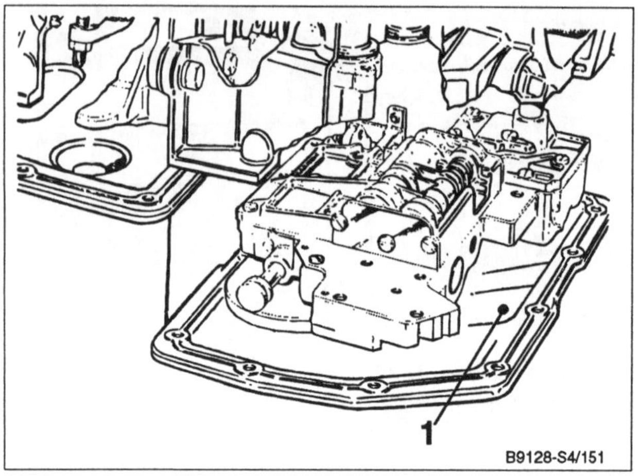

Fig. 23. Correct location of pan magnet (**1**) in front pan.

Automatic Transmissions Specifications
- Grade of ATF — Ford Specification M2C-33F (alternate: Ford Specification G)
- ATF capacity 8.5 quarts (8.0 liters)

11. Lower the car. Firmly apply the handbrake, then start the engine and shift the transmission through all gears to circulate the fluid. Check the fluid level as described earlier. Check for leaks.

Automatic Transmission Final Drive

The final drive on cars with automatic transmission is lubricated with a separate gear oil. Final drive service consists of checking the gear oil level and inspecting for leaks.

The final drive gear oil level should be checked regularly, although there is no specified replacement interval specified by Saab. The final drive oil drain plug is shown earlier in Fig. 20.

Check the lubricant level with the car level. Remove the oil filler plug, shown in Fig. 24. The level is correct when the fluid just reaches the edge of the filler hole. If the oil level is low, Add oil through the filler plug until it begins to run out of the filler hole. Install and torque the filler plug.

Tightening torque
- final drive drain or filler plug. . 39-59 Nm (29-44 ft-lb)

Final Drive Specifications (automatic transmission only)
- Grade of oil — SAE EP 80 or 75 API-GL-4 or API-GL-5
- alternate grade of oil — 10W-30 engine oil
- Fluid capacity — 1.3 quarts (1.25 liters)

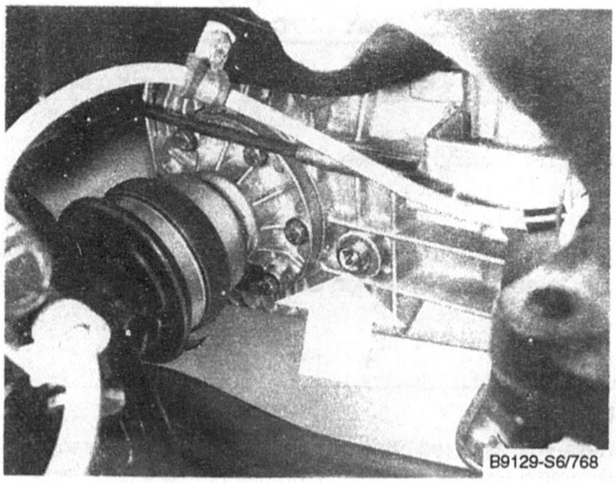

Fig. 24. Final drive fill plug (arrow).

Manual Transmission

Manual transmission service consists of checking the gear oil level and inspecting for leaks. The gear oil level should be checked regularly, although there is no specified replacement interval specified by Saab. In addition, 1990 and later models do not have a drain plug. Two types of manual transmission oil dipsticks are used on the cars covered by this manual.

On 1985 through 1990 models, the dipstick is similar to the engine oil dipstick. See Fig. 25. On 1991 and later cars, the manual transmission oil dipstick is shown in Fig. 26.

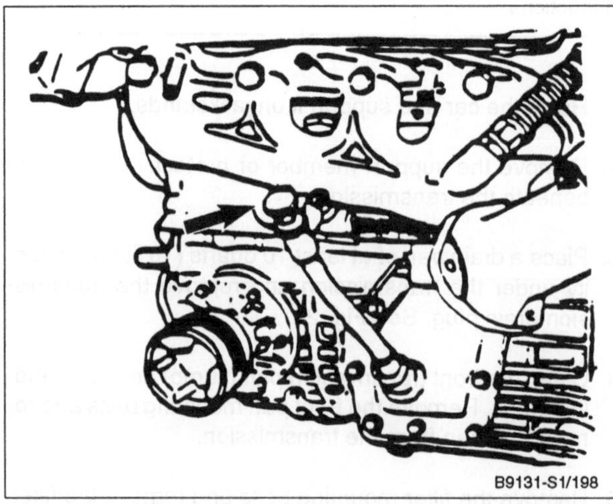

Fig. 25. Manual transmission oil dipstick (arrow) used on 1985-1990 models.

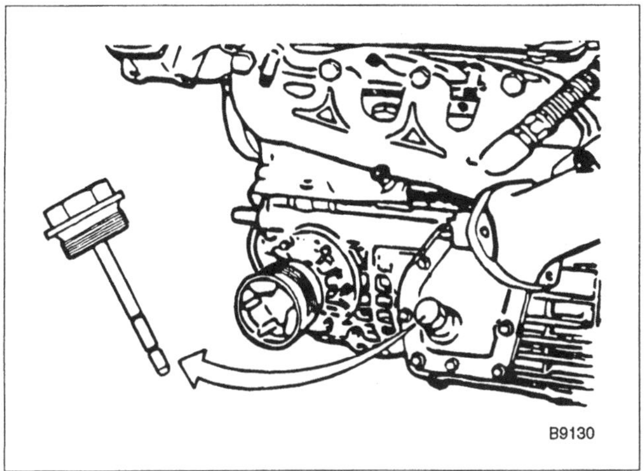

Fig. 26. Manual transmission oil dipstick (arrow) used on 1991 and later cars.

Check and fill the transmission with the car on a level surface. Remove the dipstick from the side of the transmission wipe the stick clean. Then reinsert it fully and remove it. On 1985 through 1990 models, the oil level should be between the MAX and MIN marks. On 1991 and later models, the oil level should be between the two grooves cut on the dipstick. Add oil as necessary.

Manual Transmission Oil Specifications
- Grade of oil — SAE 10W-30 SF/CC, SF/CD, SG
- alternate grade of oil — SAE EP 75 API-GL-4 or API-GL-5
- Oil capacity — 3.1 quarts (3.0 liters)

Drive Axle Joint Boots

The inner and outer boots on the front drive axles should be closely inspected for cracks and any other damage that will allow contaminants to get into the joint. See Fig. 27. If the rubber boots fail, water and dirt can enter the joint and quickly damage it. Replacement of the drive axle joint boots and inspection of the joints are described in **774 Wheel Bearings and Drive Axles.**

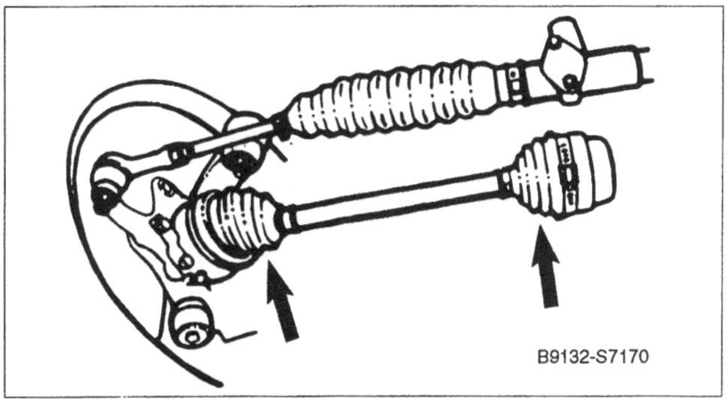

Fig. 27. Drive axle joint boots (inner and outer) should be inspected for tears and cracks. Also check that the boot clamps are tight and in good condition.

CHASSIS SERVICE

Brakes

Routine maintenance of the brake system includes maintaining an adequate level of brake fluid in the reservoir, checking brake pads for wear, checking hand brake function, and inspecting the system for fluid leaks or other damage.

WARNING
- Friction materials such as brake linings may contain asbestos fibers. Do not create dust by grinding, sanding, or cleaning the pads with compressed air. Avoid breathing asbestos fibers and asbestos dust, as it may result in serious diseases such as asbestosis and cancer, or in death.
- Brake fluid is poisonous. Do not siphon brake fluid by mouth. Wear gloves when working with brake fluid or brake pads to prevent contamination of cuts.

The level of the brake fluid will drop slightly as the brakes wear. Check the fluid level at the brake fluid reservoir, located on the driver's side, near the firewall. See Fig. 28. When filling the reservoir, use only new brake fluid from previously unopened containers.

Brake Fluid Grade
- SAE DOT 4

WARNING
Do not use SAE DOT 5 brake fluid. Brake system failure may result.

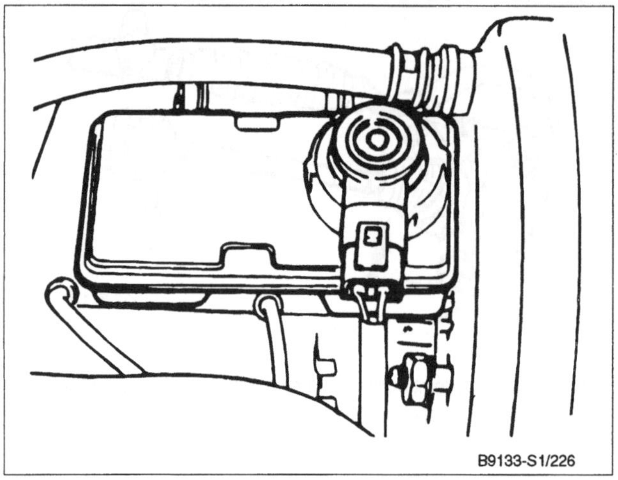

Fig. 28. Brake fluid reservoir on driver's side of engine compartment, at the firewall. Reservoir has MAX and MIN marks molded into its side.

To inspect brake hoses, gently bend them to check for cracks. Check that all hoses are correctly routed to avoid chafing or kinking. Inspect the unions and the brake calipers for signs of fluid leaks. Inspect the lines for corrosion, dents, or other damage. Replace faulty hoses or lines and bleed the brake system as described in **520 Brake Fluid and Brake Bleeding**.

WARNING

Incorrect installation or overtightening hoses, lines, and unions may cause chafing or leakage. This can lead to partial or complete brake system failure.

Brake pad thickness should be regularly checked. Brake pad thickness can be inspected by looking through an opening in the caliper after removing the wheel. See Fig. 29. For information on replacing worn pads, see **517 Brake Pads.**

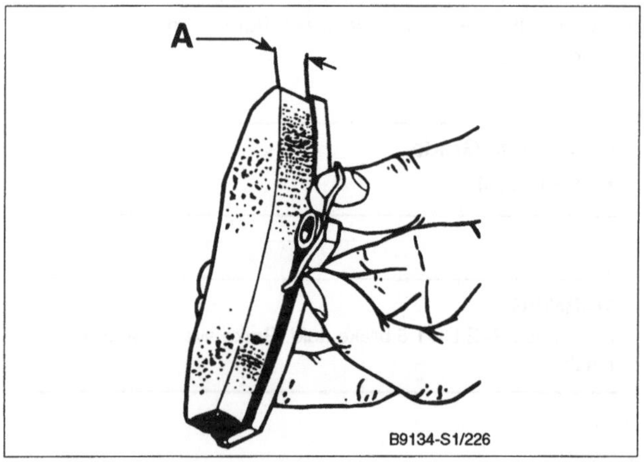

Fig. 29. Brake pad thickness (**A**)

Brake Pad Wear Limit (dimension A)
- 1985-1987 models . . . . 1.0 mm (0.04 in.)
- 1988 and later models . . . . 4.0 mm (0.16 in.)

Brake Fluid

Saab strictly recommends replacing the brake fluid every two years. Doing this will help protect against corrosion and the effects of moisture in the fluid. The procedure is described in detail in **520 Brake Fluid and Brake Bleeding**.

Parking Brake

The parking brake system is independent of the main braking system and does not require any periodic adjustment, although the function of the parking brake should be checked regularly. Check that the cable moves freely in its housing. A complete description of the parking brake is described in **551 Parking Brake.**

Power Steering Fluid

The only maintenance of the power steering system is to check the power steering fluid and add as necessary. The power steering fluid level can be easily checked through the translucent reservoir on the right side wheel housing in the engine compartment. See Fig. 30. When the car is at normal operating temperature, the fluid level should be between the HOT and COLD marks. If the level is checked when the car is cold, it should be between the COLD mark and the ADD mark.

Power Steering Fluid
- GM Power Steering Fluid (GM 9985010), Texaco TL4634 or equivalent

---

Windshield Washer Fluid

Check and add windshield washer fluid as necessary. The fluid reservoir in the front right wheel housing, directly behind the corner parking light assembly.

Ball Joints, Steering Joints (Tie Rod Ends)

The control arm ball joints and steering rack tie rod ends should be carefully checked for play with the front of the car raised. Also inspect the joint's rubber boots for damage. If the boots are torn open, water and dirt can enter the joint and quickly damage it. If the ball joint or tie rod boots are damaged, the complete joint will need to be replaced.

Fig. 30. Power steering fluid reservoir.

Replacement of the control arm ball joints is covered in **631 Ball Joints**. Replacement of the steering rack tie rod ends is covered in **643 Tie Rod Ends.**

Shock Absorbers and Bushings

Inspect the shock absorber bushings carefully, especially checking the lower bushing. See Fig. 31. If the lower bushing is worn, there is usually a constant clunking sound when traveling over rough road surfaces. The shock absorber's function should also be periodically checked. See **731 Front Suspension** for a quick check of the shock absorber condition.

NOTE
Individual shock absorber bushings are available through an authorized Saab dealer.

Tires

For stability and car control, the wheels and tires must be of the correct size and in good condition. Tires must be inflated to the recommended air pressures and the wheels must be in proper alignment. For maximum safety and best all-around handling, always install replacement radial tires having the same specifications. When possible, all four tires should be replaced at once, or at least in pairs on the front or rear. New tires do not provide maximum traction, and should be broken in gently for the first 100 miles (160 kilometers) or so.

NOTE
Be sure to also check the compact spare tire condition and inflation pressure. Keep the spare inflated to 60 psi.

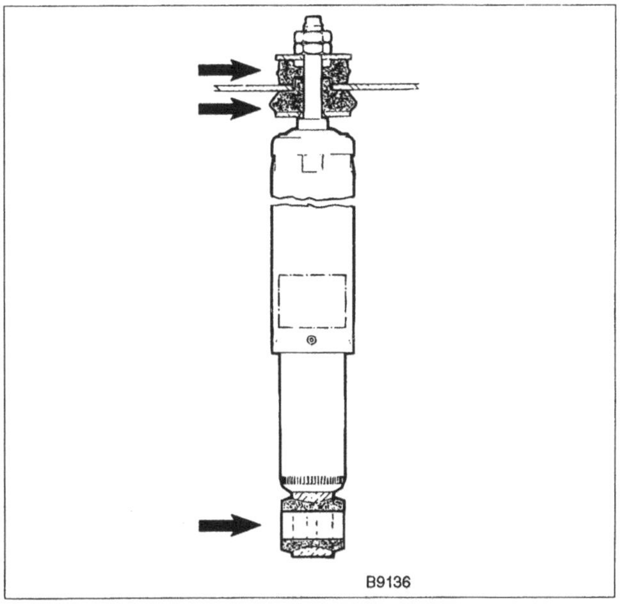

Fig. 31. Shock absorber bushings should be inspected (arrows).

Correct tire inflation pressures are important to handling and stability, fuel economy, and tire wear. Tire pressures change with temperature. Pressures should be checked often during seasonal temperature changes. The correct inflation pressures for cars covered by this manual are listed in the glovebox owner's manual. Tire pressures are also listed on a sticker located on one of the door jambs. Notice that the pressures should be higher when the car is more heavily loaded.

All inflation pressures are for cold inflation. That is, when the car has not been driven for at least three hours, or for more than one mile after sitting for at least three hours.

WARNING
Do not inflate any tire to a higher pressure than the tire's maximum inflation pressure listed on the sidewall. Use care when adding air to warm tires. Warm tire pressures can increase as much as 4 psi (0.3 bar) over their cold pressures.

To promote even wear and maximum tire life, Saab recommends rotating the tires from the front to the rear on the same side at the specified intervals listed earlier. Owing to the car's suspension design, the front tires begin to wear first at the outer shoulder and the rear tires begin to wear first at the middle of the tread or inner shoulder.

Tightening torques
- wheel lug nuts . . . . 90-110 Nm (66-81 ft-lb)
- wheel lug bolts . . . . 105-125 Nm (77-92 ft-lb)

Front Wheel Alignment

Saab recommends checking the front and rear wheel alignment at the specified interval or whenever new tires are installed. Only the front wheel alignment is adjustable. See **601 Wheel Alignment** for a more detailed discussion of alignment requirements and specifications.

BODY MAINTENANCE

Maintenance of the body includes lubricating various hinges, locks and slides.

Lubricate the seat runners with multipurpose grease. Do not apply any oil to rubber parts. Lubricate the sunroof guide rails with silicone spray. If door weatherstrips are sticking, lubricate them with silicone spray or talcum powder. The hood release cable should be lubricated as well. The radio antenna mast should be cleaned and lubricated with a product such as WD-40.

The door locks, hinges and lock cylinders can be lubricated with an oil that contains graphite. The body and door hinges, the hood latch, and the door check rods should be lubricated with SAE 30 or SAE 40 engine oil. See Fig. 32.

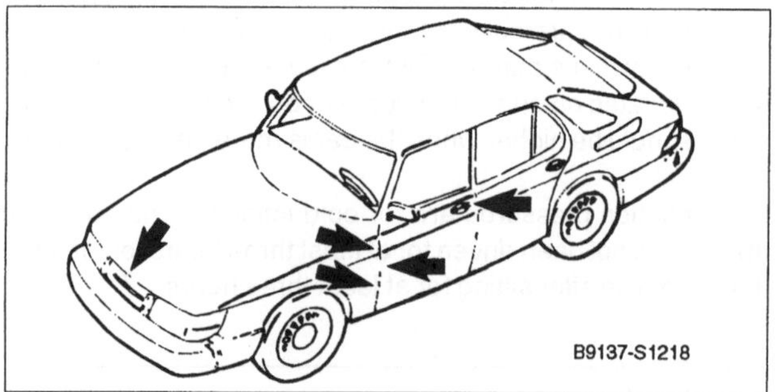

Fig. 32. Exterior body lubrication points.

Air Bag System

The air bag should be replaced every 10 years. See an authorized Saab dealer for all service relating to the SRS air bag system.

WARNING
On cars equipped with an SRS airbag system, do not attempt to service the air bag, the wiring in the steering wheel, the steering column assembly, or the airbag control unit. The airbag may ignite causing injury. See an authorized Saab dealer for all service and repairs to the SRS airbag system.

ROAD TEST

As a final check, the car should be taken for a test drive. Check the overall condition of the various systems, noting especially the function of the brakes, clutch, general engine performance and driving comfort. Check that the cruise control system and the air conditioning system are functioning correctly. See Fig. 33.

Fig. 33. Road test.

200 Engine—General

TABLES

GENERAL

Engine Assembly

Two basic engine configurations are used in the 16-valve cars covered by this manual. On all 1985 through 1990 models and on 1991 and later turbo models, a 2.0 liter engine is used. On 1991 and later non–turbo models, a 2.1 liter engine is used. See **Table a** for complete specifications.

Cylinder Head, Camshaft and Valve Train

The twin-overhead-cam cylinder head assembly is a light aluminum alloy casting. Bolts hold the cylinder head to the cylinder block. Replaceable valve guides are press-fit into the cylinder head.

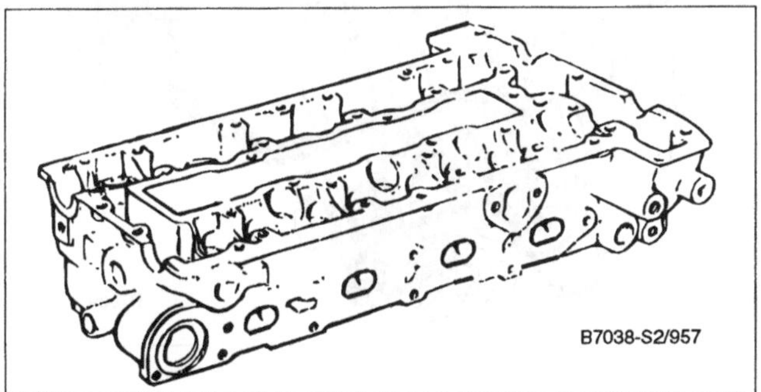

Fig. 1. Cylinder Head.

The camshafts are driven by a single chain and operate directly on the valves via hydraulic cam followers. The valves are made of steel and have chromium plated stems. The intake valve heads are induction-hardened and the exhaust valve heads are coated with stellite.

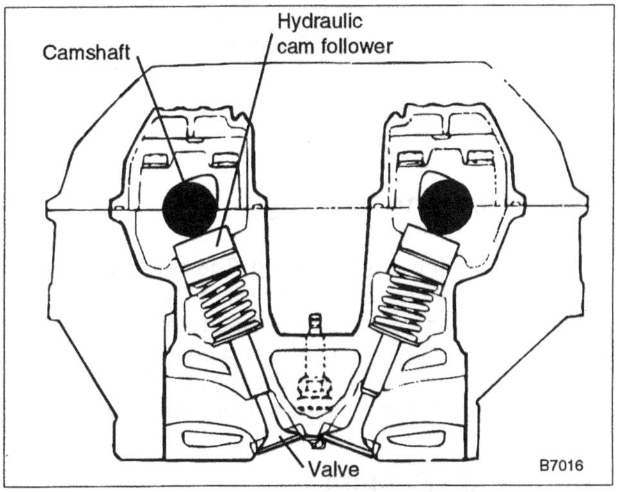

Fig. 2. Valve train.

Table a. Engine Specifications

Model	Year	Displacement cc (cu. in.)	Compression ratio	Horsepower @rpm SAE net	Torque lb-ft @rpm SAE net	Fuel System
900 900S Convertible	1986-1987	1985 (121)	10.1:1	125 @5500	128 @3000	Bosch LH 2.2
	1988	1985 (121)	10.1:1	128 @6000	128 @3000	Bosch LH 2.4
	1989-1990	1985 (121)	10.1:1	128 @6000	128 @3000	Bosch LH 2.4
	1991-on	2118 (129)	10.1:1	140 @6000	140 @2900	Bosch LH 2.4.2
900 Turbo Turbo Convertible	1985-1988	1985 (121)	9.0:1	160 @5500	188 @3000	Bosch LH 2.2
	1989-on	1985 (121)	9.0:1	160 @5500	188 @3000	Bosch LH 2.4
900 Turbo SPG	1985-1986	1985 (121)	9.0:1	160@5500	188@3000	Bosch LH 2.2
	1987-1988	1985 (121)	9.0:1	165 @5500	195 @3000	Bosch LH 2.2
	1989	1985 (121)	9.0:1	165 @5500	195 @3000	Bosch LH 2.4
	1990-1991	1985 (121)	9.0:1	175 @5500	195 @3000	Bosch LH 2.4

Cylinder Block

The cylinder block is made of cast iron. The cylinder bores are drilled straight out of the block and are surrounded by cooling jackets. The block also contains oilways, or channels, for the lubrication system.

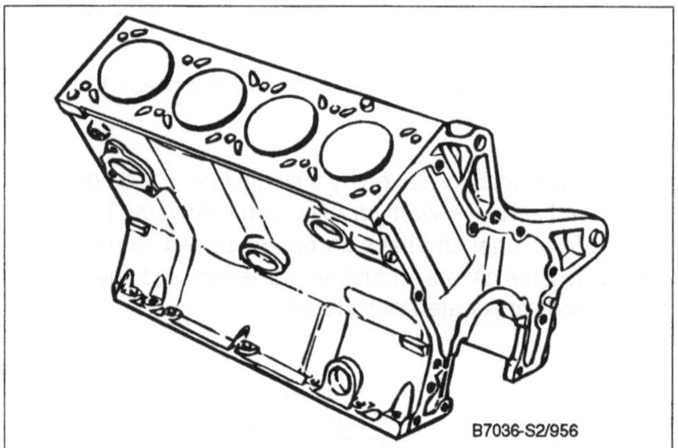

Fig. 3. Cylinder block.

Crankshaft and Bearings

The crankshaft is a steel forging. The crankshaft journals are ground, hardened, and treated with a "Tenifer" treatment. This provides a hard, non-metallic surface to help prevent bearing seizure. The crankshaft is supported by five main bearings. End float, or thrust, is controlled at the center bearing. Oilways are drilled into the shaft for bearing lubrication.

Lubrication System

The lubrication system has forced-flow circulation. The oil pressure is generated by a separate gear-type oil pump driven directly off the crankshaft. The oil is forced through a spin-

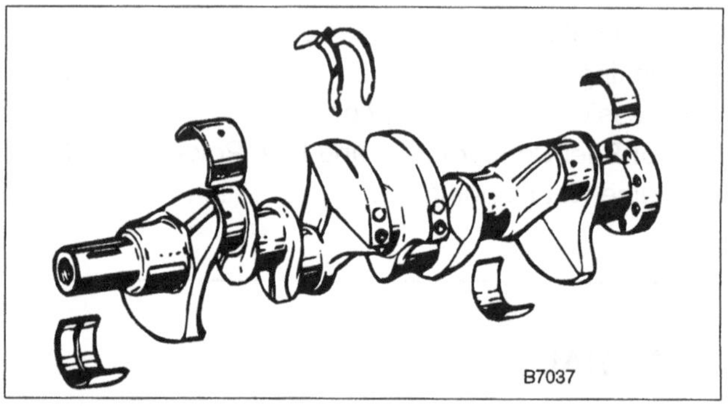

Fig. 4. Crankshaft and bearings.

on replaceable oil filter and into the engine oil passages. All models covered by this manual have a low-oil-pressure warning system. Turbo models use an engine oil cooler to help moderate engine oil temperature.

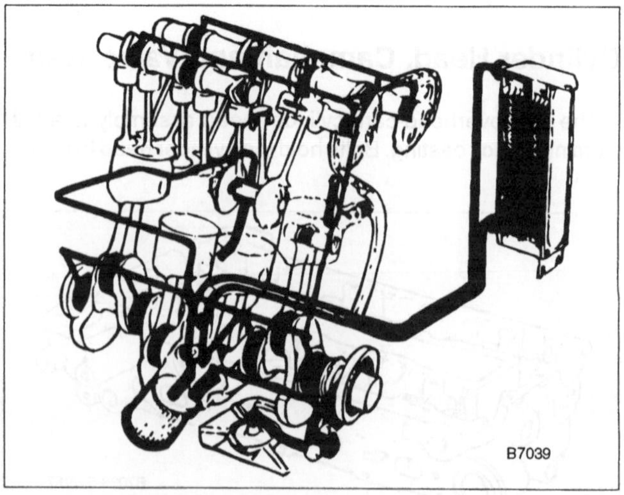

Fig. 5. Lubrication system.

Fuel Supply and Fuel Injection

The fuel supply system stores the fuel and provides the fuel injection system with a constant flow of pressurized, clean fuel. While there is some variation in components, all Saab 900 models have an electric, in-tank fuel pump and an in-line fuel filter. The injection-molded plastic fuel tank is located beneath the car, behind the rear seat. The fuel gauge sending unit and the electric fuel pump are mounted in the top of the tank.

All Saab 900 16-valve models use a Bosch LH fuel injection system. LH fuel injection is completely electronic in operation. Air flow is measured electronically, and a proportional amount of fuel is metered by electrically opening and closing the fuel injectors.

In LH fuel injection, many sensors supply information about engine operating conditions to a central electronic control unit. The control unit then calculates the amount of fuel needed for the correct air-fuel ratio and opens the fuel injectors, once for each engine revolution. The amount of fuel metered to the engine is determined solely by how long the injectors are open.

Cooling System

The engine cooling system relies on a closed system of circulating coolant to maintain an even engine temperature and help transfer heat away. The closed system becomes pressurized at normal engine operating temperature. Under pressure, the boiling point of the coolant is increased, allowing the engine and cooling system to operate at higher temperatures without boiling the coolant.

The coolant pump is driven mechanically by the engine using a V-belt. Coolant circulates through the engine to the radiator and heater core, and from the radiator back to the pump. Before the engine is up to normal operating temperature, the flow of coolant is controlled by a thermostat. When closed, the thermostat forces coolant to bypass the radiator and return directly to the coolant pump. This retains as much heat as possible until the engine is warm.

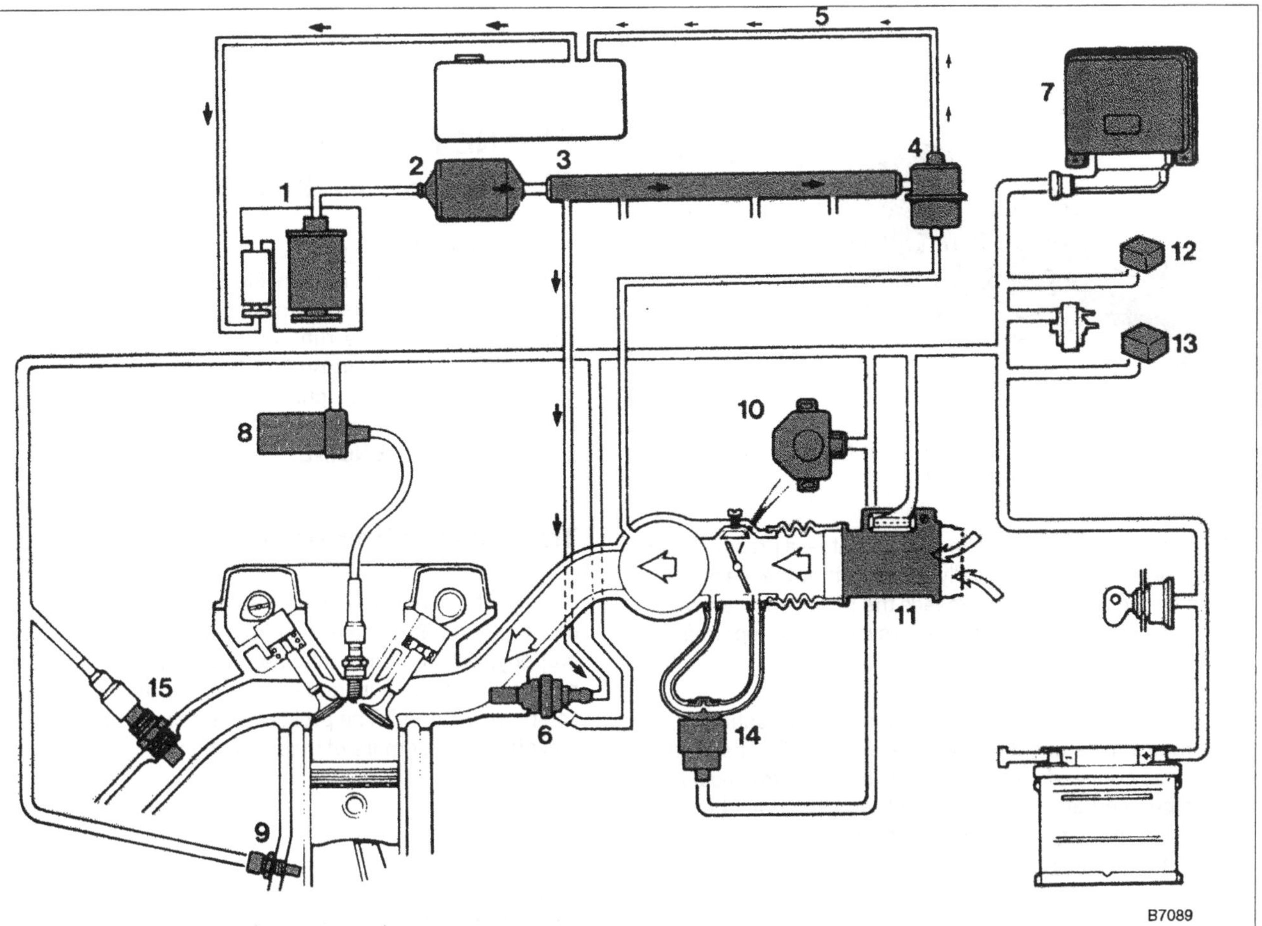

1. Fuel pump
2. Fuel filter
3. Fuel rail
4. Fuel pressure regulator
5. Fuel return line
6. Fuel injectors
7. Electronic control unit
8. Ignition coil
9. Temperature sender
10. Throttle position switch
11. Air mass meter
12. System relay
13. Fuel pump relay
14. Adaptive Idle Control (AIC) valve
15. Oxygen sensor

Fig. 6. LH fuel injection system.

An electric cooling fan, located behind the radiator, is thermostatically controlled so that it runs only when the coolant has reached a certain temperature. An auxiliary electric cooling fan is located next to the main cooling fan and operates in addition to the first fan when necessary or when the air conditioning system is on.

Turbocharger

The turbocharger is an exhaust-driven pump that forces air into the cylinders at higher than atmospheric pressure. This allows the engine to burn more fuel and therefore produce more power.

The turbocharger unit consists of two wheels mounted on a common rotating shaft. One of the wheels, called the turbine, is driven by the exhaust gas. The other wheel, called the compressor, is located in the intake air stream and compresses the intake air.

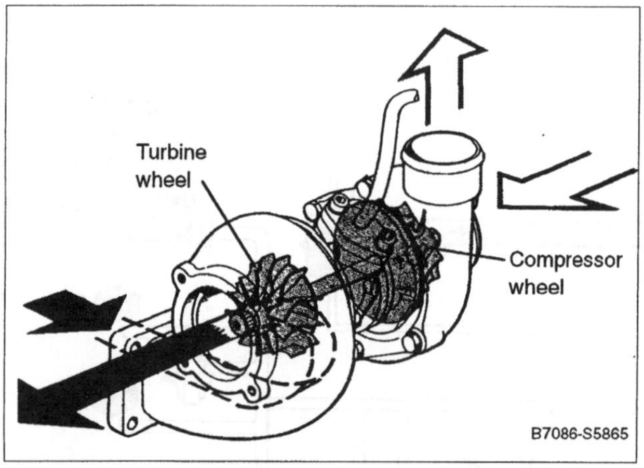

Fig. 7. Turbocharger, showing exhaust gas flow (black arrows) and intake air flow (white arrows).

The turbine is equipped with a waste gate—a pressure controlled valve that automatically bleeds off any excess pressure. If the boost pressure exceeds the predetermined value, in spite of the waste gate, a pressure switch will temporarily interrupt the fuel supply to the engine.

When the intake air is compressed by the turbocharger, its temperature rises and its density decreases. To ensure that the air is as cool and dense as possible to burn the most fuel, the compressed air is routed through an intercooler before it enters the combustion chamber. The intercooler can cool the air by as much as 110°F.

Automatic Performance Control (APC)

The APC system is used on all 16-valve turbo cars. The APC system continually monitors engine detonation or knock and adjusts the turbocharger boost pressure accordingly.

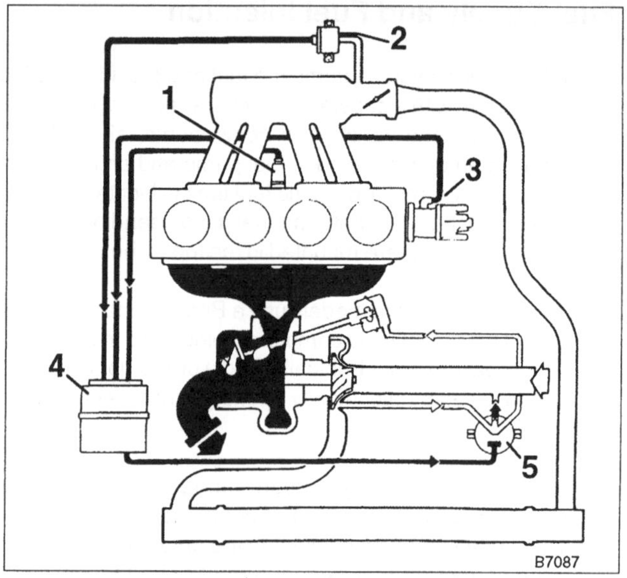

Fig. 8. The APC system control unit (**4**) receives input signals from three components: the ignition distributor (**3**), the knock sensor (**1**), and the pressure sensor (**2**). The control unit controls boost pressure via the solenoid valve (**5**).

Engine Mechanical Troubleshooting

This troubleshooting section applies to problems affecting the basic engine assembly, including the cylinder block, cylinder head, and their internal moving parts. For problems relating to how the engine runs and its overall driveability, see **Driveability Troubleshooting** below. For troubleshooting specific cooling system, ignition system, fuel system, or turbocharger/APC system problems, see the appropriate main repair group under **2 ENGINE**.

Troubleshooting Basics

Only a few basic functions are required of the engine. The block, cylinder head, and their moving parts must fit together properly, operate smoothly, seal well enough to create and maintain compression, and keep pistons, valve train, and ignition properly timed. The problems discussed below are those that affect one or more of these functions.

Noise

In order to run smoothly under harsh conditions, the internal engine parts are made to precise dimensions, assembled with precision clearances, and lubricated by a pressurized oiling system. Most unidentified engine noises result from clearances that have become too large due to worn parts, lack of adequate lubrication, or both.

Rumbling or groaning from the engine compartment may not indicate engine problems at all, but rather a worn bearing or bushing in an engine-driven accessory. They include the coolant pump, the alternator, the power steering pump, and the air conditioning compressor. To check these, run the engine briefly with the drive belt disconnected and see if the noise has stopped. Once the drive belt is removed, turning the pulley and shaft by hand may also reveal a bad bearing or bushing. A properly functioning accessory should turn smoothly.

Fluid Leaks

Fluid leaking from the engine is most likely either oil, coolant, brake fluid, or power steering fluid. Look for wet spots on the engine to help pinpoint the source. It may be helpful to start by cleaning the suspected area.

The most common areas of engine oil leaks are the timing chain cover, the engine end plate (flywheel end), the crankshaft oil seals, the cylinder head cover gasket, the transmission-to-engine mating surface, and the oil filter seal.

A pressure test of the cooling system is the best way to discover and pinpoint coolant leaks. Coolant is a mixture of water and anti-freeze, yellow-green in color or perhaps brown if the coolant is old. See **261 Radiator and Cooling System** for more information.

The brake system and power steering system are other sources of leaks near the engine. Brake fluid is clear, perhaps slightly purple, and a little slippery. Look for wet spots around the master cylinder or brake lines. Especially check the flexible hoses near the wheels. See **5 BRAKES** for repair information.

Smoking

Smoke that is visible at the tailpipe is usually either blue-gray smoke from burning oil, or white steam from the cooling system. The color of the smoke identifies the contaminant.

Blue-gray oil smoke in the exhaust usually indicates general engine wear. If smoking is most obvious under high engine vacuum, such as while coasting at high rpm, valve guide oil seals or worn valve guides are the most likely cause.

NOTE

- See **291 Turbocharger** for additional troubleshooting where applicable.

- Oil smoke or steam appearing suddenly in the exhaust, along with low compression pressure in one or more cylinders, is probably due to a failed cylinder head gasket. Look for bubbles in the coolant or coolant loss, oil in the radiator, or water in the oil (that turns the oil an opaque, creamy brown).

- Black smoke is caused by the engine getting too much fuel. See **240 Fuel Injection** for more troubleshooting information.

Excessive Oil Consumption

Some oil consumption is normal. This is why the oil level must be checked, and occasionally corrected, between oil changes. Aside from leaks, increased oil consumption will usually be accompanied by some smoking, however slight. The causes of excessive oil consumption are the same as those for oil smoke in the exhaust. As with smoking symptoms, gradual increases are caused by worn piston rings, valve guides or a failed turbocharger, where applicable. Sudden high oil consumption suggests broken piston rings.

Poor Fuel Consumption and Low Power

Poor fuel consumption and low power can suggest problems with the fuel, ignition, or turbocharger systems, particularly on a low-mileage engine. On an engine with high mileage, engine wear and low compression may be the cause. Normal wear of the valves, piston rings, and cylinder walls decreases their ability to make a good seal. The engine becomes less efficient and has to work harder, using more fuel to produce the same amount of power.

All 16-valve naturally-aspirated engines are equipped with Bosch EZK ignition. All 16-valve turbo models are equipped with an Automatic Performance Control (APC) system. These systems continuously monitor engine detonation to adjust either ignition timing (EZK) or turbocharger boost pressure (APC) to achieve maximum engine performance.

If either the EZK or APC system detects a fault, a set of predetermined values are used to prevent engine damage caused by detonation. This means the engine is slightly detuned to safely run on all gasoline octane ratings. If low power is experienced, be sure to check the operation of these systems. The EZK system is covered in **340 Ignition System.** The APC system is covered in **291 Turbocharger.**

NOTE

- In order for the EZK and APC system to operate most efficiently, the engine is allowed to knock occasionally. A limited amount of knock is considered normal.

- The EZK control unit has built-in diagnostic capabilities. A fault will show up as a series of pulses on the "**CHECK ENGINE**" light. These pulses will continually repeat themselves with a brief pause between sequences. For more information on EZK on-board diagnostics, see **340 Ignition System.**

Engine Not Running

An engine problem that affects ignition or valve timing may prevent the engine from starting or running. The camshaft timing chain and sprockets are responsible for correctly timing the actions of the valves and the ignition system. A worn timing chain may jump teeth, throwing off all the engine's timing functions, and still appear to be perfectly normal. To accurately check camshaft timing, see **215 Camshaft Timing Chain**.

Table b lists additional symptoms of engine mechanical problems, their probable causes and suggested corrective actions. The boldface numbers in the corrective action column indicate the heading in this section of the manual where the applicable test and repair procedures can be found.

DRIVEABILITY TROUBLESHOOTING

This heading covers general engine management principles and the basic requirements that allow an engine to run smoothly. Therefore, effective troubleshooting of specific running conditions can only take place after all of the common problem areas listed below have been eliminated as a source of trouble.

Most driveability problems are complex in nature. Therefore, a logical method needs to be used to isolate the trouble area. Always begin with the simplest and most fundamental engine management basics. Jumping to conclusions or

Table b. Engine Mechanical Troubleshooting

Symptom	Probable cause and corrective action
1. Pinging or rattling noise under load, uphill or accelerating, especially from low speeds. Indicates detonation or pre-ignition	**a.** Fuel octane level too low, use higher octane fuel **b.** Engine running too hot, overheating. Check for cooling system faults. See **261 Radiator and Cooling System** **c.** Spark plugs damaged or wrong plug type installed. Replace plugs. See **340 Ignition System** **d.** Air-fuel mixture too lean due to air leaks or low fuel pressure. See **240 Fuel Injection**
2. Light metallic tapping noise, varies directly with engine speed. Oil warning light **not** illuminated	**a.** Low oil pressure, defective warning light circuit. See **220 Lubrication System** **b.** Faulty hydraulic cam follower. See **214 Cylinder Head and Valve Mechanism**
3. Light metallic knock, varies directly with engine speed. Oil warning light blinking or fully illuminated (may be most noticeable on hard stops or cornering)	**a.** Low oil level, correct if necessary **b.** Oil flow restricted. Replace oil filter. See **1 LUBRICATION AND MAINTENANCE** **c.** Low oil pressure. Check oil pressure. See **220 Lubrication System**
4. Blue-gray exhaust smoke, oil fouled spark plugs. Indicates oil burning in combustion chamber	**a.** Faulty valve stem oil seals or valve guides. See **211 Cylinder Head Removal and Installation** **b.** Worn, broken, or incorrectly installed pistons or piston rings. See **212 Pistons, Connecting Rods, Cylinder Bores**
5. White steam in exhaust	**a.** Failed cylinder head gasket or warped cylinder head (probably accompanied by low compression). Replace or resurface cylinder head and gasket as necessary. See **211 Cylinder Head Removal and Installation** **b.** Cracked cylinder block. Replace engine or short block. See **212 Pistons, Connecting Rods, Cylinder Bores**
6. Black exhaust smoke	**a.** Air-fuel mixture too rich due to faulty fuel injection component. See **240 Fuel Injection**
7. Screeching or squealing noise under load. Goes away when coasting. Indicates slipping V-belt	**a.** Loose, worn, or damaged V-belt(s). Inspect, replace, or tighten belt(s). See **1 LUBRICATION AND MAINTENANCE**
8. Growling or rumbling, varies with engine rpm. Bad bearing or bushing in an engine-driven accessory	**a.** Remove belt from engine driven accessory and check for play, bearing roughness, or loose mountings
9. Check engine light illuminated or flashing	**a.** EZK ignition system on-board diagnostics fault detected. Check fault codes. See **340 Ignition System** **b.** LH fuel injection control on-board diagnostics fault detected. Check fault codes. See **240 Fuel Injection.**
10. Engine will not start or run. Starter operates, engine turns over at normal speed	**a.** No spark to spark plugs. Check ignition system. See **340 Ignition System** **b.** Incorrect valve timing. Check camshaft timing chain for wear. See **215 Camshaft Timing Chain** **c.** No fuel being delivered to engine. See **240 Fuel Injection**

searching aimlessly for the problem can be time consuming and frustrating.

NOTE

Models with EZK ignition and models with 2.4 LH and 2.4.2 LH fuel injection have built-in diagnostic capabilities which will light the "**CHECK ENGINE**" light. For more information on EZK diagnostics, see **340 Ignition System.** For more information on LH fuel injection diagnostics, see **240 Fuel Injection.**

As with any troubleshooting, careful observation of symptoms is the key to identifying and isolating driveability problems. A test drive can help by demonstrating when the problem is most pronounced, such as a hesitation that occurs only when the engine is cold, or a steady miss at high speed.

How has the symptom developed? A symptom that develops quickly is probably caused by a problem that can be corrected by simple maintenance or repair. A symptom that has developed gradually over time, especially after sixty or seventy thousand miles is more likely an indication of general wear and the need for more comprehensive work.

Warnings and Cautions

For general safety, and to protect the sensitive electronic components, the following warnings and cautions should be adhered to during any troubleshooting, maintenance, or repair work. Always follow the proper repair and working procedures in the sections that are referenced.

WARNING

- The ignition systems used on the cars covered by this manual are high-energy systems operating in a dangerous voltage range. Exposure to terminals or live parts could prove to be fatal. Use extreme caution when working on a car with the ignition on or the engine running.

- Do not touch or disconnect any of the high tension cables from the coil, distributor, or spark plugs while the engine is running or being cranked by the starter.

- Connect and disconnect ignition system wires, multiple connectors, and ignition test equipment leads only while the ignition is switched off.

- Before operating the starter without starting the engine, disable the ignition. See **340 Ignition System.** Do not disconnect terminal 4 (center terminal) from the coil or remove the distributor cap as a means of disabling the ignition.

- During any test where fuel is discharged, do not smoke or work near heaters or other fire hazards. Have a fire extinguisher handy.

CAUTION

- Do not connect test instruments with a 12-volt power supply to terminal 15 (+) of the ignition coil. The current flow will damage the ignition control unit. In general, make test connections only as specified by Saab, as described in this manual, or as described by the test instrument's manufacturer.

- Do not disconnect the battery while the engine is running.

- Do not exceed 16 volts at the battery with boosting cables attached, and do not quick-charge the battery (for boost starting) for longer than one minute. Wait at least one minute before boosting the battery a second time.

- Do not connect terminal 1 (-) of the ignition coil to ground as a means of preventing the engine from starting.

- Running the engine with a spark plug wire disconnected may damage the catalytic converter.

- Cleanliness is essential when working with fuel circuit components. Before disconnecting any fuel lines, thoroughly clean the unions. Use clean tools.

Mechanical Condition

Before troubleshooting a poorly running engine or an engine that will not start, determine the general condition of the engine, especially if it has high mileage. If the engine is severely worn or has mechanical problems, the only remedy is overhaul or repair. If a tune-up or scheduled maintenance is due, it should be done before proceeding further with troubleshooting.

Only a few basic functions are required of the engine. The parts must fit together properly, operate smoothly, and seal well enough to create and maintain compression, and keep pistons, valve train, and ignition properly timed. General engine condition can be easily assessed by performing a compression test.

NOTE

Saab does not give specific compression pressure specifications. In addition, a compression test requires special equipment. If the equipment is not available, most automotive repair shops can do these tests quickly and at a reasonable cost.

Carbon Deposits

Carbon deposits on the fuel injectors and the intake valves will affect the way the engine idles and runs. See Fig. 9. Even a ten percent decrease in the amount of fuel that the injectors deliver will cause driveability problems. These deposits normally form during the "hot soak" period immediately after the engine is turned off, at which point the engine temperature rises slightly for a short period.

Fig. 9. Examples of carbon deposits on fuel injector (left) and intake valve (right). Carbon deposits can cause a rough idle, hard cold starting, and overall poor performance.

Driving style may be the main contributor to the problem. A car that is predominantly driven on short trips around town or in city traffic seems to increase the likelihood of deposits forming.

Carbon deposits on the intake valves and injectors should be removed prior to troubleshooting driveability problems. Special fuel injector test equipment is required to accurately check for clogged injectors. If the injectors are severely clogged, they can be removed and visually inspected. Inspecting the intake valves is more difficult because the intake manifold must be removed. Check with an authorized Saab dealer for the latest information on carbon deposits and the best methods to remove them.

CAUTION

Always follow the manufacturer's directions when using fuel additives designed to remove carbon deposits and clean injectors. As a general rule, high detergency fuels should not be used together with fuel additives. The additional cleaner in the fuel can dilute engine oil and accelerate engine wear.

Tune-up and Preventive Maintenance

A tune-up is regular maintenance of the ignition and fuel system components for normal wear and contamination. The condition of the tune-up and emission control components can affect engine performance and driveability. The ignition components all carry high voltage to deliver a precisely timed spark to ignite the air-fuel mixture. If any of these components are faulty or worn, the intensity and timing of the spark will be affected. Dirt contamination of fuel system and emission control components can adversely affect fuel delivery, air-fuel mixture, and exhaust emissions.

When experiencing driveability problems, a good starting point is to do a tune-up. Many driveability problems are eliminated by simply replacing these components.

Replacement schedules and procedures for the spark plugs, spark plug wires, distributor cap, ignition rotor, fuel filter, air filter, oxygen sensor, and oil and oil filter are given in **1 LUBRICATION AND MAINTENANCE.**

NOTE

For information on inspecting ignition components, see **340 Ignition System.** For information on checking the oxygen sensor, see **254 Exhaust Emission Control.**

Basic Adjustments

In addition to tune-up component replacement, it is important that all of the basic adjustments that can be made are correctly set. Check idle speed, idle mixture (%CO), and ignition timing to be sure they are all within the specified limits. See **1 LUBRICATION AND MAINTENANCE.**

NOTE

All of the basic adjustments require the use of specialized test equipment. If any of the test equipment is not available, it is recommended that the adjustments be done by an authorized Saab dealer or other qualified repair shop. These adjustments can be made quickly and at a reasonable cost.

The systems that adapt the idle mixture (Lambda system), idle speed (AIC system), ignition timing (EZK system, where applicable), and turbo boost pressure (APC system) can only correct engine operation within a limited range. Once these limits are exceeded, driveability problems will become noticeable. For example, the oxygen (Lambda) sensor can adapt idle mixture for things such as a small vacuum leak or minor engine wear. A large vacuum leak or a severely worn engine may exceed the operating range of the sensor, causing the engine to run lean. If large adjustments are necessary, the faults that are causing these incorrect settings should be corrected prior to making any adjustments.

Oxygen Sensor

The oxygen sensor, also called the Lambda sensor, affects the air-fuel mixture by sending a varying voltage signal to the fuel injection control unit. The sensor is positioned in the exhaust stream and actually measures the amount of oxygen in the exhaust gas so that the fuel injection system can correctly adjust the air-fuel mixture. See Fig. 10. A high concentration of oxygen in the exhaust gas indicates a lean mixture and a low content indicates a rich mixture.

NOTE

The control unit ignores the signal from the oxygen sensor until the engine reaches a specified temperature. Therefore, when troubleshooting cold engine driveability problems, the oxygen sensor can be ruled out as a possible cause.

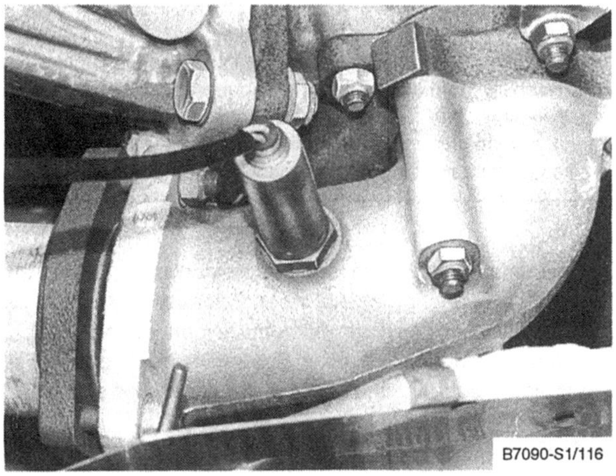

Fig. 10. Oxygen sensor reacts to oxygen content in exhaust gas to constantly monitor air-fuel mixture.

As the sensor ages, it loses its ability to react quickly to changing conditions. It may eventually cease to produce any signal at all. When this happens, idle speed may fluctuate and fuel consumption may increase. The oxygen sensor should be replaced at the specified mileage interval as described in **1 LUBRICATION AND MAINTENANCE.**

Air Flow Measurement and Vacuum Leaks

The Bosch LH fuel injection system uses an air mass meter to precisely measure incoming air. The air mass meter sends an electrical signal to the LH control unit proportional to the amount of air passing through the meter. This is the main input used to determine the amount of fuel delivered to the engine.

Because proper fuel metering depends on accurately metering the intake air, any unmeasured air entering the system will cause a lean fuel mixture and poor running. To see how air leaks can affect engine running, remove the oil filler cap while the engine is running.

There are many possible places for unmeasured air to enter the engine. Carefully inspect for cracks or looseness all hoses, fittings, duct work, and seals and gaskets. See Fig. 11. For a thorough inspection, it may be necessary to remove hoses and ducts that cannot be completely checked in their installed positions. Check that the throttle mounting nuts are tight and that the throttle plate is closed.

Use a vacuum gauge connected downstream of the throttle plate to check engine vacuum. Start the engine and let it idle. If the system is operating correctly, engine vacuum should be within the range listed below.

Engine Vacuum
- at idle (850±50 rpm) . . 12 to 23 in. Hg (-.4 to -.8 bar)

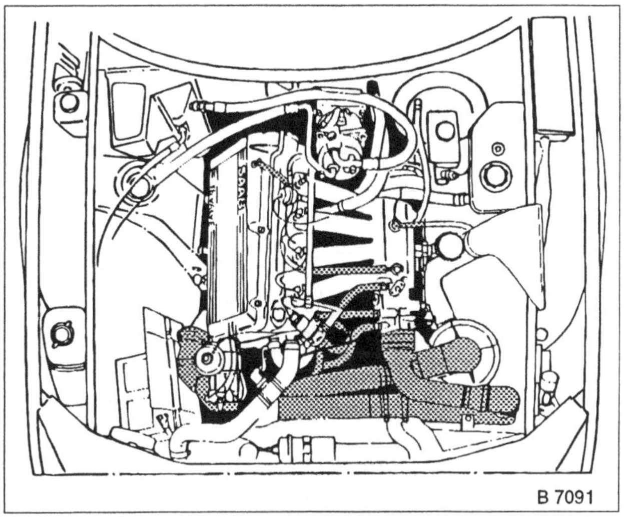

Fig. 11. Common places to check for intake air leaks (shaded areas). Intake air leaks can cause rough running, especially at idle.

In addition to air leaks, an air restriction can cause driveability problems. Remove the air filter and check that it is clean. Hold the filter up to a strong light source. If the light cannot be seen through the filter it should be replaced. Remove the air bellows from the throttle housing and check for residue around the throttle plate. Large amounts of residue on the throttle plate can cause an erratic idle. Clean the throttle housing with a carburetor cleaner and reinstall the bellows. Check the AIC idle setting as described in **240 Fuel Injection**.

NOTE
- On turbo models, be sure to check the hoses and plastic hose clamps on the turbo bypass valve. The valve is located on the compressor side of the turbocharger. See Fig. 12. If the car exhibits minor driveability problems and no other vacuum leaks can be found, check the operation of the bypass valve as described in **291 Turbocharger.** The valve may be faulty and leaking internally.
- Some early 1985 models may not have a turbo bypass valve.

Electrical System

All the cars covered by this manual use fuel injection and ignition systems that rely on precise electrical signals for proper operation. If any of these signals are distorted, incorrect, or missing, the car can develop major driveability problems.

Battery Voltage

One of the most fundamental requirements in troubleshooting engine performance problems is to make sure the battery is fully charged and in good condition. Many of the sophisticat-

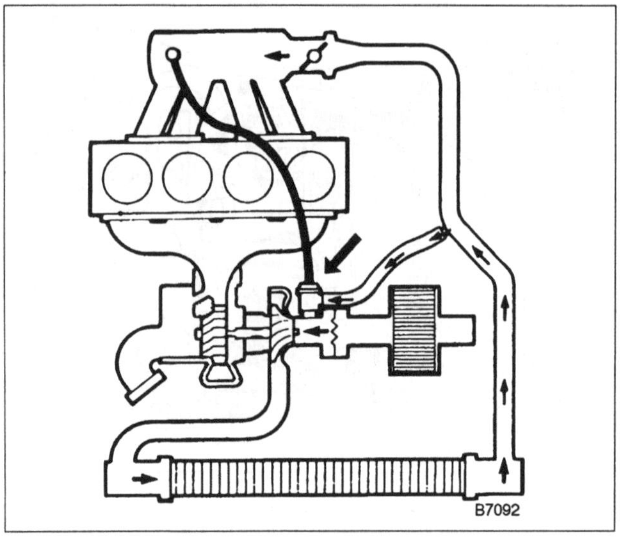

Fig. 12. Leaking turbo bypass valve (arrow) can be the source of minor driveability problems.

ed electronics used on the cars covered by this manual require a specific operating voltage to function correctly.

Battery voltage can be measured across its terminals with all cables attached. Do not eliminate the battery as a possible source of trouble until a load test has been performed, especially if starting problems are encountered. See **300 Electrical—General** for battery testing information.

NOTE

A digital voltmeter should be used to measure battery voltage. A fully charged battery will measure 12.6 volts, or more, as compared to a battery only 25% charged that measures 12.15 volts. Using an analog meter may result in inaccurate results.

For the battery to maintain its proper voltage level, the charging system must be functioning correctly. If in doubt about the condition of the charging system, have the system checked. Most automotive repair shops can test the system quickly and at a reasonable cost.

Wiring and Harness Connectors

The cars covered by this manual are equipped with electronic fuel injection and ignition systems that are controlled by central electronic control units. Many of the circuits operate on very low current and are sensitive to increased resistance due to faulty or corroded wiring or connectors.

NOTE

In most cases, faulty electronic control units are not the cause of driveability problems. These units are extremely durable and reliable. Actual failures are very rare. Driveability problems are more often caused by missing or incorrect signals to the control unit, or other faulty components.

The electrical system is subject to corrosion, vibration, roadway elements and general wear. Because of this, the integrity and freedom from corrosion in the connections, wires, and switches, including all ground connections, is one of the most important conditions for trouble-free operation of the fuel injection and ignition systems.

Always make a thorough visual inspection of all wires and connectors, switches and fuses. Loose or damaged connectors can cause intermittent problems, especially at the small terminals in each control unit connector. In most cases, a visual inspection will detect any faults. If a connector shows no visible faults, but is still suspect, perform a voltage drop test at the connector. Even a small amount of corrosion in a connector can cause a large voltage drop to the circuit's load. See **300 Electrical—General** for more electrical troubleshooting information.

Ground Connections

For any electrical circuit to work, it must make a complete path, beginning at the negative (-) battery terminal and ending at the positive (+) terminal. The negative (-) battery cable runs directly to the car's chassis. Therefore, connecting a wire to the chassis or any metal part bolted to the chassis provides a good ground source or path back to the negative (-) side of the battery.

Poor ground connections are one of the major sources of driveability problems. There are only a few main ground connections or points for the fuel injection and ignition systems. If any of these ground points are faulty, the voltage to the circuit will be reduced or even eliminated. Be sure to carefully check the main ground(s) for the LH system at the engine lifting bracket on the intake manifold. Also check the ground for the fuel pump at the center crossmember in the luggage compartment.

When checking ground wires, ground points, or ground straps, begin with a thorough visual inspection. Ground connections and wires can corrode, become loose, or break in areas that are not visible. To thoroughly check a circuit ground, check the voltage drop between the connector and a good ground source. Large voltage drops indicate too much resistance. The connection is corroded, dirty or otherwise damaged. Clean or repair the connection and retest. Also check both battery terminals and all ground straps between the engine and the body for voltage drops.

NOTE

For voltage drop tests and other general electrical troubleshooting information, see **300 Electrical—General.**

Fuel System

For the engine to start quickly when cold or hot, run correctly throughout all operating conditions, and accelerate smoothly without hesitation, the fuel system must deliver a precise amount of fuel in relation to the amount of air that is drawn in by the engine.

Fuel Supply

To start and run, the engine needs an adequate supply of fuel. Fuel from the tank is supplied to the engine via an electric fuel pump, a fuel filter, and the connecting fuel lines. If either the filter or a fuel line is restricted, the engine may not run properly. If the restriction is severe, or the main fuel pump is faulty, the engine may not start at all. If fuel delivery problems are suspected, check the fuel pump and pump relay as described in **234 Fuel Pump and Fuel Tank.**

NOTE

On some turbo models, a faulty turbocharger over-pressure switch can cause a no-start condition. The turbocharger over-pressure switch is designed to turn off the fuel pump during overboost conditions. Test this switch as described in **291 Turbocharger.**

The fuel injection system has the main function of delivering the correct amount of fuel for all engine operating conditions. Basic fuel delivery is dependent on fuel pressure and the correctly functioning injectors. Fuel pressure is often overlooked when diagnosing driveability problems.

The fuel pressure from the fuel pump is controlled by a pressure regulator. This regulator returns surplus fuel to the fuel tank. The amount of fuel delivered to the engine is varied by changing the amount of time the electric pulsed-type fuel injectors remain open. A change in fuel pressure results in a change in the amount of fuel (or fuel mixture) delivered to the engine. Fuel pressure and fuel pressure regulator tests are described in **240 Fuel Injection.**

NOTE

Fuel pressure tests require the use of a pressure gauge. If this equipment is not available, fuel pressure tests can be performed by an authorized Saab dealer or other qualified shop.

Correctly operating fuel injectors play a major role in fuel delivery. The injectors are switched on and off at the ground side of the connector. Positive (+) battery voltage is always present at the connector when the car is running. An injector can become clogged, it can completely fail or lose power and refuse to open, or it can short to ground and remain open whenever the engine is running. Checking if an injector is fundamentally working can be accomplished easily; checking an injector's spray pattern is more difficult. See **240 Fuel Injection** for additional information.

NOTE

High or low fuel pressure or a faulty injector will result in an incorrect fuel mixture and overall poor driveability. A lean mixture (too little fuel) can cause the engine to run poorly when cold or stumble upon acceleration. A rich mixture (too much fuel) can dilute the engine oil, foul the spark plugs, and cause a rough idle.

Table c lists some of the more common driveability symptoms, causes and corrective actions. The boldface type indicates the repair group where the applicable tests and repair procedures can be found.

Table c. Engine Management—Driveability Troubleshooting

SYMPTOMS

a. Engine fails to start
b. Engine starts but stops immediately
c. Erratic idling
d. Poor response on acceleration
e. Engine erratic in all speed ranges
f. Excessive fuel consumption
g. Engine fails to rev up (no power)
h. CO content too low
i. CO content too high

a	b	c	d	e	f	g	h	i	CAUSES	CHECK/REMEDY
a									Fuel pump not running	Test fuel pump. **234**
a	b								Ignition system faulty	Test ignition system components **340**
a	b								System relay faulty	Test system relay. **240**
a	b								Fuel pump relay faulty	Check fuel pump relay. **234**
a	b	c							AIC idle valve faulty	Test AIC valve. **240**
		c							EGR valve sticking or faulty	Test and clean EGR valve. **254**
	b				f				Throttle position sensor faulty or out of adjustment	Test throttle position sensor and adjust if necessary. **240**
		c	d	e	f	g		i	Air mass meter faulty	Test air mass meter. **240**
a	b	c	d	e		g	h		Induction (intake) system leaking	Checks for air leaks. Make a vacuum gauge test. **200**
a		c	d	e		g	h	i	Fuel pressure too low	Measure fuel pressure. **240**
					f			i	Fuel pressure too high	Measure fuel pressure. **240**
			d	e		g	h		Fuel pump output too low	Make fuel delivery test. **234**
a	b	c	d	e	f	g	h	i	Coolant temperature sensor faulty	Test sensor resistance, replace if necessary. **240**
		c	d	e	f				Oxygen sensor faulty	Test oxygen sensor output signal. Replace sensor if mileage interval exceeded. **254**
		c		e	f			i	Fuel injectors leaking	Check fuel injectors for leaks. Replace leaky injectors. **240**
					f	g			Binding throttle plate, or throttle plate incorrectly adjusted	Check throttle plate and adjust if necessary. **240**
a	b	c	d	e	f	g	h	i	Electrical connection(s) loose, broken, or corroded	Visually inspect electrical connectors and correct any faults found. **300**
a	b	c	d	e	f	g	h	i	Ground point(s) loose, broken, or corroded	Visually inspect ground points and correct any faults found. **300**
	b	c			f				CO content too high	Test air mass meter. **240**
	b	c	d						CO content too low	Test air mass meter. **240**
a	b	c	d	e	f	g	h	i	Input(s) signals to LH control unit missing	Make electrical tests at control unit connector. **240**
						g			Turbocharger APC system faulty	Test APC system components. **291**
a	b	c	d	e	f	g	h	i	LH control unit faulty	Test inputs to control unit. If all inputs are correct, replace control unit. **240**

201 Engine Removal and Installation

REMOVING AND INSTALLING ENGINE

The engine is removed together with the transmission as a complete assembly. The removal and installation procedure described below applies specifically to 1988 turbocharged models with manual transmission. Minor departures from the procedures may be necessary depending on model year, engine and transmission.

To remove engine and transmission assembly

1. Open the hood and disconnect the windshield washer hoses. Remove the hood hinge mounting bolts and carefully lift off the hood. See Fig. 1.

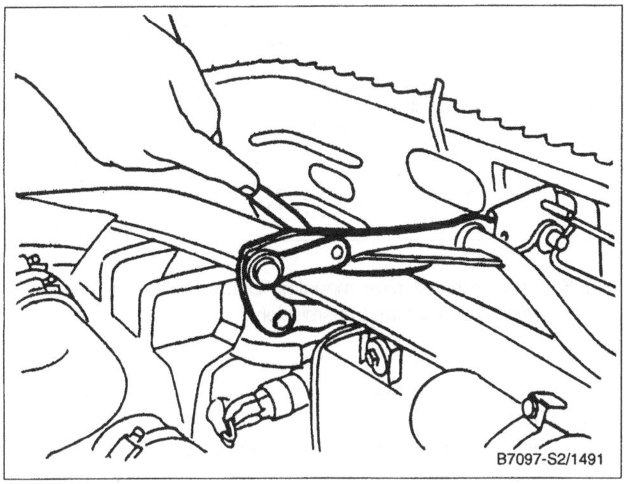

Fig. 1. Hood hinge bolt being removed.

2. Using the steering wheel, turn the right-hand wheel to its full right-lock position. Then insert Saab special tool no. 83 93 209 between the upper control arm and the body as shown in Fig. 2.

NOTE
The Saab special tool is used to hold the suspension control arm in its loaded position while the car is raised. This aids in disconnecting the drive axle inner joint from the transmission as the engine is pulled.

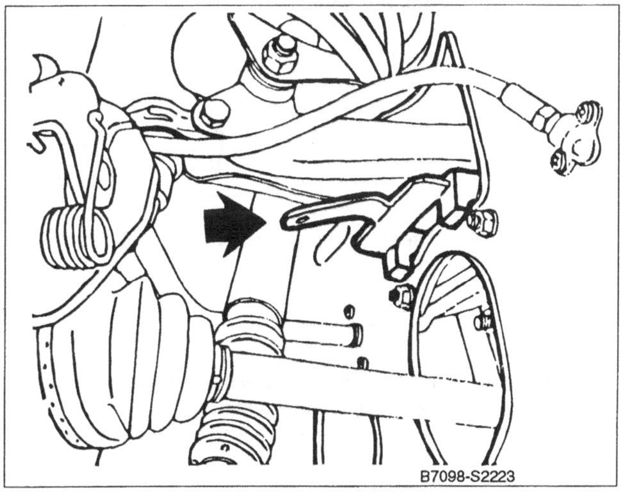

Fig. 2. Special tool (arrow) inserted between body and suspension control arm to limit suspension travel.

3. Disconnect the battery leads and remove the battery.

4. Disconnect the exhaust pipe from the exhaust manifold. See **252 Exhaust System.**

5. Loosen the right-hand wheel lug bolts.

6. Drain the power steering fluid reservoir. See **644 Power Steering.**

7. Raise the car enough to work safely under it and support it securely on jackstands.

WARNING
Never work under a lifted car unless it solidly supported on stands designed for the purpose. Do not work under a car supported solely by a jack.

8. Remove the nut from the shift selector rod taper pin and tap the pin out of the rod. See Fig. 3. Disconnect the rod from the shift linkage.

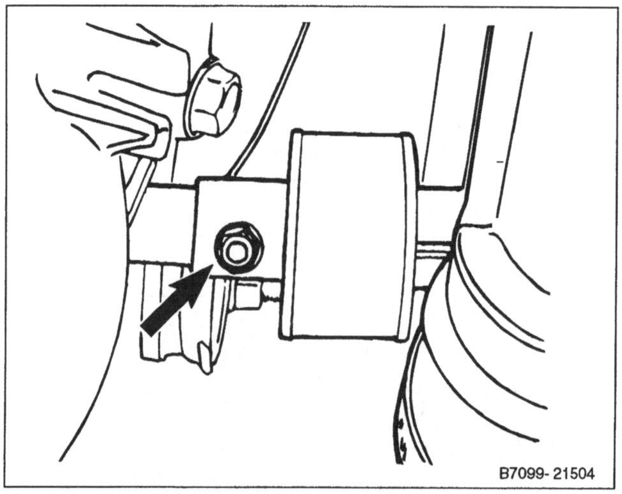

Fig. 3. Manual shift selector rod taper pin retaining nut (arrow).

NOTE
For removal of the shift cable on models with automatic transmission see **444 Automatic Transmission Controls**.

9. Thoroughly clean the area around the speedometer cable at the left-hand differential-bearing housing. Unscrew the speedometer cable from the housing. See Fig. 4.

10. Remove the mounting bolts holding the front exhaust pipe bracket to the transmission. See Fig. 5.

11. Remove the large clamps holding the inner drive axle joint boots to the drive axle drivers. Pull the boots off the drivers.

12. Remove the right-hand road wheel. Remove the two through-bolts from the lower ball joint. Separate the ball joint from the control arm and remove the inner drive axle joint from its driver. See Fig. 6. Push the drive axle assembly forward and rest it on the control arm.

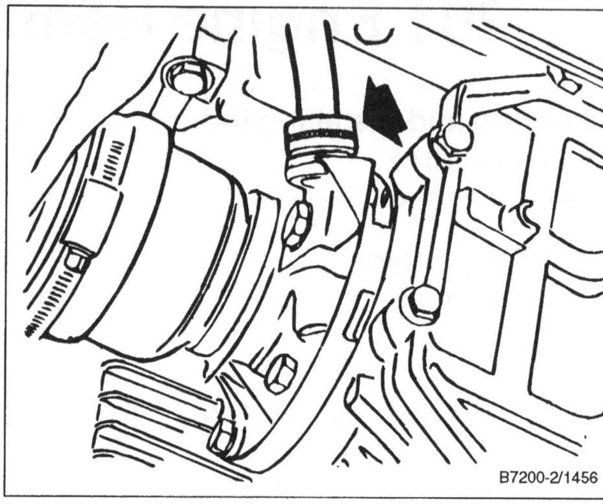

Fig. 4. Unscrew speedometer cable retaining collar (arrow) and pull cable from bearing housing.

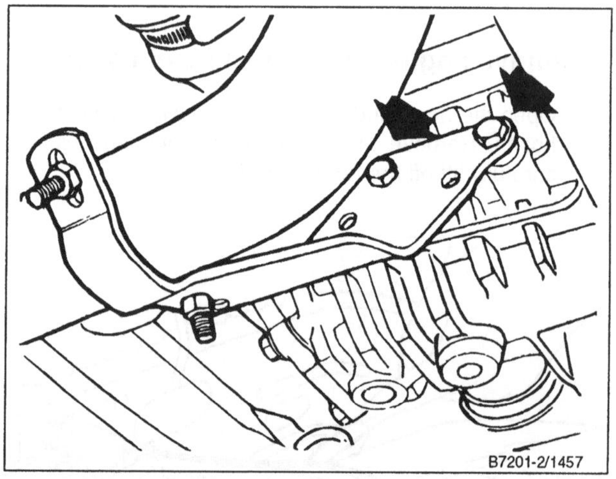

Fig. 5. Front exhaust pipe mounting bracket. Remove bracket bolts (arrows) from transmission.

CAUTION
Cleanliness is essential when working on drive axle joints. Even a small amount of sand or dirt will appreciably shorten the service life of the joint. Cover or wrap the joint and inner drivers on the transmission.

13. Disconnect the power steering pump pressure line union on the exhaust side of the engine. See Fig. 7.

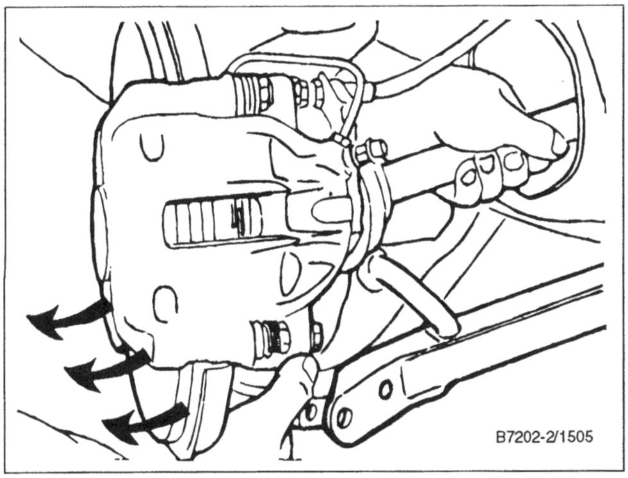

Fig. 6. Hub assembly and drive axle being pulled out to remove inner joint from drive axle driver. Rest drive axle on control arm.

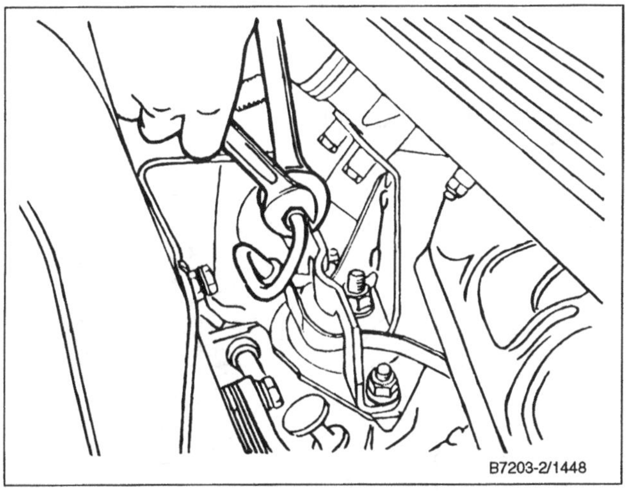

Fig. 7. Power steering pump pressure line union being disconnected.

14. Drain the coolant from the radiator and block as as described in **261 Radiator and Cooling System.** Disconnect the coolant hoses shown in Fig. 8.

15. Disconnect the wiring connectors and ground wires shown in Fig. 9. Lift the wiring loom out of the engine compartment and rest it on the A/C evaporator housing. Position all other disconnected wires and connectors so that they will not interfere with the engine's removal.

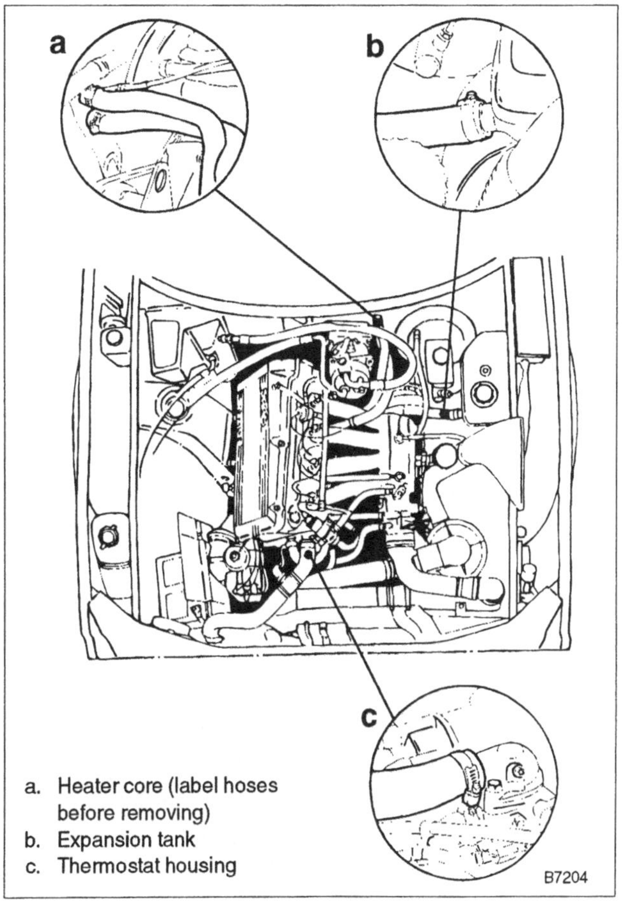

Fig. 8. Coolant hoses to be disconnected.

NOTE

Use a small screwdriver to remove the connector retaining clips where necessary.

---

16. On turbo models, disconnect the turbocharger discharge pipe.

17. Remove the air mass meter with its rubber socket connector.

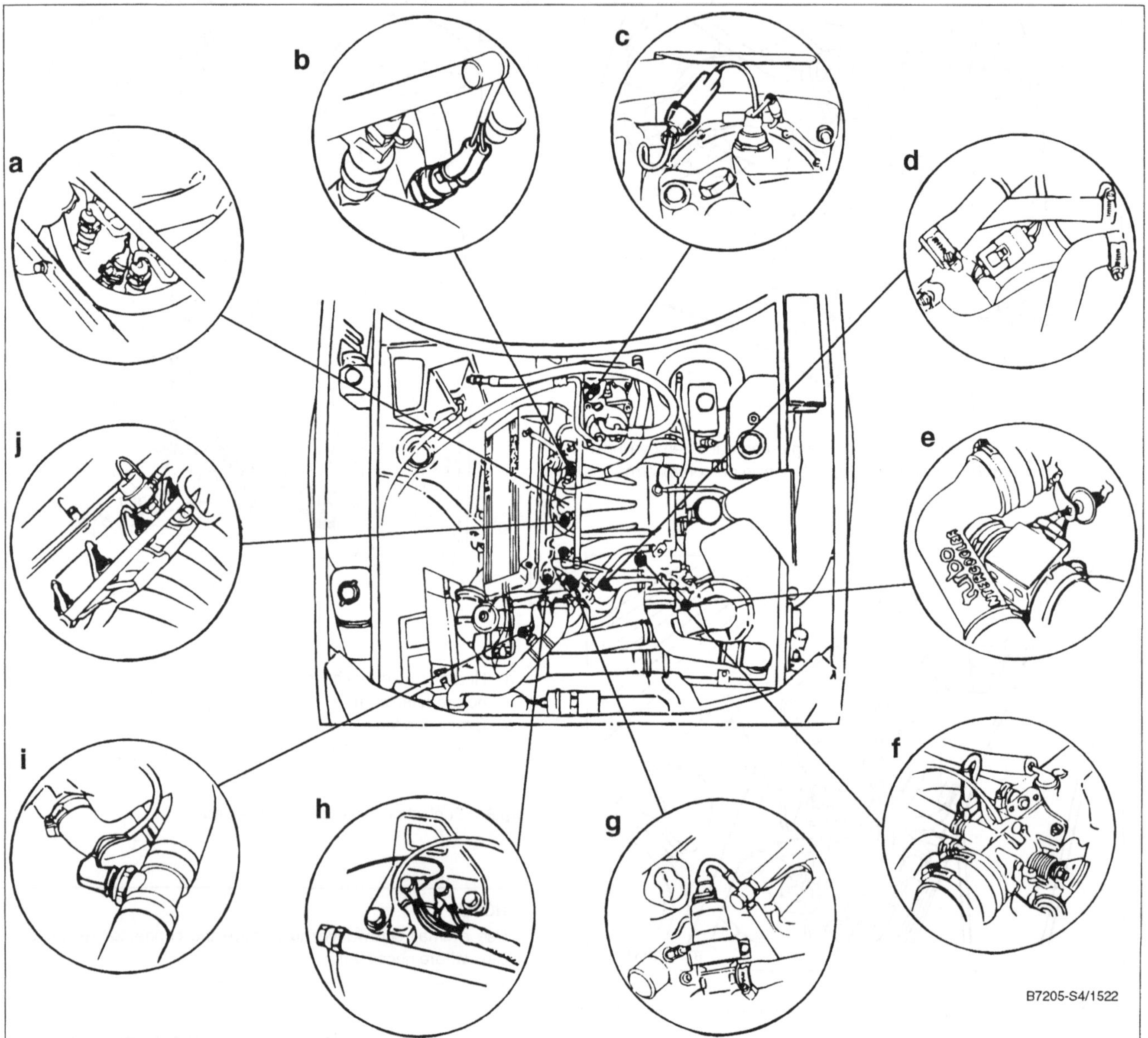

a. A/C temperature sensor
b. Coolant temperature sensor
c. A/C compressor thermo-switch
d. Shift-up indicator vacuum switch
e. Air mass meter
f. Throttle-position sensor
g. AIC valve
h. Ground wires
i. Radiator cooling fan thermo-switch
j. Injectors

Fig. 9. Wiring connectors and ground wires to be removed.

18. Remove the air cleaner together with the intake air pipe. See Fig. 10.

19. Remove the intercooler together with the air pipe. See Fig. 11. Cover or cap the inlet to the turbocharger.

20. Remove the intercooler side baffle plate and top cover. See Fig. 12.

21. Remove and label the wires at the ignition coil. Remove the coil from its bracket.

22. Remove the A/C radiator cooling fan. See **261 Radiator and Cooling System.**

23. Disconnect the lower radiator hose from the radiator.

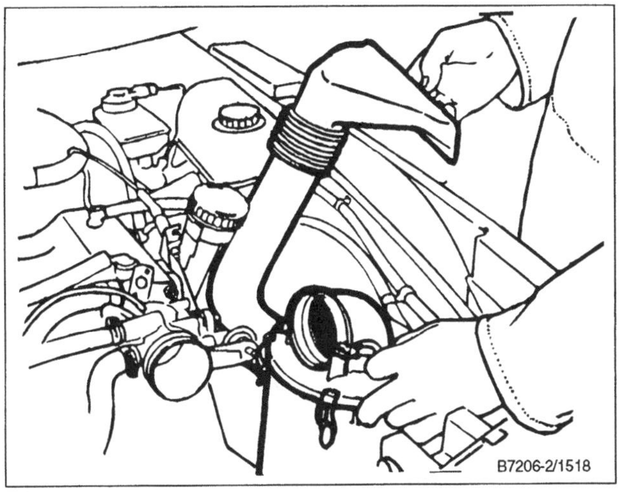

Fig. 10. Air cleaner assembly and intake air pipe being removed.

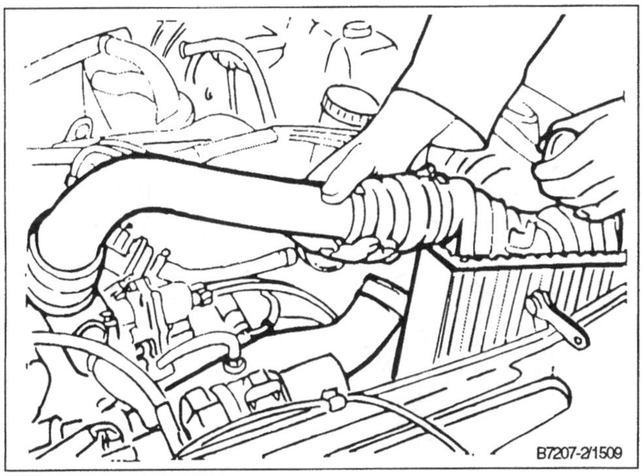

Fig. 11. Intercooler and air pipe being removed.

24. Disconnect the harness connector and hoses from the turbocharger APC solenoid. See Fig. 13. Label the hoses before removing them.

25. Disconnect the accelerator cable from the throttle housing and its bracket.

26. Disconnect the vacuum hoses for the brake booster and the charcoal canister at the intake manifold and throttle housing.

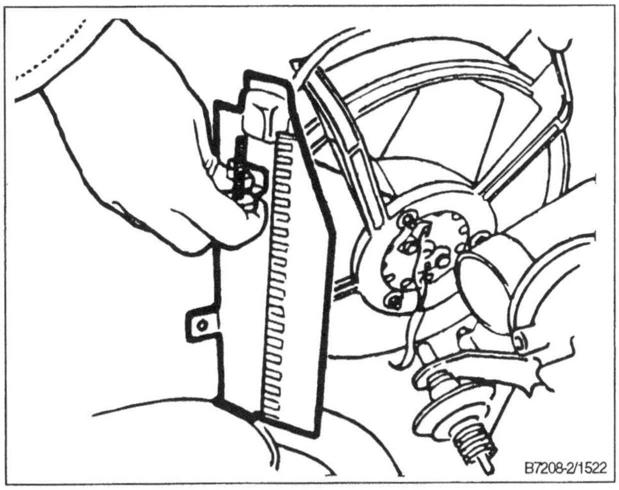

Fig. 12. Intercooler side baffle plate being removed.

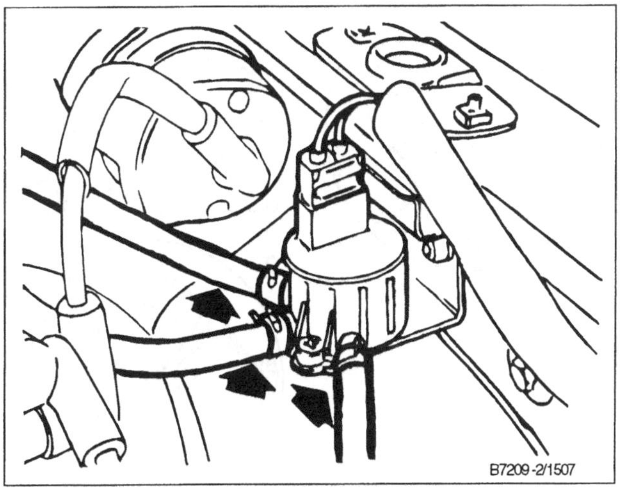

Fig. 13. APC solenoid. Remove harness connector and hoses (arrows). Label hoses before removal.

27. Disconnect the wiring connectors and wires shown in Fig. 14. Cut any wiring ties and lift the wiring loom out of the engine compartment and set aside. Position all other disconnected wires and connectors so that they will not interfere with the engine's removal.

NOTE

Use a small screwdriver to remove the connector retaining clips where necessary.

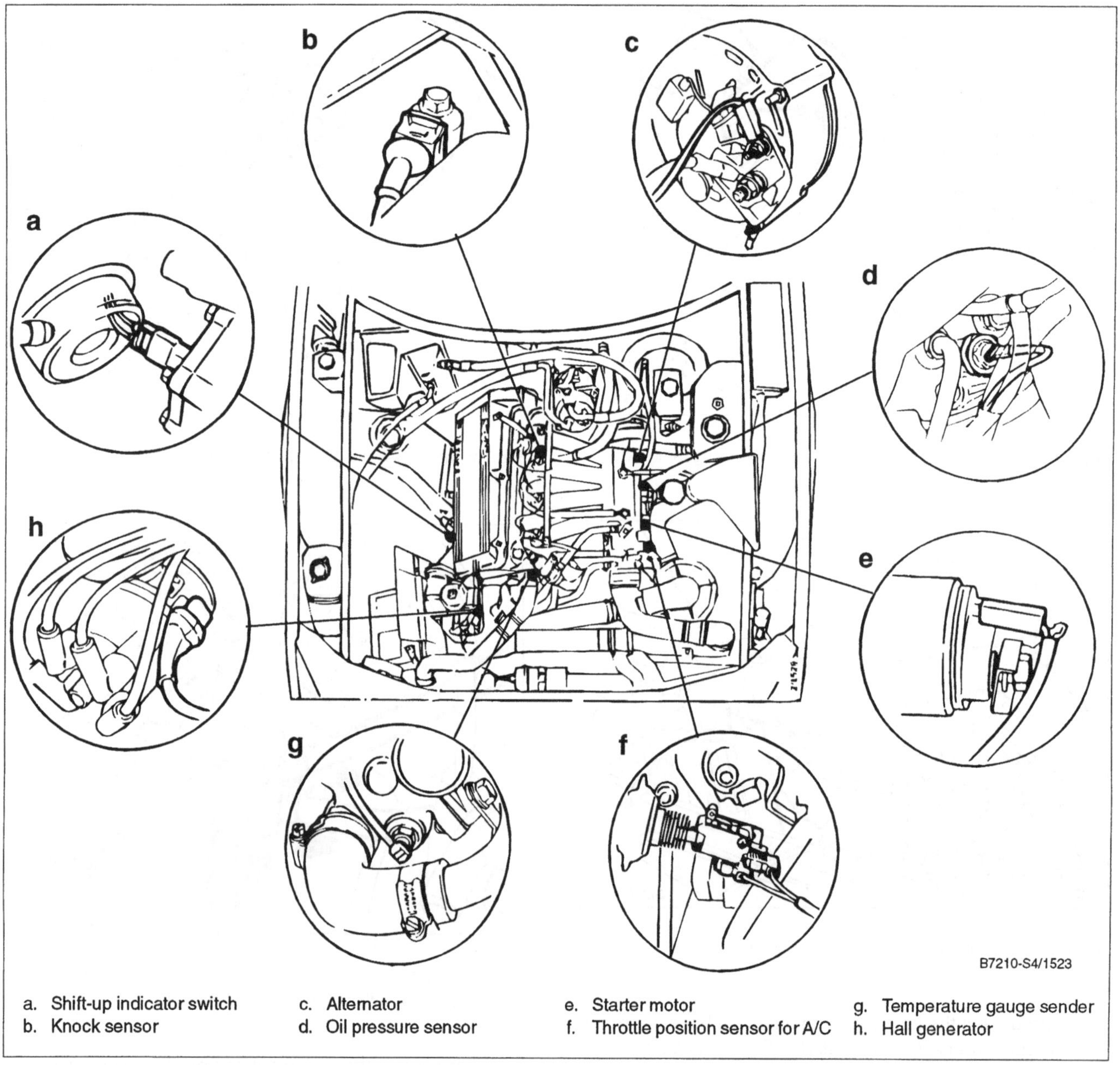

Fig. 14. Wiring connectors and ground wires to be removed.

28. Remove the positive battery cable from it brackets on the engine and on the starter motor. Set the cable on the battery tray.

29. Disconnect the ground wire at the timing chain cover and the transmission.

30. Unplug the oxygen sensor connector above the right-hand wheel well. Unhook the wiring harness from the fender and set it on the engine.

31. Disconnect the hydraulic line from the clutch slave cylinder. See Fig. 15.

32. Disconnect the fuel supply line and return line from the fuel rail and fuel pressure regulator.

33. Unbolt the A/C compressor (4 bolts) and carefully place the compressor on the A/C evaporator housing without disconnecting or loosening any refrigerant hoses.

WARNING

Loosening or disconnecting refrigerant hoses will allow the refrigerant to be discharged under high pressure, possibly causing personal injury. Recharging a discharged A/C system requires special equipment and knowledge.

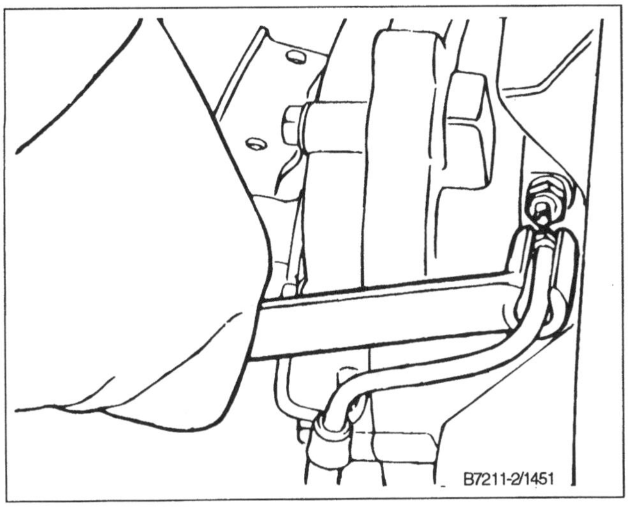

Fig. 15. Clutch slave cylinder hydraulic line being removed.

34. Attach an engine hoist to the engine. The front lifting point is at the intake manifold. The rear lifting point is at the A/C bracket. Raise the engine slightly.

NOTE

The A/C bracket has a lifting eye, which is used to attach the lifting hoist hook. If the bracket has been removed, bolt a separate lifting eye to the coolant pump cover using the existing mounting bolts.

---

35. Remove the bolts from the three engine mounts (front, left, and right). Remove the bump stop on the front and right-hand mounts. The left-hand mount is a through bolt with a spacer. See Fig. 16, Fig. 17, and Fig. 18.

Fig. 16. Front engine mount and bump stop assembly.

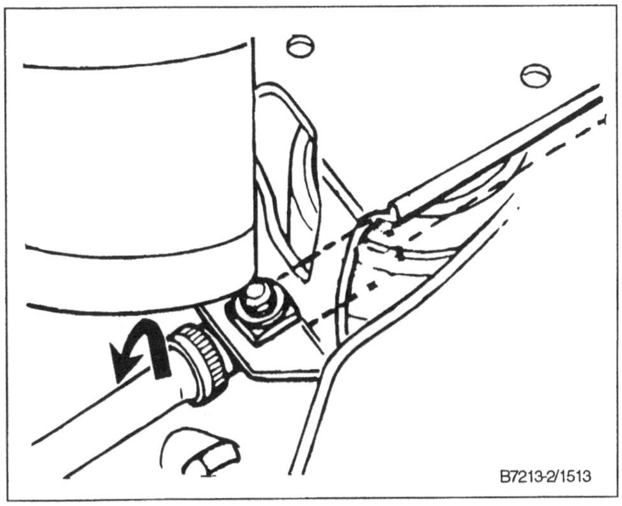

Fig. 17. Left-hand engine mount being loosened. Spacer sleeve is hidden behind bracket (dotted line).

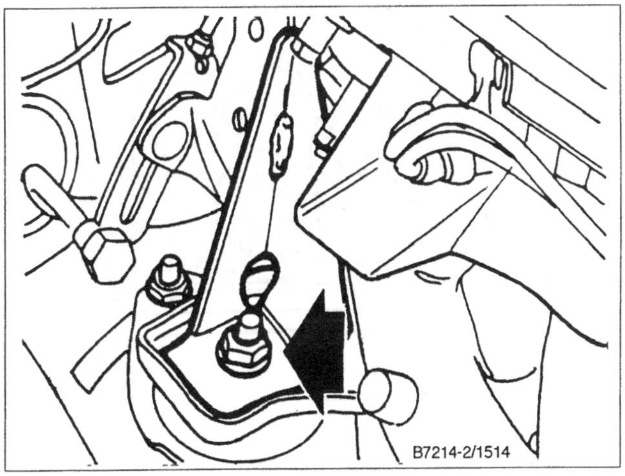

Fig. 18. Right-hand engine mount.

36. Raise the engine just enough to separate the left-hand inner drive axle joint from the driver. See Fig. 19.

CAUTION

Cleanliness is essential when working on drive axle joints. Even a small amount of sand or dirt will appreciably shorten the service life of the joint. Cover or wrap the joint and inner drivers on the transmission.

---

37. Disconnect the power steering pump return line from the bottom of the power steering pump.

38. Remove the oil pressure sensor connector and disconnect the oil cooler lines from the oil filter housing. See Fig. 20.

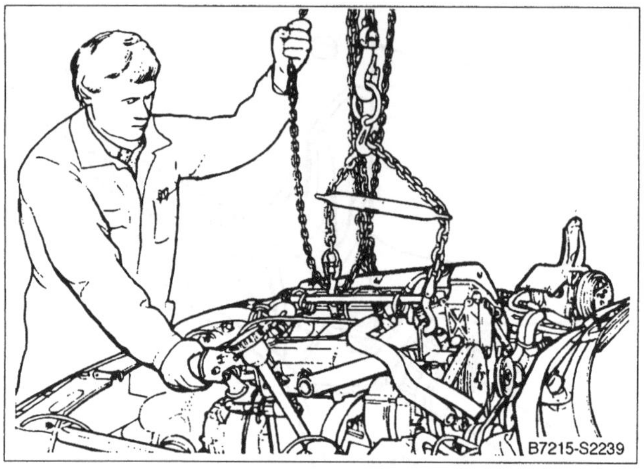

Fig. 19. Engine being raised just enough to separate inner drive axle joint from driver.

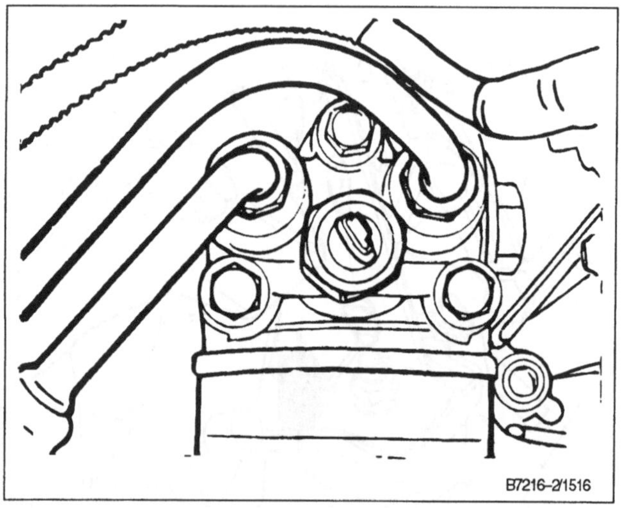

Fig. 20. Top view of oil filter housing showing oil pressure sensor and oil cooler lines.

39. Raise the engine and transmission assembly out of the car. Be careful not to damage the radiator or APC solenoid valve.

Before installing the engine make sure the drive axle drivers are packed with grease. See **774 Wheel Bearings and Drive Axles.** Install any coolant drain plugs removed earlier. When lowering the engine into the engine compartment, incline the engine slightly down in the front. See Fig. 21.

Lower the engine into position just enough to install the power steering pump return line and oil cooler lines. Align the left-hand drive axle joint and the engine mounts and lower the engine into its final position. Be sure to install the spacer sleeve on the left-hand engine mount. Reconnect all hoses, connectors and ground wires. Be sure to replace any wire ties

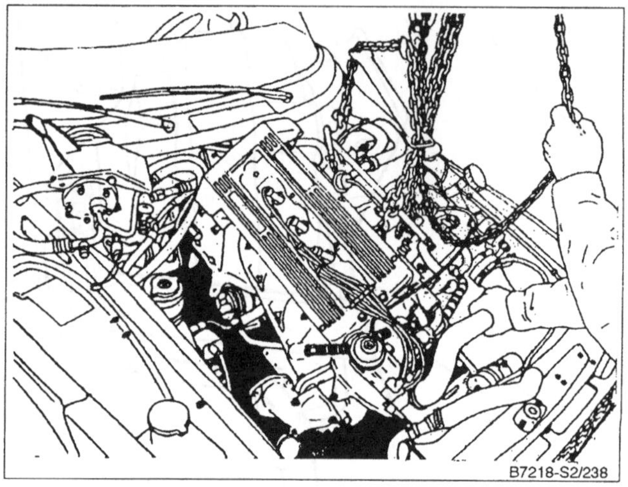

Fig. 21. Engine being lowered into engine compartment. Note engine tilted down in front during installation.

cut off during the removal procedure. Use new self-locking nuts when reinstalling the lower ball joint to the control arm.

WARNING
Always replace self-locking nuts and bolts. These fasteners are designed to be used only once. They may become unreliable and fail when used a second time.

Refill the engine with engine oil and coolant, and refill the power steering system. If applicable bleed the clutch slave cylinder as described in **412 Clutch Control.** The remainder of installation is the reverse of removal.

Tightening torques
- clutch slave cylinder to transmission case . . . . . . 6 to 10 Nm (53 to 89 in-lb)
- coolant drain plug to engine block . . 15 Nm (11 ft-lb)
- oil cooler line to oil filter housing . . . . 7 to 10 Nm (62 to 89 in-lb)
- lower ball joint to lower control arm . . . . . . 40 to 55 Nm (30 to 41 ft-lb)
- road wheel to wheel bearing hub
  - lug nuts . . . . 90 to 110 Nm (66 to 81 ft-lb)
  - lug bolts . . . . 105 to 125 Nm (77 to 92 ft-lb)
- power steering line union . . . . 20 to 34 Nm (15 to 25 ft-lb)
- all other fasteners
  - M5 bolt . . . . 5 Nm (44 in-lb)
  - M6 bolt . . . . 10 Nm (89 in-lb)
  - M8 bolt . . . . 20 Nm (15 ft-lb)
  - M10 bolt . . . . 40 Nm (30 ft-lb)

Separating Engine from Transmission

The engine must be first removed from the car to separate the transmission from the engine.

To separate engine from transmission (manual transmission)

1. Thoroughly clean the external surfaces of the engine and transmission.
2. Drain the engine oil and remove the EGR pipe, if applicable.
3. Remove the plastic clutch cover from over the flywheel.
4. Remove the transmission oil dipstick tube.
5. On turbo models, disconnect the oil return line from the turbocharger.
6. Remove the starter motor and unscrew the starter motor stay plate.
7. Remove the transmission input shaft. Remove the clutch slave cylinder mounting bolts. See **411 Clutch.**
8. Remove all of the engine to transmission mounting bolts. Remove the bracket for the oil filler pipe at the throttle housing.
9. Carefully lift the engine away from the transmission. See Fig. 22. Guide off the clutch release-bearing sleeve and slave cylinder as the engine is lifted off.

Fig. 22. Engine being lifted off transmission together with clutch release-bearing sleeve and slave cylinder.

CAUTION

If the engine does not immediately lift off the transmission, do not attempt to pry them apart until making sure that all fasteners have been removed.

---

10. Remove all old traces of oil and dirt from the flanges.
11. Before refitting the engine to the transmission, make sure the two guide sleeves are fitted in the transmission flange. Place a new gasket onto the transmission flange and apply a sealing compound (Permatex Ultrablue) to the slots shown in Fig. 23.
12. Apply a thread sealing compound (Loctite 45, Saab Part No. 3009081) to the six bolt holes shown in Fig. 24.

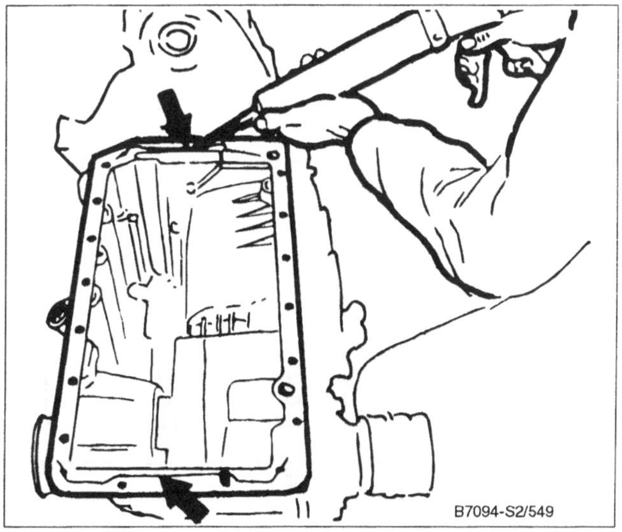

Fig. 23. Sealing compound being installed to slots (arrows) on transmission flange.

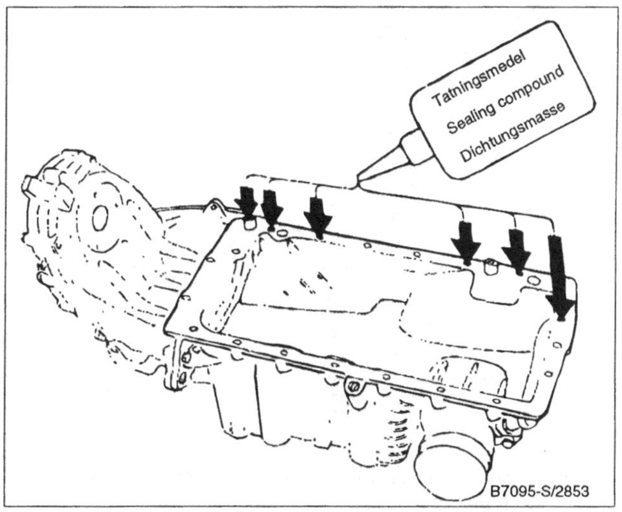

Fig. 24. Bolt holes in transmission case requiring thread sealer.

The remainder of the procedure is the reverse of removal. Fig. 25 shows engine-to-transmission bolt identification and location.

Tightening torques

- engine to transmission . . . . 25 Nm (18 ft-lb)
- all other fasteners
  - M5 bolt . . . . .5 Nm (44 in-lb)
  - M6 bolt . . . . .10 Nm (89 in-lb)
  - M8 bolt . . . . 20 Nm (15 ft-lb)
  - M10 bolt . . . . 40 Nm (30 ft-lb)

To separate engine from transmission (automatic transmission)

1. Thoroughly clean the external surfaces of the engine and transmission.

2. Drain the engine oil and remove the oil filler pipe.

3. Remove the plastic cover from above the drive plate and torque converter.

4. Disconnect the throttle (kick-down) cable at the throttle housing.

5. On turbo models, remove the turbocharger support.

6. Remove all of the engine to transmission mounting bolts. Remove the ATF cooler lines from the transmission.

7. Remove the 8 bolts holding the drive plate to the torque converter. When all bolts are removed, position the drive plate so that the two brackets on the drive plate perimeter are horizontal.

NOTE

To best access the bolts, spin the drive plate until the bolts are accessible from above.

---

8. Carefully lift the engine straight off the transmission. See Fig. 22 above for reference.

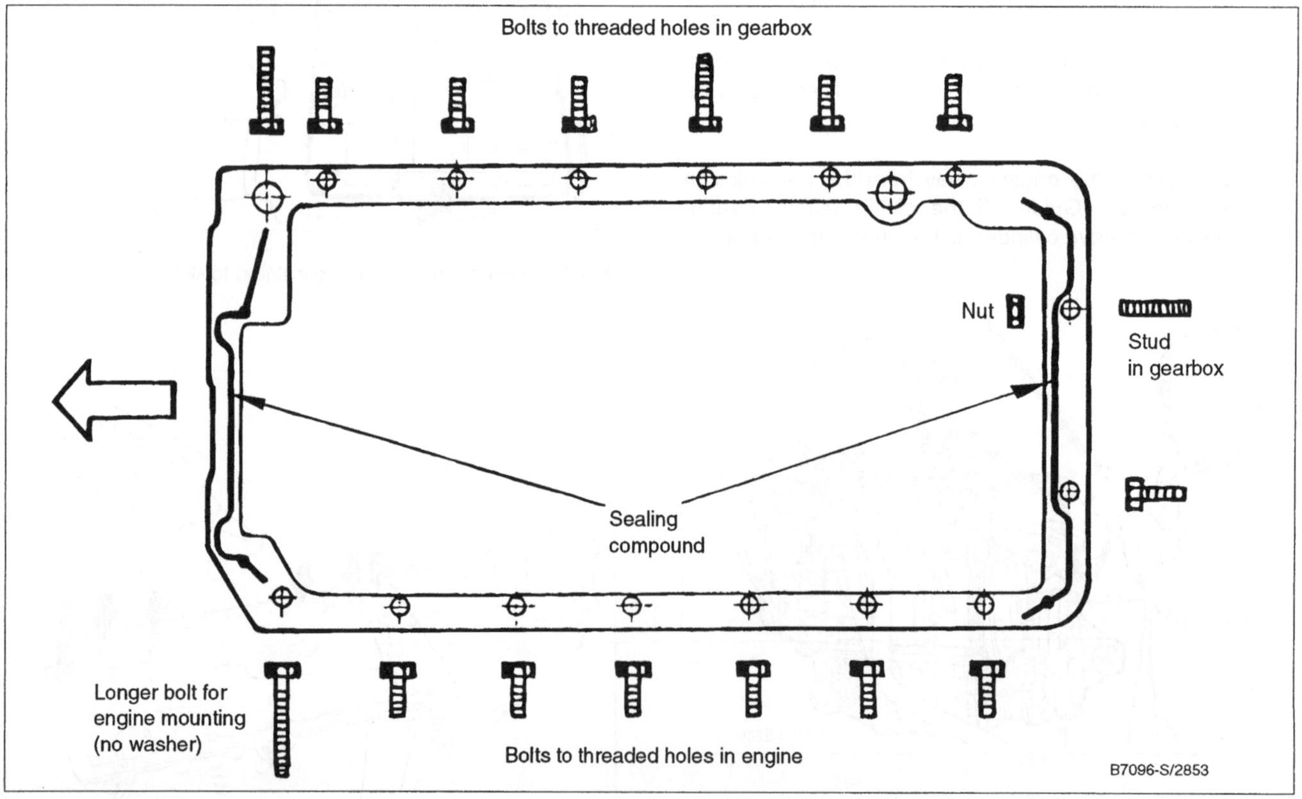

Fig. 25. Engine-to-transmission mounting bolt identification for manual transmission.

CAUTION
If the engine does not immediately lift off the transmission, do not attempt to pry them apart until making sure that all fasteners have been removed.

NOTE
Use Saab special tool 87 90 255 to support the torque converter once the engine is removed from the transmission. See Fig. 26. If the torque converter is left unsupported, the torque converter oil seal may be damaged. As an alternative, remove the torque converter and store it in its normally installed position.

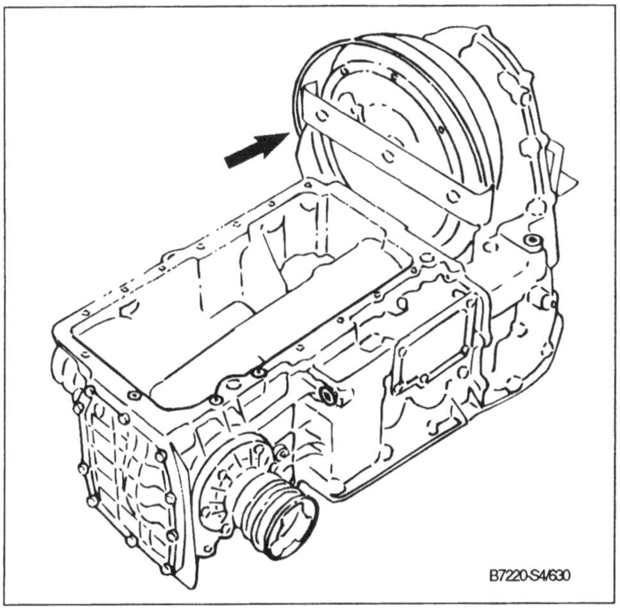

Fig. 26. Torque converter support bracket (arrow).

9. Remove all old traces of oil and dirt from the flanges.

NOTE
When cleaning the removed transmission, plug all of the openings shown in Fig. 27.

10. Before refitting the engine to the transmission, carefully inspect the drive plate for cracks. Make sure both mating surfaces are completely clean. Make sure the two brackets on the drive plate perimeter are horizontal.

11. Place a new gasket onto the transmission flange and apply a sealing compound (Permatex Ultra-blue) to the slots shown in Fig. 28.

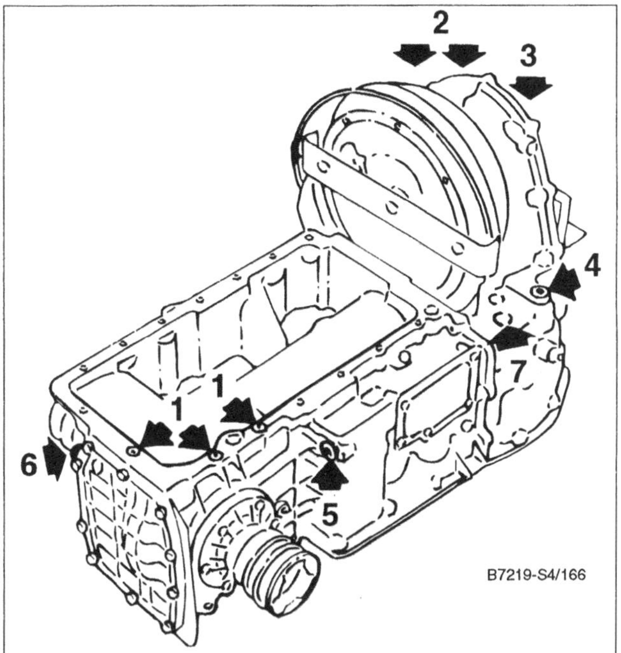

1. Mating flange (3 bolt holes)
2. Oil cooler outlets
3. Primary drive cover
4. Oil filler hole
5. Gear selector cable hole
6. Speedometer cable hole
7. Reverse gear switch hole

Fig. 27. Areas of automatic transmission which must be plugged prior to cleaning.

Fig. 28. Sealing compound being installed to slots (arrows) on transmission flange.

12. Apply a thread sealing compound (Loctite 45, Saab Part No. 45 3009081) to the three bolts holes shown in Fig. 29.

CAUTION

Saab part numbers are for reference only. Consult an authorized Saab dealer for the most up-to-date parts information.

---

13. Install the engine to the transmission. Take care not to damage the torque converter centering pin when lowering the engine onto the transmission. Tighten mounting bolts in stages.

The remainder of the procedure is the reverse of removal. Be sure to fill the engine with oil.

Tightening torques

- engine to transmission .... 25 Nm (18 ft-lb)
- all other fasteners
  - M5 bolt ....5 Nm (44 in-lb)
  - M6 bolt ....10 Nm (89 in-lb)
  - M8 bolt .... 20 Nm (15 ft-lb)
  - M10 bolt .... 40 Nm (30 ft-lb)

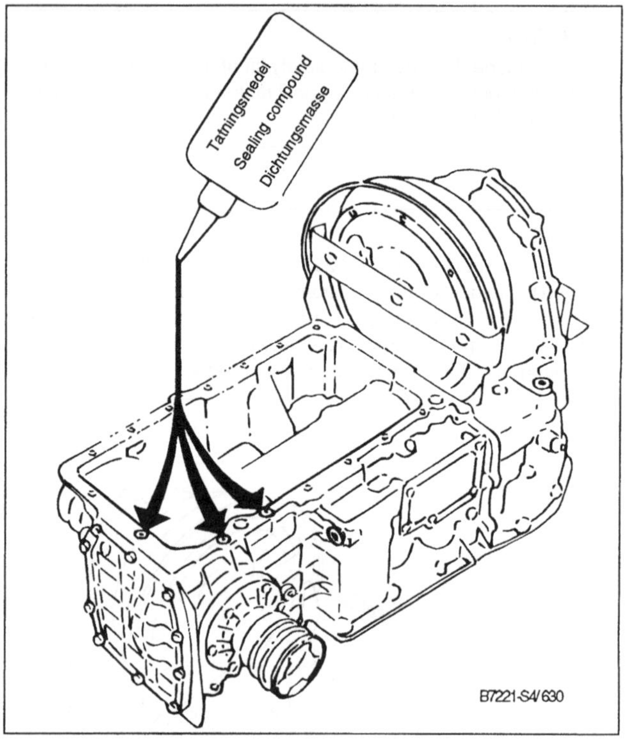

Fig. 29. Bolt holes in transmission case requiring thread sealer.

210 Engine Disassembly and Assembly

GENERAL

Fig. 1 shows the complete engine assembly. This repair group covers disassembly and assembly of the engine cylinder block only. Cylinder block reconditioning procedures and specifications can be found in **212 Pistons, Connecting Rods, Cylinder Bores**, and in **216 Crankshaft.**

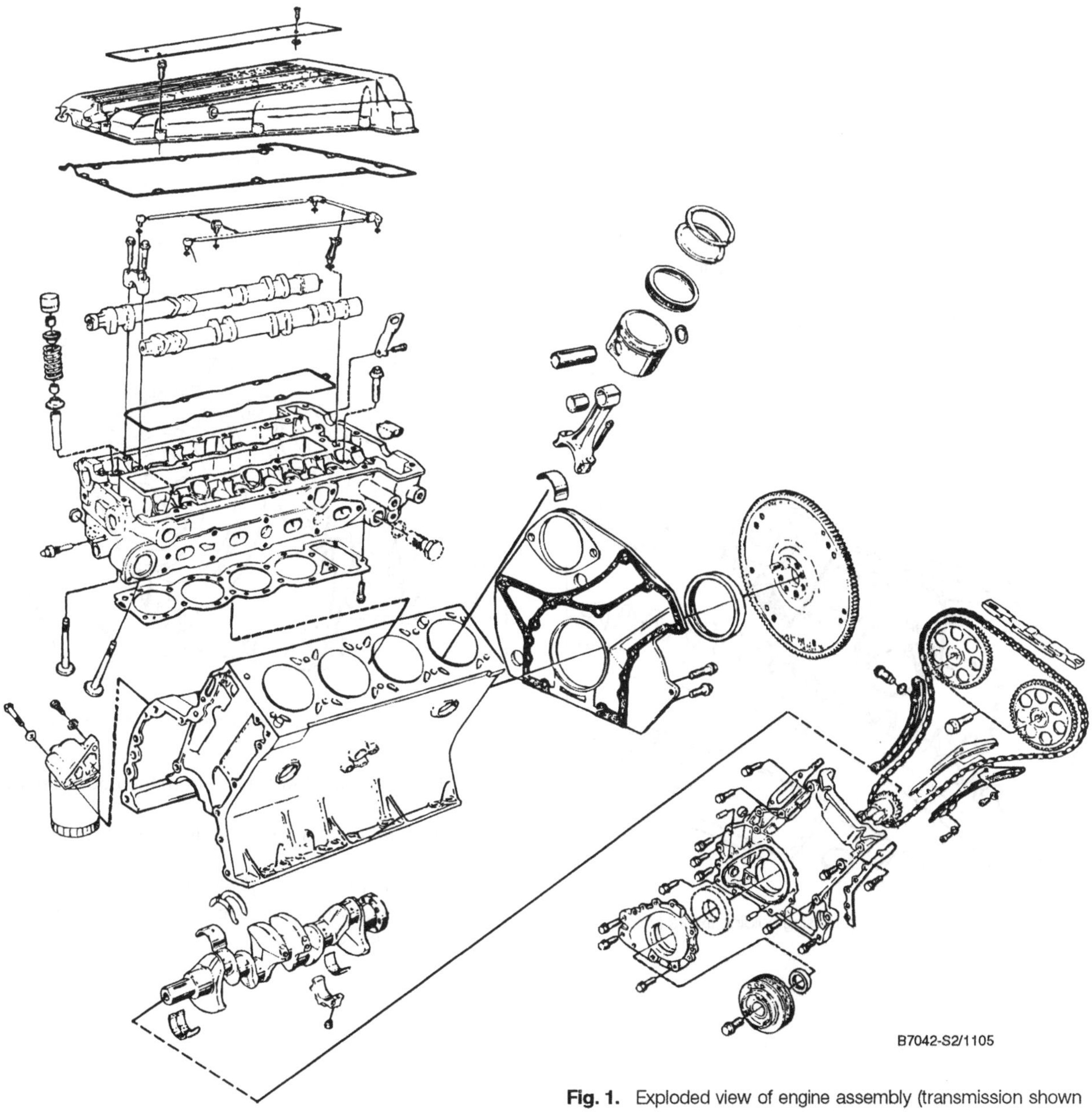

Fig. 1. Exploded view of engine assembly (transmission shown removed).

Disassembling and Assembling Cylinder Block

Disassembly and assembly procedures for the Saab cylinder block are similar to those of most other water-cooled engines.

Thoroughly mark the position and orientation of all parts as they are removed. Connecting rods, rod caps, rod bearings, pistons, main bearing caps and main bearings are assembled in an exact location and orientation. This "assembly code" is critical to proper engine operation if any of the parts are going to be reused.

To disassemble cylinder block

1. Remove the engine from the car and separate the transmission from the engine as described in **201 Engine Removal and Installation.**
2. Remove the cylinder head as described in **211 Cylinder Head Removal and Installation.**
3. Remove the plastic clutch cover from above the flywheel.
4. With the alternator V-belt installed and tensioned, loosen the bolts from the coolant pump pulley.
5. Remove the alternator V-belt.
6. Immobilize the flywheel with a holding fixture. See Fig. 2.

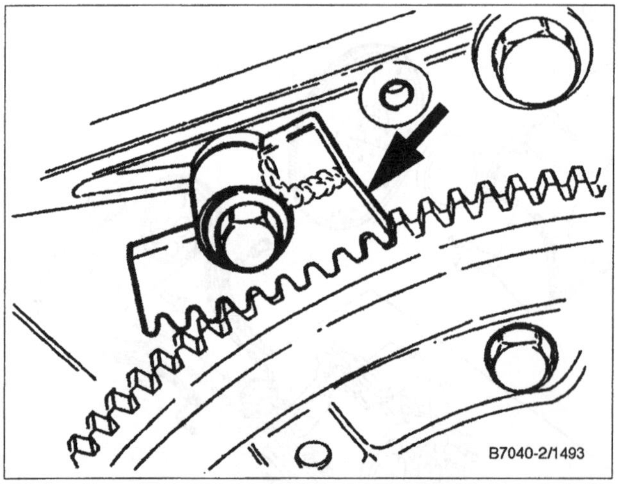

Fig. 2. Flywheel being immobilized using holding fixture (arrow).

7. Remove the center bolt from the crankshaft pulley. See Fig. 3. Remove the pulley from the crankshaft.

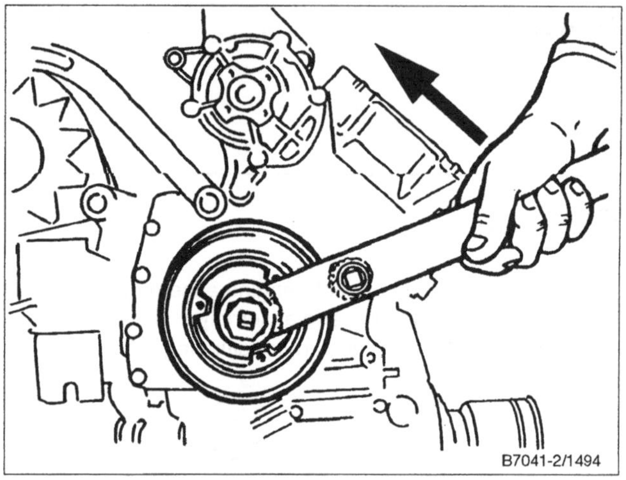

Fig. 3. Crankshaft pulley mounting bolt being removed.

8. Remove the oil pump mounting bolts and remove the pump from the timing chain cover. See Fig. 4.

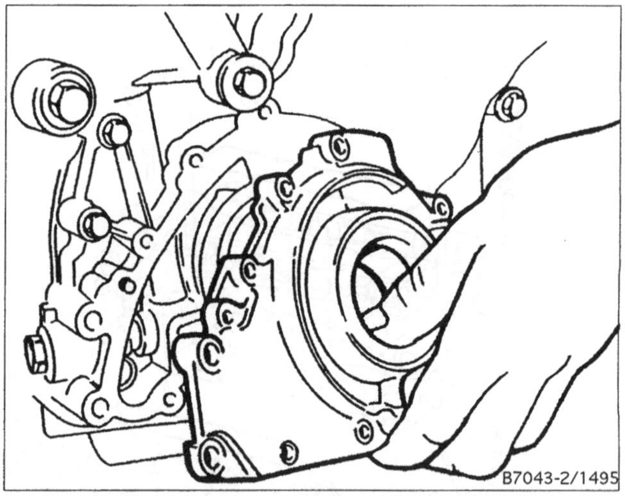

Fig. 4. Oil pump being removed from timing chain cover.

9. Disconnect the lower hose from the coolant pump.
10. Remove the alternator together with its stabilizer bar.
11. Remove the mounting bolts from the timing chain cover and remove the cover together with the water pump and hoses. See Fig. 5.

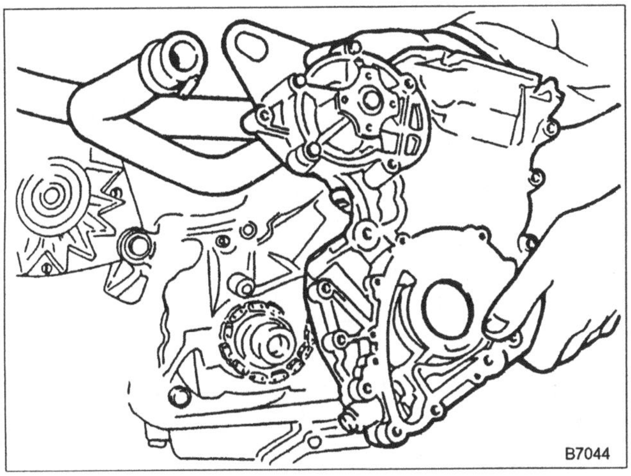

Fig. 5. Timing chain cover with coolant pump and hoses being removed.

12. Remove the pivoting timing chain guide. See Fig. 6.

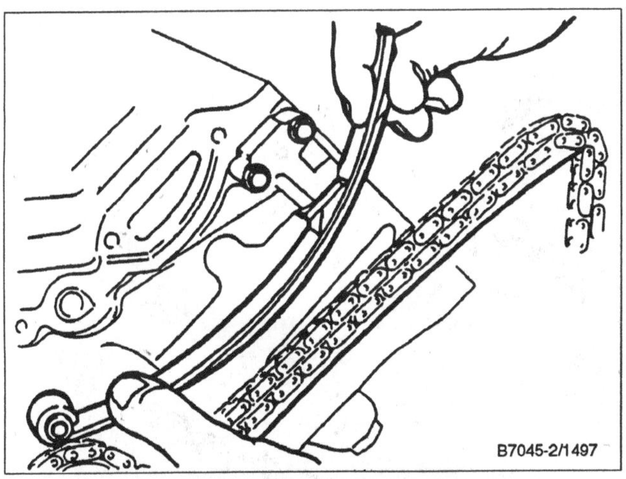

Fig. 6. Pivoting camshaft timing chain guide being removed.

13. Pull off the crankshaft sprocket with the timing chain. See Fig. 7. Remove the key from the keyway in the crankshaft.

14. Remove the fixed timing chain guide. See Fig. 8

15. Unscrew the oil filter and unbolt the oil filter housing from the cylinder block.

16. Mark the cylinder number on the top of each piston. Remove the piston and connecting rod assemblies.

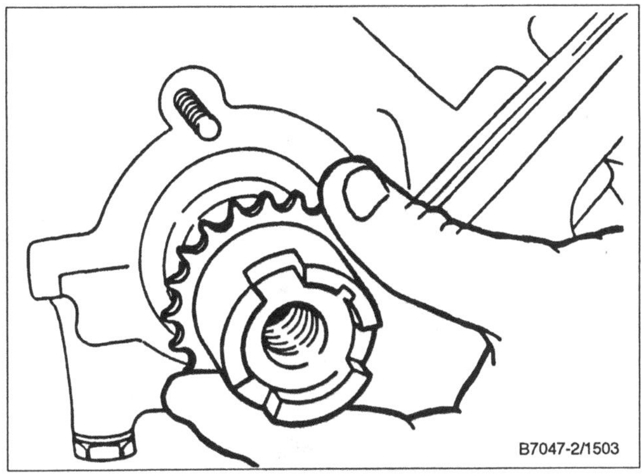

Fig. 7. Crankshaft sprocket being removed.

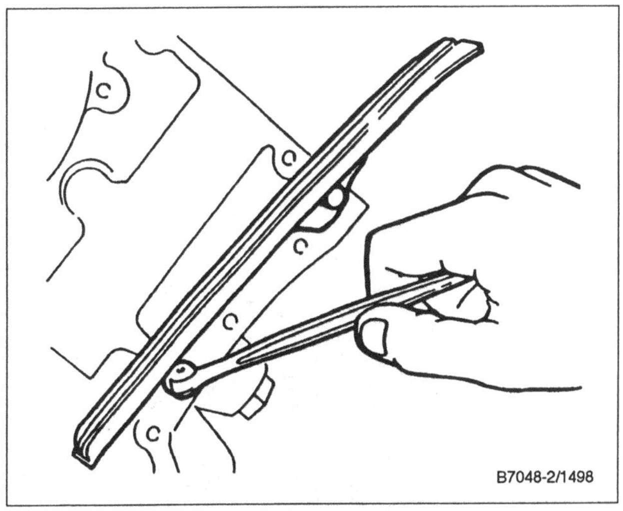

Fig. 8. Fixed timing chain guide being removed.

CAUTION

Before removing the pistons from the cylinders, remove any ridge or carbon deposits from the top of the cylinder. Install small lengths of rubber hose over the connecting rod studs to protect the cylinders during piston removal.

NOTE

Observe the markings on the connecting rod big-ends. Where necessary, mark all parts so that they can be reinstalled in their exact positions. This includes noting which bearing shells are in the upper positions and which ones are in the lower positions.

---

17. Turn the engine block on its end so that the flywheel is facing up and remove the flywheel. See Fig. 9.

Fig. 9. Flywheel being removed.

18. Remove the cylinder block end plate. See Fig. 10.

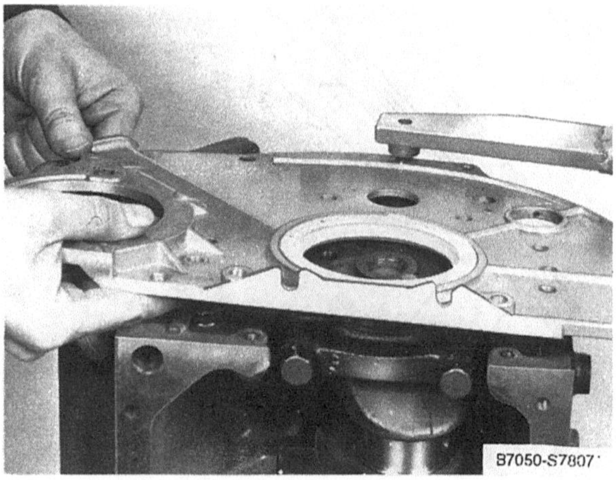

Fig. 10. Cylinder block end cover being removed.

19. Remove the crankshaft bearing caps and remove the crankshaft. See Fig. 11.

NOTE
Note the main bearing cap markings and locations for reinstallation.

To assemble cylinder block

NOTE
When reusing old parts, make sure that they are installed in their original positions.

Fig. 11. Crankshaft being removed.

1. Lubricate the bearing shell saddles and install the upper shells into position. Carefully install the crankshaft.

2. Fit the center thrust washers to the crankshaft. See Fig. 12. Check that the crankshaft end float is within specification as described in **216 Crankshaft.**

Fig. 12. Crankshaft thrust washer being installed.

3. Lubricate the lower bearing shells and install the shells and main bearing caps in their original locations.

Tightening torque
- main bearing caps
  to cylinder block . . . . 110 Nm (81 ft-lb)

NOTE
Note the markings on the bearing caps. See Fig. 13.

Fig. 13. Main bearing cap markings.

4. Apply a small amount of grease to hold the new end-plate gasket to the cylinder block. Install the plate, being careful not to damage the crankshaft oil seal. If necessary, trim the gasket at the transmission-to-engine flange. See Fig. 14.

CAUTION

- To prevent oil leaks, the crankshaft seal should be perfectly centered around the crankshaft . A special guide (Saab tool no. 83 92 540) fits over the crankshaft to center the end plate and seal. The tool also protects the crankshaft seal during the installation of the end plate.
- Do not use an adhesive sealer on the gasket.

NOTE

For procedures on replacing the crankshaft oil seal, see **216 Crankshaft.**

Tightening torque

- cylinder block end plate to engine (flywheel end) . . . . 20 Nm (15 ft-lb)

---

5. Install the flywheel using a sealant on the mounting bolts. Be sure the old bolts are cleaned before applying the fresh sealant.

Tightening torques

- flywheel to crankshaft
  17 mm bolt head . . . . 60 Nm (44 ft-lb)
  19 mm bolt head . . . . 85 Nm (63 ft-lb)

Fig. 14. Cylinder block end plate gasket being trimmed at transmission gasket surface.

6. Align the compression rings so that their end gaps are approximately aligned over the wrist pin. Make sure the double oil scraper ring joints do not align. See Fig. 15.

NOTE

When installing the second compression ring, make sure the **"TOP"** marking on the ring is facing up.

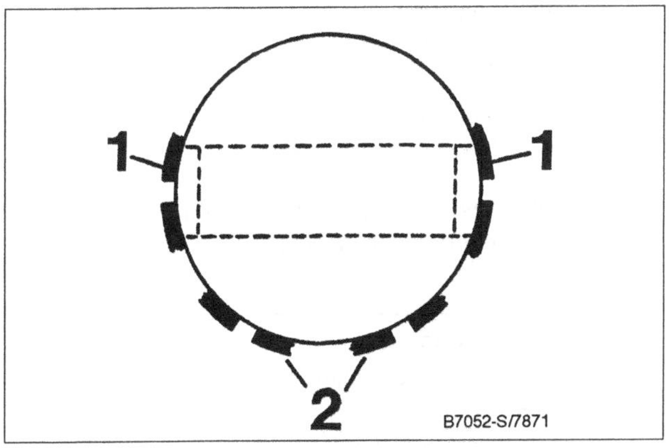

Fig. 15. Piston ring end gaps correctly aligned. Compression rings shown at **1** and double oil scraper ring shown at **2**.

7. Lightly lubricate the pistons and rings with engine oil. Install the pistons so that the notch in the top of the piston faces the camshaft timing chain and the numbers on the connecting rod big-end face the exhaust side of the engine. See Fig. 16. Install and tighten the connecting rod caps.

NOTE

Make sure the flange on the connecting rod cap nut is towards the rod cap, facing the piston.

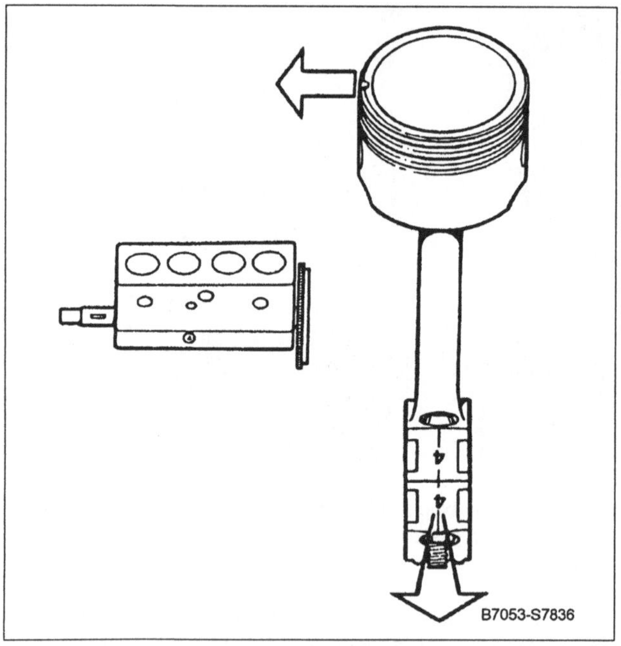

Fig. 16. Install piston so that top notch faces camshaft timing chain and connecting rod big-end numbers face the exhaust side of the block.

Tightening torque
- connecting rod cap to connecting rod . . . . 55 Nm (41 ft-lb)

8. Install the timing chain sprocket and the timing chain guides. Hang the timing chain on the sprocket.

9. Install the timing chain cover using a new gasket. Do not use sealer on this gasket.

Tightening torque
- timing chain cover to engine block . . 20 Nm (15 ft-lb)

10. Install the oil pump using a new O-ring.

Tightening torque
- oil pump to timing chain cover . . . . . . 8 Nm (71 in-lb)

11. Using engine oil, prime the cavity between the oil filter housing and the oil pump. See Fig. 17. Then install the oil filter housing and a new oil filter.

Tightening torque
- oil filter to oil filter housing . . . . 10 Nm (89 in-lb)

Fig. 17. Oil pump being primed through oil filter housing hole.

12. Install the crankshaft pulley with its key. While holding the flywheel stationary, tighten the mounting bolt.

Tightening torque
- crankshaft pulley to crankshaft
  1985-1990 . . . . 190 Nm (140 ft-lb)
  1991and later . . . . 175 Nm (129 ft-lb)

13. The remainder of the assembly is the reverse of removal.

Tightening torques
- all other fasteners
  M5 bolt . . . . 5 Nm (44 in-lb)
  M6 bolt . . . . 10 Nm (89 in-lb)
  M8 bolt . . . . 20 Nm (15 ft-lb)
  M10 bolt . . . . 40 Nm (30 ft-lb)

211 Cylinder Head Removal and Installation

GENERAL

This section covers only cylinder head removal and installation. For complete cylinder head, camshaft, and valve train repair information, see **214 Cylinder Head and Valve Mechanism.** For camshaft timing chain repair information, see **215 Camshaft Timing Chain.**

CYLINDER HEAD REMOVAL AND INSTALLATION

To remove cylinder head

1. Raise the hood and remove the hood hinge mounting bolts. See Fig. 1. Carefully remove the hood from the car.

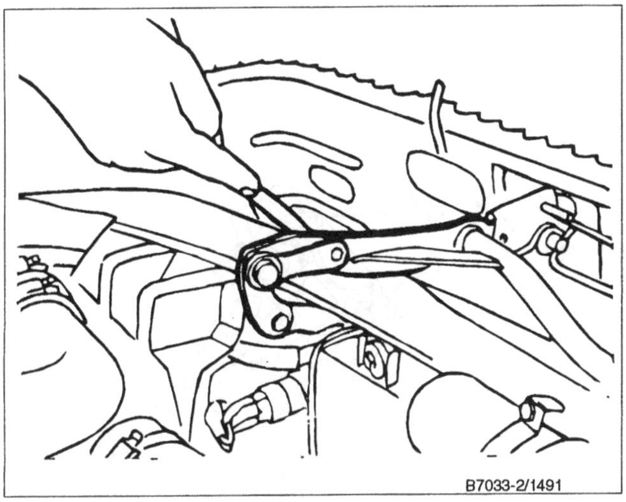

Fig. 1. Hood hinge mounting bolt being removed.

2. Disconnect the battery terminals and remove the battery from the car.
3. Drain the coolant from the radiator and from the engine block. See **261 Radiator and Cooling System.**
4. Remove the exhaust manifold.
5. If applicable, remove the turbocharger as described in **291 Turbocharger.**
6. Remove the air conditioning compressor belt and tensioning pulley.
7. Loosen the power steering pump pivot bolts. Remove the power steering pump V-belt and push the power steering pump out of the way.
8. Remove the center cover from above the spark plugs. Disconnect the spark wires from the spark plugs and remove the distributor cap with the spark plug wire assembly.
9. Remove the valve cover and the valve cover split rubber plugs as described in **214 Cylinder Head and Valve Mechanism.**
10. Remove the two bolts holding the timing chain cover to the bottom of cylinder head. See Fig. 2.

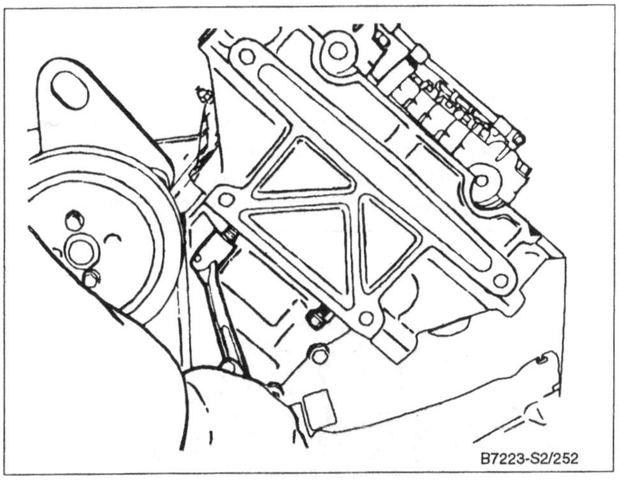

Fig. 2. Camshaft timing chain cover to cylinder head mounting bolt being removed.

11. Working from below, jack up the engine just enough to relieve the weight on the right-hand engine mount (exhaust side of engine). Then remove mounting bolts (and spacer sleeve) that are screwed into the cylinder head.

12. Disconnect the coolant hose from the thermostat housing.

13. Unbolt the fuel pressure regulator complete with bracket from the cylinder head. See Fig. 3. Do not disconnect fuel hoses.

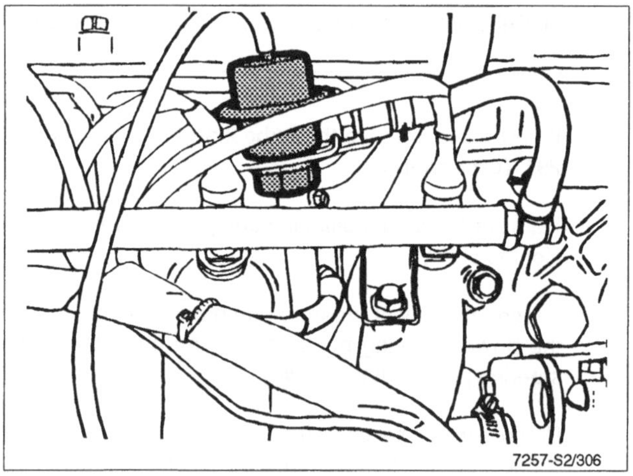

Fig. 3. Fuel pressure regulator.

14. Disconnect the fuel injection system ground wires from the cylinder head.

15. Unbolt the A/C bracket from the cylinder head.

16. Unbolt the intake manifold, complete with fuel injectors and fuel rail and set aside. See Fig. 4.

CAUTION

Cleanliness is essential when working with fuel injection components. Before removing the intake manifold, thoroughly clean the area around the injectors, the fuel rail, and the intake manifold. If possible, blow the area clean using compressed air.

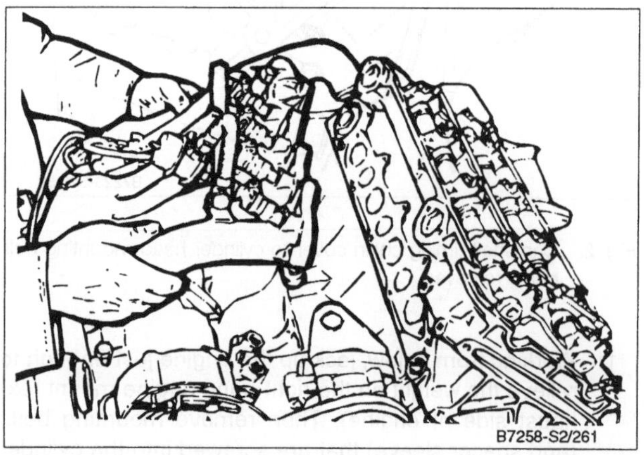

Fig. 4. Intake manifold with fuel injectors and fuel rail being removed.

17. Disconnect the wires from the temperature sensor.

18. Set the engine to TDC by rotating the crankshaft in normal engine rotation until the "0" mark on the flywheel is aligned with the timing mark on the end plate. Check that the timing marks on the camshaft are also aligned. See Fig. 5 and Fig. 6.

CAUTION

Do not turn the engine over using the camshafts. The camshafts are not designed to take this type of load. Use a wrench on the crankshaft pulley center bolt to turn the engine over.

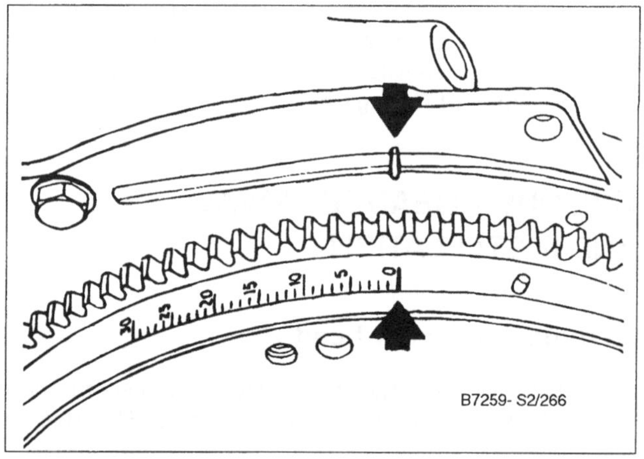

Fig. 5. Flywheel 0° TDC mark aligned with timing mark on engine end plate (arrows).

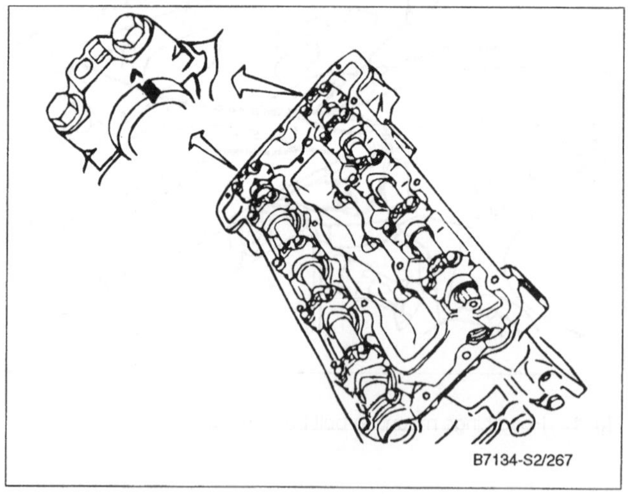

Fig. 6. Camshaft timing mark aligned with mark on camshaft bearing cap.

19. Loosen, but do not remove, the camshaft sprocket center mounting bolts. Hold the camshaft flats with an open-end wrench while loosening the bolts. See Fig. 7.

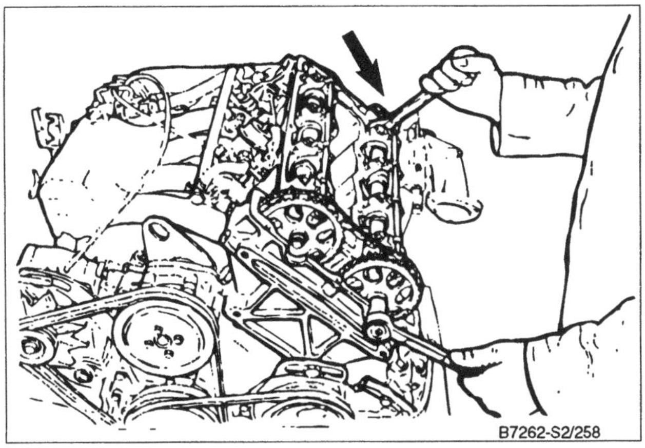

Fig. 7. Camshaft sprocket mounting bolt being loosened. Hold camshaft on flats (arrow) when loosening bolts.

20. Remove the camshaft timing chain tensioner. See Fig. 8.

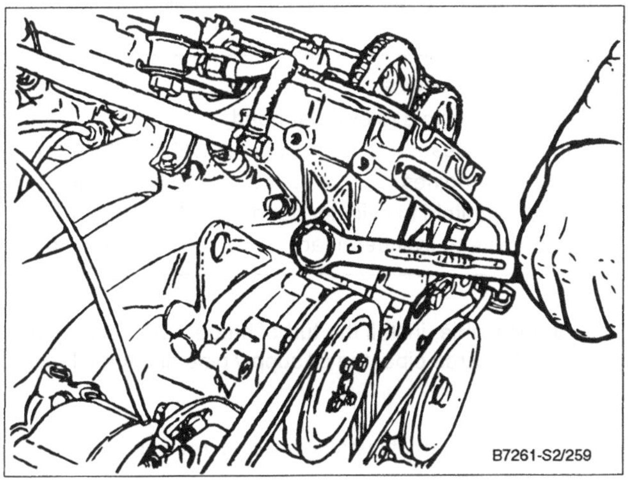

Fig. 8. Camshaft timing chain tensioner being removed from side of cylinder head.

CAUTION

Do not rotate the engine if the camshaft timing chain tension is relieved. The pistons can hit the valves and damage them.

---

21. Remove the camshaft sprockets.

NOTE

Insert a stiff wire through the chain to prevent it from falling into the timing chain cover area.

---

22. Siphon off any excess oil from the inside of the cylinder head. Then remove the cylinder head bolts in stages, working towards the center of the head starting with the outer-most bolts first.

23. Insert a threaded guide pin (Saab tool no. 83 92 128) into one of the cylinder head bolt holes. Position the timing chain and its pivoting guide so that it does not interfere with the removal of the cylinder head. Carefully lift the cylinder head off, pivoting the head as it passes the timing chain guide. See Fig. 9.

NOTE

As an alternative to the special guide pin, a wooden dowel can be used. The guide pins are used to align the head during removal. If the guide pin or dowel is not used, there is risk of damaging the chain guide when the head is lifted off.

Fig. 9. Cylinder head being removed.

24. Place the cylinder head on wooden blocks to prevent damaging the open valves.

To install cylinder head

1. Check to make sure the camshafts and the crankshaft are correctly set to 0°TDC as shown above in Fig. 5 and Fig. 6.

2. Place a new cylinder head gasket onto the cylinder block.

3. Fit the guide pin (or dowel) to the cylinder block and position the timing chain and pivoting chain guide as shown in Fig. 10.

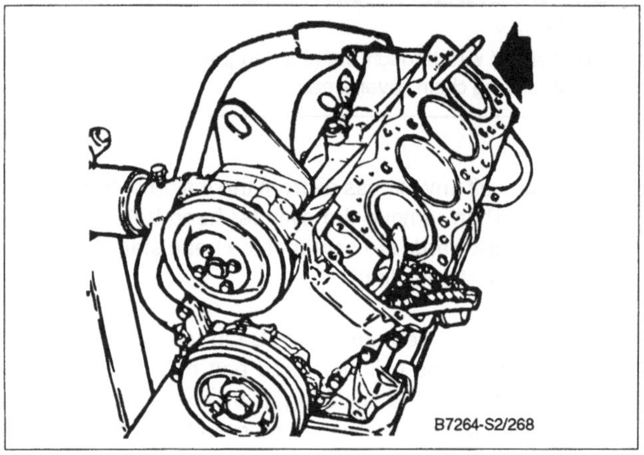

Fig. 10. Guide pin installed in cylinder block (arrow), and timing chain and chain guide correctly positioned for installation of cylinder head.

4. Slide the cylinder head into position, using the guide pin as a pivot when lowering the cylinder head over the chain guide. Install and torque the cylinder head bolts. See Fig. 11.

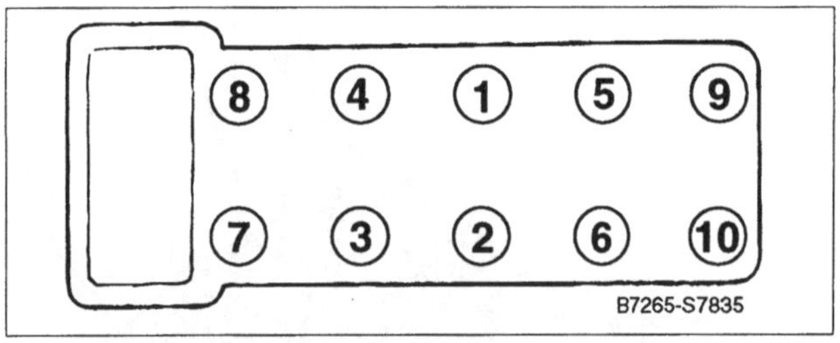

Fig. 11. Cylinder head bolt tightening sequence.

Tightening torques

- cylinder head to cylinder block
  stage 1 . . . . 60 Nm (44 ft-lb)
  stage 2 . . . . 80 Nm (59 ft-lb)
  plus and additional 90° (1/4 turn)

---

5. Install the two M8 bolts into the underside of the cylinder head as shown above in Fig. 2.

6. With all TDC (timing) marks aligned, install the exhaust-side camshaft sprocket, making sure there is no chain slack between the cam sprocket and crankshaft sprocket.

7. Install the intake-side camshaft sprocket, keeping the chain tight between the two sprockets. Hand tighten the sprocket mounting bolts. Install the timing chain tensioner.

Tightening torque

- timing chain tensioner
  to cylinder head . . . . 63 Nm (47 ft-lb)

---

8. Rotate the crankshaft by hand in the direction of normal engine rotation two full revolutions. Recheck that the TDC marks are all correctly aligned. If any faults are found reset the timing chain.

9. Tighten the camshaft sprocket mounting bolts while holding the camshafts stationary as shown earlier in Fig. 7.

Tightening torque

- camshaft sprocket to camshaft . . . . . 63 Nm (47 ft-lb)

---

10. Install the valve cover as described in **214 Cylinder Head and Valve Mechanism.**

11. Install the distributor cap, the spark plug wires, and the center cover above the spark plugs.

12. Using new nuts, install the exhaust manifold, and if applicable, the EGR pipe. Install the intake manifold.

Tightening torques

- exhaust manifold to cylinder head
  turbo engines. . . . . 25 Nm (18 ft-lb)
  all other engines . . . . 20 Nm (15 ft-lb)
- intake manifold to cylinder head. . . . 22 Nm (16 ft-lb)
- EGR valve to cylinder head . . . . 8 Nm (71 in. lb)

---

13. If applicable, install the turbocharger. Connect the turbo discharge pipe, the oil supply pipe and the oil return pipe.

14. The remainder of installation is the reverse of removal.

NOTE

Upon first start up, the hydraulic lifters will most likely be noisy. Avoid racing the engine above 3,000 rpm until the noise has dissipated.See **214 Cylinder Head and Valve Mechanism** for more information on hydraulic cam followers .

Tightening torques

- all other fasteners
  M5 bolt . . . . 5 Nm (44 in-lb)
  M6 bolt . . . . 10 Nm (89 in-lb)
  M8 bolt . . . . 20 Nm (15 ft-lb)
  M10 bolt . . . . 40 Nm (30 ft-lb)

---

212 Pistons, Connecting Rods, Cylinder Bores

TABLES

GENERAL

This repair group provides the specifications and special reconditioning information necessary to repair the Saab short block. The information contained here is intended to be used as a reconditioning guide for the professional or experienced automotive mechanic. Many of the operations described below require precision measuring equipment. If you lack the skills, tools, or a suitable workplace for reconditioning, we suggest you leave these repairs to an authorized Saab dealer or other qualified shop.

Be sure to carefully mark the position and orientation of all parts as they are removed. Connecting rods, rod caps, rod bearings, pistons, main bearing caps and main bearings are assembled in an exact location and orientation.

NOTE

Cylinder block disassembly and assembly is covered in **210 Engine Disassembly and Assembly.**

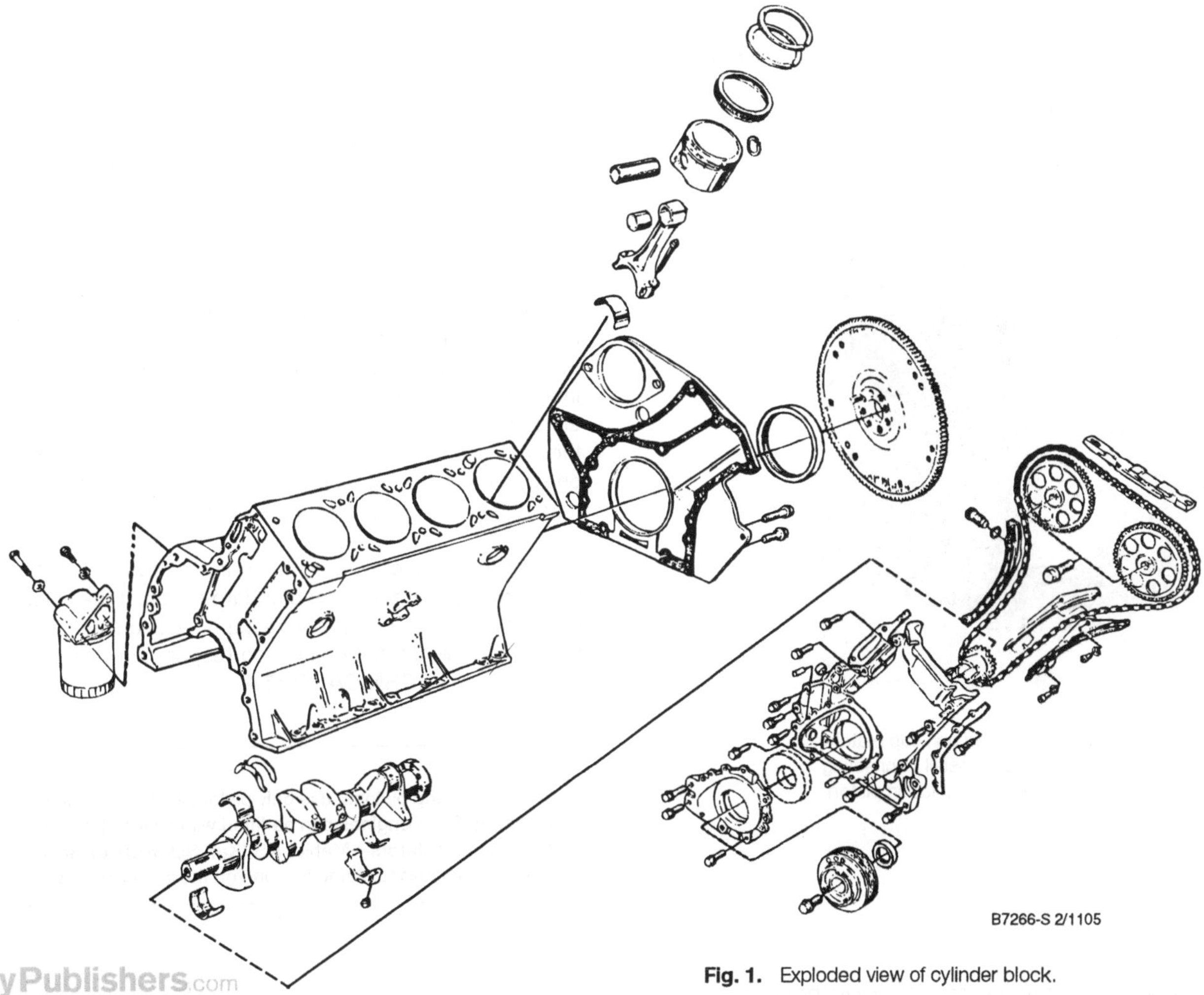

Fig. 1. Exploded view of cylinder block.

CYLINDER BLOCK AND PISTON RECONDITIONING

All of the disassembled parts should be thoroughly cleaned before inspection.

Connecting Rods and Pistons

If pistons, piston pins, piston rings, connecting rods, and bearings are to be reused, they should always be reinstalled in their exact positions. Used parts should never be interchanged between cylinders.

The piston pin should fit without any play. The fit can be quickly checked by pushing the pin through the rod by hand. There should be slight resistance as the pin moves through the rod. A more accurate check can be made using a micrometer.

Each connecting rod is matched to its cap and identified by a number. See Fig. 2. Connecting rods should be checked for bend, twist and bearing bore out-of-roundness using connecting rod alignment and measuring equipment.

CAUTION
The pistons, piston rings, and connecting rods must be installed according to their markings. See **210 Engine Disassembly and Assembly.**

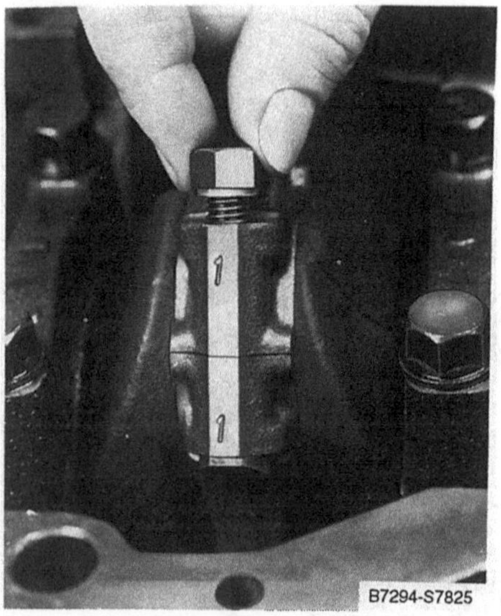

Fig. 2. Connecting rod big-end and cap matching numbers. Always install connecting rod nuts with flange on nut facing piston.

Tightening torque
- connecting rod cap to connecting rod . . . . 55 Nm (41 ft-lb)

Piston pins, connecting rod bushings, and connecting rods are all available as individual replacement parts. The piston pin circlips should be replaced with new circlips any time they are removed. **Table a** lists connecting rod and piston pin specifications.

Table a. Connecting Rod and Piston Pin Specifications

Connecting rods	
big end diameter	56.000-56.019 mm (2.2047-2.2055 in.)
small end diameter, bushing installed	24.005-24.010 mm (0.9451-0.9453 in.)
big-end oil clearance	0.026-0.062 mm (0.0010-0.0024 in.)*
maximum permissible weight variation per set	9 g (0.32 oz.)
piston pin diameter	23.996-24.000 mm (0.9447-0.9449 in.)
Piston pin to connecting rod oil clearance	
B202 engine	0.005-0.014 mm (0.0002-0.0006 in.)
B212 engine	0.002-0.011 mm (0.00008-0.0004 in.)

* See **216 Crankshaft** for more information on connecting rod big-end clearances and bearing shell color codes.

Saab supplies pistons in size ranges or classifications. These are "A", "B" and "C." The cylinder bores and the pistons are individually coded. The piston codes are on the piston domes and the cylinder block codes are stamped into the cylinder block. See Fig. 3. The Saab block is originally fitted with either an "A" class piston or a "B" class piston, depending on the final size of the bore. It is possible to find both classes in one cylinder block. Other piston classes are available as standard and oversize replacements when servicing the engine.

Small amounts of wear in the cylinders can usually be corrected by installing a larger standard piston. If the cylinder is excessively worn, it may require boring to fit an oversized piston. Piston diameters and their corresponding classifications are listed in **0 TECHNICAL DATA.**

CAUTION
When replacing pistons, use only pistons from the same manufacturer. Saab engines can be fitted with pistons from one of three manufacturers—Mahle, KS (Karl Schmidt), or Hepolite. The manufacturer's stamp is found on the inside of the piston.

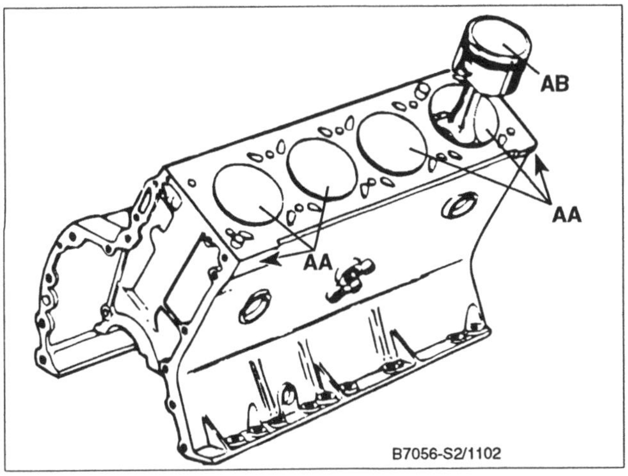

Fig. 3. Piston and cylinder classification locations.

NOTE

- The classifications for standard replacements pistons are AB, B, and C (B on B212 engines), with the C or B class pistons being the largest and most commonly used replacement piston on high-mileage engines.

- Standard piston class A is not available as a replacement part.

When measuring the piston diameter, make measurements at right angles (90°) to the wrist pin and at the height indicated in Fig. 4.

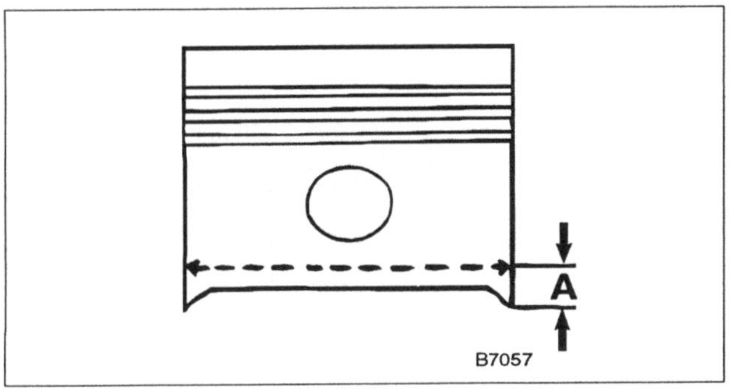

Fig. 4. Piston measuring point (**dotted line**) used when checking piston diameter. Distance varies depending on piston manu-

---

Piston measuring point (distance A)

- Manufacturer
  - Mahle . . . . 16 mm (5/8 in.)
  - KS . . . . 22 mm (7/8 in.)

Cylinders

When checking the piston-to-cylinder clearance, a set of feeler gauges with a 1/2 inch wide blade and a spring scale are needed to perform the procedure described below. A conventional piston-to-cylinder clearance measuring procedure can be used (micrometer and bore gauge), although the method described below is simpler and is the one recommended by Saab.

To check piston-to-cylinder clearance

1. Lightly oil the cylinder and the piston.

2. Insert a feeler gauge blade into the cylinder and then install the piston (piston rings removed) upside down into its matching cylinder. Select a feeler gauge that fits snugly.

NOTE

- The 1/2 in. wide feeler gauge blade is critical to accurate results.

- The feeler gauge blade should be inserted at right angles to the wrist pin.

---

3. Connect the spring scale to the feeler gauge and pull on the scale. See Fig. 5. Note the amount of force required to start the gauge moving out of the cylinder.

NOTE

Repeat the above step at several depths in the cylinder.

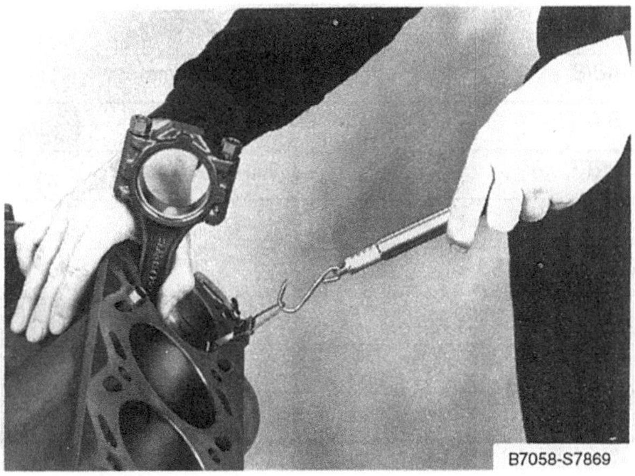

Fig. 5. Piston-to-cylinder clearance being checked using spring scale and feeler gauge with 1/2 inch blade.

4. Continue to try different feeler gauges until the pulling force is within the specified limits listed below.

Correct spring scale force
- 8-12 N (1.7-2.6 lb-ft)

CAUTION
When replacing pistons, do not mix pistons from different manufacturers. Pistons can be individually replaced, although they must be from the same manufacturer.

NOTE
- The thickness of the correct feeler gauge equals the piston-to-cylinder clearance. Piston-to-cylinder clearances vary depending on the piston manufacturer and piston/cylinder classifications. See **Table b.**
- If the piston-to-cylinder clearance is excessive, the piston should be replaced with a larger standard piston. Excessively worn cylinders can be bored to accept oversized pistons. See **0 TECHNICAL DATA** for piston classifications and diameters.

Piston Rings

Check piston ring end gaps, ring side clearances in the piston, and ring thicknesses. Piston ring end gaps are checked with the piston rings inserted one at a time into the cylinders. See Fig. 6. Use an inverted piston to position the rings squarely in the bore. Be sure to check the end gap at the bottom of the piston's travel when checking rings in a worn cylinder. Piston ring specifications are listed in **Table c.** The end gap should be checked after the final cylinder hone, if applicable.

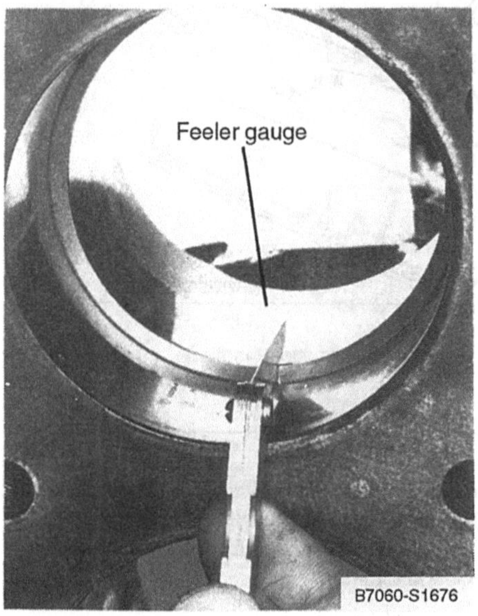

Fig. 6. Piston ring end gap being checked in cylinder block using feeler gauge.

Table b. Piston-to-Cylinder Clearances

Piston/cylinder classification	Non-turbo (KS)	Turbo	Non-turbo (Hepolite)
B202 engine (2.0 L)			
A/A	0.012-0.032 mm	0.030-0.050 mm	0.015-0.033 mm
AB/A	0.004-0.022 mm	0.022-0.040 mm	0.009-0.025 mm
AB/B	0.014-0.032 mm	0.032-0.050 mm	0.019-0.035 mm
B/A	—	0.014-0.032 mm	0.001-0.019 mm
B/B	0.006-0.024 mm	0.024-0.042 mm	0.011-0.029 mm
C/B	—	0.008-0.034 mm	—
B212 engine (2.1 L)			
A/A	—	—	0.008-0.028 mm
B/B	—	—	0.009-0.027 mm
B+/B+	—	—	0.009-0.027 mm

Table c. Piston Ring Specifications

	Top compression ring	Second compression ring	Oil scraper ring
end gap	0.35-0.55 mm (0.0138-0.0217 in.)	0.30-0.45 mm (0.0118-0.0177 in.)	0.38-1.40 mm (0.0150-0.0551 in.)*
width (thickness)	1.73-1.75 mm (0.0681-0.0689 in.)	1.98-1.99 mm (0.0780-0.0783 in.)	2.63-2.73 mm (0.1035-0.1075 in.)**
side clearance in piston groove	0.05-0.09 mm (0.0020-0.0035 in.)	0.04-0.07 mm (0.0016-0.0028 in.)	—

*Applies to segment part of ring
**Oil scraper ring segment width (thickness): 0.58-0.64 mm (0.022-0.025 in.)

When installing the rings, install the second compression ring so that the word "TOP" found on the ring faces up. See Fig. 7. Make sure the ring end gaps are offset as shown in Fig. 8. Lightly coat the cylinders and the rings with engine oil before installation.

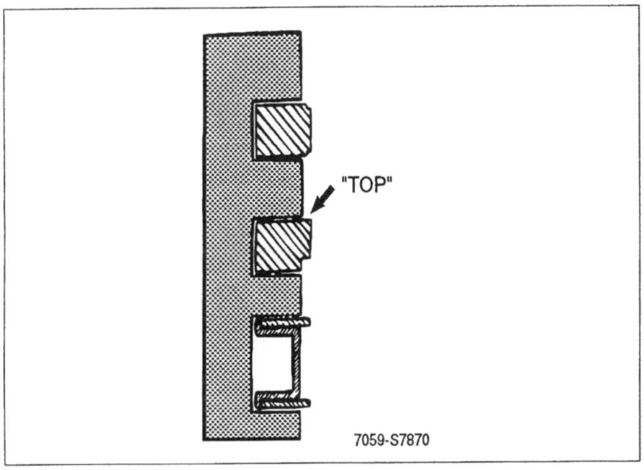

Fig. 7. "TOP" marking on second compression ring correctly installed.

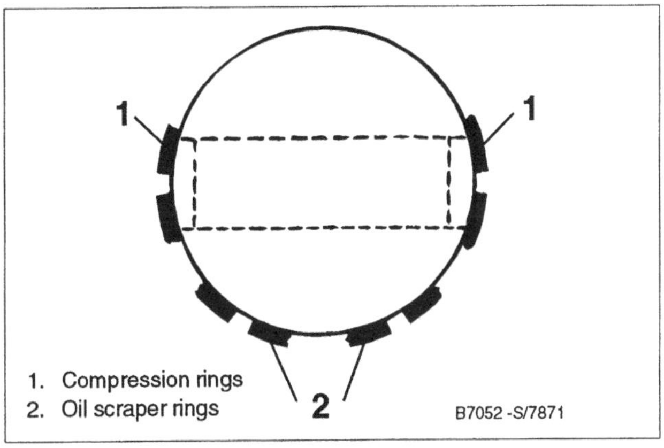

Fig. 8. Position compression ring end gaps 180° apart and directly above piston pin. Position scraper ring end gaps evenly around piston and not in line with each other.

214 Cylinder Head and Valve Mechanism

214

GENERAL

Some special tools may be required to disassemble and repair the valve mechanism owing to the cylinder head design. Read the procedures carefully before starting the job to determine what tools and equipment will be needed.

HYDRAULIC CAM FOLLOWERS

The valve gear mechanism uses hydraulic cam followers to keep the valve clearances within a limited working range of 2.055 mm (0.081 in.). See Fig. 1. The cam followers are sealed units and require no maintenance or adjustment.

Under some circumstances, such as removal and installation of the cylinder head, the cam followers may become noisy. Hydraulic cam follower noise usually makes a tapping or chattering sound. In most instances, this is considered normal as long as the noise goes away after a few minutes (maximum of 30 minutes). If the noise does not go away, either the follower is faulty, the oil pressure to the follower is too low, or there is trapped air in the system. Some of the common conditions that can cause follower noise are listed below.

1. Starting the engine after an oil change. If the noise was not present before the oil change, it should go away as soon as the oil pressure is restored.

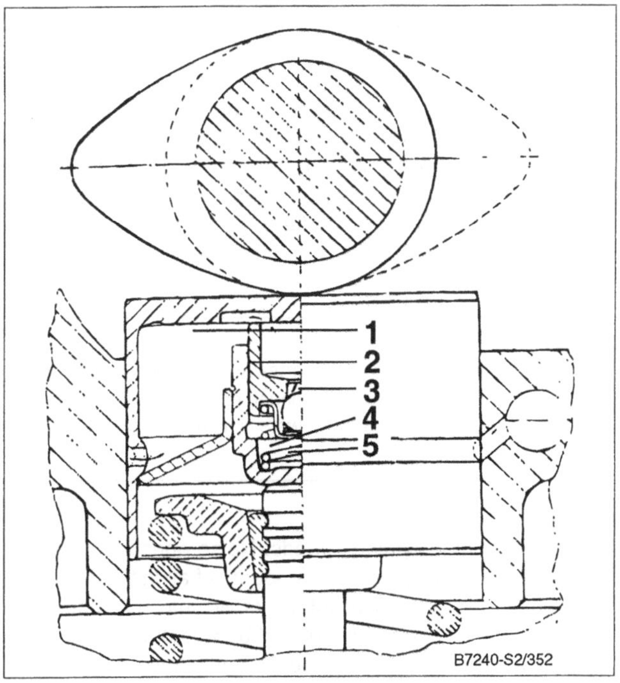

Fig. 1. Cutaway of hydraulic cam follower. Each cam follower consists of two adjacent storage chambers (**1**), a leakage passage (**2**), a check valve (**3**), a high pressure chamber (**4**), and a spring (**5**).

2. Starting a cold engine or starting an engine that has been sitting for long periods. This noise is normal and should disappear as soon as the oil reaches normal operating pressure.

3. After doing work that required cranking the engine by hand or by the starter motor. This can cause the oil in the follower to be forced out. Again, the noise should go away as soon as the oil pressure is restored. Under severe conditions where the follower has been totally drained, it may take as long as 30 minutes at engine speeds between 2,000 and 3,000 rpm to fully expel the air from the follower.

CAUTION

When trying to restore oil pressure to a hydraulic lifter, do not raise the engine speed above 3,000 rpm, as cam follower damage may result.

---

4. After installing a new replacement cam follower. The oil in a new follower may drain out if the follower is incorrectly stored. This situation is the same as the drained follower described above.

5. After a short period of idling (engine hot). Run the engine at 1,500 rpm or above. The noise should disappear. The noise is due to low oil pressure in the hydraulic cam followers when the engine is idling. Check the external oil feed pipe connectors and O-rings in the top of the cylinder head, where applicable.

NOTE

On cylinder heads used on mid-1988 and later models, the external supply pipes were discontinued owing to a new cylinder head design. The oil to the followers is now supplied via internal oil passages.

---

6. Only at high engine speeds. Noise goes away at idle. This is due to excessive amounts of air in the oil at high engine speeds. The intake of air into the lubrication system is most likely caused by a leak on the suction side of the oil pump. Check the oil pump's suction pipe O-ring and the suction pipe itself for damage.

NOTE

The oil pump suction pipe is in the engine oil sump of the transmission. To reach the suction pipe, the engine and transmission assembly needs to be removed and the engine separated from the transmission. See **201 Engine Removal and Installation.**

---

7. Noise at all times, regardless of engine speed or temperature. A noisy follower at all times usually means that the follower is faulty and should be replaced.

NOTE

On cylinder heads introduced on mid-1988 models, with internal oil passages, it is possible for air to become trapped in the system, causing cam follower noise. If more than one follower is exhibiting noise, try bleeding the lubrication system before checking and/or replacing noisy cam followers as described below.

Checking and Replacing Hydraulic Cam Followers

To check for a faulty cam follower, run the engine briefly to build up oil pressure. Then turn the engine off and remove the valve cover as described below under **Valve Mechanism.** Using a wooden stick, press down on the top of the follower. If the follower feels spongy or can be pressed down at all, it is most likely faulty and should be replaced. A faulty cam follower can be easily removed after removing the camshaft as described below. Use a magnetic tool to remove the cam followers.

NOTE

- Keep in mind that a damaged or poorly sealed oil passageway to the follower may be the cause of the noise rather than the follower itself.

- Do not mix up cam followers between cylinders. Used cam followers should always be installed in their original positions.

- Be sure to inspect the cam follower lobe for scoring or wear when replacing a faulty lifter. Camshaft specifications can found below under **Valve Mechanism Reconditioning.**

Bleeding Air Trapped in Cylinder Head (from Mid-1988 models—engine no. J082586—and later)

The cylinder head introduced on mid-1988 and later models uses small drilled oil passages to supply oil to the camshaft and cam followers. In some cases, it is possible for air to become trapped due to the tight tolerances of the passages. If more than one follower is noisy even after running the engine for more than a half-hour, try bleeding the lubrication system as described below.

NOTE
- On earlier cylinder heads (up to engine no. J082585), oil to the followers is supplied via oil pipes. Trapped air has not been a problem on cylinder heads with external oil pipes.
- See **100 Fundamentals for the Do-it-yourself Owner** for engine number location.

To bleed lubrication system

1. Remove the valve cover as described below in **Valve Mechanism.**
2. Loosen (three or four turns) the black bolt on the no. 5 intake camshaft bearing cap. See Fig. 2.

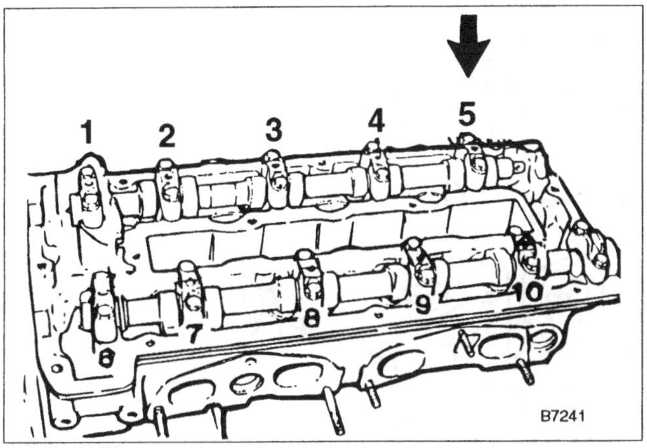

Fig. 2. Black bolt on no. 5 intake camshaft bearing cap (arrow).

3. Install the valve cover using at least one top bolt and one bottom bolt.

NOTE
Place some clean shop towels in the spark plug recesses to catch any spilled oil.

---

4. Disable the ignition by removing the harness connector from the EZK amplifier (normally-aspirated models) or the ignition control unit (turbo models).

NOTE
The EZK amplifier and the ignition control unit are mounted on the left-hand (driver's) side fender. See **340 Ignition System** for exact component location.

---

5. Make sure the car is in neutral. Then crank the starter motor for 15 seconds. Wait about 30 seconds. Repeat this procedure two or three times.
6. Remove the valve cover and inspect the loosened camshaft bearing cap bolt. There should be evidence that oil has been forced passed the bolt. Tighten the bolt.

Tightening torque
- camshaft bearing cap to cylinder head. . . . . 15 Nm (11 ft-lb)

---

7. Remove the shop towels from the spark plug area and clean up any spilled oil.
8. Thoroughly clean the valve cover gasket surface and reinstall the valve cover using a high-temperature silicone sealer as described below in **Valve Mechanism.**
9. Start the engine. The noise should go away within a few minutes.

NOTE
If the cam follower noise does not go away, it is most likely faulty and should be replaced as described earlier.

VALVE MECHANISM

Fig. 3 is an exploded view of the valve mechanism and cylinder head assembly.

To remove and install valve cover

1. Remove the two screws and center cover from above the spark plugs. See Fig. 4.
2. Remove the spark plug wires from the spark plugs and remove the wires with the grommet from the valve cover.
3. Remove the outer and inner valve cover mounting bolts and remove the cover and two cover gaskets.
4. Thoroughly clean the gasket surfaces of the cover and cylinder head. Install the new gaskets to the cylinder head. Then apply a 1/8 in. bead of high-temperature silicone sealer (Bostik Silicone 2680 or equivalent) at the three split plugs and above the distributor. See Fig. 5.
5. The remainder of installation is the reverse of removal. Tighten the mounting bolts in stages until the final torque is reached.

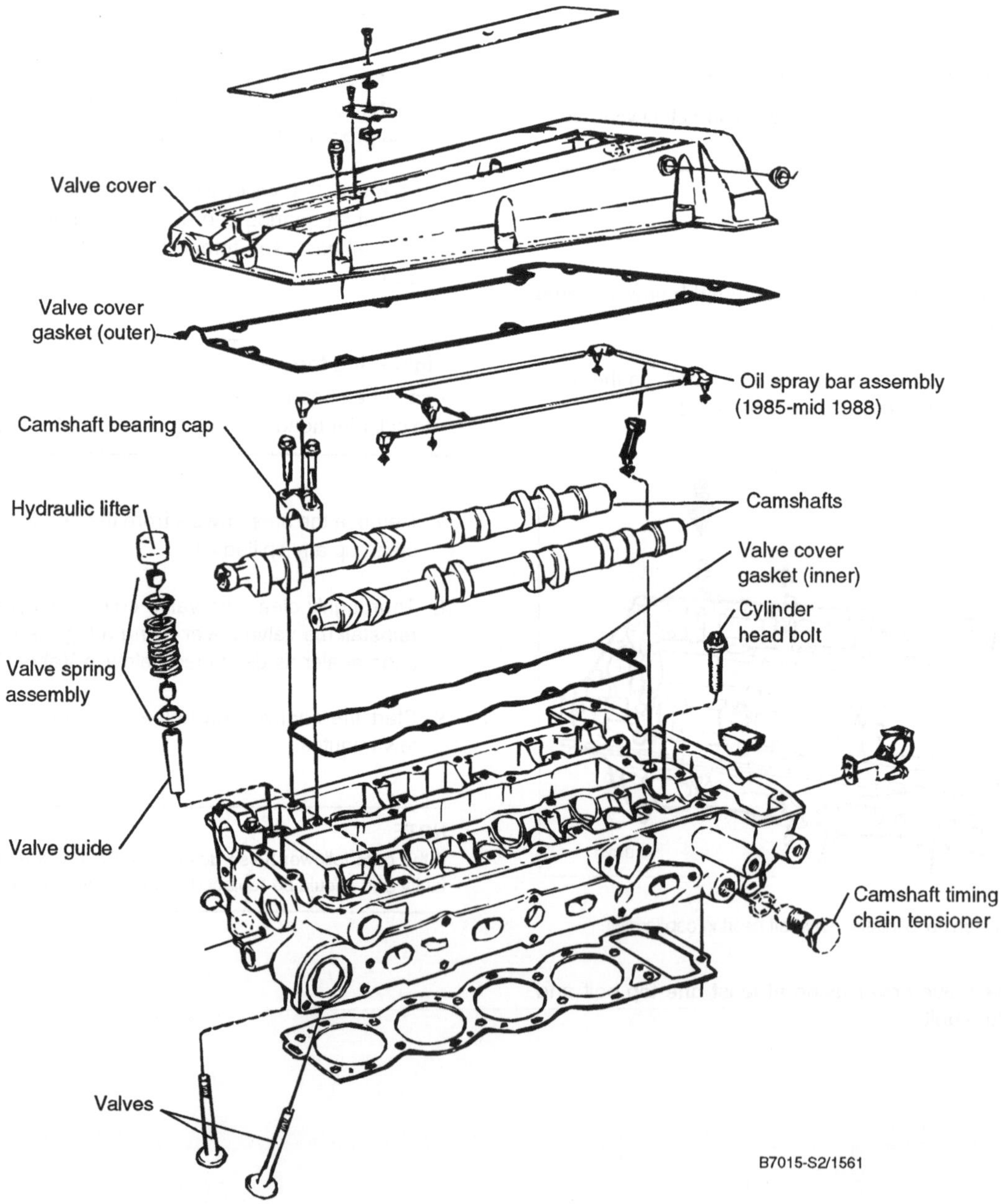

Fig. 3. Exploded view of valve mechanism and cylinder head.

Fig. 4. Center spark plug cover being removed.

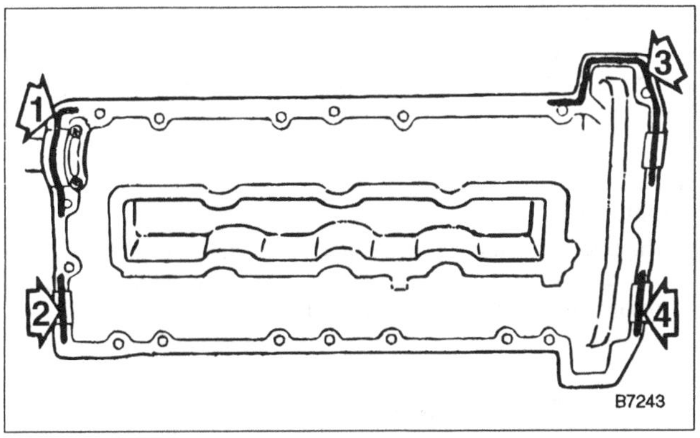

Fig. 5. Areas of valve cover gasket requiring high-temperature silicone sealer.

Tightening Torque

- valve cover to cylinder head . . . . 15 Nm (11 ft-lb)

To remove and install camshafts (cylinder head installed)

1. Remove the valve cover as described above.

2. Set the engine to TDC by rotating the crankshaft in normal engine rotation until the "0°" mark on the flywheel is aligned with the timing mark on the end plate. See Fig. 6.

NOTE

Check also that the marks on the camshafts are aligned with their matching marks on the bearing caps. See Fig. 7. If the marks are not aligned, turn the engine over an additional 360° (1 revolution).

CAUTION

If the camshaft timing marks shown below are not exactly aligned, the timing chain or one of the sprockets may be severely worn or the timing chain may have jumped out of position. See **215 Camshaft Timing Chain.** for additional information.

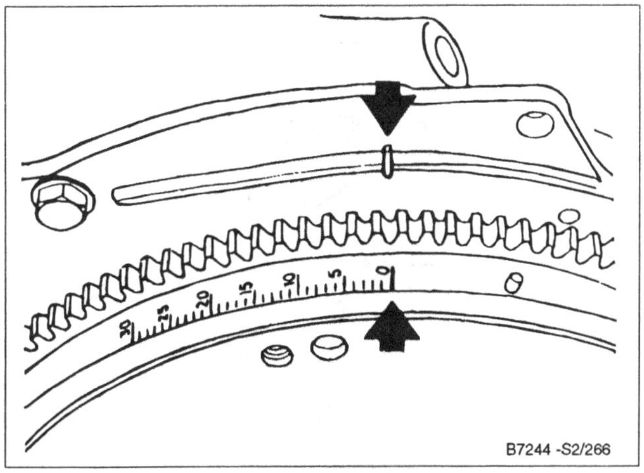

Fig. 6. Flywheel 0° TDC mark aligned with timing mark on engine end plate (arrows).

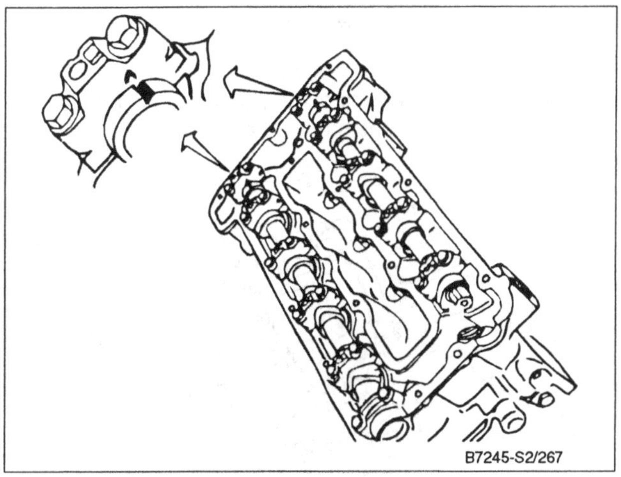

Fig. 7. Camshaft timing mark aligned with mark on camshaft bearing cap.

3. Remove the ignition distributor as described in **340 Ignition System.**

4. Remove the external oil supply pipes from the cylinder head, if applicable. See Fig. 8

CAUTION
Do not twist the pipes when removing them. This may damage the sealing surfaces between the pipes and the fittings.

NOTE
Beginning in mid-1988, a new cylinder head was used and the external oil supply pipes were discontinued.

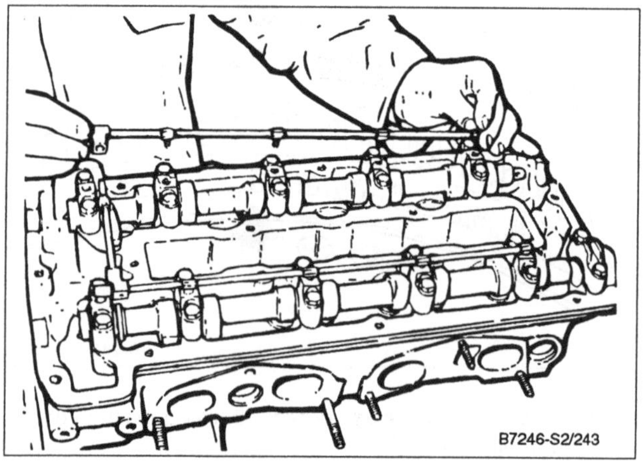

Fig. 8. Oil supply pipes for hydraulic cam followers being removed on mid-1988 and earlier cars.

5. Working from the side of the cylinder head, remove the camshaft timing chain tensioner. See Fig. 9.

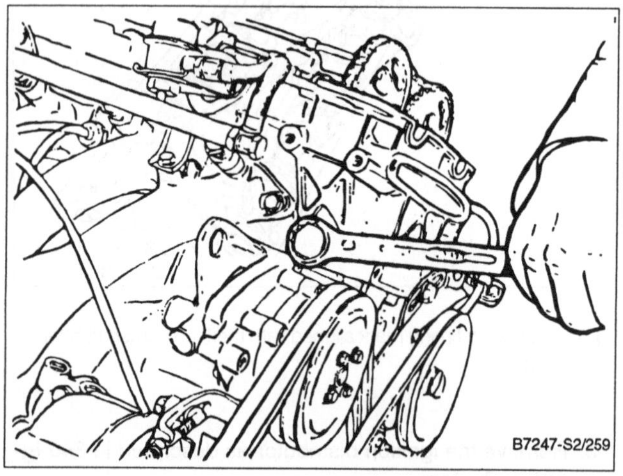

Fig. 9. Camshaft timing chain tensioner being removed.

6. Carefully unbolt the camshaft bearing caps evenly until all valve spring pressure is relieved. Remove the camshaft by lifting the chain off the sprocket.

CAUTION
Do not rotate the engine if the camshaft timing chain is removed from either camshaft. The pistons can hit the valves and damage them.

NOTE
If removing both camshafts, insert a stiff wire through the chain to prevent it from falling into the timing chain cover area.

Installation is the reverse of removal. If removed, be sure to refit the hydraulic cam followers in their original positions. Make sure the flywheel is set at the TDC mark (shown in Fig. 6 above). Install the camshafts such that the timing marks are correctly aligned (shown in Fig. 7 above) before tightening the camshaft bearing caps. When installing the camshaft bearing caps, make sure all numbers on the caps align with their numbers on the cylinder head, if applicable. Install the black camshaft bearing cap bolts on the spark plug side of the cap. See Fig. 10. Make sure the oil spray bar is correctly aligned with the bar's fittings (see Fig. 22 given later).

CAUTION
The black camshaft bearing cap bolts are drilled to provide oil passages to the hydraulic cam followers and must always be installed on the spark plug side of the cap..

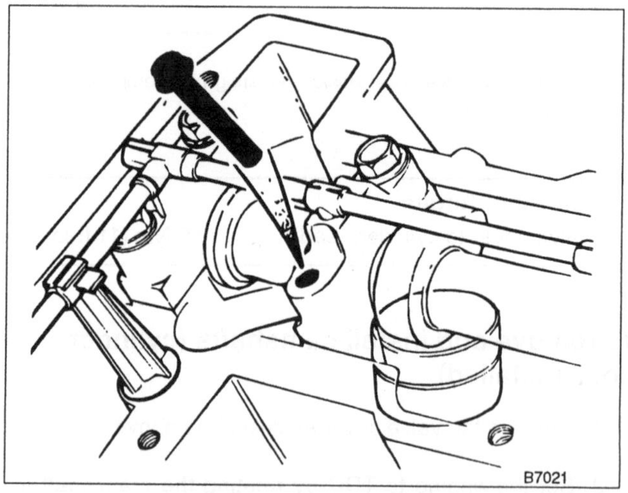

Fig. 10. Special black camshaft bearing cap bolts must be installed on spark plug side of cap.

NOTE
Bearing caps marked with numbers 1 through 5 are used on the intake camshaft and 6 through 10 are used on the exhaust camshaft.

Make sure all of the timing chain slack is on the tensioner side of the chain and install the chain tensioner. Turn the engine over several revolutions in the direction of normal rotation and bring it back to TDC. Double check all timing marks. If necessary, remove the tensioner and reposition the chain on the sprockets. Install the valve cover using new gaskets and sealer as described earlier. Install the distributor and ignition wires as described in **340 Ignition System.**

CAUTION

Correct camshaft timing is critical. If the timing chain is off by as little as two teeth on the sprocket, the pistons can contact the valves. Always check and recheck camshaft timing by turning the engine over by hand.

Tightening Torques

- camshaft bearing caps to cylinder head .... 15 Nm (11 ft-lb)
- camshaft timing chain tensioner to timing case .... 63 Nm (46 ft-lb)
- valve cover to cylinder head .... 15 Nm (11 ft-lb)

To replace valve stem oil seals (cylinder head installed)

The procedure below covers valve stem oil seal replacement with the cylinder head installed, using Saab special tools and a source of compressed air. The oil seals can also be replaced as part of the disassembly/assembly procedure described later under this repair group.

1. Set the engine to TDC and remove the camshafts as described earlier.

2. Using a magnetic tool, extract the cam followers. Label each follower so that they can be reinstalled in their original positions. See Fig. 11

CAUTION

The cam follower bores create a sealing surface for the cam followers. These surfaces must not be scratched or scored. Special plastic protector sleeves are available from an authorized Saab dealer (special tool no. 83 93 746). See Fig. 12.

---

3. Remove a spark plug from the cylinder head. Thread an appropriate adapter into the spark plug hole and connect the compressed air line to the adapter. Raise the air pressure in the cylinder.

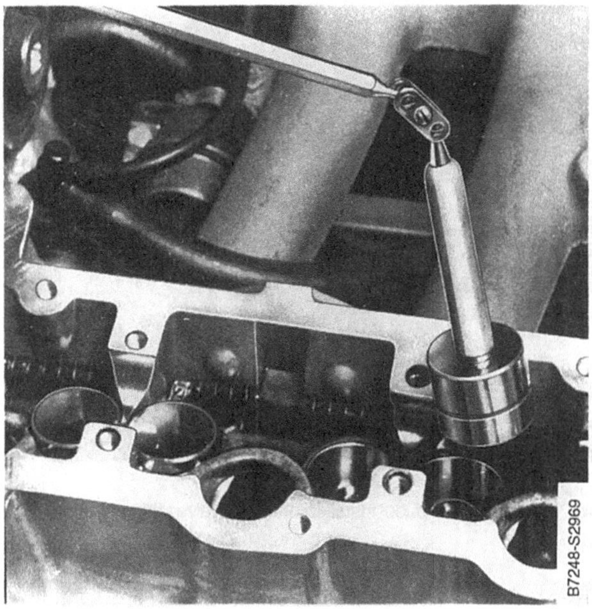

Fig. 11. Cam follower being removed from cylinder head.

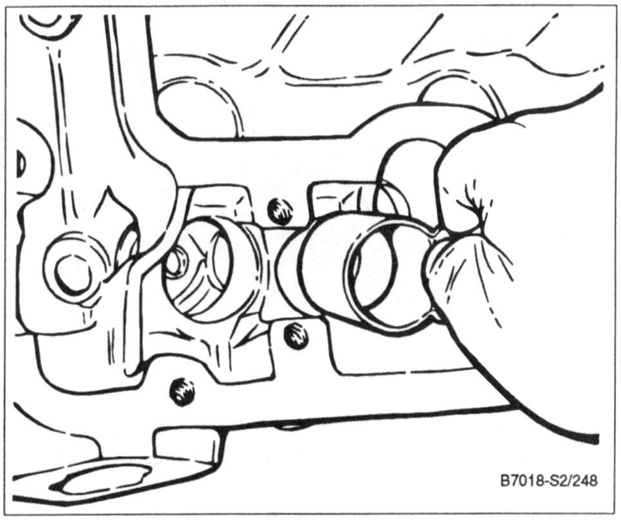

Fig. 12. Plastic sleeves being installed in cylinder head to protect cam follower bores.

NOTE

The spark plug hole adapter is available under Saab special tool no. 83 92 326.

---

4. Place Saab special tool 83 94 181 on top of the valve spring retainer. See Fig. 13.

5. Firmly tap the tool so that the valve spring keepers come free and follow the tool out. See Fig. 14.

Fig. 13. Special Saab tool installed on top of valve spring retainer.

Fig. 14. Removed valve spring keepers in special tool.

6. Lift out the valve spring retainer and the spring. Then using valve stem oil seal removal pliers, remove the seal. See Fig. 15.

7. Using Saab special tool 83 93 803, insert the new valve stem oil seal into the tool. See Fig. 16.

8. Fit the seal into position and gently tap on the tool until the seal is firmly seated. See Fig. 17.

9. Install the spring and the spring retainer. Fit the valve keepers into position on the valve spring retainer.

10. Place Saab special tools 83 94 181 and 83 94 207 onto the valve spring retainer and tap the tool firmly to seat the keepers in the keeper grooves. See Fig. 18.

11. Repeat the above procedures for the remaining valve stem oil seals.

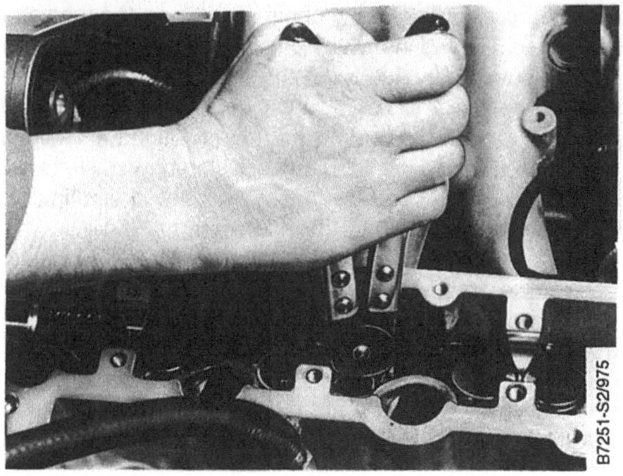

Fig. 15. Valve stem oil seal removed with special pliers.

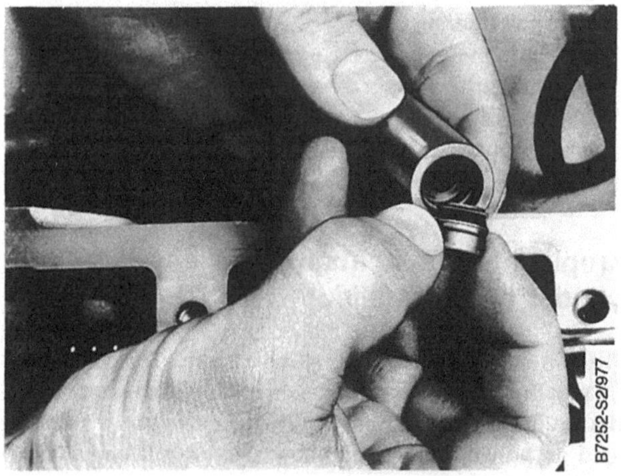

Fig. 16. Valve stem oil seal being installed into special tool.

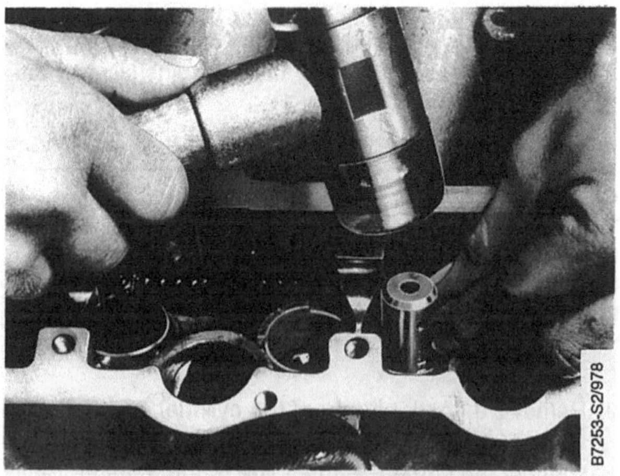

Fig. 17. Valve stem oil seal being installed on valve guide.

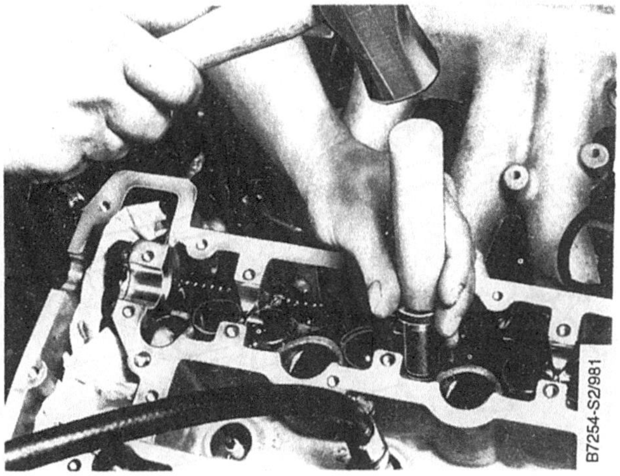

Fig. 18. Hammer and special tools being used to install valve spring keepers.

12. Remove the plastic sleeves and fit the cam followers into their original positions. Install the camshafts and valve cover as described above.

To disassemble valve mechanism (cylinder head removed)

NOTE
Cylinder head removal and installation is covered under **211 Cylinder Head Removal and Installation.**

---

1. Remove the ignition distributor as described in **340 Ignition System.**

2. Remove the intake and exhaust manifolds from the cylinder head.

3. If applicable, pull the oil spray bar assembly from the top of the cylinder head.

CAUTION
Do not twist the spray bars during removal. This can cause damage to the mating surfaces between the bars and the fittings.

---

4. Remove the camshaft bearing caps. Label each cap so that they can be reinstalled in their original positions. Remove the camshafts.

5. Using a magnetic tool, extract the cam followers. Label each follower so that they can be reinstalled in their original positions.

CAUTION
The cam follower bores create a sealing surface for the cam followers. These surfaces must not be scratched or scored. Special plastic protector sleeves are available from an authorized Saab dealer (special tool no. 83 93 746). See Fig. 12 given earlier.

---

6. Remove the valves, the valve spring retainers, the valve springs, and the spring seats. Label each complete assembly as it is removed.

NOTE
A valve spring compressor adaptor or a special valve spring compressor is needed to depress the valve springs. Saab special valve spring compressor (Saab tool no. 83 93 761 or KD tool no. 308) and special anvil (tool no. 83 93 779) are shown below in Fig. 19.

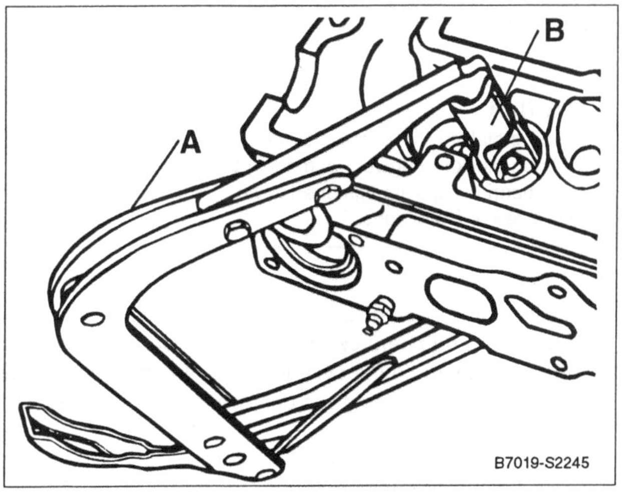

Fig. 19. Special valve spring compressor (**A**) and anvil (**B**) being used to remove valve.

WARNING
The valve springs are under considerable pressure. Use extreme care when removing the valves. Wear hand and eye protection.

---

7. Remove the valve stem oils seals using valve stem oil seal removal pliers. See Fig. 20.

Cylinder head assembly is the reverse of disassambly. Prior to installation, inspect all components for wear as described below. Be sure all parts are installed in their original positions. Lubricate the camshaft, the valve stems, and the valve stem oil seals prior to installation. Always use new valve

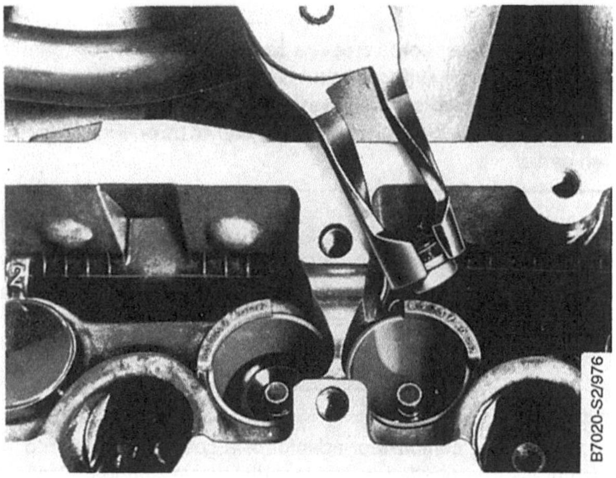

Fig. 20. Valve stem oil seals being removed using special pliers.

stem oil seals, tapping them in using an appropriate drift and a plastic hammer. Be sure to install the lower valve spring seat before installing the valve stem oil seals.

CAUTION

When installing the camshafts, place the cylinder head on wooden blocks to prevent damaging the valves as the camshaft bearing caps are tightened down. Tighten all caps evenly in three or four stages.

Always refit the cam followers in their original positions. Make sure the numbers on the camshaft bearing caps align with their numbers on the cylinder head, if applicable. Install the black camshaft bearing cap bolts on the spark plug side of the cap.

CAUTION

The black camshaft bearing cap bolts must always be fitted to the spark plug side of the bearing cap. These bolts are drilled to provide oil passages to the hydraulic cam followers. See Fig. 21.

NOTE

Bearing caps marked with numbers 1 through 5 are used on the intake camshaft and 6 through 10 are used on the exhaust camshaft.

If applicable, replace any worn or faulty O-rings for the oil spray bar fittings. To ensure proper lubrication, make sure the holes in the spray bars align with the holes in the fittings. See Fig. 22.

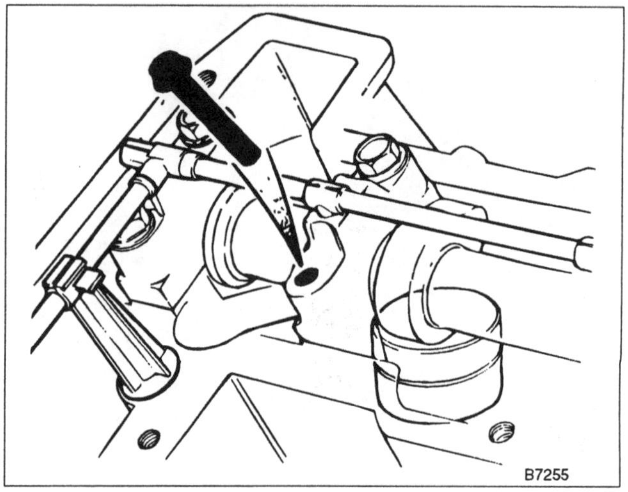

Fig. 21. Black camshaft bearing cap bolts are specially drilled and must always be installed on the spark plug side of the cap.

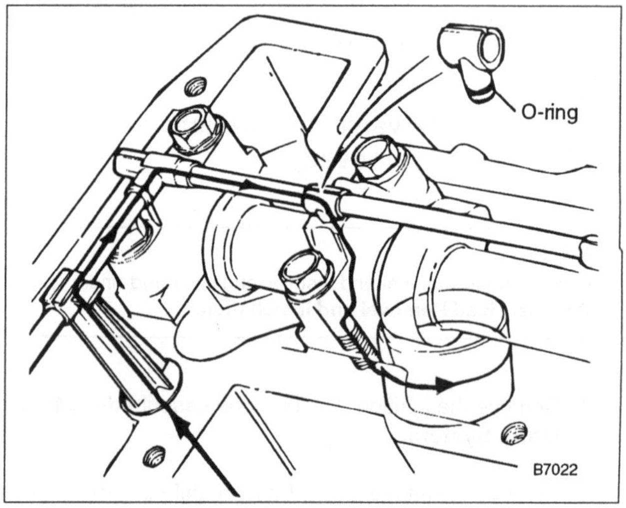

Fig. 22. Oil spray bar fittings correctly installed for proper lubrication to the camshafts and cam followers.

Tightening Torques

- camshaft bearing caps
  to cylinder head . . . . 15 Nm (11 ft-lb)
- spark plugs to cylinder head. . . . . . . 28 Nm (21 ft-lb)
- camshaft sprocket to camshaft. . . . . 63 Nm (46 ft-lb)
- intake manifold to cylinder head. . . . 18 Nm (13 ft-lb)
- exhaust manifold to cylinder head
  turbo models . . . . 25 Nm (18 ft-lb)
  normally-aspirated models . . . . 20 Nm (15 ft-lb)
- timing chain cover
  to cylinder head . . . . 20 Nm (15 ft-lb)
- valve cover to cylinder head. . . . . . . 15 Nm (11 ft-lb)
- ignition distributor to cylinder head . . 20 Nm (15 ft-lb)

VALVE MECHANISM RECONDITIONING

For anyone with the proper equipment and basic experience in cylinder head reconditioning, this section provides the necessary specifications and special reconditioning information.

Cylinder Head Assembly

The disassembled cylinder head should be carefully inspected for warpage, cracks, and wear. Check the valve guides and valve seats for wear as described below under **Valve Guides** and **Valves and Valve Seats.**

Always decarbonize and clean the head before inspecting it. A high quality straight edge can be used to check for warpage. Visually inspect the cylinder head for cracks. If a cracked cylinder head is suspected and no cracks are detected through the visual inspection, have the head further tested for cracks by an authorized Saab dealer or a qualified machine shop. A cracked cylinder head should be replaced.

A warped cylinder head can be machined provided the minimum cylinder head height is not exceeded. If further machining is required, the head should be replaced. Removing more than this amount will reduce the size of the combustion chamber and adversely affect engine performance.

CAUTION
If the cylinder head has been machined, make sure all sharp edges and burrs are removed from the combustion chamber. Lightly smooth the edges all around the combustion chamber and remove any imperfections.

Always check the total height of the cylinder head prior to machining. See Fig. 23. A maximum amount of 0.4 mm (0.016 in.) can be removed from a new cylinder head. If the cylinder head has been previously machined, it probably cannot be machined again. Cylinder head dimensions and minimum resurfacing height specifications are listed below.

Cylinder Head Height (dimension A)
- new . . . . 140.5±0.1 mm (5.531±0.004 in.)
- after machining . . . . 140.1±0.1 mm (5.516±0.004 in.)
- machining limit . . . . 0.4 mm (0.0016 in.)

Check the camshaft followers and the follower bores for wear. Replace the cam followers or the cylinder head if any signs of wear are visible or if any part does not meet the specifications listed below.

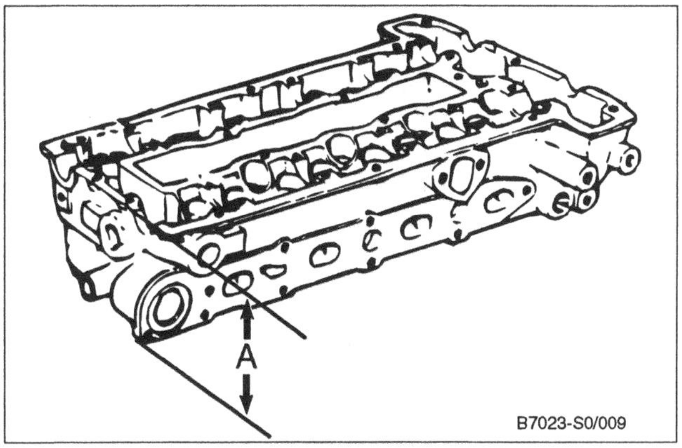

Fig. 23. Front view of cylinder head showing minimum resurfacing dimension (**A**). Measure between valve cover gasket surface and cylinder head gasket surface.

Cam followers
- diameter . . . . 32.96-32.98 mm (1.2976-1.2984 in.)
- height. . . . . .26 mm (1.02 in.)
- bore in cylinder head. . . . . 33.000-33.016 mm (1.2992-1.2998 in.)

Camshafts

Camshaft wear is usually caused by insufficient lubrication. Engine oil dilution, extended oil drain intervals, or high mileage are all additional causes of camshaft wear. Checking the camshaft for wear and runout requires precision measuring equipment.

Visually inspect the camshaft lobes and journals for wear. Lightly lubricate the camshafts and place them in the cylinder head. Install and tighten the bearing caps. Check that the camshaft turns smoothly. Using a feeler gauge, check the camshaft axial play (end float) between the camshaft flange and the thrust surface on the front of the cylinder head.

Camshaft
- axial play (end float)
  maximum permissible . . . . 0.08 to 0.35 mm (0.0031 to 0.0138 in.)
- camshaft bearing diameter . . . . . 28.922-28.935 mm (1.1387-1.1392 in.)

Valve Guides

Valve guide specifications are listed below. Check valve guide wear with a new valve as shown in Fig. 24. Inspect the valve seats to ensure that the cylinder head can be reconditioned before installing new valve guides.

Valve Guides
- length . . . . 49.00 mm (1.929 in.)
- outside diameter. . . . . 12.039-12.050 mm (0.4740-0.4744 in.)
- bore for valve guide in cylinder head . . . . 12.000-12.018 mm (0.4724-0.4731 in.)

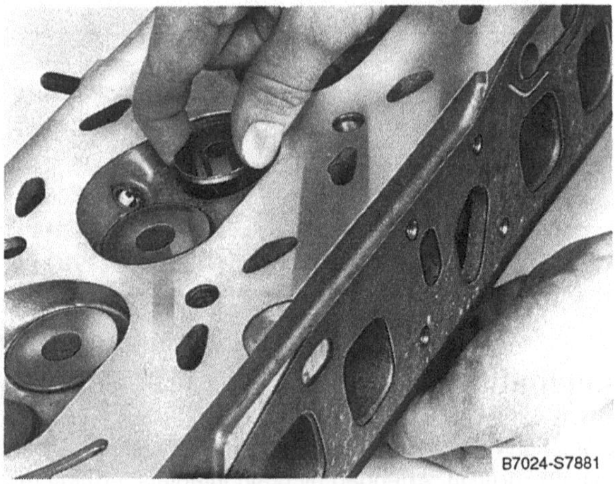

Fig. 24. Valve guide wear being checked. Raise valve 3 mm (0.12 in.) from its seat and rock the valve head to check for play. (Saab 8-valve cylinder head shown.)

Valve guide wear
(clearance between valve guide and valve stem)
- maximum permissible. . . . . 0.5 mm (0.02 in.)

Saab special tools should be used to replace worn valve guides. See Fig. 25. Always finish replacement valve guides using a 7.0 mm (H7) reamer.

NOTE
The Saab special tools are recommended to replace the valve guides, although conventional methods can also be used.

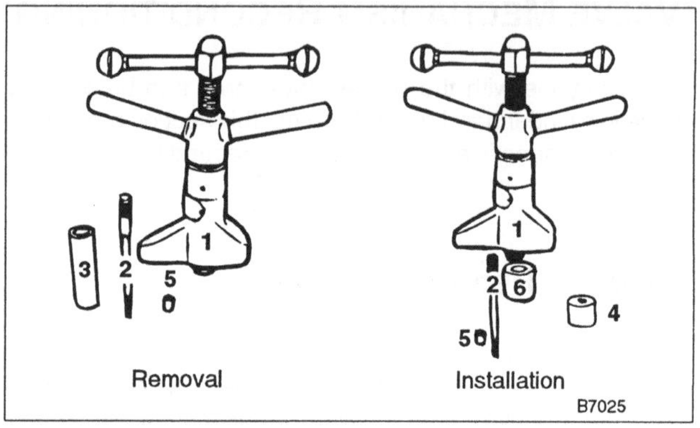

Fig. 25. Saab special tools for removing and installing valve guides.

Valve guide removal/installation special tools
- jack screw (**1**) . . . . tool no. 83 90 494
- pull rod (**2**). . . . . tool no. 83 93 811
- spacer sleeve (**3**) . . . . tool no. 83 93 829
- stop (**4**) . . . . tool no. 83 93 837
- special nut(**5**) . . . . tool no. 83 93 845
- mandrel (**6**) . . . . tool no. 83 90 379

To replace valve guides

1. Flush the cylinder head with hot water.

CAUTION
Always heat the head with hot water prior to removing and installing the valve guides. Failure to the heat the head may result in damage to the cylinder head casting.

2. Working from the camshaft side of the cylinder head, insert the special tools. See Fig. 26 and Fig. 27. Withdraw the valve guide by turning the jack screw.

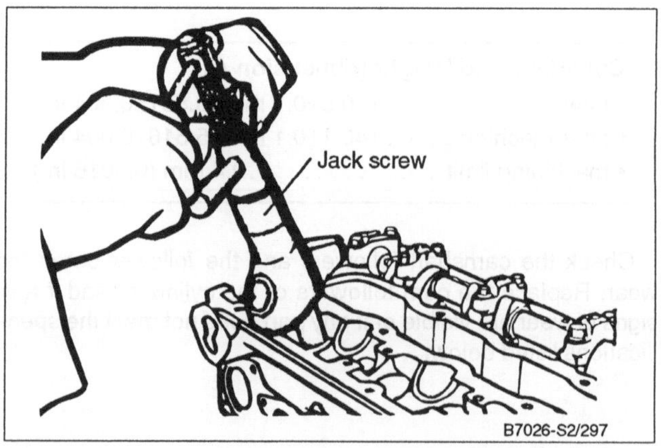

Fig. 26. Valve guide being removed using special tools.

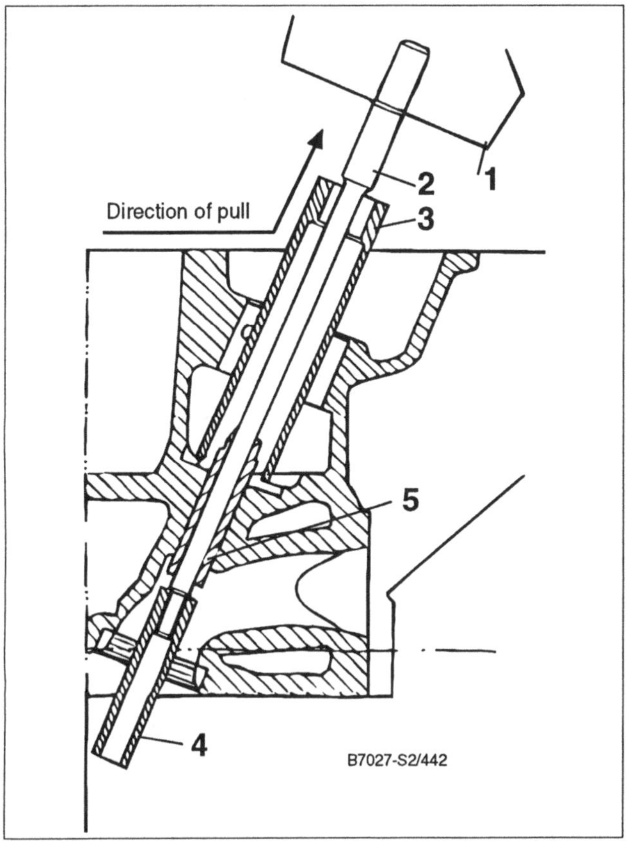

Fig. 27. Cutaway view of special valve guide removal tools. Withdraw valve guide (**5**) using jack screw (**1**), pull rod (**2**), spacer sleeve (**3**), and special nut (**4**)

3. Flush the new guide with cold water and fit the guide into the cylinder head. Working from the combustion chamber side of the cylinder head, insert the special tools. See Fig. 28. Draw the valve guide into the head until the stop is fully seated.

4. Finish the guide using a 7.0 mm undersize reamer (Saab tool no. 83 93 944). Then Follow up with a 7.0 mm H7 valve guide reamer.

Valves and Valve Seats

Valve and valve seat specifications vary between turbo and non-turbo models. Exhaust valves are sodium-filled and stellite-coated and should be hand-lapped only. Refacing exhaust valves by grinding will remove this coating and render the valve useless. Valve and valve seat dimensions are shown below in Fig. 29.

Whenever work is done on the valves or valve seats, the valve stem height must be checked. This measurement is critical to proper hydraulic cam follower operation. A special tool (Saab tool no. 83 93 753) is available to make the measurement, although precision measuring equipment can be substituted.

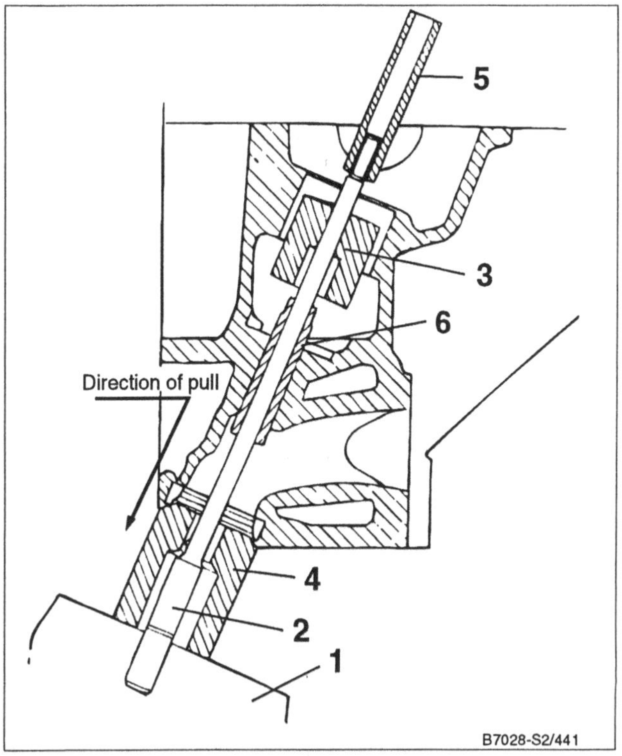

Fig. 28. Cutaway view of special tools used to install valve guide. Draw valve guide (**6**) in using jack screw (**1**), pull rod (**2**), stop (**3**), centering mandrel (**4**), and special nut (**5**).

With the valves installed and the camshafts and cam followers removed, measure the distance between the end of the valve stem and lowest part of the camshaft bearing surface as shown in Fig. 30.

NOTE

When using the special Saab stepped tool, lay the tool across the camshaft bearings so that the long and short legs are directly above the end of the valve stem. See Fig. 31. Check that the shorter step does not touch the valve (maximum valve height) and that the longer step raises the tool off the bearing surface (minimum valve height).

If any fault is found, either grind the end of the valve stem (valve stem too high) or machine the valve seat (valve stem too low). If the valve stem is worn, replacing the valve is also an alternative to valve seat grinding.

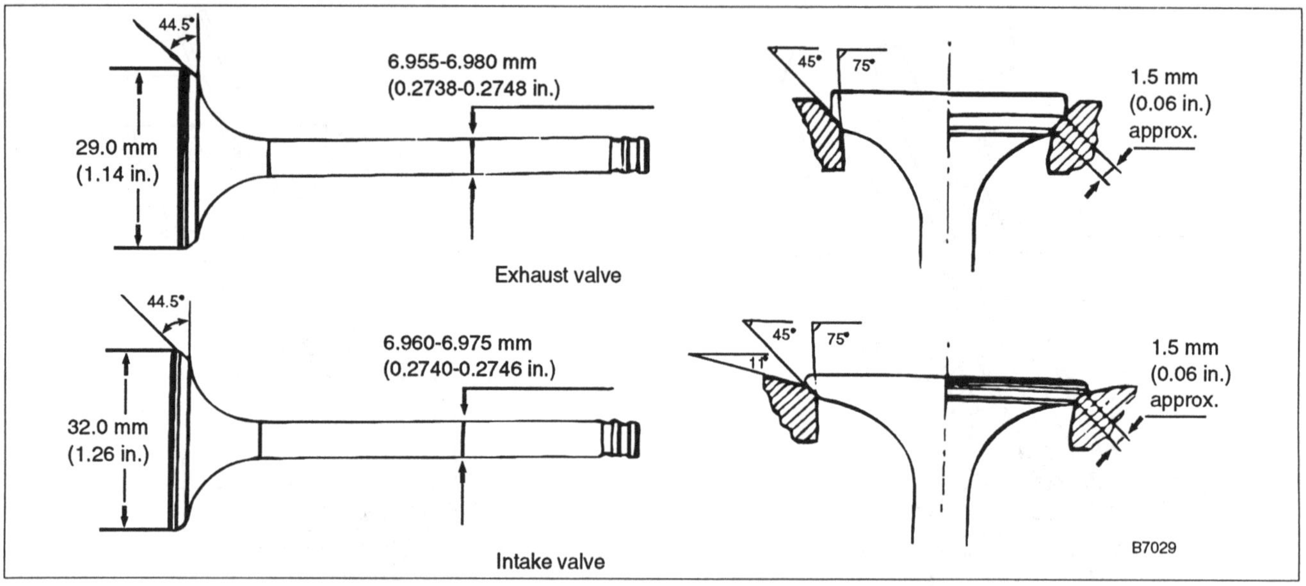

Fig. 29. Valve and valve seat dimensions.

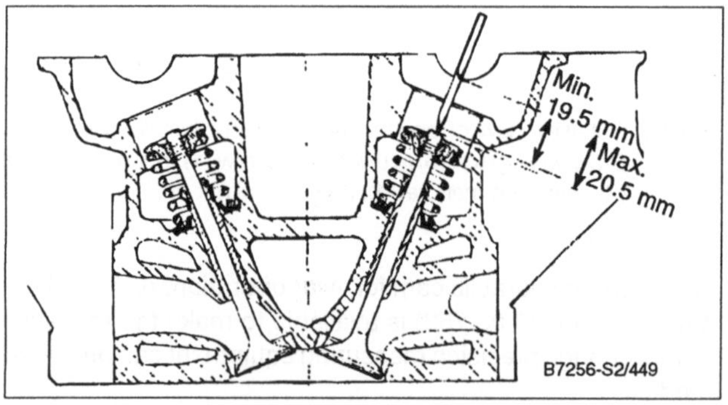

Fig. 30. Working range of hydraulic cam follower (valve stem height).

Scrapping Sodium Filled Valves

All of the fuel injected cars covered by this manual use sodium-filled exhaust valves. The sodium contained in the hollow stems of the valves is considered Hazardous Waste under the resource Conservation and Recovery Act. Disposal of used sodium-filled valves must be done in accordance with local, state, and federal regulations.

WARNING
Discarded sodium-filled exhaust valves should never, under any circumstances, be mixed with normal scrap metal as they are liable to explode when melted down.

To scrap sodium-filled exhaust valves

1. Drill a hole through the center of the valve and into the sodium compound. See Fig. 32.

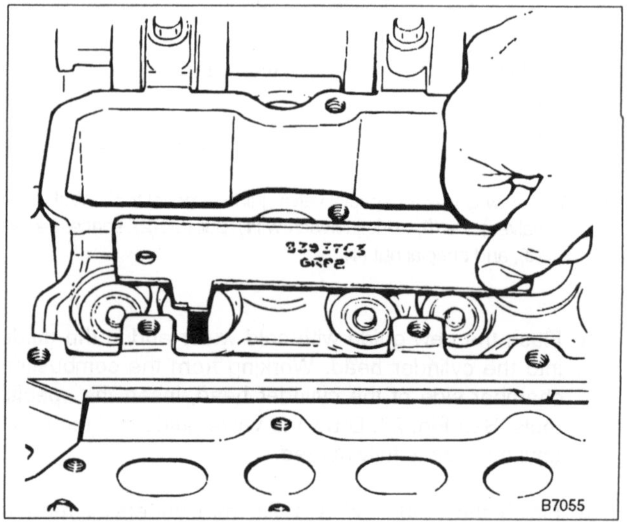

Fig. 31. Valve stem height being checked using Saab special tool no. 83 93 753.

2. Drill another hole into the stem or cut off the stem approximately 25 mm (1.0 in.) from the end of the stem.

WARNING
When drilling or cutting, keep clear of any water to avoid an explosion.

---

3. Throw the valve into a bucket of water and quickly move at least 10 feet away. See Fig. 33.

Fig. 32. Hole being drilled into sodium-filled exhaust valve.

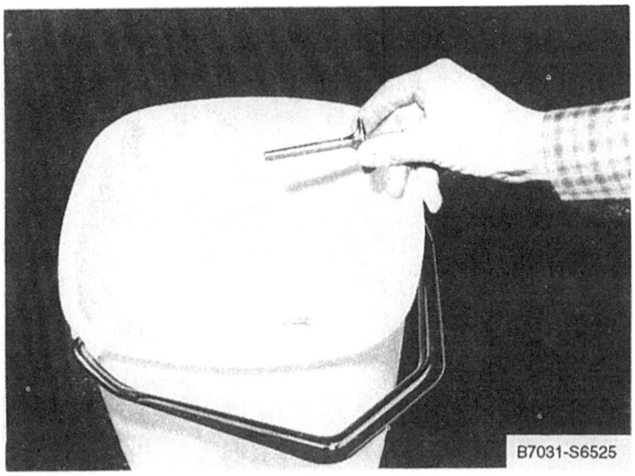

Fig. 33. Sodium-filled valve being dropped into bucket of water to neutralize sodium.

4. After about two minutes, the reaction should die down. Remove the valve from the bucket and dispose of the valve in the normal way.

Valve Springs

Valve springs should be checked for squareness and fatigue. Accurately checking for worn springs requires special spring testing equipment to compress the valve under a specified pressure. Always replace a spring that fails to meet any specification.

To quickly check the springs, line them all up in a row. Place a straight edge across the top of the springs. Any spring that is significantly shorter than the others (more than 1/16 in.) is worn and should be replaced. As a final check, measure the free height of the spring. Spring specifications are listed below.

Use a high-quality square to check the springs maximum out-of-square. Place the spring against the square and turn the spring one full rotation to obtain the largest gap. Measure the gap between the spring and the square. See Fig. 34.

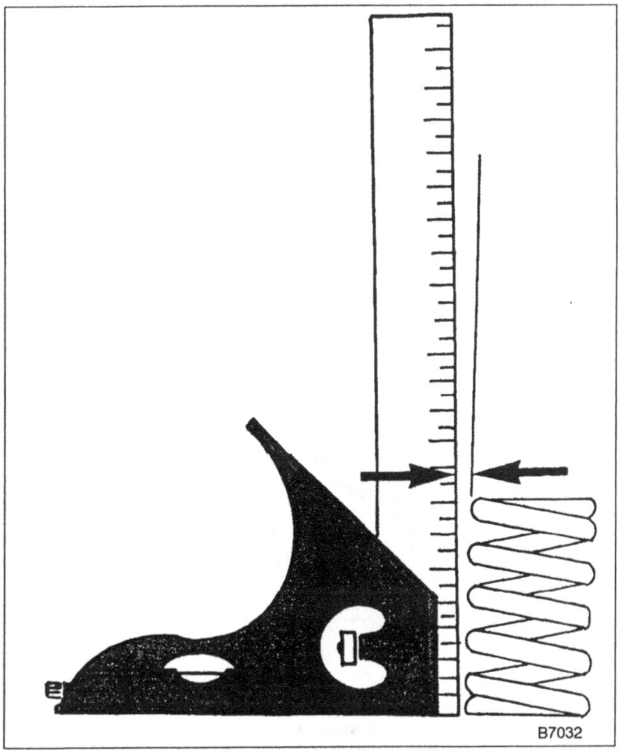

Fig. 34. Valve spring being checked for out-of-square. Make measurement at arrows.

Valve Springs

- out-of-square (maximum) .... 1.75 mm (0.07 in.)
- free height .... 45.0±1.5 mm (1.77±0.06 in.)
- height when compressed with pressure of 131-141 lbs. .... 28.4 mm (1.12 in.)

215 Camshaft Timing Chain

GENERAL

The overhead camshafts are driven by a single row chain. The chain is tensioned using a hydraulic chain tensioner system, which is mounted in the side of the cylinder head. See Fig. 1. The timing chain and its related components normally do not require maintenance. If either the chain, the lower guides or the crankshaft sprocket are worn and need to be replaced, the engine will have to be removed from the car.

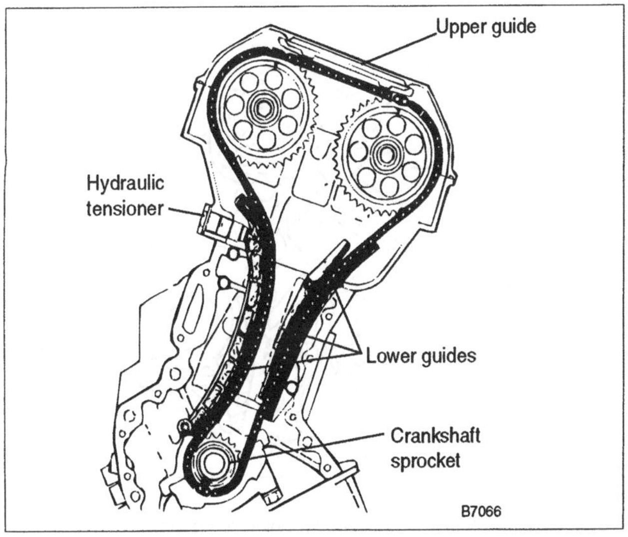

Fig. 1. Timing chain and related components used on 16-valve engines.

CAMSHAFT TIMING CHAIN

Checking Camshaft Timing Chain

The camshaft timing chain and its components can be quickly checked to determine if the chain, sprockets, or guides are worn or if the chain is incorrectly positioned on it sprockets.

CAUTION

The camshafts must always be properly timed to the crankshaft or the pistons may contact the valve heads resulting in engine damage. Do not alter the position of the crankshaft or camshafts once the timing chain is removed from its sprockets.

To check the timing chain and its components, remove the valve cover as described in **214 Cylinder Head and Valve Mechanism.** Then using a wrench on the crankshaft pulley mounting bolt, turn the engine over in the direction of normal engine rotation until the camshafts and the flywheel are set to TDC (0°TDC). See Fig. 2.

CAUTION

Do not turn the engine over using the flats on the camshaft.

NOTE

Normal engine rotation is counterclockwise when viewing the engine from the front of the car.

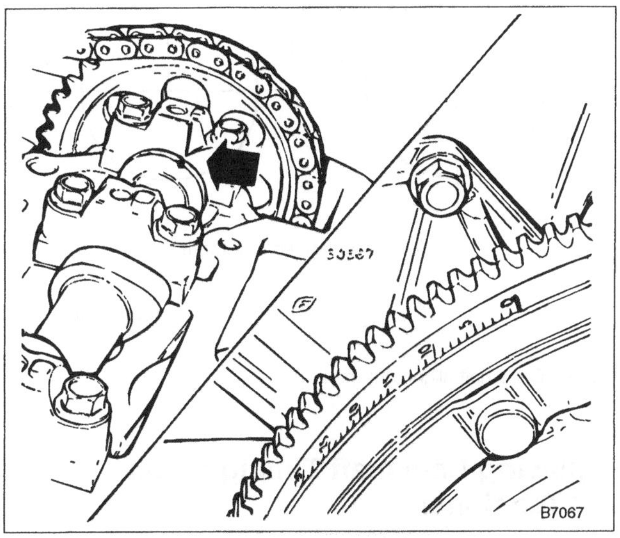

Fig. 2. Camshaft sprocket and crankshaft TDC marks set at 0°TDC. Both camshaft marks and the flywheel mark should be exactly aligned.

If all three marks are not perfectly aligned, either the timing chain is severely worn or the chain is incorrectly installed. See Fig. 3 for camshaft sprocket diagnosis. Remove and install the chain as described to below.

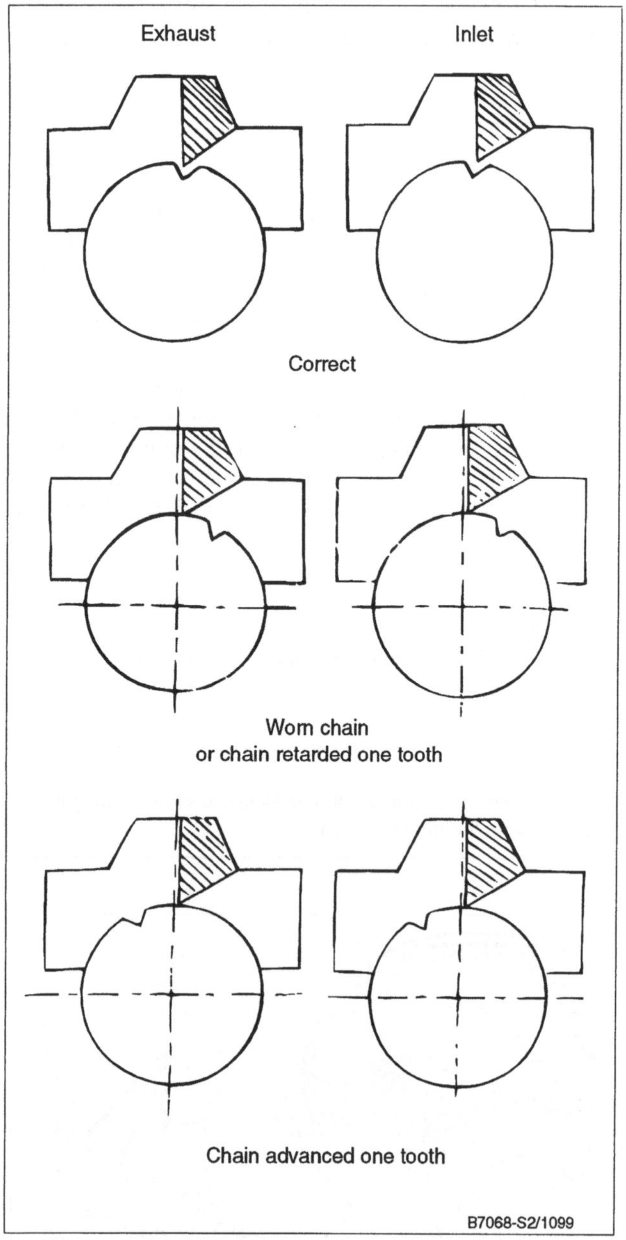

Fig. 3. Camshaft alignment marks.

Replacing Camshaft Timing Chain and Tensioner

Replacing the timing chain or any of the components behind the timing chain cover will require that the engine assembly be removed from the car. The camshaft timing chain cover cannot be properly removed with the engine installed. The camshaft sprockets, the chain tensioner, and the upper chain guide can all be replaced with the engine installed.

To replace camshaft timing chain

1. Remove the engine as described in **201 Engine Removal and Installation.**

2. Remove the valve cover and align the camshaft and the flywheel at TDC as shown above.

3. Remove the oil pump from the timing chain cover as described in **220 Lubrication System.**

4. Disconnect the lower hose from the coolant pump.

5. Remove the alternator with its support bracket.

6. Remove the bolts from the timing chain cover and remove the cover from the engine. See Fig. 4.

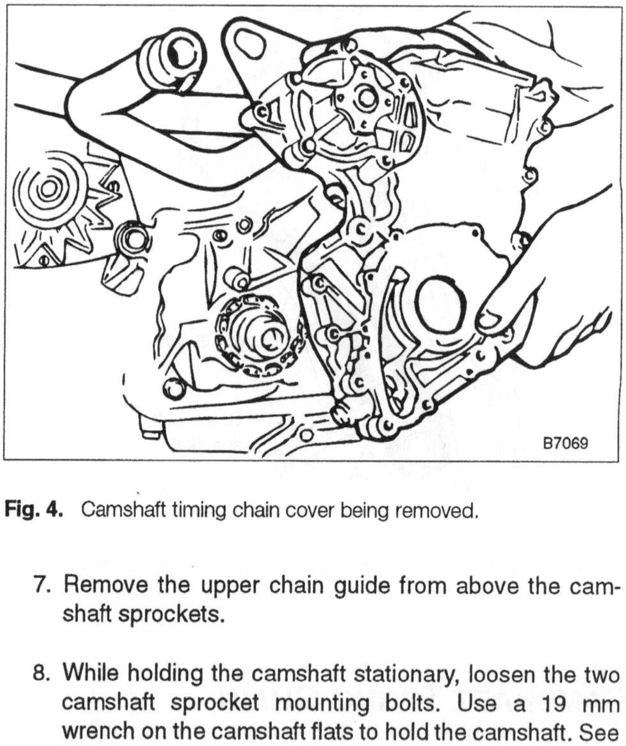

Fig. 4. Camshaft timing chain cover being removed.

7. Remove the upper chain guide from above the camshaft sprockets.

8. While holding the camshaft stationary, loosen the two camshaft sprocket mounting bolts. Use a 19 mm wrench on the camshaft flats to hold the camshaft. See Fig. 5.

9. Remove the chain tensioner from the side of the cylinder head. See Fig. 6.

10. Remove the camshaft sprockets and the chain. See Fig. 7.

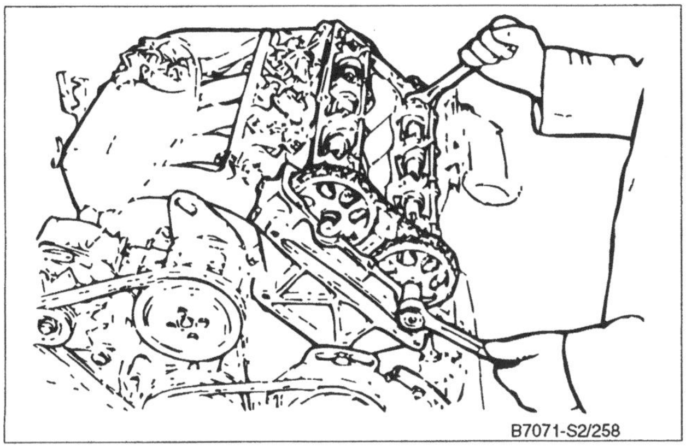

Fig. 5. Camshaft sprocket mounting bolt being loosened while camshaft is held stationary.

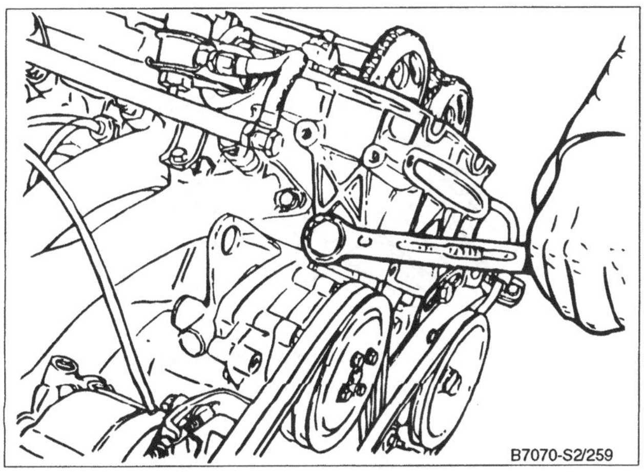

Fig. 6. Camshaft timing chain tensioner being removed.

CAUTION
The camshafts must always be correctly timed to the crankshaft or the pistons may contact the valve heads resulting in engine damage. Do not alter the position of the crankshaft or camshafts once the timing chain is removed from its sprockets.

11. Inspect the chain sprockets, the chain guides, and the chain tensioners. Replace any sprocket that is visibly worn. Replace any guide that is deeply scored.

12. With all TDC (timing) marks aligned, install the timing chain and the exhaust-side camshaft sprocket, making sure there is no chain slack between the cam sprocket and crankshaft sprocket.

13. Install the intake-side camshaft sprocket, keeping the chain tight between the sprockets. Hand tighten the sprocket mounting bolts.

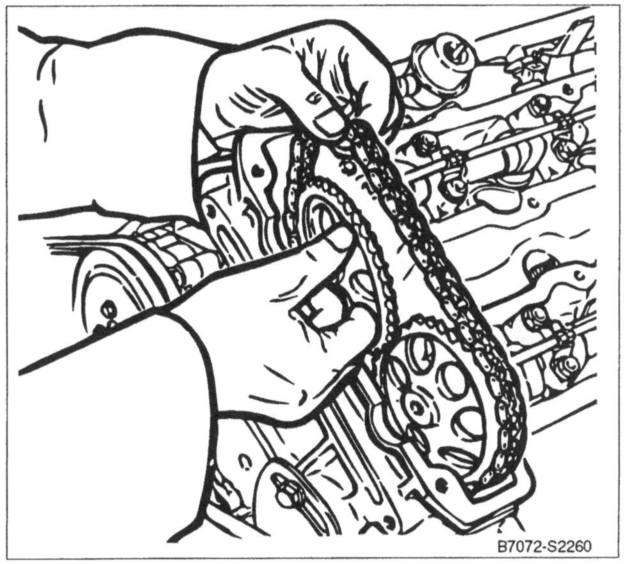

Fig. 7. Camshaft sprocket being removed from camshaft.

CAUTION
Make sure that the chain is centered on the guides.

14. Preset the tensioner so that it will not tension the chain when installed. Then install and tighten the tensioner to the cylinder head.

NOTE
- Two types of tensioners are used on 16-valve engines and are preset differently. The early-style tensioner can be preset by fully pressing the spring-loaded piston in and turning it until it locks. The late-style tensioner, which has a ratcheting mechanism, can be preset by removing the small bolt and spring in the end of the tensioner. Lift up on the ratcheting pawl and push the lever all the way in. See Fig. 8.
- The late-type ratcheting tensioner can be retrofitted on all 16-valve cars with the early-style tensioner.

Tightening torque
- timing chain tensioner to cylinder head. . . . . .63 Nm (47 ft-lb)

15. Tension the camshaft timing chain by either installing the center bolt and spring (late-style tensioner) or by releasing the tensioner piston by pressing on the guide rail. See Fig. 9.

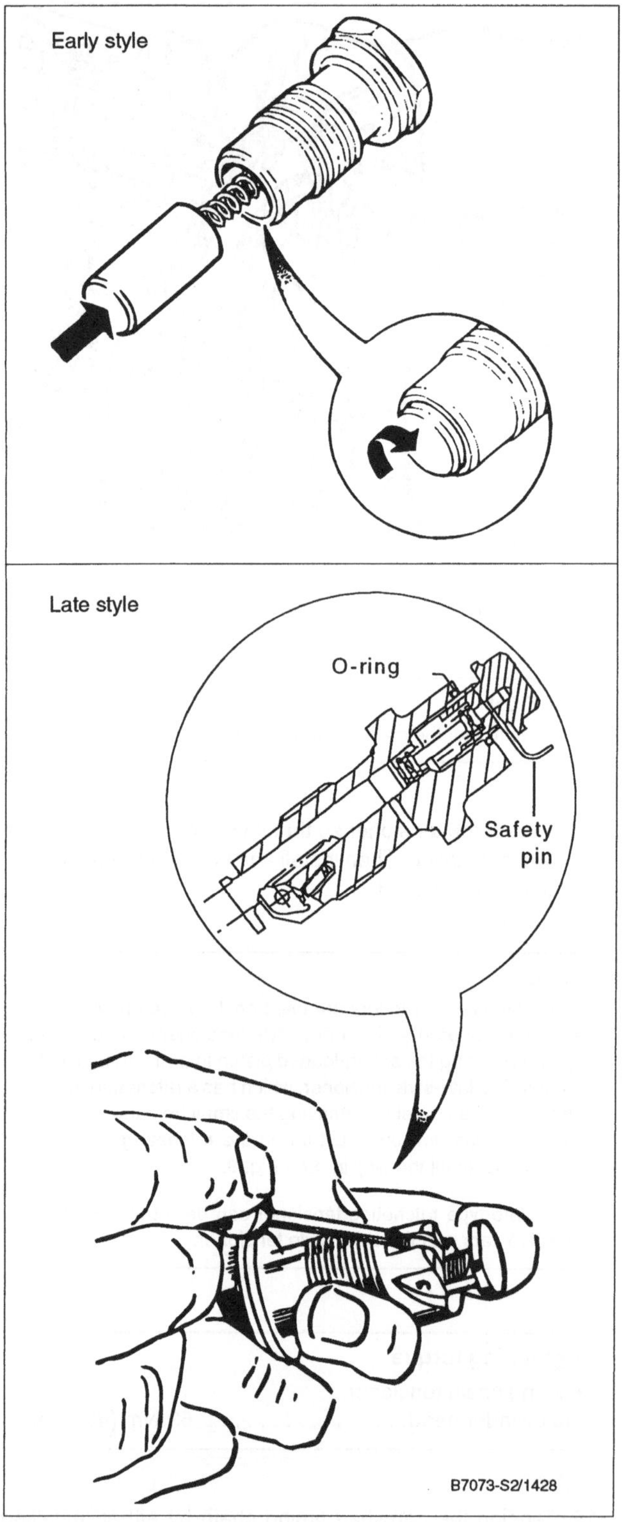

Fig. 8. Early and late style camshaft timing chain tensioners being preset prior to installation.

NOTE

New late-style tensioners come with a safety pin installed. Removing the safety pin will release the tensioner arm. Do not remove the pin until the tensioner is installed.

Tightening torque

- camshaft timing chain center tensioner plug (bolt) to tensioner . . . 22 Nm (16 ft-lb)

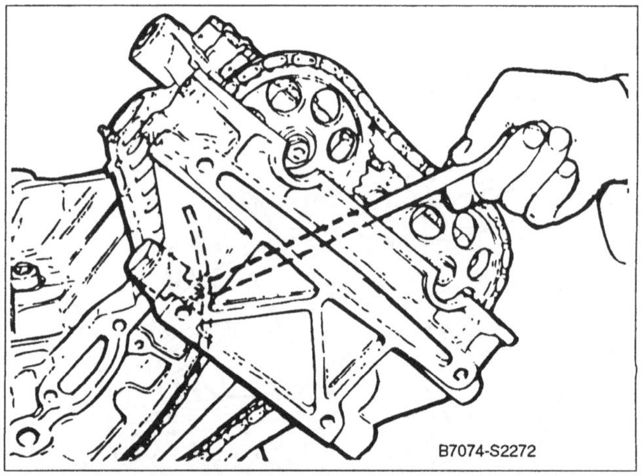

Fig. 9. Early-style camshaft timing chain tensioner being released by pressing on guide rail with blunt object.

16. Rotate the crankshaft by hand in the direction of normal engine rotation two full revolutions. Recheck that the TDC marks are all correctly aligned. If any faults are found reset the timing chain.

CAUTION

Recheck all timing marks carefully. It is very easy to be off one tooth.

---

17. Remove the camshaft sprocket bolts one at a time, coat them with Loctite, and reinstall. Tighten the bolts while holding the camshaft stationary as shown in Fig. 10.

CAUTION

Hold the camshaft steady using a 19 mm wrench on the camshaft flats when tightening the camshaft sprocket bolts. Do not allow the tensioner or the chain to take this load.

Tightening torque
- camshaft sprocket to camshaft . . . . . 63 Nm (47 ft-lb)

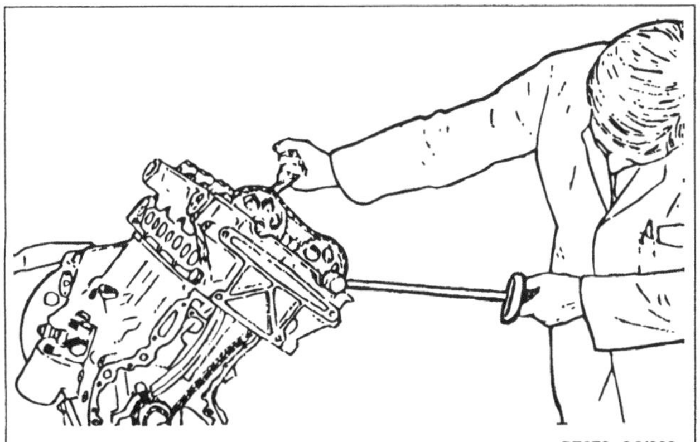

Fig. 10. Camshaft sprocket bolt being tightened.

18. Install the camshaft timing chain cover, the oil pump, and crankshaft sprocket.

CAUTION
Be careful not to damage the crankshaft oil seal when installing the oil pump.

Tightening torques
- timing chain cover to engine block and cylinder head . . . . 20 Nm (15 ft-lb)
- oil pump to timing chain cover . . . . . . 8 Nm (71 in-lb)
- crankshaft pulley to crankshaft
  1985-1990 . . . . 190 Nm (140 ft-lb)
  1991 and later . . . . 175 Nm (129 ft-lb)

19. Install the valve cover as described in **214 Cylinder Head and Valve Mechanism.**

Tightening torque
- valve cover to cylinder head . . . . 15 Nm (11 ft-lb)

20. Install the alternator as described in **321 Alternator and Charging System.**

Tightening torques
- all other fasteners
  M5 bolt . . . . 5 Nm (44 in-lb)
  M6 bolt . . . . 10 Nm (89 in-lb)
  M8 bolt . . . . 20 Nm (15 ft-lb)
  M10 bolt . . . . 40 Nm (30 ft-lb)

216 Crankshaft

CRANKSHAFT

Crankshaft main bearing caps and used bearing shells should only be installed in their original positions. Label all parts carefully.

NOTE

Crankshaft removal and installation is part of the engine disassembly procedure covered in **210 Engine Disassembly and Assembly.**

To inspect crankshaft

1. Thoroughly clean the removed crankshaft, the crankshaft bearings, and the bearing surfaces in the cylinder block. Make all parts being measured are dry and oil-free.

2. Carefully place the upper bearing shells in the cylinder block and set the crankshaft into the block. Install the center thrust washers. See Fig. 1.

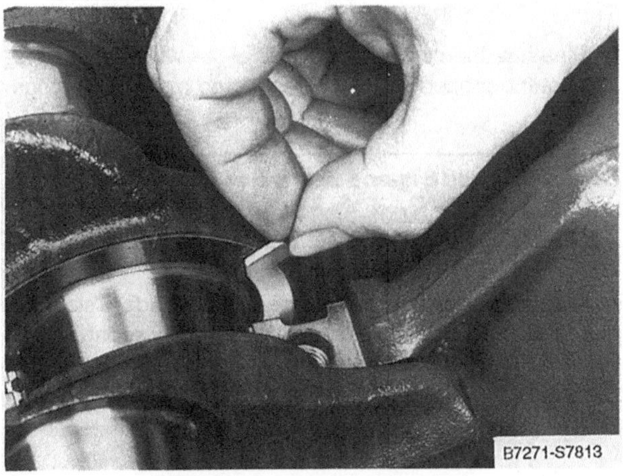

Fig. 1. Crankshaft thrust washers being installed.

3. Using a dial indicator, measure the crankshaft end float by gently levering the crankshaft back and forth using a screwdriver. See Fig. 2.

Crankshaft
• end float allowable. . . . . 0.08-0.28 mm (0.003-0.011 in.)

Fig. 2. Crankshaft end float being checked with dial indicator (arrow). Move the crankshaft back and forth with screwdriver to obtain readings.

4. Remove the crankshaft from the cylinder block and place the crankshaft in two V-blocks. Using a dial indicator at the center main journal, measure the crankshaft radial play by rolling the crankshaft in the V-blocks.

Crankshaft radial play (runout)
• maximum allowable. . . . . 0.10 mm (0.004 in.)

---

5. Measure the main journal and crankpin diameters using a micrometer. Make the measurements at several points to check for out-of-roundness. Journal diameters are listed in **Table a.**

Crankshaft journal out-of-round
- maximum allowable . . . . 0.05 mm (0.002 in.)

Table a. Crankshaft Main Journal and Crankpin Diameters

Main journals	diameter mm (in)
standard	57.981-58.000 (2.2827-2.2835)
first undersize	57.731-57.750 (2.2729-2.2736)
second undersize	57.481-57.500 (2.2630-2.2638)
third undersize	57.237-57.250 (2.2534-2.2539)
fourth undersize	56.987-57.000 (2.2436-2.2441)
Crankpin	**diameter mm (in)**
standard	51.981-52.000 (2.0465-2.0472)
first undersize	51.731-51.750 (2.0367-2.0374)
second undersize	51.481-51.500 (2.0268-2.0276)
third undersize	51.237-51.250 (2.0172-2.0177)
fourth undersize	50.987-51.000 (2.0074-2.0079)

NOTE
The crankshaft can be ground down to one undersize without affecting its hardened surface. If any more material needs to be removed, the crankshaft should be rehardened using a "Tenifer" hardening treatment. Contact your authorized Saab dealer for information on crankshaft hardening and replacement crankshafts.

6. Measure the main bearing oil clearances by placing the bearing shells and the crankshaft in the cylinder block.

7. Place a strip of plastigage about 6mm (1/4 in.) to one side of the journal's center-line. See Fig. 3. Install the bearing cap, tighten the bolts and then remove the cap.

NOTE
Plastigage comes in three thicknesses. Use the green type (PG-1) to make the measurements described here. Keep the crankshaft stationary when measuring oil clearances.

Tightening torque
- main bearing caps
  to cylinder block . . . . 110 Nm (81 ft-lb)

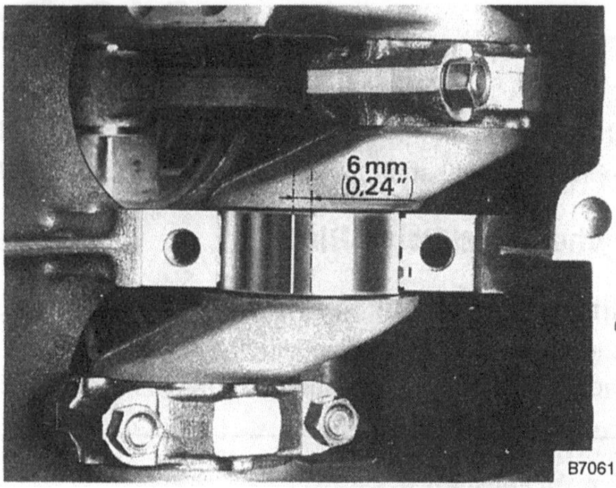

Fig. 3. Plastigage correctly positioned on crankshaft journal.

8. Measure the strip of plastigage at the widest point. Repeat the procedure for the remaining main bearings.

Crankshaft main bearing
- oil clearance . . .0.020-0.062 mm (0.0008-0.0024 in.)

9. To measure the big-end oil clearances, install the piston and connecting rod assemblies into their respective cylinders. Rotate the crankshaft so that the piston is at approximately 60° BTDC. Place a strip of plastigage about 6mm (1/4 in.) to one side of the crankpin's center-line as shown above. Install the connecting rod cap, tighten the nuts and then remove the cap.

Tightening torque
- connecting rod cap
  to connecting rod . . . . 55 Nm (41 ft-lb)

10. Measure the strip of plastigage at the widest point. Repeat the procedure for the remaining big-end bearings.

Connecting rod big-end bearing
- oil clearance . . .0.026-0.062 mm (0.0010-0.0024 in.)

Replacement bearing shells come in the standard size and four undersizes. Standard, first undersize, and second undersize shells are available in two thicknesses and are marked with a colored paint dot code, indicating the size and thickness classification.

Crankshaft Bearing Shell Color Codes
- main and big-end bearing shells
  standard
    thin-red
    thick-blue
  first undersize
    thin-yellow
    thick-green
  second undersize
    thin-white
    thick-brown

The proper way of fitting bearing shells is as follows: Begin by fitting two thin shells and measure the oil clearance using plastigage as described above. If the clearance is too big, attempt to reduce it by combining a thick and thin shell or two thick shells. If the clearance is still too large with two thick shells, the crankshaft will have to be machined to accept undersized bearing shells. See **Table a** above.

FLYWHEEL AND DRIVE PLATE

Replacement of the flywheel or the drive plate are similar operations. When removing the mounting bolts, hold the ring gear stationary using a holding fixture such as Saab tool no 83 92 987. The flywheel or driveplate should be inspected for wear and excessive runout using a dial indicator. Replace the driveplate or flywheel if it is warped, cracked, or deeply scored.

NOTE
Saab does not list wear specifications for the flywheel or drive plate.

Before installing the flywheel, check the transmission clutch shaft bearing and replace it if there is any play or if it does not roll smoothly. See Fig. 4.

Installation of the flywheel and the drive plate is the reverse of removal. Be sure all old thread sealer is cleaned from the bolts and their holes. Use a thread sealer when reinstalling the bolts and tighten them in a alternating pattern.

Tightening torques
- flywheel to crankshaft
  17 mm bolt head . . . . 60 Nm (44 ft-lb)
  19 mm bolt head . . . . 85 Nm (63 ft-lb)

CYLINDER BLOCK OIL SEALS

The cylinder block oil seal(s) at the timing chain end of the engine can be replaced with the engine removed or installed. On cars with manual transmission, the oil seal at the flywheel end of the engine can be replaced with the engine installed, but the clutch and flywheel must first be removed.

Fig. 4. Clutch shaft bearing in flywheel being removed. Special drift (Saab tool no. 83 91 997) shown.

NOTE
On cars with automatic transmission, the oil seal at the drive plate end of the engine can only be replaced after removing the transmission torque converter. Removing the torque converter requires that the transmission be separated from the engine as described in **201 Engine Removal and Installation.**

To replace crankshaft oil seal at flywheel end (cars with manual transmission only)

1. Remove the clutch and pressure plate as described in **411 Clutch.**

2. While holding the flywheel stationary, remove the flywheel mounting bolts and remove the flywheel.

NOTE
The flywheel can be held stationary using a locking tool such as Saab tool no. 83 92 987.

3. Carefully pry out the old oil seal using a screwdriver.

4. Lubricate the new seal with oil. Fit the new seal so that the spring ring side of the seal is towards the crankshaft.

5. Using a suitable seal driver install the seal. See Fig. 5.

NOTE
The old seal was originally installed flush to the cover surface. The new seal should be installed slightly deeper so that it does not ride on the same area of the shaft. This will allow for a good seal despite any wear.

Fig. 5. Crankshaft oil seal at flywheel end of engine being installed. (Saab special tool no. 83 92 540 shown).

To replace crankshaft oil seal at timing chain end

1. If necessary, remove the alternator and A/C V-belts as described **1 LUBRICATION AND MAINTENANCE.**

2. While holding the flywheel stationary, remove the crankshaft pulley mounting bolt and the pulley with its woodruff key. See Fig. 6.

Fig. 6. Crankshaft pulley being removed.

3. Pry out the old oil seal using a screwdriver. See Fig. 7.

4. Lubricate the new seal with oil. Fit the new seal and install it using the special sleeve (Saab tool no. 83 94 215), the pulley center bolt, and special wrench (Saab tool no. 33 92 185). See Fig. 8.

Fig. 7. Crankshaft oil seal at timing chain end of engine being removed.

Fig. 8. Crankshaft oil seal at timing chain end of engine being installed using sleeve **A** (Saab tool no 8394 215) and **B** wrench (Saab tool no. 83 92 185).

NOTE
The old seal was originally installed flush to the cover surface. The new seal should be installed slightly deeper so that it does not ride on the same area of the shaft. This will allow for a good seal despite any wear.

---

5. Install the crankshaft pulley with its woodruff key. Install and tension the engine V-belts as described in **1 LUBRICATION AND MAINTENANCE.**

Tightening torque
- Crankshaft pulley to crankshaft
  - 1985-1990 . . . . 190 Nm (140 ft-lb)
  - 1991 and later . . . . 175 Nm (129 ft-lb)

---

220 Lubrication System

GENERAL

This section covers inspection and repair of the lubrication system, which includes the oil pump, the oil pressure warning system, and the oil cooler.

Oil pressure is generated by a gear-type pump mounted externally to the front of the cylinder block. The pump is driven directly by the crankshaft. When the engine is running, oil is drawn from the oil pickup tube in the oil sump and circulated through the engine, the turbocharger and the oil cooler (where applicable).See Fig. 1.

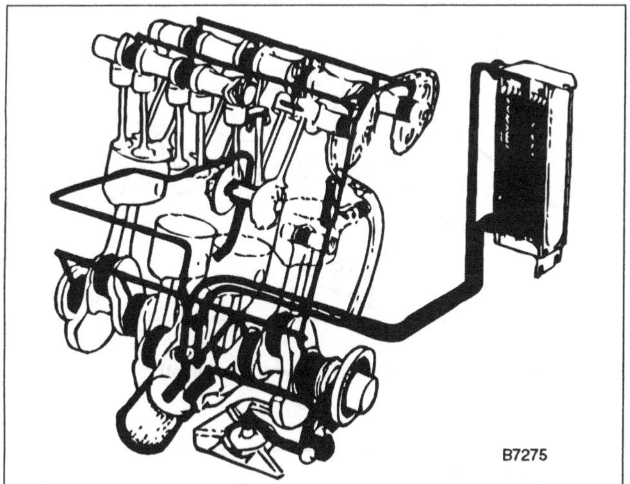

Fig. 1. Schematic of lubrication system.

The oil sump is an integral part of the transmission housing, although the engine oil and the transmission oil are separated from each other.

NOTE
Oil and oil filter change are covered in **1 LUBRICATION AND MAINTENANCE.**

All engines covered by this manual have an oil pressure warning system that warns the driver of insufficient oil pressure immediately after startup. Other safety features include a filter bypass, to guard against bursting the filter due to over pressure, and an oil pump pressure relief valve to prevent excessive system pressure.

OIL PUMP AND OIL PICKUP TUBE

There is normally no need to remove the oil pump or pickup tube unless oil pressure is inadequate. Check the oil pressure as described below under **Oil Pressure Warning Systems.**

The oil pump can be removed with the engine installed or removed. Removal of the oil pickup tube requires that the engine be removed and separated from the transmission.

To remove, inspect, and install oil pump

1. Remove the V-belts from the crankshaft pulley and drain the engine oil as described in **1 LUBRICATION AND MAINTENANCE.**

2. While holding the flywheel steady, remove the crankshaft pulley mounting bolt and remove the pulley from the crankshaft. See Fig. 2.

NOTE
Immobilize the crankshaft by locking the flywheel using a locking tool, such as Saab special tool no. 83 92 987.

3. Thoroughly clean the area around the oil pump.

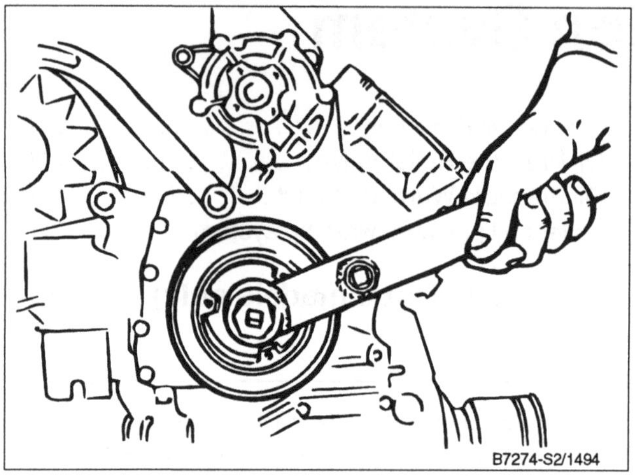

Fig. 2. Crankshaft sprocket mounting bolt being removed.

4. Remove the oil pump mounting bolts and remove the pump. See Fig. 3.

NOTE
Before removing the pump, inspect for any signs of oil around the crankshaft oil seal in the oil pump housing. If the seal is leaking, it should be replaced at this time as described in **216 Crankshaft**.

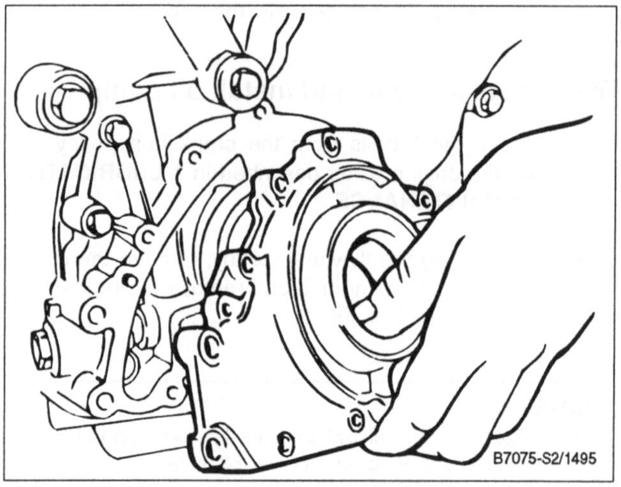

Fig. 3. Oil pump being removed.

5. Disassemble and clean the oil pump gears. Inspect the inside of the oil pump housing for scoring or wear. Inspect the oil pump O-ring and the crankshaft seal. Replace any worn or faulty parts

6. Reassemble the pump making sure the small punch mark in the outer gear is facing up. See Fig. 4.

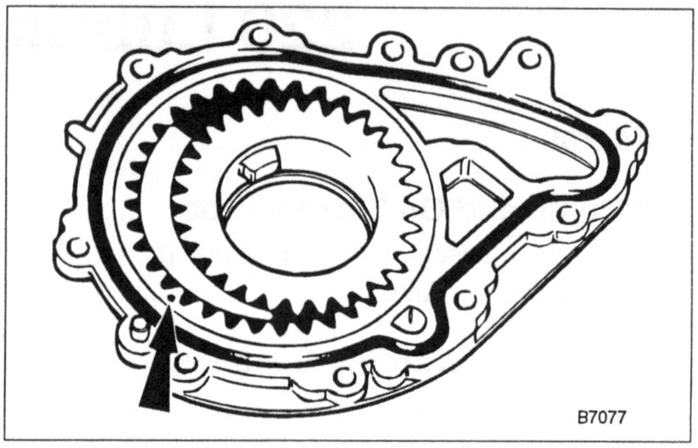

Fig. 4. Oil pump outer gear correctly installed with punch mark facing up (arrow).

7. Using a straight edge and a feeler gauge, check the end float between the pump body and the gear wheel. See Fig. 5.

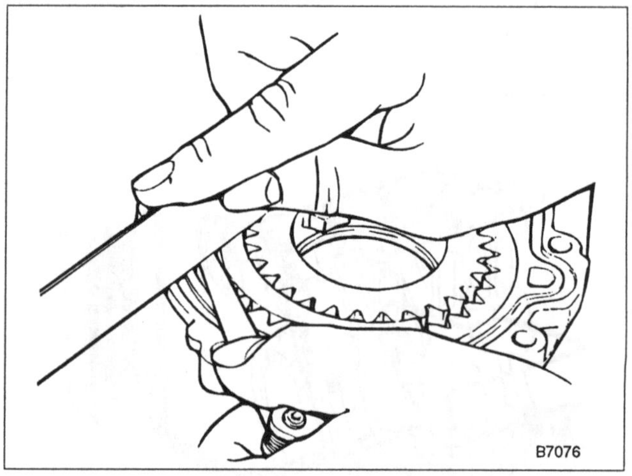

Fig. 5. Oil pump being checked for wear (end-float).

Oil pump end float
- 0.03-0.08 mm (0.0012-0.0031 in.)

---

8. Fit the O-ring in the oil pump groove. Lubricate the crankshaft oil seal and the pump gears with engine oil.

9. Pull the inner gear out of the pump slightly and install the pump, making sure that the dog on pump gear engages the slot on the crankshaft hub. Push the pump on while aligning the dowel. See Fig. 6.

Fig. 6. Oil pump being installed. Pull gear out of housing to align dog on gear with slot on crankshaft hub.

NOTE
The oil pump locating dowel aids in centering the pump to the crankshaft. Some models may not be equipped with the dowel. If the dowel is missing, a special tool (Saab tool no. 83 93 589) is available to center the pump and protect the oil seal during oil pump installation. See Fig. 7.

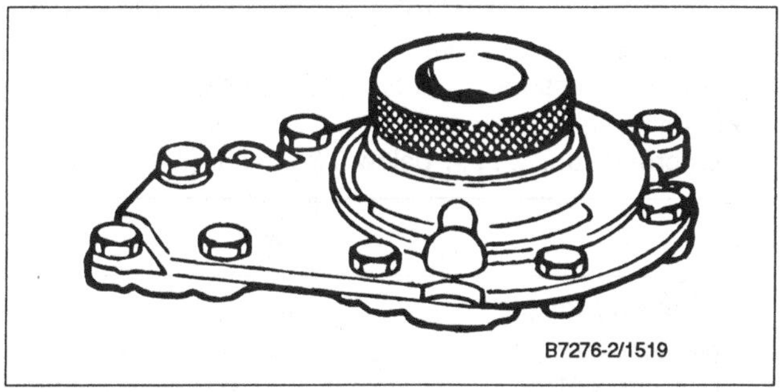

Fig. 7. Saab special tool installed in oil pump prior to pump installation.

10. Install and tighten the pump mounting bolts in an alternating pattern. Carefully install the crankshaft pulley and tighten the mounting bolt. Install the V-belts.

CAUTION
Be careful not to damage the crankshaft oil seal when installing the crankshaft pulley. Thoroughly clean and oil the pulley flange before installation.

Tightening torques
- oil pump to timing chain cover...... 8 Nm (71 in-lb)
- crankshaft pulley to crankshaft
  - 1985-1990 ....190 Nm (140 ft-lb)
  - 1991 and later ....175 Nm (129 ft-lb)

11. Remove the oil filter housing from the cylinder block. Then fill the oil pump pressure cavity with clean engine oil to prime the pump. See Fig. 8.

CAUTION
The oil pump must be primed any time it is removed.

Fig. 8. Oil pump being primed through oil pump pressure cavity.

12. Using a new gasket, install the filter housing and refill the engine with oil.

Oil Pump Pickup Tube

The oil pickup tube is mounted to bottom of the engine block. Because the oil sump is an integral part of the transmission case, the engine needs to be removed and the transmission separated from the engine to reach the pickup tube. See **201 Engine Removal and Installation** for these procedures.

CAUTION
Different oil pickup tubes are used depending on the type of transmission installed. Cars with manual transmission use a longer tube than cars with automatic transmission. If the wrong tube is installed, inadequate lubrication and engine damage can result.

OIL PRESSURE WARNING SYSTEMS

When diagnosing the cause of an illuminated oil pressure warning light, certain basic checks should be made before assuming that engine damage has occurred.

CAUTION

If the oil pressure warning indicator stays on after the engine is started, or flashes on while driving, always assume that low oil pressure is the cause.

First, check that the warning light electrical circuit is functioning correctly by making the quick checks described below in **Oil Pressure Warning Circuit**. Check that the engine oil level and oil type is correct.

Second, if the engine has been run for any length of time with the warning light on, it should be determined if any engine damaged has occurred. If abnormal noises are heard with engine running, turn the engine off and remove the valve cover and the camshaft bearing cap from the no. 4 cylinder (cylinder closest to the radiator). This is the cylinder farthest from the oil pump. Inspect the bearings surface for damage. If the surfaces are badly worn, chances are that other parts of the engine are also damaged.

Third, check the oil pressure as described below. If the oil pressure is below the specification listed, remove and clean the oil pressure relief valve (also described below). Then prime the oil system (see Fig. 8 above) and reinstall the oil pressure relief valve. Recheck the oil pressure and the oil pressure warning light.

NOTE

Occasionally, a small particle of metal or other contaminant may become lodged in the valve, causing it to stick open. In some cases when the valve is removed for inspection, the oil that runs out of the bore carries the contaminants away, leaving the impression that the valve was not the cause of the low pressure. Do not disassemble the engine until first cleaning the pressure relief valve. If this corrects the problem, be sure to change the oil and the filter to remove any additional debris.

Testing Oil Pressure

Test the oil pressure by removing the pressure switch and installing a pressure gauge. See Fig. 9. Run the engine just long enough to get an accurate reading. If testing shows that the oil pressure is too low or too high, either the oil pump is worn or faulty, the relief valve is stuck, or the engine is severely worn. If the pressure is as specified and the warning light still stays on, check the light's electrical circuit as described below in **Oil Pressure Waring Circuit**.

NOTE

For the most accurate pressure readings, change the oil and the oil filter before making the pressure test.

Fig. 9. Oil pressure switch (arrow).

Lubrication system test pressures

- oil pressure
  at 2000 rpm,
  engine temp. 80°C (176°F) . . . . at least 2.7 (39 psi)
- oil pressure relief valve
  opening pressure . . . . 3.6-5.2 bar (52-75 psi)
- warning light
  on pressure. . . . . below 0.3-0.5 bar (4.4-7.2 psi)

Tightening torque

- oil pressure switch
  (1/4"-18 NPTF) . . . . 10 Nm (89 in-lb)

Oil Pressure Relief Valve

The oil pressure relief valve is located in the side of the camshaft timing chain cover. See Fig. 10. The valve is designed to open and route oil back to the inlet side of the oil pump when the oil pressure exceeds a predetermined level. If the pressure is abnormally high or low, remove and clean the valve. Change the oil and the oil filter and recheck the oil pressure. If the pressure is still low, remove and inspect the oil pump as described above in **Oil Pump and Oil Pickup Tube**.

NOTE

Low oil pressure can also be caused by general engine wear, such as worn engine bearings on a high-mileage car.

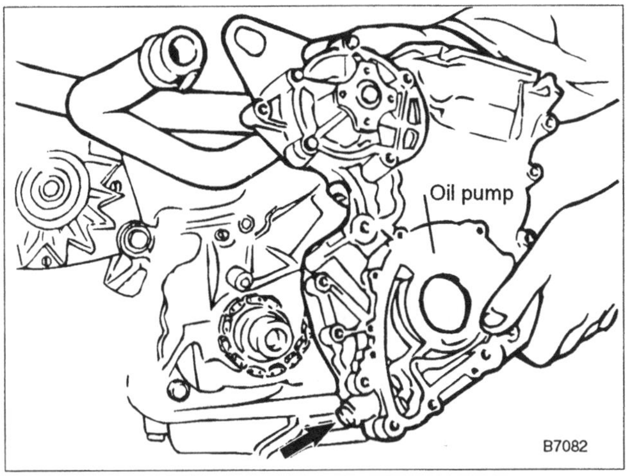

Fig. 10. Oil pressure relief valve (arrow).

Oil Pressure Warning Circuit

The oil pressure warning electrical circuit consists of an oil pressure switch mounted in the oil circuit and an instrument panel warning light. When the ignition switch is turned on, a warning light comes on. When the engine is started and the oil pressure rises, the oil pressure switch opens and the warning light goes out. If the light doesn't come on at all or stays on all the time, check to make sure the oil level and the oil pressure are correct before doing the tests listed below.

NOTE

For information on center console gauges (oil pressure and oil temperature gauges), see **3 ELECTRICAL SYSTEM.**

To test oil pressure warning system

1. Turn the ignition switch on. The warning light on the instrument panel should light up. Remove the wire from the oil pressure switch (See Fig. 9 above). The light should go out.

NOTE

If the light does not go out, the wiring to the switch is most likely shorted to ground between the terminal and the warning light. See **3 ELECTRICAL SYSTEM** for electrical wiring troubleshooting information.

---

2. If the warning light does not light when the ignition is on, remove the terminal from the oil pressure switch and ground the wire to a clean metal surface.

NOTE

If the warning light comes on, check the switch as described in the next step. If the warning light does not come on, the wiring circuit to the dash light or the light itself is faulty.

---

3. To test the switch, connect an ohmmeter between the switch terminal and ground. With the engine off, there should be continuity. With the engine running there should be no continuity. If any faults are found, the switch is faulty and should be replaced.

Tightening torque

- oil pressure switch to oil filter housing
  small (1/4"-18 NPTF) . . . . 10 Nm (89 in-lb)
  large (M14X1.5) . . . . .30 Nm (22 ft-lb)

CAUTION

If the oil pressure switch is not faulty and the light remains on while the engine is running, the oil pressure may be dangerously low. Do not operate the engine until the problem is corrected. The engine may be severely damaged.

OIL COOLER (TURBO MODELS ONLY)

Turbocharged engines uses an oil-to-air cooler to aid in engine oil cooling. A thermostat in the oil filter housing prevents oil flow to the cooler until the oil temperature reaches 167°F (75°C).

The engine oil cooler is fitted beneath the front left-hand head light. The oil cooler thermostat is screwed into the side of the oil filter housing. Fig. 11 shows the oil cooler and its hoses.

To remove oil cooler

1. Remove the left-hand turn signal/parking light assembly.
2. Remove the baffle plate from behind the light assembly.
3. Remove the left-hand headlight.
4. Working from below, remove the spoiler grille from beneath the car.
5. Working from in the engine compartment, remove the oil cooler mounting nut from the radiator member. See Fig. 12.

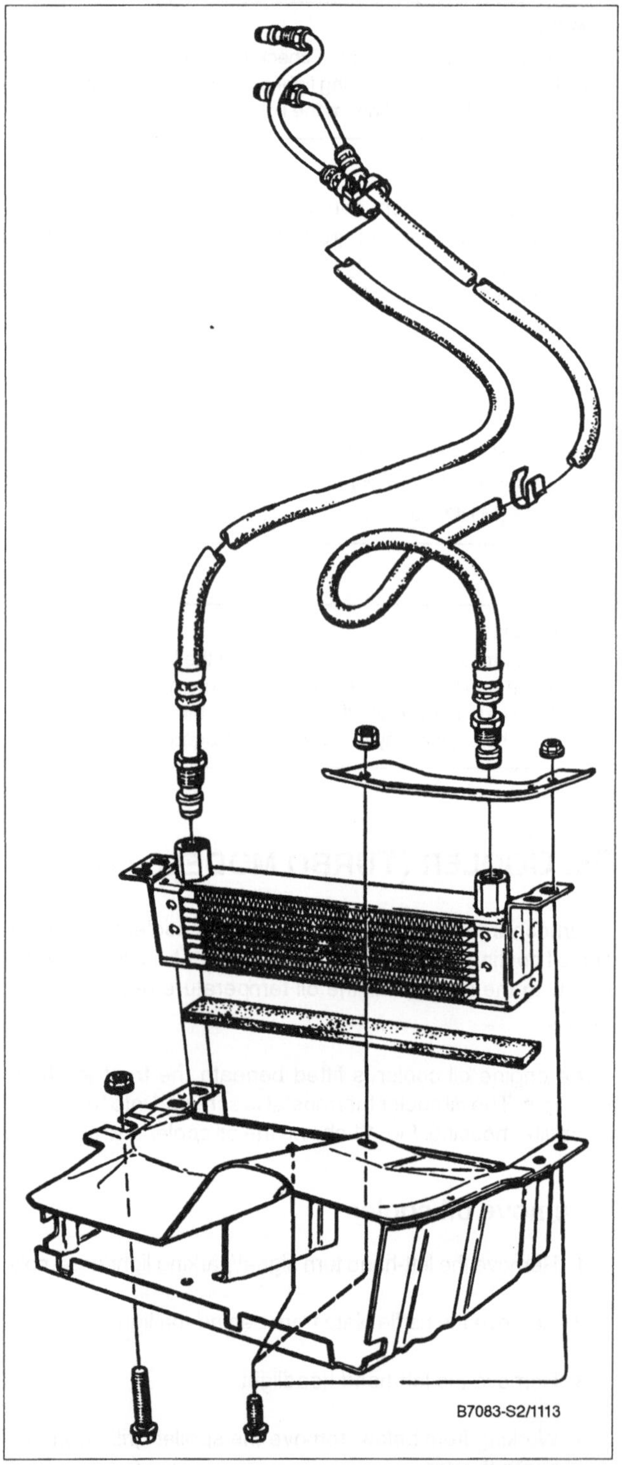

Fig. 11. Oil cooler and related components for turbo engines.

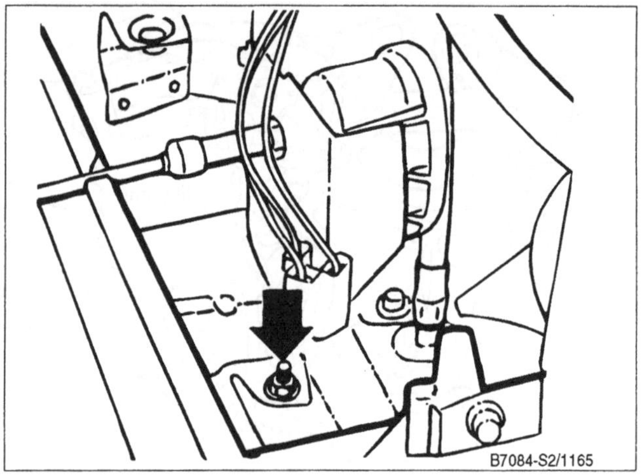

Fig. 12. Oil cooler mounting nut in engine compartment near radiator.

6. Unbolt the cooler together with its shroud from below. Carefully pull the cooler and shroud down. See Fig. 13. Unbolt the shroud from the cooler and remove the lines from the cooler.

Installation is the reverse of removal.

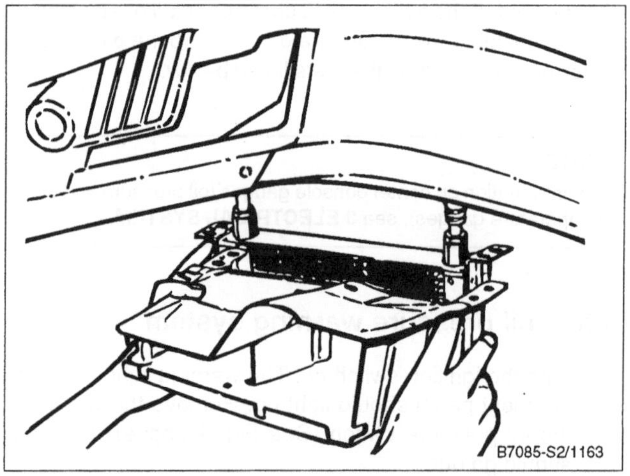

Fig. 13. Oil cooler being removed from recess behind headlight.

234 Fuel Pump and Fuel Tank

TABLES

234

GENERAL

The injection-molded plastic fuel tank is located beneath the car, behind the rear seat. The fuel gauge sender and the electric fuel pump are mounted in the top of the tank. See Fig. 1. The fuel tank is designed to prevent overfilling and allow for fuel expansion. The filler cap contains a valve to prevent a vacuum from occurring in the tank. To prevent a backflow of fuel from the tank when the car is stopped, there is a check valve in the fuel return line from the fuel injection system.

The fuel pump is an electric rotary pump mounted inside the fuel tank. The fuel pump constantly circulates fuel throughout the system. Fuel is drawn from the tank, filtered, routed to the fuel injectors, regulated to maintain a constant pressure, and then recirculated back into the fuel tank.

NOTE

Fuel filter replacement is covered in **1 LUBRICATION AND MAINTENANCE.**

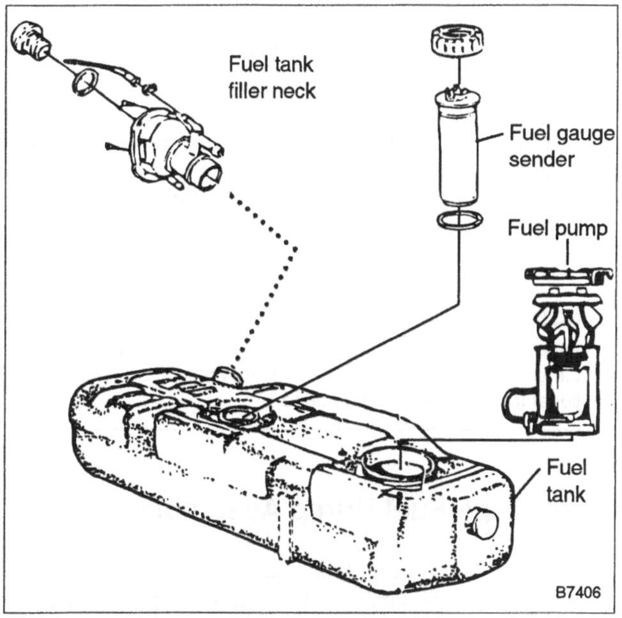

Fig. 1. Fuel tank and related components.

Fuel flows through the pump and over the pump's electric motor. This helps cool the pump during normal operation. An inlet strainer helps prevent pump clogging. The main pump is mounted in a reservoir to help prevent fuel starvation. The fuel pump contains a bypass valve to prevent high pressure from damaging the system if there is a kinked or blocked fuel line. Also, there is a check valve at the pump outlet to hold pressure in the system after the pump is turned off.

The electrical current that operates the fuel pump is controlled by a relay. This ensures that the pump will not flood the engine, or continue running in the event of an accident or if the engine stalls. Turbo models have an additional pump safety feature that turns the pump off during overboost situations.

Two different fuel pump combinations and mountings are installed on Saab 900 models depending on the model year. See **Table a.** The two pump types are not interchangeable, owing to a different fuel tank design.

Table a. Fuel Pump Applications

Model	Pump type
1986-1988 Non-turbo *	Roller
1985-1989 Turbo models*	
1989- and later Non-turbo	Rotor
1990 and later, Turbo	

*Some early models may be equipped with a special replacement rotor-type fuel pump. This dealer-installed service pump was designed to eliminate a complaint of poor driveability or noisy pump operation and can be identified by its rubber mounting collar—other rotor-type pumps are mounted with a threaded plastic collar.

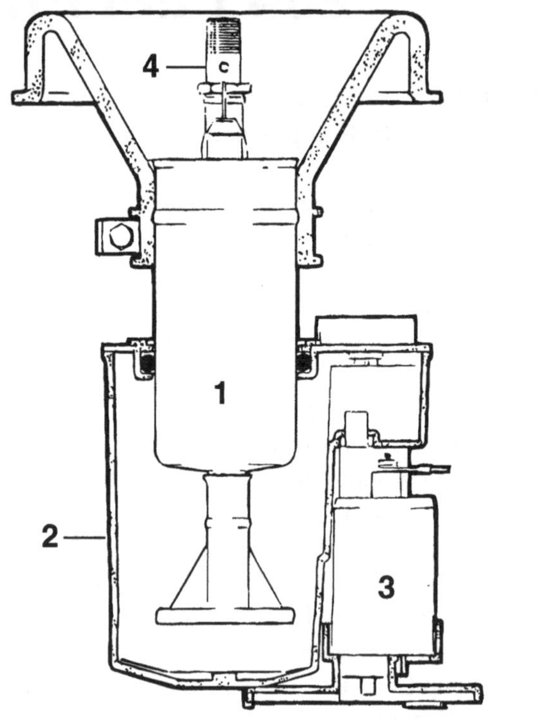

1. Fuel pump
2. Reservoir (pressure vessel)
3. Supply pump
4. Pump outlet check valve

Fig. 2. Roller-type fuel pump assembly.

Roller-type Fuel Pump Assembly with Supply Pump

Fig. 2 shows the roller-type fuel pump assembly. The pump is mounted in a fuel reservoir that is fed with fuel by the electric supply pump. In addition, the fuel return from the fuel injection system also feeds the fuel reservoir. This design helps prevent fuel starvation during hard cornering. The reservoir has a safety valve that opens at 0.1 bar (1.4 psi). The pump is mounted to the tank by a rubber collar.

NOTE
It may require at least two gallons of fuel to prime the pump if the car ran out of gas. If the car is not on level ground, additional fuel may need to be added.

Rotor-type Fuel Pump Assembly

Fig. 3 shows the rotor-type “ejector-style” fuel pump. The rotor-type pump is similar to the roller-type pump, although the separate supply pump is not used. The ejector picks up additional fuel from the bottom of the tank and forces it into the reservoir.

Safety Precautions

Please read the following warnings before doing any work the fuel pump, fuel tank or fuel lines.

WARNING
- Always disconnect the negative (–) battery terminal and cover the terminal with an insulated material whenever working on any fuel related component.
- Gasoline is dangerous to your health. Wear hand and skin protection when working on the fuel system. Do not breathe fuel vapors and always work in a well-ventilated area.
- Fuel and fuel vapors will be present during many of the operations described in this repair group. Do not smoke or create sparks. Have an approved fire extinguisher handy.
- The fuel injection system is designed to maintain pressure in the system after the engine is turned off. Fuel will be expelled under pressure as fuel lines are disconnected. This can be a fire hazard, especially if the engine is warm. Always wrap a clean shop rag around the fitting before loosening or disconnecting any fuel line.

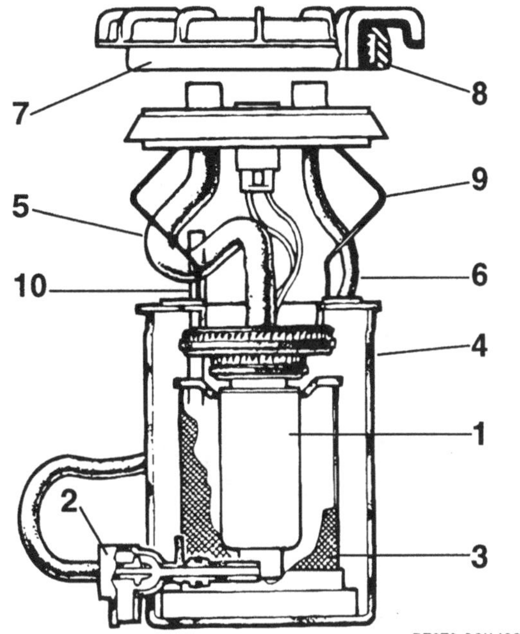

1. Main fuel pump
2. Ejector pump
3. Filter
4. Reservoir
5. Fuel delivery line
6. Fuel return line
7. Threaded retaining collar
8. Seal
9. Spring
10. Filter breather pipe

Fig. 3. Rotor-type fuel pump assembly.

FUEL PUMP TROUBLESHOOTING

The fuel supply system is an integral part of the operation of the fuel injection system. Problems such as a no-start condition, hesitation or stalling may be due to poor fuel delivery. There are some preliminary tests to quickly determine if the fuel pump or its electrical circuit are causing the problem. Some of the tests described below require special test equipment, such as a fuel pressure gauge.

NOTE

General engine management and driveability troubleshooting should begin by consulting **200 Engine—General.**

Cleanliness is essential when working with the fuel system. Even a tiny particle of dirt can cause trouble if it reaches an injector.

CAUTION

• Thoroughly clean fuel line unions before loosening or disconnecting them. Use only clean tools.

• Keep removed components clean, and seal or cover them with a clean, lint-free cloth, especially if the repair cannot be finished immediately.

CAUTION

• Avoid the use of compressed air nearby, and do not move the car while the fuel system is open.

• When replacing parts, install only new, clean components. Seals and O-rings should always be replaced rather than reused.

NOTE

To help prevent poor fuel delivery due to pump cavitation, Saab recommends avoiding the use of alcohol-enhanced fuel above ambient temperatures of 75°F (24°C).

Basic Troubleshooting Principles

The basic function of the fuel pump is to deliver an adequate amount pressurized fuel. Problems caused by a fault with fuel pump range from a no-start condition to intermittent poor performance and driveability. These problems can be the result of a fault in the pump electrical circuit or a worn or faulty pump.

Begin troubleshooting with a simple check of the fuel pump electrical circuit. The pump should run while cranking the engine with the starter. If necessary, remove the access cover under the trunk floor and listen or feel to see whether the pump is running. If the pump does not run, see **Fuel Pump Electrical Tests** below.

If fuel delivery problems are suspected, begin troubleshooting with a check of the fuel pump delivery rate. See **Fuel Pump Delivery Rate.** The test will indicate whether further tests are necessary. This is especially important on high-mileage cars, where normal pump wear may decrease delivery volume.

Also check for correct pump installation, a clogged pump strainer, or a bad supply/ejector pump. Any combination of these may cause the problems listed above.

NOTE

• On Turbo models, a faulty over-pressure switch may cause momentary fuel pump cut-out or a no-start condition. See **Testing Turbo Over-pressure Switch (1985-1988 models only)** for more information.

• On models with rotor-type pumps, poor performance may be caused by a ruptured or ballooned delivery hose in the pump assembly. This hose is available as a replacement part. See **Replacing Rotor-type Pump** for additional information.

• Poor driveability or performance on 1988 and some 1989 models with roller-type or rotor-type pumps may be due to a blocked return line check valve at the fuel tank, causing excessive line pressure. The problem is covered under Saab's vehicle warranty. For more information see an authorized Saab dealer.

NOTE
During 1990 model production, a modified rotor-type pump was installed as a running change. The new design helps eliminate pump cavitation, pump noise, and pump overheating. The new pump can be retrofitted to earlier models with the rotor-type (threaded collar) pump.

Fuel Pump Noise

Fuel pump noise does not always indicate a bad pump. The amount of noise that is transmitted to the passenger compartment varies with the car. Humming or buzzing at various levels is generally considered normal, while clicking, ratcheting or grinding may signal imminent pump failure.

On roller-type pumps, improper pump or fuel tank installation can increase pump noise, as can a faulty supply pump. On rotor-type pumps, a clogged ejector pump can often cause noise. Many times the main fuel pump is replaced needlessly when the supply or ejector pump is at fault.

NOTE
Supply pump test procedures are covered below under **Supply Pump Quick Check (roller-type pump only).**

Operating Fuel Pump for Tests

The electric fuel pump only runs when the cars is running or being started. This is to prevent running the fuel pump indefinitely if the engine stalls. Testing the fuel pump and the fuel injection system requires running the pump with the engine off.

The procedure for operating the pump uses a temporary wiring connection on the fuse/relay panel to bypass the fuel pump relay. The preferred method is to use a remote switch, such as the Saab special tool, Part No. 83 93 886, but you can accomplish the same thing with a homemade jumper wire and an in-line switch. See Fig. 4.

CAUTION
- A homemade jumper wire with a switch should be at least 1.5 mm metric wire size (14 gauge-AWG) and, for safety, should include a 20 amp in-line fuse.
- Avoid damaging the fuse sockets. The ends of the jumper wire should be flat-blade connectors that are the same size as the sockets in the fuse panel.
- Connect and disconnect the remote switch or jumper wire only with the switch in the off position.

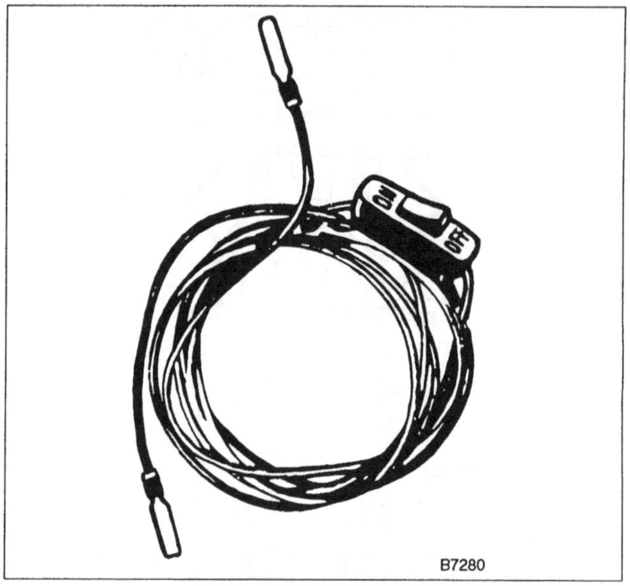

Fig. 4. Jumper wire with flat-blade connectors and switch for running fuel pump without running engine.

To run the pump, remove fuses number 27 and 30. With the ignition off, connect the jumper wire and switch as shown in Fig. 5. Turn the switch ON to run the pump.

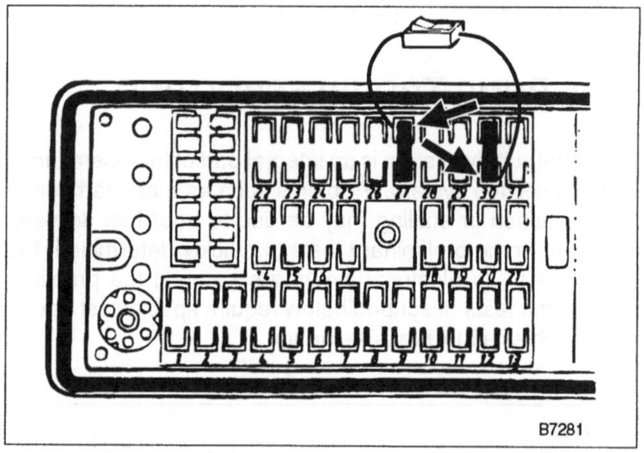

Fig. 5. Schematic view of jumper wire and switch connected at fuse sockets (fuses removed) for running fuel pump. Fuse/relay panel is in engine compartment, on top of left fender.

NOTE
If the pump runs only when the jumper is connected, the relay or wiring is faulty. If the pump doesn't run with the jumper installed, the problem is in the wiring to the pump or the pump itself. See **Fuel Pump Electrical Tests** for more information

Supply Pump Quick Check
(roller-type pump only)

The following test will help determine whether a faulty supply pump is the cause of fuel delivery problems. A faulty supply pump usually manifests itself through a noisy fuel pump.

WARNING
- The fuel tank must be no more than 3/4 full when making this check or fuel may be spilled.
- Fuel will be expelled. Do not disconnect wires that could cause sparks. Have a fire extinguisher handy. Perform the test outside if possible. Have a supply of rags handy to clean up any spills.

To test roller-type supply pump

1. Working in the trunk, remove the pump access cover. Disconnect the main pump terminals and slacken and remove the mounting boot clamp. See Fig. 6.

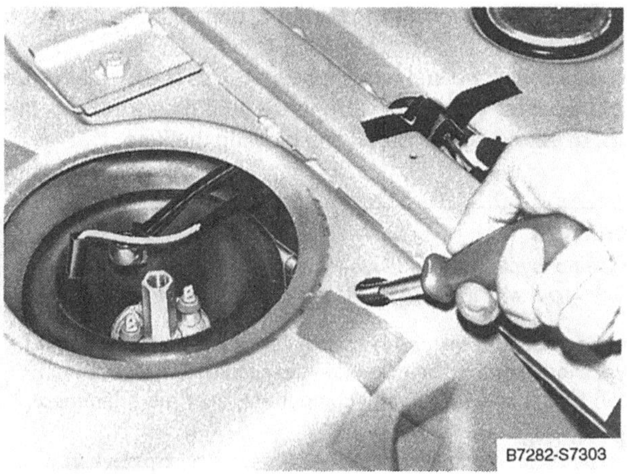

Fig. 6. Fuel pump mounting boot clamp being loosened using flexible socket driver.

2. Carefully peel back the pump mounting boot so that the top of the fuel pump reservoir is visible.

3. Run the fuel pump as described above under **Operating Fuel Pump for Tests** As the supply pump runs, fuel should be forced out of the reservoir relief valve. See Fig. 7.

WARNING
The fuel forced out of the relief valve will be partially vaporized and is extremely flammable. Do not smoke or create sparks. Have an approved fire extinguisher handy.

NOTE
If fuel comes out of the relief valve, the supply pump is working correctly. If no fuel comes out, the supply pump is most likely faulty and should be replaced. The supply pump is available as a replacement part. See **Replacing Roller-type Pump** below.

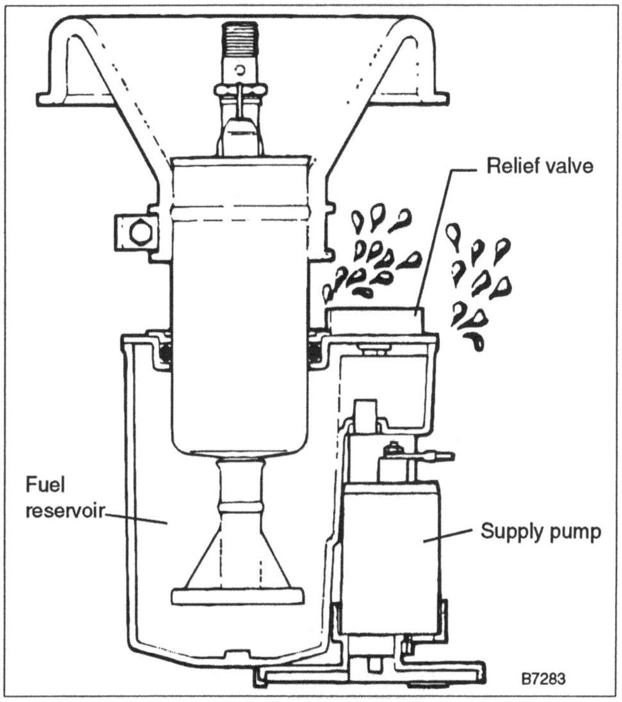

Fig. 7. Cutaway view of roller-type fuel pump showing fuel forced out of relief valve as supply pump operates.

4. When finished, clean up all spilled fuel and reinstall the pump mounting boot.

FUEL PUMP

Fuel Pump Delivery Rate

This test checks the fuel pump output. Insufficient fuel delivery can result in poor performance, hard starts, hesitation, and stalling.

To test pump delivery

1. Working in the engine compartment, disconnect the fuel return line from the pressure regulator. See Fig. 8. Attach a length of fuel line to the pressure regulator and place the hose into a 2 liter (2 quart) container.

WARNING
Fuel will be expelled as the line is disconnected.

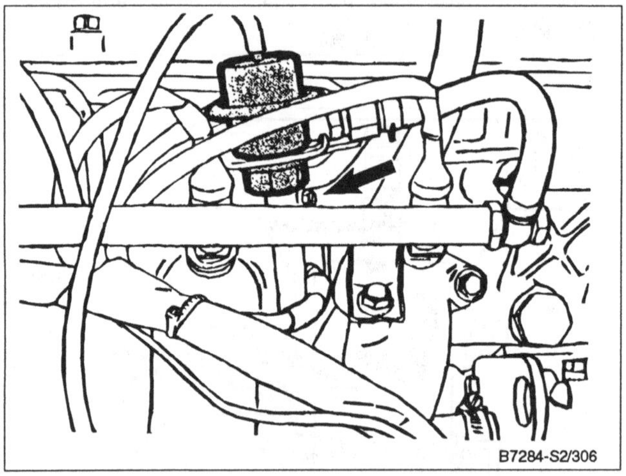

Fig. 8. Disconnect fuel return line at pressure regulator (arrow). Attach length of hose to regulator and place hose end into container.

2. Install the switched jumper (see Fig. 5) and run the fuel pump for exactly 30 seconds. Check the amount of fuel in the container.

Fuel Pump Delivery Volume
- minimum of 900 ml (30 oz.) in 30 seconds

Causes of insufficient delivery:
- system pressure incorrect. Check fuel pressure as described in **240 Fuel Injection**
- clogged pump strainer/fuel filter/supply line
- voltage at fuel pump incorrect. See **Fuel Pump Electrical Tests**
- worn fuel pump

If fuel delivery volume is low, reconnect the fuel return line to the pressure regulator. Install a pressure gauge as described in **240 Fuel Injection.** Run the fuel pump and slowly clamp shut the fuel return line at the pressure regulator. Immediately release the pinched hose as soon as the pressure approaches 6 bar (83 psi). When the return line is clamped shut, the gauge needle should rapidly rise to about 6 bar (83 psi). If not then the pump is probably faulty.

CAUTION
Allow the fuel pressure to rise slowly. Do not allow the pressure to exceed 6 bar (83 psi). Damage to the fuel pump or lines can result.

Fuel Pump Electrical Tests

The fuel pump electrical circuit is relay controlled, which means that power to the pump must pass through the relay before the pump can run. The LH control unit controls the relay.To keep the relay closed and the pump running, the LH control unit must get a signal that the engine is turning over. This pulse signal comes from the ignition system. 1985 through 1988 Turbo models use an additional over-pressure switch to prevent engine damage as a result of turbocharger wastegate malfunction. The over-pressure switch cuts off power to the fuel pump relay whenever boost pressure is above a safe level.

NOTE
- Testing of the turbo over-pressure switch is covered below.
- An additional engine over-rev protection feature turns off the fuel injectors if the engine speed goes above 6,000 rpm. This feature is handled by the LH control unit.

If the pump fails to run at all when the starter is cranked, the problem can be due to one of the following: the fuel pump relay, the electrical circuit, the LH control unit, or the pump itself.

Always check the fuel pump fuse first. This is especially true in cold weather when moisture in the fuel may freeze in the pump, overloading the circuit and blowing the fuse.

CAUTION
- Do not bypass a failed fuse to power the circuit directly until finding out why the fuse has failed.
- Repeated fuse failures indicate an electrical problem, usually a short to ground or an overloaded circuit. A circuit overload can be caused by a faulty component, such as a jammed pump.

NOTE
For the best troubleshooting results, perform all tests listed below in the sequence in which they appear.

Testing Fuel Pump Circuit

Check the circuit that powers the pump by bypassing the relay to run the pump. Remove the fuel pump relay, shown in Fig. 9. Install a jumper wire between sockets 30 and 87b in the relay panel. The pump should run.

If the pump runs with the relay bypassed, you can assume that the pump and the wiring to the pump is OK. It is either the relay, or the circuit that controls the relay that is faulty. See **Testing Relay Circuit** below.

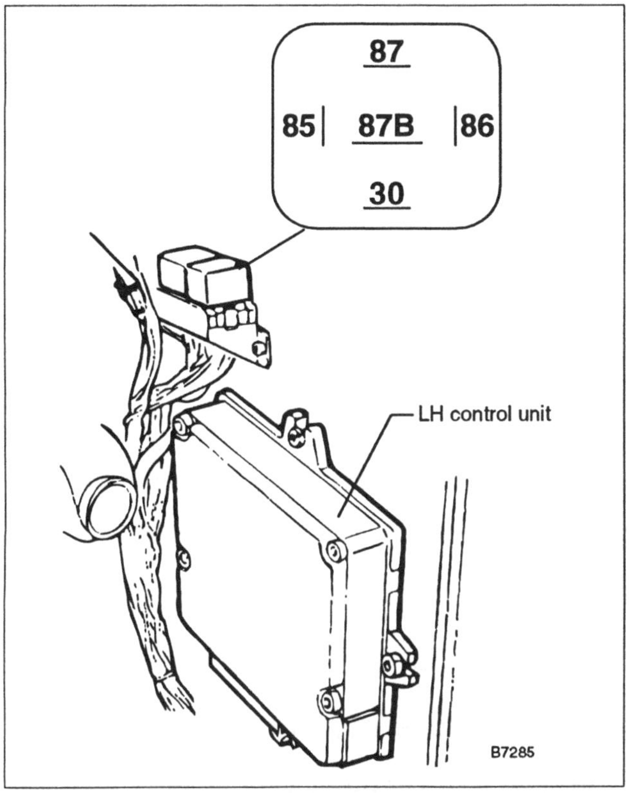

Fig. 9. Fuel pump relay and LH control unit located behind right side trim panel in front passenger footwell. Inset shows relay terminal designations on relay panel with relay removed.

If the pump does not run with the relay bypassed, then there is a fault in the fuel pump circuit. Power may not be getting from the relay panel to the pump, or even from the battery to the relay panel. Use the wiring diagrams in **371 Wiring Diagrams, Fuses and Relays** to check power and ground in the circuit.

To test fuel pump relay control circuit

1. Remove the fuel pump relay. See Fig. 9 above.
2. Turn the ignition on and check for power at terminal 86 of the fuel pump relay socket. If there is no power then either the fuel-injection main relay or the wiring from the main relay to socket 86 of the fuel pump relay panel is faulty.

NOTE
See **240 Fuel Injection** for information on checking the LH main relay. See **371 Wiring Diagrams, Fuses and Relays** for all LH wiring diagrams.

---

3. Connect a voltmeter across relay socket terminals 86 (+) and 85 (–). Briefly crank the starter and check that there is battery voltage. If not, the ground signal from the LH control unit to the fuel pump relay is not present.

CAUTION
Use only a digital voltmeter with high input impedance when testing solid-state components.

NOTE
The LH control unit must get an ignition pulse telling it that the engine is being cranked or running. This signal comes from the ignition system. If this signal is not present at the LH control unit, the fuel pump relay will not operate. See **240 Fuel Injection** for LH control unit electrical tests.

---

4. If there is power at relay sockets as described above, test the relay itself. When battery power is applied to terminals 85 (–) and 86 (+), the relay should click closed (pull-down) and there should be continuity between terminals 30 and 87 and between 30 and 87b. If not, the relay is faulty.

NOTE
The above test in step 4 does not guarantee that the relay is working correctly under all operating conditions. Sometimes the relay may work intermittently only when it has been operating for some time.

Checking for Voltage at the Fuel Pump

This test checks to see if the correct voltage is reaching the pump. To make the test, remove the pump access cover in the floor of the luggage compartment. With the engine off, run the fuel pump as described above under **Operating Fuel Pump for Tests.** Then measure the voltage across the pump terminals. See Fig. 10.

Voltage at fuel pump with pump running
• at least 11.5 volts

If the voltage is below 11.5 volts, check all wiring connections in the circuit for looseness or corrosion. Clean or tighten as necessary. If there is no voltage at the pump, check the wire between the pump and the relay and the ground wire from the pump to the center crossmember in the trunk.

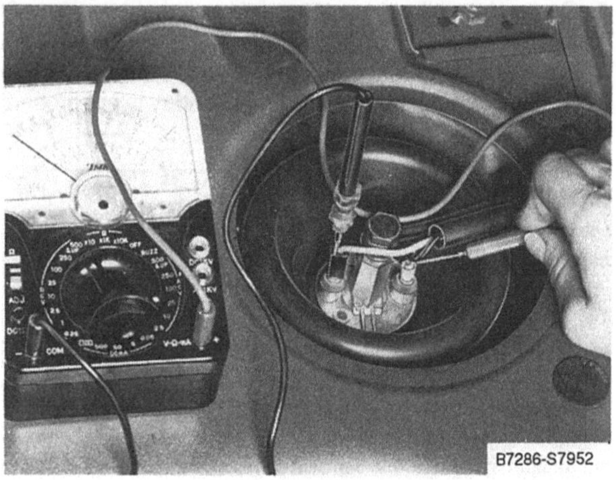

Fig. 10. Checking for voltage across fuel pump terminals with pump running and engine off. Roller-type pump shown. On models with rotor-type pump probe from back of wiring connector.

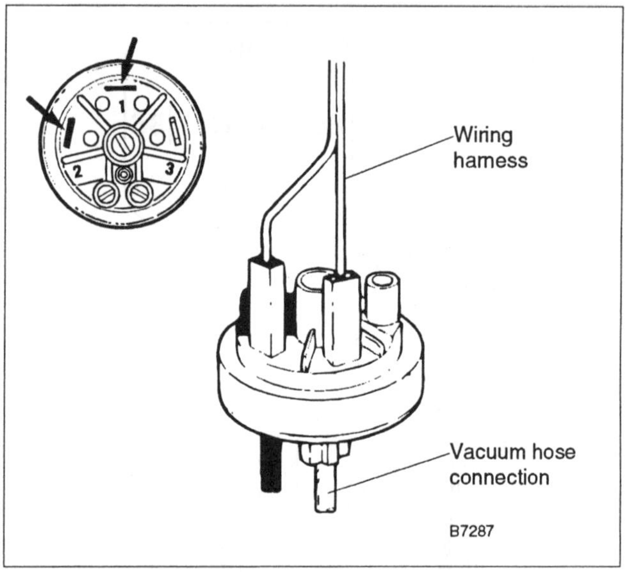

Fig. 11. Turbo over-pressure switch used on 1985-1988 Turbo models. Be sure to connect switch wires to terminals 1 and 2, as shown at arrows.

Testing Turbo Over-pressure Switch (1985-1988 models only)

The over-pressure switch is mounted under the instrument panel to the left of the steering wheel, just behind the fuse/relay panel. Fig. 11 shows the switch. The switch is used as a backup safety device in the event the turbocharger wastegate malfunctions. If the turbocharger boost pressure exceeds a safe level, the over-pressure switch interrupts power to the fuel pump. The tests described below require the use of hand-held vacuum/pressure pump.

NOTE
All 1989 and later Turbo models also have an over-pressure safety feature, although it is handled by the 2.4 LH control unit instead of a separate pressure switch.

To test turbo over-pressure switch

1. Working beneath the left-hand side of the instrument panel, disconnect the vacuum hose from the switch and plug the hose. Attach the pump to the switch using the piece of hose.

2. With the engine running, slowly increase pressure. The engine should shut off when the pressure reaches approximately 1 bar (14.5 psi).

3. If the engine does not shut off, turn off the engine and disconnect the harness connectors from the switch. Connect an ohmmeter across switch terminals 1 and 2. With no pressure applied to the switch, there should be continuity. With 1 bar (14.5 psi) of pressure applied to the switch, there should be no continuity. If any faults are found, replace the switch.

Replacing Roller-type Pump

When replacing a roller-type pump, pay close attention to the installed height and orientation of the pump. Incorrect installation can lead to fuel starvation or pump failure. Be sure to have new fuel line sealing washers on hand.

To replace roller-type pump assembly

1. Disconnect the battery negative (–) cable and cover the terminal pole on the battery. Remove the fuel pump access cover in floor of luggage compartment. Disconnect the pump wiring.

NOTE
On cars with LH 2.4 fuel injection, disconnecting the battery will erase the control unit's adaptive memory. When the engine is restarted it may operate erratically. The engine must be driven for at least 30 minutes for the adaptive memory to be reset.

---

2. Disconnect the fuel line from the pump. Using an open-ended wrench, hold the pump still while removing the dome nut. See Fig. 12.

WARNING
Fuel will be expelled when the fuel lines are disconnected and the pump is removed. Do not smoke or create sparks. Have an approved fire extinguisher handy.

CAUTION
Be careful not to damage the fuel pump terminals when loosening the center banjo bolt. If a terminal is broken off it is not repairable and the entire pump must be replaced. A special wrench is available to reduce the possibility of breaking a pump terminal. The wrench is Saab Part No. 83 94 330

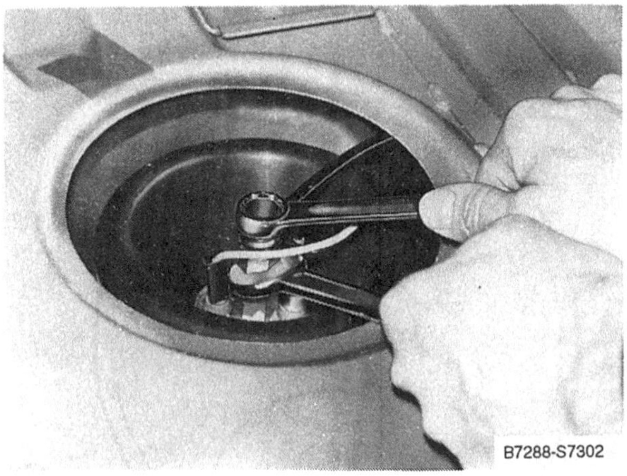

Fig. 12. Fuel line being removed from fuel pump.

3. Lift the pump up just enough to disconnect the fuel return line from the fuel reservoir and the wiring from the supply pump.

4. Using a flexible nut driver, loosen the collar mounting strap. See Fig. 13. Carefully lift out the pump.

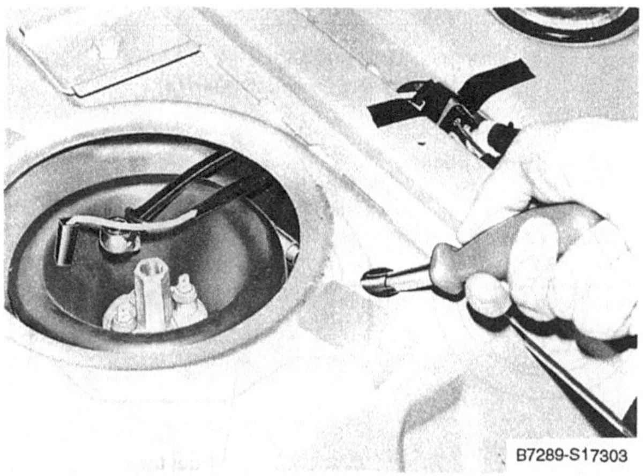

Fig. 13. Fuel pump mounting collar strap being loosened.

5. To disassemble the pump, pull the main pump out of the reservoir. The supply pump is held with a bracket and two screws.

6. To assemble the pump, install the wiring to the supply pump and install the supply pump into the reservoir, pressing it against the stop.

NOTE
On the supply pump, the white wire goes to the positive (+) terminal and the black wire to the negative (–) terminal.

7. Install the main pump into the collar and adjust it to the height shown in Fig. 14. Install the pump with collar into the reservoir and adjust the total height of the pump assembly to the specification shown in the figure.

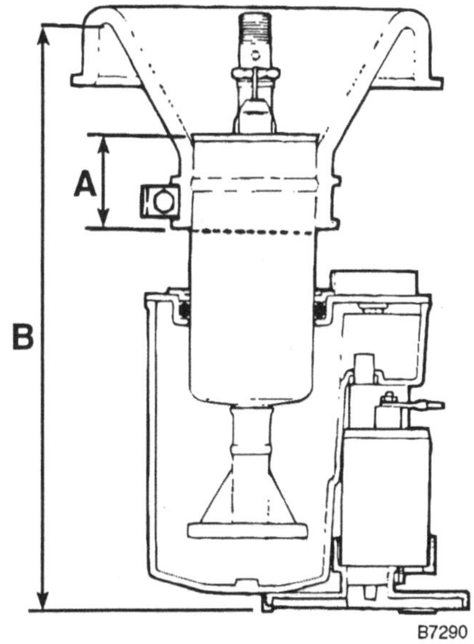

Main Fuel Pump

- Dimension **A**
  All. . . . . 42±2 mm (1.65±.08 in.)
- Dimension **B**
  1985-early 1988
  (up to VIN J3001527
  and J7004073) . . . . 225 mm (8 7/8 in.)
  Early 1988 and later
  (from VIN J3001528
  and J7004074). 242 mm (9 1/2 in.)

Fig. 14. Assembled dimension of main fuel pump in reservoir and rubber collar.

NOTE

- Proper pump height and pump orientation in the tank is very important. Scribe a mark 42 mm (1 21/32 or 1.65 in.) from the top of the pump to accurately position the collar to the pump during assembly.
- See **1 LUBRICATION AND MAINTENANCE** for information on VIN numbers and their locations.
- The supply pump on some early models may be equipped with a "bow-tie" style strainer. For better fuel pick-up and delivery, a modified strainer should be installed. See Fig. 15. The new strainer is Saab Part No. 93 86 491.

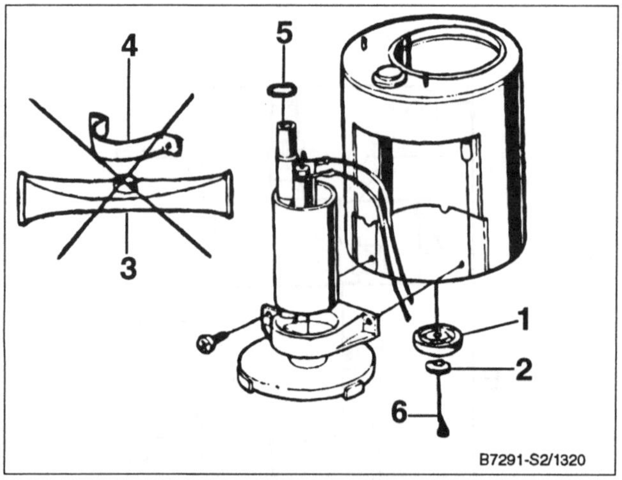

Fig. 15. Where installed, always replace bow-tie strainer (**3**) with modified strainer shown. Replace supports (**1** and **2**), clamp (**4**), and O-ring (**5**) when using new strainer. Screw (**6**) must be reinstalled for proper pressurization of reservoir.

8. Install the pump into the tank, making sure to reconnect the supply pump connector and the fuel return line before fully installing the pump. Orient the pump as shown in Fig. 16 or Fig. 17. Tighten the mounting collar clamp.
9. Using new sealing washers, reinstall the fuel line and banjo bolt/dome nut. Reconnect the wiring.

NOTE

The direction of pump rotation is determined by the polarity of the wiring, so + and – signs are cast in the pump housing to ensure proper wiring connection.

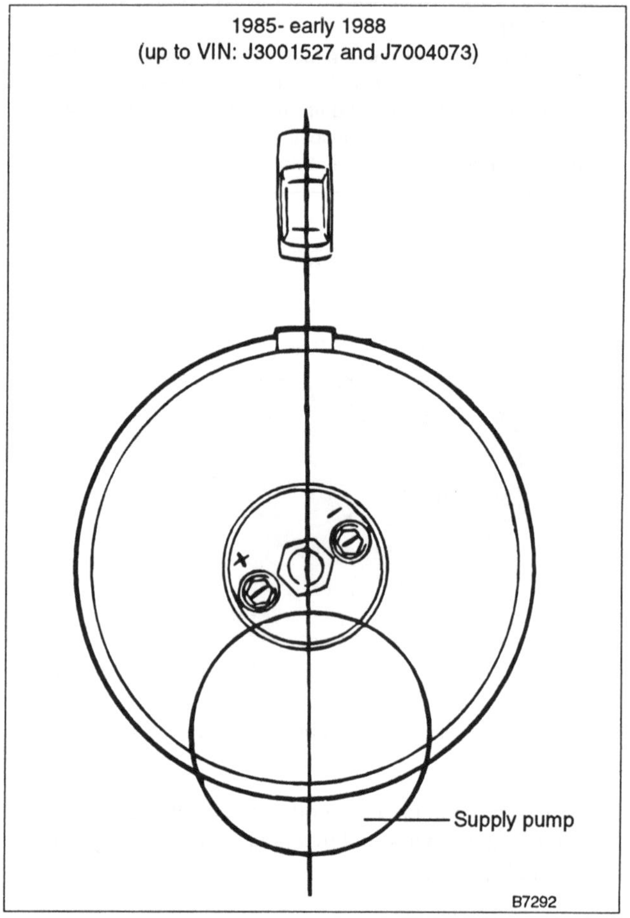

Fig. 16. 1985 to early 1988 fuel pump orientation relative to car.

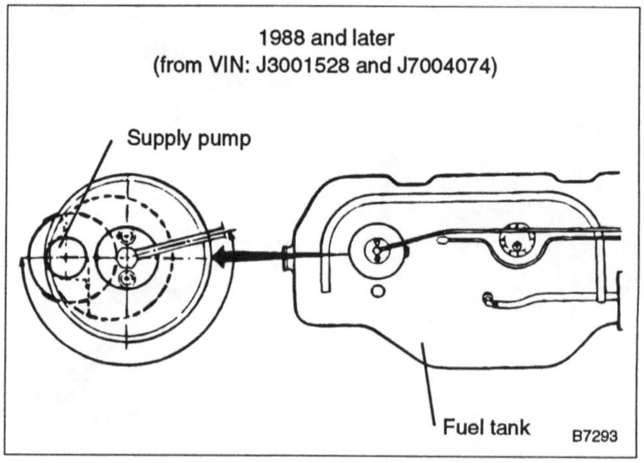

Fig. 17. Early 1988 and later fuel pump orientation relative to fuel tank.

Replacing Rotor-type Pump

When removing and installing the rotor-type fuel pump, always replace the O-rings that seal the fuel lines and the pump to the tank. This is especially important on 1990 and later models because the O-rings are designed to swell when exposed to fuel vapor. These O-rings cannot be reinstalled unless they are completely dry.

There were two styles of rotor-type pumps installed. The early version had the supply check valve installed in the pump assembly. The later pump (identified by the internal T-fitting in the pump's supply line) had the supply check valve fitted in the supply line fitting. If the early pump is replaced, a new check valve must be inserted into the end of the supply line using special tools, as the early-style pump is no longer available. See **Replacing Fuel Pump Check Valve** below for replacement procedures.

To replace rotor-type pump assembly

1. Disconnect the battery negative (–) cable and cover the terminal pole on the battery.

NOTE
On cars with LH 2.4 fuel injection, disconnecting the battery cable will erase the control unit's adaptive memory. When the engine is restarted it may operate erratically for up to 30 minutes.

2. Working in the luggage compartment, remove the access panel from above the pump.

3. Remove the clamp that secures the fuel lines and wiring. See Fig. 18. Disconnect the wiring from the pump.

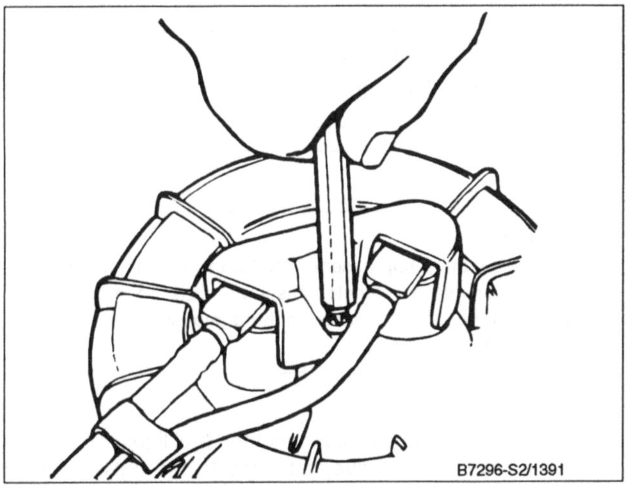

Fig. 18. Fuel line and pump wiring hold-down clamp.

4. Using a round-jawed pliers, disconnect the fuel lines by pulling them straight up. See Fig. 19. Tie the fuel lines back.

WARNING
Fuel will be expelled under pressure. Wrap clean rags around the fuel line fittings before removing them from the pump.

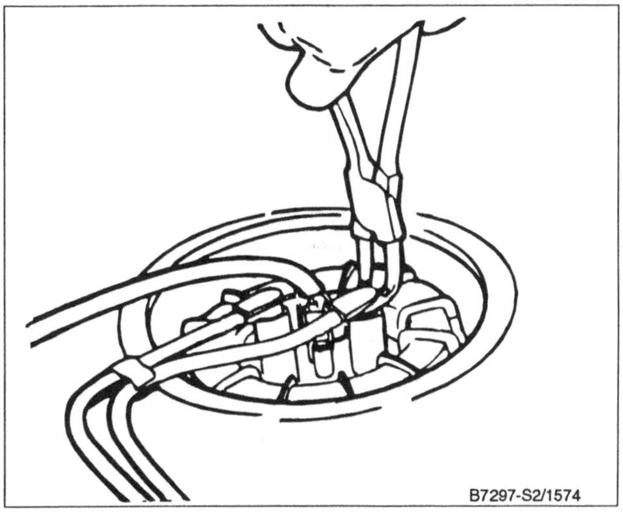

Fig. 19. Fuel lines being removed.

5. Loosen the large threaded retaining collar while holding the pump stationary.

CAUTION
The pump must not turn when loosening the retaining collar or the ejector assembly could be damaged. A special tool is available to hold the pump stationary as the collar is loosened. See Fig. 20.

6. Lift the pump assembly up to allow excess fuel to drain, then tilt it slightly to remove it from the tank. Place it in a receptacle to catch any remaining fuel.

NOTE
Carefully inspect the fuel delivery line . If the hose is distorted or ballooned, it should be replaced. This hose is available as a replacement part through and authorized Saab dealer. Be sure to check the fuel system pressure if the hose is distorted. Excessive fuel pressure due to a blocked check valve will cause the new hose to balloon.

7. Carefully insert the new pump assembly (ejector pump side first) into the tank so that the fuel line fittings on the pump are parallel with the length of the car and that the matching mark on the pump is aligned with the mark on the tank. See Fig. 21.

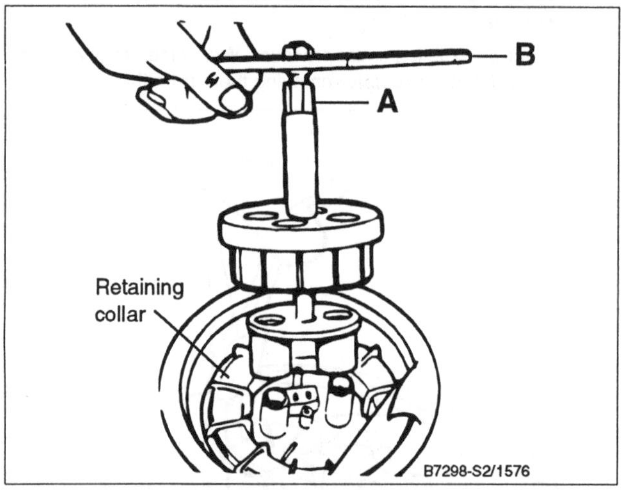

Fig. 20. Special tool for fuel pump removal. Loosen collar by turning outer part of tool with open-end wrench at **A** while holding pump stationary with handle (**B**) on inner tool.

NOTE

Make sure that the pump fits between the locating ribs in the bottom of the tank and that the matching marks on the pump and tank are exactly aligned.

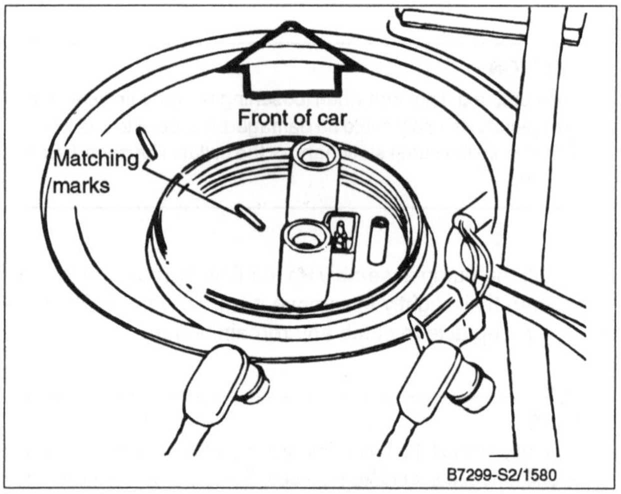

Fig. 21. Correct alignment of pump assembly relative to tank.

8. On 1989 models, lubricate the new collar seal with petroleum jelly (Vaseline) and insert it into the retaining collar. Install the collar while aligning the matching marks. See Fig. 22. Press down firmly on the collar until it engages the tank. Tighten the collar until it stops.

CAUTION

Do not allow the pump assembly to turn when tightening the collar. Otherwise the ejector assembly may be damaged.

NOTE

The locating marks on the retaining collar must be within 30° to either side of the marks on pump and tank.

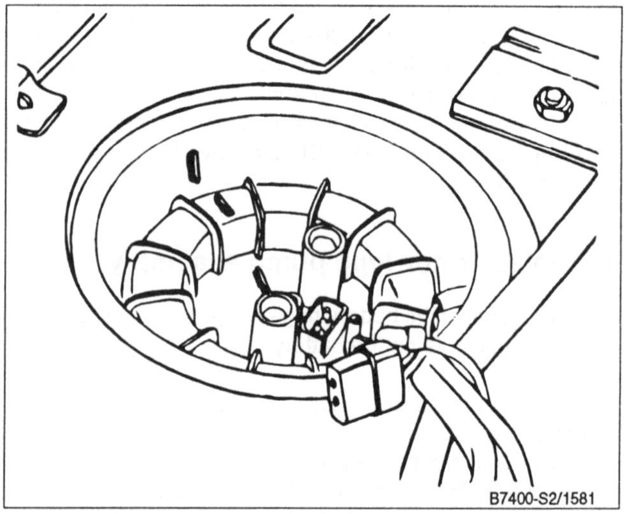

Fig. 22. Matching marks on retaining collar and fuel tank correctly aligned.

9. On 1990 and later models, fit the new collar seal in the groove in the fuel tank. Install the retaining collar, tightening it using Saab special tool no. 83 94 462 or an equivalent. See Fig. 23.

Tightening torque

- Fuel pump to fuel tank (threaded collar)
  1990 and later . . . . 55 Nm (40 ft-lb)

CAUTION

Do not allow the pump assembly to turn when tightening the collar. Otherwise the ejector assembly may be damaged. Check that the locating marks on the pump and tank are still aligned after tighening.

---

10. Using new O-rings, reconnect the fuel lines. The return line (with check-valve) attaches to the outlet nearest the front of the car. The supply line attaches to the rear outlet. Reconnect the wiring and fit the clamp.

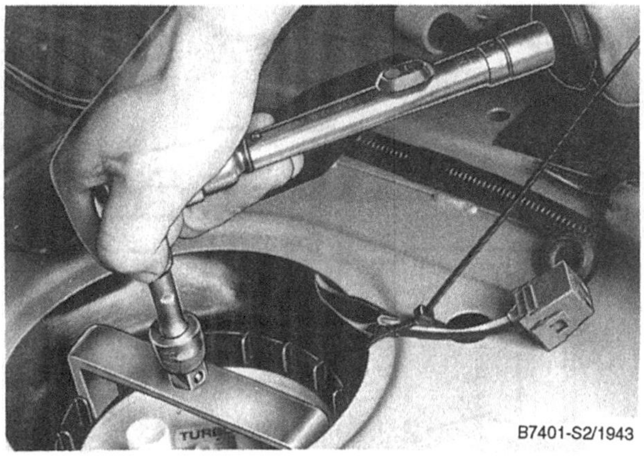

Fig. 23. Threaded collar being tightened using Saab special tool no. 83 94 462 on 1990 and later models.

NOTE
During 1990 production, the supply line check valve was moved from the supply-line socket in the pump to the end of the supply line itself. Make sure you identify each fuel line correctly before connecting it.

To replace fuel pump screen (models with rotor-type pump only)

1. Carefully ease back the lugs that secure the ejector pump, turn the ejector 90° (1/4 turn) clockwise and remove it from the reservoir. See Fig. 24.

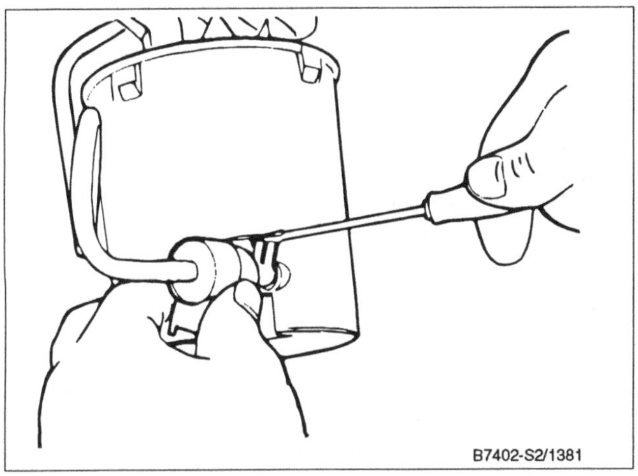

Fig. 24. Ejector pump retaining lugs being eased back. Turn ejector 1/4 turn and remove from reservoir assembly.

2. Separate the reservoir and filter unit from the top section. See Fig. 25.

3. Install the new reservoir and filter assembly to the pump upper part using a new O-ring.

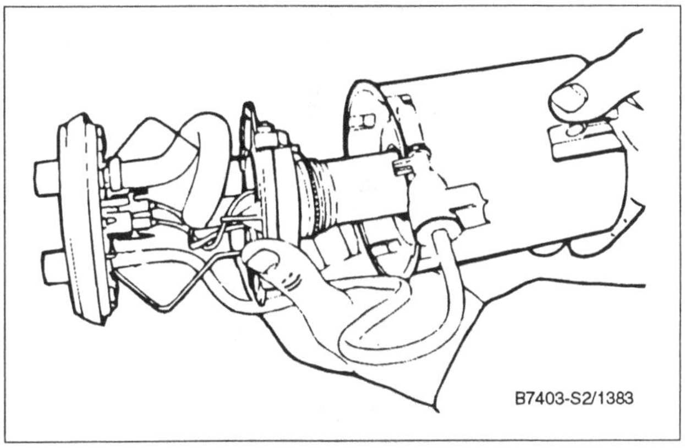

Fig. 25. Reservoir and filter unit being separated from pump top section.

4. Fit a new O-ring to the ejector pump and reinstall it into the reservoir, turning it 90° counter-clockwise to lock it to the dog on the reservoir.

NOTE
Check that the ejector pump is clean. If necessary blow through using compressed air to remove any foreign material.

Replacing Fuel Pump Check Valve

A one-way check valve is used at the fuel pump supply line to hold fuel pressure in the system after the pump shuts off. A faulty valve can cause longer engine cranking on restart because of the loss of fuel system rest pressure. The check valve is tested as part of a fuel system pressure test as described **240 Fuel Injection**

WARNING
- Fuel under pressure will be expelled. Wipe up any spilled fuel immediately. Wrap fuel lines with a clean rag before removing them. Do not smoke or create sparks. Have an approved fire extinguisher handy.
- Always remove the battery negative (–) cable before doing any work on any fuel system component.

To replace the check valve on models with roller-type pump, remove the banjo fitting while holding the pump steady. Then carefully remove the check valve from the pump. Install the new check valve and the banjo fitting, using new sealing washers. See Fig. 26.

CAUTION
Do not grip the check valve too tightly, as it may become damaged.

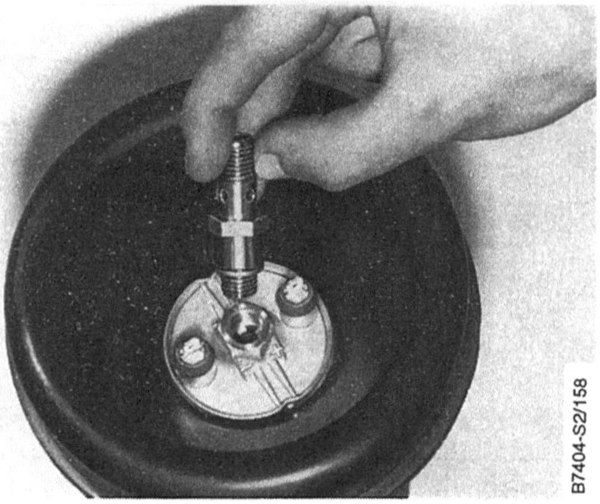

Fig. 26. Fuel pump check valve shown removed from roller-type fuel pump.

On models with the rotor-type pump, two types of check valves were used on the cars covered by this manual. On 1989 models, the return line check valve was part of the fuel pump. On 1990 and later models, the check valve was integrated into the supply line fitting (elbow). If a 1989 pump is being replaced with the newer superseded pump or if the new style check valve is faulty, installation of a new check valve will require the use of a special tool.

To install the check valve in the supply line, first cut the old fitting from the end of the line as close as possible to the fitting. Using the special tool (Saab tool no. 83 94 546) and a channel-lock pliers, carefully press the new check valve fitting into the fuel supply line. See Fig. 27. Use new O-rings when reconnecting the fuel supply line to the pump.

NOTE

Do not cut more than 12–14 mm (1/2 in.) of the supply line when removing the fitting/check valve. If more than this is removed, a new fuel-pump-to-fuel-filter fuel line will have to be installed.

FUEL TANK AND LINES

Fuel Tank

The fuel tank is held in place with two straps. Removal and installation is fairly straightforward as long as the rear of the car can be raised and supported securely. The following information is a general guide to the necessary steps for tank removal and installation.

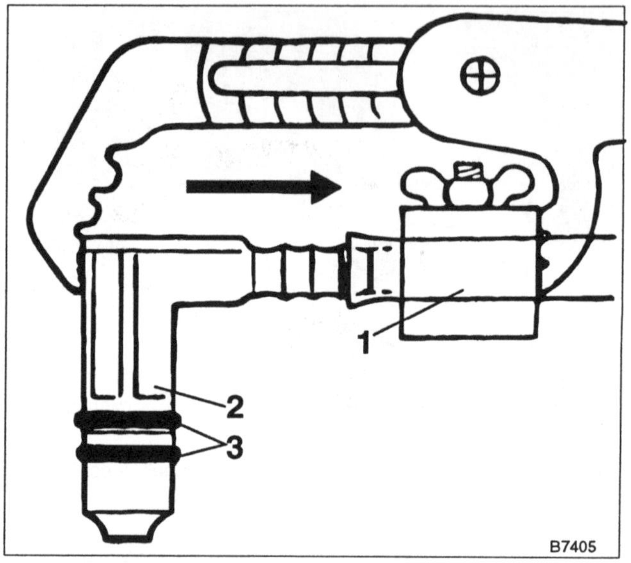

Fig. 27. New supply line check valve (**2**) being installed with Saab special tool (**1**). Always use new O-rings (**3**).

WARNING

Fuel will be expelled when the fuel lines are disconnected. Do not smoke or create sparks. Have an approved fire extinguisher handy.

To remove and install fuel tank

1. Drain the fuel tank using the fuel pump. Run a hose from the pump outlet on the tank into a suitable container. See Fig. 28. Then run the fuel pump without running the engine as described earlier under **Fuel Pump Troubleshooting.**

NOTE

The fuel tank is not equipped with a drain plug.

---

2. Disconnect the battery negative (–) cable.

NOTE

On cars with LH 2.4 fuel injection, disconnecting the battery cable will erase the control unit's adaptive memory. The engine must be driven for about 30 minutes for the adaptive memory to be reset.

---

3. Working from the luggage compartment, remove the floor covers and disconnect the electrical leads from the pump and fuel gauge sender.

4. Raise the rear of the car and support it securely.

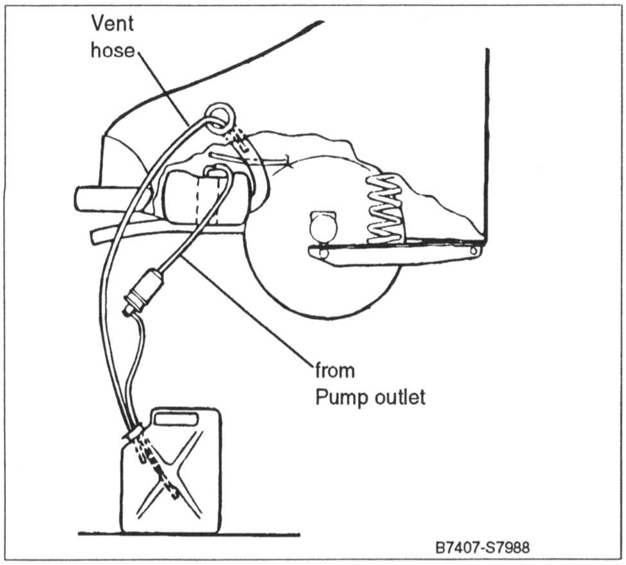

Fig. 28. Setup used to drain fuel tank. A vent hose should be used to prevent emission of hydrocarbons.

5. Disconnect the filler pipe, breather lines, and fuel return line from the tank. Undo the fuel line clips.

6. Remove the strap nuts and lower the tank.

NOTE
Before installing the tank, check that the rubber seal around the fuel sender opening is positioned correctly and undamaged. Cover the tank openings with tape.

---

7. Raise the tank into position and support it with the straps. Adjust the tank's lateral position, then tighten the straps and remove the tape.

8. Reconnect the filler pipe, making sure the rubber grommet is in position.

9. Reconnect all fuel lines, clips, and electrical connections.

10. Lower the car and reconnect the battery negative (–) cable.

To remove and install fuel gauge sender unit

WARNING
Fuel will be exposed. Do not smoke or create sparks. Have an approved fire extinguisher handy.

NOTE
There are different fuel sender gaskets depending on the sender installed. For the most correct replacement, bring the old gasket to an authorized Saab dealer to match up with the replacement gasket.

---

1. Disconnect the battery negative (–) cable. Cover the battery terminal.

2. Remove the cover from above the sender in the luggage compartment. Disconnect the sender wiring.

3. Remove the sender's threaded retaining ring.

NOTE
A Saab special tool (tool no. 83 93 365) is available to remove and install the threaded retaining ring. See Fig. 29.

Fig. 29. Fuel gauge sender retaining ring being removed using socket wrench and Saab special tool no. 83 93 365. (Saab Part No. 83 93 365).

4. Withdraw the sender and let it drain. Remove the old seal and discard.

5. Fit a new sender seal, insert the sender into the tank, and tighten the retaining ring. Reconnect the wiring and fit the cover.

NOTE
New senders are shipped with a retaining wire at the bottom to prevent float damage. Remove the wire before installing the sender, or else the gauge will only read EMPTY.

To replace fuel tank rollover valve

1. Working in the luggage compartment, remove the trim from the right-hand side.

2. Remove the rollover valve mounting screws and pull the valve into the luggage compartment. See Fig. 30.

Fig. 30. Rollover valve in luggage compartment being removed.

3. Disconnect the hose from the valve and remove the valve.

Installation is the reverse of removal. Make sure the long outlet on the valve is facing up when installing.

Fuel Lines

The fuel lines that run from the fuel tank to the engine compartment are routed along the left-hand sill inside the car.

To replace fuel lines inside car

1. Remove the left-hand sill scuff plate and roll the carpet back.

2. Peel back the tape holding the fuel lines to the body.

3. Carefully remove the insulation felt from the bulkhead.

4. Disconnect the fuel return and supply lines in the engine compartment and pull the lines through the bulkhead grommets and into the passenger compartment.

5. Release the lines from the retaining clips at the fuel tank and remove the lines from the car.

Installation is the reverse of removal. Clean the fuel lines using low pressure compressed air. Tape the open ends before reinstalling them. Always replace fuel fitting sealing washers and any crimp-type hose clamps.

To replace banjo fitting on fuel line

WARNING
The procedure below requires the use of a soldering iron with a knife tip. Completely remove the hose and its fitting from the car. Move a safe distance away from the car and any open fuel when making the repair.

---

1. Using a soldering iron, melt a notch into the into the end of the line. See Fig. 31.

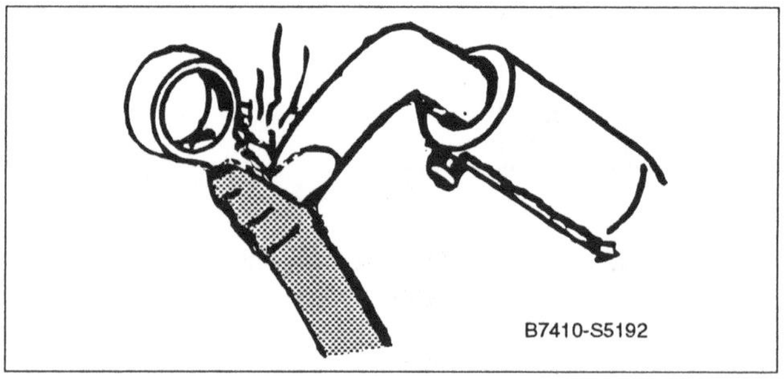

Fig. 31. Soldering iron being used to melt notch in fuel line.

2. Pull the fitting out of the fuel line. Cut the fuel line back just enough to remove the melted notch made earlier.

3. Using a section of thick rubber hose, slice it open and place it over the fuel line. Place the fuel line in a vice. Allow the end of the line to overhang the vise jaws slightly. See Fig. 32.

NOTE
As a better alternative to the split rubber hose, use Saab special tool no. 83 94 546 and a channel-lock pliers to install the fitting to the fuel line. See Fig. 27 shown earlier.

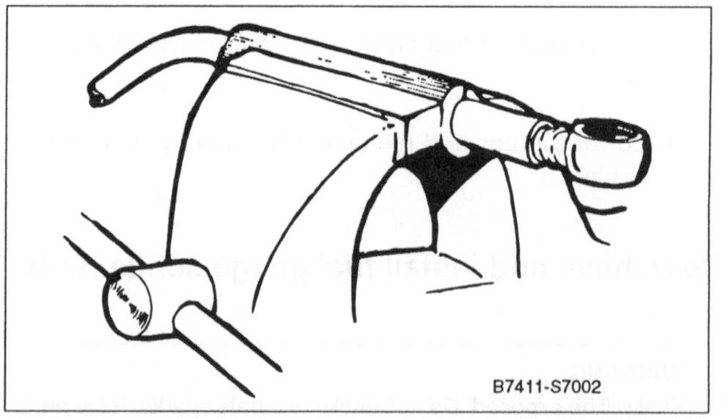

Fig. 32. Fuel line (protected with rubber hose) installed in vice.

4. Tap the fitting into the fuel line until it is fully seated.

240 Fuel Injection

TABLES

GENERAL

All Saab 16-valve models feature Bosch LH Fuel Injection as standard equipment. This system is also know as Hot-wire fuel injection, owing to the operation of the air mass meter, which uses a heated wire to help measure the incoming air.

Fuel injection component troubleshooting and repair are covered in detail within this repair group. Special equipment may be necessary for some of the procedures included here. If you do not have the equipment or the experience required to accurately do the job, we suggest leaving those tests or repairs to an authorized Saab dealer. The Saab dealer is equipped with special diagnostic test equipment; the ISAT (Intelligent Saab Tester) and the LH Tester. Both of these electronic tools are capable of quickly pinpointing hard-to-find LH fuel injection problems.

NOTE

- Fuel supply—the system that pressurizes the fuel and delivers it to the injection system—is covered in **234 Fuel Pump and Fuel Tank.**

- Exhaust emission systems, such as Evaporative Loss Control (ELCD), and Exhaust Gas Recirculation (EGR) are covered separately in **254 Exhaust Emission Control.**

Principle of Operation

Bosch LH fuel injection is completely electronic in operation. Air flow is measured electronically, and a proportional amount of fuel is metered by electrically opening and closing the fuel injectors.

In LH fuel injection, many sensors supply information about engine operating conditions to a central electronic control unit (ECU). The control unit then calculates the amount of fuel needed for the correct air-fuel ratio and opens the fuel injectors, once for each engine revolution. The amount of fuel metered to the engine is determined solely by how long the injectors are open.

There are three versions of LH fuel injection on 16-valve models. Each has the same basic components and operating principles. The differences among the three are mainly in refinements of certain operations and additional functions, introduced on newer models. See **Table a.**

Air Intake. All air entering the engine passes through a pleated paper air filter in the air cleaner. Air flow is controlled by the throttle valve in the throttle housing. Along with the throttle valve, the housing contains idle air passages, connec-

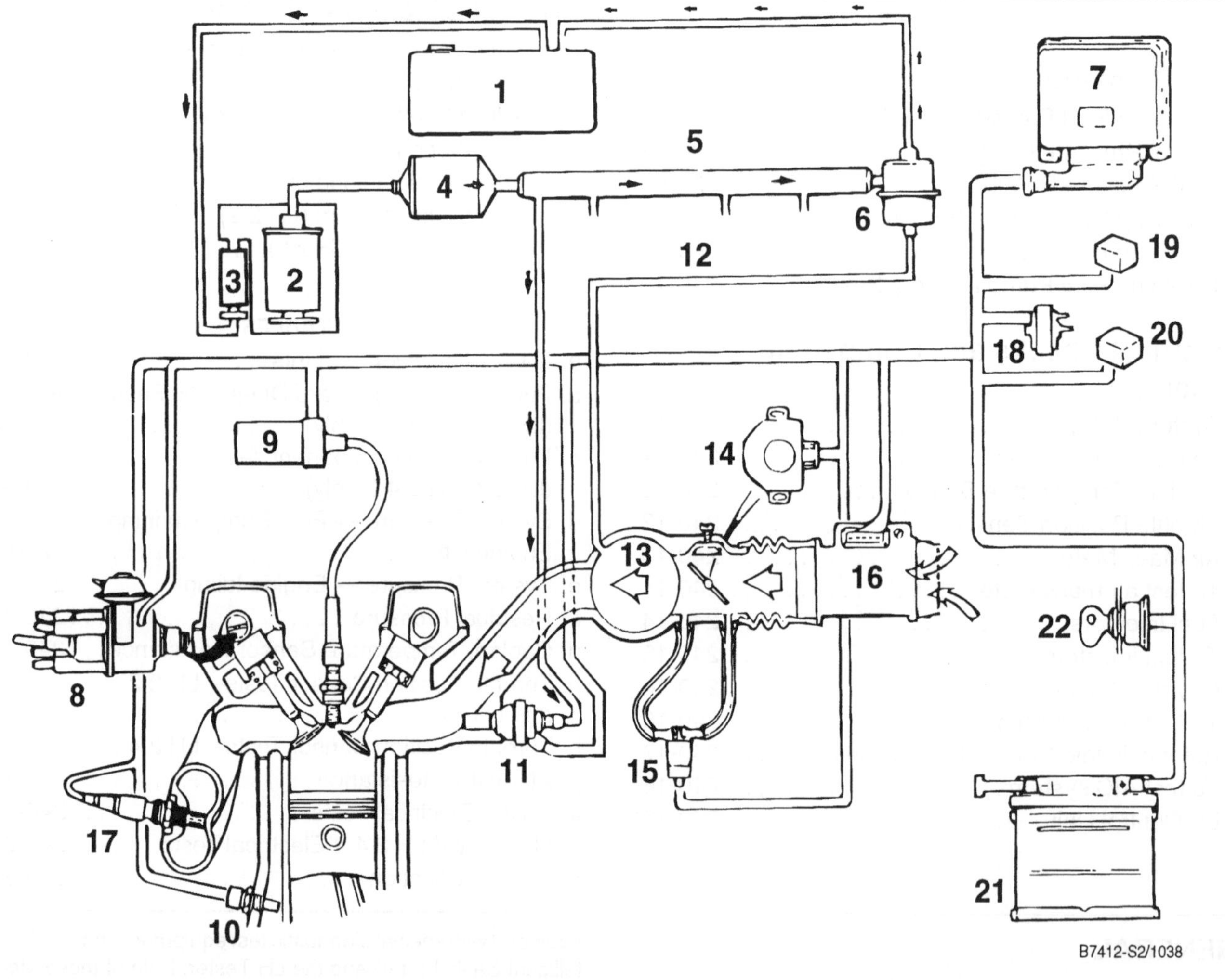

1. Fuel tank
2. Fuel pump
3. Supply pump (Bosch pump only)
4. Fuel filter
5. Fuel rail
6. Fuel pressure regulator
7. LH control unit
8. Ignition distributor
9. Ignition coil
10. Coolant temperature sensor
11. Fuel injector
12. Vacuum line
13. Intake manifold
14. Throttle position sensor
15. AIC (Automatic Idle Control) valve
16. Air mass meter
17. Oxygen sensor
18. Over-pressure switch (1985-1988 turbo only)
19. System relay
20. Fuel pump relay
21. Battery
22. Ignition switch

Fig. 1. Schematic view of LH fuel injection.

Table a. LH Fuel Injection Variants

System	Year/Model	Identification
LH 2.2	1985-1988 Turbo	Three-wire AIC idle valve, metal air mass meter with a sealed mixture adjustment screw, and a throttle dashpot
	1986-1987 Non-turbo	
LH 2.4	1989-1992 Turbo	Two-wire AIC idle valve, plastic air mass meter without a mixture adjustment screw
	1988-1990 Non-turbo	
LH 2.4.2	1991and later Non-turbo	Three-wire AIC idle valve, similar plastic air mass meter without a mixture adjustment screw, and a throttle potentiometer.

tions for the Adaptive Idle Control (AIC) valve, and vacuum connections. The housing is connected to the hot-wire air mass meter by flexible rubber ducts. Air entering the engine is measured by the air mass meter. The meter has no moving parts. Instead, the meter measures air flow electronically and generates a voltage signal. This signal is then used by the control unit to determine how much fuel to inject.

Fuel Metering. The control unit meters fuel by changing the opening time (pulse time) of the fuel injectors. To ensure that injector pulse time is the only factor that determines fuel metering, fuel pressure is precisely controlled by a fuel pressure regulator. The injectors are mounted to a common fuel supply called the fuel rail. The control unit monitors engine speed (ignition pulse) to determine the rate of injector openings. Other signals to the control unit help determine injector pulse time for different operating conditions. A temperature sensor signals engine temperature for cold-start and warm-up enrichment. A throttle position sensor signals throttle position for full-throttle and idle. An oxygen sensor signals information about combustion efficiency for control of the air-fuel mixture.

Automatic Idle Control (AIC). Idle speed is electronically controlled. The LH control unit controls an idle valve that allows air to bypass the closed throttle valve. This bypass control is known as the AIC system. The control unit makes continual changes to the valve opening to adjust idle speed based on engine operating conditions. The bypass air passage is upstream of the air mass meter, so mixture is not affected by changing the idle speed.

LH 2.4 Fuel Injection. Both the AIC and mixture control systems are adaptive on this system. That means that each system compensates automatically for changes in the engine due to age or small problems. As a result, idle speed and mixture do not need to be adjusted. A deceleration fuel shut-off function in the LH 2.4 control unit replaces the deceleration dashpot used on LH 2.2. An integrated Evaporative Loss Control Device (ELCD) system is used on LH 2.4. Operation of the ELCD valve is controlled by the LH control unit. Another main feature of the LH 2.4 system is that it has on-board diagnostics. See **On-board Diagnostics** for more information.

LH 2.4.2 Fuel Injection This fuel injection system is used only on 900 models with the 2.1 liter engine. LH 2.4.2 has all of the features of the earlier LH 2.4 system, plus additional refinements. The system uses a new AIC valve. A throttle potentiometer replaces the throttle switch, and the control unit has been reprogrammed to modify the control of many other functions, including additional diagnostic codes.

NOTE

Unless otherwise noted, all procedures and specifications in this chapter apply to all LH fuel injection variants. Where distinction is important, the systems will be referred to by their alpha-numeric codes—LH 2.2, LH 2.4, LH 2.4.2.

Safety Precautions

The following warnings and caution should be adhered to whenever doing work on the fuel injection system.

WARNING

- Fuel will be discharged during many fuel system test procedures. Do not smoke or work near heaters or other fire hazards. Have a fire extinguisher handy. Work only in a well-ventilated area.
- Wear suitable hand protection, as prolonged contact with fuel can cause illnesses and skin disorders.

CAUTION

- Connect and disconnect wires and test equipment only with the ignition off.
- Before making any electrical tests that require the engine to be cranked using the starter, disable the ignition system as described in **340 Ignition System**
- On models with LH 2.4 and LH 2.4.2, always wait at least 40 seconds after turning off the ignition before removing the control unit connector. If the connector is removed before this time, residual power in the system relay may damage the control unit.
- Cleanliness is essential when working with parts of the fuel system open. Thoroughly clean fuel line connections and surrounding areas before loosening. Avoid the use of compressed air, and avoid moving the car. Only install clean parts.

CAUTION

Fuel system cleaners and other chemical additives other than those specifically recommended by Saab may seriously damage the catalytic converter, the oxygen sensor and the plastic gas tank.

FAULT DIAGNOSIS

The management of engine functions is controlled by a number of systems—ignition, fuel injection, and emission control. Because these functions are interrelated, it is difficult if not impossible to isolate general driveability problems by examining components of the fuel injection system alone. For this reason, engine management and driveability troubleshooting information can found in **200 Engine—General.** This information is organized to help isolate problems and suggest more specific troubleshooting steps by taking all of the interrelated systems into consideration.

Fig. 2 shows an overall view of the engine compartment to help you identify component locations for testing and replacement.

Basics Requirements for LH Fuel Injection

The following list contains basic checks that should be made when experiencing fuel injection problems. Again, if in doubt of whether the fault is caused by the fuel system or

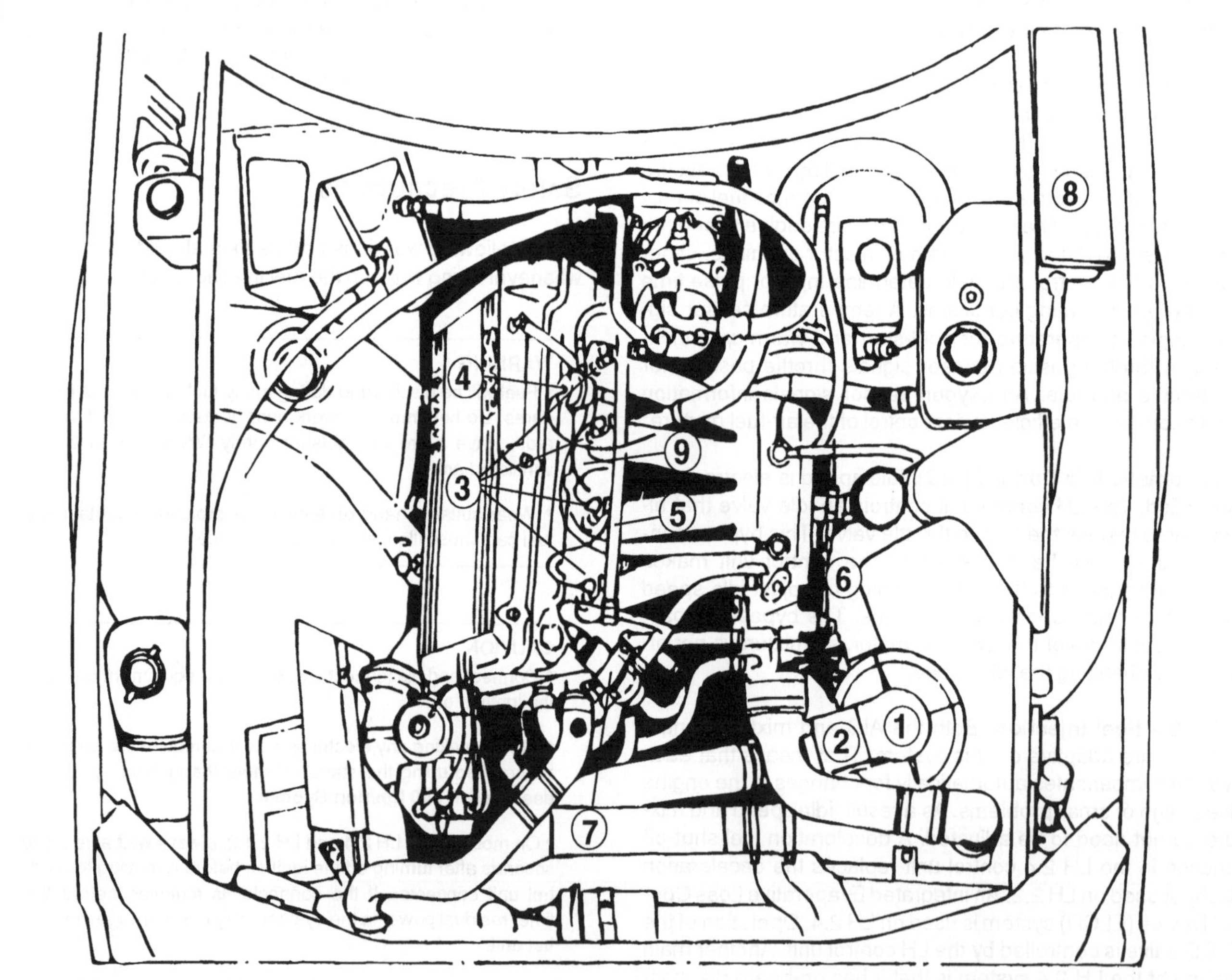

1. Air cleaner
2. Air mass meter
3. Fuel injectors
4. Fuel pressure regulator
5. Fuel rail
6. Throttle housing
7. AIC idle valve
8. Fuse/relay panel
9. Intake manifold

Fig. 2. Typical view of engine compartment from front of car.

some other system such as the ignition system, consult **200 Engine—General** first.

1. Check the intake (induction) system for leaks. Check for cracked, loose, or disconnected hoses and duct work. Check that all hose clamps are tight. On Turbo cars, be sure to check the hoses and hose clamps at the turbocharger by-pass valve. See **291 Turbocharger.**

NOTE

An air leak allows unmeasured air to enter the engine, often resulting in an in overly lean fuel mixture and causing driveability problems.

---

2. Check that the battery is in good condition. Check that the battery cables are tight and free of corrosion. Check that all related ground points are firmly connected and in good condition. Don't forget to check the main ground at the battery. Check all of the harness connectors for damage and corrosion. Check for power and ground at the LH control unit. See **371 Wiring Diagrams, Fuses and Relays.**
3. Check the fuses.
4. Check for sufficient fuel in the tank. If the engine ran out of fuel it takes up to two gallons to prime the pump. See **234 Fuel Pump and Fuel Tank.**
5. Check for spark at the spark plugs. If the tachometer needle bounces while the engine is cranked by the starter then the ignition system is probably working correctly. See **340 Ignition System.**
6. On cars with LH 2.4 and LH 2.4.2, check for any faults through the on-board diagnostics system (LH 2.4 and LH 2.4.2 only) as described below.

On-board Diagnostics (LH 2.4 and LH 2.4.2 only)

LH 2.4 and LH 2.4.2 fuel injection systems feature built-in diagnostic circuitry that detects and stores coded fault information in the LH control unit's memory. When the system compensates for values that are outside the permitted limits, a fault code is generated. The fault codes can be "read-out" through the **Check Engine** light.

The system has the ability to store up to three fault codes, including intermittent faults. The system also has a built-in test cycle that can actively test all major fuel injection components and their signals.

NOTE

- As a general rule, a steady-on Check Engine light means the LH control unit has detected a fault. A flashing Check Engine light means that the EZK Ignition system has detected a fault. As an exception to the rule, the check engine light on 1992 models does not flash. Therefore, a steady-on Check Engine light on these models could be either an EZK system or an LH system fault.

- LH 2.2 is not equipped with LH on-board diagnostics. Although these models are equipped with a check engine light, it is primarily used for EZK ignition system fault detection. On LH 2.2, the Check Engine light will come on in the event of an air mass meter fault and the engine will go into a "limp-home" mode.

The fault code display is triggered by grounding a pin at the LH control unit test connector. A special Saab test harness (tool no. 83 93 886), available from an authorized Saab dealer, should be used to make the test. As an alternative, a switched test lead can be constructed using simple electrical components. Be sure to make the lead long enough to reach the driver's seat.

The following on-board diagnostic codes are broken into two parts—one for LH faults and the other for component and signal testing.

To display LH fault codes

1. With the ignition key off and the test lead switch off, connect one end of the lead to ground and the other end to the connector terminal shown in Fig. 3.

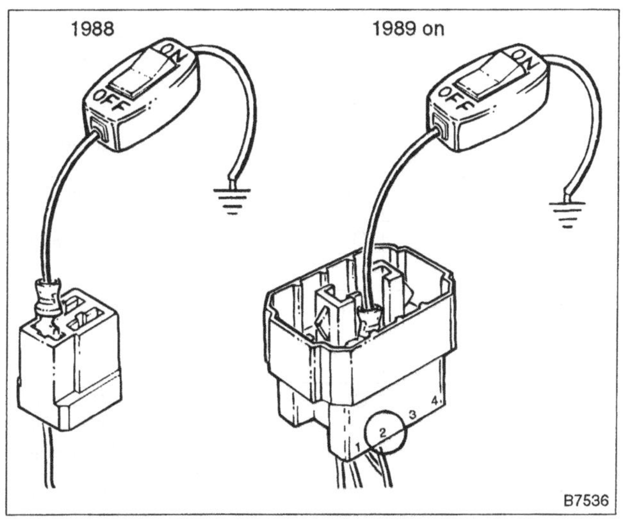

Fig. 3. Switch lead shown connected to test connector. On 1988 models the 3-pin connector is in the engine compartment, near the fresh air intake. On 1989 and later sedan models the 10-pin connector is beneath the rear seat. On 1989 and later convertible models, the 10-pin connector is under the bellows ahead of the shifter.

CAUTION
Carry out the tests exactly as shown. An extra few seconds at the wrong time can clear the fault memory.

2. With the engine off and the test lead switch OFF, turn the ignition key to the ON position.

3. When the Check Engine light comes on, turn the test lead switch ON. The light should immediately go out.

4. Wait until the Check Engine light flashes once and then immediately turn the test lead switch OFF.

5. A fault code (if present) will begin to flash through the Check Engine light. There is a long flash at the beginning and end of each code. The code itself is a series of short flashes. Once the code is fully displayed it will begin to repeat itself until the next code is called up. **Table b** lists fault codes, their probable faults, and corrective actions.

Table b. Fuel Injection On-board Diagnostic Fault Codes (LH 2.4, LH 2.4.2 only)

Error code	Check Engine light in normal driving	Probable fault	Corrective action, first check
12231	Off	No rpm signal (this code will always be present if the test is run with the engine off). Crank the engine for 5 seconds. If the code disappears, the rpm signal is O.K.	Check fuse no. 3. Check the ignition system input at pin 1 of the LH control unit
12221	On	No signal from air mass meter (LH system is in limp-home mode)	Check air mass meter connections. Check for air mass meter burn-off. (Do not interchange air mass meter from LH 2.2—component damage may result)
12214	On	Coolant temperature signal incorrect (temperature signal out of range: below -90°C or above 160°C)	Test coolant temperature sensor. Make resistance test at pin 13 of LH control unit and ground. (Resistance should be approx. 2280-2720 ohms at 68°F (20°C) and 290-365 ohms at 176°F (80°C))
12211	Off	Incorrect battery voltage (below 10 volts or over 16 volts with engine running)	Check charging system, battery connections, grounds, etc.
12225	On	Incorrect oxygen sensor output. Sensor pre-heater circuit may be open. (Engine temp. must be above 176°F for 1988 model, or above 158°F for 1989 and later model)	Check oxygen sensor output voltage (voltage fluctuating between 0 and 1.0 volt). Check oxygen sensor pre-heater circuit and fuse
12223	On	Air-fuel mixture too lean	Tighten all boot clamps and carefully check all hose connections. Check injector O-rings. Check ELCD valve.Check fuel pressure. (Also see fault code 12225 above)
12241	On	Air-fuel mixture fault	
12224	On	Air-fuel mixture too rich	
12233	On	LH control unit fault (ROM mixture fault)	Replace LH control unit
12242	On	Air mass meter burn-off function fault	Check air mass meter wiring and burn-off function
12243	Off	No signal from road speed sensor	Check signal output from road speed sensor on speedometer. Check for power and ground supply at sensor
12244	Off	A/T "Drive" signal faulty. (signal absent while driving or constantly present even when stopped)	Check the fuse for the reverse lights, the reverse light switch, the wiring connections, etc
12245	On	EGR temperature too low (where applicable)	Check function of EGR system. See **254 Exhaust Emission Control**
12232	Off	Memory voltage below 1 volt	Check for battery voltage at pin 4 of the LH control unit with the ignition off
12212	Off	Idle contacts in throttle position switch shorted, closed throttle switch while driving (LH 2.4 only)	Check and adjust throttle position switch. Check wiring harness for short to ground

continued on next page

Table b. Fuel Injection On-board Diagnostic Fault Codes (LH 2.4, LH 2.4.2 only) (continued)

Error code	Check Engine light in normal driving	Probable fault	Corrective action, first check
12251	Off	Incorrect or missing signal from throttle potentiometer (LH 2.4.2 only)	Check throttle potentiometer function (harness disconnected at potentiometer)
12213	Off	Full throttle contacts in throttle position switch shorted, full throttle signal at idle (LH 2.4 only)	Check and adjust throttle position switch. Check wiring harness for short to ground
12222	Off	AIC valve circuit fault	Check electrical connections between LH control unit and AIC valve
12111	Off	Oxygen sensor self-compensating circuit problem (incorrect air-fuel ratio while driving)	Check for air fuel leaks. Check the oxygen sensor pre-heater circuit and sensor output. Check ELCD system. Check fuel pressures
12112	Off	Oxygen sensor self-compensating circuit problem (incorrect air-fuel ratio at idle)	
12113	Off	AIC self-compensating circuit problem (system unable to reduce idle speed to correct level)	Check throttle stop screw setting. Check throttle plate for an over-center condition. Check for vacuum leaks
12114	Off	AIC self-compensating fault. (system unable to increase idle to an acceptable level)	Check for binding AIC valve. Check for a mechanical engine problem causing low idle problem
00000	Off	No more faults	No corrective action necessary. This code must be present before fault memory can be erased

NOTE

As an example, if the test is run with the engine off, the first code displayed would be for "no RPM signal" (fault code 1-2-2-3-1). This would be viewed as a long flash followed by a single flash, two flashes, two flashes, three flashes, and ending with a single flash. The light would then come one for a long flash and the code would begin to repeat itself. See Fig. 4.

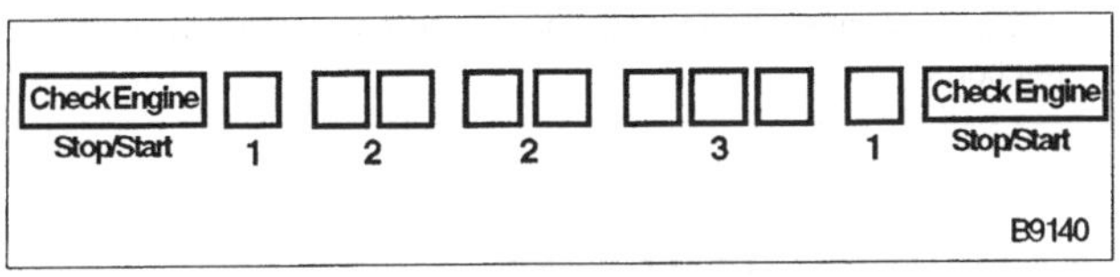

Fig. 4. Graphic representation of flashing Check Engine light.

6. To cycle to the next fault code, simply turn the test lead switch ON, wait until the Check Engine light flashes once, and then turn the switch OFF. Any new codes will begin to flash.

NOTE

The test sequence can be restarted by allowing the Check Engine light to flash twice before turning the switch OFF. Do not let the light flash three times, as this can cancel the fault memory.

To erase fault code memory

Before erasing the memory, the error code 00000 ("no more faults") listed in **Table b** must be present.

NOTE

After making repairs, it takes a certain amount of time for the LH system to re-adapt or compensate for the new conditions. Usually ten minutes of driving, allowing the engine to reach operating temperature, should be sufficient. Recheck for the "no more faults" code only after making a test drive to prevent the same error codes from reoccurring.

1. With the engine off, the test lead switch OFF, and no faults codes present as described above, turn the ignition key to the on position.
2. As soon as the Check Engine light comes on, turn the test lead switch ON. The light should immediately go out.
3. Wait until the Check Engine light flashes three times and then turn the test lead switch OFF. The fault memory will now be erased.

To test components and signals

The LH 2.4 and LH 2.4.2 control unit has a built-in component test that actively tests all major fuel injection components. The test will also display a code through the Check Engine light.

1. On cars with automatic transmission, place the transmission in "D".

2. With the ignition key off and the test lead switch off, connect the switched test lead as shown above in Fig. 3.

3. Turn the test lead switch ON.

4. Turn the ignition key ON and wait for a short flash of the Check Engine light, then immediately turn the test lead switch OFF.

NOTE

As soon as the check engine light flashes, the fuel pump should run briefly (provided the pump is functioning correctly). This is the first component tested, verifying that the component and signal diagnosis program has been accessed. The Check Engine light will not flash a test code during the fuel pump test.

---

5. To move on to the next component test (and test code), turn the test lead switch on. After a short flash, turn the switch OFF. The fuel injectors should be clicking on and off.

6. Proceed through the remaining tests using the same procedure of turning the test lead switch on, waiting for a short flash and turning the switch OFF. **Table c** lists the test sequence, the test code, the signal or component being tested, and the operational check being made.

FUEL PRESSURE TESTS

Checking the fuel pressure is a fundamental part of troubleshooting and diagnosing the system. An accurate fuel pressure gauge will be needed to make the tests described below.

There are two significant fuel pressure values: 1) System pressure—created by the main fuel pump and maintained by the pressure regulator, and 2) Residual pressure—the pressure maintained in the closed system after the engine (and fuel pump) are shut off.

Relieving Fuel Pressure and Connecting Fuel Pressure Gauge

Much of the work on fuel injection requires the disconnecting of fuel lines. To prevent fuel from spraying all over, especially on a hot engine, it is a good idea to relieve the pressure in the lines before fully disconnecting them.

One basic method of relieving the pressure is to remove the fuses for the fuel pump and then run the engine until it stalls. Another method is to connect a hand vacuum pump to the vacuum connection on the fuel pressure regulator. When vacuum is applied to the regulator it will open and dump fuel back to the tank.

WARNING

Fuel will be discharged during many fuel system tests described below. Do not smoke or work near heaters or other fire hazards. Have a fire extinguisher handy.

Basic system troubleshooting requires a pressure gauge to measure fuel pressure under different operating conditions. To install the pressure gauge, remove the banjo bolt holding the fuel supply line to the fuel rail. See Fig. 5. Then reinstall the supply line and the gauge fitting.

NOTE

Use Saab pressure gauge Part No. 83 93 852, or an equivalent that has a range of at least 0 to 5 bar (approximately 0 to 100 psi). Always use new sealing washers when making connections.

Table c. Fuel Injection Component Test (LH 2.4, LH 2.4.2 only)

Test sequence	Test Code	Component/signal being checked	Operational check
1	No light display	Fuel pump circuit	Listen for pump running or fuel flow in the engine compartment
2	12411	Injector circuit	Listen for clicking injectors
3	12412	AIC actuator circuit	AIC valve will audibly cycle between open and closed once a second
4	12413	ELCD (charcoal canister purge valve)	ELCD valve audibly clicks once a second
5	12421	Automatic transmission drive position signal	Flashing stops when shifting from drive to neutral
6	12424	Closed throttle signal from throttle position switch	Flashing stops when throttle is opened
7	12431	Full throttle signal from throttle position switch	Flashing stops as accelerator pedal approaches full throttle

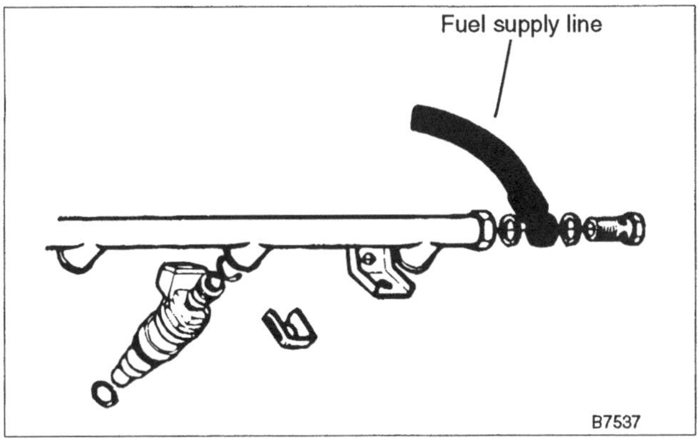

Fig. 5. Fuel pressure gauge installation location at fuel supply line on front of fuel rail.

CAUTION
Absolute cleanliness is essential when working with fuel circuit components. Even a minute particle of dirt can cause trouble if it reaches an injector. Thoroughly clean the unions before disconnecting any fuel lines. Use clean tools.

Fuel Pressure Regulator

The fuel pressure regulator is mounted to the end of the fuel rail and is spring operated. See Fig.6 and Fig. 7. Fuel pump pressure presses on the regulator valve diaphragm. The valve opens at a set pressure and returns excess fuel to the tank. When the pump shuts off, the regulator valve closes, holding residual pressure in the fuel system. This helps prevent fuel vaporization in the lines and possible vapor lock.

The pressure regulator is connected to the intake manifold through a vacuum hose. As manifold pressure changes, it

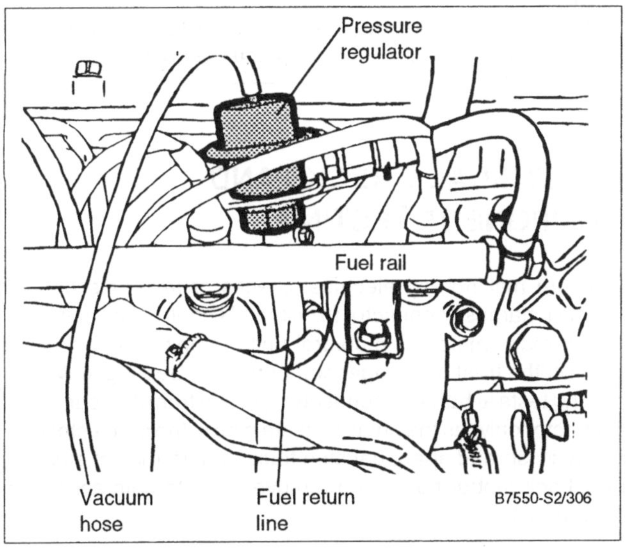

Fig. 6. Fuel pressure regulator.

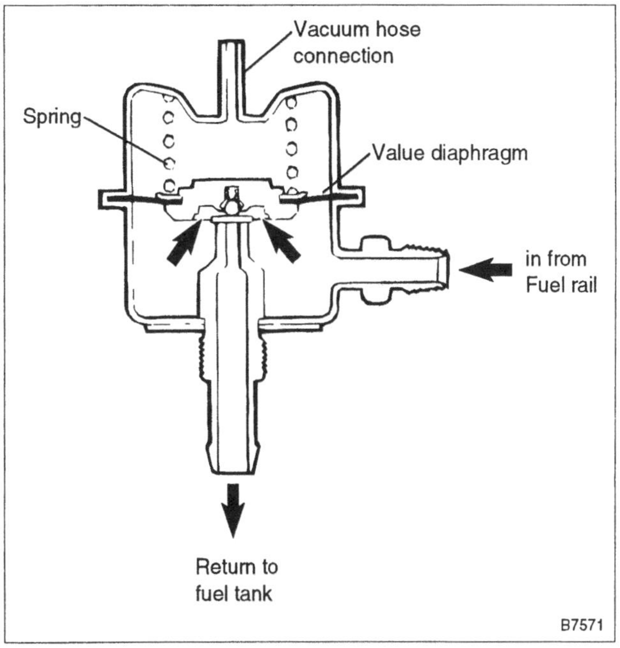

Fig. 7. Cutaway view of LH fuel pressure regulator.

acts on the pressure regulator diaphragm to increase or decrease fuel pressure in the same ratio. Thus the pressure differential between fuel pressure in the system and manifold pressure is constant. This ensures that the amount of fuel injected is influenced only by injector opening time, and not by intake manifold pressure.

Checking System Line Pressure

System pressure is checked in two steps. The first step checks basic operation of the pressure regulator. The second step checks the regulator's response to manifold pressure.

With the fuel gauge connected as described above and the engine off, run the fuel pump by bypassing the fuel pump relay as described in **234 Fuel Pump and Fuel Tank.** System pressure specifications are given in **Table d.**

Table d. System Pressure — Fuel Pump Running, Engine Off

Model	Fuel Pressure
Non-Turbo models	3.0 bar (43.5 psi)
Turbo models	25±.05 bar (36.3±0.7 psi)

NOTE
Pressure specifications give fuel pressure relative to normal atmospheric pressure (1 bar/14.5 psi) in the intake manifold.

If system pressure is too high, first check for a blocked or damaged fuel return line from the pressure regulator. Also check the return line check valve. If there are no obstructions, replace the pressure regulator.

NOTE
The fuel return line check valve is either near the fuel tank or in the top of the fuel tank depending on the type of tank installed. See **234 Fuel Pump and Fuel Tank.**

If system pressure is too low, first check for fuel leaks. Test the fuel pump as described in **234 Fuel Pump and Fuel Tank.** A restricted fuel filter or a damaged fuel line are also potential causes. If there are no other faults, replace the pressure regulator.

To check pressure regulator response to manifold pressure, reinstall the fuel pump fuse and then start the engine. With the engine idling, system pressure should be slightly lower than with the engine off. See **Table e.** If not, check the regulator vacuum hose and its connection to the intake manifold. If there are no faults, the regulator is faulty and should be replaced.

Table e. System Pressure — Engine Idling

Model	Fuel Pressure
Non-Turbo models	2.4 bar (34.8 psi)
Turbo models	1.9 bar (27.6 psi)

NOTE
- These pressures are approximate, and based on a manifold pressure of –0.6 bar (–8.7 psi) at idle. Allowances should be made for altitude and other factors influencing manifold pressure.
- The most accurate test of pressure regulator response is made by using a hand vacuum pump and pressure pump (Turbo models) to apply vacuum/pressure to the regulator with the fuel pump running. Fuel pressure should always be lesser/greater than system pressure by the same amount of vacuum/pressure that is applied.

Checking Residual Pressure

Hot-start problems are usually the only reason to suspect a residual pressure problem, but it deserves a routine check as long as the gauge is connected.

With the engine off, run the fuel pump for about 30 seconds or so to pressurize the system. Turn the pump off and watch the pressure gauge. **Table f** gives the specification.

If pressure drops beyond specification, check for leaks in the fuel lines, at the fuel pump, and at the fuel injectors. If there are no leaks, then either the pressure regulator or the fuel pump check valve is faulty.

Table f. Residual Pressure

All models	0.1–0.2 bar (1.5–3 psi) below system pressure, maximum after 10 minutes

To determine which, turn the pump on and off again, then immediately clamp shut the fuel return line at the pressure regulator. If pressure still drops, then the fuel pump check valve is faulty. See **234 Fuel Pump and Fuel Tank** for replacement. If residual pressure is now OK, then the pressure regulator is faulty.

Checking Fuel Pump Delivery Pressure

The fuel pump is capable of developing a much higher pressure than that regulated by the pressure regulator. The fuel pump has an internal check-valve that will not allow the system pressure to exceed 6.0 bar (87 psi).

To check fuel pump delivery pressure, run the fuel pump with the engine off. Then briefly pinch shut the return line at the pressure regulator. The fuel pressure should quickly rise. If the pressure does not respond quickly, and no leaks can be found, the fuel pump is worn and should be replaced.

CAUTION
In the event the fuel pump check valve is faulty (stuck closed), make sure the fuel pressure does not rise above 6.0 bar (87 psi). Damage to the fuel lines or fuel system components could result.

NOTE
Additional fuel pump tests, including testing the fuel pump relay, are covered in **234 Fuel Pump and Fuel Tank.**

ELECTRICAL CHECKS AND COMPONENT TESTING

Before making any electrical checks or tests, disconnect the main harness connector from the LH control unit and from the air mass meter. The control unit is behind the side kick panel in the front passenger compartment. See Fig. 8. When making tests at harness connectors, probe from the rear of the connector only, as the small delicate pins in the connector can be easily spread apart, causing faulty or intermittent signals. Peel back rubber boots and remove seals to gain access to the rear of the connector.

CAUTION
- Avoid damaging harness connectors or relay panel sockets. Jumper wires should end with flat-blade connectors that are the same size as the connector or relay terminals.
- Always switch a meter to the appropriate function and range before making test connections.
- Connect and disconnect test equipment only with the ignition switched off.

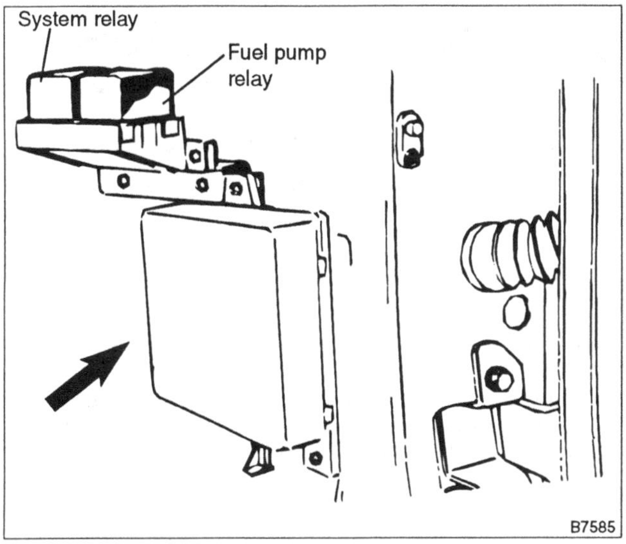

Fig. 8. LH control unit (arrow) behind kick panel in passenger compartment ahead of front door.

System Relay

The system relay supplies power to the LH control unit and the fuel pump relay when the ignition key is in the run position. On later models with LH 2.4, the system relay also supplies power to the fuel injectors, the air mass meter, the AIC idle valve, and the charcoal canister purge valve (ELCD). If this relay is faulty the engine will not start.

To test system relay

1. With the ignition key off, remove the harness connector from the air mass meter and the LH control unit.
2. Working at the LH control unit connector, check for ground at pin 11 on LH 2.2 or at pin 17 on LH 2.4 and LH 2.4.2. Repair any wiring faults using the appropriate wiring diagram shown in **371 Wiring Diagrams, Fuses and Relays.**

CAUTION
On models with LH 2.4 and LH 2.4.2, always wait at least 40 seconds after turning off the ignition before removing the control unit connector. If the connector is removed before this time, residual power in the system relay may damage the control unit.

3. Connect pin 21 in the LH control unit connector to ground. Turn the ignition key on and check for voltage at pin 9 of the LH control unit connector. See Fig. 9. Check for voltage at the air mass meter connector; on cars with LH 2.2 check for voltage at pin 2 and on LH 2.4 and LH 2.4.2, check for voltage at pin 5.

NOTE
If no faults are found, the system relay is working correctly. If the voltage is not as specified, continue testing as described below.

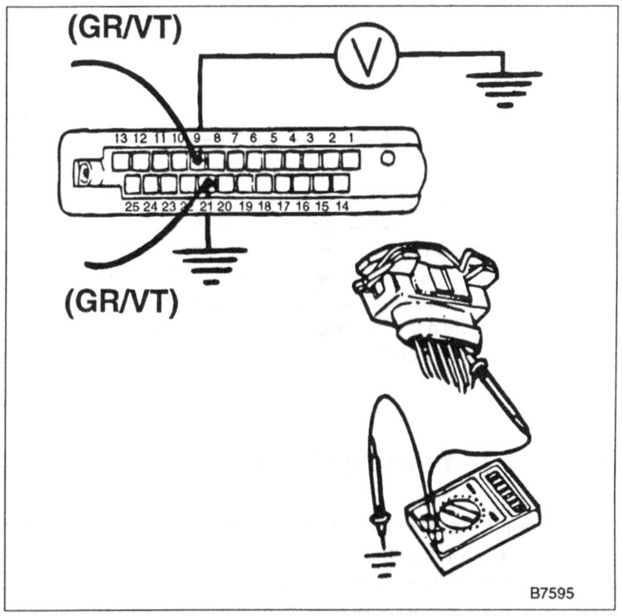

Fig. 9. Voltage being measured at air mass meter connector (pin 2 on LH 2.2 or pin 5 on LH 2.4 and LH 2.4.2) and LH control unit connector (pin 9).

4. Remove the system relay and check for voltage at terminal 30 and terminal 86 of the relay socket. See Fig. 10. If any faults are found, check the wiring between the battery and the relay sockets using the appropriate wiring diagram.

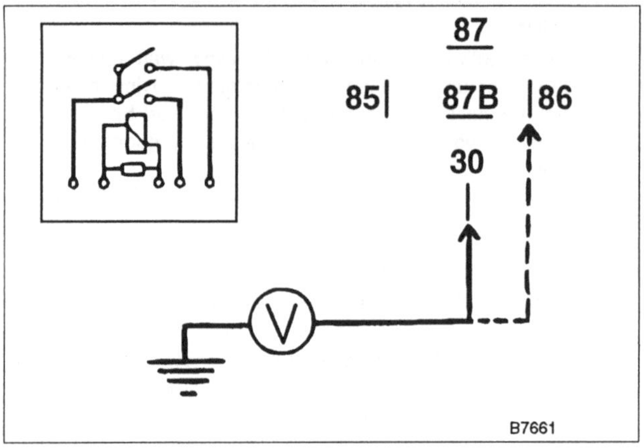

Fig. 10. Voltage being checked at system relay terminals 30 and 86.

5. If voltage is as specified in step 3, check the wiring between the system relay sockets and the following components; the LH control unit connector (pin 21 and pin 9), the air mass meter connector (pin 2 on LH 2.2 or pin 5 on all others), and the fuel pump relay (terminal 86). If no wiring faults can be found, the system relay is probably faulty.

NOTE

On 1985 through 1988 turbo models, a turbo over-pressure switch interrupts power to the fuel pump relay (terminal 86) when the turbocharger's boost pressure gets too high. Check for continuity in the wire between terminal 87 of the system relay socket and terminal 86 of the fuel pump relay socket. If there is no continuity, test the turbo over-pressure switch as described in **234 Fuel Pump and Fuel Tank.**

Coolant Temperature Sensor

The coolant temperature sensor sends continuous engine temperature information to the LH control unit. If there is break in the signal from the temperature sensor, the ECU will simulate a fixed signal, and the system will continue to operate based on an engine temperature of 68°F (20°C) on cars with LH 2.2 or 113°F (45°C) on cars with LH 2.4 and LH 2.4.2.

The sensor is mounted in the intake manifold. See Fig. 11. It projects into the cylinder head to measure engine coolant temperature. The sensor has a negative temperature coefficient (NTC), which means that its resistance goes down as its temperature goes up.

To check the sensor, disconnect the harness connector and check the resistance across the sensor terminals. **Table g** lists sensor resistances based on engine temperature. If any faults are found, the sensor should be replaced as described later under **Component Replacement.**

Fig. 11. Coolant temperature sensor.

NOTE

A faulty coolant temperature sensor will cause a noticeable increase in idle speed, as well as possibly affecting driveability. On LH 2.4 and LH 2.4.2 the Check Engine light should illuminate.

Table g. Coolant Temperature Sensor Resistance

Temperature	Resistance (ohms, ±10%)
–4°F (–20°C)	14,000
14°F (–10°C)	9,000
32°F (0°C)	5,800
50°F (10°C)	3,800
59°F (15°C)	3,000
68°F (20°C)	2,600
76°F (25°C)	2,000
86°F (30 °C)	1,700
176°F (80°C)	320

Throttle Position Sensor

The throttle position sensor is mounted on the side of the throttle housing and is directly connected to the throttle valve shaft. See Fig. 12. The sensor provides a signal to the control unit based on the position of the throttle valve.

On models with LH 2.2 and LH 2.4 fuel injection, there are two switches in the housing that provide a ground signal whenever the throttle is fully closed or fully open.

On models with LH 2.4.2 fuel injection, there is a potentiometer in the housing that changes resistance as the throttle moves. The control unit sends a voltage signal to the position sensor and monitors the voltage that comes back. Resistance decreases (voltage increases) as the throttle opens.

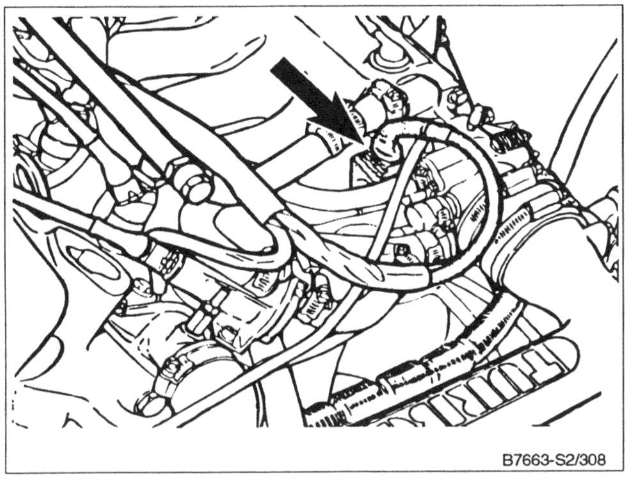

Fig. 12. Throttle position sensor (arrow) mounted to side of throttle housing.

Check throttle position sensor function by disconnecting the harness connector and checking continuity across the terminals while changing the throttle position. See Fig. 13. **Table h** and **Table i** list the correct test results. If the results are incorrect, check sensor adjustment as described below under **LH Fuel Injection Adjustments.** If adjustment does not correct the problem, replace the throttle position sensor.

NOTE

On models with LH 2.4.2, the throttle potentiometer is non-adjustable. If test results are incorrect, replace the throttle position sensor.

Table h. Throttle Position Sensor Tests—LH 2.2 and LH 2.4

Throttle position	Terminals checked	Test results
closed	1 and 2	continuity
	2 and 3	no continuity
half open	1 and 2	no continuity
	2 and 3	no continuity
wide open	1 and 2	no continuity
	2 and 3	continuity

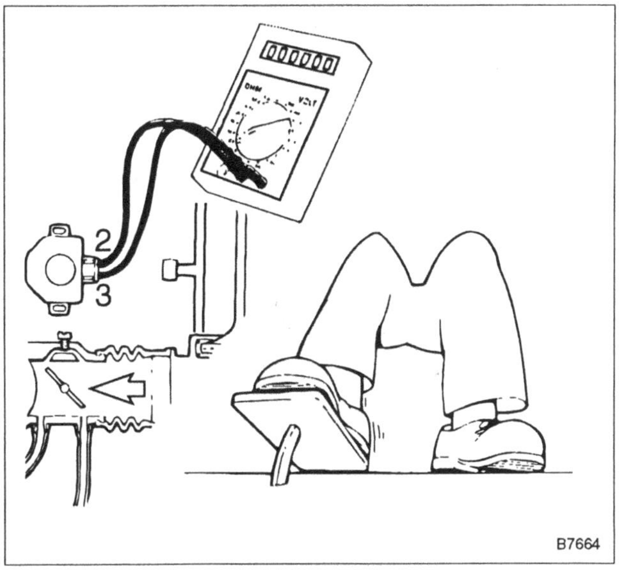

Fig. 13. Throttle position sensor being checked on cars with LH 2.2 and LH 2.4.

Table i. Throttle Position Sensor Tests—LH 2.4.2

Throttle position	Terminals checked	Test results
closed	1 and 3	2600–3000 ohms
wide open	1 and 3	1100–1500 ohms
between closed and wide open	1 and 3	continuously variable between 1100–3000 ohms, with no dropouts

Air Mass Meter

The heart of the LH fuel injection system is the air mass meter. It is located between the air filter and the intake manifold. The air mass meter electronically measures the air entering the engine. The air mass meter has no moving parts.

When the engine is running, a current is used to heat a thin wire in the center of the meter. See Fig. 14. The current flowing through the wire is regulated to maintain a wire temperature that is always 100°C more than that of the cooler air passing over it. The current used to heat the wire is then electronically converted into a precise voltage measurement corresponding to the air entering the engine.

Because the hot wire measures air mass, no other inputs to the control unit are needed for air temperature or altitude. A screen across the meter intake helps protect the wire and also breaks up air flow for more accurate measurements. To keep the wire free from dirt, it is heated to a temperature of about 1000°C (1830°F) for one second. This "burn-off" cycle takes place automatically, four seconds after the engine is turned off.

If the hot wire breaks or if there is no output from the air mass meter, the control unit automatically switches to a "limp-home" mode and turns on the Check Engine light. The engine

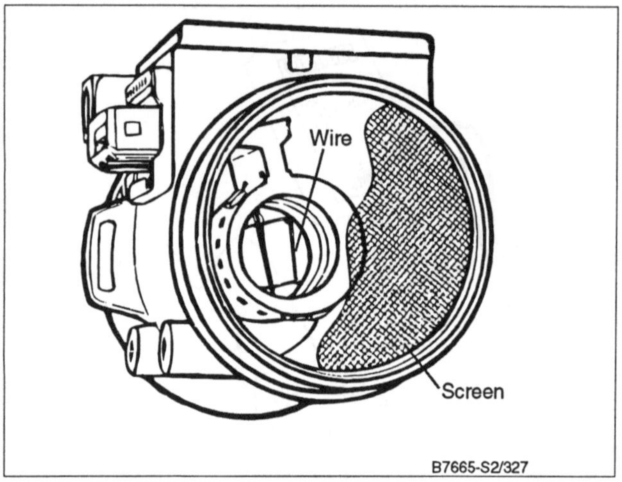

Fig. 14. Air mass meter.

can usually be started and driven. The air mass meter has no internal moving parts and cannot be serviced.

To test air mass meter

1. Disconnect the air mass meter from the air cleaner only. Leave it connected to the duct leading to the intake manifold and leave the wiring harness connected.
2. Start engine and run it to normal operating temperature
3. Rev the engine to at least 2500 rpm, then shut it off. Look through the front of the meter at the hot wire. After approximately four seconds the wire should glow brightly for about one second.

NOTE

If the wire glows as specified, then the air mass meter and control unit are probably operating correctly. If the wire does not glow, continue testing.

---

4. Remove the air mass meter and look through it to see if the wire is broken. It's very small, so check carefully. If the wire is broken, the meter will have to be replaced as described below. If the wire is intact, continue testing.
5. Reinstall the air mass meter and the harness connector. Peel back the rubber boot from the harness connector. Working from the rear of the connector, connect a digital voltmeter across terminals 1 and 4. See Fig. 15.

CAUTION

Use only a digital voltmeter for the above test. An analog meter can damage the air mass meter.

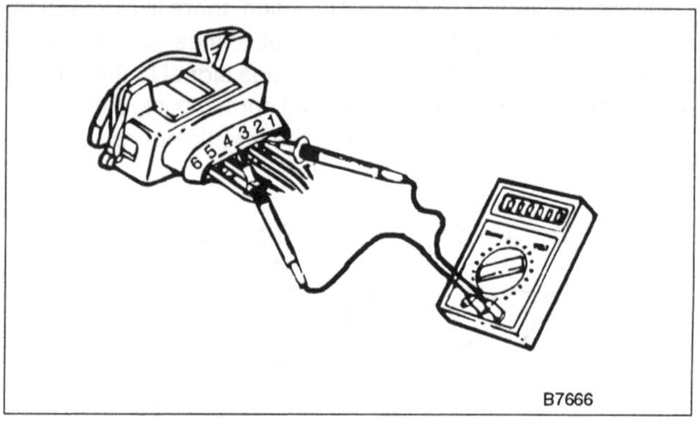

Fig. 15. Voltmeter connected to air mass meter terminals 1 (–) and 4 (+) to check for meter burn-off voltage.

6. Start and rev the engine to at least 2500 rpm, then shut it off. After about 4 seconds, the voltage should rise to about 4 volts for a duration of one second. If the voltage is as specified, but the wire does not glow as described in step 3 above, the air mass meter is faulty and should be replaced.

NOTE

• On models with LH 2.2, a new replacement air mass meter should be adjusted as described below under **LH Fuel Injection Adjustments.** Air mass meters used on LH 2.4 and 2.4.2 do not require any adjustments.

• The air mass meter used on LH 2.2 (aluminum body) is not the same meter used on LH 2.4 and LH 2.4.2 (black plastic body). **Do not interchange these meters.**

---

7. If voltage is not present in step 6, turn the ignition key on and check for voltage and ground at the meter's connector. There should be ground at pin 4. There should be positive (+) battery voltage at pin 2 (LH 2.2) or at pin 5 (LH 2.2, LH2.4.2).

If any faults are found, check for wiring breaks between the air mass meter and the LH control unit and between the air mass meter and the system relay. Also check the operation of the system relay as described above.

Fuel Injectors

The fuel injectors are electrically operated solenoid valves. They are switched on and off (open and closed) by the LH control unit. The injectors are connected to a common fuel supply, called the fuel rail. See Fig. 16.

There are a couple of common injector problems. Carbon deposits that form at the tips of injectors will clog them and reduce fuel flow. This often causes rough running, stalling, and power loss. This problem can usually be reduced and controlled through the use of quality gasolines containing deter-

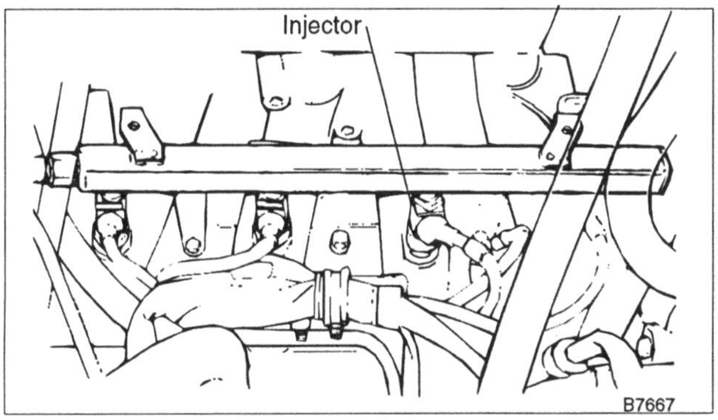

Fig. 16. Fuel rail is common fuel supply and mounting for all four injectors.

gents. If the problem is serious, a Saab dealer or other qualified repair shop has specialized injector cleaning equipment.

A second problem is old and cracked O-rings. They allow unmeasured air to enter the engine. Since the air is unmeasured, additional fuel is not injected and the mixture is leaned. The LH control unit will usually adapt itself to this extra air, but a symptom of this problem is increased fuel consumption. The injectors must be removed as later described under **Component Replacement.**

NOTE

Damaged injector O-rings are just one of many possible causes of increased fuel consumption. For more information see **200 Engine—General.**

To test injectors

NOTE

The information below is an electrical test only. The test does not check the spray patterns or fuel supply to the injectors. A Saab dealer or other qualified repair shop can perform such a test as part of the injector cleaning procedure.

---

1. Disconnect a harness connector from an injector and connect to it an LED test light, an injector tester, or a digital voltmeter across the connector. See Fig. 17. When the engine is cranked, the LED should flash or the following voltage should be present:

Injector cranking voltage (approximate VDC)

- engine temp. 32°F .... 1 volt
- engine temp. 68°F .... 0.6 volt
- engine temp. 176°F .... 0.3 volt

CAUTION

Use only a digital voltmeter, an LED test light or an LED injector tester. Use of an analog VOM or incandescent test light may damage the control unit.

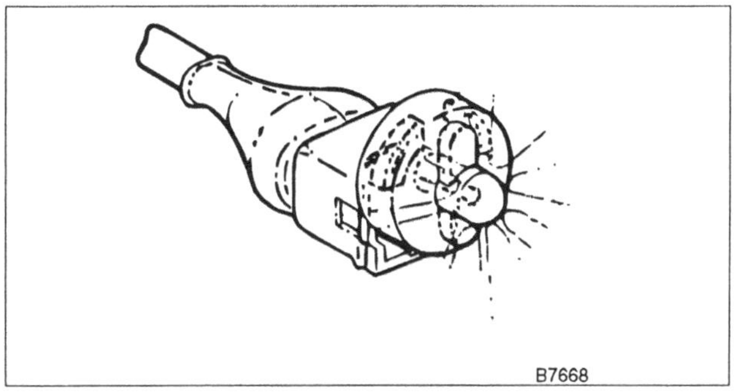

Fig. 17. Special LED test light connected across fuel injector connector for voltage test. Simple LED test light or digital VOM can also be used.

NOTE

If the engine runs but you think one injector is bad, place a screwdriver or an automotive stethoscope on the injector. There should be a buzzing or vibration as the engine runs. If the injector is not buzzing, check for voltage to the injector and check its resistance as described below. If the injector is buzzing, very briefly unplug the harness connector from the injector with the engine running. If the injector is working correctly, there should be a noticeable rpm drop.

---

2. If the light doesn't flash or there is no voltage, check for power to the injector. There should be battery voltage (+) at the blue/red wire of each injector connector with the ignition key on. See Fig. 18. If not, check the wiring to the injector using the wiring diagrams listed in **371 Wiring Diagrams, Fuses and Relays.**

NOTE

If there is positive (+) battery voltage as described in step 2, but there was no response through the LED or voltmeter in step 1, check the wire from the control unit to the injectors. If no wiring faults can be found, the ground signal from the control unit is missing. See **Electrical Checks and Component Testing** for additional electrical tests.

---

3. If power is present as described in the above step, unplug the injector connectors and check the injector resistance. Replace the injector if the resistance is incorrect.

Injectors

- resistance .... 16 ohms at 68°F (20°C)

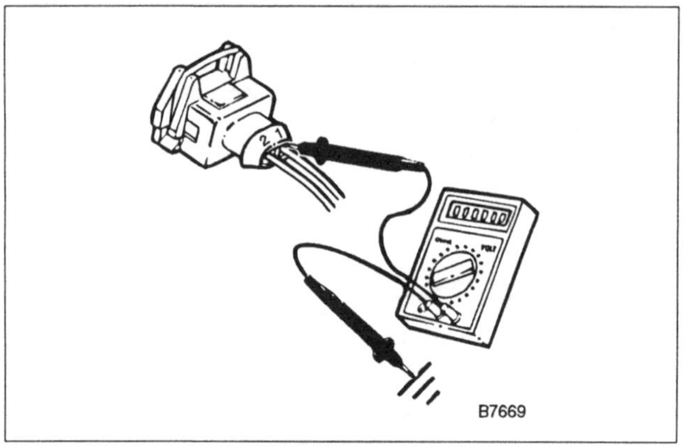

Fig. 18. Checking power to injector with ignition on. Check at red-blue wire of connector.

NOTE

Injector resistance will vary slightly depending on temperature. In general, the range should be between 10 and 20 ohms.

---

4. If no faults are found up to this point, check the pulse-time regulation function of the control unit. Peel back the rubber boot from the injector connector and connect a voltmeter to the wires in the connector. Start the engine. Check that the voltage decreases as the engine warms up. The values given below are approximate.

Injector warm-up enrichment voltage (VDC)

- engine temp. 32°F .... 1 volt
- engine temp. 68°F .... 0.6 volt
- engine temp. 176°F .... 0.3 volt

If the voltage readings are incorrect, check the temperature sensor as described above. Also check the CO reading and the air mass meter burn off function. If these are OK, then either the wiring to the control unit or the control unit itself is faulty. Check the control unit inputs and wiring by making the electrical checks described below under **LH Control Unit.**

Oxygen Sensor

The oxygen sensor monitors the exhaust gas and provides the LH control unit with feedback about the air-fuel ratio and combustion efficiency. Using this information, the control unit continuously adjusts the air-fuel mixture to ensure optimum driveability and exhaust emissions.

The oxygen sensor is mounted in the exhaust system. See Fig. 19. The oxygen sensor produces a small voltage (0-1 volt) based on the oxygen content in the exhaust gas as compared to the oxygen outside the exhaust pipe. The bigger the differential, the greater the output voltage. When the mixture is rich, there is very little oxygen in the exhaust (high output voltage). When the mixture is lean, there is an excess of oxygen in the exhaust (low output voltage).

Fig. 19. Location of oxygen sensor (arrow) on turbo model (in exhaust manifold). On non-turbo model, the oxygen sensor is mounted in the front exhaust pipe, near the steering rack.

The oxygen sensor must be at a temperature of at least 600°F to generate voltage, so it is electrically heated to help it reach operating temperature more quickly. The oxygen sensor is checked by seeing if it generates a voltage when the engine is idling at normal operating temperature.

To test oxygen sensor

1. Disconnect the sensor's single black wire (sensor signal wire). See Fig. 20.

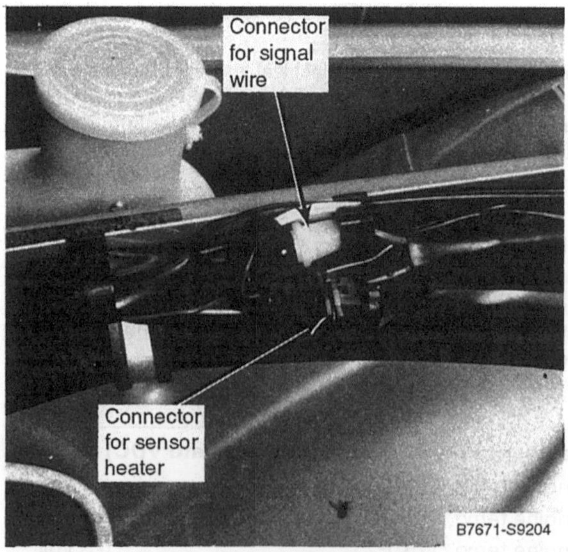

Fig. 20. Oxygen sensor wiring connectors are located on right (passenger) inner fender well, just above battery.

2. Connect the positive lead of a voltmeter to the black wire coming from the sensor and connect the negative meter lead to chassis ground. See Fig. 21. Start the car and let it idle. The oxygen sensor should start to produce a fluctuating voltage within a short period.

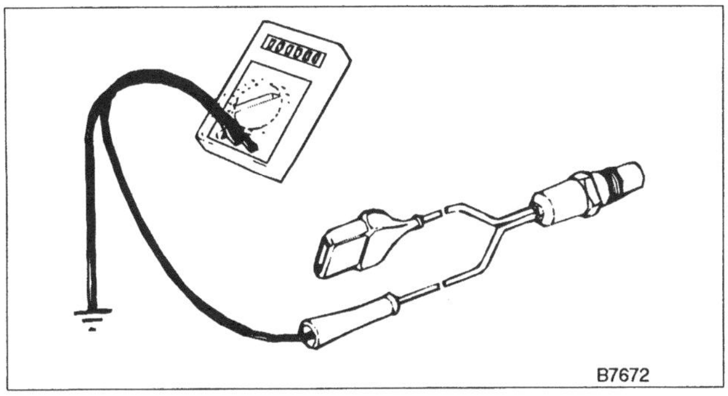

Fig. 21. Oxygen sensor output voltage being checked (shown schematically). Sensor should be installed and the engine should be running.

Oxygen sensor
- voltage at idle ....0.4 to 1 volt

NOTE
To further check sensor response to lean and rich mixtures, slightly open the oil filler cap to create an air leak, or pull the vacuum hose off of the fuel pressure regulator to increase fuel pressure.

3. If the sensor does not emit a fluctuating voltage, turn the engine off and check the sensor's preheater circuit. Disconnect the two-wire connector and check for battery voltage at the connector with the engine running. If voltage is not present, check the wiring to the preheater and check the preheater fuse. If voltage is present, check the preheater coil resistance. See Fig. 22. If any faults are found, replace the sensor.

NOTE
On 1986 through 1988 models, the preheater fuse is in-line, in the engine compartment on the right-hand side fender. On 1989 and later cars, the fuse is in the main fuse/relay panel (position no. 1). On 1985 models, the preheater is not fuse protected.

Oxygen sensor
- preheater resistance ....approx. 4 ohms

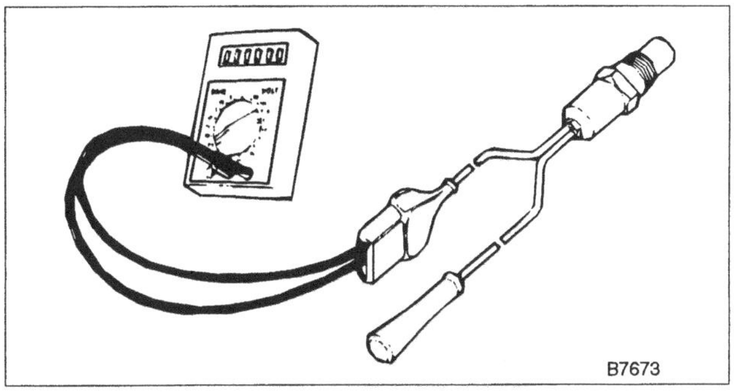

Fig. 22. Oxygen sensor heating coil resistance being measured (shown schematically).

If the oxygen sensor doesn't produce voltage, and the preheater circuit is OK, replace the sensor.

Tightening Torque
- Oxygen sensor .... 40 Nm (29 ft-lb)

NOTE
Coat the oxygen sensor threads with an anti-seize compound before installation. Do not get the compound on the sensor tip. This can damage the sensor.

Automatic Idle Control (AIC) System

The AIC system maintains idle speed through the AIC idle control valve. See Fig. 23. The system compensates for engine load and engine operating conditions, so periodic adjustment of the idle speed is unnecessary. The AIC valve is controlled by the LH control unit.

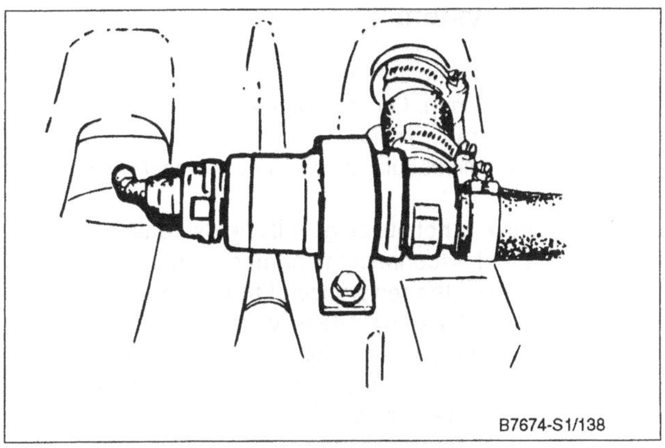

Fig. 23. AIC valve.

The LH control unit monitors engine rpm and other operating conditions such as air conditioning and automatic transmission loads, and then sends an electrical signal to open or close the valve. This changes the amount of air that bypasses the closed throttle valve.

NOTE

There are three different idle control valves used on the cars covered by this manual, depending on the fuel system. These valves operate differently and are not interchangeable between fuel systems.

On LH 2.4 and LH 2.4.2 fuel injection, the AIC function is completely adaptive. The control unit remembers how far the valve was open the last time the engine was at idle and reverts to this as a base setting. In addition, the valves used on these models have a limp-home feature that maintains a fixed idle speed in the event of an AIC valve failure. This idle speed is approximately 1,200 rpm on cars with LH 2.4 and 850 rpm on cars with LH 2.4.2.

Before proceeding with troubleshooting an idle problem, check that the throttle position sensor is working correctly as described earlier under **Throttle Position Sensor**. On cars with high mileage, enough residue and dirt may collect on the throttle valve plate and bore to restrict air flow and affect the AIC system. remove the AIC valve and the air intake duct from the throttle housing. Clean both the AIC valve and the throttle plate and bore. Reinstall the valve and duct and run the engine to check idle.

NOTE

On models with LH 2.2 fuel injection, the base idle may need to be reset after this procedure. See **LH Fuel Injection Adjustments** as described below.

To test AIC system

1. With the engine running, check that the idle control valve is buzzing. On cars with air conditioning (A/C) or automatic transmission, turn on the A/C or shift the car into drive. Idle should remain steady or increase slightly.

2. If the valve is not buzzing, or if idle decreases, stop the engine and disconnect the harness connector from the valve. Check the resistance of the valve across its terminals. Fig. 24 shows the valve terminals. **Table j** lists the test terminals and results. If resistance is incorrect, replace the valve. If resistance is OK, test for power to the valve as described below.

NOTE

If you suspect an intermittent fault, lightly tap the valve while you are testing resistance.

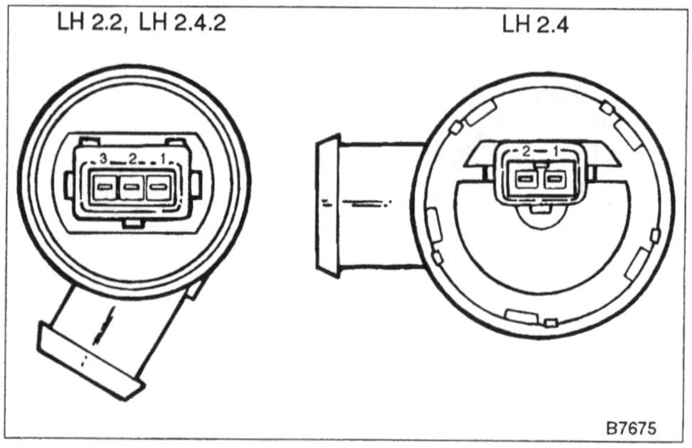

Fig. 24. Idle control valve terminals.

Table j. AIC Valve Resistance

Fuel System	Test terminals	Resistance (ohms)
LH 2.2	1 and 2	20±2
	2 and 3	20±2
	1 and 3	40±4
LH 2.4	1 and 2	8±2
LH 2.4.2	1 and 2	10–15
	2 and 3	10–15

3. With the valve's harness connector disconnected, check for voltage at the blue/red wire in the connector. When the ignition is turned on there should be approximately battery voltage.

NOTE

On models with LH 2.2 fuel injection, the voltage will be momentary—only for as long as the fuel pump relay is closed. To get a constant voltage supply, run the engine.

---

4. If there is no voltage at the connector, check the wiring to the valve's connector. See **371 Wiring Diagrams, Fuses and Relays.**

NOTE

On LH 2.2, the AIC valve receives positive (+) battery voltage from the fuel pump relay. On LH 2.4 and LH 2.4.2, the valve receives voltage from the system relay.

---

5. If voltage is present as described in step 3, check the wiring between the control unit and the valve and fix any breaks. If no wiring faults are found, check the LH control unit signal to the AIC valve. If the signal to the valve is as specified, the AIC valve is probably faulty and should be replaced.

NOTE

• On cars with LH 2.2, the AIC valve control signal can be checked using a duty cycle meter (or dwell meter) as described below under **LH Fuel Injection Adjustments.**

• On cars with LH 2.4 and LH 2.4.2, the AIC valve control signal can be checked by making the electrical tests outline below under **LH Control Unit.**

There are some additional inputs to the control unit that affect idle speed (i.e. throttle position, A/C-on signals, and A/T Drive position). To check these signals, make the electrical tests described below under **LH Control Unit.**

LH Control Unit

The following voltage and continuity tests will help determine whether there are faults in the wiring or components that provide information to the LH control unit. If all inputs and wiring are OK but operational problems still exist, the control unit itself may be faulty.

Generally, a complete absence of voltage or continuity means there is a wiring or connector problem. See **371 Wiring Diagrams, Fuses and Relays** for help in diagnosis. Test results with different values than those specified do not necessarily mean that a component is faulty. Check for loose connections or connections that are inadequate due to corrosion or contamination. If the results are still incorrect, test the component itself.

Make these tests at the LH control unit connector using a digital VOM. The control unit is mounted in the passenger compartment, in the passenger-side footwell. See Fig. 8 given earlier. Remove the control unit, unplug the connector, and make tests from the back of the connector to avoid damaging connector terminals. See Fig. 25. Control unit connector terminals are shown in Fig. 26. **Table k** and **Table l** list the tests and correct test results.

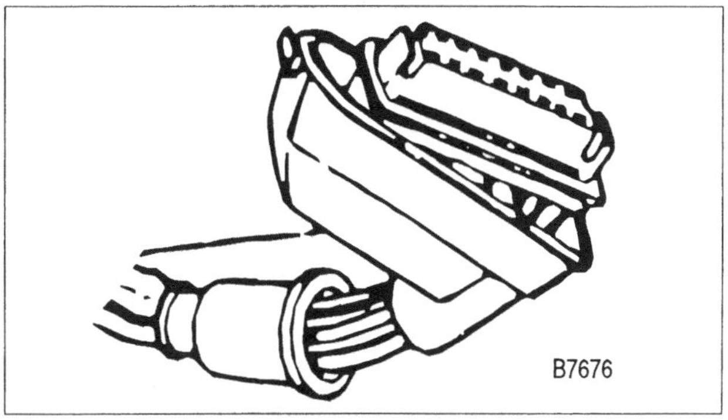

Fig. 25. For access to back of connector terminals when testing, disconnect connector and remove rubber gasket.

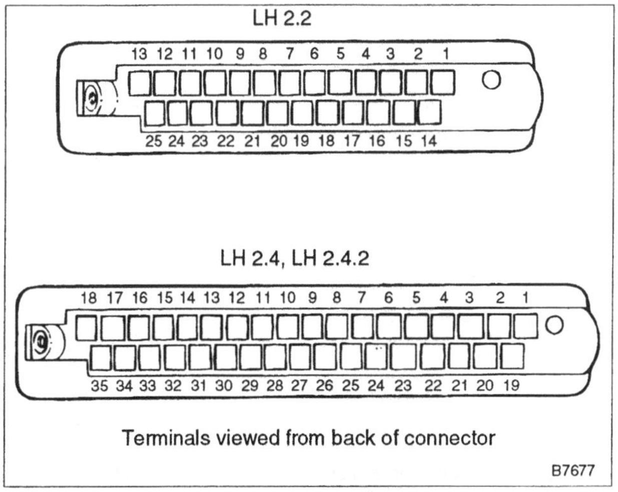

Fig. 26. Terminal identification for LH fuel injection control unit connectors. Take care not to damage connector terminals with test probes.

CAUTION

- Always connect or disconnect the control unit connector and meter probes with the ignition off to avoid damage to electronic components.
- On models with LH 2.4 and LH 2.4.2, always wait at least 40 seconds after turning off the ignition before removing the control unit connector. If the connector is removed before this time, residual power in the system relay may damage the control unit.

NOTE

- On cars with LH 2.4 and LH 2.4.2 fuel injection, disconnecting the control unit connector will erase the control unit's adaptive and diagnostic memory. When the engine is restarted it may operate erratically. The engine must be driven for at least 30 minutes for the adaptive memory to be reset.
- The electrical tests below call for measuring voltage, current or resistance using a digital volt-ohm meter (DVOM) or a multimeter. Using an analog meter may damage the control unit or other solid state components

Table k. LH 2.2 Electrical Tests

Component or circuit	Test terminals (Test conditions at LH connector unless otherwise noted)	Test conditions	Test value
Control unit power (switched), fuse 22	18 and ground	Ignition on	Approximately 12 volts (battery voltage)
Control unit grounds	18 and 5, 18 and 11, 18 and 25	Ignition on	Approximately 12 volts (battery voltage)
System relay	Jumper terminals 5 and 21		System relay closes, power to pin 9, power to pin 2 of air mass meter
Fuel pump relay (see **234 Fuel Pump and Fuel Tank**)	Jumper terminals 5 and 17, and 21 and 25		Fuel pump runs, power to oxygen sensor preheater (blue/red wire), power to fuel injectors (blue/red wire)
Ignition pulse signal	1 and 25	Crank engine	2–3 volts (VAC)
Coolant temperature sensor	2 and 25		Resistance according to **Table g** shown earlier
Throttle position sensor	3 and ground	Throttle closed	Continuity
		Throttle 1/2 open	No continuity
	12 and ground	Throttle wide open	Continuity
Automatic transmission switch (see **444 Automatic Transmission Controls**)	4 and 25	Ignition on, transmission in Drive	Approximately 12 volts (battery voltage)
AIC valve	10 and 23		40±4 ohms
Fuel injectors	13 and socket 87 of fuel pump relay		4 ohms
Air conditioning ON signal	12 and 16	Jumper sockets 16 and TK of A/C compressor relay	Continuity
Oxygen sensor	20 and 25	Oxygen sensor connected	No continuity
		Separate connector and connect green wire to ground	Continuity
Signal to EZK ignition system (ex. turbo models) (see **340 Ignition System**)	3 and 24	Disconnect EZK control unit connector, bridge terminals 7 and 8 of connector	Continuity

Table I. LH 2.4 and LH 2.4.2 Electrical Tests

Component or circuit	Test terminals (at disconnected LH connector unless otherwise noted)	Test conditions	Test value
Control unit power (switched), fuse 22	35 and ground	Ignition on	Approximately 12 volts (battery voltage)
Control unit power (constant)	4 and ground		Approximately 12 volts (battery voltage)
Control unit grounds	4 and 5, 4 and 17		Approximately 12 volts (battery voltage)
System relay	Jumper terminals 17 and 21		Relay closes, power to pin 9, power to pin 5 of air mass meter, power to fuel injectors (blue/red wire)
Fuel pump relay (see **234 Fuel Pump and Fuel Tank**), power from system relay, oxygen sensor heater power	Jumper terminals 5 and 20, and 17 and 21		Fuel pump runs, power to blue-red wire at oxygen sensor connector
Ignition pulse signal (LH 2.4)	1 and 17	Crank engine	2–3 volts (VAC)
Ignition pulse signal (LH 2.4.2)	1 and 17	Crank engine	6.5 volts (VAC)
Coolant temperature sensor	13 and 17		Resistance according to **Table g** shown earlier
Throttle position sensor (LH 2.4)	2 and ground	Throttle closed	Continuity
		Throttle 1/2 open	No continuity
	3 and ground	Throttle wide open	Continuity
Throttle potentiometer (LH 2.4.2)	2 and 10		Resistance according to **Table i** shown earlier
Automatic transmission switch. (see **444 Automatic Transmission Controls)**	30 and 17	Ignition on, transmission in Drive	Approximately 12 volts (battery voltage)
AIC valve (LH 2.4)	33 and 17	LH control unit connector connected, engine running, probe terminals from rear of connector	Approximately 8 volts at idle. Voltage should drop as load increases
AIC valve (LH 2.4.2)	33 and 17	LH control unit connector connected, engine running, probe terminals from rear of connector	Approximately 7 volts at idle. Voltage should drop as load increases
Fuel injectors	18 and socket 87 of system relay		4 ohms
Air conditioning ON signal (LH 2.4)	14 and ground	LH control unit connector connected, engine running, probe terminals from rear of connector	less than 0.1 volt with A/C off and battery voltage with A/C on.
Air conditioning ON signal (LH 2.4)	14 and ground11 and ground	LH control unit connector connected, engine running, probe terminals from rear of connector	less than 0.1 volt with A/C off and battery voltage with A/C on.
Oxygen sensor	24 and 17	Oxygen sensor connected	No continuity
		Separate connector and connect green wire to ground	Continuity
EGR modulating valve (where applicable—see **254 Exhaust Emission Control)**	19 and 17	Ignition on	Approximately 12 volts (battery voltage)

continued on next page

Table I. LH 2.4 and LH 2.4.2 Electrical Tests (continued)

Component or circuit	Test terminals (at disconnected LH connector unless otherwise noted)	Test conditions	Test value
EGR valve temperature sensor (where applicable—see **254 Exhaust Emission Control)**	23 and 17	Run engine at operating temperature. Turn off engine and disconnect control unit connector (after waiting 45 seconds)	Resistance should measure several megohms
Charcoal canister purge valve(ELCD) (see **254 Exhaust Emission Control**)	27 and 17	Ignition on	Approximately 12 volts (battery voltage)
Vehicle speed sensor	34 and ground	Ignition on, raise front of car and rotate front left wheel slowly	Voltage alternates between 0.5 and 5.0 volts
Upshift light	Jumper 26 and 17	Ignition on	Light illuminates
Check Engine light	Jumper 22 and 17	Ignition on	Light illuminates

COMPONENT REPLACEMENT

To remove LH control unit

1. Disconnect the negative (–) battery cable.
2. Remove the door sill plate from the passenger side of the car. Peel up the lower front edge of the door weather stripping.
3. Remove the trim plate holding the carpet to the front side panel and peel the carpet until the control unit and relays are visible.
4. Unplug the control unit harness connector by releasing the fastener and then pivoting the connector up and off of the unit.
5. Remove the mounting screws holding the control unit to the body.

Installation is the reverse of removal.

To remove coolant temperature sensor

1. Detach the crankcase ventilation hose from the valve cover.
2. Unbolt the fuel pressure regulator and its mounting bracket from the cylinder head and push it to one side.

NOTE

On some models, it may be necessary to remove the regulator completely for access to the temperature sensor.

---

3. Drain approximately 4 quarts of coolant from the cooling system. This is so that coolant will not leak out when the sensor is removed. See **261 Radiator and Cooling System.**
4. Disconnect the sensor wiring and remove the sensor using a 3/4 in. deep-well socket.
5. Before installing the new sensor, check the mating surface on the intake manifold. Use a new copper sealing washer.

Tightening torque

• Coolant temperature sensor. . . . . . . 20 Nm (15 ft-lb)

Installation is the reverse of removal. Refit the pressure regulator and crankcase ventilation hose. Make sure the LH system grounds are properly replaced on the pressure regulator mounting.

To remove throttle position sensor

1. Unplug the harness connector from the sensor. See Fig. 12 given earlier.
2. Remove the mounting screws holding the sensor to the side of the throttle housing.

Installation is the reverse of removal. On LH 2.2 and LH 2.4, the switch must be adjusted as described below under **LH Fuel Injection Adjustments.**

To remove air mass meter

1. Loosen the large hose clamp securing the rubber duct to the air mass meter.

2. Disconnect the harness connector from the meter.

3. Release the two spring clip fasteners from the air cleaner housing cover. See Fig. 27. Remove the air mass meter from the duct and cover.

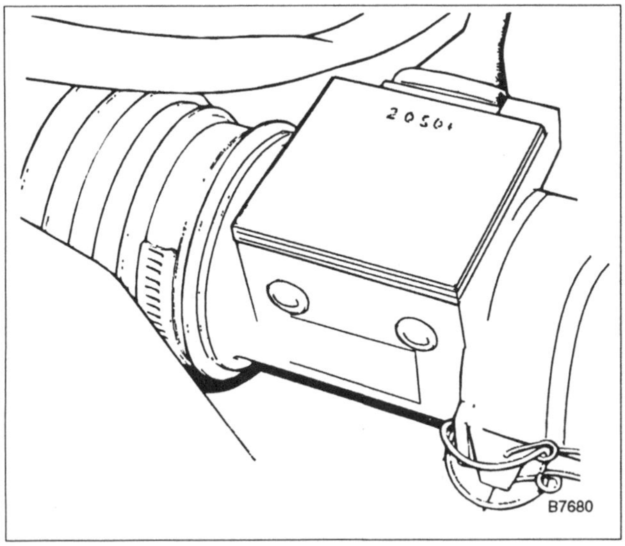

Fig. 27. Air mass meter and spring clip.

Installation is the reverse of removal. Where applicable, make sure the ridge in the air mass meter and the recess in the air cleaner cover are aligned before fastening the spring clips.

To remove fuel injectors

WARNING
Fuel will be expelled when fuel lines are disconnected. Do not smoke or work near heaters or other fire hazards. Keep a fire extinguisher handy.

CAUTION
Absolute cleanliness is essential when working with fuel circuit components. Even a minute particle of dirt can cause trouble if it reaches an injector. Thoroughly clean the unions before disconnecting any fuel lines. Use clean tools.

1. Detach the crankcase ventilation hose from the valve cover.

2. Unplug the wiring harness connectors from the injectors. Free the wiring harness from the fuel rail by undoing the clips. Carefully pull the harness to the side.

3. Disconnect the fuel inlet and outlet hoses on the fuel rail by removing the banjo bolts. See Fig. 28.

WARNING
Wrap a clean shop towel around the banjo bolts before loosening them.

CAUTION
To prevent damaging the fuel rail assembly, use an open-end wrench on the flats on the rail to hold it steady while loosening the fuel lines.

Fig. 28. Fuel rail assembly and fuel lines (arrows).

4. Remove the bolts that hold the fuel rail to the intake manifold and pull the fuel rail away complete with injectors.

5. To remove the injectors from the fuel rail, pry off the retaining clips, then twist the injectors slightly and pull them off.

When installing the injectors, fit new O-rings to the injectors. For ease of installation, lightly lubricate the O-rings with petroleum jelly. Install the injectors to the fuel rail but do not install the retaining clips. Position the fuel rail and injectors on the intake manifold. Make sure that injectors and the fuel rail are fully seated and then install the retaining clips and mounting bolts. Reconnect the fuel hoses to the fuel rail. Reconnect the crankcase ventilation hose.

To remove fuel pressure regulator

WARNING
Fuel under pressure will be expelled when fuel lines are disconnected. Do not smoke or work near heaters or other fire hazards. Keep a fire extinguisher handy.

1. Disconnect the vacuum hose from the regulator.

2. Disconnect from the regulator the fuel hose coming from the fuel rail.

3. Remove the fuel pressure regulator mounting bolts and lift out the regulator and its bracket. Disconnect the fuel return hose from the regulator. Remove the regulator from the bracket.

Installation is the reverse of removal. Remember to reconnect the ground wires to the fuel pressure regulator mounting bolt.

To remove AIC valve

1. Disconnect the harness connector from the AIC valve.

2. Loosen the valve's hose clamps and disconnect the hoses from the valve. See Fig. 29.

Fig. 29. AIC valve hose clamp being loosened.

3. Remove the valve bracket mounting bolt and remove the valve from the bracket.

Installation is reverse of removal. On models with LH 2.2, adjust the base setting of the valve as described below.

LH Fuel Injection Adjustments

Except for the adjustment of the throttle position sensor, all of the adjustment described below are for cars with LH 2.2 only. Cars with LH 2.4 and LH 2.4.2 are fully adaptive and do not require any adjustments.

Setting the base idle speed ensures that the AIC idle control valve is operating in about the middle of its range under normal conditions. Base idle speed is set at the factory and does not normally need to be checked or adjusted, unless repairs are made to the throttle valve or housing.

Base idle speed is set by reading the control unit signal to the valve. This signal is known as the duty cycle. It is the ratio of on-time to off-time of voltage and is expressed as a percentage. It is measured using a duty cycle or dwell meter.

NOTE

- Remember that 100% on a duty cycle meter corresponds to 90° on a dwell meter.

- The test meter must be a quality digital dwell meter, with an accuracy of at least 0.5%. Using an inexpensive analog meter that has an error of 2° would result in a setting 25% out of calibration.

To adjust AIC valve and basic idle speed (LH 2.2)

NOTE

For the most accurate adjustment, the engine should be at operating temperature, the ignition timing and idle mixture should be checked and adjusted if necessary.

1. Back off the dashpot so that the plunger no longer contacts the throttle lever. See Fig. 30. Check that there is slack in the accelerator cable. If necessary, loosen the cable.

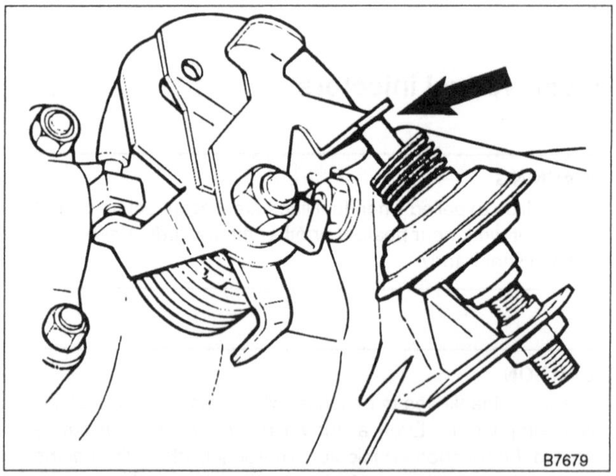

Fig. 30. Dashpot plunger should be backed off throttle lever (arrow).

2. Loosen the screws for the throttle position sensor so that it doesn't affect the throttle valve. Disconnect the sensor harness connector and connect a jumper wire as shown in Fig. 31. This will simulate a closed-throttle signal.

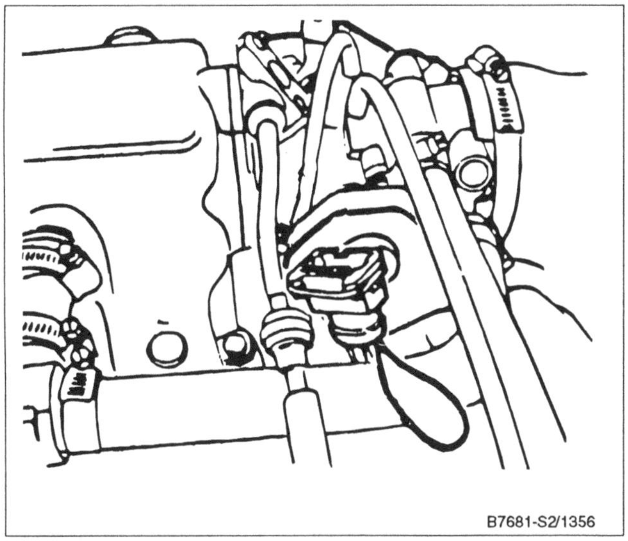

Fig. 31. Jumper terminals 1 (gray wire) and 2 (black-white wire) of throttle position sensor connector. Connect wire from back of connector to prevent damage to terminals.

3. Loosen the throttle stop screw locknut and back out the screw until it is no longer contacting the throttle lever. See Fig. 32. Thread the screw in until it touches the lever then turn it an additional 3/4 turn. Tighten the locknut.

Fig. 32. Throttle stop screw and locknut (arrow).

4. Working from the rear of the AIC valve's harness connector, connect the dwell meter to terminals 2 and 3. See Fig. 33. Connect the tachometer.

5. Start the engine and let it run until the radiator fan cycles at least once.

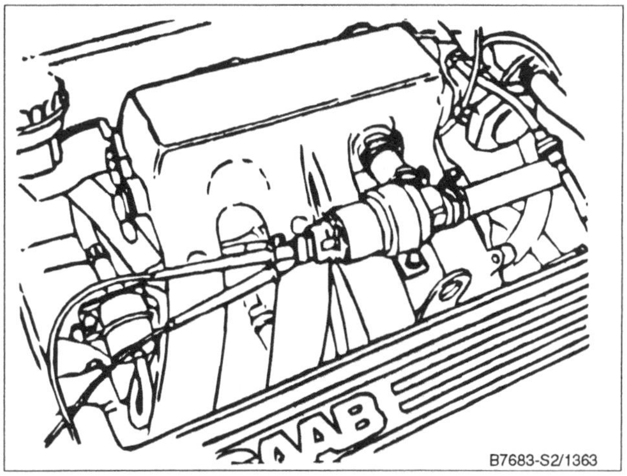

Fig. 33. Meter leads connected to idle control valve terminals 2 and 3 (valve terminal 4 and 5).

NOTE

Make sure all other electrical accessories on the car are off. Make all adjustments to idle only when the fan is off.

---

6. Depending on the throttle housing, turn either the throttle valve stop screw (Fig. 32) or the throttle valve air bleed screw (Fig. 34) out until the duty cycle is 38% (34° dwell).

NOTE

If the correct reading cannot be obtained, check for intake air leaks or other engine problems.

Fig. 34. Throttle valve air bleed screw (arrow) used on early models. Later models with LH 2.2 do not have this screw.

7. Next, turn the screw back in until the meter reading is as specified in the table below.

NOTE

When the correct reading is obtained, the base idle speed should be 850±50 rpm. If it isn't, check for intake air leaks or other engine problems.

Base Idle Speed (at 850±50 rpm)

- **Duty cycle** 33%
- **Dwell** .... 30°

8. Lock the adjusting screw and recheck the reading.

9. Turn off the engine and remove all test equipment. Reset the dashpot and throttle switch as described below

To adjust throttle position sensor (LH 2.2 and LH 2.4)

The throttle position sensor has two switches in the housing that provide a ground signal whenever the throttle is fully closed or fully open.

1. Loosen the switch mounting screws.

2. Attach a VOM to terminals 1 and 2 of the switch to measure continuity. See Fig. 35.

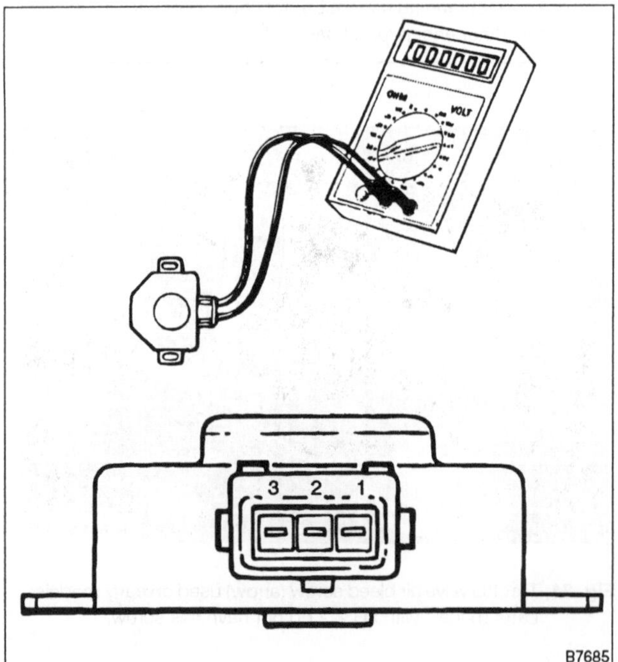

Fig. 35. Ohmmeter shown connected to throttle position sensor.

3. Rotate the switch so that the switch is just open (no continuity), then rotate the switch in the opposite direction until the switch just closes (continuity). Rotate the switch a bit further (approximately.5 mm/.020 in.) in the same direction and tighten the screws.

NOTE

Rotation of the throttle position sensor too far can cause the switch to hold the throttle open. This may lead to excessive idle speed.

---

4. Check that the switch opens (no continuity) as the throttle is just opened from idle.

5. Connect the ohmmeter to terminals 2 and 3 to check the full throttle switch contacts. There should be no continuity until the switch is almost fully open. At 72° of open throttle, the switch should close (continuity). If the contacts do not close, the switch should be replaced.

To check basic setting of fuel injection system (LH 2.2)

The procedure below should only be carried out when the engine function has been disturbed, such as for the replacement of the air mass meter or after major engine work has been completed. Saab does not recommend any specified interval for this check. For the most accurate results, a high-quality digital multimeter should be used.

1. Run the engine until it reaches operating temperature.

2. Connect the positive lead of a voltmeter to the black wire coming from the oxygen sensor. Connect the negative lead to chassis ground.

NOTE

Do not disconnect the wires from the oxygen sensor.

---

3. With the engine running, check the voltage signal coming from the oxygen sensor. The voltage should be fluctuating around the middle of its operating range.

Oxygen sensor

- voltage at idle, engine fully warmed .... approx. 0.5 volts (fluctuating)

NOTE

If the sensor is not putting out a signal or if the signal is not fluctuating, make sure the oxygen sensor system is functioning correctly as described earlier under **Oxygen Sensor.**

---

4. If the voltage is too high or too low, turn the engine off and remove the harness connector from the air mass meter. Connect a digital ohmmeter across terminals 3 and 6 of the meter and measure the resistance.

Air Mass Meter Base Setting

• resistance measured at meter terminals 3 and 6 . . . . 380 ohms

5. If the resistance reading is incorrect, remove the anti-tamper plug from the side of the air mass meter. See Fig. 36. Turn the screw until the correct reading is obtained.

NOTE
If a reading of 380 ohms cannot be obtained, the air mass meter is faulty and should be replaced.

Fig. 36. Air mass meter anti-tamper plug (arrow).

6. Reconnect the harness connector and start the car. Recheck the voltage signal at the oxygen sensor as described above. The voltage should now be fluctuating around the middle of its operating range (approximately 0.5 volts)

7. If the sensor voltage is incorrect, adjust the screw in the side of the air mass meter until the oxygen sensor voltage is correct.

8. Install a new anti-tamper plug anti-tamper plug from the side of the air mass meter.

If the correct oxygen sensor reading still cannot be obtained, check the wiring between the air mass meter and the control unit. Test the oxygen sensor and its circuit. Try substituting a known good air mass meter or oxygen sensor. If no faults can be found, the LH control unit may be faulty.

To adjust dashpot (LH 2.2)

The dashpot is mounted to the throttle housing and holds the throttle valve open for a moment when the throttle is closed suddenly. This helps reduce the exhaust emission of unburned fuel. A defective or misadjusted dashpot may be the cause of a high idle or the cause of an idle that takes a long time to return to normal when the throttle is released.

NOTE
Before making the adjustment below, check that the basic throttle adjustment is correct.

1. With the engine at operating temperature, rev the engine up to at least 3000 rpm, then release the throttle and check that the engine speed returns to normal within 3 to 6 seconds. If the dashpot fails this test, proceed with dashpot adjustment.

2. Loosen the dashpot locknut and turn the dashpot so that its operating rod moves away from the throttle lever. See Fig. 29 above.

3. Rev the engine up to the speed given below and hold the engine speed steady. Turn the dashpot in until its rod just touches the throttle lever. See **Table n.**Tighten the locknut.

Table n. Dashpot Adjustment Specifications

Model	Speed at which dashpot rod touches throttle lever
Non-turbo	2500±100 rpm
Turbo	2600±100 rpm

4. Recheck dashpot operation. If it still fails the test, replace it.

252 Exhaust System

General

Fig. 1 and Fig. 2 are exploded views of the exhaust systems used on the cars covered by this manual. Proper exhaust and emission system functions depend on each component being free from holes, with air-tight seals at all joints. Exhaust systems components are subjected to vibration, extreme temperatures and road hazards. Although the exhaust system is designed to be maintenance free, regular inspection is warranted due to these harsh operating conditions.

The catalytic converter is similar in appearance to a small muffler and is mounted to the front pipe. Its chemically-reactive core contains surfaces coated with precious metal catalysts. These catalysts promote chemical reactions in the exhaust gases to reduce the quantity of harmful pollutants in the gases. Under normal conditions, the catalytic converter does not require replacement unless it damaged.

The catalytic converter reduces emissions most efficiently when the percentage of oxygen in the exhaust system falls within a narrow range. The oxygen sensor monitors the exhaust gas and provides feedback about combustion efficiency to the fuel injection control unit. For information on testing and removing the oxygen sensor, see **240 Fuel Injection.** For information on testing other emission control systems, including the EGR system, see **254 Exhaust Emission Control**.

WARNING

Toxic exhaust gases are colorless and odorless. Do not run the engine in a non-ventilated area. Repair any structural damaged to the body that would allow the entrance of exhaust gases.

Exhaust System

To remove and install exhaust system

New fasteners, rubber hangers and clamps are always recommended. Gaskets should be replaced whenever the flanged joints are disconnected.

1. Raise the car to a safe level.

WARNING

Never work under a lifted car unless it is solidly supported on stands designed for the purpose. Do not work on a car that is supported solely by a jack.

---

2. Remove the bolts securing the front exhaust pipe to the exhaust manifold or turbocharger.
3. Loosen all clamps and flange bolts.
4. On turbo models, remove the front pipe-to-transmission bracket.
5. Detach the rubber mountings. Separate and remove the pipes and mufflers as necessary. Be sure to inspect the heats shields while the system is removed.

CAUTION

Inspect the heat shields carefully whenever the exhaust system is removed, especially the front heat shields. See Fig. 3. The front shields protect the ignition switch wiring harness from the heat of the hot catalytic converter. If these shields are deteriorated or damaged, they should be replaced

---

Installation is the reverse of removal. Loosely install and assemble the complete system. Then evenly tighten the clamps and mounting bolts. Make sure that no part of the exhaust system contacts any part of the body.

Make sure the rear retaining rings are slightly preloaded toward the front of the car. See Fig. 4. Install the clamps 9 mm (1/3 in.) from the ends of the pipes. See Fig. 5. When installation is complete, start the engine and check for leaks by briefly placing a clean rag over the tailpipe and listening for hissing sounds. If any leaks are present, loosen the system fasteners and readjust the system. Check the system alignment by driving the car over rough road surfaces to check if the system contacts the underbody.

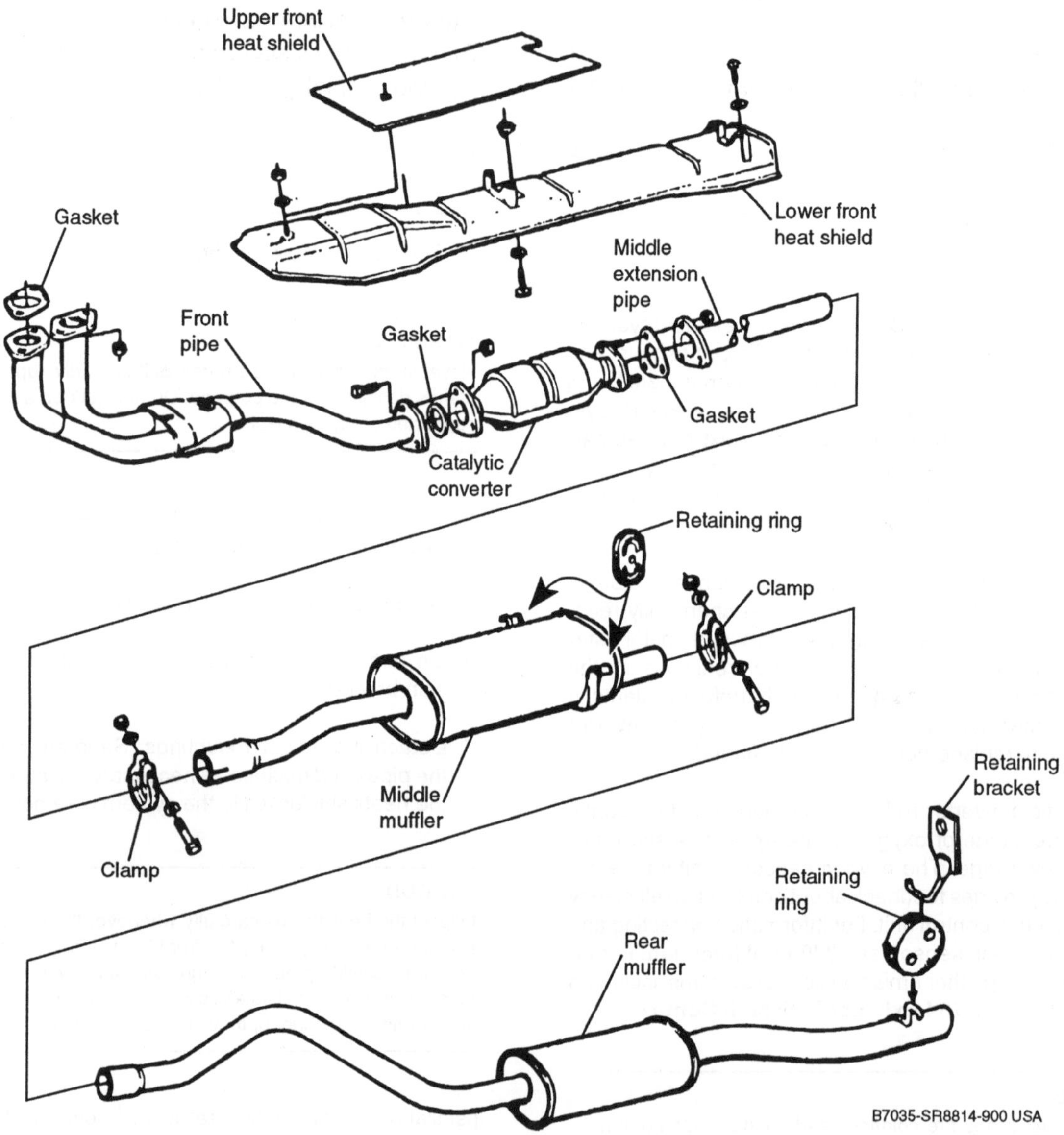

Fig. 1. Exploded view of exhaust system used on cars with normally-aspirated engines.

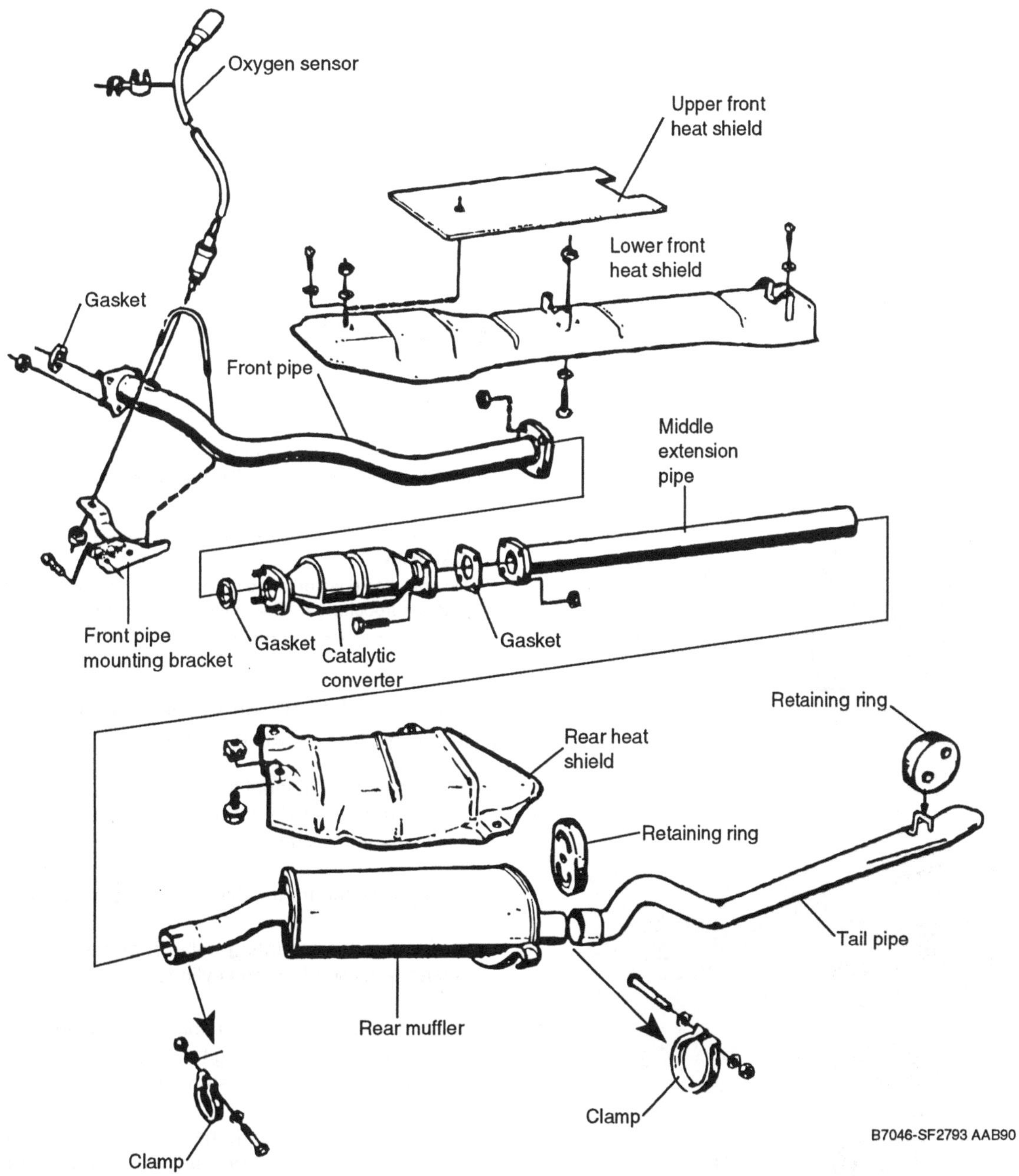

Fig. 2. Exploded view of exhaust system used on cars with turbo-charged engines.

NOTE

Anti-seize compound used on threaded fasteners and exhaust pipe joints will make future replacement easier.

CAUTION

Do not let anti-seize compound or penetrating oil come in contact with the slit portion of the oxygen sensor body. The oxygen sensor will be destroyed.

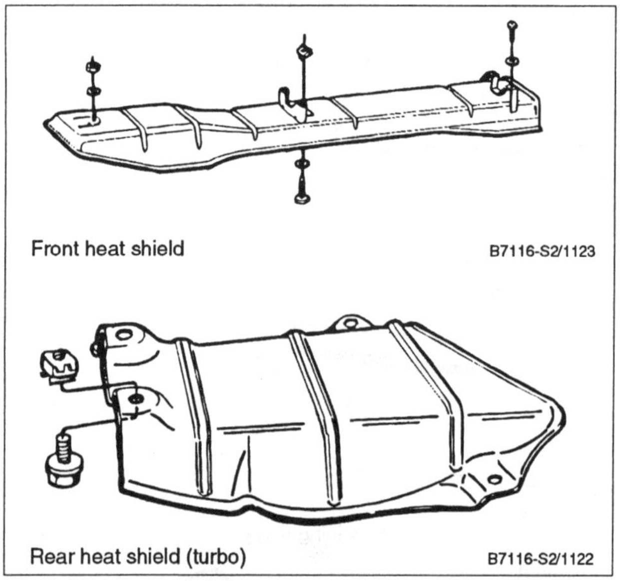

Fig. 3. Exhaust system heat shields.

Fig. 4. Rear muffler retaining ring shown preloaded for exhaust system expansion.

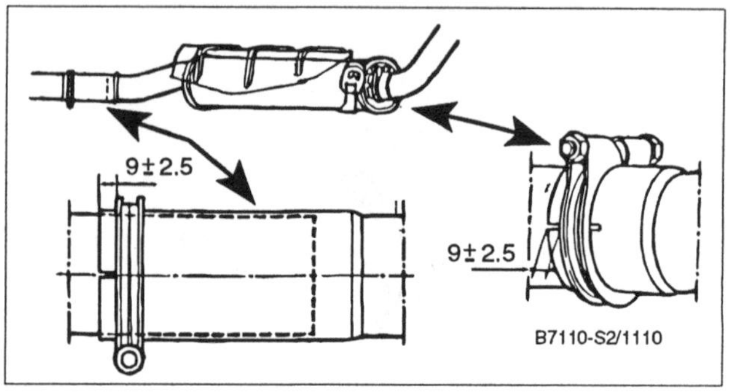

Fig. 5. Exhaust system clamps should be installed 9 mm (1/3 in.) from end of pipe.

EXHAUST MANIFOLD

The exhaust manifold is made of cast iron. Normally-aspirated models use a two-piece manifold assembly. Turbo models use a one-piece manifold. See Fig. 6. Always use a new manifold gasket when installing the manifold.

CAUTION

Do not let cold water come in contact with a hot exhaust manifold or turbocharger. The manifold could crack as a result.

To remove and install exhaust manifold (normally-aspirated engine)

1. On cars with manual transmission, remove the oil filler pipe from the side of the transmission. Plug the opening in the transmission.
2. Separate the front exhaust pipe from the manifold.
3. Remove the mounting bolts from the center manifold and remove it from the cylinder head.
4. Remove the upper-left manifold mounting stud from the cylinder head by locking two nuts on the stud. Use the inner nut to thread the stud out.
5. Remove the remaining mounting bolts from the outer manifold. Remove the manifold by pivoting the front end up while lifting it off the studs. See Fig. 7.

Installation is the reverse of removal. Use new mounting nuts, tightening them evenly to their specified tightening torque.

Tightening torque
normally-aspirated engine
- exhaust manifold to cylinder head . . 20 Nm (15 ft-lb)

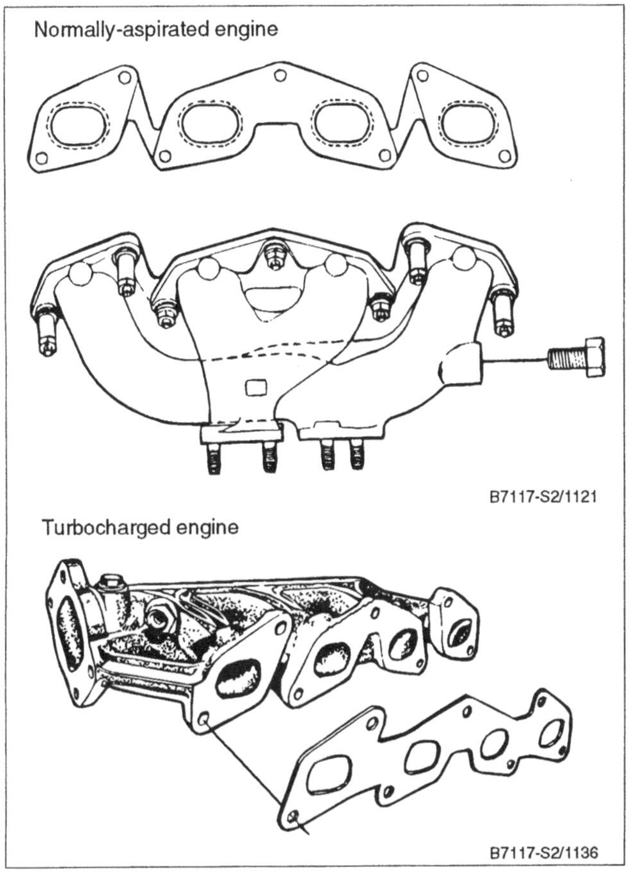

Fig. 6. Exhaust manifolds.

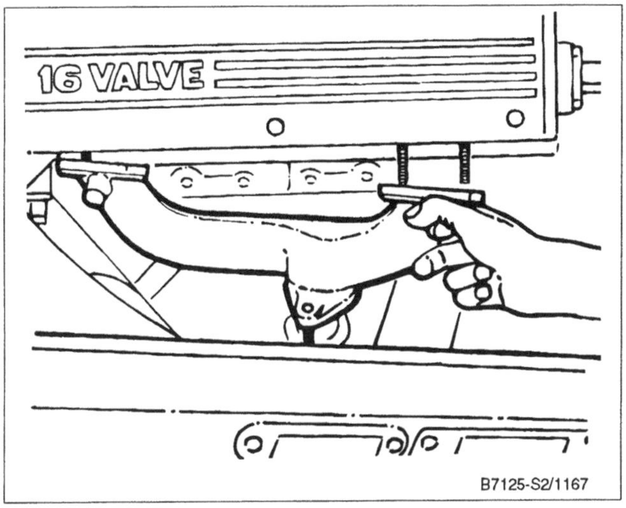

Fig. 7. Exhaust manifold being removed.

To remove and install exhaust manifold (turbo engine)

1. Remove the battery from the car.
2. Remove the ignition distributor as described in **340 Ignition System.**
3. Remove the distributor heat shield from the cylinder head.
4. Disconnect the turbocharger air intake pipe, air outlet pipe and oil supply pipe from the turbocharger. Disconnect the turbocharger oil supply pipe from the engine.
5. On 1988 and later models, drain the cooling system as described in **261 Radiator and Cooling System.** Then remove the coolant lines from the turbocharger.
6. Remove and label the small pressure hoses from the APC solenoid on the radiator crossmember.
7. Where applicable, disconnect the EGR pipe from the exhaust manifold and from the EGR valve.
8. Remove the mounting bolts from the turbo unit bracket at the transmission.
9. On cars with manual transmission, remove the oil filler tube from the side of the transmission. Plug the opening in the transmission.
10. Disconnect the oil return pipe from the turbocharger.
11. Unbolt the front exhaust pipe from the turbocharger.
12. Remove the exhaust manifold mounting nuts, spacers, and washers. Slide the manifold with turbocharger off the studs and out of the car. Separate the turbocharger from the exhaust manifold, if necessary.

Installation is the reverse of removal. Use new mounting nuts, tightening them evenly to their specified tightening torque. Replace the manifold gasket. Replace all turbocharger oil line seals and gaskets. For additional information on the turbocharger see **291 Turbocharger.**

Tightening torque
turbo engine
- exhaust manifold to cylinder head . . 25 Nm (18 ft-lb)

254 Exhaust Emission Control

TABLES

GENERAL

This repair groups covers only the Evaporative Loss Control Device (ELCD) and the Exhaust Gas Recirculation (EGR) system. For repair and troubleshooting information of other emissions-related systems, such as the oxygen sensor system, the throttle dashpot, and the Automatic Idle Control (AIC) system, see **240 Fuel Injection.**

For proper exhaust emissions, a number of engine systems (ignition, fuel injection, and emission controls) must be functioning correctly. Because of the interrelation of these systems, it is difficult to isolate general driveability problems by examining a single emission control system. For this reason, engine management and driveability troubleshooting information can found in **200 Engine—General.** This information is organized to help isolate problems and suggest more specific troubleshooting steps by taking all of the interrelated systems into consideration.

NOTE
Cars with LH 2.4 and LH 2.4.2 fuel injection systems feature built-in diagnostic circuitry that detects and stores system faults in the LH control unit's memory, including emissions-related faults. When a system compensates for values that are outside the permitted limits, a fault code is generated in the control unit's memory. It is highly recommended that checking for any faults codes be the first step in diagnosing an emissions-related fault. See **240 Fuel Injection** for instructions on viewing the fault codes.

EVAPORATIVE LOSS CONTROL DEVICE (ELCD)

All 900 models use a charcoal-filter canister to absorb vapors from the fuel tank. This system is known as the Evaporative Loss Control Device (ELCD). See Fig. 1. The fuel tank is vented to the canister and the canister is in turn vented to the intake manifold via the ELCD valve. When the engine is running, the accumulated vapors in the canister are drawn through the open ELCD valve and into the engine to be burned. The charcoal canister is mounted in the driver's side wheel well, behind the headlight. The ELCD valve is mounted to the top of the canister. Two types of ELCD systems are used, depending on the fuel system installed. See **Table a**.

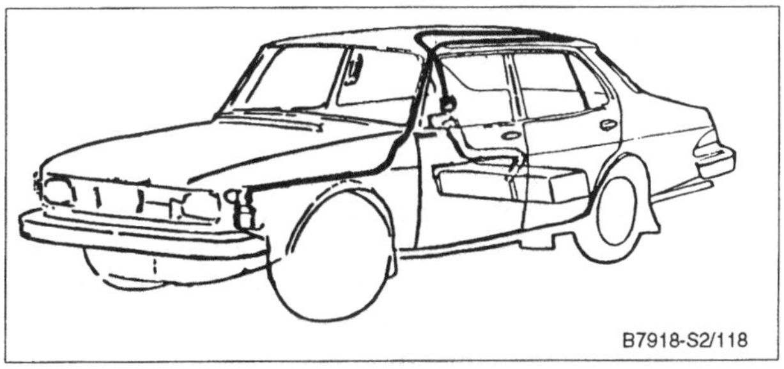

Fig. 1. Evaporative Loss Control Device (ELCD) system. Note how hoses run through headliner.

Table a. ELCD Systems Applications

Fuel Injection Systems*	ELCD System
LH 2.2	Vacuum operated
LH 2.4, LH 2.4.2	Electronically controlled

*see 240 Fuel Injection for system identification

The ELCD valve controls the amount of fuel vapors drawn into the engine and therefore can affect the fuel mixture and exhaust emissions. A fault in the ELCD system can often result in an overly lean or an overly rich mixture, causing driveability problems, especially at idle.

NOTE
On most models, charcoal canister replacement is a maintenance procedure scheduled at a specified mileage interval. See **1 LUBRICATION AND MAINTENANCE** for the recommended service interval.

On models with LH 2.2 fuel injection, a vacuum-operated diaphragm valve on the canister controls when fuel vapors are drawn into the engine. See Fig. 2. Vacuum from the intake manifold opens the valve during cruising.

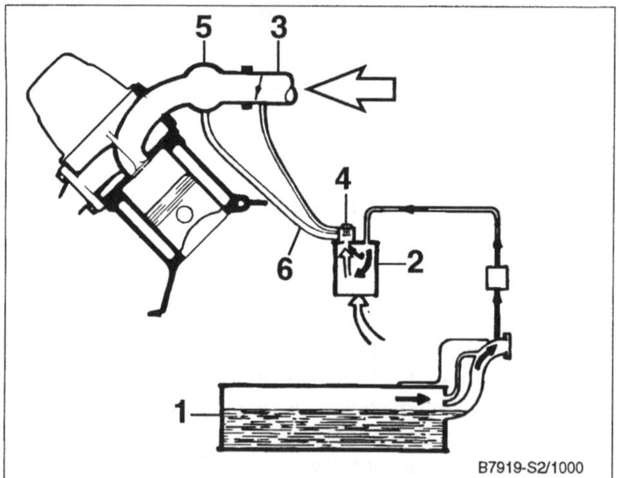

1. Fuel tank
2. Activated-charcoal canister
3. Intake manifold
4. ELCD valve
5 Throttle plate
6. Hose

Fig. 2. Schematic view of ELCD system.

To test the vacuum-operated ELCD valve, disconnect the vacuum hoses from the valve. If necessary, remove the canister from the car. Blow into the lower (larger) hose fitting on the valve. Air should not pass through the valve. Connect a hand-held vacuum pump to the top hose fitting on the valve. With vacuum applied to the top fitting, air should now pass through the lower fitting. See Fig. 3. If any faults are found, the valve is faulty and should be replaced.

On models with LH 2.4 and LH 2.4.2 fuel injection, an electric solenoid valve controls canister purge. The valve is opened and closed electronically by the LH control unit. With the engine running, the valve should be audibly buzzing, and the buzzing noise should vary in intensity depending on engine speed.

The electric ELCD valve can be tested by disconnecting its harness connector and supplying battery voltage to the valve terminals. The valve should click open when voltage is supplied. If no faults are found, check for voltage to the valve's connector with the ignition in the run position. There should be battery voltage between the gray/white wire and ground. If voltage is not present, check for faults using the appropriate wiring diagram shown in **371 Wiring Diagrams, Fuses and Relays**. If voltage is present, check the electrical resistance of the valve. See Fig. 4. If the resistance is not as specified, the valve should be replaced.

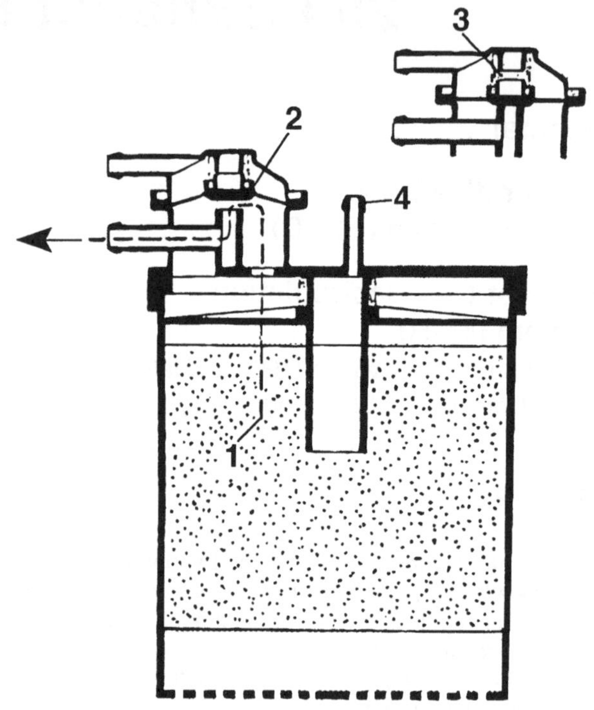

1. to Inlet manifold
2. ELCD valve shown open
3. ELCD valve shown closed
4. from Fuel tank

Fig. 3. Vacuum-operated ELCD valve and charcoal canister.

ELCD valve

• valve resistance . . . . 40 to 60 ohms

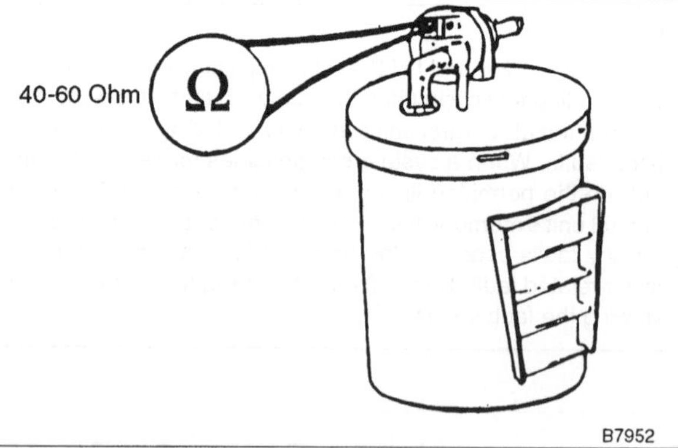

Fig. 4. Electric ELCD valve resistance being measured (shown schematically).

EXHAUST GAS RECIRCULATION SYSTEM (EGR)

The exhaust gas recirculation system functions to recirculate a small amount of the exhaust gases into the intake manifold. This lowers combustion chamber temperatures and reduces exhaust emissions.

Electronic EGR System (1990 and later Californian and SPG models only)

This electronic system is controlled by the LH control unit and uses engine temperature, engine load, and road speed inputs to determine the optimum amount of EGR needed. The main components of the system are the valve, the temperature sensor, and the modulating valve. See Fig. 5

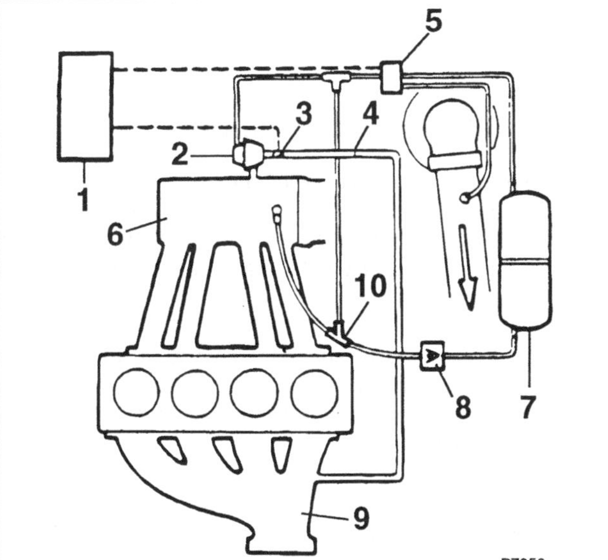

1. LH control unit
2. EGR valve
3. Temperature sensor (NTC)
4. EGR pipe
5. Modulating valve
6. Intake manifold
7. Vacuum tank
8. Check valve
9. Exhaust manifold
10. Overflow valve (turbo only)

Fig. 5. Schematic of Electronic EGR system and hose routing.

The LH control unit sends a signal to the modulating valve, which allows engine vacuum to open the EGR valve. The temperature sensor, which is mounted to the EGR valve, monitors when the valve is open and closed. When troubleshooting the EGR system, check that all hoses are properly connected. Check the hoses for cracks or deterioration. Check the wiring to the EGR modulating valve and the temperature sensor for any visible damage.

NOTE

- If a fault occurs in the EGR system, the Check Engine light will come on and a fault code will be stored in the memory of the LH control unit.
- On cars equipped with the electronic EGR system, the Check Engine light may come on intermittently, even though there are no emissions or fuel system-related faults. To remedy this problem, a special wiring harness has been made. See an authorized Saab dealer for additional information.

To test EGR valve

The EGR valve is shown in Fig. 6.

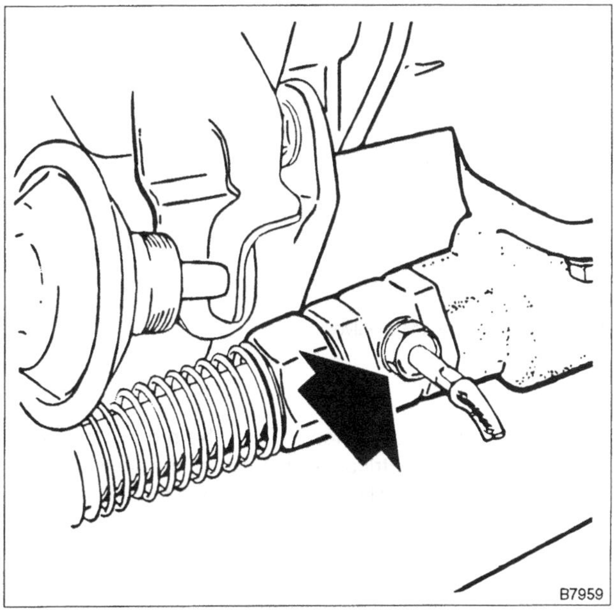

Fig. 6. EGR valve on underside of intake manifold. Temperature sensor shown at arrow.

1. Start and run the engine until it is at operating temperature. Disconnect the hose leading to the EGR valve.
2. Apply vacuum to the disconnected hose. Use a hand-held vacuum pump or suck on the end of the hose to create the vacuum. The engine idle speed should drop noticeably, and the engine may even stall. Reconnect the hose to the EGR valve.

If any faults are found, remove and clean the EGR valve. Be sure to also clean the EGR valve opening in the bottom of the intake manifold. Install the EGR valve and recheck. If any faults are found, the EGR valve should be replaced.

To test modulating valve and signal to modulating valve

The EGR modulating valve is mounted to the driver's side of the fender. See Fig. 7.

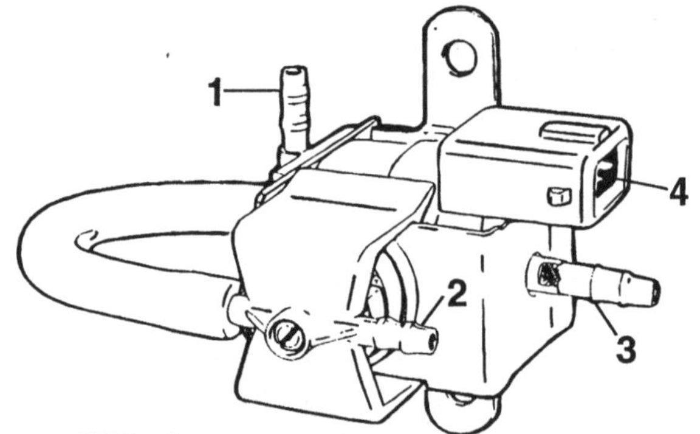

1. to EGR valve
2. to Intake manifold via vacuum tank
3. Turbo: to air hose between air mass meter and turbocharger Non-turbo: to air hose between air mass meter and throttle housing
4. Two pin connector (signal from LH control unit)

Fig. 7. EGR modulating valve.

1. Disconnect the connector from the modulating valve. Check the electrical resistance of the valve with an ohmmeter or apply battery voltage to the valve's terminals. The valve should audibly click when voltage is applied. If the resistance is not as specified or the valve doesn't click, it should be replaced.

EGR modulating valve

• valve resistance . . . . approximately 30 ohms

NOTE

It may be easier to apply voltage to the valve by first removing it from its mountings.

2. Connect an LED test light or a fuel injector test light to the valve's harness connector. See Fig. 8.

3. Start and run the engine until it is at operating temperature.

4. Raise the idle speed above 2500 rpm. The LED should flicker.

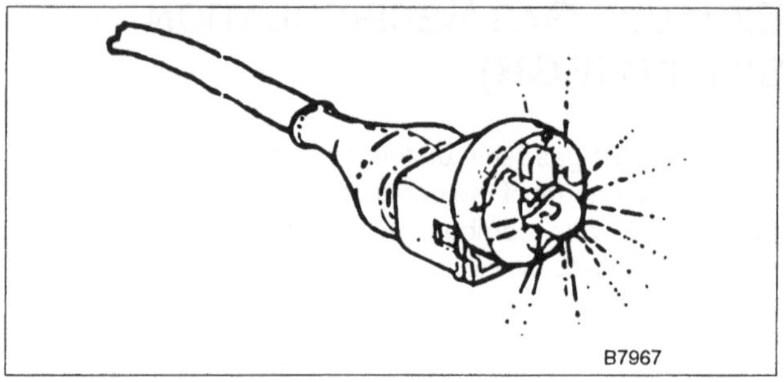

Fig. 8. LED test light connected to modulating valve connector.

If the light does not flicker as described above, turn the engine off and check for power at pin 1 of the connector with the ignition key on. If battery voltage is not present, check for wiring faults to the valve using the appropriate wiring diagram shown in **371 Wiring Diagrams, Fuses and Relays.** If battery voltage is present, check for continuity in the wire between the valve and the LH control unit (pin 19). If no wiring faults can be found, the LH control unit may be faulty.

To test EGR temperature sensor

1. With the engine cold, disconnect the 2-pin connector from the temperature sensor on the EGR valve (see Fig. 6 shown above). Connect an ohmmeter to the sensor terminals. Check that there are several megohms of resistance.

2. Run the engine at speeds above 2500 rpm. After a short period, the sensor's resistance should gradually drop to several thousand ohms or less as the EGR valve heats up.

If the EGR temperature sensor does not respond as specified, it should be replaced. Be sure to check for a good ground at pin 2 (black wire) of the sensor.

261 Radiator and Cooling System

GENERAL

This section covers repairs and troubleshooting information for the engine cooling system only. For heater core and related heating and air conditioning components, see **854 Heating and Air Conditioning.** For information on the engine oil cooler used on turbo models, see **220 Lubrication System.** For information on the ATF cooler used on models with automatic transmission, see **4 TRANSMISSION.**

Additional cooling system components and accessories, such as a winter thermostat for extremely cold climates, an air diverter for extra air flow across the radiator for warm climates, and engine block heaters for easier winter starting, are available through an authorized Saab dealer. Check with your authorized dealer for application and availability information.

Fig. 1 is a schematic view of the cooling system.

Coolant Pump and Thermostat

The coolant pump is crankshaft-driven by a V-belt. A three-stage thermostat controls the flow of coolant. When the engine is cold the thermostat is closed and coolant bypasses the radiator to circulate coolant through the engine and heater core only. When the engine warms up, the thermostat opens and coolant circulates through the complete system. When the engine reaches full operating temperature, the thermostat closes off the lower outlet in the thermostat housing to substantially reduce the coolant flow through the heater core. See Fig. 2.

NOTE

On certain 1985 through 1989 models, the heater core may not receive sufficient coolant flow at low engine speeds during stop-and-go driving in cold weather. If low heater output is experienced during these conditions only, see **854 Heating and Air Conditioning** for information on installing a special circulation pump.

Radiator and Expansion Tank

The radiator is a crossflow design. A translucent expansion tank, or overflow reservoir provides for coolant expansion at higher temperatures and easy monitoring of coolant level.

Radiator Cooling Fans

The cooling fans are mounted to the radiator and are electrically operated. The main fan is switched on and off via a coolant thermoswitch. The main fan also runs whenever the A/C

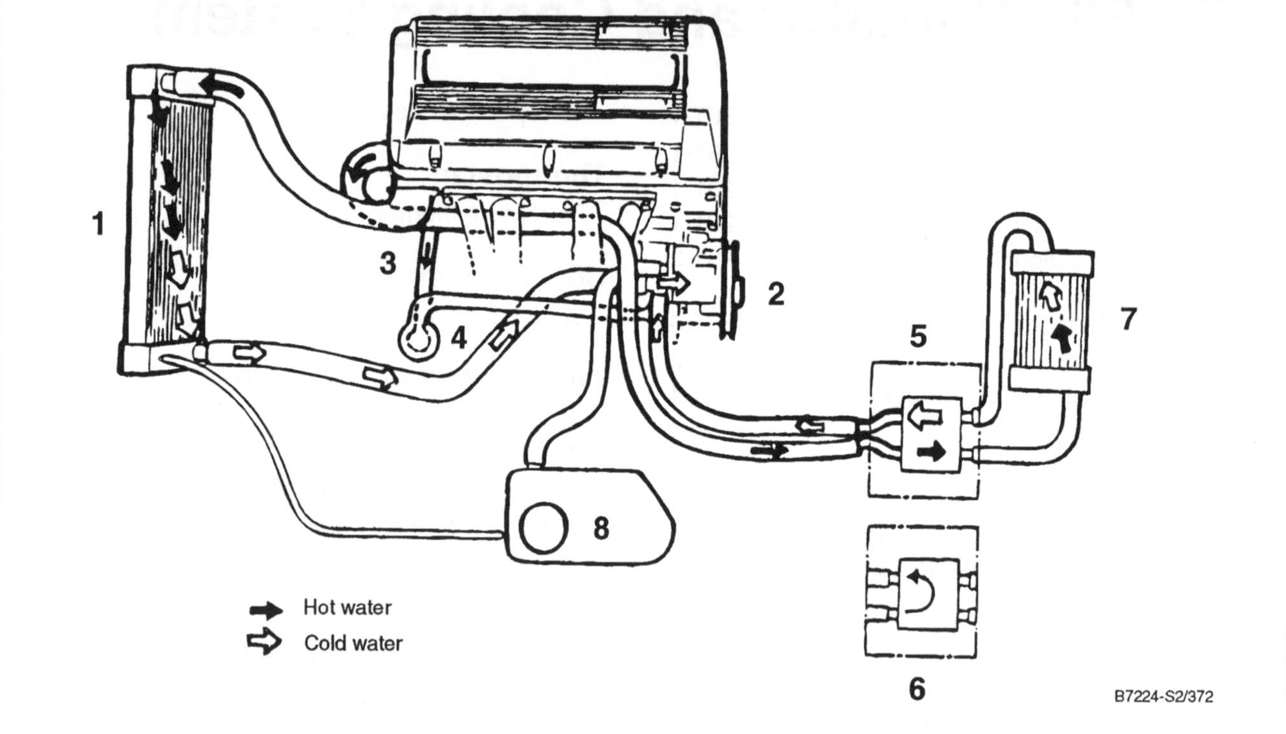

1. Radiator
2. Coolant pump
3. Thermostat housing
4. Preheater, intake manifold
5. Heater valve in open position
6. Heater valve in closed position
7. Heater core
8. Coolant expansion tank

Fig. 1. Schematic view of cooling system.

switch is on. The auxiliary cooling fan runs anytime the A/C compressor is cycling.

Warnings and Cautions

The following warnings and cautions should be followed when working on the cooling system.

WARNING

- At normal operating temperature the cooling system is pressurized. Allow the system to cool as long as possible before opening—a minimum of an hour—then release the cap very slowly to allow safe release of pressure.
- Releasing the cooling system pressure lowers the coolant's boiling point, and the coolant may boil suddenly. Use heavy gloves and wear eye and face protection to guard against scalding.
- Use extreme care when draining and disposing of engine coolant. Coolant is poisonous and lethal to pets. Pets are attracted to coolant because of its sweet smell and taste. See a veterinarian immediately if any amount of coolant is ingested by the animal.

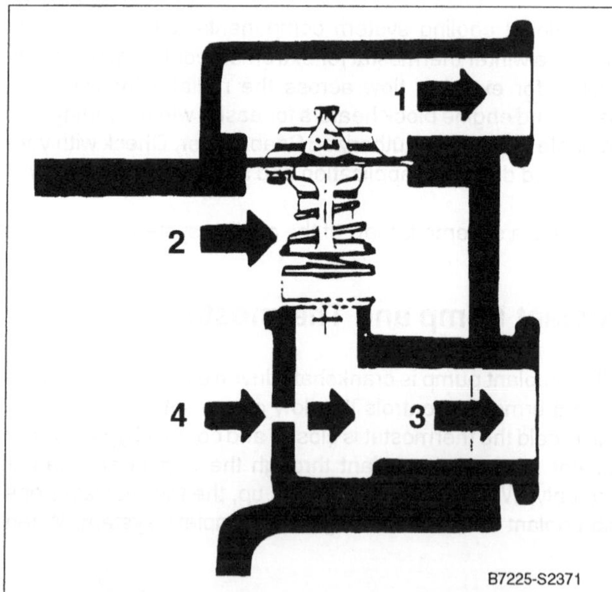

1. to Radiator
2. From engine block/cylinder head
3. to Heater
4. By-pass port

Fig. 2. Schematic view of three-way thermostat.

CAUTION
Avoid adding cold water to the coolant while the engine is hot or overheated. If it is necessary to add coolant to a hot system, do so only with the engine running and coolant pump turning.

TROUBLESHOOTING

When investigating the cause of overheating or coolant loss, begin with a visual inspection of the system. Carefully check for a broken or loose coolant pump V-belt. Check the coolant level and check for evidence of coolant leaks.

Check that the radiator fins are not blocked with dirt or debris. Clean the radiator fins using low-pressure water or compressed air. Blow from the engine side out.

Inspect the coolant pump by first removing the V-belt from the pump. Firmly grasp opposite sides of the pulley and check for play in all directions. Rotate the pulley and check that the shaft runs smoothly. If any faults are found the pump should be replaced.

NOTE
The coolant provides some lubrication for the pump shaft, so an occasional drop of coolant leaking from the pump is acceptable. If coolant drips from the vent hole, the pump should be replaced.

Most cooling system faults can be grouped into one of three main categories: 1) cooling system leaks, 2) poor coolant circulation, or 3) radiator cooling fan electrical faults.

The cooling system becomes pressurized at normal operating temperature, which raises the boiling point of the coolant. Leaks may prevent the system from becoming pressurized, allowing the coolant to boil at a lower temperature. If visual evidence is inconclusive, a cooling system pressure test, as described below, will help to pin point hard to find leaks.

If the cooling system is full of coolant and holds pressure, the next most probable cause of overheating is the result of poor coolant circulation due to a failed thermostat or coolant pump, a pinched or restricted hose, or a clogged system.

To quickly check if the thermostat is opening and if coolant is circulating through the radiator, allow a cold engine to reach operating temperature (temperature gauge needle approximately centered). Feel the radiator hoses. If the hoses are hot to the touch, the coolant is probably circulating correctly. If there are any cool areas in the hoses or radiator, coolant flow to the radiator is restricted. Check for a faulty thermostat or a plugged radiator.

WARNING
Be careful when working near the cooling fans when the engine is hot. The fan can come on at anytime—even if the ignition key is off.

NOTE
A thermostat that is stuck open will cause the engine to warm up slowly and run below normal temperature at highway speed. A thermostat that is stuck closed will restrict coolant flow to the radiator and cause overheating.

An otherwise sound cooling system will overheat if the electric cooling fans are not operating correctly. The fans should switch on and off according to engine coolant temperature or whenever the A/C is on. Below are a few quick checks to help diagnose cooling system faults.

If the engine overheats and no other cooling system tests indicate trouble, the radiator may have some plugged passages that are restricting coolant flow. This does not necessarily mean that the radiator must be replaced. In many cases the radiator can be chemically cleaned by a qualified radiator repair shop.

Pressure Test

A pressure test uses a special tester to pressurize the system and simulate normal running conditions. If the system is unable to hold pressure, the engine will overheat more easily. Aside from locating external leaks, a pressure test can be used to check for internal leaks. Some of the common sources of internal leaks are a faulty cylinder head gasket, a cracked cylinder head, or a cracked cylinder block.

Requirements:
- cooling system pressure tester
- engine warm

With the engine at normal operating temperature, pressurize the system. See Fig. 3. Loss of pressure indicates leaks that may also be detected by the seepage of coolant. If the pressure drops rapidly and there is no sign of coolant leakage, the cylinder head gasket may be faulty. See **211 Cylinder Head Removal and Installation** for gasket replacement procedures.

The screw-on type expansion tank cap should also be tested using a pressure tester and the correct tester adapter. Carefully inspect the cap for damage. A faulty cap or a damaged cap gasket should be replaced.

Cooling system test pressures
- Maximum system pressure . . . . . . 1.2 bar (17.5 psi)
- Expansion tank cap opening pressure . . . . 0.9 to 1.2 bar (13 to 17.5 psi)

CAUTION
Exceeding the specified test pressure could damage the radiator or other cooling system components.

Fig. 3. Cooling system being checked for leaks using pressure tester.

Temperature Gauge Quick-Check

A quick-check of the coolant temperature sender can be made to determine the cause of a faulty coolant temperature gauge. The checks should be made at the coolant temperature sender on the front of the cylinder head, below the distributor. See Fig. 4.

Disconnect the wire from the sender and hold it to ground. Turn the ignition on. If the gauge needle moves upward, the sending unit is faulty. If the gauge does not respond, the wiring to the gauge is broken (open circuit) or the gauge itself is faulty. To check the resistance of the sender measure between ground and the sender terminal with the wire disconnected.

Coolant temperature sender specification
- sender resistance at 90°C (194°F) . . . 51.2±4.3 ohms

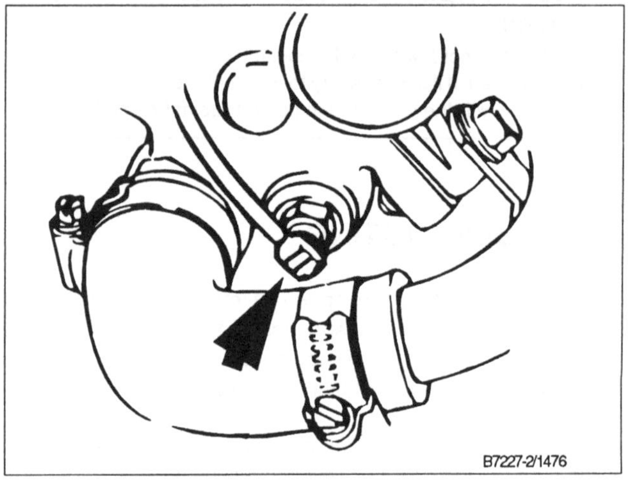

Fig. 4. Coolant temperature sender (arrow) below distributor on front of cylinder head.

When replacing a faulty coolant temperature sender, drain the cooling system down below the level of the sender and then unscrew the sender.

Radiator Fan Quick-Check

If the main fan does not come on, a quick-check can be made to determine whether the fault is in the fan, the thermoswitch, or the fan's electrical circuit. Be sure to check fuse no. 25 (30 amp) and fuse no. 6 (30 amp) before making any tests.

To test the fan, turn the ignition on and remove the rubber boot from the fan's thermoswitch on the left side of the radiator. Using a screwdriver, short the two switch terminals together and check that the fan runs. See Fig. 5.

Fig. 5. Main cooling fan thermoswitch being shorted using screwdriver.

If the fan runs only when the switch is bypassed, the thermoswitch is faulty and should be replaced. If the fan does not run, check for battery voltage between the green wire and ground at the thermoswitch. If voltage is present, the fan is faulty and should be replaced. If voltage is not present, check the wiring and the fan relay using the diagrams shown in **371 Wiring Digrams, Fuses and Relays.** Additional information on replacing the thermoswitch and the radiator cooling fan is given below.

NOTE
The auxiliary fan located beside the main fan on the radiator is part of the A/C system. See **8 BODY** for additional information on this fan.

Thermostat Quick-Check

To quickly check if the thermostat is opening and if coolant is circulating through the radiator, allow a cold engine to reach operating temperature (temperature gauge needle approximately centered). Feel the radiator hoses. If the hoses are hot to the touch, the coolant is circulating. If the hoses are not hot, coolant flow to the radiator is restricted. Check for a faulty thermostat or a plugged radiator.

COOLANT AND HOSES

To drain and fill cooling system

1. Disconnect the negative (–) battery cable, remove the expansion tank cap, and set the temperature control knob in the passenger compartment to warm.

2. Loosen the radiator drain plug at the bottom of the radiator next to the battery. See Fig. 6.

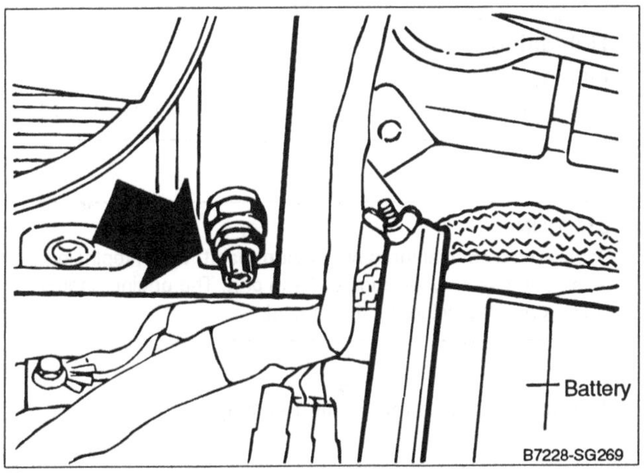

Fig. 6. Radiator drain plug (arrow).

3. Remove the engine block drain plug. See Fig. 7. The drain plug is located behind the exhaust manifold.

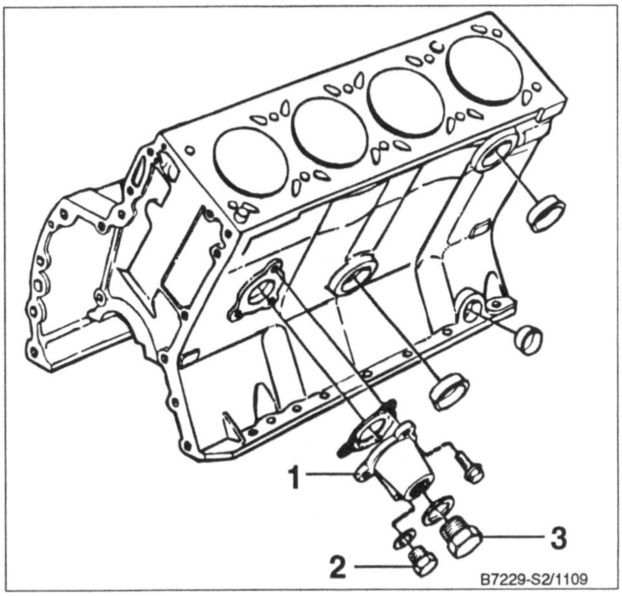

Fig. 7. Cylinder block showing location of coolant drain plug **2**. Do not remove cover **1** or plug **3**.

4. Reinstall the drain plugs using new gaskets.

Tightening torque
- Engine block drain plug (with new gasket) . . . . .15 Nm (11 ft-lb)

5. With the engine off, open the bleeder screw on the thermostat housing. See Fig. 8. Using a coolant mixture of 50% anti-freeze and 50% water, fill the expansion tank slowly so that air is allowed to escape. Close the bleeder screw when coolant begins to come out of the bleeder. Run the engine and check for leaks.

Cooling system capacity
- 10 liters (10.5 qts.)

Fig. 8. Exploded veiw of thermostat housing and bleeder screw (arrow). 16 valve model shown.

CAUTION

- Always use Saab original coolant to avoid the formation of harmful, clogging deposits in the cooling system. Use of other anti-freeze solutions may be harmful to the cooling system.
- The final coolant level is best checked when the cooling system is fully cold.
- Do not reuse the coolant when replacing damaged engine parts. Contaminated coolant may damage the engine or cooling system.
- If coolant comes in contact with paint or bodywork rinse with water. Coolant will damage paint.

To replace a hose

1. Drain the coolant as described above.
2. Remove the hose. Using a screwdriver, loosen each hose clamp and slide the clamps away from the hose ends.

NOTE
If a radiator hose is stuck to the radiator connection, cut the old hose off the connection as shown in Fig. 9. Prying the hose loose may damage the connection or the radiator.

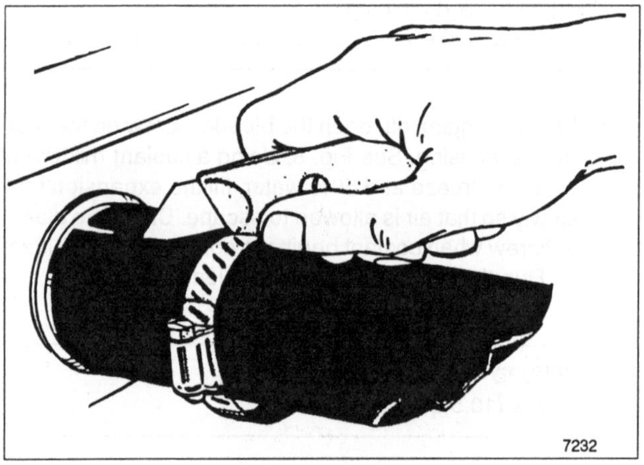

Fig. 9. Stuck hose being removed by cutting.

3. Clean the hose connections and remove any bits of old hose and sealer.
4. Install the new hose. Slide the loose clamps over the hose and slide the hose ends over the connections.
5. Position and tighten the clamps. Place the clamp as near the bead as possible and at least 4 mm (5/32 in.) from the hose end, as shown in Fig. 10. Tighten the clamps enough to compress the hose firmly around the connections.

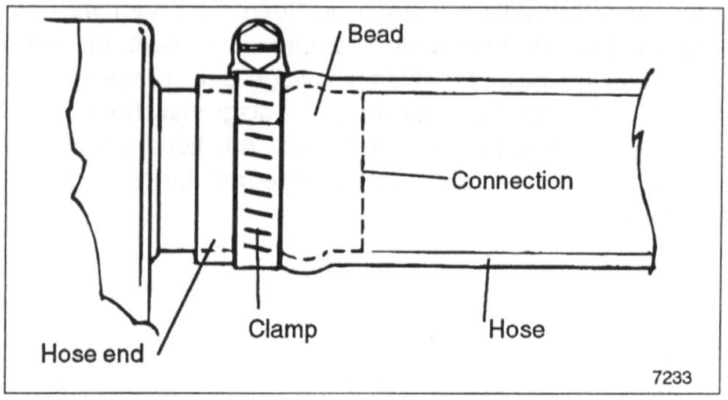

Fig. 10. Hose clamp correctly installed on hose end.

CAUTION
Do not overtighten clamps. Tighten just enough to seal. Overtightening may damage the hose and cause premature failure.

6. Refill and bleed the radiator as described above. Run the engine until warm. Check for leaks.

THERMOSTAT

To replace thermostat

1. Drain the coolant as described above.
2. Remove the thermostat housing bolts. Remove the housing and thermostat. See Fig. 8 above.
3. Position the new thermostat and gasket in the outlet. Install the thermostat housing.

Tightening torque
- thermostat housing to cylinder head . . . . 18 Nm (13 ft-lb)

CAUTION
Saab uses a special three stage thermostat. When replacing, be sure to use only an original Saab part. The engine should not be operated without a thermostat.

4. Refill and bleed the cooling system.

COOLANT PUMP

Fig. 11 is an exploded view of the coolant pump. Always plan to replace gaskets and seals and to have them on hand before beginning repairs.

To replace coolant pump

1. Disconnect the negative (–) battery cable.
2. Drain the cooling system as described above.

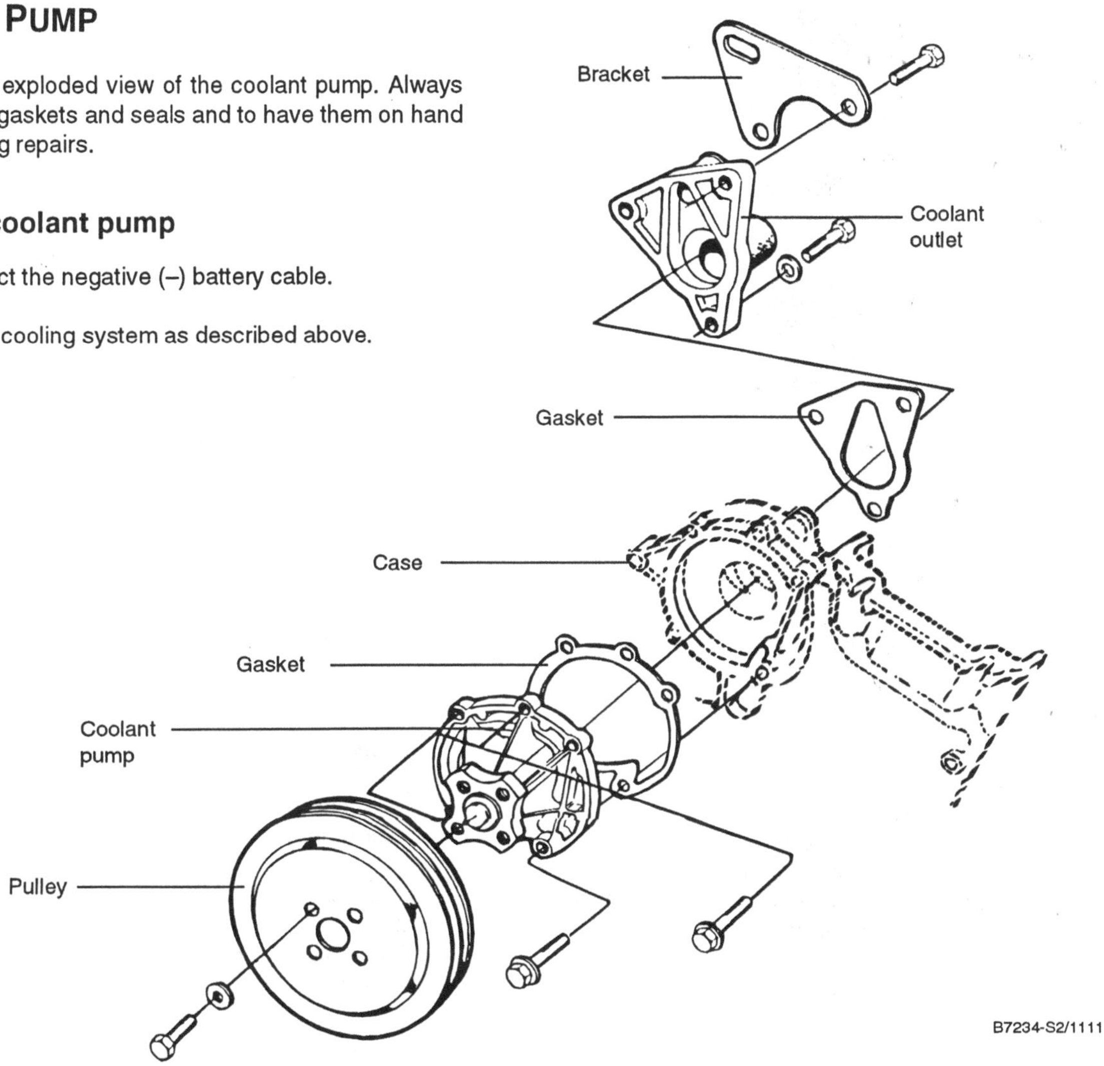

Fig. 11. Coolant pump exploded view.

3. Loosen the four bolts securing the pulley to the pump.
4. Remove the V-belts from the pump as described in **1 LUBRICATION AND MAINTENANCE**. Then remove the pulley mounting bolts and the pulley. See Fig. 12.
5. Remove the coolant pump bolts and remove the coolant pump from the case. See Fig. 13. Note the position of the two longer bolts.

Installation is the reverse of removal. Clean the pump mating surface before installing a new gasket. Install the V-belts as described in **1 LUBRICATION AND MAINTENANCE**. Fill and bleed the cooling system as described above. Start the engine and check for leaks.

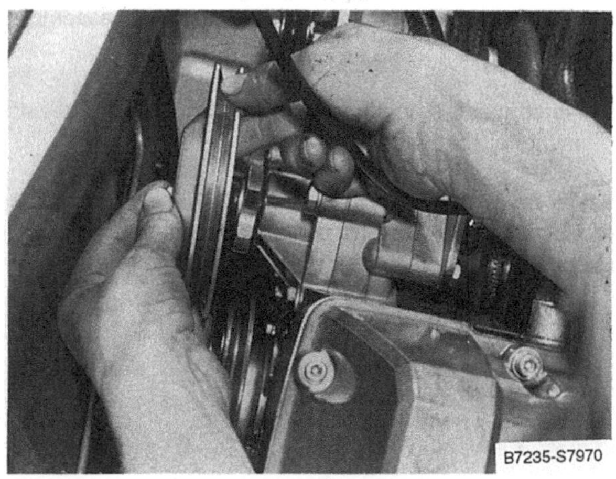

Fig. 12. Coolant pump pulley being removed.

Fig. 13. Coolant pump being removed.

RADIATOR AND COOLING FANS

Replacement radiators from Saab may not fit into some of 1989 and early cars without installing a special radiator mounting kit and making some minor modifications. See your Saab dealer for details.

To remove radiator

1. Disconnect the negative (–) battery cable. Disconnect the harness connectors from the fans, the ground straps, and the thermoswitch wires.

2. Drain the cooling system as described above. Disconnect all coolant hoses from the radiator.

3. Remove the air filter intake shroud and the distributor cap.

4. Remove the two bolts securing the radiator and remove the radiator and cooling fans as an assembly. See Fig. 14.

5. Working at the bench, remove the fan assemblies and any other hardware.

Installation is the reverse of removal. Inspect the rubber lower radiator mountings and replace any that are damaged. Fill and bleed the cooling system as described above.

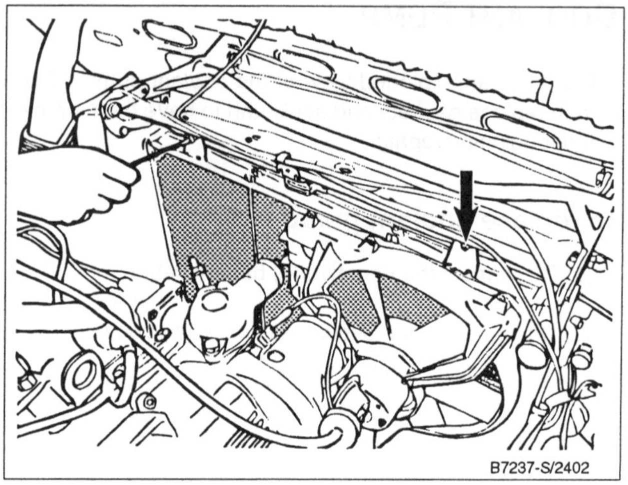

Fig. 14. Radiator mounting screw being removed. Also remove screw (arrow) on other side. Tilt radiator back to remove.

To replace radiator cooling fan thermoswitch

1. Disconnect the negative (–) battery cable.

2. Drain the cooling system as described earlier.

3. Disconnect the wires from the thermoswitch and unscrew the switch. Install the new switch. See Fig. 15.

NOTE

- Always the replace the sealing ring when installing a new thermoswitch.

- To test the thermoswitch, heat it in a pan of water with an ohmmeter connected to it terminals. Check the temperature of the water when the switch closes.

Radiator Cooling Fan Thermoswitch

- closes (fan on) . . . . 194–203°F (90–95°C)
- opens (fan off). . . . . 185–194°F (85–90°C)

---

4. Fill and bleed the cooling system. Start the engine and check for leaks. Always verify that both fans operate by allowing the engine to run at temperature and by switching on the A/C.

Fig. 15. Main cooling fan thermoswitch in left-side of radiator (arrow).

To remove radiator cooling fans

1. Disconnect the negative (–) battery cable. Disconnect the fan harness connector and the ground strap.

2. Remove the screws securing the fan shroud to the radiator. Remove the fan and shroud as a unit. See Fig. 16.

3. Working at the bench, remove the screws or nuts holding the fan motor to the shroud assembly.

Installation is reverse of removal. Use a thread locking compound such as Loctite® on the screws or nuts that secure the fan motor to the shroud. Always test fan operation after reinstalling the fan/shroud assembly.

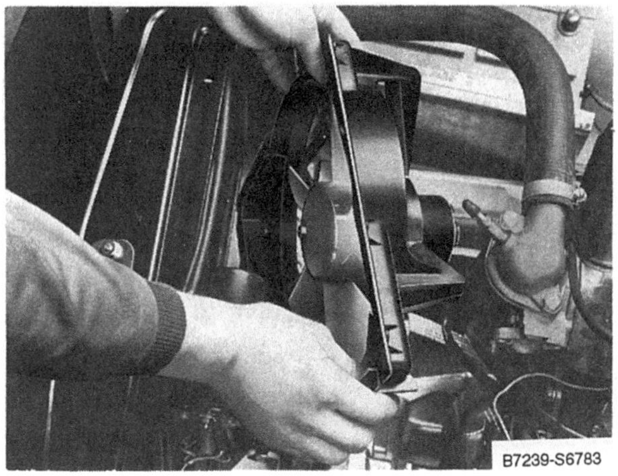

Fig. 16. Fan and shroud assembly being removed.

NOTE

- If the engine overheats even though the fan cycles on and off as specified, the fan may be running in reverse. Check that the fan is drawing air through the radiator and not blowing air out. Some fans were wired improperly at the connector. The wires in the fan connector should mate with the same color wires in the harness connector.

- Always check the tightness of the fan to the shroud. These fasteners can vibrate loose. Use a thread locking compound on all screws or nuts.

---

291 Turbocharger

TABLES

GENERAL

This section covers turbocharger and Automatic Performance Control (APC) system service and repair. Special test equipment is required for some of the test procedures. If this equipment is not available, the tests should be left to an authorized Saab dealer.

Turbocharger

A schematic of the turbocharger with intercooler is shown in Fig. 1. The exhaust-driven turbocharger rotates at very high speeds and is precisely balanced. The turbocharger is cooled and lubricated with engine oil. On 1988 and later models, the turbocharger is also water-cooled. The turbocharger's boost pressure is limited by a mechanical wastegate. When the boost pressure exceeds a predetermined level, the wastegate opens to bypass some of the exhaust gases around the turbine.

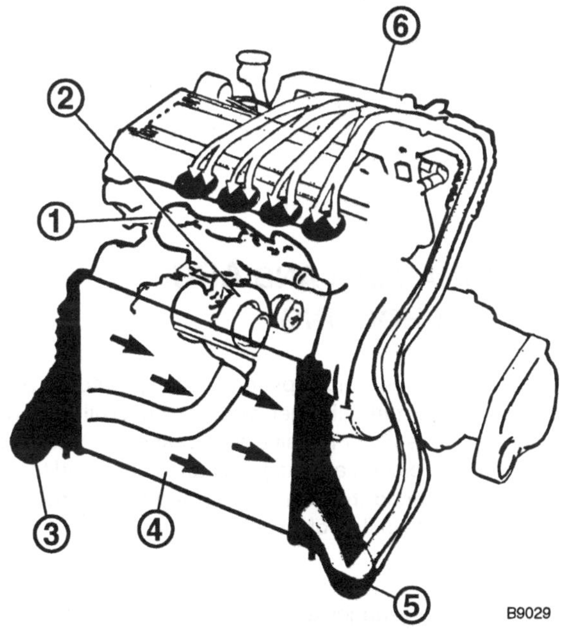

1. Exhaust manifold
2. Turbocharger
3. Intercooler (intake air)
4. Airflow
5. Intercooler outlet
6. Intake manifold

Fig. 1. Turbocharger system.

Automatic Performance Control (APC)

The Automatic Performance Control (APC) system enables the engine to achieve optimum performance and good fuel economy, regardless of the grade of fuel being used at any given time. The APC system consists of simple electronic components. See Fig. 2.

A knock sensor on the engine block senses the onset and degree of engine knock and transmits an electronic signal to the control unit. The control unit also receives intake manifold pressure information from a pressure sensor and engine speed information from the ignition system. Using this information, the APC control unit electrically modulates the APC solenoid valve to regulate boost pressure via the wastegate.

Through the APC system, the boost pressure is allowed to increase (up to a limit) until the engine begins to knock. It does this by venting some of the manifold pressure away from the wastegate and in effect "tricking" the wastegate that the boost pressure is lower than it really is. When the engine begins to knock, the solenoid valve closes and true boost pressure is applied to the wastegate diaphragm and the wastegate valve is pushed off its seat.

NOTE

In the event the APC system or the wastegate malfunction, allowing the boost pressure to exceed a safe level, the fuel pump will be turned off. On cars with LH 2.2, this feature is handled by an overpressure switch. On 1989 and later cars with LH 2.4, the LH control unit handles this feature. See **234 Fuel Pump and Fuel Tank** for more information on the overpressure switch.

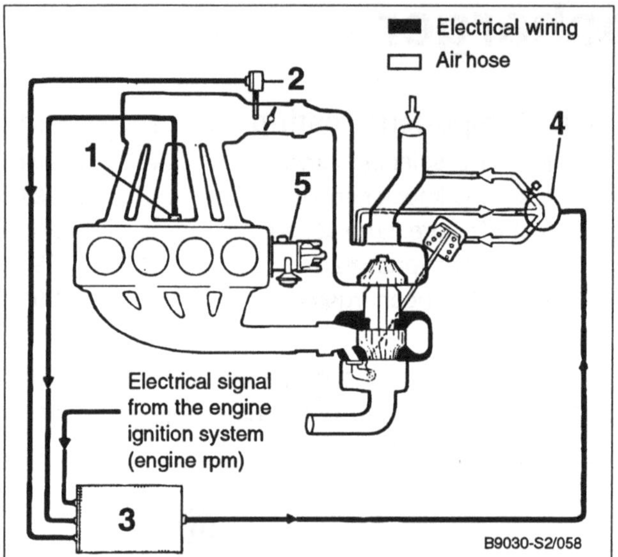

1. Knock sensor
2. Pressure sensor (transducer)
3. APC control unit
4. Solenoid valve
5. Ignition distributor

Fig. 2. Schematic of APC system.

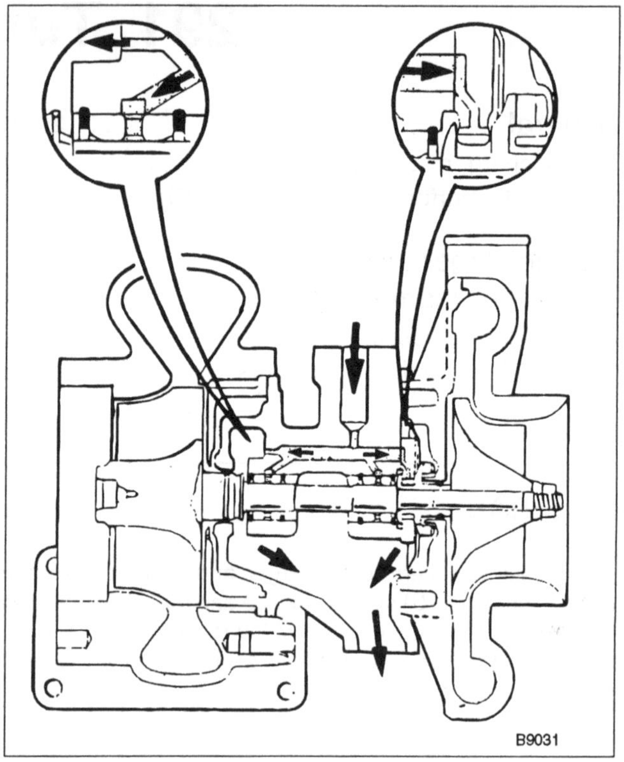

Fig. 3. Cutaway view of turbocharger showing lubrication circuit.

TURBOCHARGER AND APC TROUBLESHOOTING

Due to the extremely high speeds and temperatures, proper lubrication of the turbocharger is critical. Small oil passages are drilled into the turbo housing and bearings. Piston-type sealing rings are placed around the shaft at each end to prevent oil from entering the compressor and turbine housing. There is a high rate of flow of oil to ensure the shaft floats on a film of oil. See Fig. 3.

When the turbo bearings and seals become worn, the lubricating oil can slip past the seals and into the combustion chamber. The result is blue-gray oil smoke exiting through the tailpipe. When the bearings become severely worn, the turbocharger compressor and turbine impeller may contact the turbocharger housing, making loud screeching or thumping noises.

The major cause of turbocharger bearing and seal failure is due to carbon deposits caused by "coking" or baking of the oil in the turbocharger's oil passages. During operation, the temperature of the turbocharger housing can exceed the boiling point of the oil running through it. Because the oil is constantly circulating, it stays cool enough to prevent coking. If the oil pipes are clogged or if the engine is turned off before the turbocharger has had time to cool down, the oil can quickly coke. When coking deposits form, the oil flow to the turbocharger is reduced. In addition, this substance is abrasive and can wear the seals and bearings prematurely.

If none of the above faults are present, but engine performance and the turbocharger's output is low, first check for any loose or broken hoses. Check the small vacuum/pressure hoses as well as the larger intake air hoses. Be sure to check the hoses at the turbo bypass valve, which is directly below the turbocharger. Check the electrical connectors to the APC control unit (under rear seat on 1985 cars; at driver's side wheel housing on all 1986 and later cars), the solenoid valve (at radiator crossmember) and the pressure sensor (under left-hand side of instrument panel). If there are any faults in the APC system, the boost pressure will only boost to the "basic" pressure as described below.

In addition to the APC system components, there are two other components that can reduce the turbocharger's boost pressure. The cruise control vacuum switch and the brake light switch (1986 and later cars) will revert the system to the "basic" boost setting automatically whenever the brakes are applied or the cruise control system is activated. If either one of these components is faulty, the wastegate will open at a low boost level. Testing and adjusting information for the brake light switch is given in **364 Electric Controls and Switches.** Testing information for the cruise control vacuum switch is given in **368 Cruise Control.**

Besides low boost pressure, the boost pressure can also be too high, and this can result in complete engine failure. **Table a** lists turbocharger and APC system faults, their probable causes, and corrective actions.

> **NOTE**
> It is a good idea to measure the boost pressure as a first step in turbocharger and APC troubleshooting.

Turbo Boost Pressure

There are two boost pressure measurements that should be made to check the performance of the turbocharger and to check that the APC system is operating.

The first part of the pressure check measures basic boost. The basic boost pressure is the base or "basic" setting of the wastegate and is adjustable. This pressure setting is a safety feature to prevent engine damage caused by detonation. In the event of an APC component malfunction, the system reverts to the basic setting of the wastegate. The basic setting is also selected whenever the cruise control system is selected or the brake pedal is depressed (1986 and later models only).

The second part of the pressure check measures the efficiency of the turbocharger and also checks the operation of the APC system. This setting is not adjustable and is regulated based on the quality of the fuel in the tank. The better the quality of fuel in the tank, the higher the boost level.

Table a. Turbocharger and APC Troubleshooting

Symptom	Probable cause	Corrective action
1. Noise or vibration from turbo unit	**a.** Insufficient lubrication to turbo shaft and bearing	**a.** Remove and check the turbo oil supply pipe for clogging or deposits. Check the oil pressure and oil flow to the turbocharger
	b. Leakage in the intake or exhaust system	**b.** Check for leaking seals (O-rings) or pipe connections. Tighten all connections and replace defective seals
	c. Turbo shaft unbalanced	**c.** Replace turbo unit
2. Boost pressure too low	**a.** Leakage between compressor and cylinder head or turbine and exhaust system	**a.** Check for leaking pipe connections. Tighten all connections
	b. Incorrectly adjusted charge pressure	**b.** Check and adjust basic boost and maximum boost
	c. Wastegate sticking open	**c.** Clean and lubricate wastegate. Replace the wastegate assembly if it cannot be made to move freely
	d. Partially clogged exhaust system	**d.** Check exhaust system for blockage, especially in the catalytic converter. Replace faulty exhaust components
	e. Clogged air cleaner	**e.** Replace air cleaner
	f. Binding turbo shaft	**f.** Replace turbo unit
	g. Faulty brake light switch (1986 and later) or faulty cruise control vacuum switch	**g.** Test switches and replace if faulty
	h. Faulty APC component	**h.** Test the knock sensor, the pressure sensor, and the solenoid valve. If no faults are found, the APC control unit may be faulty
3. Boost pressure too high	**a.** Leak in hose at APC solenoid valve or faulty solenoid valve	**a.** Check for leaks and replace any faulty hoses. Test the APC solenoid and replace if faulty
	b. Leaky wastegate diaphragm unit	**b.** Replace diaphragm unit
	c. Leaking hose to pressure sensor or faulty pressure sensor	**c.** Check for leaks and replace any faulty hoses. Test the pressure sensor and replace if faulty
	d. Wastegate sticking closed or binding	**d.** Clean and lubricate wastegate. Replace the wastegate assembly if it cannot be made to move freely
	e. Incorrectly adjusted charge pressure	**e.** Check and adjust basic boost and maximum boost
	f. Faulty pressure sensor or APC control unit	**f.** Test pressure sensor. If no faults are found the APC control unit may be faulty
4. Normal boost pressure despite intensive knocking	**a.** APC system not working	**a.** Test APC system components
	b. Knock sensor or wiring to knock deter faulty	**b.** Test knock sensor and replace if faulty
	c. Knock sensor not correctly tightened to block	**c.** Check that sensor is torqued to 14 Nm (10 ft-lb)
	d. Wastegate sticking closed	**d.** Clean and lubricate wastegate. Replace the wastegate assembly if it can not be made to move freely

CAUTION
Do not measure the maximum turbocharger boost pressure by eliminating the knock sensor from the APC system. Although the system is designed to limit boost pressure based on the input from the pressure sensor (and at even higher levels, the fuel pump is turned off), the pressure at which this occurs may be too high if the fuel quality is low, resulting in engine failure.

To check and adjust boost pressure

An accurate vacuum/pressure gauge with a long extension hose is required to measure the boost pressure. See Fig. 4. Hang the gauge from the rear view mirror for ease of viewing to obtain the most accurate results. In addition, a helper is required to make the test described below.

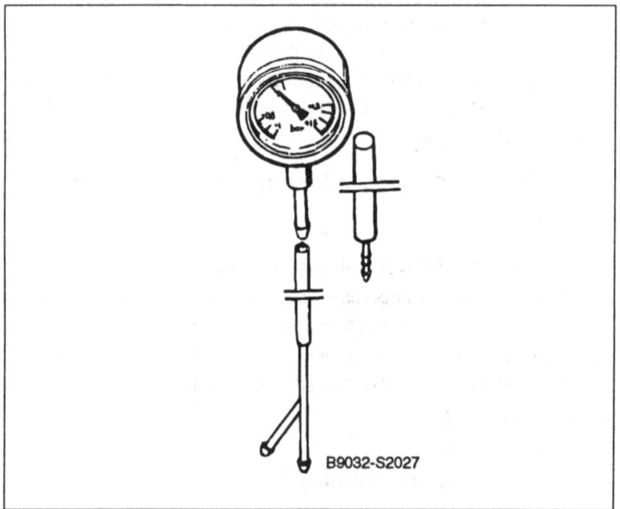

Fig. 4. Vacuum/pressure gauge (Saab tool no. 83 93 514) for measuring turbo boost pressure and for testing APC components.

WARNING
The following test requires that the car be driven on the road while checking the boost pressure on the gauge. This should always be done using a helper. Do not make the test if a helper is not available to read and record the gauge results.

CAUTION
It is recommended that the highest octane-rated fuel available be used when making this test. Otherwise the engine may knock continually during the test, possibly causing engine damage.

1. Run the engine until it is fully warmed up.

2. Working at the radiator crossmember, unplug the wiring connectors from the solenoid valve. See Fig. 5.

Fig. 5. APC solenoid valve wiring connectors being disconnected.

NOTE
When using Saab tool no. 83 94 074, the APC solenoid valve can be isolated from the system by turning the solenoid-valve switch off. Make sure the knock detector switch is also off. See Fig. 6.

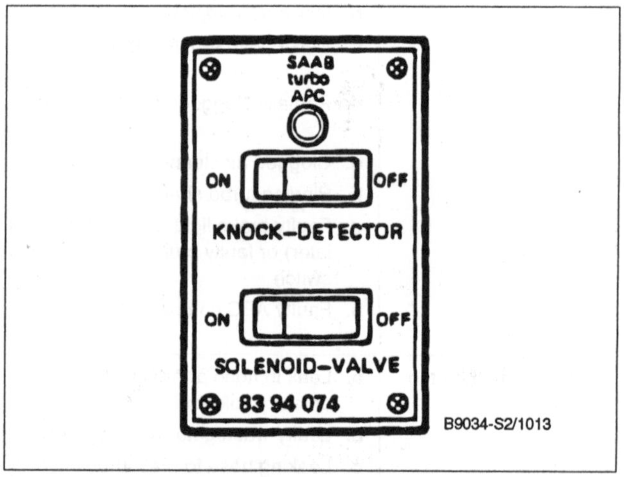

Fig. 6. Optional Saab APC test box (Saab tool no. 83 94 074). Switches should be set to off position.

3. Using a pressure/vacuum gauge with a long hose, connect the gauge hose between the fitting on the intake manifold and the hose leading to the pressure sensor. See Fig. 7. Place the gauge in the passenger compartment, preferably hanging vertically from the rear view mirror.

4. While driving the car on a long open stretch of road, place the transmission in 3rd gear with the engine speed at or below 1,500 rpm.

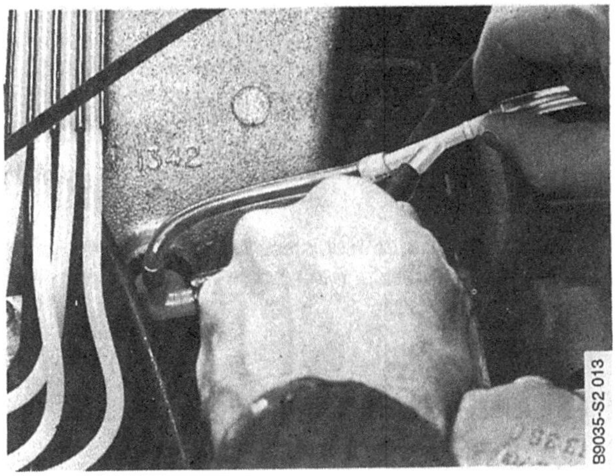

Fig. 7. Pressure/vacuum gauge hose being connected to fitting on intake manifold. Use a long hose so gauge can reach into passenger compartment.

NOTE
On cars with automatic transmission, place the transmission in the "1" drive position.

---

5. Depress the accelerator pedal to the floor until the engine speed approaches 3,000 rpm. Apply the brakes hard as necessary to keep rpm at 3,000 while continuing to hold the accelerator pedal to the floor. Within a short time, the gauge reading will stabilize. Have the helper record the gauge reading at this point.

CAUTION
Make the test as quickly as possible (within three to five seconds) to avoid overheating the brakes. If the test is to be made a second time, drive the car about one mile before repeating the test to allow the brakes to cool.

Turbocharger Basic Boost Pressure

- All ex. 1987 and later SPG models . . . . 0.35±0.03 bar (5.0±0.4 psi)
- 1987 and later SPG models . . . . 0.40±0.03 bar (5.8±0.4 psi)

---

6. If the basic boost setting is incorrect, allow the turbocharger to cool. Then remove the anti-tamper seal (if installed) and the circlip from the end of the wastegate actuator rod. Remove the rod from the wastegate lever. See Fig. 8.

CAUTION
Use extreme care when working near a hot turbocharger.

NOTE
Some late models may not have the wastegate seal installed, as the seal has been discontinued in production and does not need to be reinstalled.

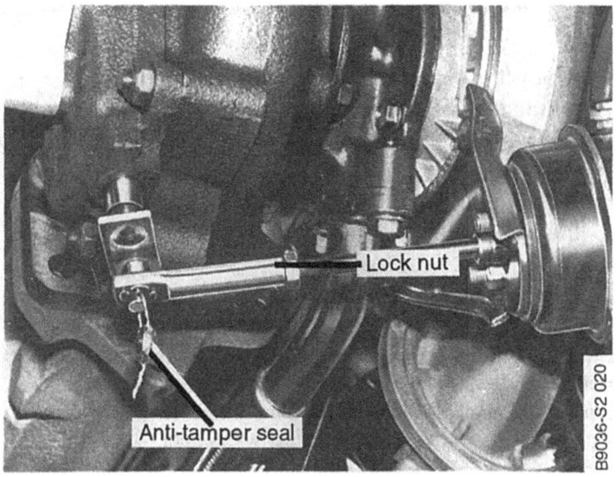

Fig. 8. Wastegate actuator (diaphragm) rod.

7. While holding the rod stationary, loosen the locknut and thread the end piece in or out to readjust the length of the rod. See Fig. 9. Then tighten the locknut. Reinstall the rod to the lever. Recheck the basic boost pressure as described above. If necessary, readjust the length of the rod.

CAUTION
Do not allow the wastegate actuator rod to turn the diaphragm or the diaphragm may be damaged.

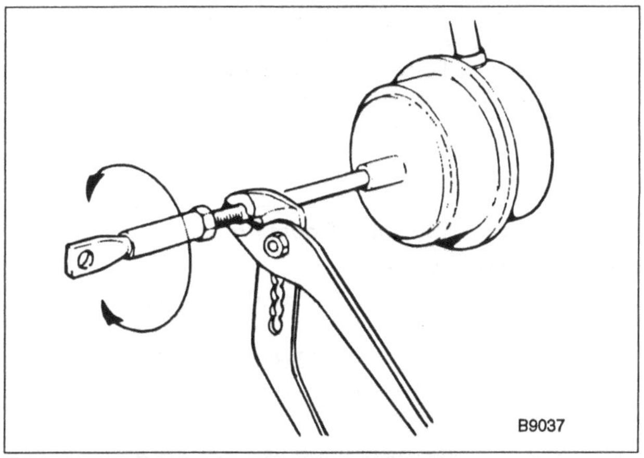

Fig. 9. Hold wastegate actuator rod stationary with pliers when adjusting the end piece or tightening the locknut.

NOTE
If the charging pressure is too low, turn the end piece clockwise to shorten the rod. If the pressure is too high, turn the end piece counter clockwise to lengthen the rod. Adjust the rod in small increments. One complete turn of the end piece roughly equates to a 0.04 bar (0.6 psi) pressure change.

8. When the basic boost pressure setting is correct, reconnect the wiring connectors to the solenoid valve. Keep the vacuum/ pressure gauge connected.

9. On 1986 and later cars, disconnect the white wire from the brake light switch above the brake pedal.

NOTE
On 1986 and later cars, a voltage signal via the brake light switch (white wire) is applied to the APC control unit when the brakes are applied. This reduces the maximum boost pressure to the basic boost setting of the wastegate.

10. Again, drive the car on a long open stretch of road and repeat the above test. In a short time, the gauge reading should stabilize. Have the helper record the reading at this point.

CAUTION
Make the test as quickly as possible. (in three to five seconds) to avoid overheating the brakes. If the test needs to be made a second time, drive the car about one mile before repeating the test to allow the brakes to cool.

NOTE
The specifications listed below should be used only as a guide. Do not expect the gauge reading to reach this level, as this maximum pressure is controlled by the signal from the knock sensor. When engine knocks, the wastegate will open, resulting in gauge reading considerably lower than that specified.

Turbocharger Maximum Boost Pressure
- All ex. 1987 and later SPG models . . . . 0.75±0.05 bar (10.9±0.7 psi)
- 1987 and later SPG models . . . . 0.85±0.05 bar (12.3±0.7 psi)

11. If the pressure stabilizes at the basic setting or goes above the maximum setting, test the turbocharger APC components as described below. If the pressure reading is only slightly above the basic setting even though the gas is of high-quality, the turbocharger may be faulty. See the troubleshooting table given earlier.

CAUTION
Do not attempt to adjust the maximum boost pressure through the wastegate actuator rod. This will not increase the boost pressure or engine performance, as maximum boost pressure is controlled by the APC system.

12. When the pressure test is complete, remove the pressure gauge and reconnect the wiring at the brake light switch, if applicable.

APC COMPONENT TESTING

To test APC solenoid valve

The APC solenoid valve is mounted in the engine compartment, at the radiator crossmember. Based on the signal from the APC control unit, the valve regulates the control pressure to the wastegate. See Fig. 10. When the solenoid is closed (connection between port C and W only), the wastegate will open at a low (basic) boost pressure. When the solenoid is open (all three ports open), there is not enough pressure to open the wastegate and the boost pressure is allowed to increase until engine knock is detected.

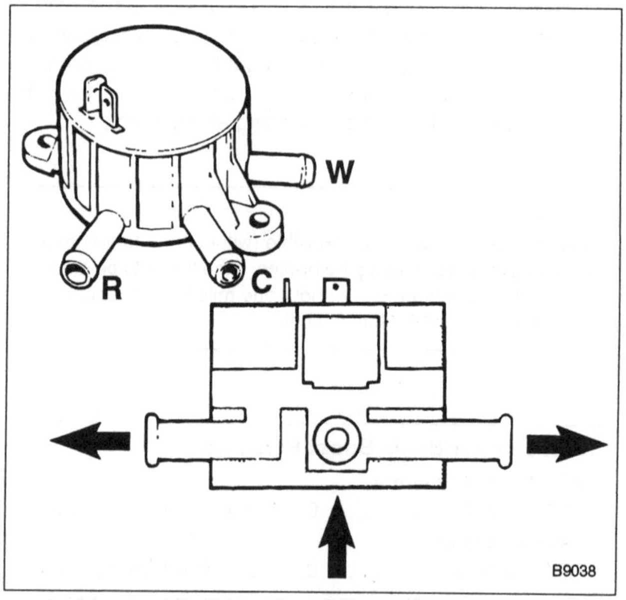

Fig. 10. APC solenoid valve. Port **C** is the pressure line from the turbo compressor (true manifold pressure), **W** is the pressure line to the wastegate, and **R** is a bleed or return line back to the turbo compressor.

1. Disconnect the electrical connectors from the APC solenoid valve. See Fig. 5 given above.

2. Disconnect the hose between the turbo compressor inlet and the connection marked "R" on the solenoid.

3. Using jumper wires, connect battery voltage to the solenoid terminals. Blow through the disconnected hose (port R) and check that air passes freely through the valve. See Fig. 11.

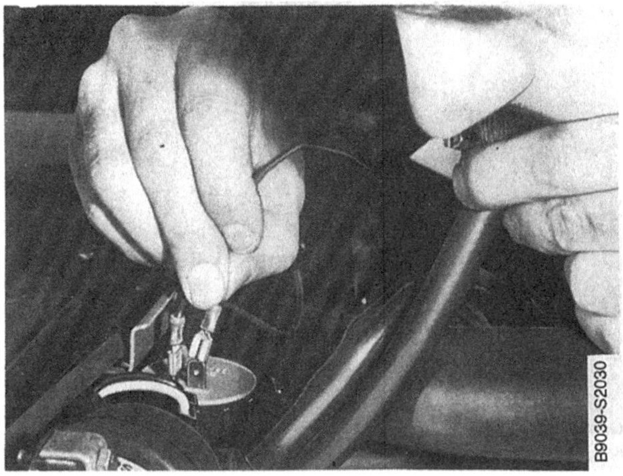

Fig. 11. APC solenoid being tested. When battery voltage is applied to terminals, air should pass through valve.

4. Remove the voltage and check air does not pass through the valve. If any faults are found, the valve is faulty and should be replaced.

To test APC knock sensor

The knock sensor is mounted to the engine block between cylinders no. 2 and 3, below the intake manifold. See Fig. 12. The sensor functions like a microphone and is able to convert mechanical vibration (knock) into a small AC voltage signal. The APC control unit is programmed to react to the signals whose frequencies are characteristic of combustion chamber noise.

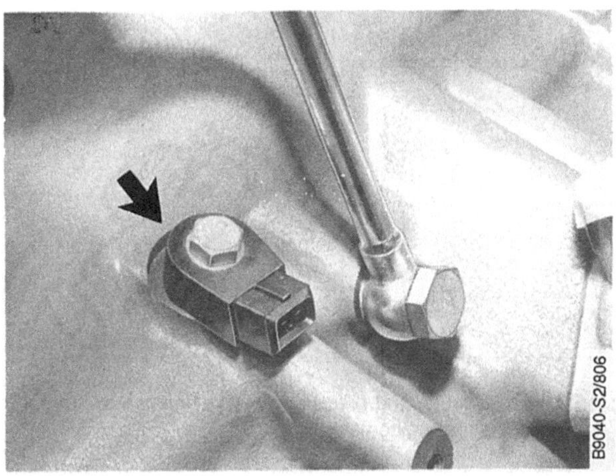

Fig. 12. APC knock sensor (arrow). Sensor is bolted to engine block beneath intake manifold.

1. Disconnect the harness connector from the APC control unit. Using a digital voltmeter (set to VAC scale), connect it to control unit terminals 2 and 3 on 1985 models or 16 and 17 on all 1986 and later models. See Fig. 13.

NOTE

On 1985 models, the APC control unit is under the rear seat. On 1986 and later models, the control is mounted to the left-hand side wheel well in the engine compartment.

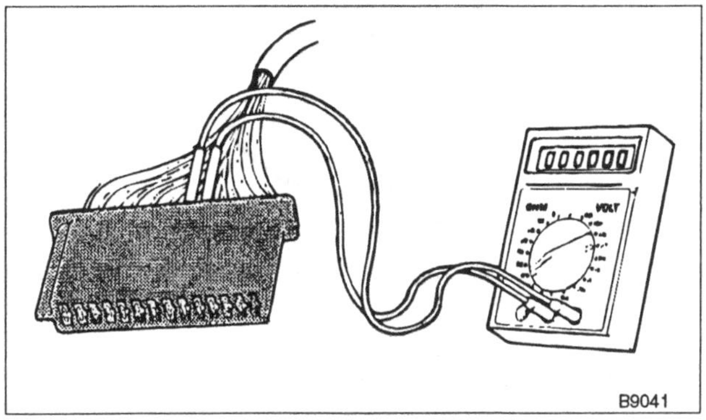

Fig. 13. Voltmeter connected to knock sensor terminals in APC control unit connector. Back-probe connector to prevent damaging connector terminals.

2. Gently tap on the knock sensor mounting bolt. The voltmeter should register a small voltage and the reading should vary depending on the intensity of the tap.

3. If the knock sensor does not respond to the tapping, remove the sensor from the engine. Then connect the voltmeter directly to the sensor terminals and repeat the above tapping test. If the sensor now tests good, check the wiring from the sensor connector to the control unit.

If the sensor does not respond as specified, it is faulty and should be replaced. When replacing the sensor, thoroughly clean the mounting surface on the engine and the threads on the mounting bolt. Install the new sensor with the electrical connector towards the front, offset to the left about 20°. See Fig. 14. Torque the bolt to the specification given below while holding the sensor steady.

Tightening torque

- APC knock sensor to engine block .... 14 Nm (10 ft-lb)

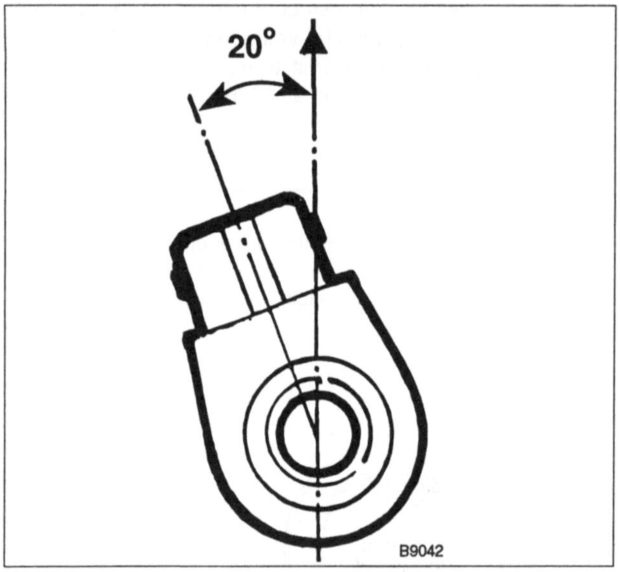

Fig. 14. Knock sensor installation angle. Hold sensor steady when tightening bolt.

To test APC pressure sensor

The sensor monitors the turbo boost pressure in the manifold downstream of the throttle. The pressure is converted into an electrical resistance signal and is monitored by the control unit.

The pressure sensor is mounted beneath the left-hand (driver's) side of the dashboard. A hand-pump and pressure gauge is required to accurately check the sensor.

1. Disconnect the harness connector from the APC control unit. Using a digital voltmeter (set to resistance scale), connect it to control unit terminals 8 and 9 on 1985 models or 10 and 23 on all 1986 and later models. Check the sensor's resistance. If any faults are found, the sensor should be replaced.

Pressure Sensor Test Specifications
- resistance at atmospheric pressure . . . . 5 to 13 ohms

NOTE
On 1985 models, the APC control unit is under the rear seat. On 1986 and later models, the control is mounted to the left-hand side wheel well in the engine compartment.

2. Remove the center storage console and the lower dashboard pad as described in **853 Dashboard and Consoles.**

3. Unscrew the pressure sensor mounting bracket from the relay panel member to access the sensor. See Fig. 15.

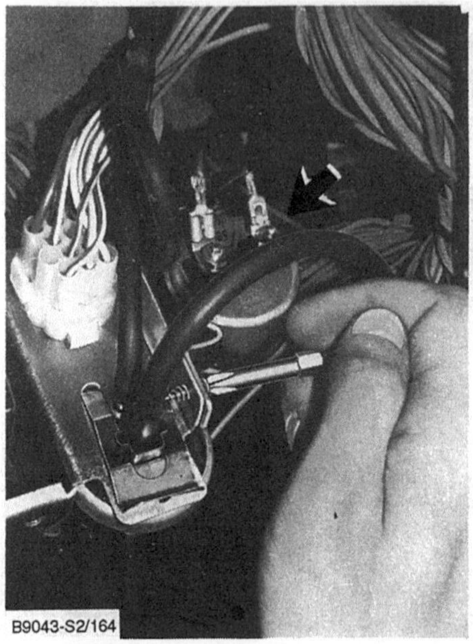

Fig. 15. Pressure sensor mounting bracket beneath left hand side of dashboard. Pressure sensor shown at arrow.

4. Disconnect the hose leading to the sensor and connect a pressure gauge and hand-pump. See Fig. 16. Pump the pressure up to 1 bar (14.5 psi), then reduce the pressure to 0.6 bar (8.7 psi) while tapping lightly on the side of the sensor. Check the sensor's resistance. If any faults are found, the sensor should be replaced.

Pressure Sensor Test Specifications
- resistance at 0.6 bar (8.7 psi) pressure . . . . 83 to 92 ohms

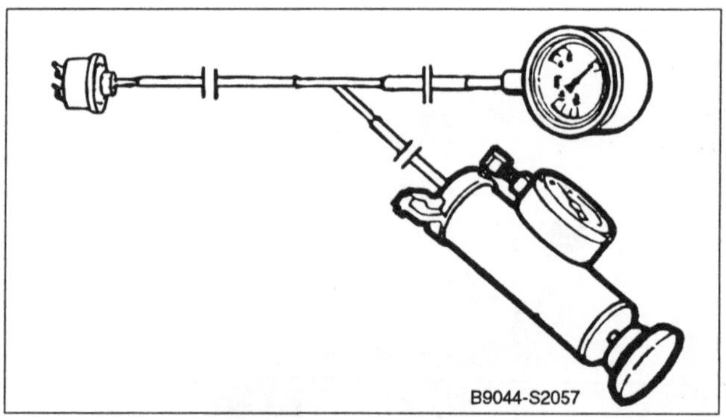

Fig. 16. Pressure gauge and pressure pump connected to pressure sensor (shown schematically).

To test APC bypass valve

The vacuum controlled bypass valve redirects turbocharger output (boost) to the suction side of the compressor when the throttle is snapped shut. This dampens the distribution of intake air flow that results when full boost is suddenly bounced *off the closed* throttle plate. See Fig. 17.

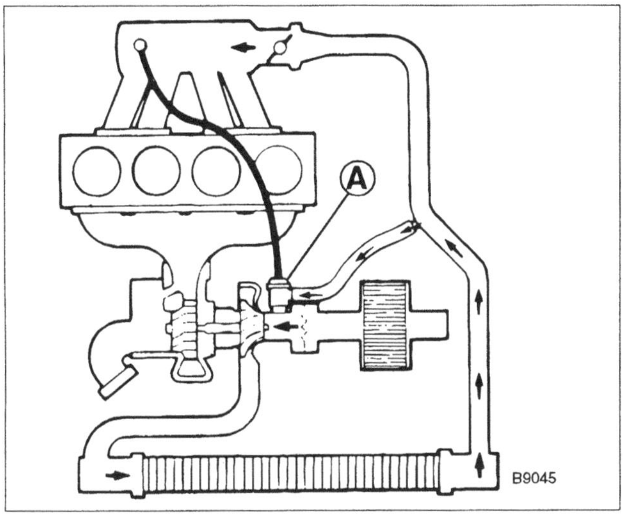

Fig. 17. Schematic of APC system showing the operation of the bypass valve (**A**).

The symptoms of a faulty bypass valve are usually minor driveability problems (slight surging when opening the throttle on cars with manual transmission), and noises (hooting sounds). The bypass valve is mounted below the center of the turbocharger. Use a hand-held vacuum pump to accurately test the valve.

NOTE
Some early 1985 models with automatic transmission may not have a bypass valve.

---

1. With the engine cold, remove the bypass valve below the turbocharger.

WARNING
Do not remove the bypass valve when the engine is hot. The turbocharger operates at very high temperatures.

---

2. Using a hand-pump, apply vacuum to the fitting on the valve and check that it lifts off its seat. See Fig. 18. With the vacuum applied, check that the valve stays open. If the valve slowly closes, the diaphragm is leaking and the valve should be replaced.

NOTE
Inspect the hose clamps on the bypass valve. If the old style plastic clamps are installed, they should be replaced with metal screw-type clamps. The plastic clamps are known to break easily, allowing the pressure from the turbocharger to blow the hoses off the valve.

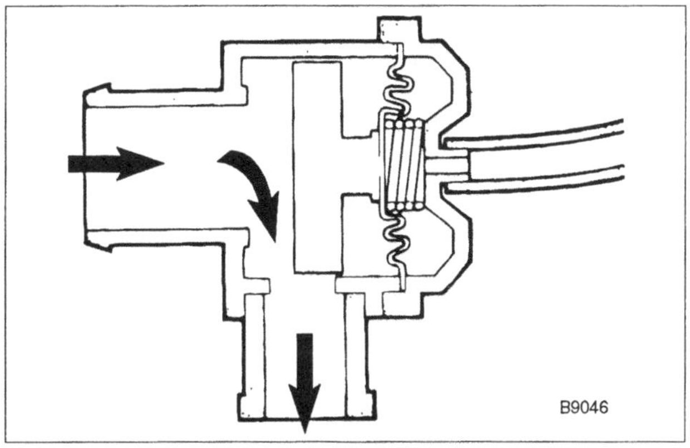

Fig. 18. APC bypass valve shown with vacuum applied to diaphragm. The valve diaphragm should hold valve stay off its seat.

Replacing Turbocharger

Turbocharger removal and installation is done by first removing the exhaust manifold with the turbocharger as a complete unit. with the manifold removed, unbolt the turbocharger from the manifold. See **252 Exhaust System** for removal and installation of the exhaust manifold.

NOTE
The turbocharger-to-exhaust manifold mounting nuts are self-locking and should always be replaced.

When replacing the turbocharger, the oil return and supply pipes should be removed and thoroughly cleaned using a stiff brush and solvent. See Fig. 19. If the pipes cannot be completely cleaned, they should be replaced. The engine oil and oil filter should also be changed.

NOTE
If the turbocharger is being replaced because of excessive exhaust smoke, there will be some oil remaining in the exhaust system that will most likely cause smoking upon start-up. Be sure to test drive the car long enough to clear the leftover oil from the exhaust system before suspecting that the replacement turbocharger is faulty.

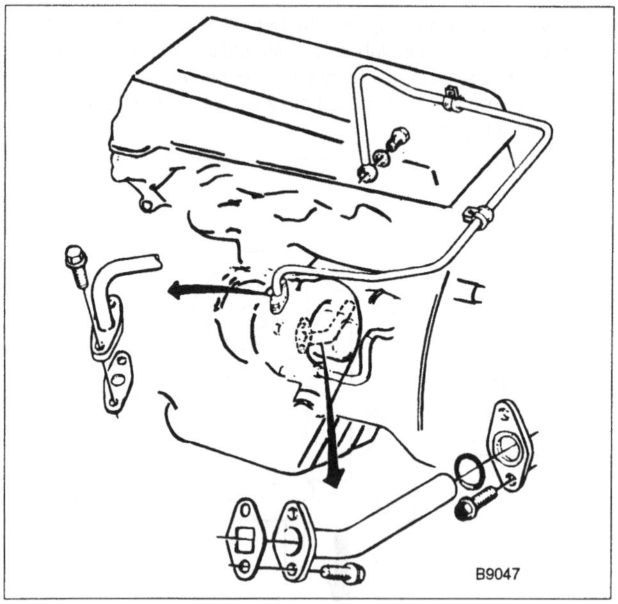

Fig. 19. Turbocharger oil supply and return pipes.

300 Electrical System —General

TABLES

300

GENERAL

A brief description of the principal parts of the electrical system is presented here. Also covered here are general electrical system troubleshooting tips as well as instructions on using the Saab wiring diagrams.

Voltage and Polarity

Saab electrical systems are 12-volt direct current (DC) negative-ground systems. A voltage regulator controls the system voltage at approximately the 12-volt rating of the battery. All circuits are grounded by direct or indirect connection to the negative (–) terminal of the battery. A number of ground connections throughout the car connect the wiring harness to chassis ground. These circuits are completed by the battery cable or ground strap between the body and the battery negative (–) terminal.

Wiring, Fuses, and Relays

Nearly all parts of the wiring harness connect to components of the electrical system with keyed, push-on connectors that lock into place. Notable exceptions are the heavy battery cables and the alternator wiring. The wiring is color-coded for circuit identification.

With the exception of the battery charging system, all electrical power is routed from the ignition switch or the battery through the fuse/relay panel, located in the right side of the engine compartment. Fuses prevent excessive current from damaging components and wiring. Fuses are color coded to indicate their different current capacities.

Fuse rating and color

Color	Rating
Red	10 amp
Blue	15 amp
Yellow	20 amp
Green	30 amp

The relays are electromagnetic switches that operate on low current to switch a high-current circuit on and off. Many of the relays are mounted on the fuse/relay panel in the engine compartment, although later models are equipped with additional auxiliary relay panels. For information on relay and fuse locations, see **371 Wiring Diagrams, Fuses and Relays.**

Electrical System Safety Precautions

Please read the following warnings and cautions before doing any work on your electrical system.

WARNING
- Some of the cars covered by this manual are equipped with a Supplemental Restraint System (SRS) that automatically deploys an airbag. The airbag unit uses an explosive device to electrically ignite a powerful gas. On cars so equipped, any work involving the steering wheel and SRS system should only be performed by an authorized Saab dealer. Performing repairs without disarming the SRS may cause serious personal injury.

- On cars equipped with SRS (airbag) the ignition switch must be in the lock position before replacing fuses. If fuse no. 7 is removed with the ignition switch on, an SRS fault will be detected. If the SRS warning light comes on, see an authorized Saab dealer.

WARNING

- The ignition system of the car operates at lethal voltages. People with pacemakers or weak hearts should not expose themselves to the ignition system. Extra caution must be taken when working on the ignition system or when servicing the engine while it is running or the key is on. See **340 Ignition System** for additional ignition system warnings and cautions.

- Before operating the starter without starting the engine (as when making a compression test), disable the ignition system as described in **340 Ignition System.**

- Keep hands, clothing and other objects clear of the radiator cooling fan when working on a warm engine. The fan may start at any time, even when the ignition is switched off.

CAUTION

- Always turn off the engine and remove the negative (–) battery cable before removing any electrical components.

- Connect and disconnect ignition system wires, multiple connectors, and ignition test equipment leads only while the ignition is off.

- Do not disconnect the battery while the engine is running.

- Do not quick-charge the battery (for boost starting) for longer than one minute, and do not exceed 16.5 volts at the battery with the boosting cables attached. Wait at least one minute before boosting the battery a second time.

- Do not use a test lamp that has a normal incandescent bulb to test circuits containing electronic components. The high electrical consumption of these test lamps may damage the components.

- Many of the solid-state modules are static sensitive. Static discharge will permanently damage them. Always handle the modules using proper static prevention equipment and techniques.

- To avoid damaging harness connectors or relay panel sockets, use jumper wires with flat-blade connectors that are the same size as the connector or relay terminals.

- Always switch a test meter to the appropriate function and range before making test connections.

- Do not try to start the engine of a car which has been heated above 176°F (80°C), (for example, in a paint drying booth) until allowing it to cool to normal temperature.

- Disconnect the battery before doing any electric welding on the car.

- Do not tow a car suspected of having a defective ignition system without first disconnecting the ignition control unit.

- Do not wash the engine while it is running, or anytime the ignition is switched on.

Electrical Test Equipment

Many of the electrical tests described in this manual call for measuring voltage, current or resistance using a digital multimeter (DMM). Digital meters are preferred for precise measurements and for electronics work because they are generally more accurate than analog meters. The numerical display is also less likely to be misread, since there is no needle position to be misinterpreted by reading at an angle.

An LED test light is a safe, inexpensive tool that can be used to perform many simple electrical tests that would otherwise require a multimeter. The LED indicates when voltage is present between any two test-points in a circuit.

CAUTION

- Choose test equipment carefully. Use a meter with at least 10 megohm input impedance, or an LED test light. An analog meter (swing-needle) or a test light with a normal incandescent bulb may draw enough current to damage sensitive electronic components.

- An ohmmeter must not be used to measure resistance on solid state components such as control units or time delay relays.

- Always disconnect the battery before making resistance (ohm) measurements on the circuit.

Wiring Diagrams

The wiring diagrams shown in **371 Wiring Diagrams, Fuses and Relays** have been specially designed to enable quick and efficient diagnosis and troubleshooting of electrical malfunctions.

Each electrical sub-system, such as the ignition system, the hazard warning lights, etc., is described individually and a separate wiring diagram is shown for each sub-system on the car. Each wiring sub-system consists of a spread that includes a brief description of the circuit operation, fault tracing, a list of all the components used in the circuit along with their location in the car, and the wiring diagram(s) applicable to that circuit.

Wiring Codes and Abbreviations

A tremendous amount of information is included in each wiring diagram if you know how to read them. For example, you will notice that all electrical components, connectors, fuses, and ground locations are identified using a unique number. Each of these numbers corresponds to a particular part in the circuit. A complete listing of these component numbers, their identifications and locations is given in **371 Wiring Diagrams, Fuses and Relays.**

All wire colors in the diagrams are abbreviated. Combined color codes indicate a multi-colored wire. For example the code BL/RD indicates a Blue wire with a Red stripe. Wire color abbreviation are listed below. Immediately following the wire color is the nominal cross-sectional area of the wire given in mm^2.

NOTE

Sometimes the color of an installed wire may be different than the one on the wiring diagram. Don't be concerned, just be sure to confirm that the wire connects to the proper terminals.

Wire color codes

- BL or BU . . . . . . Blue
- BR or BN . . . . . . Brown
- GL or YE . . . . . . Yellow
- GN . . . . . . Green
- GR or GY . . . . . . Gray
- OG . . . . . . Orange
- RA or PK . . . . . . Pink
- RD . . . . . . Red
- SV or BK . . . . . . Black
- VL . . . . . . Violet
- VT or WH . . . . . . White

Most terminals are identified by numbers on the components and harness connectors. The terminal numbers for major electrical connections are shown in the diagrams. Though many terminal numbers appear only once, several other numbers appear in numerous places throughout the electrical system and identify certain types of circuits. Several of the most common circuit numbers are listed below in **Table a.**

Table a. Terminal and Circuit Numbers

Number	Circuit description
1	Low voltage switched (1) terminal of coil
4	High voltage center terminal of coil
+X	Originates at ignition switch. Supplies power when the ignition switch is in the PARK, RUN, or START position
S	Originates at ignition switch. Supplies power whenever the key is in the ignition switch. (power in LOCK, PARK, RUN, or START position)
15 or +15	Originates at ignition switch. Supplies power when ignition switch is in RUN or START position
30 or +30	Battery positive (+) voltage. Supplies power whenever battery is connected. (Not dependent on ignition switch position, unfused)
31	Ground, battery negative (–) terminal

continued

Table a. Terminal and Circuit Numbers (continued)

Number	Circuit description
50 or +50	Supplies power from battery to starter solenoid when ignition switch is in START position only
+54	Originates at ignition switch. Supplies power when ignition switch is in the RUN position only
85	Ground side (–) of relay coil
86	Power-in side (+) of relay coil
87	Relay change-over contact
D	Alternator warning light and field energizing circuit

Additional abbreviations shown in the wiring diagrams are given below.

Abbreviations

- ABS . . . . . . antilock brakes
- AC . . . . . . air conditioning
- AIC . . . . . . automatic idle control
- APC . . . . . . automatic performance control (turbo models)
- AUT . . . . . . automatic transmission
- CAB . . . . . . cabriolet (convertible)
- CC . . . . . . cruise control
- CONV . . . . . . convertible
- ECU . . . . . . electronic control unit
- EZK . . . . . . electronic ignition with knock control
- I . . . . . . fuel injection engine
- LH . . . . . . fuel injection system
- LHD . . . . . . left hand drive
- LHF . . . . . . left hand, front
- LHR . . . . . . left hand, rear
- LHS . . . . . . left hand, side
- M__ . . . . . . designation for model year, i.e. M89 for 1989 model
- MAN . . . . . . manual transmission
- P . . . . . . passenger
- RHD . . . . . . right hand drive
- RHF . . . . . . right hand, front
- RHR . . . . . . right hand, rear
- RHS . . . . . . right hand, side
- SRS . . . . . . supplemental restraint system Airbag (see warnings in text)
- T . . . . . . turbo
- TSI . . . . . . timing service instrument (test socket)
- 2-D . . . . . . 2-door sedan
- 3-D . . . . . . 3-door hatchback
- 4-D . . . . . . 4-door sedan
- US . . . . . . Cars for sale in the United States
- CA . . . . . . Cars for sale in Canada
- UC . . . . . . Cars for sale in California

CAUTION

Many of the wiring diagrams are applicable to cars delivered to other world markets served by Saab. At times, there may be certain parts of the diagram that are applicable only to these markets. This manual covers only Saab cars delivered to the United States (US), California (UC), and Canada (CA). Abbreviations other than US, UC, and CA found on a particular diagram should be ignored. For example the abbreviation "ME" is for cars sold in the Middle East. **Table b** is a complete list of the market codes.

Table b. Market Codes

Market codes	Market	Market codes	Market
AT	Austria	FR	France
AU	Australia	GB	Great Britain
BE	Belgium	GR	Greece
CA	Canada	IS	Iceland
CH	Switzerland	IT	Italy
DE	Germany	JP	Japan
DK	Denmark	ME	Middle East
ES	Spain	NL	Netherlands
EU	Europe	NO	Norway
FE	Far East	SE	Sweden
FI	Finland	US	USA

Using the Wiring Diagrams

An example of a wiring diagram is shown below together with an explanation of the symbols used in the diagram. See Fig. 1.

NOTE

- Unless specified otherwise, switches and relays are shown in the rest or de-energized position.

- Throughout the wiring diagrams, a fuse is shown together with the designation **22A.** This 22A code is the component number for the fuse holder in the fuse/relay panel and should not be confused with the fuse rating or the fuse position.

- Designations such as **+30**, **+15**, **+54** are positive voltage supplies, usually found at the top of each circuit. Each number represents when voltage is present in the circuit. For example, +30 is live at all times (unfused) and +15 is live when the key is in the run or start position. See above for a listing of the common terminal and circuit numbers.

ELECTRICAL TROUBLESHOOTING

Four things are required for current to flow in any electrical circuit: a voltage source, wires or connections to transport the voltage, a consumer or device that uses the electricity, and a connection to ground. Most problems can be found using only a digital multimeter (volt/ohm/amp meter) to check for voltage supply, for breaks in the wiring (infinite resistance/no continuity), or for a path to ground that completes the circuit.

Electric current is logical in its flow, always moving from the voltage source toward ground. Keeping this in mind, electrical faults can be located through a process of elimination. When troubleshooting a complex circuit, separate the circuit into smaller parts. The general tests outlined below may be helpful in finding electrical problems. The information is most helpful when used with the wiring diagrams.

Be sure to analyze the problem. Use the wiring diagrams to determine the most likely cause of the problem. Get an understanding of how the circuit works by following the circuit from ground back to the power source.

You will find the problem if you follow a simple and logical step-by-step procedure. Test portions of the circuit at one time, starting with the area or component most likely to be at fault. Test first at points that you can reach most easily. When you find the cause of the problem, make the repair. Use appropriate tools and procedures. As a final check, test the functions of the circuit that you worked on.

When making test connections at connectors and components, use care to avoid spreading or damaging the connectors or terminals. Some electrical tests may require jumper wires to temporarily bypass components or connections in the wiring harness. When connecting jumper wires, use blade connectors at the wire ends that match the size of the terminal being tested. The delicate internal contacts are easily spread apart, and this can cause intermittent or faulty connections that can lead to more problems.

Checking for Voltage and Ground

Checking for the presence of voltage or ground is usually the first step in troubleshooting a problem circuit. When checking for voltage, a digital voltmeter or LED test light should be used. for example, if a parking light does not work, check for voltage at the bulb socket will quickly determine if the circuit is functioning properly or if the bulb itself is faulty. If voltage and ground are found at the socket, then the bulb is most likely faulty.

Another valuable troubleshooting technique is a voltage drop test. This is a good test to make if current is flowing through the circuit, but the circuit is not operating correctly. Sluggish wipers or dim headlights are examples of this. A voltage drop test will help to pinpoint a corroded ground strap or a faulty switch. Normally, there should be less than 1 volt drop across most wires or closed switches. A voltage drop across a connector or short cable should not exceed 0.5 volts.

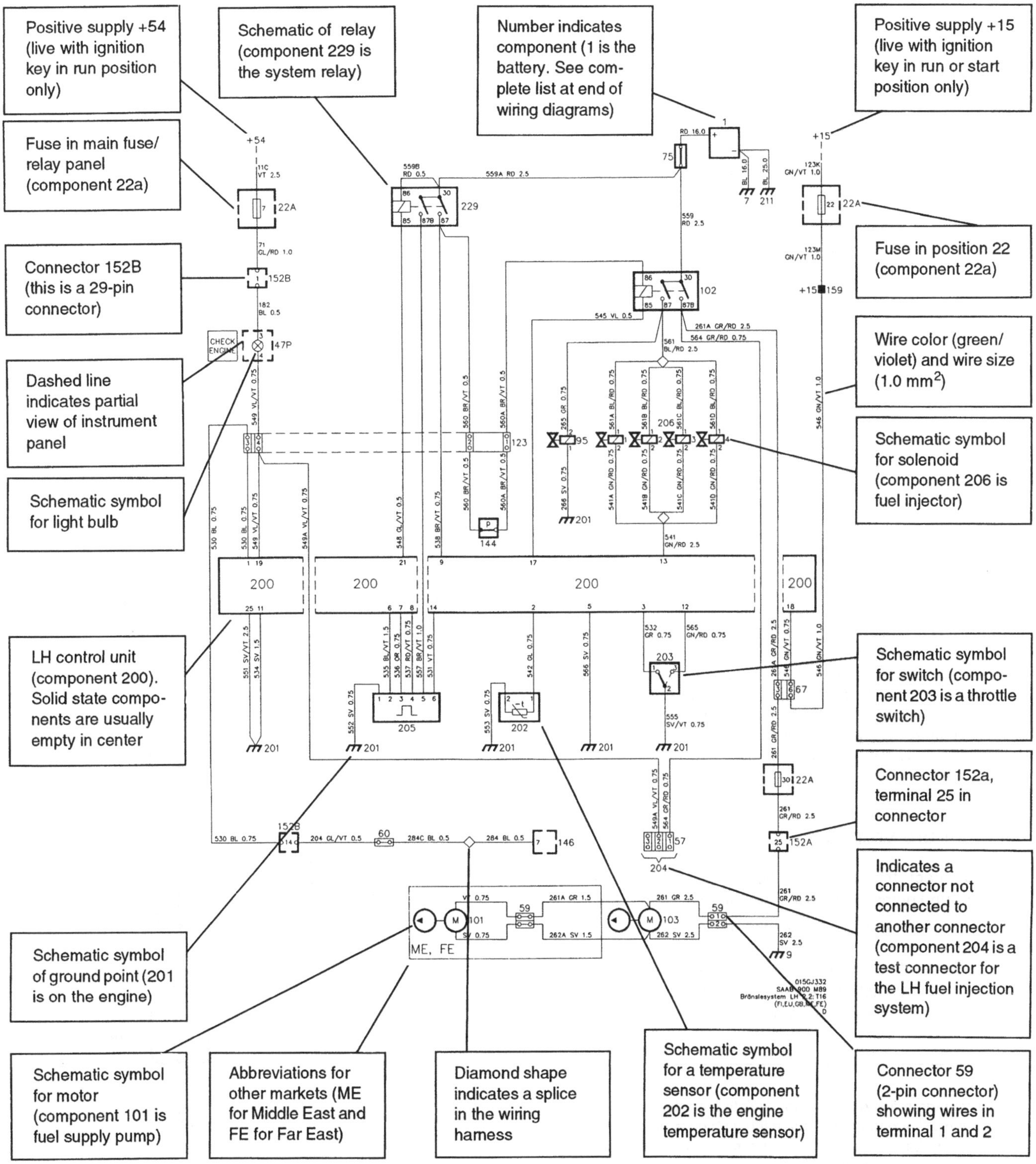

Fig. 1. Example of how to read Saab wiring diagrams.

The wires, connectors, and switches that carry current are designed with very low resistance so that current flows with a minimum loss of voltage. A voltage drop is caused by higher than normal resistance in a circuit. This additional resistance actually decreases or stops the flow of current. A voltage drop can be noticed by problems ranging from dim headlights to sluggish wipers. Some common sources of voltage drops are faulty wires or switches, dirty or corroded connections or contacts, and loose or corroded ground wires and ground connections.

Voltage drop can only be checked when current is running through the circuit, such as by operating the starter motor or turning on the headlights. Making a voltage drop test requires measuring the voltage in the circuit and comparing it to what the voltage should be. Since these measurements are usually small, a digital voltmeter should be used to ensure accurate readings. If a voltage drop is suspected, turn the circuit on and measure the voltage at the circuit's load.

NOTE

- A voltage drop test is generally more accurate than a simple resistance check because the resistances involved are often too small to measure with most ohmmeters. For example, a resistance as small as 0.02 ohms would results in a 3 Volt drop in a typical 150 amp starter circuit. (150 amps x 0.02 ohms =3 volts).

- Keep in mind that voltage with the key on and voltage with the engine running are not the same. With the ignition on and the engine off (battery voltage), voltage should be approximately 12.6 volts. With the engine running (charging voltage), voltage should be approximately 14.0 volts. Measure voltage at the battery with the ignition on and then with the engine running to get exact measurements.

To measure voltage

1. Connect the voltmeter negative lead to a reliable ground point on the car. The negative (–) battery terminal is always a good a ground point.

2. Connect the voltmeter positive lead to the point in the circuit you wish to measure. See Fig. 2. If a reading is obtained, current is flowing through the circuit.

NOTE

The voltage reading should not deviate more than 1 volt less than the voltage at the battery. If the voltage is less than this, there is probably a fault in the circuit, such as a corroded connector or a loose ground wire.

To check for a voltage drop

1. Connect the voltmeter positive lead to the positive (+) battery terminal or a positive power supply close to the battery source.

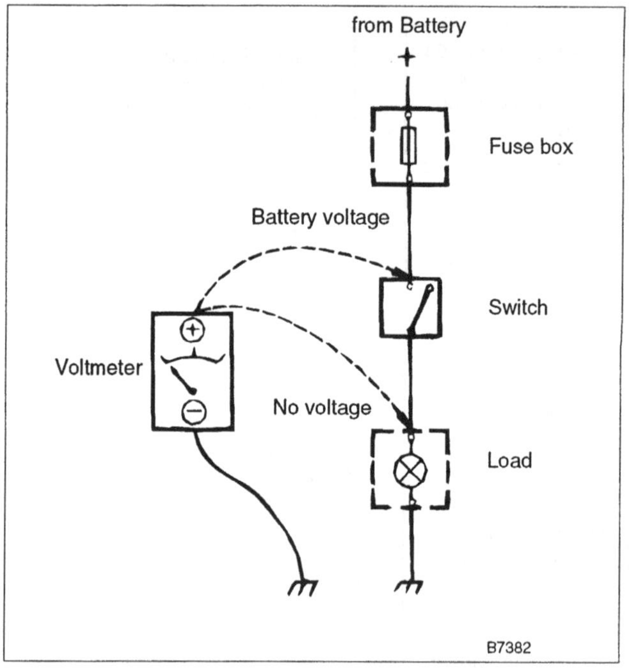

Fig. 2. Voltmeter being used to check for voltage.

2. Connect the voltmeter negative lead to the other end of the cable or switch being tested. See Fig. 3.

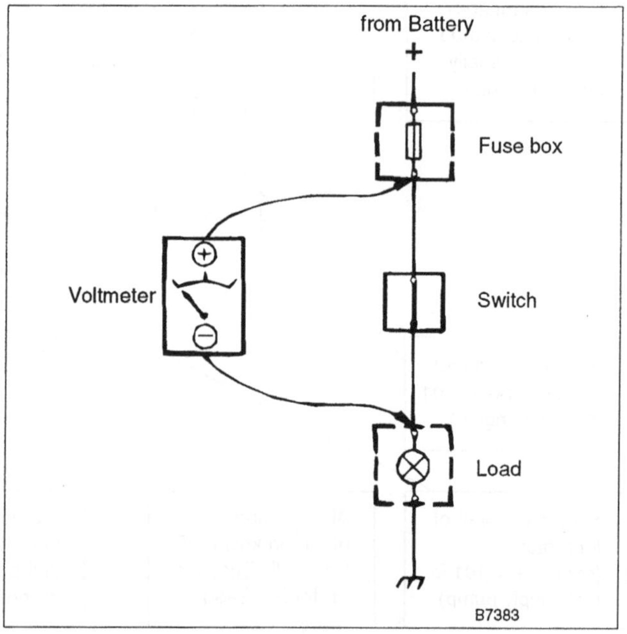

Fig. 3. Voltmeter being used to check for voltage drop.

3. With the power on and the circuit working, the meter shows the voltage drop (the difference between the two points). This value should not exceed 1 volt.

NOTE

The maximum voltage drop in an automotive circuit, as recommended by the Society of Automotive Engineers (SAE), is as follows: 0 Volts for small wire connections; 0.1 Volts for high current connections; 0.2 Volts for high current cables; and 0.3 Volts for switch or solenoid contacts. On longer wires or cables, the drop may be slightly higher. In any case, a voltage drop of more than 1.0 Volt usually indicates a problem.

Checking for Continuity

The continuity test can be used to check a circuit or switch. Because most automotive circuits are designed to have little or no resistance, a circuit or part of a circuit can be easily checked for faults using an ohmmeter. An open circuit or a circuit with high resistance will not allow current to flow. A circuit with little or no resistance allows current to flow easily.

CAUTION

Do not use an analog (swing-needle) ohmmeter to check circuit resistance or continuity on any electronic (solid-state) components. The internal power source used in most analog meters can damage solid state components. Use only a high quality digital ohmmeter having high input impedance when checking electronic components.

When checking continuity, the ignition should be off. On circuits that are powered at all times, the battery should be disconnected. Using the appropriate wiring diagram, a circuit can be easily tested for faulty connections, wires, switches, relays, and engine sensors by checking for continuity. Fig. 4 shows continuity test being made on a brake light switch.

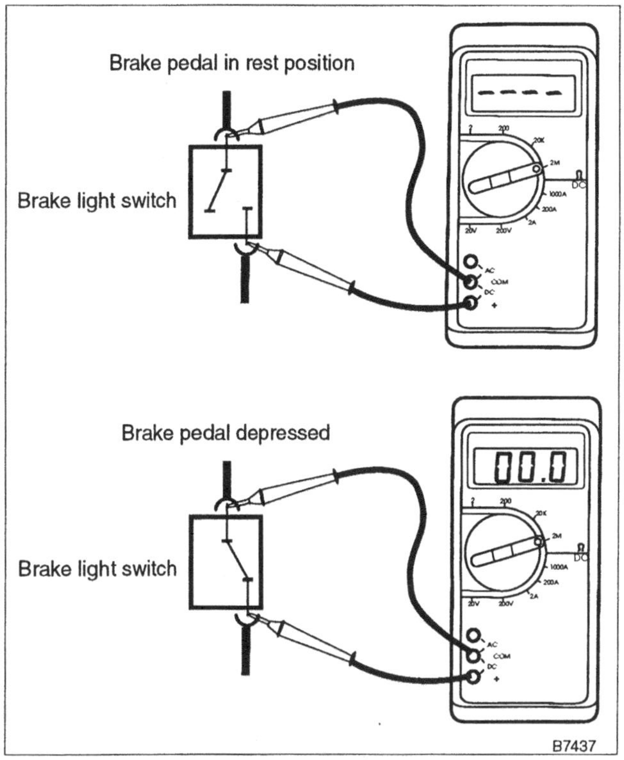

Fig. 4. Brake light switch being tested for continuity. With brake pedal in rest position (switch open) there is no continuity (infinite ohms). With the pedal depressed (switch closed) there is continuity (zero ohms).

Checking for Short Circuits

A short circuit is exactly what the name implies. The circuit takes a shorter path than it was designed to take. The most common short that causes problems is a short to ground where the insulation on a positive (+) wire wears away and the metal wire is exposed. When the wire rubs against a metal part of the car or other ground source, the circuit is shorted to ground. If the exposed wire is live (positive battery voltage), a fuse will blow and the circuit may possibly be damaged.

Shorts to ground can be located with an ohmmeter or a voltmeter. Short circuits are often difficult to locate and may vary in nature. Short circuits can be found using a logical approach based on the current path.

CAUTION

• On circuits protected with large fuses (25 amp and greater), the wires or circuit components may be damaged before the fuse blows. Always check for damage before replacing fuses of this rating.

• When replacing blown fuses, use only fuses having the correct rating. Always confirm the correct fuse rating printed on the fuse/relay panel cover.

To check for a short circuit using an ohmmeter

1. Remove the blown fuse from the circuit and disconnect the cables from the battery.
2. Disconnect the harness connector from the circuit's load or consumer.
3. Using an ohmmeter, connect one test lead to the load side of the fuse terminal (terminal leading to the circuit) and the other test lead to ground. See Fig. 5.
4. If there is continuity to ground, there is a short to ground.

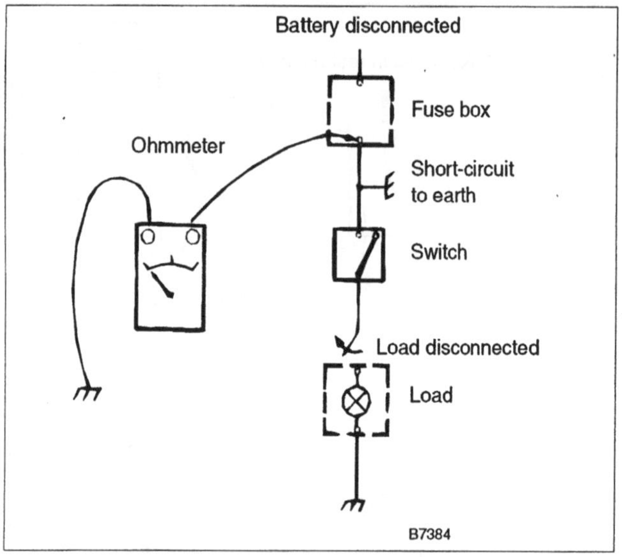

Fig. 5. Ohmmeter being used to find short circuit.

5. If there is no continuity, work from the wire harness nearest to the fuse/relay panel and move or wiggle the wires while observing the meter. Continue to move down the harness until the meter displays a reading. This is the location of the short to ground.

Visually inspect the wire harness at this point for any faults. If no faults are visible, carefully slice open the harness cover or the wire insulation for further inspection. Repair any faults found.

To check for a short circuit using a voltmeter

1. Remove the blown fuse from the circuit.

2. Disconnect the harness connector from the circuit's load or consumer.

3. Using a voltmeter, connect the test leads across the fuse terminals. See Fig. 6. Make sure power is present in the circuit, if necessary turn the key on.

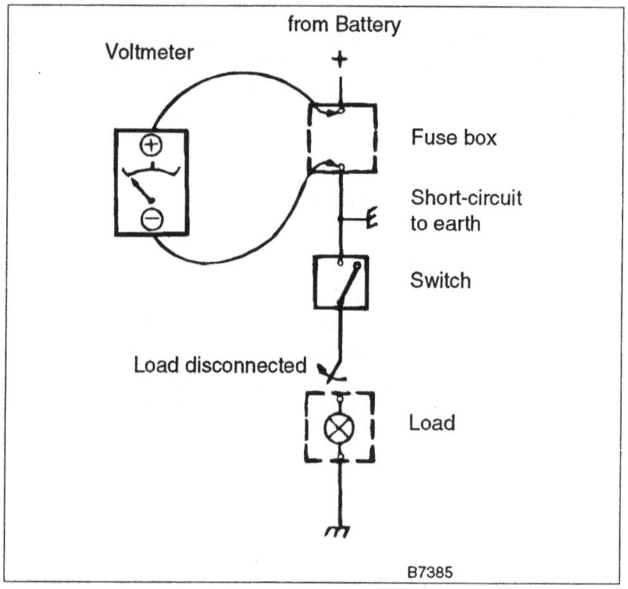

Fig. 6. Voltmeter being used to find short circuit.

4. If voltage is present at the voltmeter, there is a short to ground.

5. If voltage is not present, work from the wire harness nearest to the fuse/relay panel and move or wiggle the wires while observing the meter. Continue to move down the harness until the meter displays a reading. This is the location of the short to ground.

Visually inspect the wire harness at this point for any faults. If no faults are visible, carefully slice open the harness cover or the wire insulation for further inspection. Repair any faults found.

311 Battery

TABLES

GENERAL

The six-cell, 12-volt lead-acid battery capacity is rated by ampere/hours (Ah) and cold cranking amps (CCA). The Ah rating is determined by the average amount of current the battery can deliver over time without dropping below a specified voltage. The CCA rating is determined by the battery's ability to deliver starting current at 0° (–18°C). The original battery installed is rated at 62 Ah and is located in the engine compartment, behind the right headlight assembly.

Maintenance consists of keeping the battery and terminals clean and keeping the connections tight. After cleaning, a light coating of petroleum jelly should be spread on the terminals. Inspect the battery cables for chafing and deterioration of the insulation due to high heat.

For general information on troubleshooting electrical systems see **300 Electrical System—General**. For information on battery wiring see **371 Wiring Diagrams, Fuses and Relays**.

WARNING
- Wear goggles, rubber gloves, and a rubber apron when working with battery electrolyte. Electrolyte contains sulfuric acid and can cause skin irritation and burning. If electrolyte is spilled on your skin or clothing, flush the area at once with large quantities of water. If electrolyte gets into your eyes, bathe them with large quantities of clean water for several minutes and call a physician.
- Batteries that are being charged or are fully charged give off explosive hydrogen gas. Keep sparks and open flames away. Do not smoke.

CAUTION
- Do not disconnect the battery when the engine is running. The alternator will be damaged.
- Replace batteries if the case is cracked or leaking. Electrolyte can damage the car. If electrolyte is spilled, clean the area with a solution of baking soda and water.
- Always allow a frozen battery to thaw before attempting to recharge it.
- Always disconnect the battery terminals during battery recharging. This will prevent damage to solid-state components.

CAUTION
- Do not quick-charge the battery (for boost starting) for longer than one minute, and do not exceed 16.5 volts at the battery with the boosting cables attached. Wait at least one minute before boosting the battery a second time.
- Always disconnect the negative (–) battery cable first and reconnect it last. Cover the battery post with an insulating material whenever the terminal is removed.
- When boost starting the car using jumper cables, be careful not to reverse the battery connections. Even a momentary connection will damage the alternator.

BATTERY TESTING

Battery testing determines the state of battery charge. The most common methods are open-circuit and load voltage testing. Batteries that have filler caps can also be tested by checking the specific gravity of the electrolyte. The specific gravity test checks the amount of acid in the electrolyte as an indication of battery charge. Inexpensive specific gravity testers are available at most auto supply stores.

Open-Circuit Voltage Test

An open-circuit voltage test checks battery voltage by connecting an accurate digital voltmeter to the battery posts after disconnecting the battery ground cable. Before making an open-circuit voltage test on a battery, first load the battery with 15 amps for one minute, for example by turning on the headlights without the engine running. Open-circuit voltage levels and their corresponding percentages of charge are in **Table a.**

Table a. Open-Circuit Voltage and Battery Charge

Open-circuit voltage	State of charge
12.6 V or more	Fully charged
12.4 V	75% charged
12.2 V	50% charged
12.0 V	25% charged
11.7 V or less	Fully discharged

311

The battery is in satisfactory condition if the open-circuit voltage is at least 12.4 volts. If the open-circuit voltage is at this level or above, but the battery still lacks power for starting, make a load voltage test to determine the battery's service condition. If the open-circuit voltage is below 12.4 volts, recharge the battery. If the battery cannot be recharged to at least 75%, it should be replaced.

Load Voltage Testing

A load voltage battery test is made by connecting a specific resistive load to the battery terminals and then measuring voltage. The test requires a special tester and can generally be performed quickly and inexpensively by an authorized Saab dealer or qualified repair facility. The battery should be fully charged for the most accurate results. If the equipment is available, disconnect the negative (-) battery cable. Then apply a 200 Amp load for 15 seconds and measure the battery's voltage. If the voltage is below that listed, the battery should be replaced.

WARNING
Always wear protective goggles and clothing when performing a load test.

Battery load test minimum voltage
- (200 amp load for 15 seconds)
  - 9.6 V at 80°F (27°C)
  - 9.5 V at 60°F (16°C)
  - 9.3 V at 40°F (4°C)
  - 8.9 V at 20°F (-7°C)
  - 8.5 V at 0°F (-18°C)

If the battery quickly becomes discharged, there may be a fault in the charging system or in the electrical system. See **321 Alternator and Charging System** for charging system tests. On all models, be certain that the appropriate heat shields are securely attached to the battery.

NOTE
If the key is left in the ignition switch, a battery draw of about 40 milliamps will discharge the battery over a period of time.

BATTERY CHARGING

Discharged batteries can be recharged using a battery charger, but a battery can never be charged to a voltage in excess of that which it is capable of producing electro-chemically. Prolonged charging causes gassing that will evaporate the electrolyte to a level that can damage the battery.

Always read and follow the instructions provided by the battery charger's manufacturer. Do not use a charger if the instructions are [illegible] available. **Table b** lists charging rates and times that should be followed when charging batteries.

WARNING
The gasses given off by the battery during charging are explosive. Do not smoke. Keep open flames away from the top of the battery, and prevent electrical sparks by turning off the battery charger before connecting or disconnecting it.

CAUTION
Always disconnect a battery cable when using a battery charger. This will prevent damage to solid-state components.

Table b. Battery Charging Specifications

Charging rate (low-maintenance batteries)	Specific gravity	Approximate charging time
Fast charge (at 80% to 90% of battery's capacity, example: 44 to 50 amperes for a 55-ampere hour battery)	1.150 or less	1 hour
	1.150 to 1.175	3/4 hour
	1.175 to 1.200	1/2 hour
	1.200 to 1.225	1/4 hour
Slow charge (at 10% of battery's capacity, example: 5.5 amperes for a 55-ampere hour battery)	Above 1.225	Slow charge only, to a specific gravity of 1.250 to 1.265

321 Alternator and Charging System

GENERAL

The charging system consists of a belt-driven 14-volt alternator and a voltage regulator. The voltage regulator, which is mounted on the alternator, also serves as the alternator brush holder. The charging system provides the current necessary to keep the battery charged and to operate the car electrical accessories.

To prevent damage to the alternator or regulator when making tests or repairs, make all connections with negative (–) to negative, and positive (+) to positive. Even momentary contact with a conductor of the wrong polarity can damage the alternator's diodes.

Please read the following cautions before doing any work on the charging system or alternator.

CAUTION

- Never operate the engine with the battery disconnected. Never operate the alternator with its output terminal (B+ or 30) disconnected and the other terminals connected. Never short, bridge, or ground any terminals of the charging system.

- Always disconnect the negative (–) battery cable before removing the alternator. This will prevent an accidental short from one of the alternator terminals.

- Do not reverse the polarity of the battery. Damage to the alternator will result.

- Always disconnect both battery cables before doing any electric (arc) welding to the car.

CHARGING SYSTEM TROUBLESHOOTING

Charging system trouble is indicated by an illuminated alternator warning light on the instrument panel, or by an under- or overcharged battery.

The alternator generates electrical current by electrical induction. When the engine is running, part of the current it produces energizes its electromagnetic field. When starting, some other current must be provided to initially energize the field and begin the current generating process. This current is provided by the battery through the alternator warning light in the instrument cluster. If the lamp burns out, the alternator will not charge the battery properly.

Charging System Quick-check

As a quick-check, measure the voltage across the battery terminal with the key off and then again with the engine running. The battery voltage should be approximately 12.6 with key off and 14.0 with the engine running. If the voltage does not increase when the engine is running, there is fault in the charging system.

321

To test charging system

1. Inspect the V-belts and make sure they are tight. Make sure the battery is fully charged, as described in **311 Battery.**

2. Check the charge warning lamp. On 1987 and later cars, check that the ignition switch relay closes when the key is turned on. See **371 Wiring Diagrams, Fuses and Relays** for a complete listing of fuse and relay locations.

NOTE

- The charging system warning light should come on when the key is on and go out when the car is started. Turning the key on is a test that checks the warning light circuit to the alternator. If the light does not come on when the key is turned on, check the light bulb and the wiring to the bulb using the appropriate wiring diagram before making any more tests. Wiring diagrams can be found in **371 Wiring Diagrams, Fuses and Relays.**

- When replacing the charging system warning bulb, be sure to use the correct 2.0 watt replacement bulb. A similar 1.2 watt bulb (with black socket base) will also fit in this location. If the smaller bulb is installed, the battery will not charge correctly at idle. The light will also go out at a higher rpm. See an authorized Saab parts department for additional information.

3. Check for battery voltage between ground and terminal **B+** at the back of the alternator. Then check for battery voltage between terminal **D+** and ground with the ignition key on.

NOTE

If the voltage is not present at either terminal, check the alternator wiring and electrical circuit using the appropriate wiring diagram.

---

4. With the engine running, carefully measure the voltage between terminal D+ and ground and then between B+ and ground. The difference should not exceed 0.7 volts.

NOTE

If battery voltage was present at terminal D+ and B+ as described in step 3, but the voltage difference in step 4 is not as specified, the alternator is faulty and should be replaced.

---

5. If no faults are found up to this point, the alternator should be tested using special test equipment, such as the Sun VAT40. Alternator test specifications are shown below.

Alternator output specifications

• Bosch 80 Amp . . . . .54 Amps @ 1900 rpm

NOTE

If a load tester is not available, an output test can be done by running the engine at 1900 rpm and turning on all electrical loads such as fans, lights and heated window. Voltage at the battery should be approximately 12 volts or higher.

To remove and install alternator

1. Disconnect the negative (–) battery cable.

2. Loosen the alternator mounting bolts, pivot the alternator and remove the alternator V-belts.

3. Disconnect the wires from the rear of the alternator.

4. Remove the alternator mounting bolts and the alternator. See Fig. 1

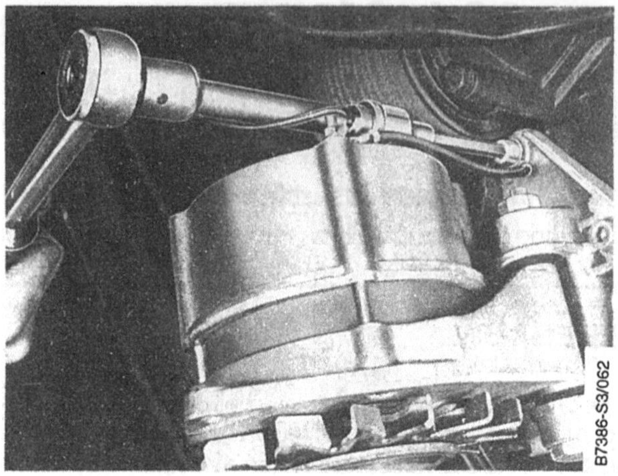

Fig. 1. Alternator mounting bolt being removed.

5. Install the new alternator, leaving the mounting bolts slightly loose. Install the V-belt and tension it as described in **1 LUBRICATION AND MAINTENANCE.**

CAUTION

V-belt tension is critical to the life of alternator and proper alternator operation. A loose belt will cause undercharging (and often a loud screeching sound). A tight belt will quickly destroy the alternator bearings and the bearings of other component driven off the same belt.

---

6. Tighten all mounting bolts. Install the wiring to the alternator. Reconnect the battery terminal and check the operation of the charging system warning lamp.

NOTE

Be sure to reinstall the insulating boots to the wire terminals on the back of the alternator.

Regulator/Brush Assembly

The voltage regulator is easily removed from the back of the alternator by removing the two mounting screws. See Fig. 2. The regulator is available as replacement part. In addition, the carbon brushes can also be separately replaced.

With the regulator removed measure the brush length. See Fig. 3. If the brush length is less than the minimum length specified in the figure, they should be replaced. Replacing the brushes requires a soldering iron and electric/electronic solder.

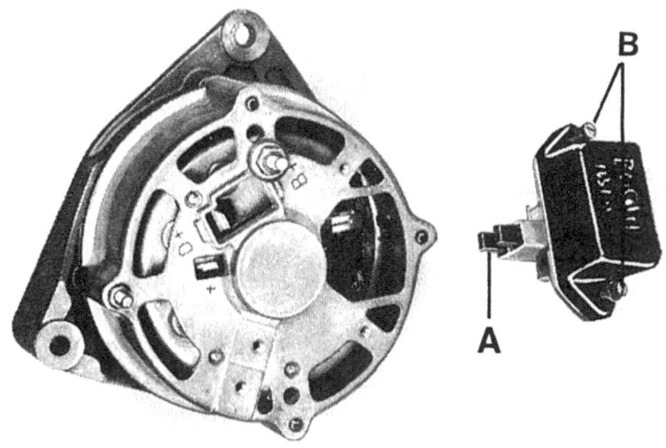

Fig. 2. Voltage regulator and brush holder removed from back of alternator. The brushes are at **A**. Remove the holder screws at **B**.

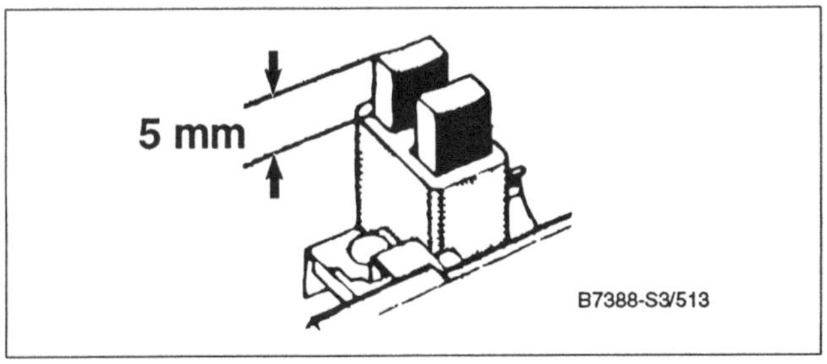

Fig. 3. Minimum regulator brush length. If brushes are shorter than 5 mm (0.2 in.), they should be replaced.

To replace regulator brushes

1. Remove the voltage regulator from the alternator.

2. Carefully unsolder the brush lead from the brush holder terminal, withdrawing the brush from the holder at the same time.

CAUTION
Work quickly to prevent overheating the solid state regulator.

---

3. Remove any traces of solder from the brush holder terminal.

4. Fit the new brush into the holder, insert the brush lead into the terminal, and solder it into place.

5. Reinstall the regulator.

331 Starter

TABLES

STARTING SYSTEM

The starter is an electric motor that drives a ring gear on the flywheel via a pinion on the end of the starter. When the ignition key is turned to the start position, a solenoid moves the pinion into engagement with the ring gear. The solenoid then closes the main electrical contacts to the starter motor and the motor turns to start the car.

When the engine has started, the ring gear speed exceeds that of the pinion, and the pinion is disengaged from the pinion shaft. When the key is released from the start position, a spring returns the pinion to the rest position. Fig. 1 shows the starter components.

The starter is located at the front of the engine under the intake manifold. The starter and solenoid are removed together as an assembly. The solenoid can be separated from the starter motor once removed. Although the starter is generally replaced as an exchanged unit, the solenoid and other parts are available from an authorized Saab dealer.

Starter Troubleshooting

Before troubleshooting the starter, make sure the battery is fully charged and the battery cables and ground connections are free of corrosion and in good condition. Troubleshooting information for the starting system appear in **Table a.**

NOTE

- For information on starter wiring and component location see **371 Wiring Diagrams, Fuses and Relays**.
- Starter efficiency is affected by engine oil viscosity. This is especially true in cold weather. Make sure the correct oil is in the engine. See **1 LUBRICATION AND MAINTENANCE.**

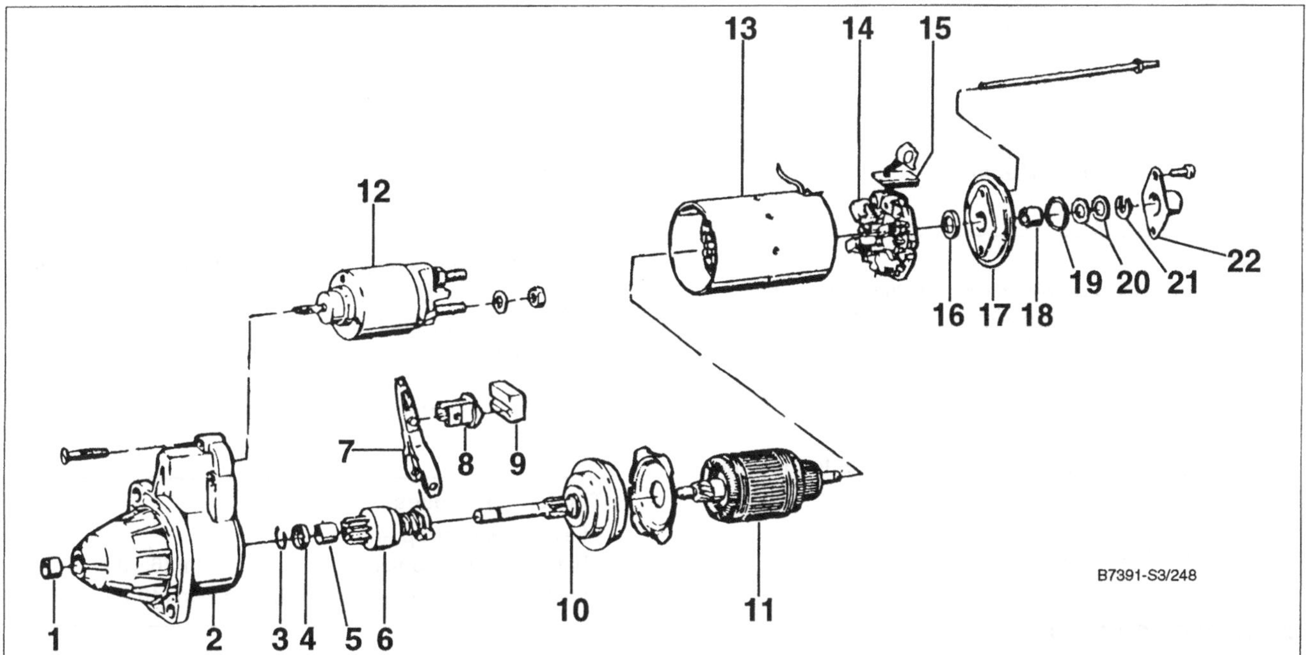

1. Starter bushing
2. Pinion housing
3. Circlip
4. Stop ring
5. Pinion bushing
6. Starter pinion
7. Pinion engaging lever
8. Bearing bracket
9. Seal
10. Planetary gear train
11. Armature
12. Solenoid
13. Stator frame
14. Brush holder assembly
15. Seal
16. Seal
17. Commutator end bracket
18. Commutator end bushing
19. Seal
20. Shims
21. Spring washer
22. End cover

Fig. 1. Exploded view of starter motor.

Table a. Battery, Starter and Charging System Troubleshooting

Symptom	Probable cause	Corrective action
1. Starter does not operate when ignition switch is turned to START	a. Ignition switch or wire leading from ignition switch to solenoid faulty (less than 8 volts to solenoid switch)	a. Test for voltage at terminal 50 of solenoid switch with ignition switch at START. If not at least 8 volts, test for voltage at terminal 50 of ignition switch with switch at START. Replace ignition switch (See **432 Manual Transmission Control**) or eliminate open circuit between ignition switch and solenoid switch
	b. Solenoid switch faulty (less than 8 volts to starter motor)	b. Test for voltage at field-winding strap with ignition at START. If not at least 8 volts, replace solenoid
	c. Starter motor faulty	c. Test for voltage at field-winding strap with ignition at START. If 8 volts or more, repair or replace starter motor
	d. Automatic transmission neutral safety switch faulty (models with automatic transmission only)	d. Test switch. See **371 Wiring Diagrams, Fuses and Relays**. See **364 Electric Controls and Switches** for switch adjustment
2. Starter turns slowly or fails to turn engine	a. Dirty, loose, or corroded starter connections	a. Remove, clean, and tighten connections
	b. Dirty, loose, or corroded ground strap between engine and body	b. Remove and clean or replace strap
	c. Starter worn or faulty	c. Repair or replace starter
3. Starter makes unusual noise, turns erratically, or fails to turn	a. Drive pinion defective	a. Repair or replace starter
	b. Flywheel or driveplate ring gear damaged	b. Replace flywheel or driveplate. See **216 Crankshaft**
4. Starter operates, but does not turn engine	a. Starter drive pinion or armature shaft faulty	a. Repair or replace starter
	b. Solenoid switch mechanism faulty	b. Replace starter solenoid switch

To remove starter

1. Disconnect the negative (–) battery cable. Remove any hoses that may be in the way, such as the air intake pipe.

2. Disconnect the large battery cable and the wire at terminal 50 from the starter motor.

NOTE

Terminal 50 has a small wire leading into the terminal. Terminal 30 has a large cable coming from the battery. The field-winding strap is a braided wire running from the solenoid to the starter housing.

---

3. Remove the two bolts holding the starter to the engine end plate and remove the starter.

Installation is the reverse of removal.

To replace solenoid

1. Remove the starter.

2. Remove the nut from the field-winding strap and disconnect the strap from the solenoid.

3. Remove the solenoid retaining screws and withdraw the solenoid, lifting it off the pinion engaging lever.

NOTE

The mounting screws are tightly fitted and may be difficult to remove. To help loosen the screws, use an impact driver with a slotted bit.

---

4. Install the new solenoid, hooking it over the pinion engaging lever. Make sure the terminals are correctly orientated and install the retaining screws and the field-winding strap.

5. Install the starter motor.

Tightening torques

- field-winding strap and battery cable to solenoid . . . . . . 7-9 nm (62-80 in-lb)
- solenoid to starter. . . . . 4.5-5.5 Nm (40-49 in-lb)

---

340 Ignition System

TABLES

GENERAL

Two versions of Hall-effect electronic ignition systems are used on the cars covered by this manual. The term Hall-effect refers to the Hall Sensor, an electronic device in the distributor that replaces conventional mechanical breaker points.

Ignition System Applications
- 1985 and later Turbo . . . . . basic Hall-effect ignition
- 1986 and later Non-turbo . . . . EZK ignition—Hall-effect with knock sensor

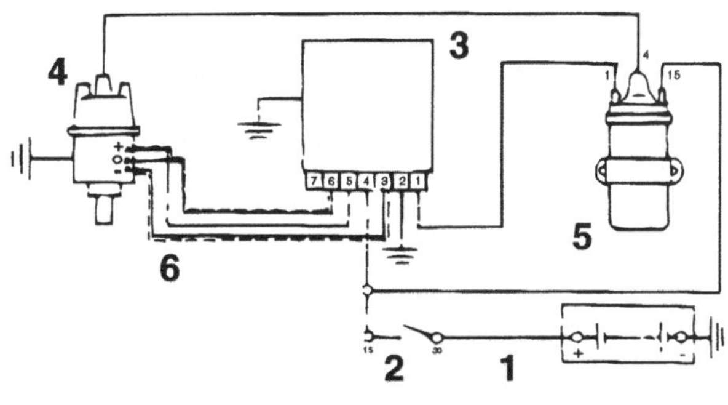

1. Battery
2. Ignition switch
3. Ignition control unit
4. Ignition distributor
5. Ignition coil
6. Shielded cable

Fig. 1. Schematic of basic Hall-effect ignition system.

Basic Hall Ignition System (1985 and later Turbo)

Fig. 1 shows a schematic of the basic Hall-effect ignition system. The firing of the spark is triggered by the Hall sensor in the distributor. The Hall sensor produces a voltage, which provides the ignition control unit with information about engine speed and crankshaft position. With this input, the control unit operates the ignition coil secondary circuit to precisely time the firing of the spark plugs.

EZK Ignition System (1986 and later Non-Turbo)

Fig. 2 shows a schematic of the EZK ignition system. This system is more sophisticated than the basic Hall system in that it includes a knock sensor to monitor and control ignition knock or ping. A knock sensor functions like a microphone and is able to convert mechanical vibration (knock) into electrical signals. The EZK control unit is programmed to react to the signals whose frequencies are characteristic of combustion chamber noise and change ignition timing accordingly.

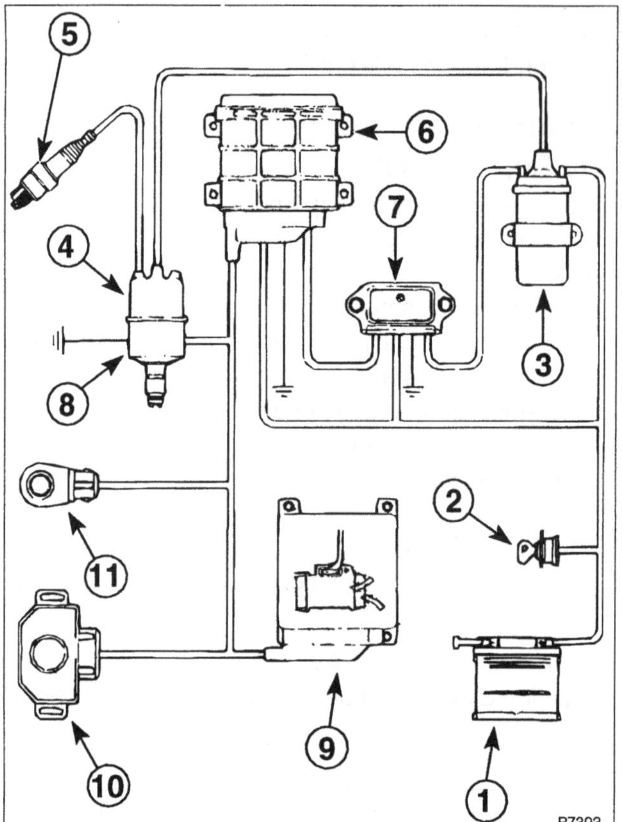

1. Battery
2. Ignition switch
3. Ignition coil
4. Distributor
5. Spark plug
6. EZK control unit
7. Ignition amplifier
8. Hall sensor
9. LH control unit
10. Throttle position sensor
11. Knock sensor

Fig. 2. Schematic of EZK ignition system with knock sensor.

The throttle switch (also used in the LH fuel injection system) signals the EZK control unit when the engine is idling. This allows the EZK control unit to set a base timing when the throttle is closed. The EZK control unit also receives an "engine load signal" from the LH control unit.

NOTE

- For more information on the throttle switch, including adjusting and testing procedures, see **240 Fuel Injection.**

- Turbo models also use a knock sensor to detect ignition knock as part of the APC (Automatic Performance Control) system. For more information, see **291 Turbocharger.**

EZK Knock Sensor Operation and Ignition Timing

During cranking, the ignition timing is locked at 5°BTDC. This setting will be held until engine speed exceeds 700 rpm. The timing will then change to a fixed 14°BTDC setting until the throttle is opened.

When the throttle is opened, the knock sensing system takes over and adjusts ignition timing according to engine conditions. The timing can vary from 5° to 22° BTDC in response to engine speed, engine load, or detonation. If engine knock occurs, the ECU determines which cylinder(s) are knocking and retards just the affected cylinder by 3°. If the knocking continues, the timing is further retarded in steps of 3°. The timing position will be maintained for a short period and will then begin to advance the timing by 1/3° increments until normal ignition is resumed.

NOTE

An occasional knocking may be heard on engines with EZK. This is normal and is necessary for proper operation of the EZK system.

IGNITION SYSTEM TROUBLESHOOTING

Poor driveability may have a variety of causes. The fault may lie with the ignition system, the fuel system, parts of the emission control system, or a combination of the three. Because of the interrelated functions of these systems, it is often difficult to know where to begin looking for problems. For this reason, effective troubleshooting should always consider these systems in unison, as one major system.

A complete failure of the ignition system to produce spark at the spark plugs is self-evident. For other problems such as rough idle, misfiring, or poor starting, however, the cause is not so clear. For troubleshooting engine management and the way the car runs, see **200 Engine—General.**

Ignition system service and repair work must be carried out carefully. The ignition system contains sensitive electronic components. To protect the system, and for general safety, the following warnings should be observed during any ignition system troubleshooting, maintenance, or repair work.

WARNING

- The ignition systems installed on the cars covered by this manual are high-energy systems operating in a dangerous voltage range which could prove to be fatal if exposed terminals or live parts are contacted. Use extreme caution when working on a vehicle with the ignition on or the engine running.

WARNING

- Connect and disconnect ignition system wires, multiple connectors, and ignition test equipment leads only while the ignition is off. Do not touch or disconnect any of the high tension cables from the coil, distributor, or spark plugs while the engine is running or being cranked by the starter.

- Before operating the starter without starting the engine (as when making a compression test), always disable the ignition as described below.

- Inefficient combustion can overload the catalytic converter with raw fuel, leading to converter overheating or plugging. An overheated catalytic converter can be a fire hazard.

CAUTION

- Do not conduct ignition system tests with a test lamp that uses a normal incandescent bulb. The high electrical consumption of these test lamps may damage the electronic components.

- Do not connect test instruments with a 12-volt supply to terminal 15 (+) of the ignition coil. The voltage backflow can damage the ignition control unit.

- Do not connect terminal 1 of the coil to ground as a means of preventing the engine from starting (for example, when installing or servicing anti-theft devices).

- Do not disconnect terminal 4 (center terminal) from the coil or remove the distributor cap to disable the ignition.

- Do not disconnect the battery while the engine is running.

On-board Diagnostics (EZK ignition only)

The EZK ignition system features built-in diagnostic circuitry that detects and stores in memory coded information about system faults. These fault codes can be "read-out" through flashes of the **Check Engine** light. Before troubleshooting an EZK fault and a flashing Check Engine light, check for broken or loose engine mounts or brackets.

NOTE

- As a general rule, a steady-on Check Engine light means the LH control unit has detected a fault. A flashing Check Engine light means that the EZK Ignition system has detected a fault. As an exception to the rule, the Check Engine light on 1992 models does not flash. Therefore, a steady-on Check Engine light on these models could be either a EZK system fault or an LH fuel system fault.

- On 1989 models with EZK only, the Check Engine light may flash intermittently due to a faulty EZK control unit. Check for a control unit date code of "945" or lower (see Fig. 13 for control unit location). The date code is stamped on the unit next to the connector housing. If the control unit is within these dates codes, see an authorized Saab dealer for additional information on correcting this fault.

On 1986-1991 models, start the engine and allow it to reach operating temperature. Briefly raise engine speed above 2000 rpm. Count the number of times the check engine light flashes before repeating the flashing code. The number of times the Check Engine light flashes determines the fault in the system.

EZK Check Engine light fault diagnosis

- three flashes—EZK control unit internal fault*
- four flashes — fault in the knock sensor, sensor wiring, or EZK control unit
- five flashes — faulty input signal from LH control unit.

*This code is applicable to 1989 and later models only.

NOTE

If the wiring harness to the EZK knock sensor is defective, a replacement wire harness is available from an authorized Saab dealer. This harness eliminates the need to replace the entire engine harness.

Test Equipment

Many of the tests of ignition system components require the use of high-impedance test equipment to prevent damage to the electrical components. A high-impedance digital multimeter should be used for all voltage and resistance tests. An LED test light should be used in place of an incandescent-type test lamp.

Many tests require checking for voltage, continuity, or resistance at the terminals of the components' harness connectors. The blunt tips of a multimeter's probes can spread open the small connector terminals and cause poor connections. To prevent damage, use flat male connectors to probe the harness connector terminals.

Disabling Ignition System

To prevent the engine from starting and to prevent the presence of dangerous voltages, the ignition system should be disabled anytime work is done on the ignition system. The ignition system should also be disabled any time work is being done in the engine compartment with the ignition key on or if the starter needs to be operated without running the engine, such as when making an engine compression test.

Disabling the ignition system is easily accomplished by disconnecting the Hall sensor connector. On all 1985-1988 models and 1989 and later Turbo models, disconnect the Hall sensor connector from the side of the distributor. On 1989 and later models with EZK ignition, the Hall sensor is mounted behind the crankshaft pulley. To disconnect the sensor connector, follow the wiring harness up from the sensor until you reach the connector in the engine compartment. See Fig. 3.

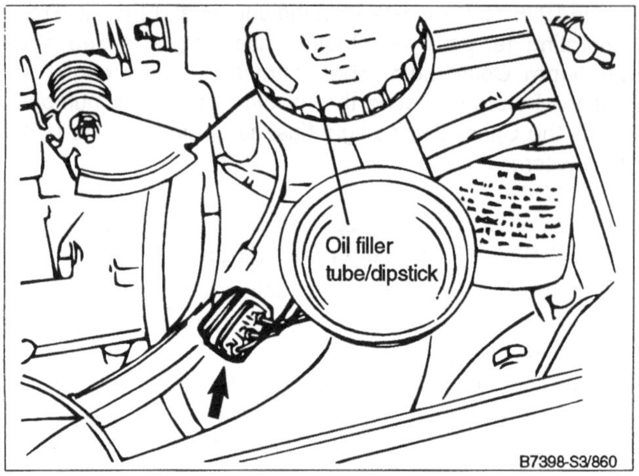

Fig. 3. Hall sensor connector on 1989 and later cars with EZK ignition. The sensor on these cars is mounted behind crankshaft pulley.

Basic Troubleshooting Principles

An engine that starts and runs indicates that the ignition system is fundamentally working—delivering voltage to the spark plugs. A hard-starting or poor-running engine, however, may indicate a problem with how well the spark is delivered. A faulty coil or control unit, cracked or deteriorated spark plug wires, a worn or cracked distributor cap or rotor, and worn or fouled spark plugs are all causes of reduced spark intensity and inefficient combustion.

An engine that has good cranking speed but will not start may indicate a complete failure of the system to produce spark. Inspect the ignition system visually. Make sure the spark plug wires have not been interchanged. Ignition firing order is described below under **Firing Order.**

If no visual faults are located, make a basic quick-check to see if spark is being produced as described below. This will be the most important first troubleshooting step. If a strong spark is observed, then the failure to start is due to another cause, perhaps no fuel being delivered to the engine.

Quick-Check of Ignition System

The first step in troubleshooting a no-start condition is to determine whether the problem is caused by the ignition system or some other system, such as a fuel delivery problem. This done by checking that the spark plugs are firing. If no spark is present, then more detailed testing of the ignition system is necessary.

To make the check, turn the ignition off and remove a spark plug wire from a spark plug. Connect to the plug wire a spark-tester or a known good spark plug gapped to 4 mm (0.16 or 5/32 in.) Position the plug so that the outer electrode is grounded on the engine.

NOTE

For the most accurate test results, the battery should be fully charged.

While a helper actuates the starter, look and listen for spark in the spark plug gap. A bright blue spark indicates a healthy ignition system. A yellow-orange spark is weaker and indicates that, while spark is present and the system is functioning, it is not operating at peak efficiency. Check the condition of the ignition system components as described below **Visual Inspection.** If there is no spark, test the system components as described below under **Electrical Tests and Fault Tracing.**

WARNING

- Do not hold the spark plug or its connector, even if using insulated pliers.
- If ignition system failure is not the problem, the engine may start during this test. Be prepared to turn off the ignition immediately. Running the engine with a spark plug wire disconnected may damage the catalytic converter.

If no faults have been detected up to this point but there is still no spark or a weak spark, refer to **Table a** for more troubleshooting information. If a strong spark is observed but the engine still will not start, refer to **200 Engine—General** for more troubleshooting information.

Firing Order

The firing order of the cylinders is 1-3-4-2. Cylinder no. 1 is next to the V-belts by the firewall and cylinder no. 4 is at the front of the car. The engine rotates counter-clockwise when viewed from the front of the car. See Fig. 4.

Table a. Ignition System Troubleshooting

Symptom	Probable cause	Corrective action
1. No spark or weak spark observed during spark test	a. Wet or damp distributor cap and/or spark plug wires	a. Remove cap and wires. Dry and reinstall
	b. Faulty wires or connectors (primary circuit).	b. Test and replace as needed
	c. Weak or faulty coil	c. Test and replace as needed
	d. Defective spark plug wires	d. Test and replace as needed
	e. Worn or fouled spark plugs	e. Replace spark plugs. See **1 LUBRICATION AND MAINTENANCE**
	f. Faulty Hall sensor	f. Test and replace as needed
	g. Faulty ignition control unit or amplifier	g. Test and replace as needed

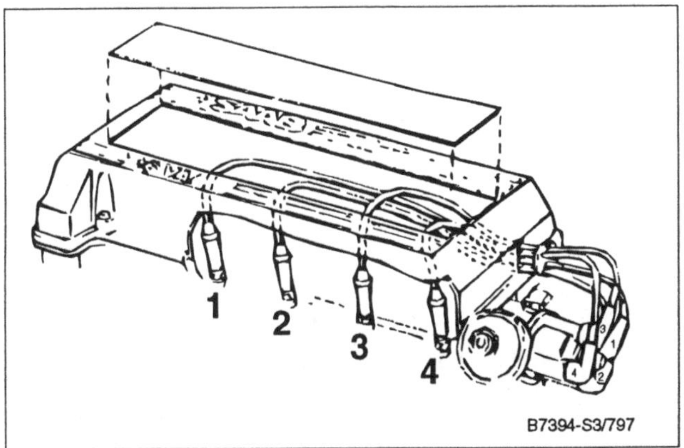

Fig. 4. Firing order and proper orientation of high tension leads.

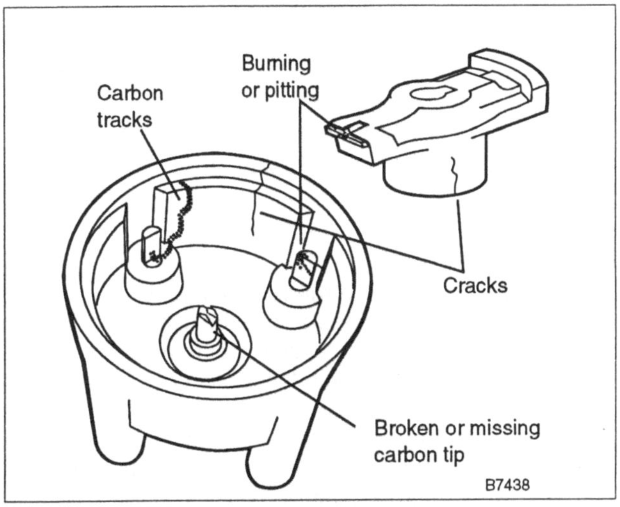

Fig. 5. Distributor cap and rotor. Inspect cap and rotor at contact points.

Visual Inspection

The spark plug wires, the distributor cap, and the distributor rotor are subject to wear and electrical breakdown, which will impair their ability to deliver a crisply timed and powerful spark. Dirt and moisture on any of these components are also potential causes of poor spark at the spark plugs.

To inspect the distributor cap and rotor, first unclip and remove the cap. Inspect the contacts inside the distributor cap and at the tip of the rotor for carbon tracks, wear, or pitting. See Fig. 5. Cracks or carbon tracks in the distributor cap may cause shorts to ground. The cracks may be fine and difficult to see. Check that the center black carbon brush inside the cap springs back when compressed. If any faults are found, replace the cap and/or rotor.

To visually check the spark plug wires, gently bend them in several places. This will expose cracks in the insulation which may cause spark "leaks". Peel back the rubber boots and check them for pliancy and the ability to seal out dirt and moisture. Wires that are cracked, oil-soaked or dry and brittle should be replaced.

The coil should be closely examined for cracks, burns, carbon tracks, or any leaking fluid. The coil tower, center terminal 4, should be clean and dry. If necessary, remove the coil for cleaning and closer examination. Check that the wiring at the coil top is not contacting the metal coil housing. Loosen the nuts and reposition the wires if necessary.

Inspect all primary wires and connections for any corrosion or damage. Clean or repair any faults found. In these sensitive electronic ignition systems, corroded or loose connections may interfere with the ignition function.

ELECTRICAL TESTS AND FAULT TRACING

The following electrical test should be done in the sequence given. Review the information given earlier before making the tests described here. On cars with EZK ignition, be sure to check for any EZK fault codes as described earlier.

For the most accurate test results, check that the battery is fully charged, there is fuel in the tank, and that engine and ambient temperature are between 32°F and 104°F (0°C and 40°C).

CAUTION
Do not unplug components or disconnect grounds when the ignition key is on. Damage to electronic components could result.

To test basic Hall ignition system (turbo models only)

1. **Check the basic components subject to wear and electrical breakdown.** Test the condition of the spark plug wires and the ignition rotor by testing the resistance of each component.

Spark Plug Wire and Rotor Resistances
- Ignition rotor .... 1,000 Ohms
- Coil high tension lead (complete with end connectors) .. 500-1,500 Ohms
- Spark plug leads (complete with end connectors) .... 2,000-4,000 Ohms

2. **Check for power to the Hall sensor.** Unplug the wiring connector from the side of the distributor. Connect a voltmeter between the two outer terminals in the connector. See Fig. 6. Turn the ignition on. The voltmeter should read at least 10 volts (VDC). Turn the ignition off and reconnect the connector. If no faults are found, proceed to the next step. If voltage is not present, check the wires between the distributor connector and the control unit. Repair any wiring faults before proceeding.

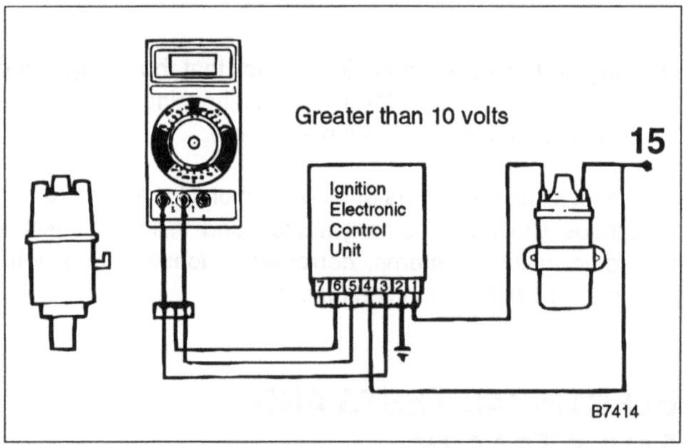

Fig. 6. Voltmeter being used to check for voltage to Hall sensor.

3. **Check for power to the ignition control unit.** Disconnect the connector from the ignition control unit. See Fig. 7. Connect a voltmeter to terminal 4 (+) and terminal 2 (–) of the control unit connector. See Fig. 8. Turn the ignition on and check that the voltage is at least 10 volts (DC). Turn the ignition off and reconnect the connector. If voltage is as specified, proceed to the next step. If voltage is not present, use the diagrams shown in **371 Wiring Diagrams, Fuses and Relays** to check for wiring faults.

NOTE
It may be necessary to remove the control unit from its mountings to gain access to the connector.

Fig. 7. Ignition control unit in left front (driver's side) fender well.

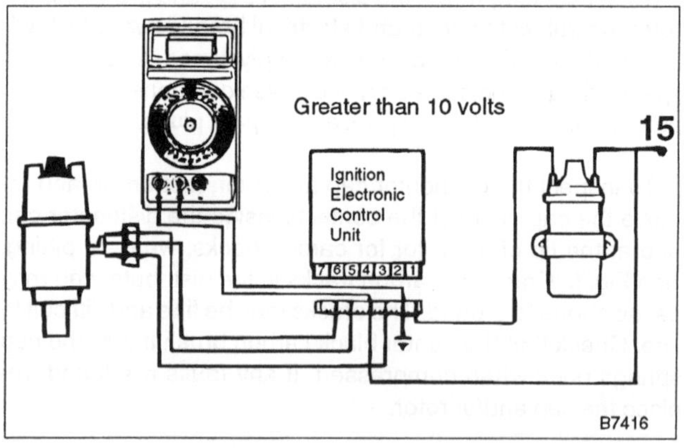

Fig. 8. Voltmeter being used to check for voltage to control unit.

4. **Check the Hall sensor output signal.** Peel back the rubber boot on the control unit connector. Connect a voltmeter between terminal 6 and 3, working from rear of the connector. See Fig. 9. Turn the ignition on and slowly turn the engine over by hand. The voltage should oscillate between 0.4 (DC)volts (or less) and 3.0 volts (or higher). If no faults are found, proceed to the next step. If any faults are found, check the wires between the control unit and the Hall sensor. If no wiring faults are found, the Hall sensor is faulty and should be replaced.

CAUTION

Turn the engine over only in the direction of normal rotation. Otherwise damage to the timing chain and its components may result. To determine normal rotation, put the car in gear and rock the car forward, noting the rotation of the ignition rotor, the flywheel, or a belt-driven component. As an alternative, briefly operate the starter.

NOTE

The engine can be turned over using a wrench on the crankshaft pulley.

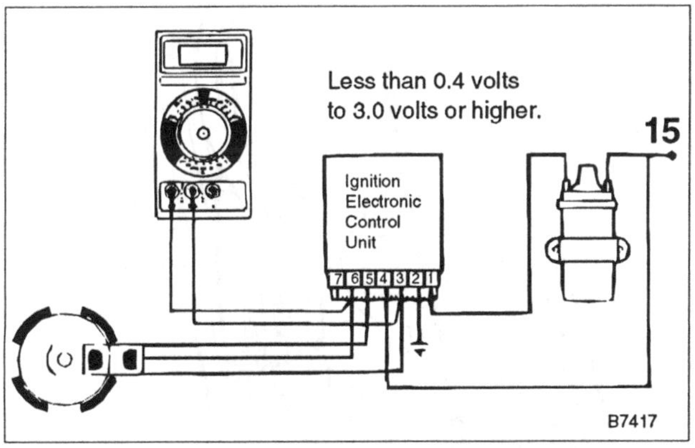

Fig. 9. Voltmeter being used to check Hall sensor operation.

5. **Check for power to the coil.** Connect a voltmeter between terminal 15 (+) of the coil and ground. Turn the ignition on and check that the voltage is at least 10 volts (DC). Turn the ignition off. If voltage is as specified, proceed to the next step. If voltage is not present, use the diagrams shown in **371 Wiring Diagrams, Fuses and Relays** to check for wiring faults.

6. **Check the coil primary and secondary circuit resistance.** Remove all wires from the coil. Measure the coil primary and secondary circuit resistance as shown below in Fig. 10. Reconnect the wires to the coil. If the coil resistance is not as specified, the coil is faulty and should be replaced. If no faults are found, proceed to the next step.

Ignition coil resistance*

- between terminal 1 and 15 .... 0.5 to 0.9 Ohms
- between terminal 1 and center tower .... 7200 to 8200 Ohms

* measure with all wires disconnected from coil

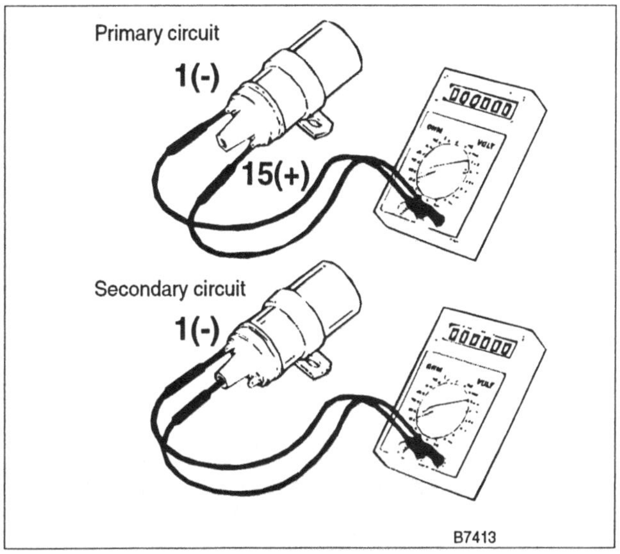

Fig. 10. Coil circuits being tested using ohmmeter. Coil terminal numbers are found on coil.

7. **Check the ignition control unit switching function to the coil.** Disconnect the Hall sensor connector from the distributor. Connect the voltmeter between coil terminals 15 and 1. See Fig. 11. Turn the ignition on. The voltage should rise to about 5 volts (DC) and then drop to 0 volts within 1 to 2 seconds. Reconnect the connector. If any faults are found, the control unit is faulty and should be replaced. If no faults are found, the ignition system is functioning correctly.

NOTE

If on 1986 and 1987 models, all electrical checks are as specified but the car still does not start, check the ignition pulse amplifier on the main fuse relay panel. See Fig. 12. This relay-like amplifier is designed to "amplify" the tachometer signal for more reliable control of electronic components affected by the ignition system (LH fuel injection, fuel pump control, etc.). To check the amplifier, simply remove it from its socket. Then using a jumper wire, jumper the fuse/relay panel sockets corresponding to relay terminals TD-I and TD-O. Try to start the engine. If the engine now runs, replace the amplifier.

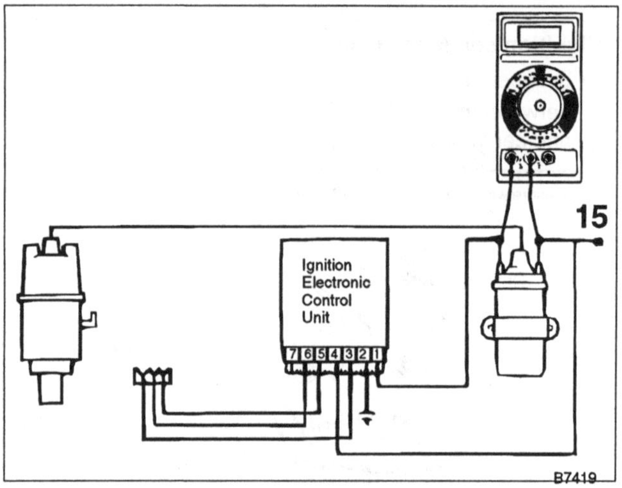

Fig. 11. Voltmeter being used to check control unit switching function.

Fig. 12. Ignition pulse amplifier used on 1986 and 1987 models. A faulty pulse amplifier will cause a no-start condition.

To test EZK ignition system (non-turbo models only)

1. **Check the basic components subject to wear and electrical breakdown.** Test the condition of the spark plug wires and the ignition rotor by testing the resistance of each component.

Spark Plug Wire and Rotor Resistances

- Ignition rotor. . . . . 1,000 ohms
- Coil high tension lead (complete with end connectors). . . 500-1,500 ohms
- Spark plug leads (complete with end connectors) . . . . 2,000-3,000 ohms

---

2. **Test the knock sensor output signal.** Remove the harness connector from the EZK control unit. See Fig. 13. Using a digital voltmeter (set to VAC scale), connect it to terminal 12 and 13 of the control unit connector. Gently tap on the knock sensor mounting bolt. The voltmeter should register a small voltage and the reading should vary depending on the intensity of the tap.

NOTE

- The knock sensor is mounted to the engine block between cylinders no. 2 and 3, below the intake manifold.

- Procedure step 1 checks both the knock sensor and the wiring to the control unit. If the knock sensor does not respond to the tapping, remove the sensor and its connector. Then connect the voltmeter directly to the sensor terminals and repeat the above test. If the sensor now tests good, check the wiring from the sensor connector to the EZK control unit. If the sensor does not respond, it is faulty and should be replaced.

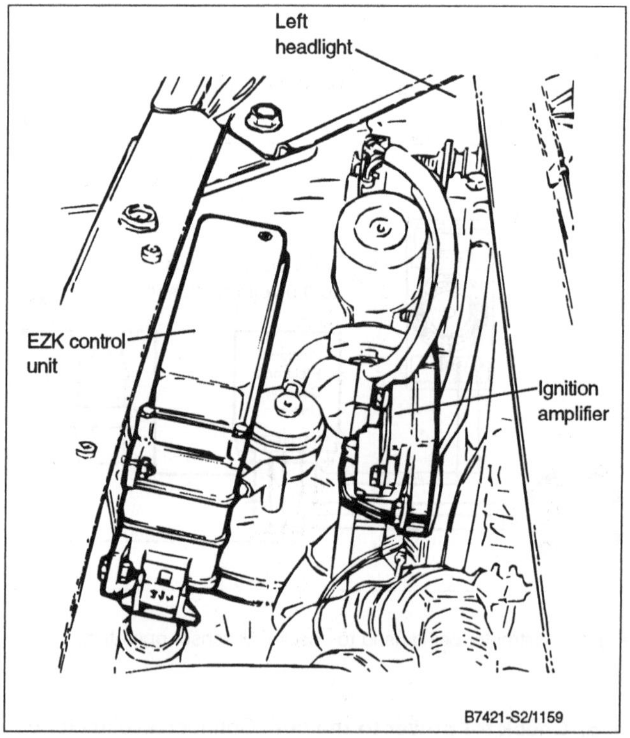

Fig. 13. EZK control unit and amplifier locations in left (driver's) side fender well. On 1988 and 1989 cars, amplifier is on side of ignition coil.

NOTE

On all cars except 1988 and 1989 models with EZK ignition, a conventional coil with a separate amplifier is used. On 1988 and 1989 cars with EZK ignition, a special coil with integrated amplifier is used. See Fig. 14. Testing procedures and replacement parts vary depending on coil style.

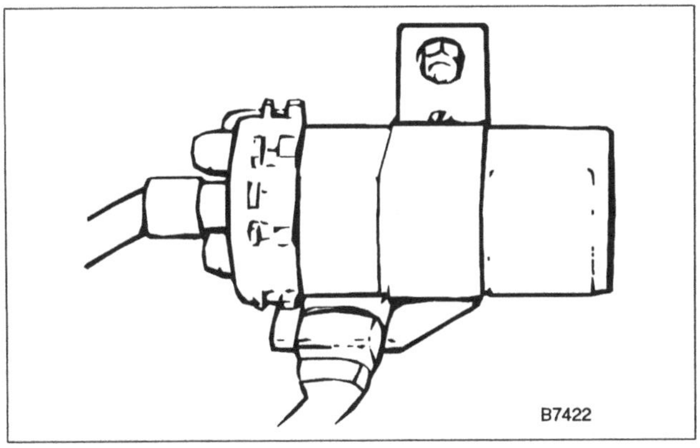

Fig. 14. Ignition coil with integrated amplifier (power stage) used on 1988 and 1989 cars. Coil is mounted on right-hand side of engine compartment, near the battery.

3. **Check for power to the EZK control unit.** Unplug the connector from the EZK control unit and connect a voltmeter (VDC scale) between terminal 6 (+) and 20 (–) of the connector. With the ignition on, there should be battery voltage. Turn the ignition off and reconnect the connector. If no faults are found proceed to the next step. If voltage is not present, use the diagrams shown in **371 Wiring Diagrams, Fuses and Relays** to check for wiring faults.

4. **Check for power to the amplifier.** Peel back the boot on the amplifier harness connector. Connect a voltmeter between the terminal 4 (+) and 2 (–) on fender-mounted amplifiers (Fig. 15) or between terminals 3 (+) and 2 (–) on coil-mounted amplifiers. With the ignition on, there should be battery voltage. If no faults are found proceed to the next step. If voltage is not present, use the wiring diagrams to check for wiring faults from the ignition switch.

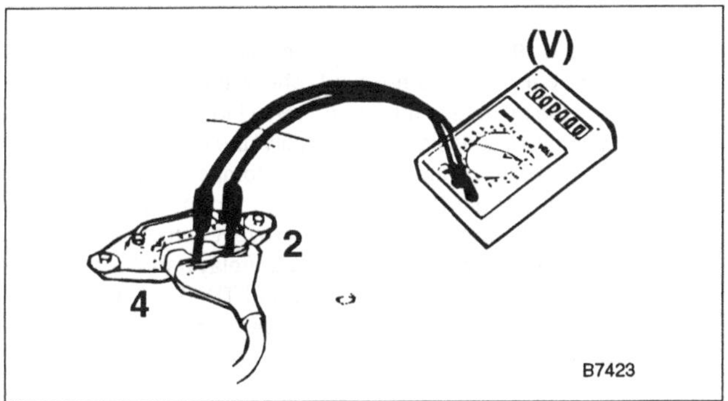

Fig. 15. Voltmeter being used to check for voltage to EZK amplifier (fender-mounter amplifier shown).

NOTE

On fender-mounted amplifiers, the amplifier should be removed from its mountings to access the harness connector.

---

5. **Check for power to the coil.** Connect a voltmeter between terminal 15 and ground on conventional coils or between terminal 3 and ground on coils with amplifier. With the ignition on, there should be battery voltage. If no faults are found proceed to the next step. If voltage is not present, use the wiring diagrams to check for wiring faults from the ignition switch.

6. **Check the coil primary and secondary circuit resistance.** With the ignition off, remove all wires from the coil. Measure the resistance of the coil circuits as shown in Fig. 16. If any faults are found the coil is faulty and should be replaced. If no faults are found, proceed to the next step. Reconnect the wires to the coil.

Ignition coil resistance*

Model	Test terminals and resistance
except 1988-1989	1 and 15 — 0.5 to 0.9 Ohms 1 and center tower — 7200 to 8200
1988-1989	4 and 3 — 0.5 to 0.9 Ohms 4 and center tower — 7200 to 8200 Ohms

* measure with all wires disconnected from coil

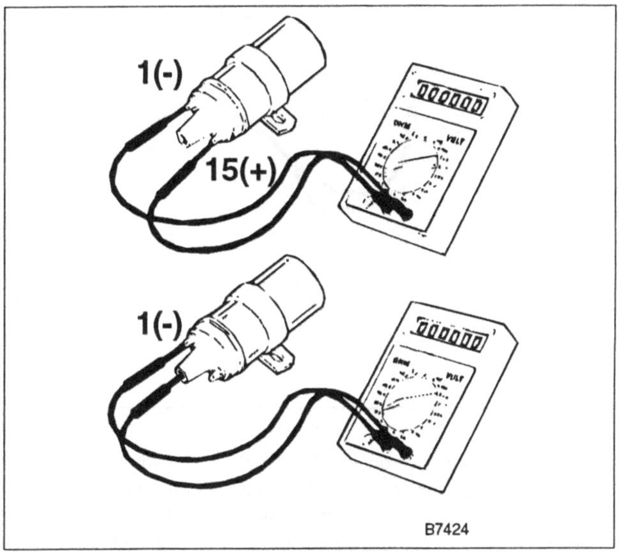

Fig. 16. Coil circuits being tested using ohmmeter (conventional coil shown).

7. **Check the amplifier switching function.** Connect the voltmeter between coil terminals 15 (+) and 1 (–) on conventional coils or between terminal 3 (+) and 4 (–) on coils with integrated amplifier. Turn the ignition on. The voltage should rise to about 6 volts (DC) and then drop to 0 volts within 1 to 2 seconds. If any faults are found, replace the amplifier. If no faults are found proceed to the next step.

8. **Check the pulse signal from the EZK control unit to the amplifier.** Peel back the rubber boot on the control unit connector. Connect a voltmeter between the terminals 5 (+) and 2 (–) on fender-mounted amplifiers or between terminals 4 (+) and 2 (–) on coil mounted amplifiers. While operating the starter, there should be an oscillating voltage (pulses between 0 and 8 volts (DC), average of 2.8 volts). If no faults are found, the amplifier is faulty and should be replaced. If there is no pulse, proceed to the next step.

9. **Check the Hall sensor signal to the EZK control unit.** Peel back the rubber boot on the EZK control unit connector. Connect a voltmeter between terminal 24 (+) and 20 (–). See Fig. 17. Turn the ignition on and slowly turn the engine over by hand. The voltage should oscillate between 0.4 volts (DC) (or less) and 3.0 volts (or higher). If no faults are found, the EZK control unit is probably faulty and should be replaced. If any faults are found, proceed to the next step.

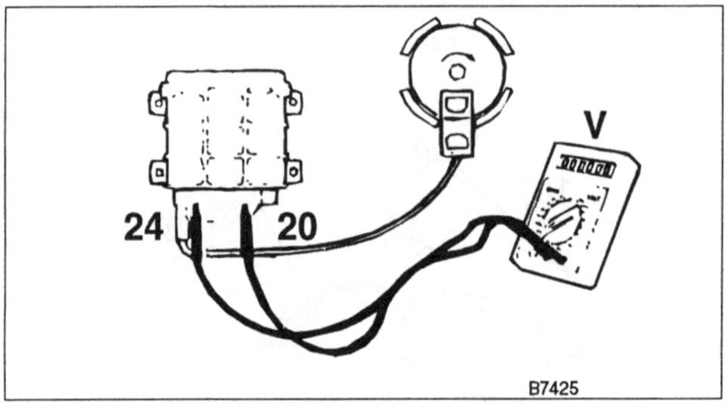

Fig. 17. Voltmeter being used to check Hall sensor operation at EZK control unit.

CAUTION

Turn the engine over only in the direction of normal rotation. Otherwise damage to the timing chain and its components may result. To determine normal rotation, put the car in gear and rock the car forward, noting the rotation of the ignition rotor, the flywheel, or a belt-driven component. As an alternative, briefly operate the starter.

NOTE

The engine can be turned over using a wrench on the crankshaft pulley.

10. **Check for voltage to the Hall sensor.** Working from the rear of the EZK control unit connector, connect a voltmeter between terminal 4 (+) and 10 (–). See Fig. 18. With the ignition on, there should be battery voltage. If no faults are found, the Hall sensor is faulty and should be replaced. If voltage is not present, the EZK control unit is faulty.

NOTE

On all 1986-1988 cars, the Hall sensor is in the distributor. On 1989 and later cars, the Hall sensor is mounted behind the crankshaft pulley. Hall sensor replacement procedures are described below.

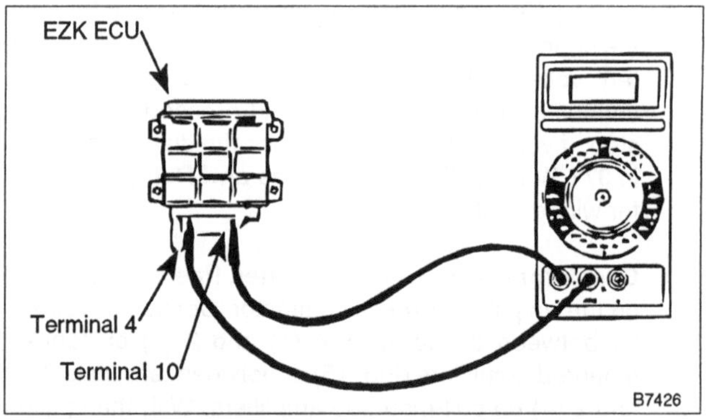

Fig. 18. Voltmeter being used to check for voltage supply to Hall sensor from EZK control unit.

NOTE

If on 1986 and 1987 models, all electrical checks are as specified but the car still does not start, check the ignition pulse amplifier on the main fuse relay panel. See Fig. 12 above. This relay-like amplifier is designed to "amplify" the tachometer signal for more reliable control of electronic components affected by the ignition system (LH fuel injection, fuel pump control, etc.). To check the amplifier, simply remove it from its socket. Then using a jumper wire, jumper the fuse/relay panel sockets corresponding to relay terminals TD-I and TD-O. Try to start the engine. If the engine now runs, replace the pulse amplifier.

Component Removal and Replacement

NOTE
Spark plug replacement is covered in **1 LUBRICATION AND MAINTENANCE.** Spark plugs are replaced at regular intervals and should not be cleaned or regapped.

To replace distributor cap and rotor

1. Unhook the spring clips from the distributor cap and pull off the cap. Leave the wires in the cap.

2. Remove the rotor by cracking it off the distributor shaft using a slip-joint pliers. See Fig. 19.

NOTE
Replacement rotors may not always be glued in place. Check to see if the rotor can be pulled off first.

Fig. 19. Ignition rotor being crushed to remove it from shaft.

3. Install a new rotor with a locking compound (Loctite 601 locking compound or equivalent).

4. To replace the distributor cap, clip the new cap into position. Then transfer one ignition wire at a time to maintain proper firing order of the spark plugs.

To remove and install distributor

1. Make matching marks on the base of the distributor and the cylinder head.

2. Unhook the spring clips from the distributor cap and pull off the cap. Leave the wires in the cap.

3. Remove the vacuum hose from the vacuum diaphragm and the harness connector from the Hall sensor, where applicable.

NOTE
Turbo models have no mechanical advance in the distributor. Models with EZK have no vacuum or mechanical advance.

4. Remove the distributor clamp-down bolt and bracket. See Fig. 20.

Fig. 20. Distributor clamp-down bolt being removed.

5. Pull the distributor from the cylinder head. Inspect the O-ring on the distributor shaft and replace if faulty.

6. Install the distributor into the cylinder head while rotating the rotor until the dog on the shaft engages the offset cutout in the camshaft. See Fig. 21.

7. Install the clamp-down bracket and its mounting bolt. Do not fully tighten the bolt.

8. Rotate the distributor housing until the matching marks made earlier are aligned. Check and adjust ignition timing as described below.

Tightening torque
- Ignition distributor hold-down bolt . . 15 Nm (11 ft-lb)

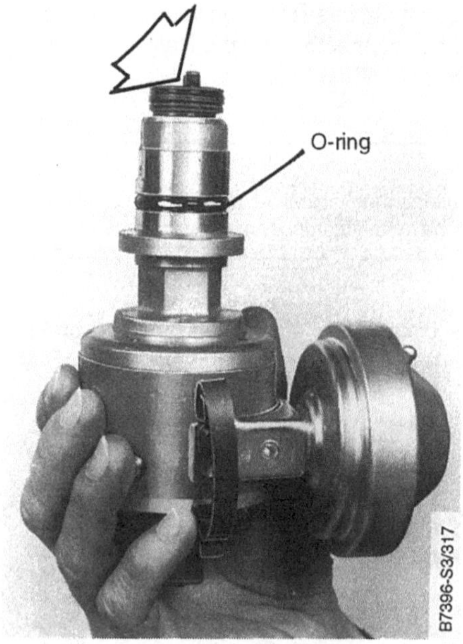

Fig. 21. Offset dog on distributor shaft (arrow) must engage cutout in camshaft.

Hall Sensor

Two different Hall sensors are used in the cars covered by this manual. Replacement procedures vary between styles.

Hall sensor location	
1986-1988 Non-Turbo	in distributor
1985 and later Turbo	
1989-and later Non-turbo	behind crankshaft

NOTE
Distributor replacement parts, including the a rebuild kit for the distributor-mounted Hall sensor, are available from an authorized Saab dealer.

To replace Hall sensor (distributor-mounted sensor)

1. Remove the distributor as described above.
2. Thoroughly clean the distributor shaft, removing any glue or other residues from the shaft.
3. Remove the field rotor circlip. See Fig. 22.
4. Using a small 2-jaw puller, carefully remove the field rotor. See Fig. 23. Remove and save the field rotor locating pin. See Fig. 24.

Fig. 22. Field rotor circlip being removed.

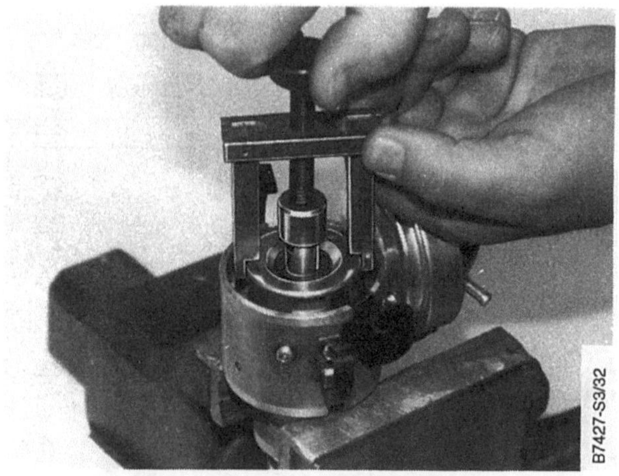

Fig. 23. Hall sensor field rotor being removed.

Fig. 24. Field rotor locating pin (arrow).

5. Remove from the distributor shaft the two circlips and spacer from above the Hall sensor. Remove the plastic cover.

6. Carefully pry off the plastic connector housing from the side of the distributor. See Fig. 25. Lift the sensor out of the distributor.

NOTE
Make sure the spacer underneath the Hall sensor does not come out with the sensor.

Fig. 25. Hall sensor connector housing being removed.

7. Install the new sensor into the distributor. Where applicable, make sure the locating pin on the sensor engages the hole in the diaphragm arm. See Fig. 26.

8. The remainder of installation is the reverse of removal. Be sure to install the field rotor locating pin (see Fig. 24).

Fig. 26. When installing sensor, engage pin on sensor with hole in vacuum diaphragm arm (arrows).

To replace Hall sensor (crankshaft-mounted sensor)

1. Disconnect the negative (–) battery cable.

2. Remove all the V-belts from the crankshaft pulley. See **1 LUBRICATION AND MAINTENANCE** for more information on removing V-belts.

3. While holding the flywheel stationary, remove the crankshaft pulley center mounting bolt. Remove the pulley and its woodruff key.

NOTE
A special flywheel holding fixture (Saab tool no. 83 92 987) is available from an authorized Saab dealer.

---

4. If necessary remove the Hall sensor field rotor from the pulley by removing the three bolts. See Fig. 27.

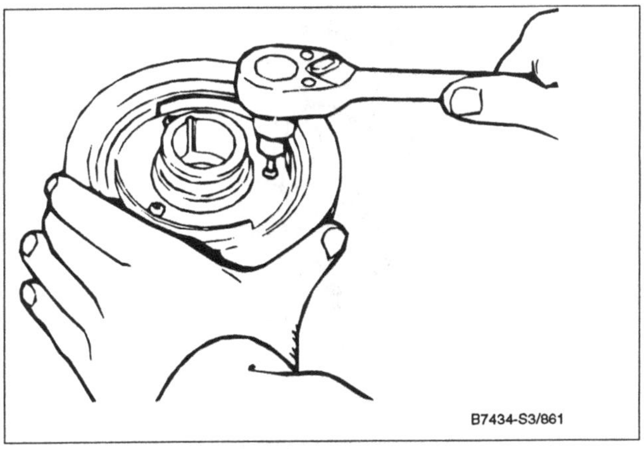

Fig. 27. Ignition field rotor being removed from rear of crankshaft pulley.

5. Disconnect the Hall sensor connector. The connector is located below the oil dipstick tube. See Fig. 28.

6. Remove the Hall sensor bolts and remove the sensor. See Fig. 29.

Installation is the reverse of removal. Use Loctite®270 on the sensor mounting bolts. Reinstall the pulley and woodruff key. Tighten the bolt while holding the flywheel steady.

Tightening torque
- crankshaft pulley bolt to crankshaft
  - 1989 . . . . 190 Nm (140 ft-lb)
  - 1991and later . . . . 175 Nm (128 ft-lb)

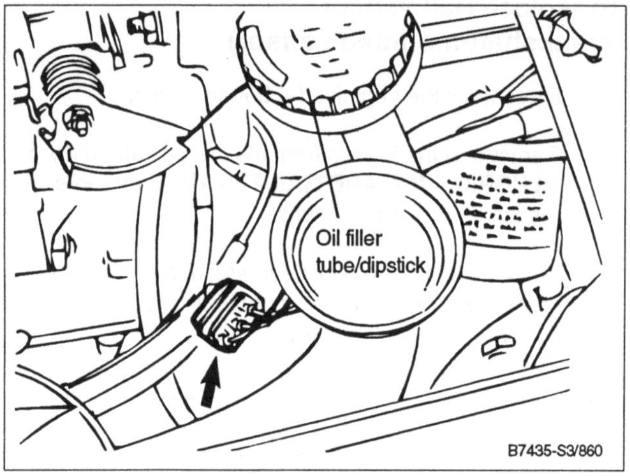

Fig. 28. Hall sensor harness connector (arrow).

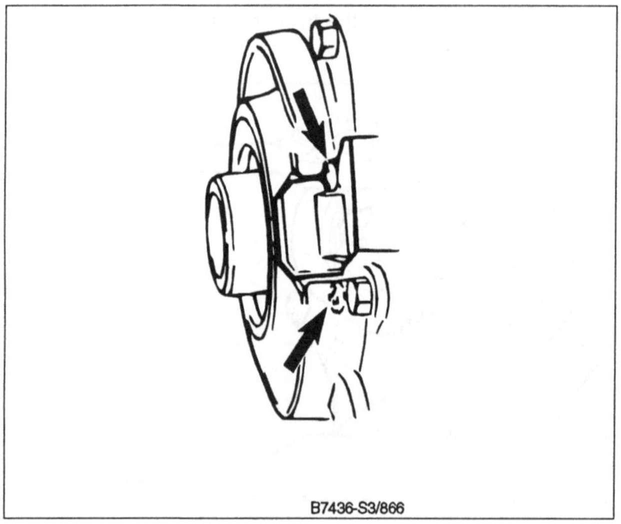

Fig. 29. Hall sensor mounting bolts (arrows).

To replace knock sensor (EZK ignition only)

1. Disconnect the knock sensor harness from the sensor. See Fig. 30.

NOTE
The knock sensor is mounted to the engine block between cylinders no. 2 and 3, below the intake manifold.

---

2. Remove the knock sensor mounting bolt and remove the sensor. See Fig. 31.

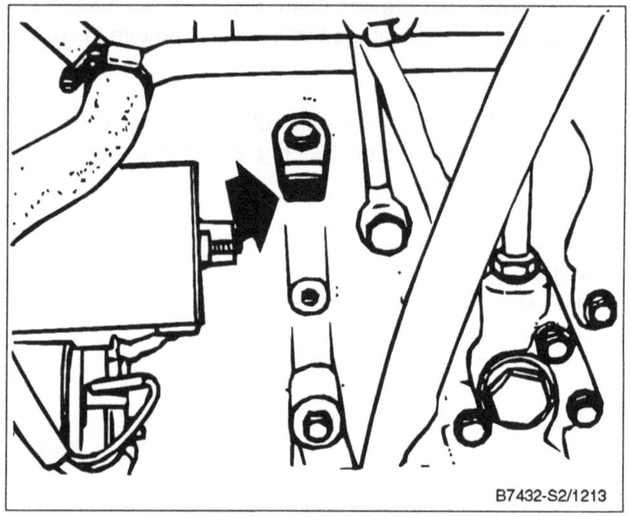

Fig. 30. Knock sensor connector removed from sensor (arrow).

Fig. 31. Knock sensor being removed.

3. Thoroughly clean the knock sensor mounting surface on the engine. Clean the mounting bolt threads. Hold the sensor steady and install and torque the bolt. Reconnect the harness connector.

Tightening torque
- Knock sensor to engine block . . . . . 20 Nm (15 ft-lb)

IGNITION TIMING

On most models, ignition timing should be checked at regular service intervals as listed in **1 LUBRICATION AND MAINTENANCE.** Timing should also be checked any time the distributor position has been altered.

NOTE
On 1989 and later models with a crankshaft-mounted sensor, checking and adjusting ignition timing at regular intervals is no longer necessary.

To check and adjust ignition timing

WARNING
The ignition systems installed on the cars covered by this manual are high-energy systems operating in a dangerous voltage range which could prove to be fatal if exposed terminals or live parts are contacted. Use extreme caution when working on a vehicle with the ignition on or the engine running.

1. Connect a timing light and a tachometer according to the manufacturer's instructions. Start the engine and allow it to reach operating temperature.

2. If applicable, disconnect the vacuum advance hose from the throttle housing. Check that the idle speed is as specified. See Fig. 32. If idle speed is incorrect, adjust it as described in **240 Fuel Injection.**

3. While pointing the timing light at the flywheel, check that the timing marks are correctly aligned. See Fig. 33. If the marks are not aligned, loosen the distributor clamp-down bolt and slowly rotate the distributor until the marks are aligned. Tighten the bolt.

Ignition Timing
- Turbo models (vacuum hose disconnected) . . . . 16°BTDC @ 850 rpm
- Non-turbo models . . . . 14°BTDC @ 850 rpm

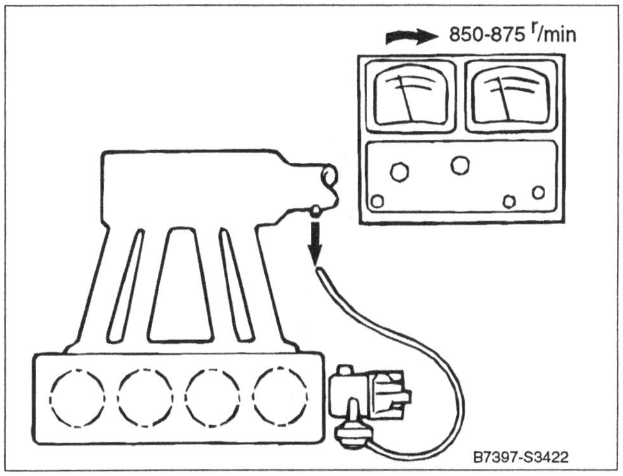

Fig. 32. When checking timing, idle speed must correct. Where applicable, disconnect the vacuum hose (arrow).

Fig. 33. Ignition timing marks on flywheel and clutch cover (arrows). Clutch cover is in front of engine.

Tightening torque
- Ignition distributor hold-down bolt . . . 15 Nm (11ft-lb)

4. Turn the engine off. Disconnect the test equipment. Reconnect the vacuum hose, if applicable.

351 Lighting

GENERAL

This repair group covers procedures for replacing exterior and interior bulbs.

NOTE
Also see **853 Dashboard and Consoles** for additional cluster bulb replacement information.

Lighting faults can usually be traced by observing whether one light is out, or other lights on the same circuit are out. If only one bulb is out, chances are the bulb has failed. If more than one bulb is out, there is probably an electrical fault in the circuit, such as a poor ground connection. Always begin by checking fuses and connectors. For wiring diagrams and additional lighting information, see **371 Wiring Diagrams, Fuses and Relays.**

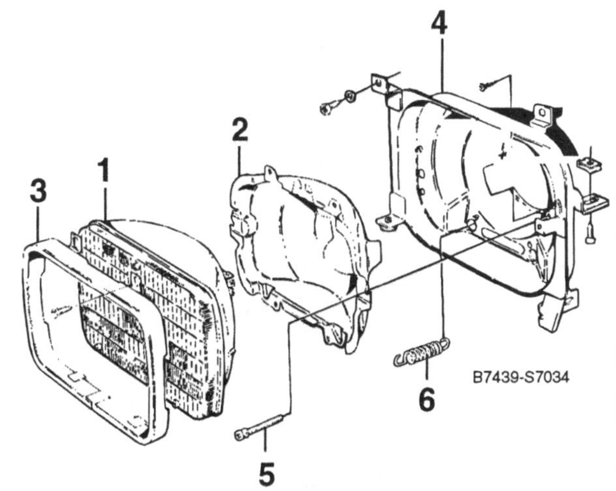

1. Sealed beam
2. Holder
3. Bezel and screw
4. Headlight frame
5. Aiming screw (1 of 2 shown)
6. Spring

Fig. 1. Sealed beam headlight on 1985 and 1986 models.

To replace sealed beam headlight (1985 and 1986 models)

1. Remove the headlight trim mounting screws. Remove the trim and disconnect the wiring harness for the park light. See Fig. 2.

CAUTION
Do not disturb the headlight aiming screws when changing the bulb. If the headlight aim is changed, special adjusting equipment will be required to correctly re-aim the headlights.

2. Remove the four sheet metal screws holding the sealed beam bezel to the headlight frame.

3. Lift out the headlight and disconnect the harness connector.

Installation is the reverse of removal. Be sure to install the new sealed beam so that the glass tabs on the beam correctly engage the cutouts of the headlight holder.

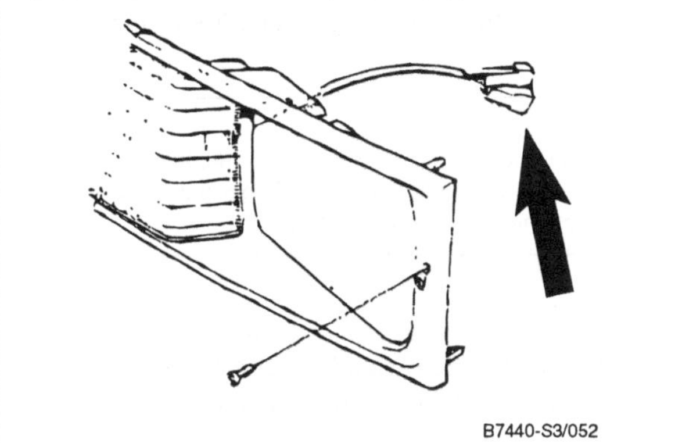

Fig. 2. Headlight trim with integrate park light. Disconnect connector for inner park light (arrow).

To replace headlight bulb (1987 and later models)

1. Open the hood and remove the bulb holder from behind by turning the locking ring counterclockwise. See Fig. 3 Remove the bulb.

CAUTION
Do not disturb the headlight aiming screws when changing the bulb. If the headlight aim is changed, special adjusting equipment will be required to correctly re-aim the headlights.

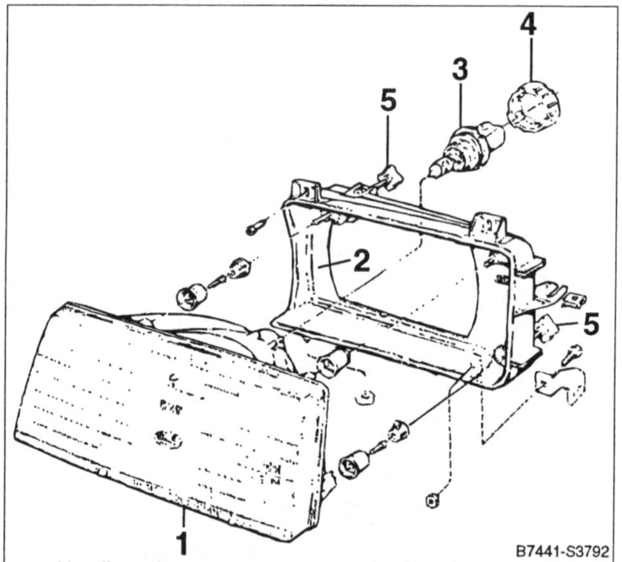

1. Headlamp lens
2. Headlight frame
3. Bulb
4. Locking ring
5. Adjusting screw

Fig. 3. Headlight with replaceable bulb on 1987 and later models.

2. Disconnect the harness connector from the bulb.

CAUTION
Do not touch the bulb glass with your hands. If the bulb was handled, use a clean rag to clean it. Finger prints can cause hot spots on the bulb, shortening its life.

---

3. Connect the harness connector to the new bulb and install it into its holder. See Fig. 4. Make sure the bulb is squarely seated. Install and tighten the locking ring.

Fig. 4. Headlight bulb being installed on 1987 and later models

To replace headlight lens assembly (1987 and later models)

1. Open the hood and remove the side light assembly mounting screws. See Fig. 5. Pull the front edge of the light assembly out slightly, disconnect the harness connector and remove the assembly.

Fig. 5. Front side light assembly mounting screw being removed.

2. Remove the three headlight assembly mounting screws and the remove the assembly. Two screws are removed from the front. See Fig. 6. One screw is removed from the rear, lower corner.

3. Carefully pry the headlight lens assembly out of the headlight frame. See Fig. 7.

4. Installation is the reverse of removal. Be sure to have the headlight aim checked after replacing the lens assembly.

Fig. 6. Headlight mounting screw being removed. One mounting screw is removed from rear of assembly.

Fig. 7. Headlight lens assembly being removed from headlight holder.

Replacing front park, direction indicator, stop, side marker and side reversing light bulbs

On all 1985 and 1986 models, the bulbs in the front corner cluster are accessed after removing the lens from the front of the cluster. See Fig. 8. On 1987 and later models, the bayonet-mounted bulbs are removed from the rear of the cluster. See Fig. 9.

To replace the parking light bulb in the headlight trim (inner side) first remove the trim frame as shown above in Fig. 2. Then remove the bulb holder from the rear of the light. Replace the bulb, reinstall the bulb holder and the headlight trim.

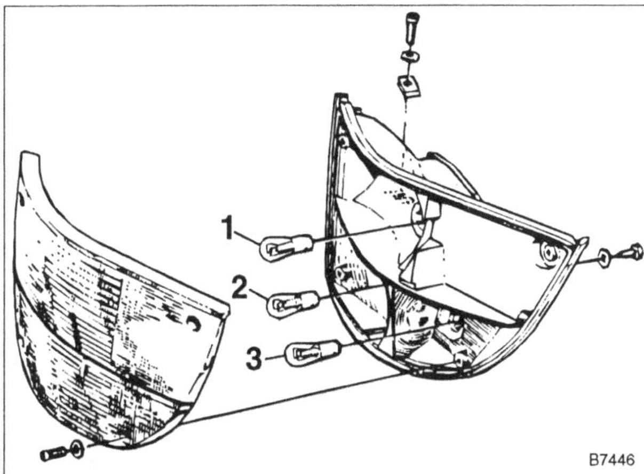

1. Direction indicator/side marker (21/5 watt)
2. Cornering light/parking light (21/5 watt)
3. Side reversing light (21 watt)

Fig. 8. Front lamp cluster on 1985 and 1986 models.

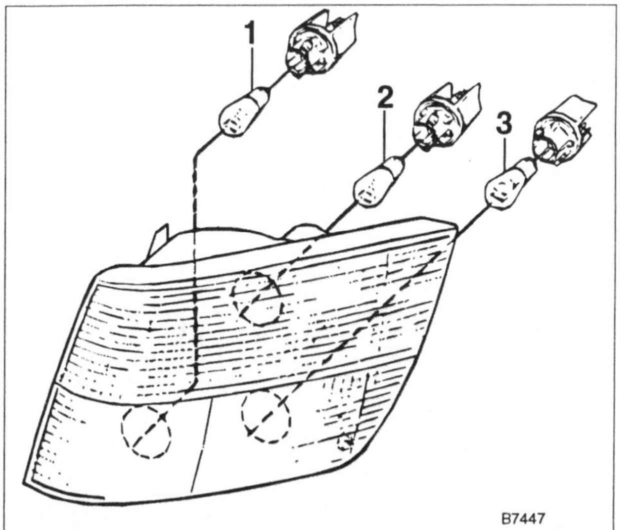

1. Direction indicator/side marker (21/5 watt)
2. Cornering light/parking light (21/5 watt)
3. Side reversing light (21 watt)

Fig. 9. Front lamp cluster on 1987 and later models.

To replace the fender-mounted direction indicator bulbs (1986 and later models), slide the light assembly toward the rear of the car while carefully prying it out. See Fig. 10. Replace the bulb and reinstall the light.

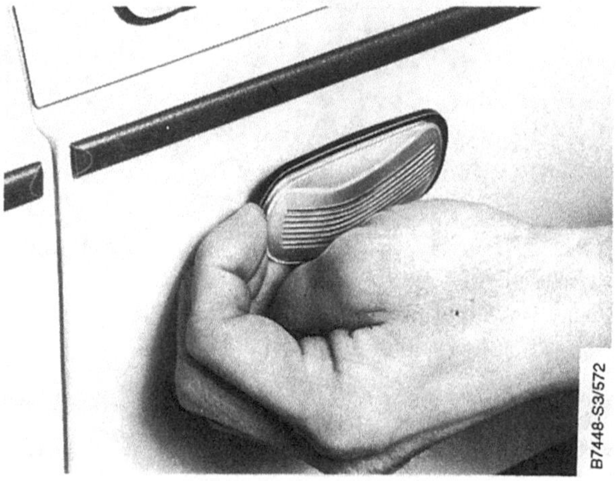

Fig. 10. Fender-mounted direction indicator lens being removed.

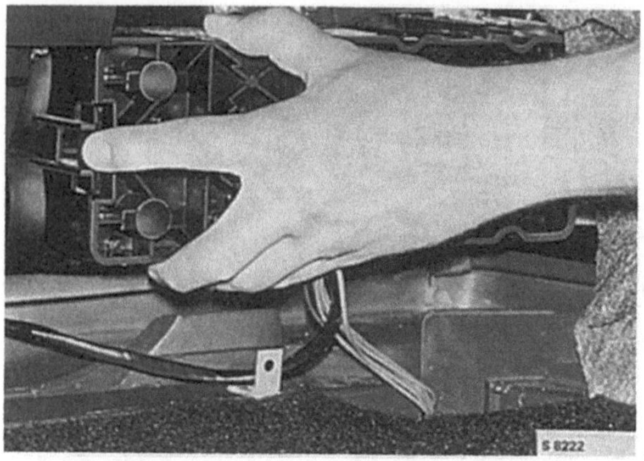

Fig. 12. Rear light cluster on sedan models being removed.

Replacing rear park, direction indicator, stop, reversing light, and license plate illumination bulbs

On all 3-door hatchback models, the bulbs in the rear light cluster are accessed after removing the lens from the front of the cluster. See Fig. 11. On all sedan models, the bulbs are accessed through the luggage compartment. Remove the bulb holder by depressing the plastic locking tab and swinging the assembly out. See Fig. 12.

1. Direction indicator (21 watt)
2. Backup light (21 watt)
3. Parking light/brake light (5/21 watt)
4. Brake light (21 watt)
5. Parking light (5 watt)

Fig. 11. Rear light cluster on hatchback models.

To replace the center (3rd) brake light bulb, remove the brake light cover by squeezing the two ribbed markings on the cover while sliding the cover off. Replace the bulb and reinstall the cover. To access the center brake light bulbs on convertible models, remove the two screws and remove the lens. See Fig. 13.

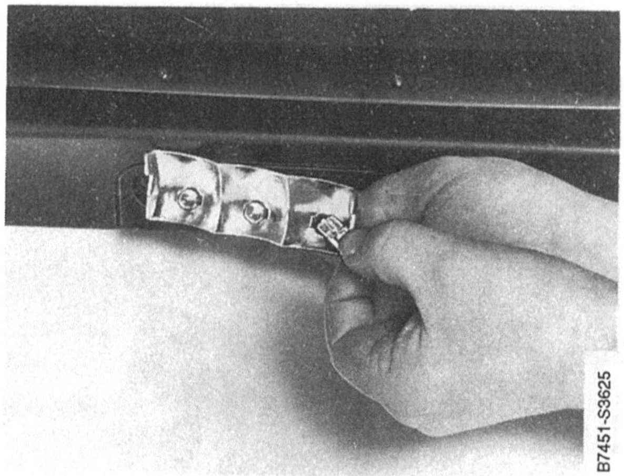

Fig. 13. Center brake light bulb being removed on convertible model.

The license plate illumination bulbs can be replaced after removing the light assembly see Fig. 14.

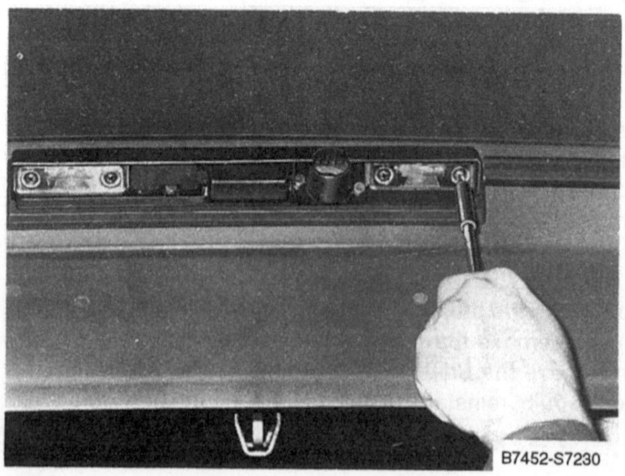

Fig. 14. License plate light assembly being removed.

Replacing Interior Bulbs

The interior dome lights and the glovebox light can be changed by carefully prying away the clear plastic cover. Some models have a time delay feature controlled by the time-delay relay. This relay is located under the back seat on the left side.

The ignition key light can be changed by peeling away the shift lever boot on cars with manual transmission or removing the plastic trim piece around the shift lever on cars with automatic. Then remove the front console and rear console screws (rear screws are behind rear passenger ashtray). Carefully lift up the console and pull the bulb from its holder.

NOTE
Remove the ignition key before attempting to remove the console.

All instrument cluster lamps, except the fuel level warning lamp, can be replaced without removing the cluster. These bulbs can be accessed after removing the left-front speaker grille. See Fig. 15.

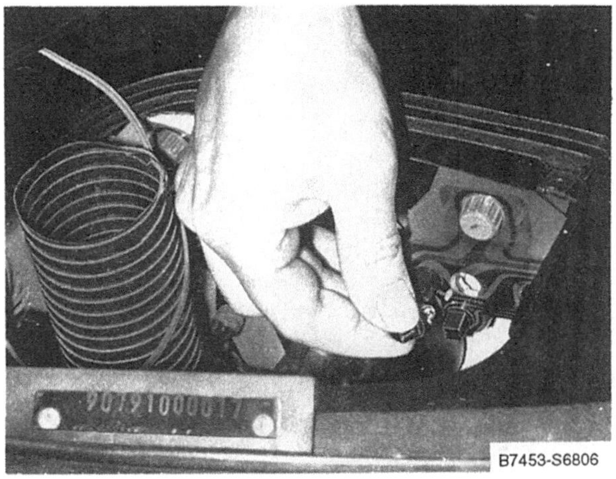

Fig. 15. Instrument cluster illumination bulbs being replaced through left speaker grille access hole.

On most 1985 dash-mounted switches, the illumination bulbs can be replaced once the switch is carefully pried out of the dashboard. On 1986 and later models, the switch illumination is integral with the switch and the bulbs cannot be separately replaced.

363 Wipers and Washers

Windshield Wipers and Washers

The windshield wiper assembly is located in the engine compartment on the firewall. The wiper motor and the linkage assembly are removed as a complete assembly. See Fig. 1. Windshield wiper wiring diagrams can be found **371 Wiring Diagrams, Fuses and Relays**.

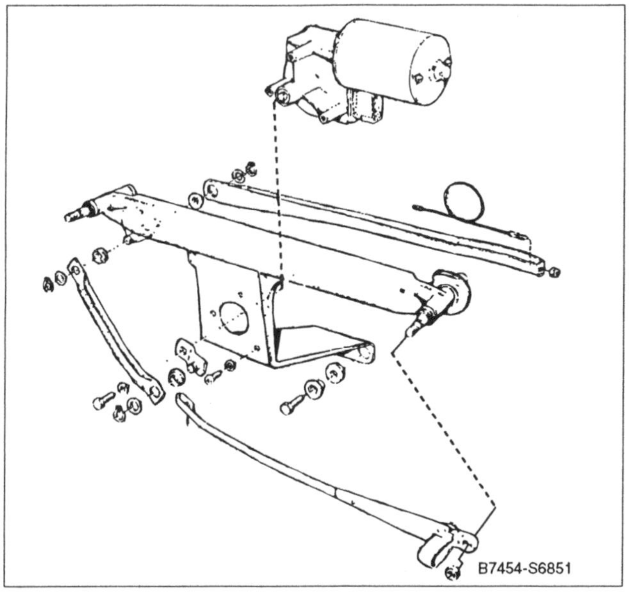

Fig. 1. Exploded view of windshield wiper assembly.

The windshield washer pump is located under the washer fluid reservoir next to the battery. To remove the washer pump, remove the reservoir bottle hold-down bracket and lift up while disconnecting the fluid hose and the pump wiring. Pull the washer pump out of the bottle. Replace the O-ring when installing a new pump.

CAUTION

Do not operate the pump unless there is fluid in the bottle. Damage to the pump may result.

To remove the windshield wiper assembly

1. Disconnect the negative (–) battery cable.

2. Lift the wiper arms off the glass. Pry up the retaining nut cover and remove the retaining nut. See Fig. 2.

NOTE

Before removing the wiper arms, note the parked position of the arms on the glass.

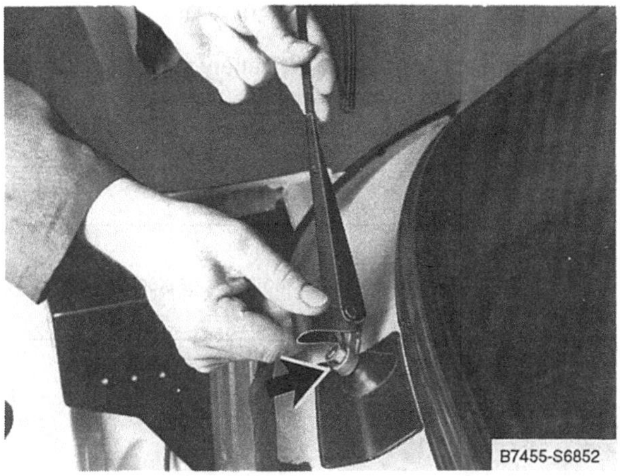

Fig. 2. Wiper arm cover being lifted to expose arm mounting nut (arrow).

3. Remove the rubber grommets covering the wiper shafts.

4. Remove the four bolts holding the wiper assembly in position. Disconnect the wiring and remove the assembly from the car. See Fig. 3.

Fig. 3. Wiper assembly being removed.

Installation is the reverse of removal. To make sure the wiper motor is stopped in the park position, operate the motor through several cycles before reattaching the wiper arms.

HEADLIGHT WIPERS AND WASHERS (1991 AND LATER MODELS)

All 1991 and later models covered by this manual are equipped with headlight wipers. The headlight wipers and washer pump will operate if the stalk switch is pulled toward the steering wheel. After five cycles, the wipers will park.

To remove headlight wiper motor

1. Disconnect the negative (–) battery cable.
2. Lift the wiper arm off the headlight. Pry up the retaining nut cover and remove the retaining nut. See Fig. 4.

NOTE

Before removing the wiper arms, note the parked position of the arms on the headlight.

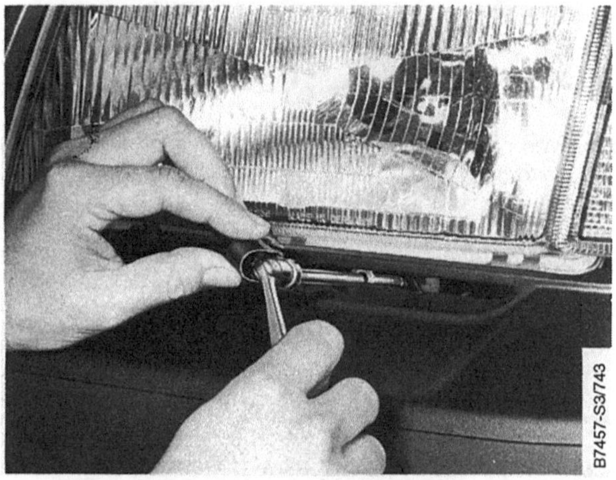

Fig. 4. Headlight wiper arm nut being removed.

3. Remove the headlight assembly as described in **351 Lighting**. Remove the bolts holding the wiper motor assembly to the body. See Fig. 5.

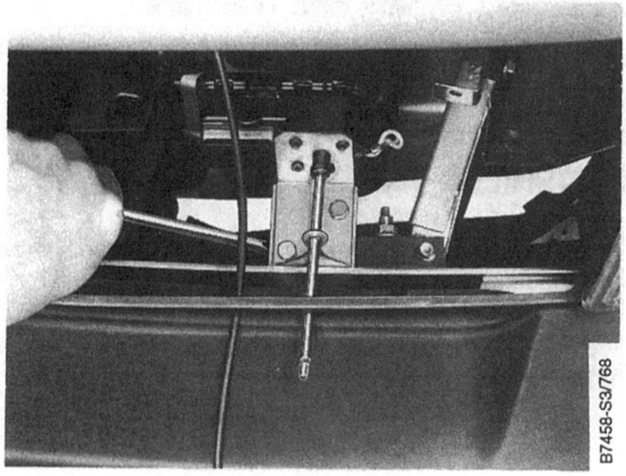

Fig. 5. Headlight wiper motor being removed.

Installation is the reverse of removal. To make sure the wiper motor is stopped in the park position, operate the motor through several cycles before reattaching the wiper arms.

364 Electric Controls and Switches

GENERAL

This repair group covers mainly the switches and controls used throughout the car's interior. Some engine control switches and instruments are also covered here. For switch testing, use the wiring diagram schematics found under **371 Wiring Diagrams, Fuses and Relays.**

WARNING
1990 and later models are equipped with an airbag or SRS system. The airbag is a explosive device. Do not attempt to remove the steering wheel or service the wiring in the steering column assembly. You could accidentally ignite the airbag, causing serious personal injury. See an authorized Saab dealer for any repairs to the SRS system or to disarm the system prior to removing the steering wheel.

SWITCHES AND CONTROLS

Replacing switches is straightforward and easy to do. Some switches may need to be adjusted for proper operation after they are installed. Usually the switches needing adjustment have a threaded body or have mounting screw adjusting holes. Some of the fasteners are Torx-type bolts and screws. Plan to have Torx drivers on hand before starting the job. If you are replacing a switch with many wires, it helps to draw a picture of the terminals and wire colors. This will prevent hooking up the wires improperly during installation. See **371 Wiring Diagrams, Fuses and Relays** for more information on identifying components and their locations.

WARNING
Always check the function of the circuit after replacing a switch to verify proper and safe operation.

CAUTION
Always disconnect the negative (–) battery cable before replacing any switches.

Instrument Panel Switches

Most of the switches located in the instrument panel can be replaced without removing the instrument panel. An exception to this is the air distribution switch, the heater control knob, and the extra dimmer switch (rheostat)—which all require removal of the instrument panel as described below.

Fig. 1 shows the headlight switch being removed . After removing the switch, disconnect the wires as shown. The panel switches can be carefully pried away from the panel. See Fig. 2.

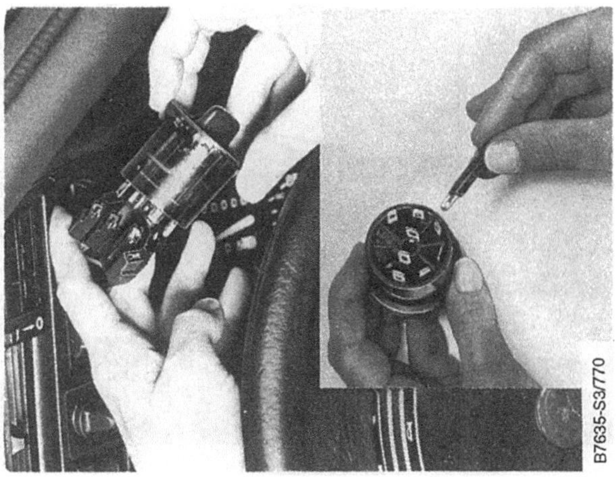

Fig. 1. Headlight switch being replaced. Inset shows bulb in back of switch being removed.

364

Fig. 2. A/C switch being removed. All square switches pull away from the instrument panel.

To remove instrument panel (1985-1989 models only)

1. Disconnect the negative (–) battery cable.

2. Remove the steering wheel as described in **641 Steering Column.**

3. Working from below the instrument panel, remove and label the four long mounting screws. See Fig. 3. and Fig. 4.

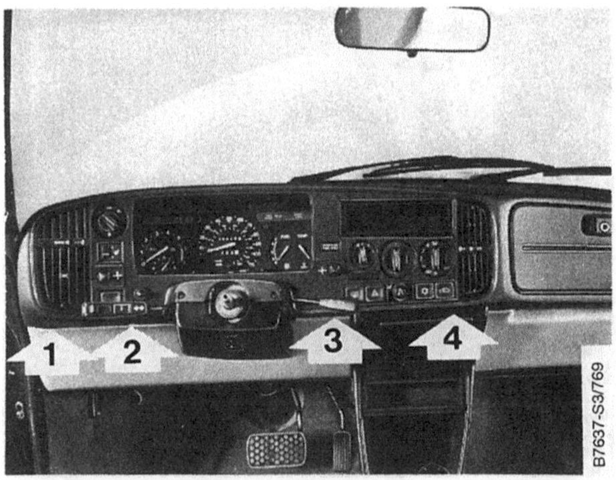

Fig. 3. Instrument panel mounting screw locations (arrows).

4. Carefully pull the panel out. Label any vacuum hoses or wires before removing them. Disconnect the control rod to the heater knob.

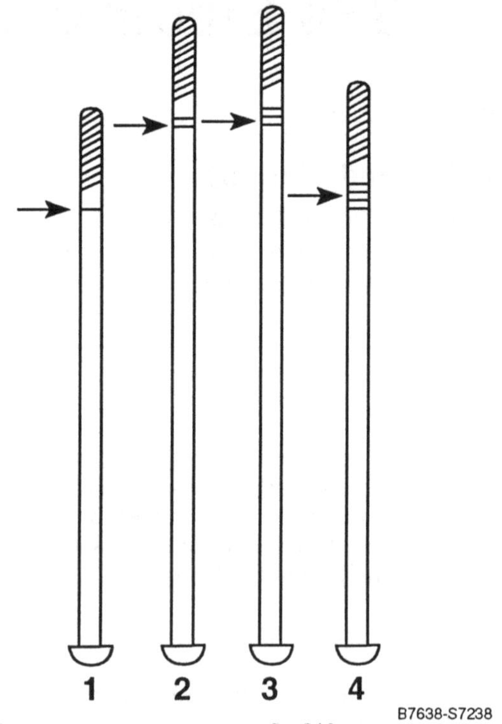

1. 176 mm
2. 205 mm
3. 210 mm
4. 189 mm

Fig. 4. Each instrument panel screw is a different length and marked accordingly.

NOTE

The extra rheostat for instrument panel lighting is removed after removing the knob and locking nut from the front of the panel. The air distribution switch and the heater control knob can be removed after removing the mounting screws from the rear of the panel. See Fig. 5.

Fig. 5. Air distribution switch mounting screw being removed. Switch is held in place with two screws (arrows). Heater control knob removal is similar.

Installation is the reverse of removal. Be sure the panel mounting screws are installed in their correct positions. If the screws are interchanged, the padded dashboard could be damaged.

Steering Column Stalk Switches and Horn Switch (1985-1989 models)

The direction indicator and windshield washer switches are mounted to the steering column. To replace a switch, remove the lower steering column trim. Then remove the screws holding the switch and disconnect the electrical connector. See Fig. 6.

WARNING

1990 and later models are equipped with an airbag or SRS system. The airbag is an explosive device. Do not attempt to remove the steering wheel or service the wiring in the steering column assembly. You could accidentally ignite the airbag, causing serious personal injury. See an authorized Saab dealer for any repairs to the SRS system or to disarm the system prior to removing the steering wheel, stalk switches, or steering column wiring.

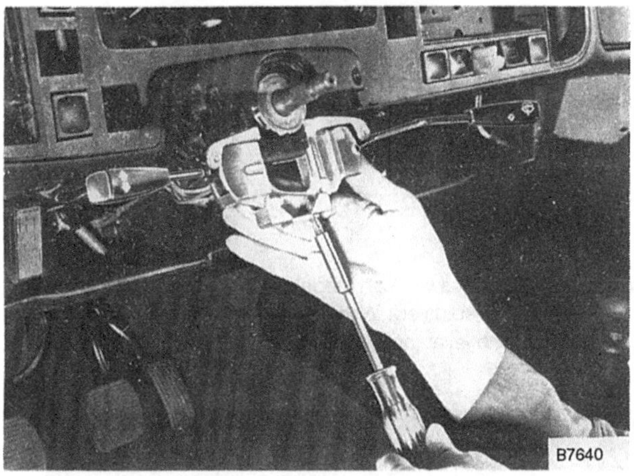

Fig. 6. Steering column stalk switches being removed.

To remove the horn switches, carefully pry them away from the steering wheel. See Fig. 7.

Interior Switches

On 1985 through 1990 models, the power window switches, sunroof switch, and interior light switch can all be changed by carefully prying the assembly from the center console. See Fig. 8.

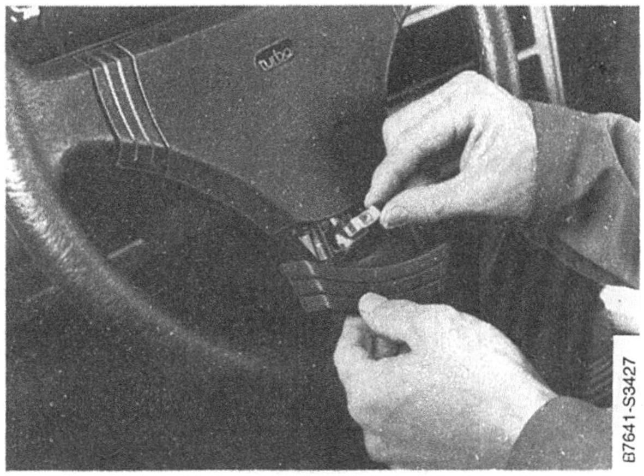

Fig. 7. Horn button switch being removed. Turbo model shown, other models are similar.

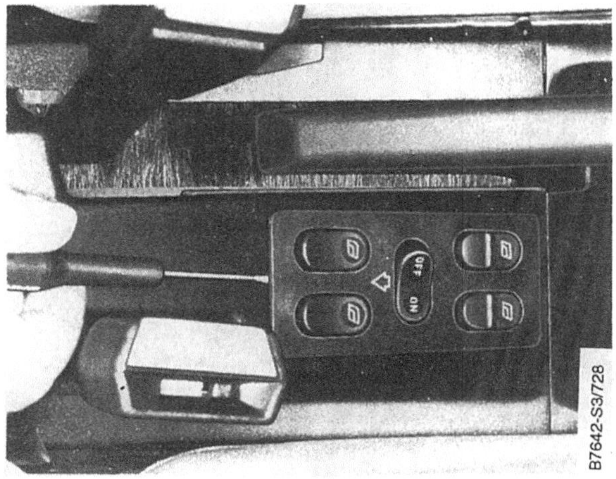

Fig. 8. Power window switch assembly in center console being pried up with a screwdriver.

NOTE

- On 1985 through 1990 models, the individual switches in the switch assembly can be separately replaced.

- On 1991 and later models, the entire power window/sunroof switch assembly must be replaced if a single switch is faulty. The bulb for the switch assembly can be replaced by turning the bulb holder 90° and removing the bulb. See Fig. 9.

The brake light switch is located on the upper part of the brake pedal assembly. See Fig. 10. Adjust the switch so the brake lights come on as soon as the pedal is depressed and go out when the pedal is in the rest position. For the most accurate adjustment, use the specification shown below.

Brake light switch adjustment

- brake lights come on when brake pedal is depressed. . . . . 0.4 in. (10 mm)

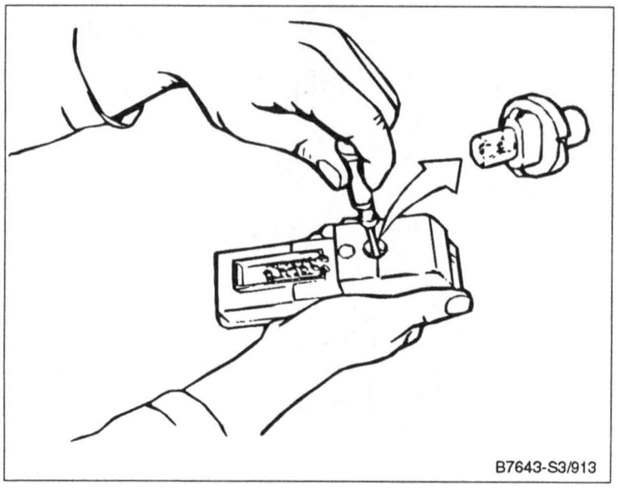

Fig. 9. Power window switch light bulb being replaced.

Fig. 10. Brake light switch (arrow) on brake pedal cluster. Switch should always be adjusted after installation.

NOTE
The cruise control brake switch and clutch switch are covered in **368 Cruise Control.**

The back-up light switch, used only on cars with manual transmission, screws into the left side of gear shift housing. See Fig. 11. To access the switch, remove the center console as described in **432 Manual Transmission Controls.**

On models with automatic transmission, the back-up light switch is part of the neutral/safety (start inhibit) switch. To adjust this switch, move the lever to N (neutral) then loosen the screws on the side and rotate the switch until the mark on the switch aligns with the shift lever. See Fig. 12.

The handbrake warning lamp switch is located between the seats under the handbrake bracket. Remove the center console as described in **432 Manual Transmission Controls** then remove the switch. See Fig. 13.

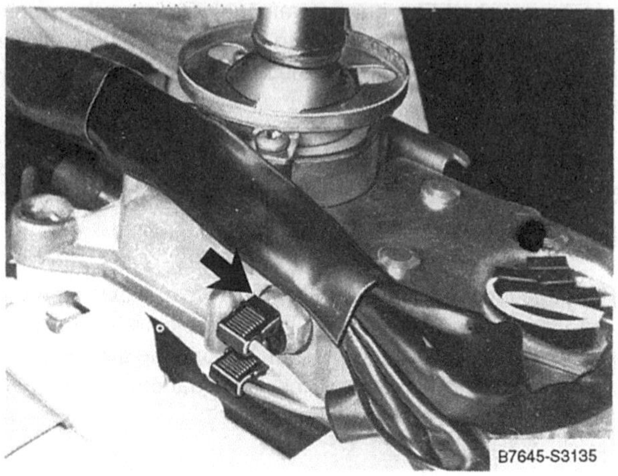

Fig. 11. Back-up light switch (arrow) on gear shift housing (manual transmission). Gear shift lever shown at top.

Fig. 12. Back-up light switch on models with automatic transmission being adjusted. With shift lever in **N**, align mark on switch with lever (arrows).

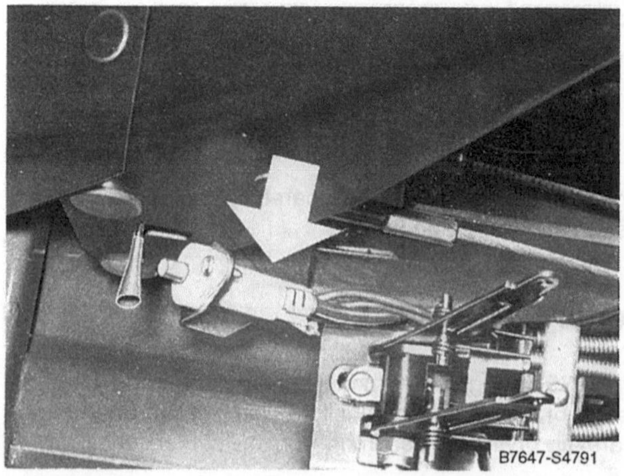

Fig. 13. Handbrake warning lamp switch (arrow) can be removed after removing center console.

The mercury switch on convertible models is used to prevent the rear window defogger from turning on when the roof is in the down position. The switch can be removed after removing the screws. See Fig. 14. The switch is located at the top of the frame member for the convertible top.

14. Mercury switch (arrow) interrupts power to heated rear window when convertible top is down.

Engine Compartment Switches

Most of the switches located in the engine compartment are used to monitor engine conditions and are covered in the various repair groups under **2 ENGINE.**

The temperature switch for the shift-up indicator prevents the shift-up system from operating until the engine reaches a certain temperature. The switch screws into the intake manifold. See Fig. 15.

Fig. 15. Temperature switch (arrow) for shift-up indicator on intake manifold. Saab 8-valve engine shown.

The fifth-gear lockout switch for shift-up indicator is shown in Fig. 16.

Fig. 16. 5th gear lockout switch (arrow) for shift-up indicator. Switch screws into right front transmission side cover, near battery.

The throttle switch for the shift-up indicator is used on cars with LH 2.2 Fuel Injection. On these models, a throttle dashpot holds the throttle open for a short period of time. The throttle switch is used to tell the shift-up relay that the accelerator pedal is released, even though the engine speed is still high. This switch is mounted on the throttle housing. See Fig. 17. To remove the switch, simply remove the screws.

Fig. 17. Throttle switch (arrow) for shift-up indicator on throttle housing.

The brake fluid warning lamp switch is part of the fluid reservoir cap. A float switch turns on the lamp when the fluid is low. To replace the switch, remove the connector and unscrew the cap. See Fig. 18.

Fig. 18. Location of brake fluid level warning lamp switch on master cylinder.

Extra Gauges

Extra gauges, available from Saab, can be located in the center console storage area. See Fig. 19. When installing new gauges, always follow the instructions that come with the instrument. The wiring shown in Fig. 20, Fig. 21, Fig. 22, and Fig. 23 can be used as a guide in installing the gauges. All of the gauges require a 52 mm diameter space in the console.

The outside temperature gauge can be adjusted as shown in Fig. 24. To calibrate the gauge, use a good quality temperature probe such as the Fluke® 80 multimeter.

NOTE

The calibration of the outside temperature gauge should be done over a period of 20 minutes to prevent misleading temperature readings.

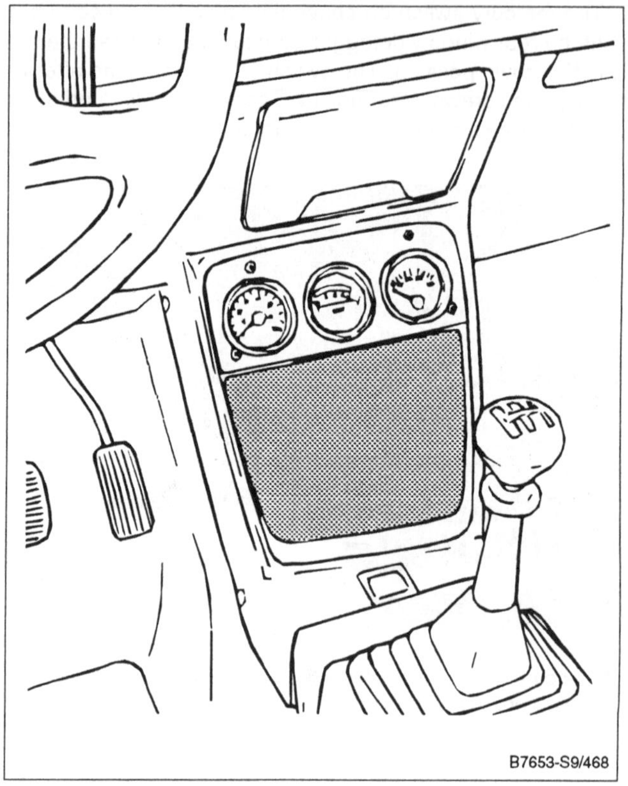

Fig. 19. Extra gauges in center console.

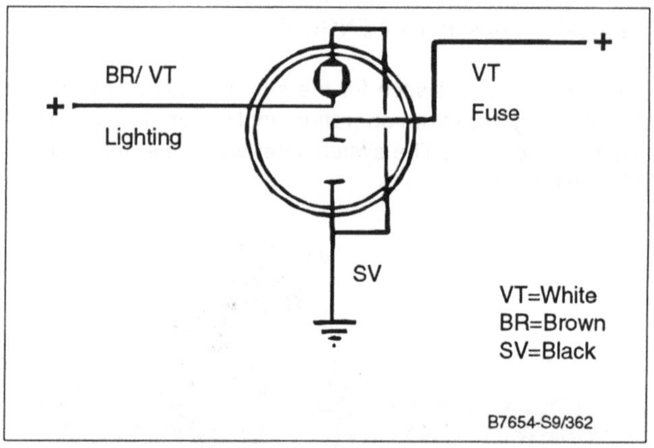

Fig. 20. Wiring diagram for extra voltmeter.

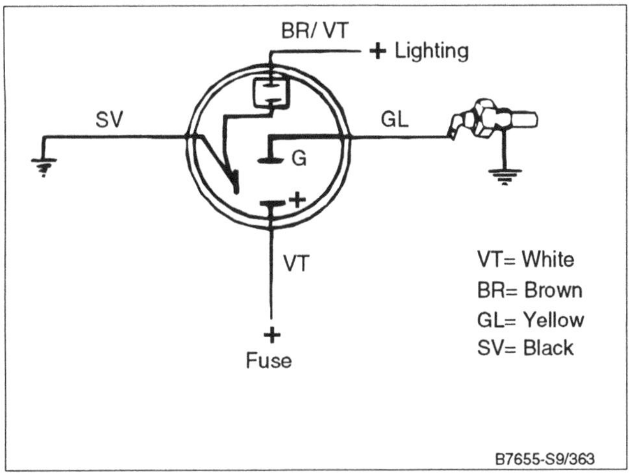

Fig. 21. Wiring diagram for extra oil temperature gauge.

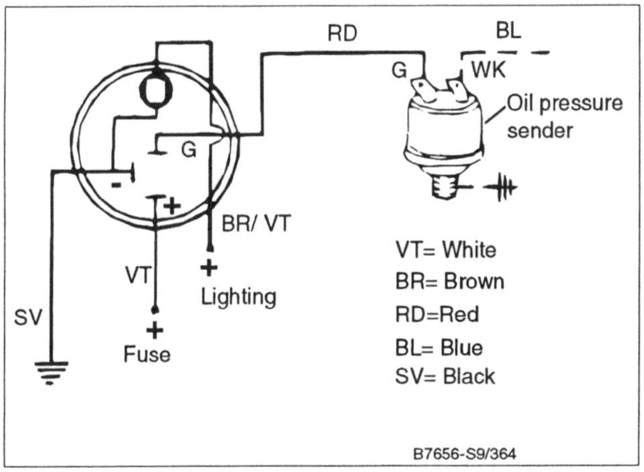

Fig. 22. Wiring diagram for extra oil pressure gauge.

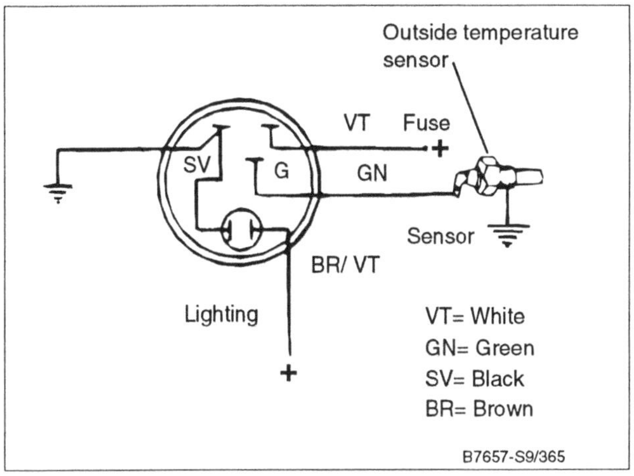

Fig. 23. Wiring diagram for extra outside temperature gauge.

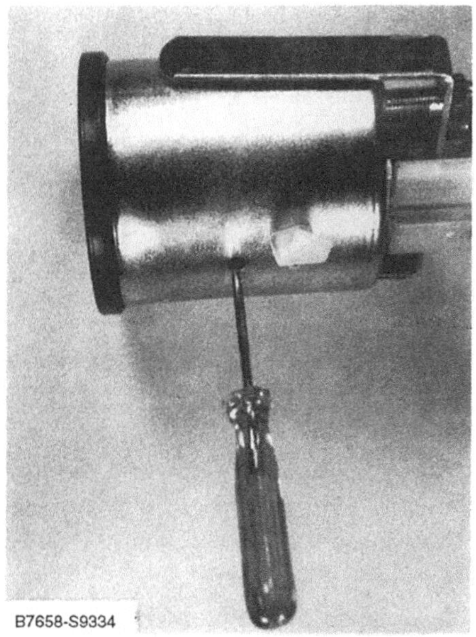

Fig. 24. Outside temperature gauge being calibrated.

368 Cruise Control

368

GENERAL

WARNING
1990 and later models are equipped with an airbag or SRS system. The airbag is a explosive device. Do not attempt to remove the steering wheel or service the wiring in the steering column assembly. You could accidentally ignite the airbag, causing serious personal injury. See an authorized Saab dealer for any repairs to the SRS system or to disarm the system prior to removing the steering wheel.

The cruise control system enables the driver to select a cruising speed that will be maintained without the use of the accelerator pedal. The system can be turned on and off at the steering column stalk switch. The system will automatically disengage when either the clutch pedal or the brake pedal is depressed. Fig. 1 shows the components of the cruise control system.

When the Set button is depressed and the road speed is above 23 mph, the speed transmitter sends an electronic road-speed signal to the control unit. The control unit then operates the vacuum pump to increase or reduce vacuum to the vacuum regulator, changing the position of the accelerator pedal and throttle plate.

When the system is in operation, temporary acceleration can be accomplished by depressing the accelerator pedal. As soon as the pressure on the pedal is released, the car will resume the preselected cruising speed.

When the brake pedal or the clutch pedal is depressed, the pedal switch releases vacuum in systems, allowing the throttle plate to return to the idle position. The vacuum pump is also switched off.

Provided the road speed has not fallen below 23 mph, the previously selected road speed can be resumed using the Resume function at the control switch. If the road speed has gone below 23 mph, if the stalk switch is moved to the off position, or if the ignition is turned off, this memory function is canceled.

On turbo models, engaging the cruise control system causes the vacuum switch to interrupt power to the APC solenoid. This automatically drops the turbocharger boost pressure to a safe preset value. On 1989 and later models, an instrument cluster lamp is fitted that lights up when the cruise control is engaged.

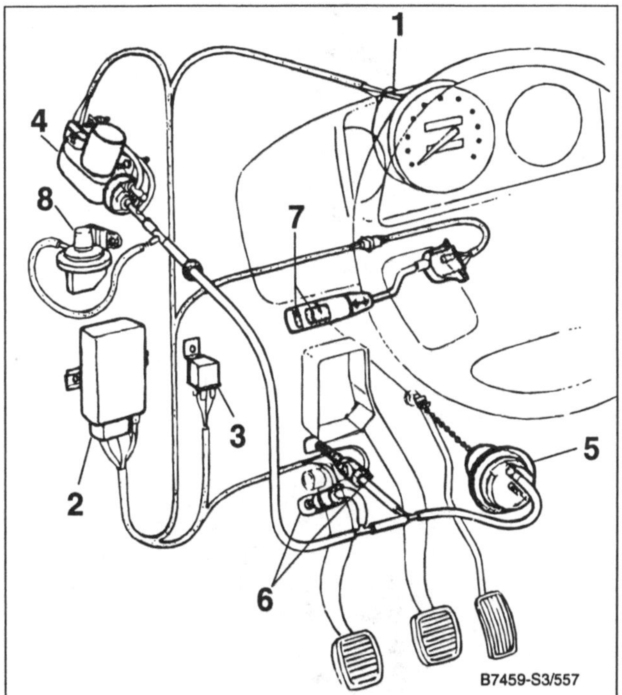

1. Speed transmitter
2. Electronic control unit
3. Relay (1985-1988 models only)
4. Vacuum pump
5. Vacuum regulator
6. Pedal switches
7. Control switch
8. Vacuum switch (turbo models only)

Fig. 1. Schematic view of cruise control system.

NOTE
Replacement of the cruise control indicator lamp on 1989 and later models is covered in **351 Lighting.**

COMPONENT LOCATIONS

The cruise control switch is located on the steering column. For information on removing the switch, see **364 Electric Controls and Switches.**

The speed transmitter is fitted behind the speedometer. The transmitter can be accessed through left speaker grille access hole. See Fig. 2.

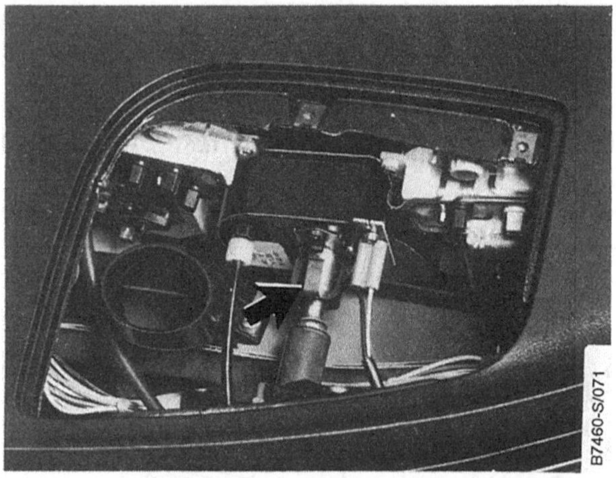

Fig. 2. Speed transmitter (arrow) on rear of speedometer. Access transmitter through left speaker grille.

The electronic control unit and the cruise control relay (where applicable) are located under the left-hand side of the dashboard. See Fig. 3.

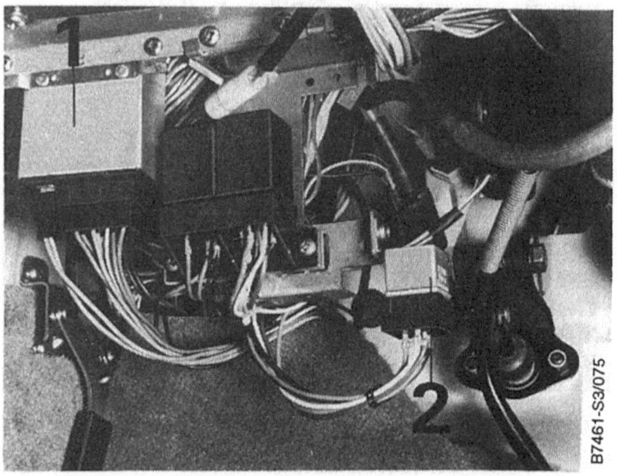

Fig. 3. Cruise control electronic control unit (**1**) and relay (**2**) as viewed from underneath left-hand side of dashboard (lower trim panel removed).

The vacuum regulator is mounted on a bracket in the heater housing area, under the center of the dashboard. The brake switch and the clutch switch (manual transmission only) are mounted above the pedals. See Fig. 4.

The vacuum pump is mounted in the engine compartment, forward of the left-hand wheel housing. On turbo models, the vacuum switch is mounted in the engine compartment on the left front fender.

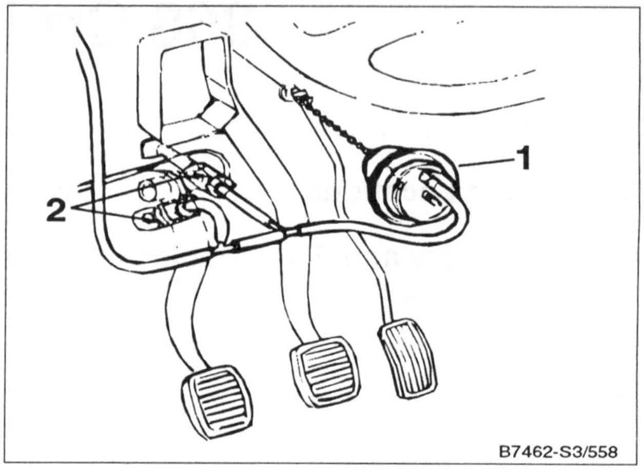

Fig. 4. Cruise control vacuum regulator (**1**) and pedal switches (**2**). Remove lower trim panel to reach components.

Electrical Tests and Fault Tracing

There are some basic fault tracing and electrical tests that can be made to test the cruise control components. Be sure to check the fuses. If the brake light bulbs are burned out, the system will not engage. Complete cruise control wiring diagrams are shown in **371 Wiring Diagrams, Fuses and Relays.**

Checking Pedal Switches

If the system is not working correctly, the first thing to check is the vacuum lines at the pedal switches. Remove the lower trim and inspect the hoses for cracks or deterioration. If no faults are found, disconnect the vacuum hose at the switch and connect in its place a length of rubber hose. While blowing through the hose with the pedal in the rest position, air should not pass through the switch. When the pedal is depressed, air should pass through the switch.

NOTE

The brake pedal switch is not the same switch used for the brake lights. On 1989 and later models, the brake light switch is also used in the cruise control circuit. Be sure to check that the brake lights are operating correctly. If necessary, adjust the brake light switch as described in **364 Electric Controls and Switches.**

To check the electrical part of the switch, disconnect the wires from the switch and connect an ohmmeter across the switch terminals. With pedal in the rest position, there should be continuity. There should be no continuity when the pedal is depressed.

If any faults are found, check that the switch is correctly adjusted as shown in Fig. 5. If no faults are found, the switch is faulty and should be replaced.

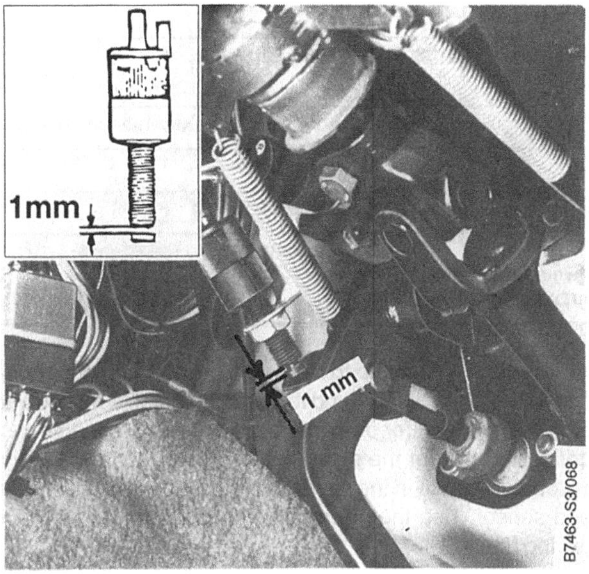

Fig. 5. Pedal switch adjustment. Adjust switch so that plunger is exposed 1 mm (0.04 in.) as shown in inset

Checking Vacuum Pump and Regulator

Remove the Harness connector from the vacuum pump. Connect a 12-volt source to terminal 3 (+) of the pump connector. Connect terminals 1 and 2 to ground. See Fig. 6. The vacuum pump should run and draw the vacuum regulator in. Check that the accelerator throttle plate is open.

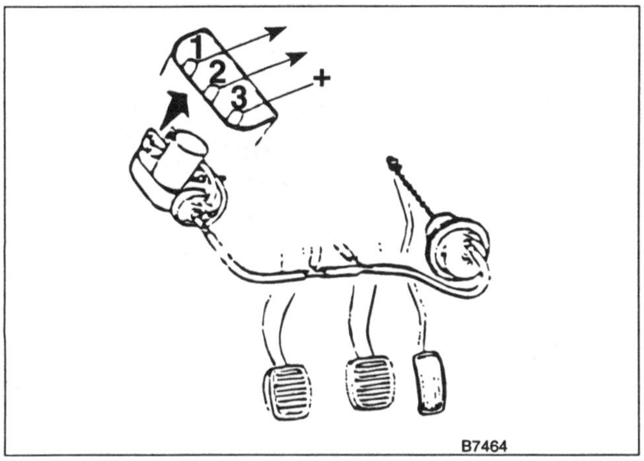

Fig. 6. Vacuum pump and terminal designation.

If the pump does not run, it is faulty and should be replaced. If the pump runs, but the accelerator does not pull in and open the throttle, check that the ball chain on the regulator is connected to the accelerator linkage. Check for vacuum to the regulator with the pump running. If the pump operates as specified and the regulator does not respond correctly, the regulator is faulty and should be replaced.

Remove the ground wire from terminal 2 only. The pump should stop and the accelerator should remain in the open position. If the accelerator slowly returns the rest position, either the regulator is leaking or the pump is faulty. To check for a leaking regulator, connect to it a length of hose. Suck on the hose and then block off then end. If the regulator slowly returns to its rest position, it is leaking and should be replaced.

Remove the wire from terminal 1 and check that the regulator returns to its rest position. If any faults are found, the pump is faulty and should be replaced.

NOTE

When replacing the regulator, use care not to damage the rubber diaphragm during installation. Install the chain so that it is as tight as possible without tensioning the accelerator.

Checking Cruise Control Switch

WARNING

1990 and later models are equipped with an airbag or SRS system. The airbag is a explosive device. Do not attempt to remove the steering wheel or service the wiring in the steering column assembly. You could accidentally ignite the airbag, causing serious personal injury. See an authorized Saab dealer for any repairs to the SRS system or to disarm the system prior to removing the steering wheel.

Remove the lower steering column trim and disconnect the switch's 4-pin harness connector. Cut any wire ties if necessary. Fig. 7 shows the connector terminals. Test the switch by making the checks listed in **Table a.** If any faults are found, the switch is faulty and should be replaced as described in **364 Electric Controls and Switches.**

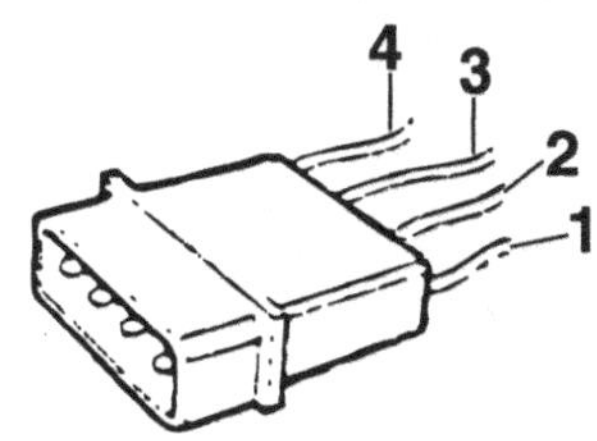

1. Red wire
2. Brown wire
3. Green wire
4. Yellow wire

Fig. 7. Cruise control switch harness connector used on 1985 through 1988 cars. Connector used on 1989 and later cars is different, although wire colors and their respective terminals numbers are the same.

Table a. Cruise Control Switch Continuity Test

Switch position	1985–1987 (up to chassis no. H2013308	1987 (from chassis no. H2013309) 1988	1989 and later
OFF	no continuity between any terminals	2 and 3	no continuity between any terminals
ON	1 and 2	1 and 2	1 and 2
RESUME	1 and 2, 2 and 4	1 and 2, 2 and 4	1 and 2, 2 and 4
SET SPEED ON	2 and 3	2 and 4 (continuity between 2 and 3 is removed when button is depressed)	2 and 3

NOTE

On 1985 through 1988 models, the original design cruise control switch is no longer available. The superceded switch kit requires replacing the harness connector to match the connector on the new switch, removing and discarding the cruise control relay and splicing the green wire of the relay connector (terminal 86) into the red/white wire in the relay connector
(terminal 30).

If no faults are found up to this point, use the wiring diagram to check for power and ground to the control unit. On 1989 and later models, check for power and ground to the speed transmitter on the rear of the speedometer. If no faults are found the problem may lie in the circuit wiring, the control unit, or the speed sensor. Use the appropriate wiring diagram to check for wiring faults.

371 Wiring Diagrams, Fuses and Relays

TABLES

MAIN FUSE AND RELAY LOCATIONS

With the exception of the battery charging system, all electrical power is routed from the ignition switch or the battery through the fuse/relay panel.

The main fuse/relay panel is located in the engine compartment, on the driver's side wheel housing. See Fig. 1. Always check the cover on the panel for exact descriptions of fuses and relays. Three main connectors run from the fuse/relay panel to the interior of the car. These connectors are accessible from inside the car, behind the instrument panel. Note that the connectors vary depending on model variant. See Fig. 2.

CAUTION

On cars equipped with SRS (airbag) the ignition switch must be in the lock position before replacing any fuses. If fuse no. 7 is removed with the ignition switch in the drive or accessory position, an SRS fault will register causing the SRS warning light to come on. Only an authorized Saab dealer can erase the fault and turn off the light.

Main Fuse Positions

Fuses prevent excessive current from damaging components and wiring. Fuses are color coded to indicate their different current capacities. **Table a** through **Table d** list the fuse positions in the main fuse/relay panel.

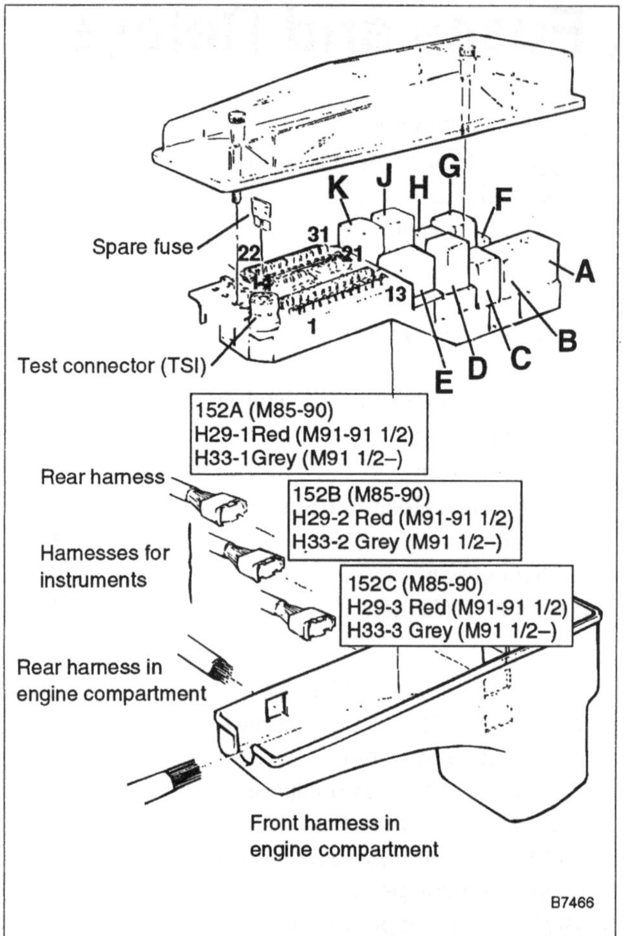

Fig. 1. Exploded view of main fuse/relay panel. Main connectors vary between years.

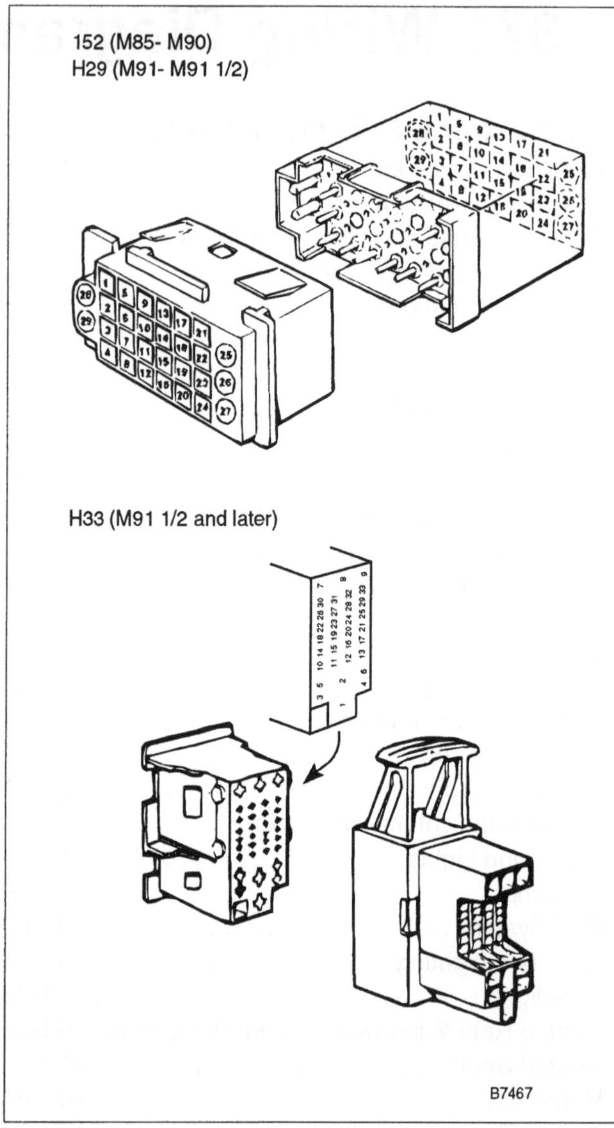

Fig. 2. Terminal identification for main bulkhead connectors at side of fuse/relay panel.

Fuse color and rating	
• Red	10 amp
• Blue	15 amp
• Yellow	20 amp
• Transparent	25 amp
• Green	30 amp

Table a. 1985 and 1986 Fuse positions—main fuse/relay panel

Position	Rating	Function
1	—	spare
2	—	spare
3	—	spare
4	10 A	shift-up indicator
5	15 A	brake fluid warning, wipers, seatbelt warning, exhaust warning (where applicable)
6	30 A	Air conditioner (A/C)
7	15 A	direction indicators , tachometer (1986 models), alternator warning light (1986 models)
8	10 A	power mirrors, cruise control
9	30 A	blower fan
10	10 A	APC system (turbo models)
11	30 A	power windows, sunroof
12	15 A	heated seats, time-delay for interior lights
13	20 A	backup lights, cigarette lighter
14	15 A	right high beam
15	15 A	left high beam, high beam warning
16	15 A	right low beam
17	15 A	left low beam
18	10 A	right front park and right rear lights, license plate light, glovebox and ashtray lights, instrument lights
19	10 A	left front park and left rear lights, burglar alarm (where applicable)
20	15 A	radio, front cornering lights
21	15 A	—
22	10 A	fuel system, ignition system, temperature gauge, handbrake warning light, tachometer (1985 models), alternator warning light (1985 models)
23	—	spare
24	10 A	central locking
25	30 A	radiator fan
26	25 A	horn
27	15 A	hazard warning lights
28	15 A	interior lighting, clock, radio, power antenna
29	20 A	heated rear window
30	20 A	fuel pump
31	15 A	brake lights

Table b. 1987 and 1988 Fuse positions—main fuse/relay panel

Position	Rating	Function
1	—	spare
2	30 A	rear windows (convertible) (1988 models)
3	30 A	rear windows (convertible) (1987 models)
4	10 A	shift-up indicator
5	15 A	brake fluid warning, wipers, seatbelt warning, exhaust warning
6	30 A	Air conditioner (A/C)
7	15 A	direction indicators, tachometer, alternator warning light
8	10 A	power mirrors, cruise control
9	30 A	blower fan
10	10 A	APC system, lighting for controls and glovebox
11	30 A	power windows, sunroof
12	20 A	heated seats, time-delay for interior lights
13	20 A	backup lights, cigarette lighter
14	15 A	right high beam
15	15 A	left high beam, high beam warning light
16	15 A	right low beam
17	15 A	left low beam
18	10 A	right front park and right rear light, license plate light
19	10 A	left front park and left rear light, burglar alarm (where applicable)
20	15 A	radio, front cornering lights
21	15 A	extra fog lights
22	10 A	fuel system, ignition system, temperature gauge, handbrake warning light
23	—	spare
24	10 A	central locking
25	30 A	radiator fan
26	25 A	horn
27	15 A	hazard warning lights
28	15 A	interior lighting, clock, radio, power antenna
29	20 A	heated rear window
30	20 A	fuel pump
31	15 A	brake lights

Table c. 1989 and 1990 Fuse positions—main fuse/relay panel

Position	Rating	Function
1	10 A	lambda sensor preheater
2	30 A	rear windows (convertible)
3	15 A	ignition system
4	20 A	daytime driving lights (Canada only)
5	15 A	wipers, seatbelt reminder lamp
6	30 A	Air conditioner (A/C)
7	15 A	direction indicators, tachometer, warning lights for alternator, check engine, oil pressure, and SRS airbag (1990 models)
8	10 A	power mirrors, cruise control
9	30 A	blower fan
10	10 A	APC system, ABS (where applicable)
11	30 A	power windows, sunroof
12	20 A	heated seats, time-delay for interior lights, convertible top actuation, seatbelt warning buzzer
13	20 A	backup lights, cigarette lighter, passive seatbelts (1989 models)
14	15 A	right high beam
15	15 A	left high beam, high beam warning light
16	15 A	right low beam
17	15 A	left low beam
18	10 A	right front park, sidemarker lights, right rear lights, license plate light
19	10 A	left front park, sidemarker lights, left rear lights, burglar alarm (ex. convertible models)
20	15 A	radio, cornering lights, burglar alarm (ex. 1989 sedan models)
21	15 A	extra fog lights
22	10 A	fuel system, temperature gauge, fuel level gauge, speed sensor (cruise control), warning lights for handbrake, ABS and footbrake
23	10 A	lighting for controls and glovebox
24	10 A	central locking, burglar alarm
25	30 A	radiator fan
26	25 A	horn
27	15 A	hazard warning lights
28	15 A	interior lighting, clock, radio, antenna, luggage compartment light
29	20 A	heated rear window
30	20 A	fuel pump
31	15 A	brake lights, ABS (1990 models)

Table d. 1991-1993 Fuse positions—main fuse/relay panel

Position	Rating	Function
1	10 A	lambda sensor preheater
2	30 A	rear windows (convertible)
3	15 A	ignition system, EGR modulating valve (California only)
4	20 A	daytime driving lights (Canada only)
5	15 A	wipers, seatbelt reminder light
6	30 A	Air conditioner (A/C)
7	15 A	direction indicators, tachometer, warning lights for alternator, check engine, oil pressure and SRS airbag, shift-up indicator
8	10 A	headlight wipers, power mirrors, cruise control, EZK test connector
9	30 A	blower fan
10	10 A	APC system
11	—	spare
12	—	spare
13	20 A	backup lights, cigarette lighter
14	15 A	right high beam
15	15 A	left high beam, high beam warning light
16	15 A	right low beam
17	15 A	left low beam
18	10 A	right front park, sidemarker lights right rear lights, license plate light
19	10 A	left front park, sidemarker lights, left rear lights
20	15 A	cornering lights
21	15 A	extra fog lights
22	10 A	fuel system, temperature gauge, fuel level gauge, speed sensor (cruise control), warning lights for handbrake, ABS and footbrake
23	10 A	lighting for controls and glovebox
24	—	spare
25	30 A	radiator fan
26	25 A	horn
27	15 A	hazard warning lights
28	15 A	clock
29	—	spare
30	20 A	fuel pump
31	15 A	brake lights, ABS

Main Relay Positions

The relays are electromagnetic switches that operate on low current to switch a high-current circuit on and off. Many of the relays are mounted on the main fuse/relay panel in the engine compartment (see in Fig. 1), although later models are equipped with additional auxiliary relay panels. **Table e** through **Table g** list the relay positions in the main fuse/relay panel.

CAUTION

- Relays positions are subject to change and can vary from car to car, depending on options and new equipment installed. If questions arise, consult the appropriate wiring diagrams. Please remember that an authorized Saab dealer is the best source for the most accurate and up-to-date information.

- Most relays incorporate several different electrical functions. Although another relay may plug into the same panel position, do not interchange relays.

Table e. 1985 Relay positions—main fuse/relay panel

Position	Function
A/B	headlights
C	heated rear window
D	A/C compressor
E	ignition switch
F	—
G	—
H	—
J	A/C radiator fan
K	horn

Table f. 1986–1988 Relay positions—main fuse/relay panel

Position	Function
A/B	headlights
C	heated rear window
D	ignition pulse amplifier (1986-1987 models)
E	ignition switch
F	shift up indicator
G	extra fog lights
H	A/C compressor
J	A/C radiator fan
K	horn

Table g. 1989-1993 Relay positions—main fuse/relay panel

Position	Function
A/B	headlights
C	heated rear window (1989-1990)
D	extra fog lights
E	ignition switch
F	daytime driving lights (Canada only)
G	radiator cooling fan
H	A/C compressor
J	A/C radiator fan
K	horn

Additional Fuse and Relay Locations

Several other fuse and relay locations are used depending on model year and equipment.

On 1989 models, the ABS relay panel is located in the engine compartment on the driver's side wheel housing. On 1990 and later models, the ABS relay panel is located in the engine compartment on the passenger side wheel housing by the door hinge. Fig. 3 shows the fuse/relay panel used for ABS functions.

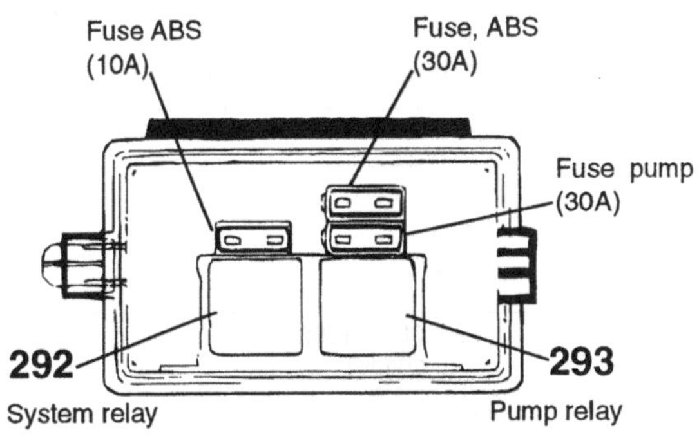

Fig. 3. ABS fuse/relay panel located in engine compartment.

On models with motorized passive seat belts, the fuses and relays for the seatbelt system are located under the rear seat on the driver side.

On 1989 and 1990 convertible models and all 1991 and later models, an extra fuse/relay panel (component 401A) can be found under the rear seat. See Fig. 4. **Table h** gives a list of fuse and relay functions for this fuse/relay panel.

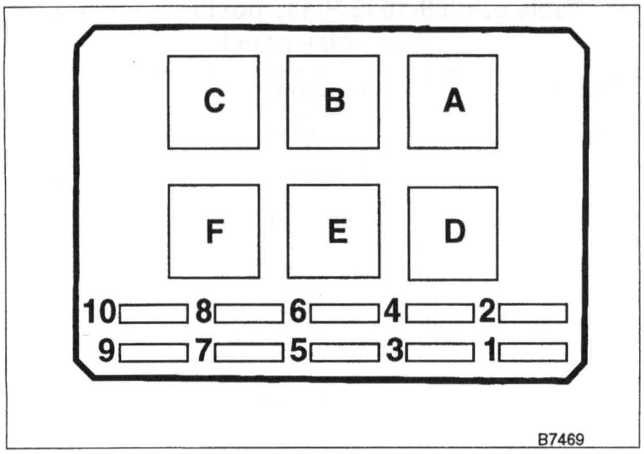

Fig. 4. Extra fuse/relay panel under rear seat used on 1989 and 1990 convertible and all 1991 and later models.

NOTE

Always check the panel cover for correct fuse ratings and relay functions. The fuses are numbered 1 through 10. The relays are lettered A through F.

NOTE

On some convertible models, there are in-line fuses located in the engine compartment for raising and lowering the convertible top. These fuses are on the passenger side bulkhead near the exhaust manifold.

Components, Connectors and Ground Locations

All of the components, connectors, and ground locations are uniquely numbered throughout the wiring diagrams. Each number corresponds to a specific component, connector, or ground point. As a general rule, the first page of each wiring topic contains a complete list of component, connector, and ground locations.

NOTE

A complete listing of component, connector, and ground locations can also be found at the end of this repair group. This master list is presented twice; first in a numerical order and then alphabetically.

Ground points are an important part of the electrical system. There are only a small number of main ground locations used throughout the car. See Fig. 5.

Table h. Fuse/relay panel under rear seat

Position	Rating	1989-1990 Convertibles	All 1991 and later models
1	20 A		convertible top
2	20 A		heated seats, interior lighting, seatbelt and key warning
3	25 A		heated rear window, power mirrors
4	10 A (15 A 1992 convertible)		interior lighting, burglar alarm, central locking on convertible
5	20 A		radio, burglar alarm
6	20 A		power antenna
7	30 A		power windows (front)
8	30 A		power windows (rear), sunroof
9	30 A		power seats (driver's side)
10	30 A		power seats (passenger's side)
A	—	relay for power windows (convertible models)	relay for seatbelt warning
B	—	relay for interior lighting time delay	relay for heated rear window
C	—	relay for raising convertible top	relay for raising convertible top
D	—	relay for lowering convertible top	relay for lowering convertible top
E	—	relay for seatbelt warning	relay for interior lighting time delay
F	—	relay for burglar alarm	relay for ignition switch

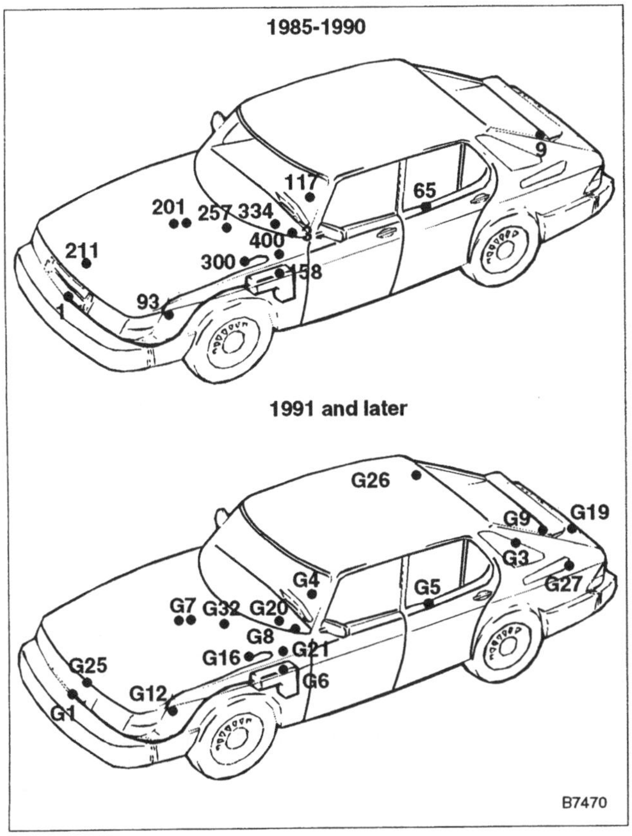

Fig. 5. Main chassis ground distribution.

Ignition Switch and Positive Supply

With the exception of the battery charging system, all electrical power is routed from the ignition switch or the battery through the main fuse/relay panel.

Battery voltage supplied directly to components is designated terminal **30** in the wiring diagrams. This supply is always live. All switched battery voltage is supplied via ignition switch terminals **15**, **50**, **X** and **54**, depending on the position of the ignition key. Additionally, as long as the key is in the ignition switch, terminal **S** of the switch will be live to the warning relay. See Fig. 6.

Circuit closed between ...	Key Position			
	Lock	Park	Run	Start
30-X		■	■	■
30-15			■	■
30-54			■	
30-50				■
30-S+	■	■	■	■

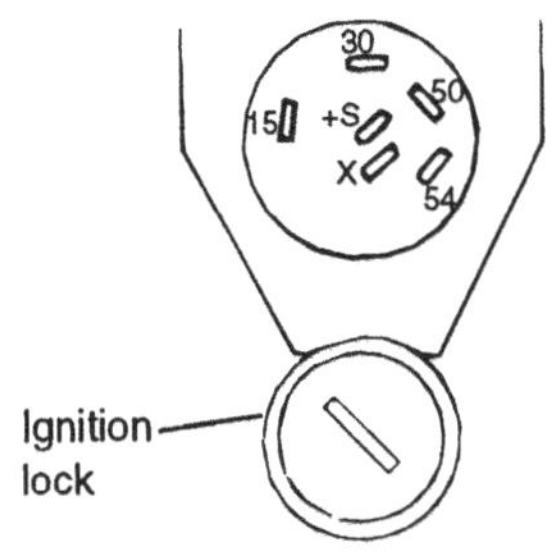

Fig. 6. Ignition switch terminals and terminal distribution.

Wiring Diagrams

The following wiring diagrams have been specially designed to enable quick and efficient diagnosis and troubleshooting of electrical malfunctions. Each wiring sub-system or topic, such as Ignition System, or Power Mirrors, is described individually. Most sub-systems include a brief circuit operation description, fault tracing, a list of component, connector and ground locations, and the wiring diagram(s).

NOTE

Many of the wiring diagrams are applicable to cars delivered to other world markets served by Saab. At times, there may be certain parts of the diagram that are applicable only to these markets. This manual covers only Saab cars delivered to the United States (US), California (UC), and Canada (CA). Abbreviations other than US, UC, and CA found on a particular diagram should be ignored. For example the abbreviation "ME" is for cars sold in the Middle East.

Anti-lock Braking System (ABS)

Operation

The Anti-lock Braking System or ABS is controlled and monitored electrically. For information on the mechanical components of the braking system see **5 BRAKES**.

A sensor at each wheel supplies the ABS control unit with information on the wheel rotation speed. If any of the wheels should begin to lock, the control unit will adjust the braking pressure to the wheel(s). ABS braking on the front wheels is individually controlled, whereas on the rear wheels it is controlled jointly.

A hydraulic pump (component 297) controls the servo action of the brakes. The hydraulic fluid pressure is controlled at the required value by means of a pressure switch (component 294) which controls the pump motor.

The ANTI-LOCK and the BRAKE FLUID warning lamps in the instrument cluster monitor the system. The ANTI-LOCK warning lamp will light up if a problem or fault occurs in the Anti-lock Braking System. The BRAKE FLUID warning lamp will light up if the brake fluid level in the reservoir falls too low.

WARNING
If either the ANTI-LOCK or the BRAKE FLUID warning lamp should come on, the car must not be driven. Brake failure may occur. Have the problem corrected promptly by an authorized Saab dealer.

Fault Tracing

The following fault tracing should be used as a guide only. Comprehensive fault tracing is best carried out using special ABS test equipment available through an authorized Saab dealer.

1. Check the fuses. The control unit (component 291), the system relay (component 292), and the pump relay (component 293) are all protected by fuses. These fuses are located in the ABS electrical distribution box (component 302).
2. Test the components shown in the table using a DMM (digital multi-meter). Use the wiring diagram(s) for terminal identification.

ABS component	Test value	Comments
wheel sensors	800 to 1400 ohms	
wheel sensors	at least 0.1 VAC	with wheel rotating at 1 revolution per second
master valve	2 to 5 ohms	measure between terminals 1 and 2 at master valve (295) on hydraulic unit
inlet valves	5 to 7 ohms	measure inlet valve coil resistances (IFL, IFR, IR) at ABS valve block (296)
outlet valves	3 to 5 ohms	measure outlet valve coil resistances (OFL, OFR, OR) at ABS valve block (296)
brake fluid level sensor	10 ohms at contacts 1 and 2	contacts closed, float at bottom
brake fluid level sensor	1 ohm at contacts 3 and 4	contacts closed, float at top

Component Locations

- 20 ignition switch in center console
- 21A Ignition switch relay, in relay panel
- 22A main fuse/relay panel in left wheel housing
- 29 brake light switch at brake pedal
- 47F brake warning light in instrument cluster
- 47Q ABS warning lamp in instrument cluster
- 47T SRS airbag warning lamp in instrument cluster. See SRS warnings
- 158 negative distribution in main fuse/relay panel
- 291 ABS control unit (1989) at ABS distribution box, engine compartment left wheel housing
- 291 ABS control unit (1990 and later), under rear seat, right side
- 292 ABS system relay (1989) in ABS distribution box, engine compartment left wheel housing
- 292 ABS system relay (1990 and later) in ABS distribution box, behind right wheel housing
- 293 ABS pump relay (1989) at ABS distribution box, left wheel housing
- 293 ABS pump relay (1990 and later) in ABS distribution box, behind right wheel housing
- 294 ABS pressure switch on the ABS hydraulic unit
- 295 ABS master valve on the hydraulic unit
- 296 ABS valve block in front of left wheel housing
- 297 ABS hydraulic pump motor in engine compartment at hydraulic unit
- 298A ABS wheel sensor in left front wheel
- 298B ABS wheel sensor in right front wheel
- 298C ABS wheel sensor in left rear wheel
- 298D ABS wheel sensor in right rear wheel
- 299 ABS brake fluid level sensor in brake fluid reservoir
- 302 ABS electrical distribution box (1989) engine compartment, left wheel housing
- 302 ABS electrical distribution box (1990 and later) behind right wheel housing
- 302A ABS fuse holder (1989) engine compartment, left wheel housing
- 302A ABS fuse holder (1990-and later) behind right wheel housing
- 303B ABS diode (1989) engine compartment, left wheel housing
- 303B ABS diode (1990 and later) behind right wheel housing
- 397 ABS diagnostic 2-pin test socket near the hydraulic unit (in wiring harness)

Anti-lock Braking System (ABS)

Connector Locations

1989-1990 models

57 3-pin connector in right front wheel sensor
57 3-pin connector in left front wheel sensor
58 12-pin connector behind instrument panel, to left of steering column
59 2-pin connector at the main fuse/relay panel (22)
59 2-pin connector under rear seat (1990 models)
59 2-pin connector at the ABS hydraulic unit in engine compartment
59 2-pin connector in engine compartment at right side of the bulkhead partition
60 1-pin connector in main fuse/relay panel (1990 models)
98 10-pin connector at the ABS control unit
98 10-pin connector behind the instrument panel to the left of the steering column
123 4-pin connector (2) at the ABS control unit
123 4-pin connector (black) left wheel sensor
123 4-pin connector (gray) right wheel sensor
152A 29-pin bulkhead connector in main fuse/relay panel
152B 29-pin bulkhead connector in main fuse/relay panel
152C 29-pin bulkhead connector in main fuse/relay panel

1991and later

H2-12 2-pin connector under rear seat, left side
H2-22 2-pin connector in engine compartment near ABS hydraulic unit
H2-23 2-pin connector in engine compartment right side partition
H2-24 2-pin connector under rear seat, left side
H2-25 2-pin connector under rear seat, right side
H2-27 2-pin connector in main fuse/relay panel
H3-4 3-pin connector in main fuse/relay panel
H29-1 29-pin bulkhead connector behind main fuse/relay panel (1991-1991 1/2)
H29-3 29-pin bulkhead connector behind main fuse/relay panel (1991-1991 1/2)
H33-1 33-pin bulkhead connector behind main fuse/relay panel (1991 1/2 and later)
H33-3 33-pin bulkhead connector behind main fuse/relay panel (1991 1/2 and later)

Ground Locations

1990

3 ground point behind instrument panel
65 ground point at the handbrake
93 ground point at left wheel housing
211 ground point at gearbox
257 ground point at alternator bracket
300 ABS ground point at the front of the hydraulic unit

1991 and later

G1 ground point at radiator cross member
G5 ground point under rear seat
G8 ground point behind instrument panel
G12 ground point at left wheel housing
G16 ground point in front of the ABS hydraulic unit
G25 ground point on gearbox

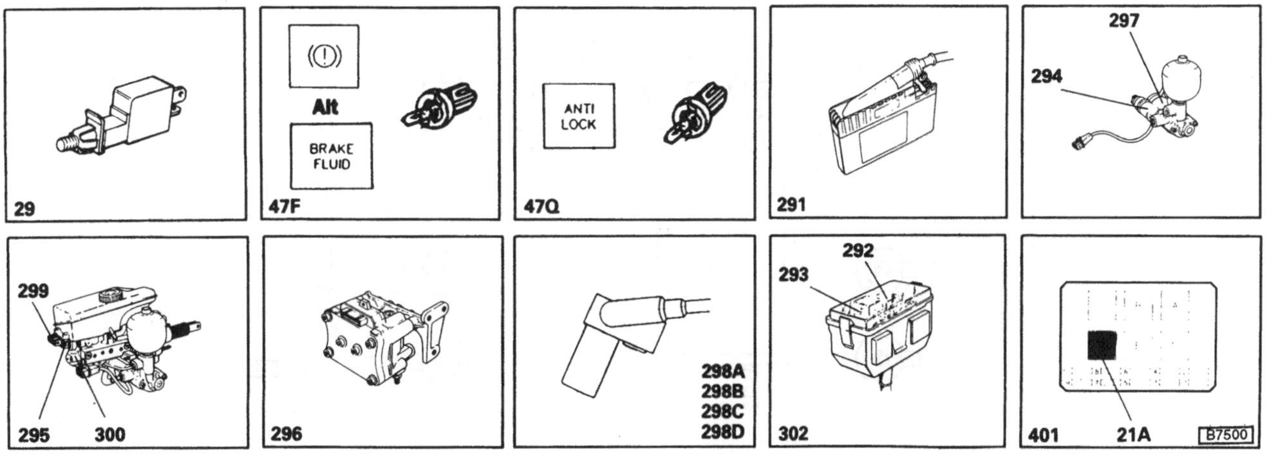

Anti-lock Braking System (ABS)

1989-1990

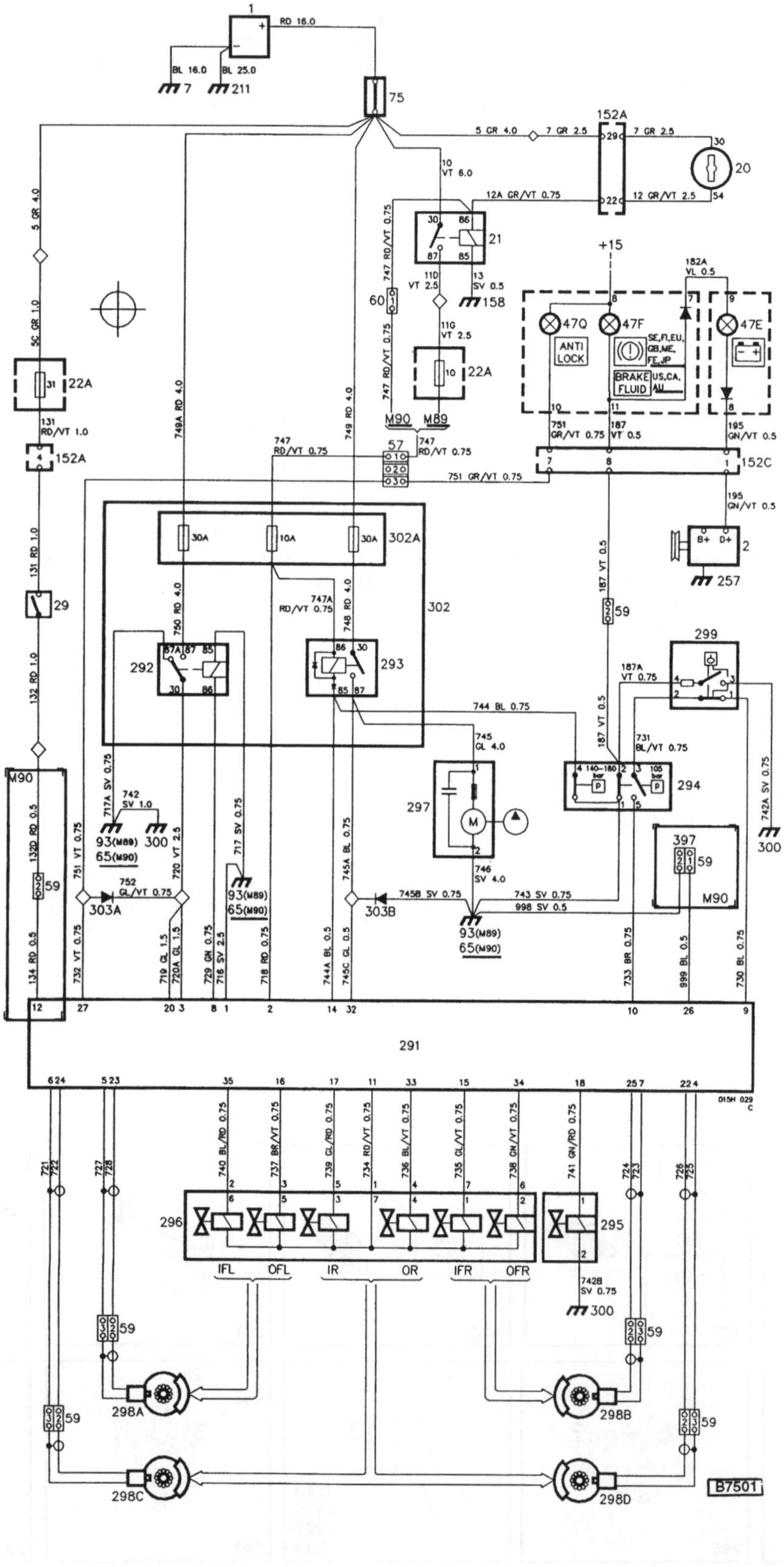

Anti-lock Braking System (ABS)

1991-1993 (page 1 of 2)

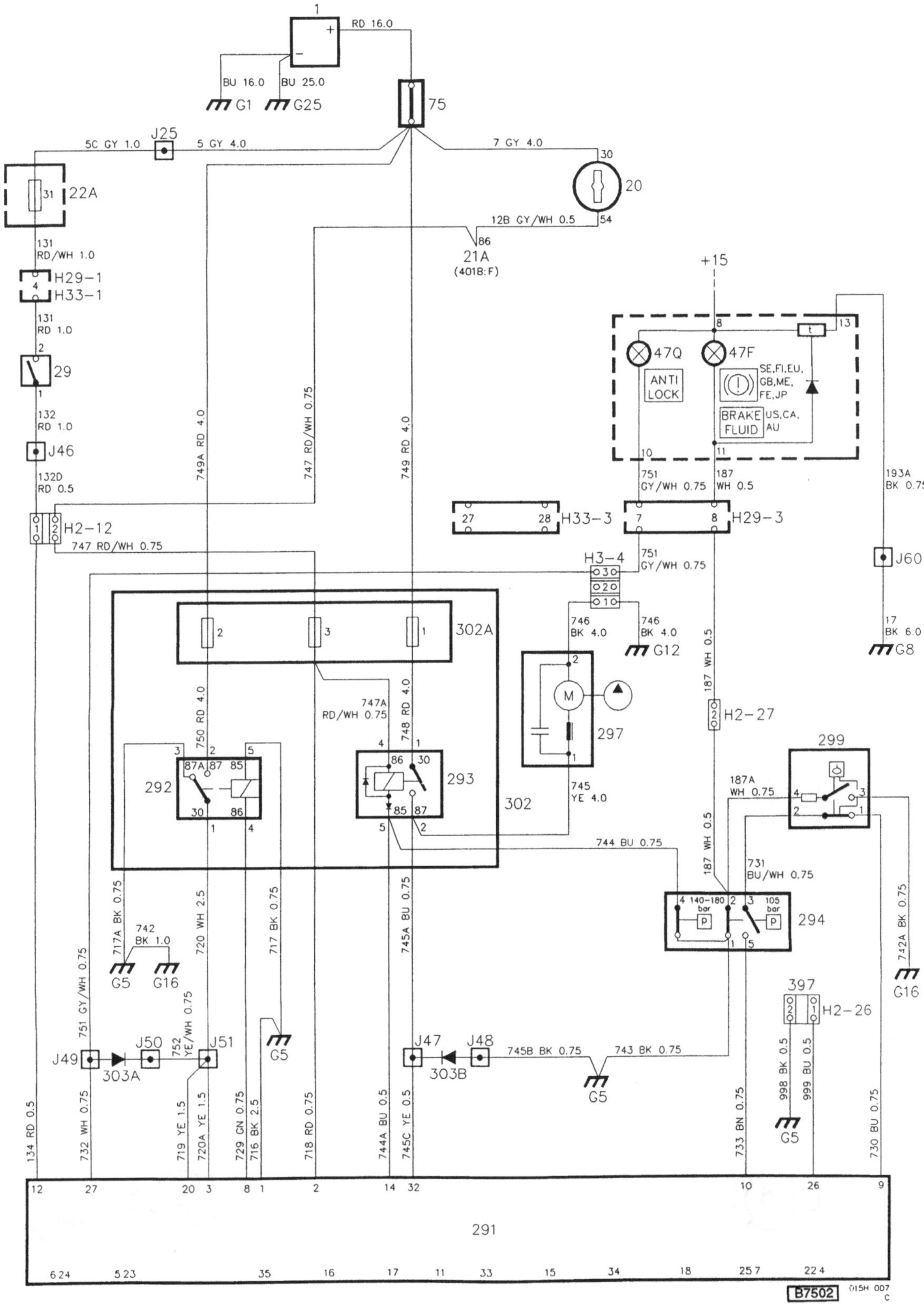

Anti-lock Braking System (ABS)

1991-1993 (page 2 of 2)

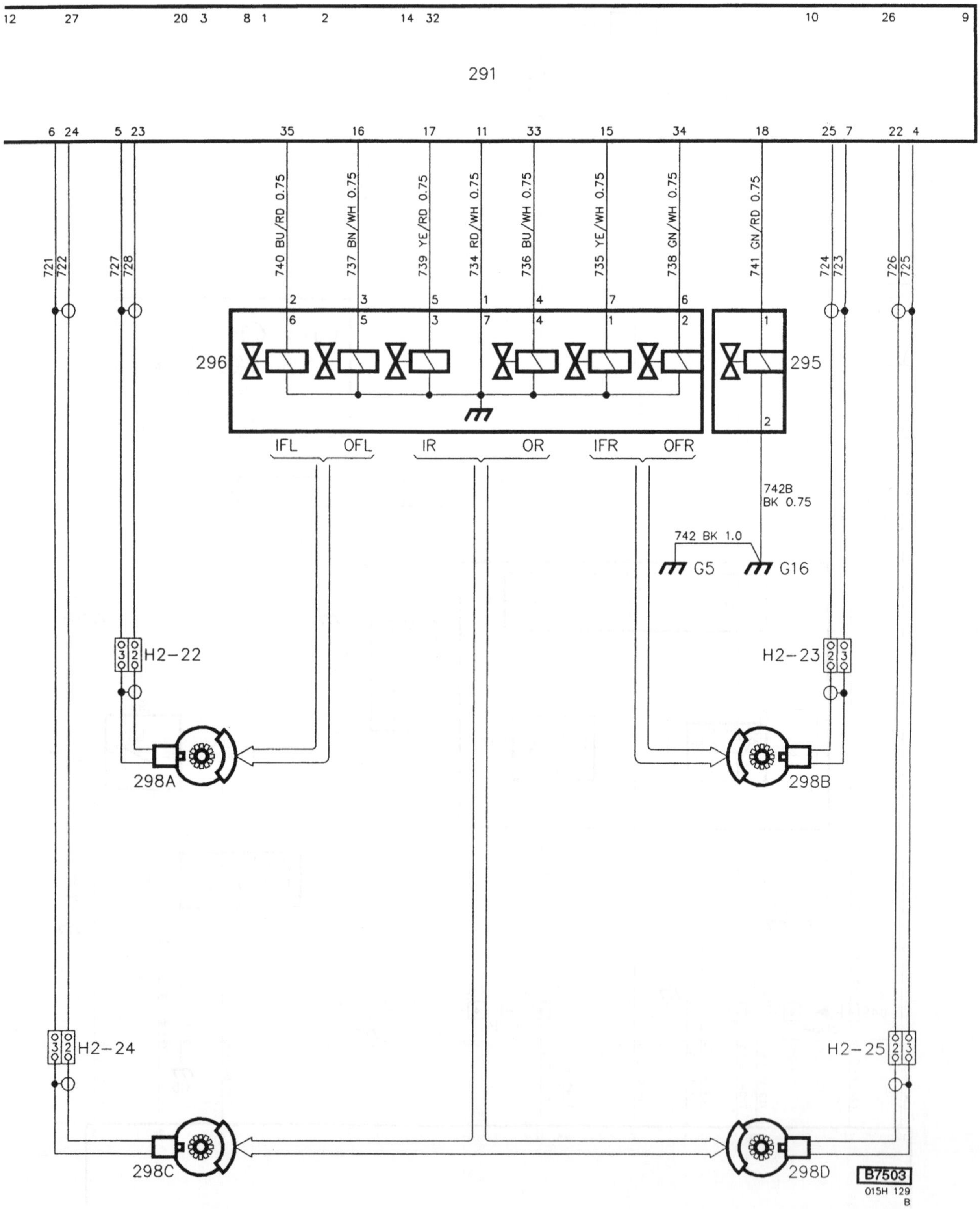

Air Conditioner (A/C)

Operation

The air conditioner is controlled by the A/C switch (component 169) and the recirculation switch (component 143) on the instrument panel. The A/C compressor (component 170) and evaporator are protected by pressure and temperature safety switches. The system also has an A/C cooling fan (component 172) behind the radiator. Idle speed compensation is done by the AIC valve.

For A/C repair information see **854 Heating and Air Conditioning.** For repair information on the cooling system see **261 Radiator and Cooling System.**

WARNING
Air conditioner repair and service requires special tools and knowledge. Pressures in excess of 300 psi are created in the system. The refrigerant (R12) can be poisonous and can immediately freeze anything it contacts including eyes and skin. Always wear eye protection and gloves when working on the A/C system.

Fault Tracing

1. Check the fuse(s).
2. Listen for an audible click of the compressor clutch when pushing the A/C switch on with the key on.
3. Check for voltage at the A/C compressor (component 170) and the A/C cooling fan (172).

Component Locations

20 ignition switch
22A main fuse/relay panel in left wheel housing
35 ventilation fan switch in instrument panel
37 radiator cooling fan motor behind radiator
38 recirculation valve (solenoid)
1985–1987 in engine compartment, right wheel housing
1988 and later– behind instrument panel, near right side A pillar
74 resistor pack for ventilation fan under the left speaker grille
143 recirculation switch in dashboard
149 main switch for ventilation fan integrated in the air distribution switch
150 air distribution switch in instrument panel
155 relay for A/C cooling fan in main fuse/relay panel
156 A/C compressor relay in main fuse/relay panel
158 negative distribution in main fuse/relay panel
166 A/C pressure switch for cooling fan in receiver/dryer unit, near battery
167 A/C throttle switch in throttle housing
168 A/C coolant temperature switch in radiator hose near distributor
169 A/C switch in instrument panel
170 A/C compressor at top of engine near firewall
171 A/C antifreeze thermostat in instrument panel, right side ahead of A pillar
172 A/C cooling fan behind radiator
173 A/C compressor diode in engine wiring harness, rear of valve cover
200 LH control unit, forward of right front, door behind trim
203 throttle angle transmitter on throttle housing
396 radiator fan relay in main fuse/relay panel

Connector Locations

1985-1990

59 2-pin connector next to radiator cooling fan
59 2-pin connector next to A/C cooling fan
60 1-pin connector at A/C compressor, top rear of engine
67 6-pin connector in engine compartment near fresh air intake
152A 29-pin bulkhead connector in main fuse/relay panel
152B 29-pin bulkhead connector in main fuse/relay panel
152C 29-pin bulkhead connector in main fuse/relay panel

1991 and later

H1-10 1-pin connector near A/C compressor in engine compartment
H2-1 2-pin connector at the radiator fan motor
H2-2 2-pin connector next to A/C cooling fan (172)
H2-10 2-pin connector behind cigarette lighter
H6-2 6-pin connector in engine compartment right wheel housing
H29-3 29-pin bulkhead connector behind main fuse/relay panel (1991-1991 1/2)
H33-3 33-pin bulkhead connector behind main fuse/relay panel (1991 1/2 and later)

Ground Locations

1985-1990

3 ground point behind instrument panel
7 ground point at radiator cross member
201 ground point for fuel injection at engine lifting lug, 2 points

1991 and later

G1 ground point at radiator cross member
G6 ground distribution in main fuse/relay panel
G7 ground point in engine lifting lug
G8 ground point behind instrument panel

37, 172

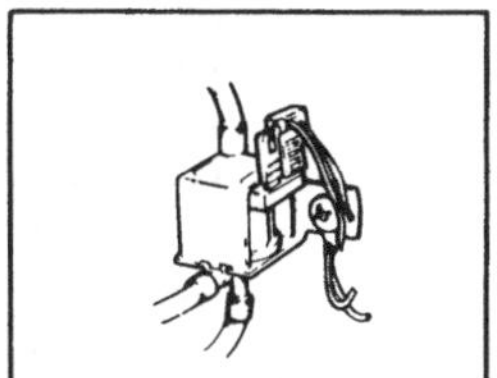
38

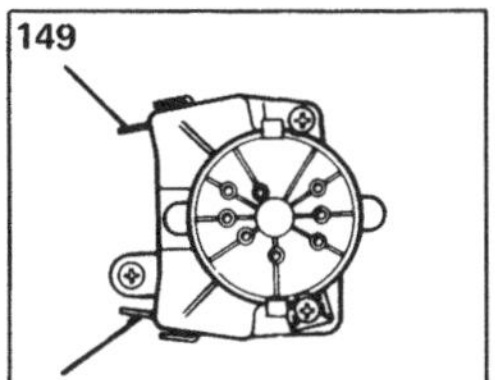
149
150

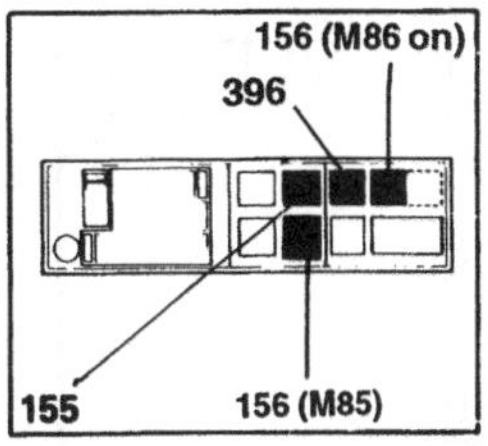

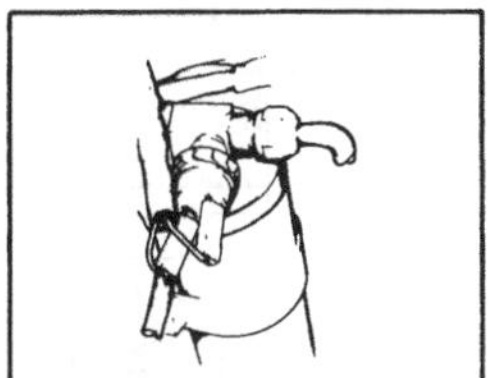
166

167

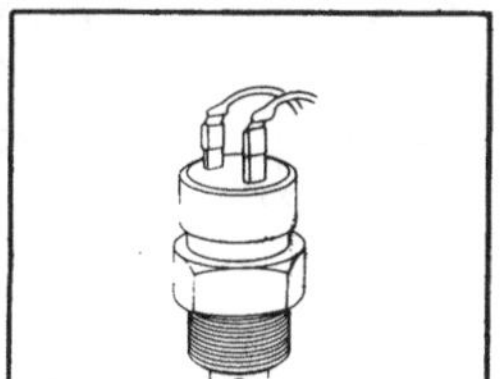
168

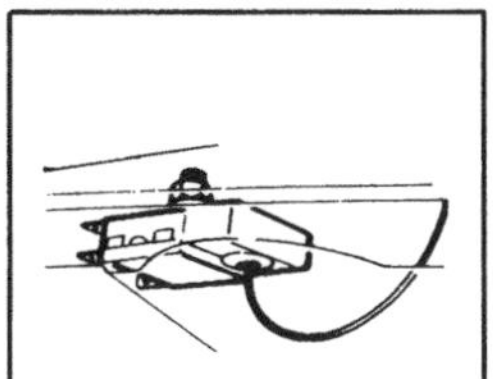
171

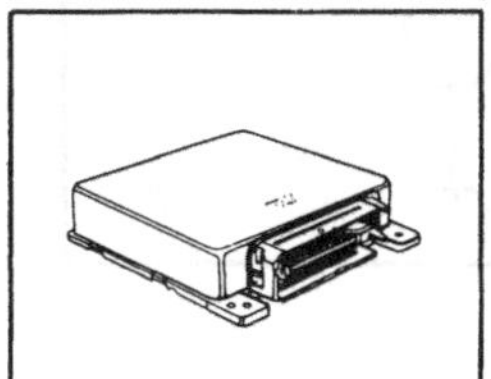
200

203 (EZK, LH 2.4) B7504

Air Conditioner (A/C)

1985

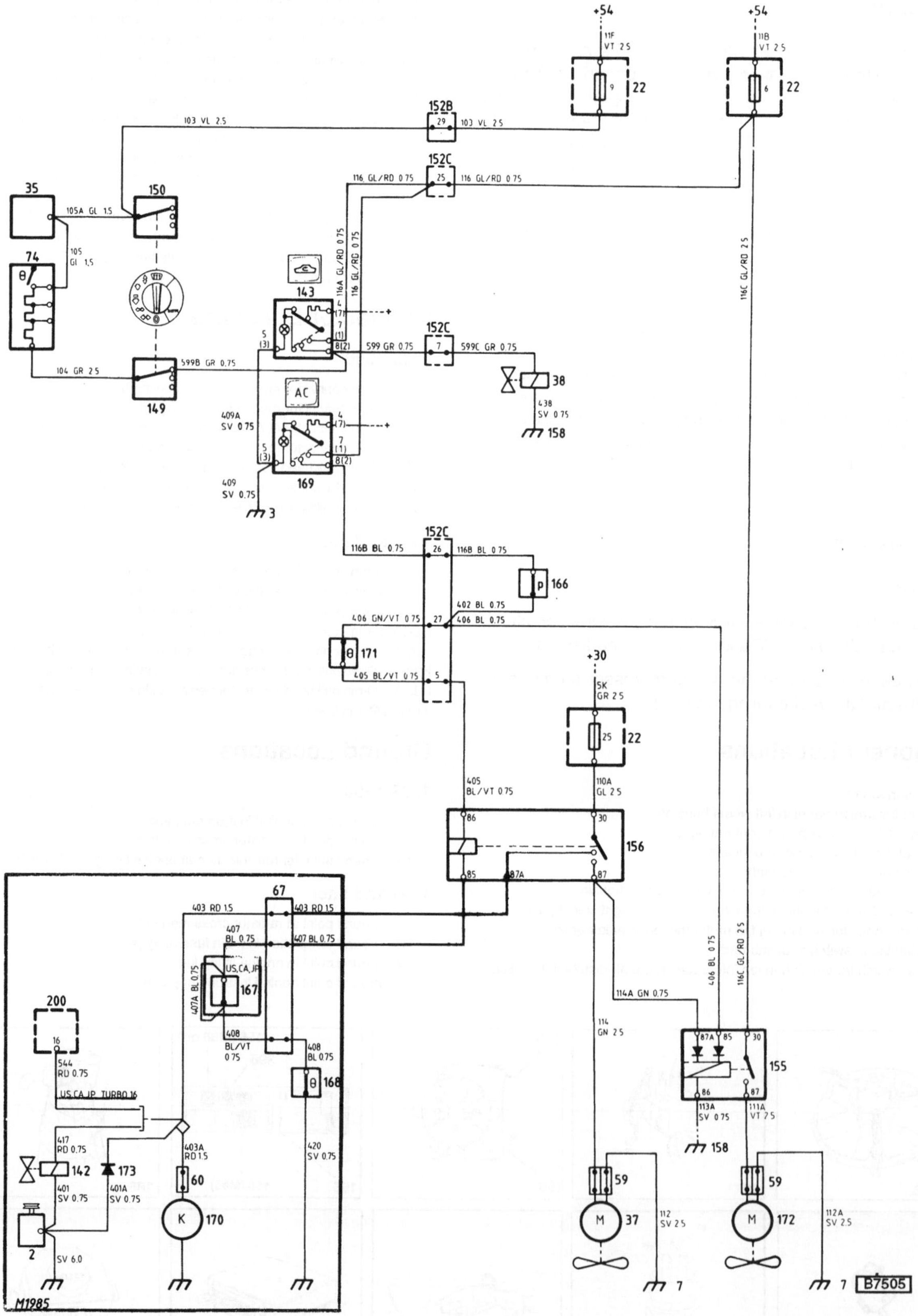

Air Conditioner (A/C)

1986-1988

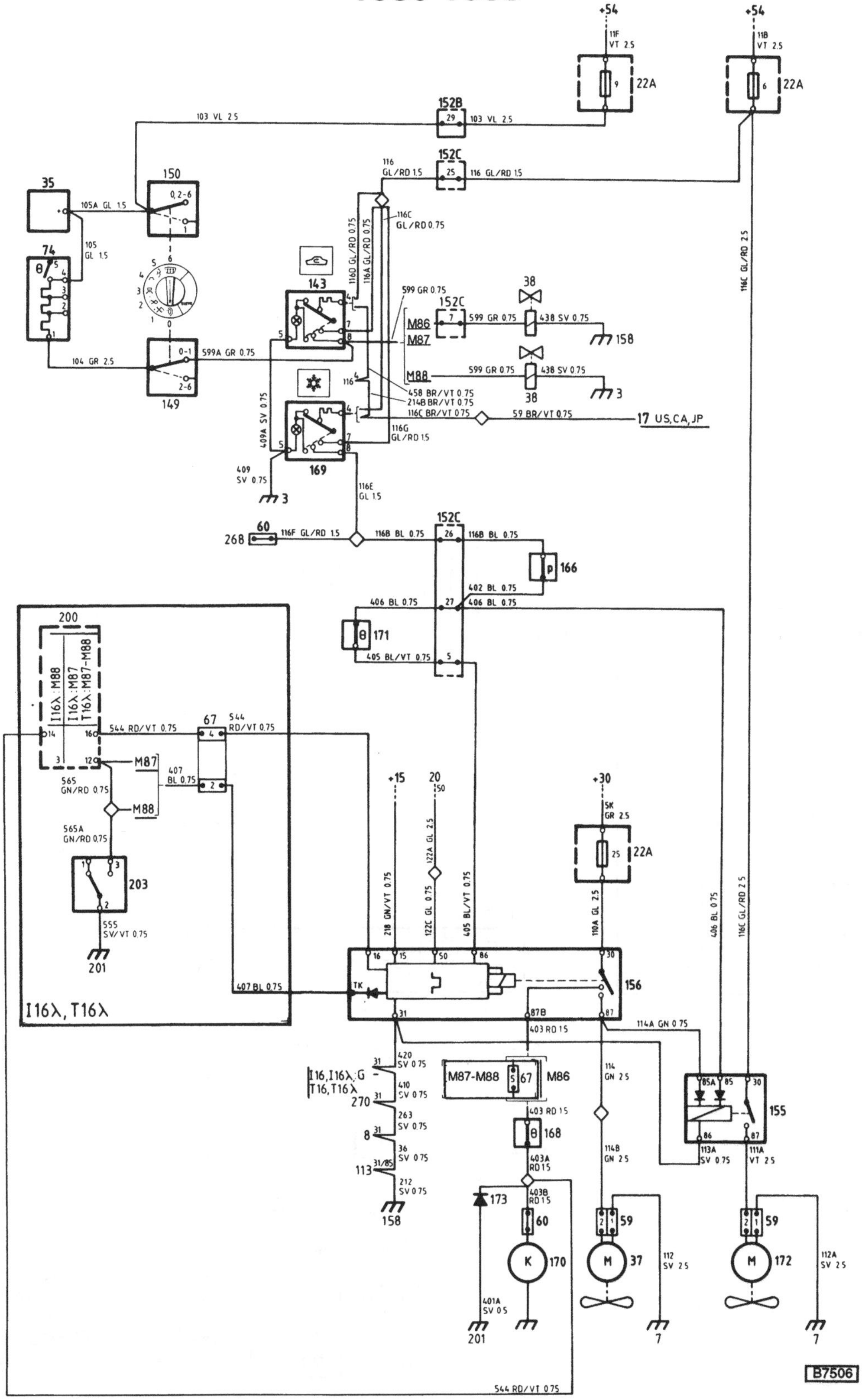

Air Conditioner (A/C)

1989-1990

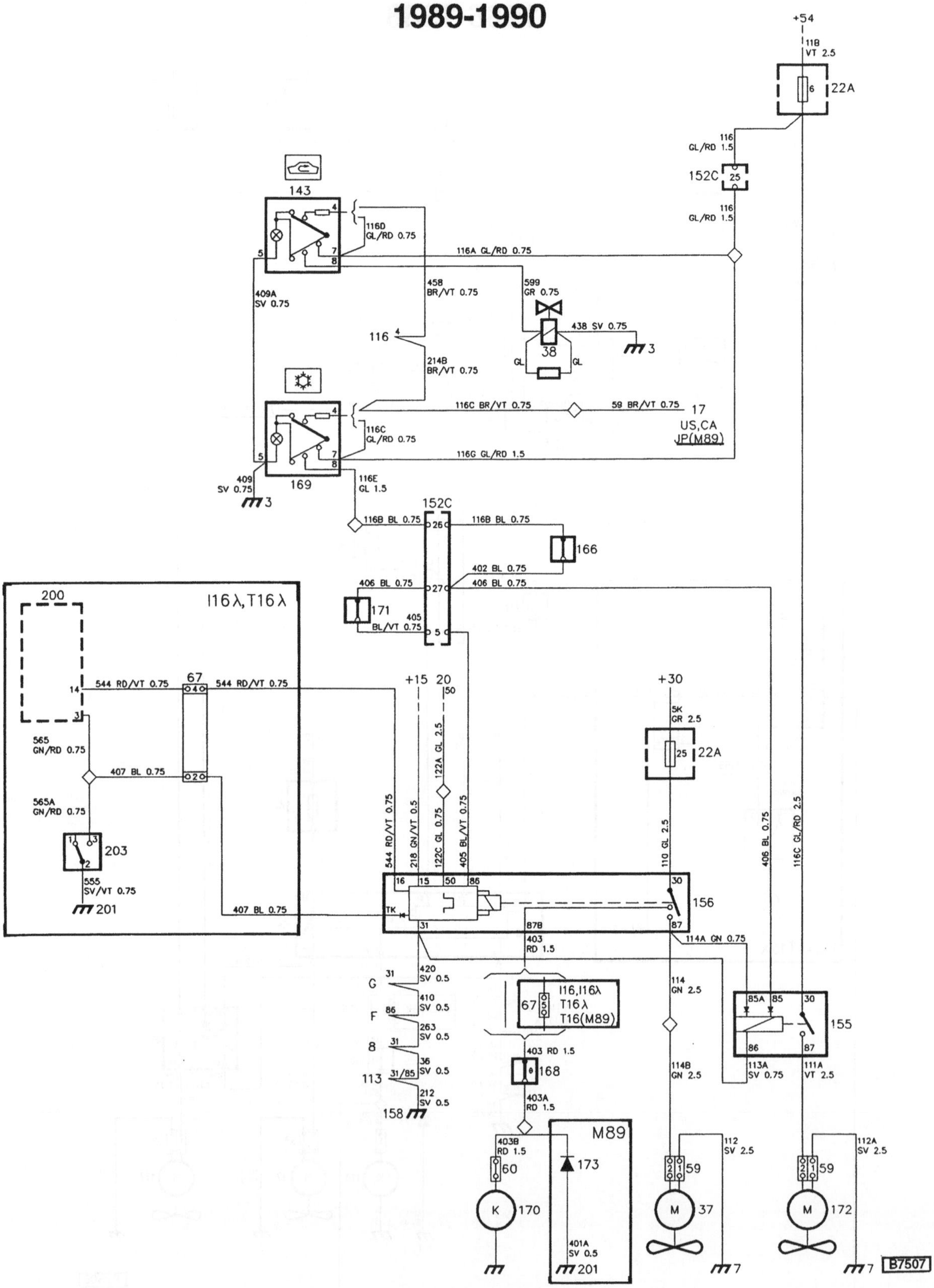

Air Conditioner (A/C)

1991-1993 non-turbo

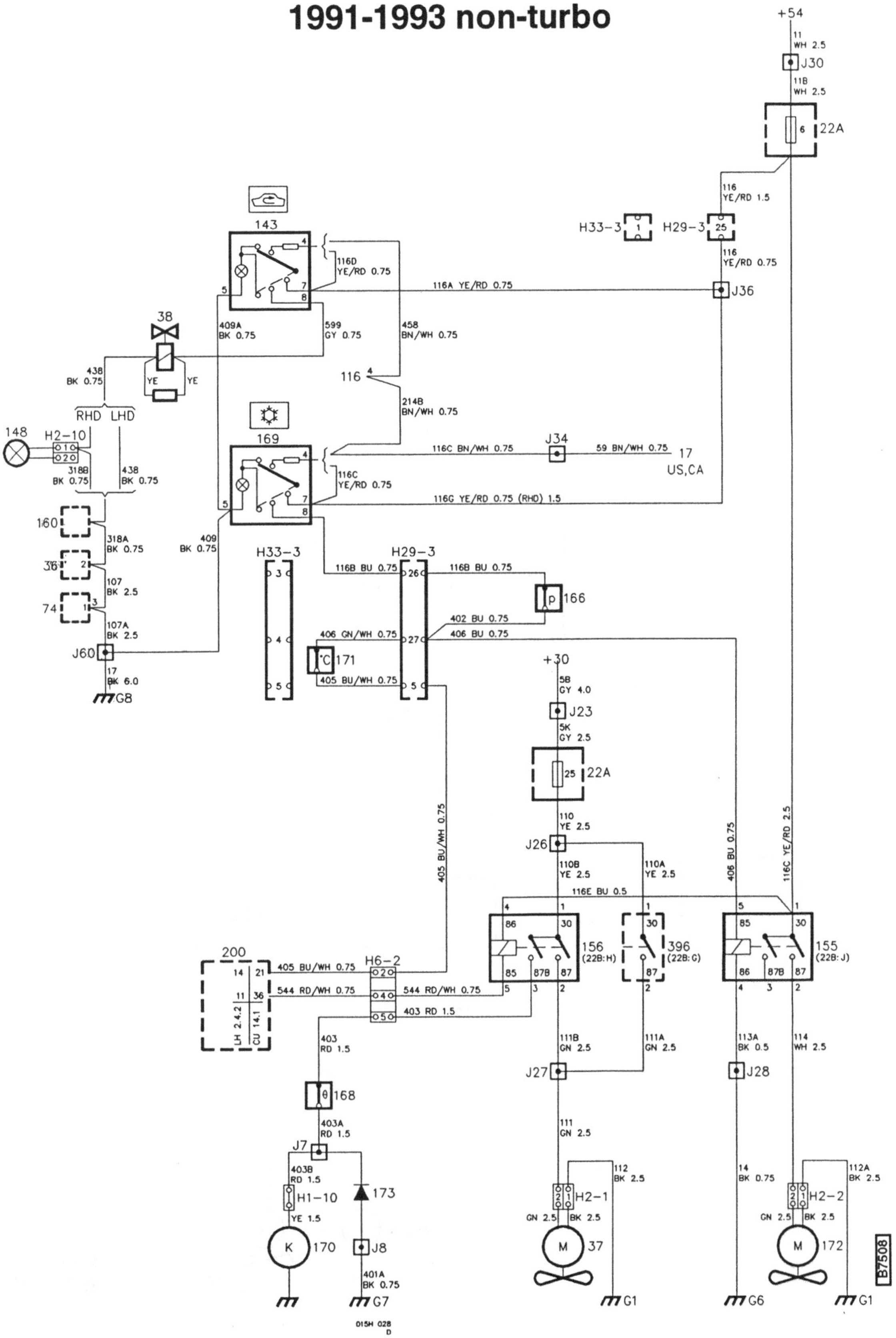

Air Conditioner (A/C)

1991-1993 turbo

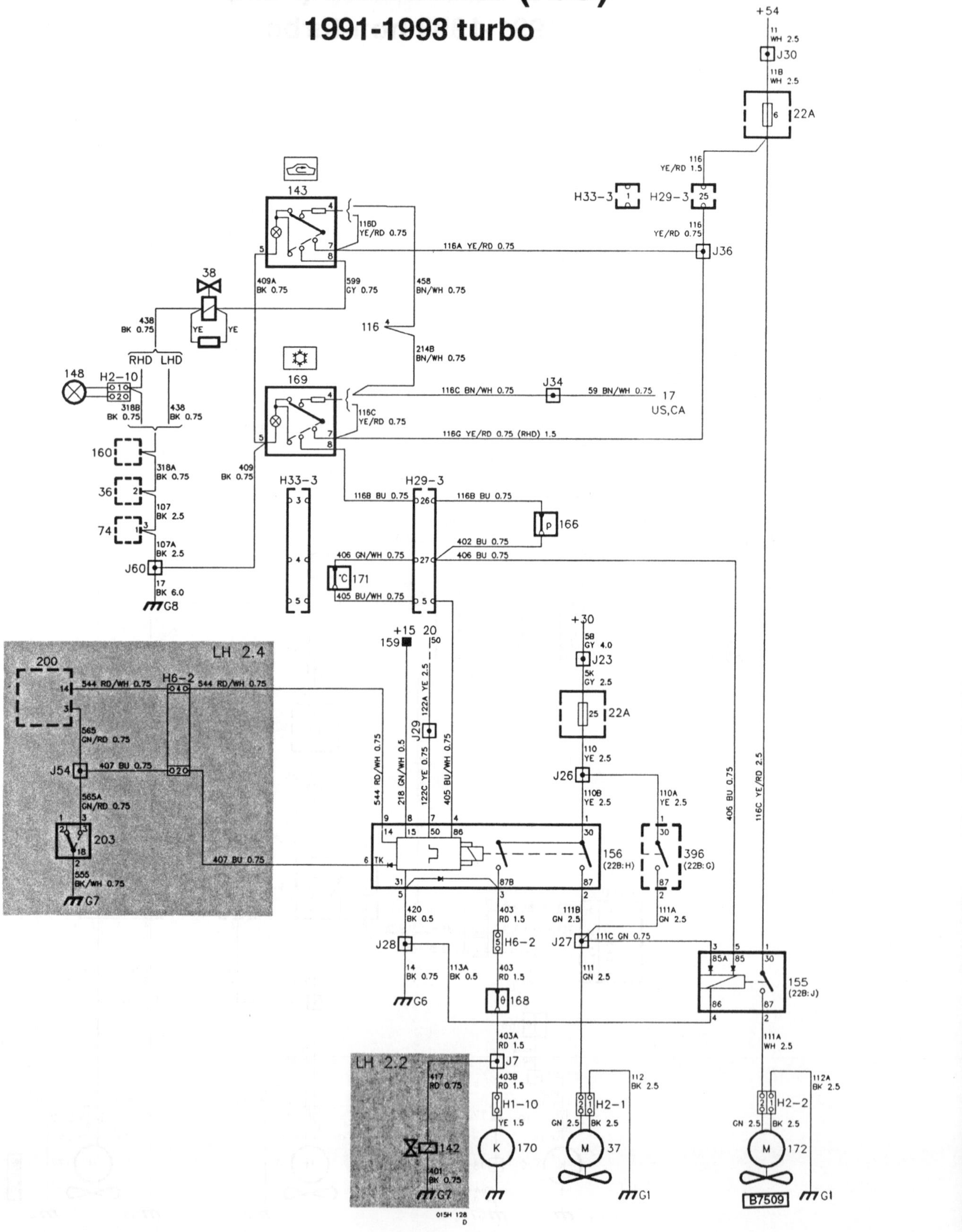

Alternator and Charging System

Operation

The alternator supplies the electrical components and recharges the battery. The alternator is not fused. Voltage regulation is controlled internally by a regulator. The alternator circuit includes a warning lamp in the instrument cluster. If the lamp burns out, the alternator will not charge the battery properly. See **853 Dashboard and Consoles** and **351 Lighting** for information on changing the lamp.

See **321 Alternator and Charging System** for alternator and charging system repair information and additional fault tracing.

CAUTION

- Do not disconnect the battery or any wire when the engine is running. Do not reverse the polarity of the battery. Do not ground any alternator terminals. This will damage the alternator.

- Always disconnect the negative (–) battery cable before removing the alternator. Unfused battery voltage is always present at the rear of the alternator.

- Always disconnect the negative (–) battery cable and all wires to the alternator before doing any electric (arc) welding to the car.

Fault Tracing

1. Inspect the V-belts and make sure they are tight. Test for a fully charged battery as described in **311 Battery.**
2. Check for faulty connections and broken wires.
3. If no visible faults can be found, test the alternator output using an alternator load tester.

Alternator output specification

- Bosch type 80 A . . . . .54 Amps @ 1900 rpm

NOTE

If a load tester is not available, an output test can be done by running the engine at 2000 rpm and turning on all electrical loads such as fans, lights and heated window. Voltage at the battery should be approximately 12 volts or higher.

Component Locations

1 battery
2 alternator in back of engine compartment
4 starter motor under intake manifold
21 ignition switch relay in main fuse/relay panel
22A main fuse/relay panel in left wheel housing
47E charging lamp in instrument cluster
75 distribution block in right wheel housing
159 +15 distribution in main fuse/relay panel

Connector Locations

1985-1990

59 2-pin connector in main fuse/relay panel
152A 29-pin bulkhead connector behind main fuse/relay panel
152B 29-pin bulkhead connector behind main fuse/relay panel
152C 29-pin bulkhead connector behind main fuse/relay panel

1991 and later

H29-1 29-pin bulkhead connector behind main fuse/relay pane l (1991-1991 1/2)
H29-2 29-pin bulkhead connector behind main fuse/relay panel (1991-1991 1/2)
H29-3 29-pin bulkhead connector behind main fuse/relay panel (1991-1991 1/2)
H33-1 33-pin bulkhead connector behind main fuse/relay panel (1991 1/2 and later)
H33-2 33-pin bulkhead connector behind main fuse/relay panel (1991 1/2 and later)
H33-3 33-pin bulkhead connector behind main fuse/relay panel (1991 1/2 and later)

Ground Locations

1985-1990

7 ground point at radiator cross member
158 ground distribution in main fuse/relay panel
211 ground point on gearbox

1991 and later

G1 ground point at radiator cross member
G6 ground distribution in main fuse/relay panel
G25 ground point on gearbox

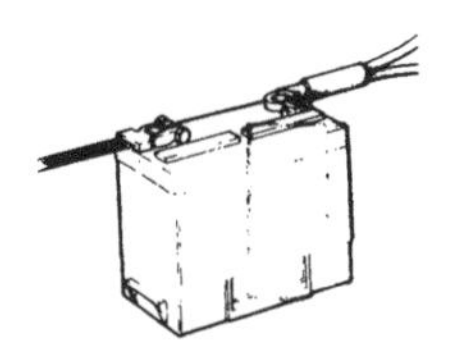
1

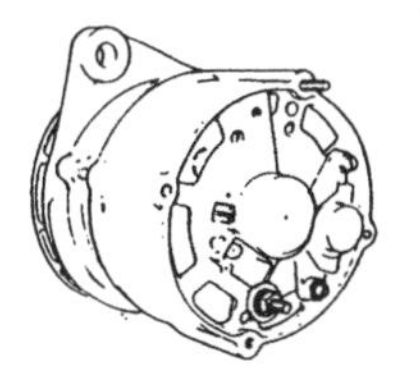
2

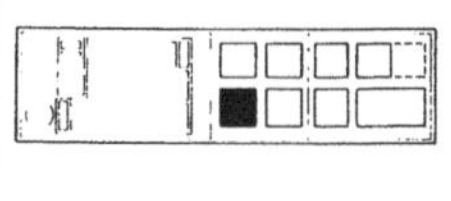
21, 22a

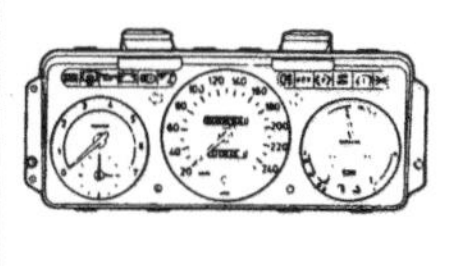
47E, 110.

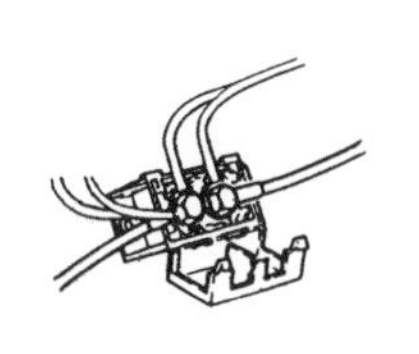
75

B7533

Alternator and Charging System

1985-1986

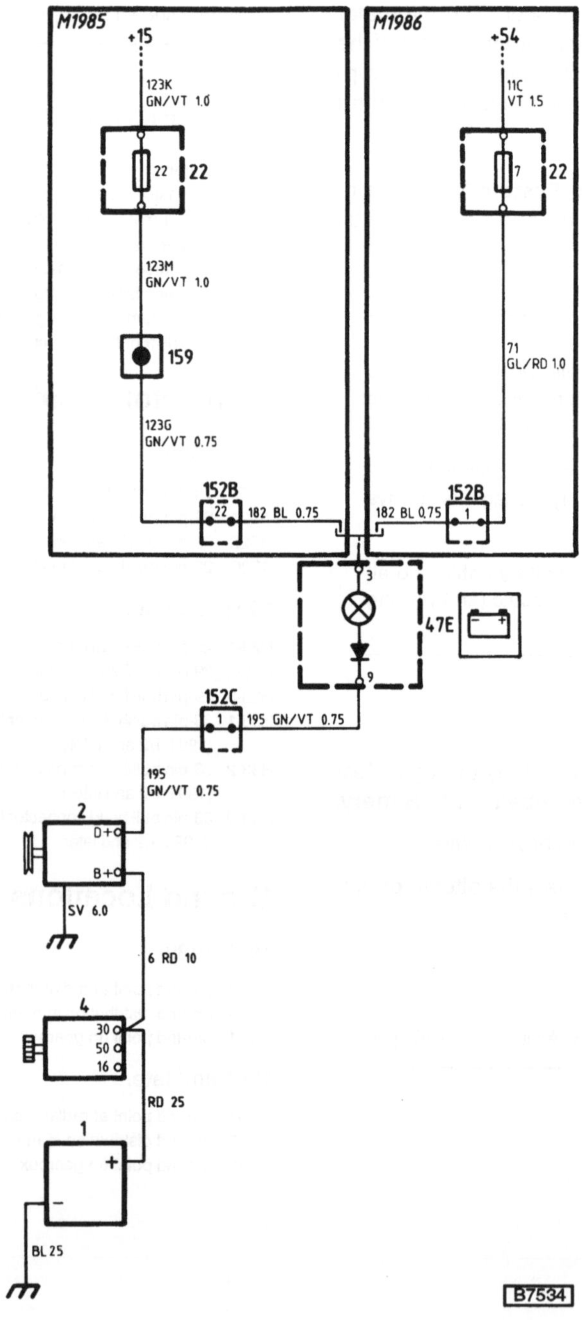

Alternator and Charging System

1987-1990

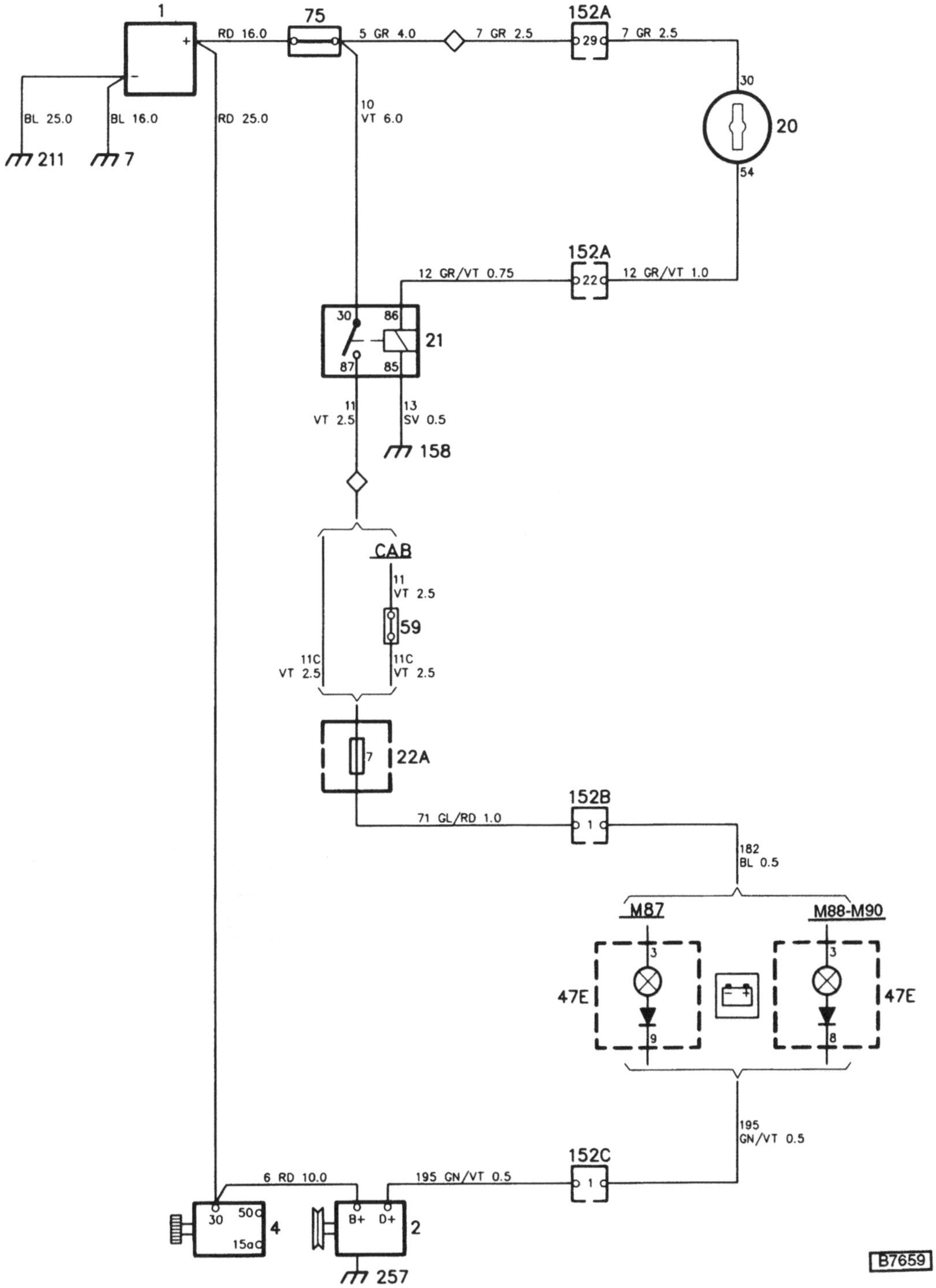

Alternator and Charging System

1991-1993

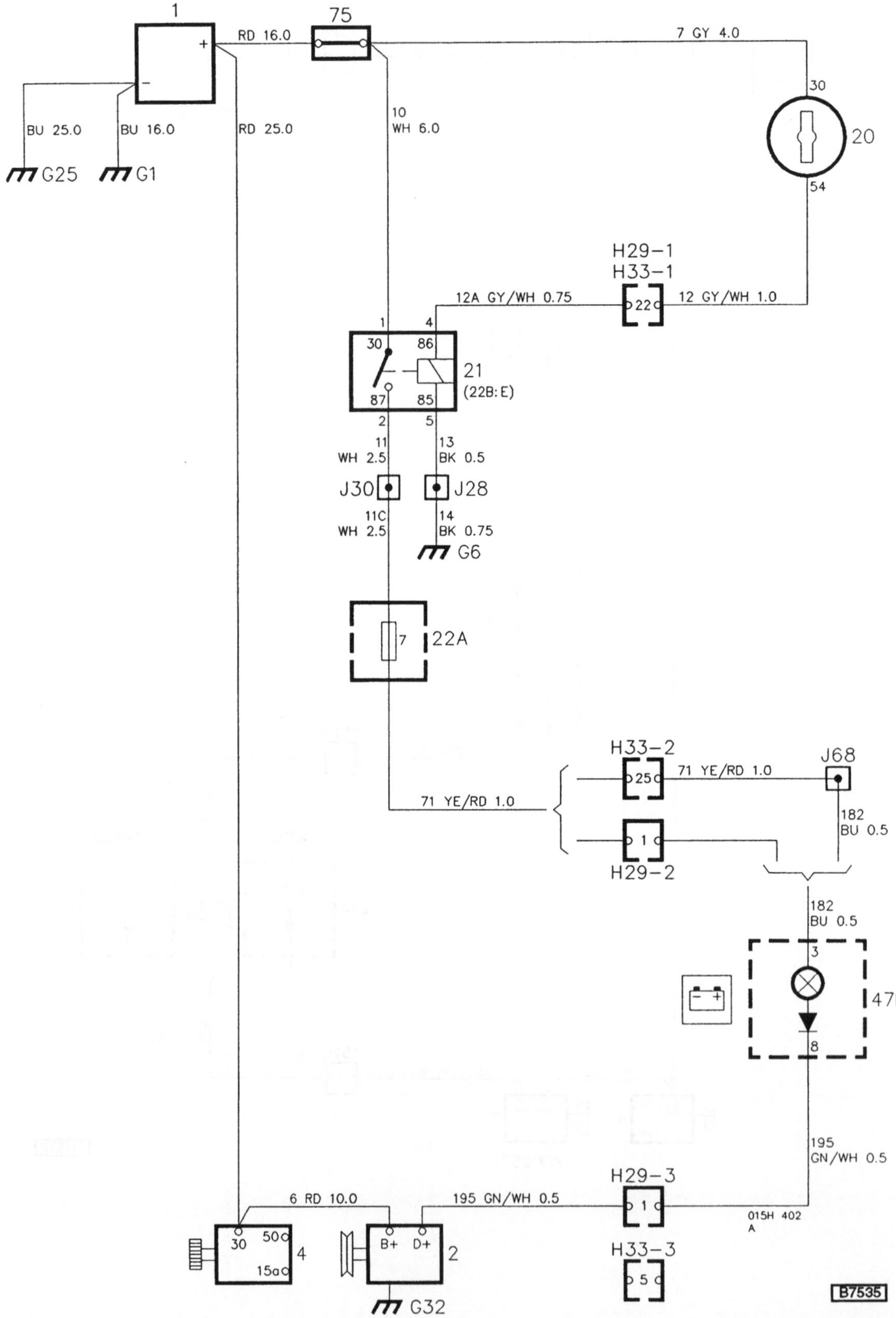

APC System

Operation

APC or Automatic Performance Control is used on turbo models only. The APC system continuously monitors engine detonation and adjusts the turbocharger boost pressure accordingly. This allows the engine to be run at a higher compression ratio for better fuel economy and throttle response. See **291 Turbocharger** for a complete system description and repair information.

Fault Tracing

1. Check fuse 10 and fuse 31 (where applicable).
2. On 1986 and later models, check that the brake light switch is functioning correctly.

NOTE
On 1986 and later cars, the brake light switch cancels the APC system when the brakes are applied, allowing the system to operate at a reduced boost pressure.

---

3. Measure the resistance at the intake manifold pressure sensor connector.

Intake manifold pressure sensor (180)
• resistance . . . . .5 to 13 ohms

Component Locations

5 ignition coil on radiator
29 brake light switch at the brake pedal
73 TSI connector at main fuse/relay panel
146 ignition amplifier in engine compartment, left bulkhead
147 ignition pulse amplifier in main fuse/relay panel (1986-1987)
177 APC control unit under back seat (1985), ahead of left wheel housing (1986 and later)
178 knock sensor on engine block, below intake manifold
179 pressure boost solenoid valve on the radiator fan casing
180 boost pressure sensor behind instrument panel to the left of the steering column
187 vacuum pump for cruise control
233 vacuum switch in engine compartment, left wheel housing

Connector Locations

1985-1990

59 2-pin connector in the engine compartment left wheel housing
60 1-pin connector in the engine compartment
152A 29 pole white bulkhead connector
152C 29 pole black bulkhead connector

1991 and later

H2-34 2-pin connector on the left wheel housing
H29-1 29-pin white bulkhead connector (1991-1991 1/2)
H29-3 29-pin black bulkhead connector (1991-1991 1/2)
H33-1 33-pin black bulkhead connector (1991 1/2 and later)
H33-3 33-pin blue bulkhead connector (1991 1/2 and later)

Ground Locations

1985-1990

7 ground point at radiator cross member
9 ground point in luggage compartment
93 ground point in left wheel housing

1991 and later

G1 ground point at radiator cross member
G6 ground distribution in main fuse/relay panel
G12 ground point in left wheel housing

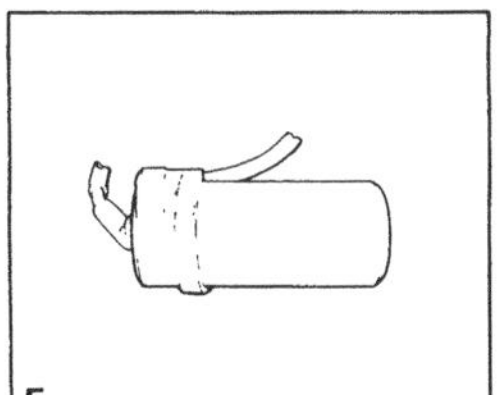
5

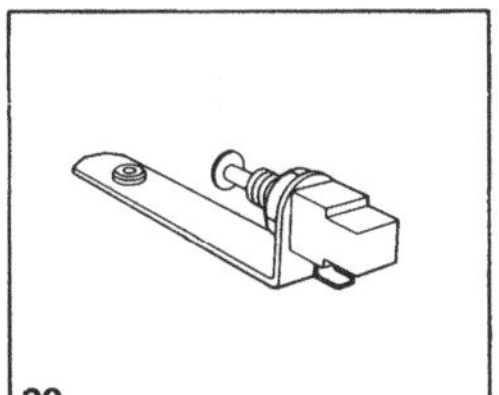
29

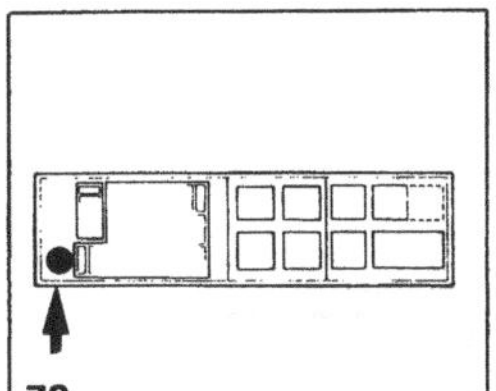
73

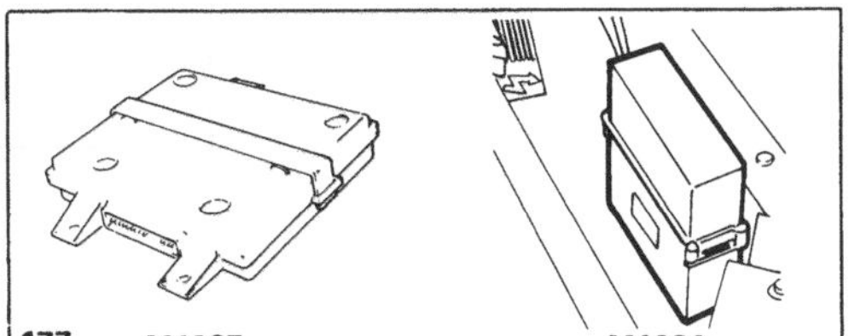
177 M1985 M1986

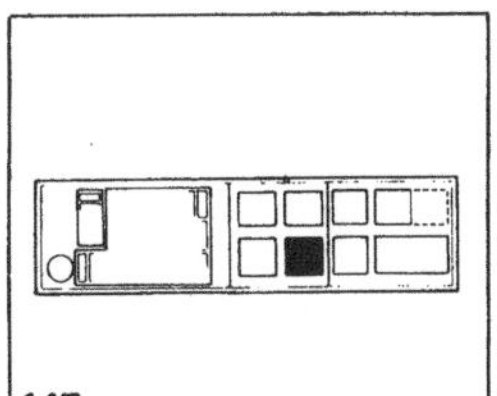
147

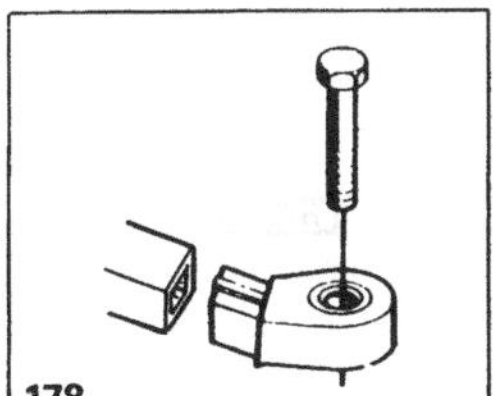
178

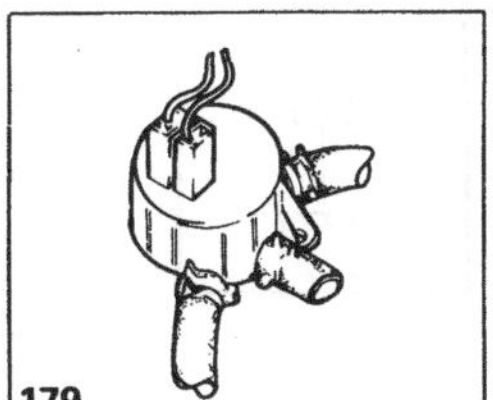
179

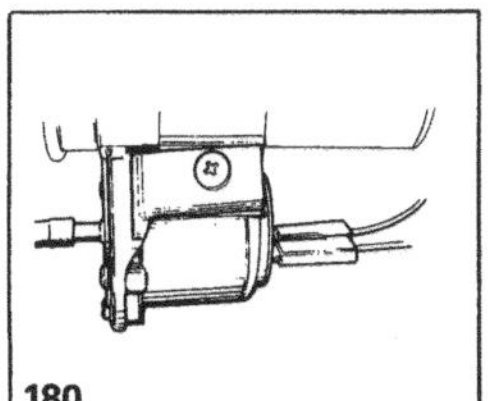
180

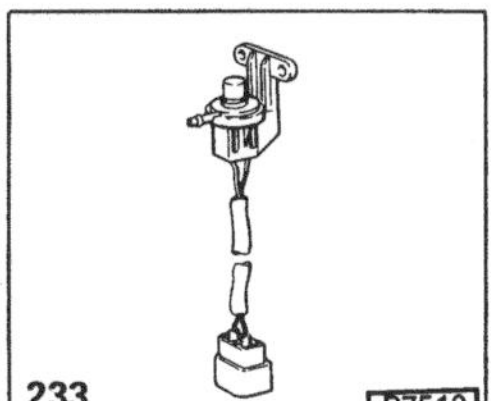
233 B7510

APC System

1985

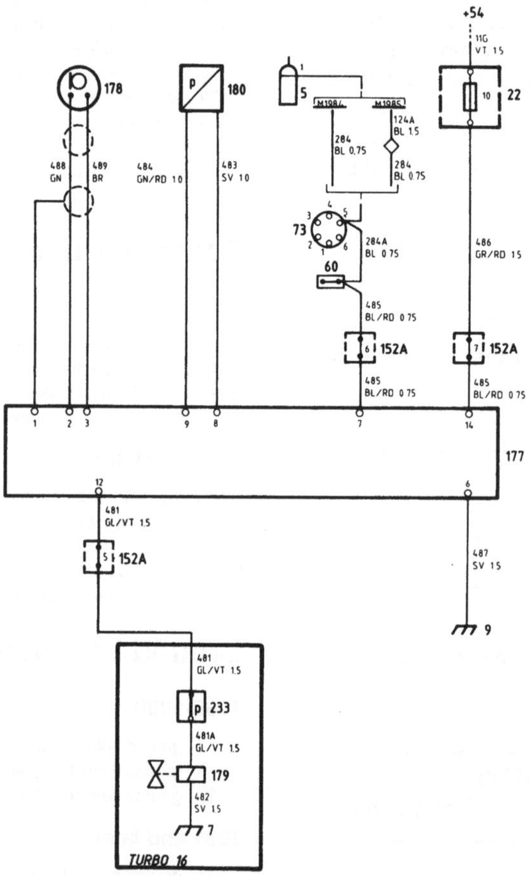

1986

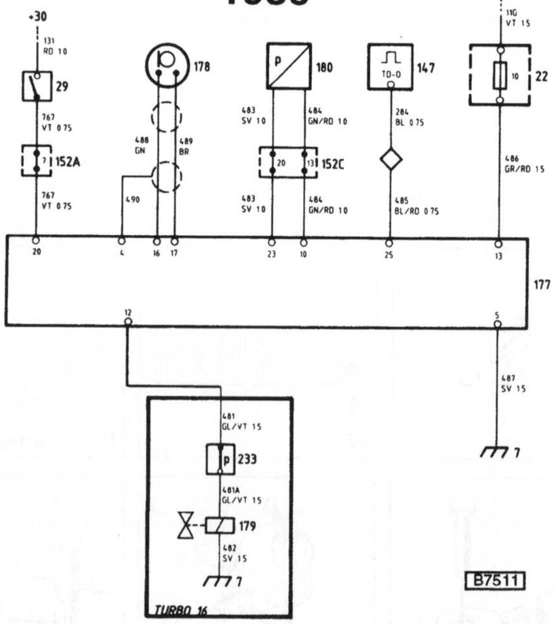

APC System

1987-1993

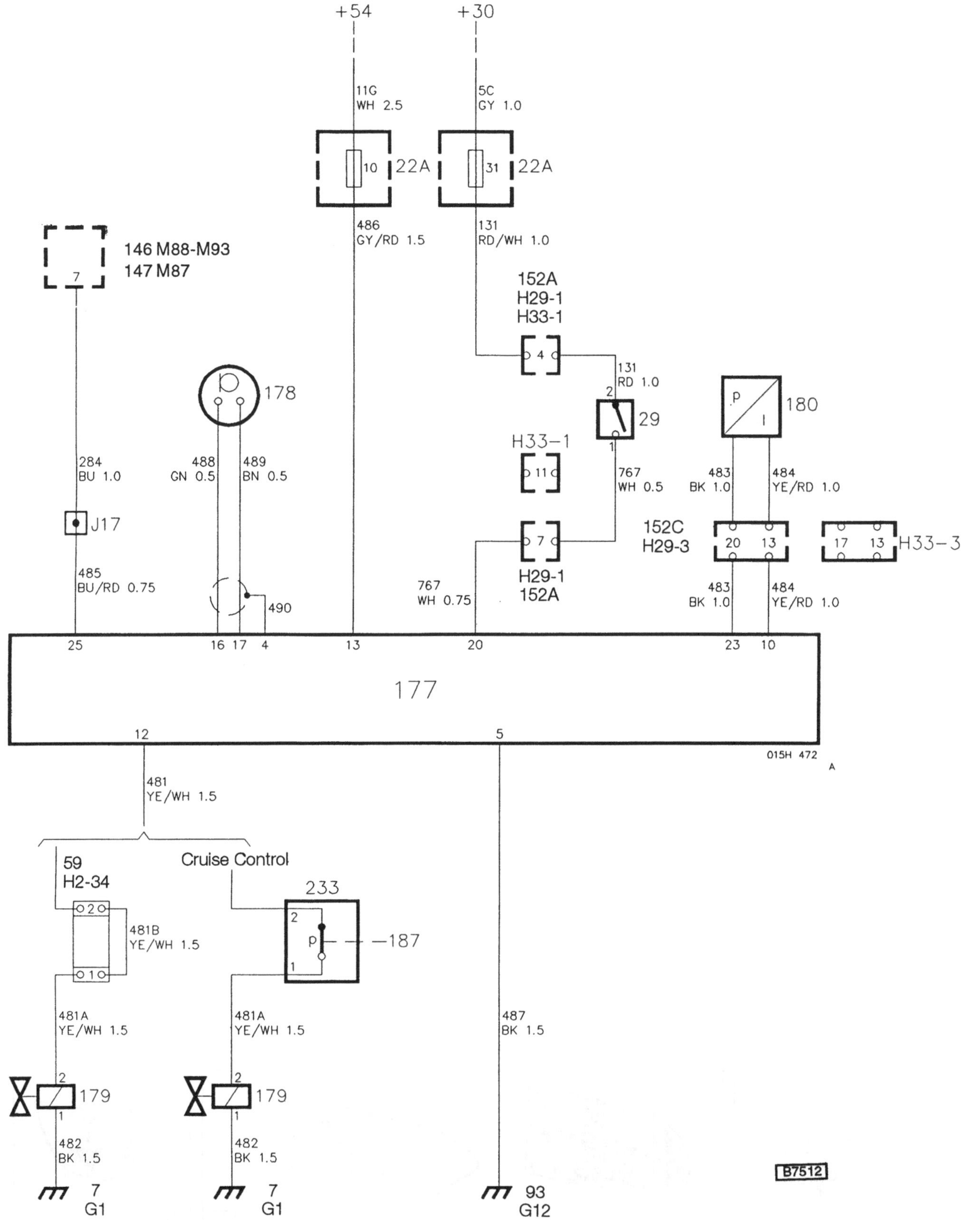

Backup Lights

Operation

Voltage is supplied to the backup light switch whenever the key is on. When the transmission is in reverse and the key is on, the switch closes to supply voltage to the backup lights.

Models with automatic transmission have a selector lever switch that can be adjusted as described in **364 Electric Controls and Switches.** Models with manual transmission have a non-adjustable switch that screws into the side of the shift-lever housing. See **432 Manual Transmission Controls.**

Fault Tracing

1. Check fuse 13.
2. With the ignition key in the off position, check for resistance across the switch terminals. There should be continuity with the transmission in reverse and no continuity when the transmission is taken out of reverse.
3. Check for voltage at the reverse lights with transmission in reverse and the key on.

Component Locations

22 main fuse/relay panel
31 backup light switch under the center console
32 backup lights in rear light cluster
119 side reversing lights in the front light cluster

Connector Locations

1985-1990

152A 29-pin bulkhead connector in main fuse/relay panel

1991 and later

H29-1 29-pin bulkhead connector behind main fuse/relay panel (1991-1991 1/2)
H33-1 33-pin bulkhead connector behind main fuse/relay panel (1991 1/2 and later)

Ground Locations

1985-1990

7 ground point at radiator cross member
9 ground point in luggage compartment
93 ground point at left wheel housing

1991 and later

G1 ground point at radiator cross member
G3 ground point in luggage compartment
G12 ground point at left wheel housing

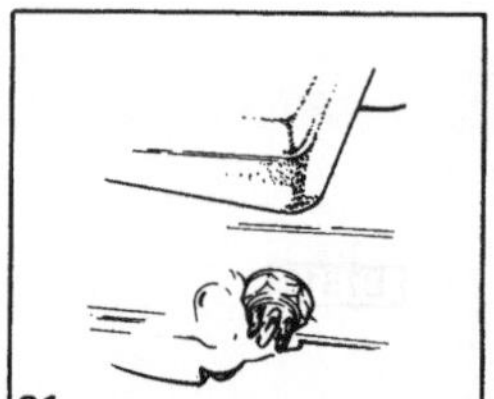
31

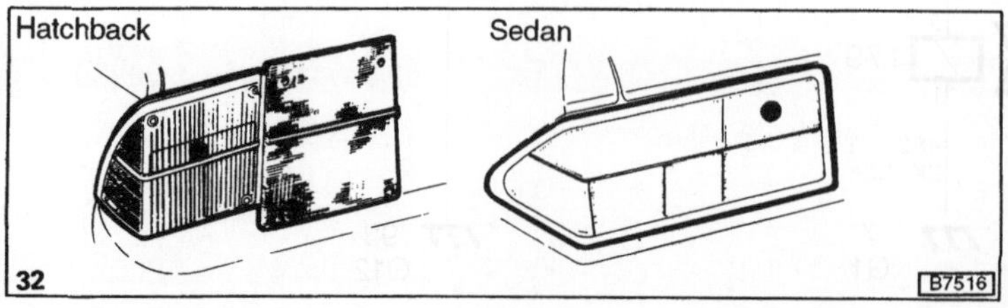

32

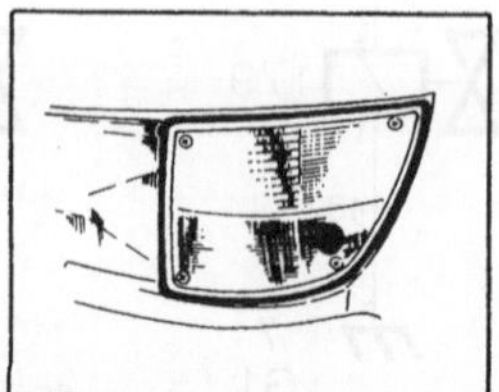
119

Backup Lights

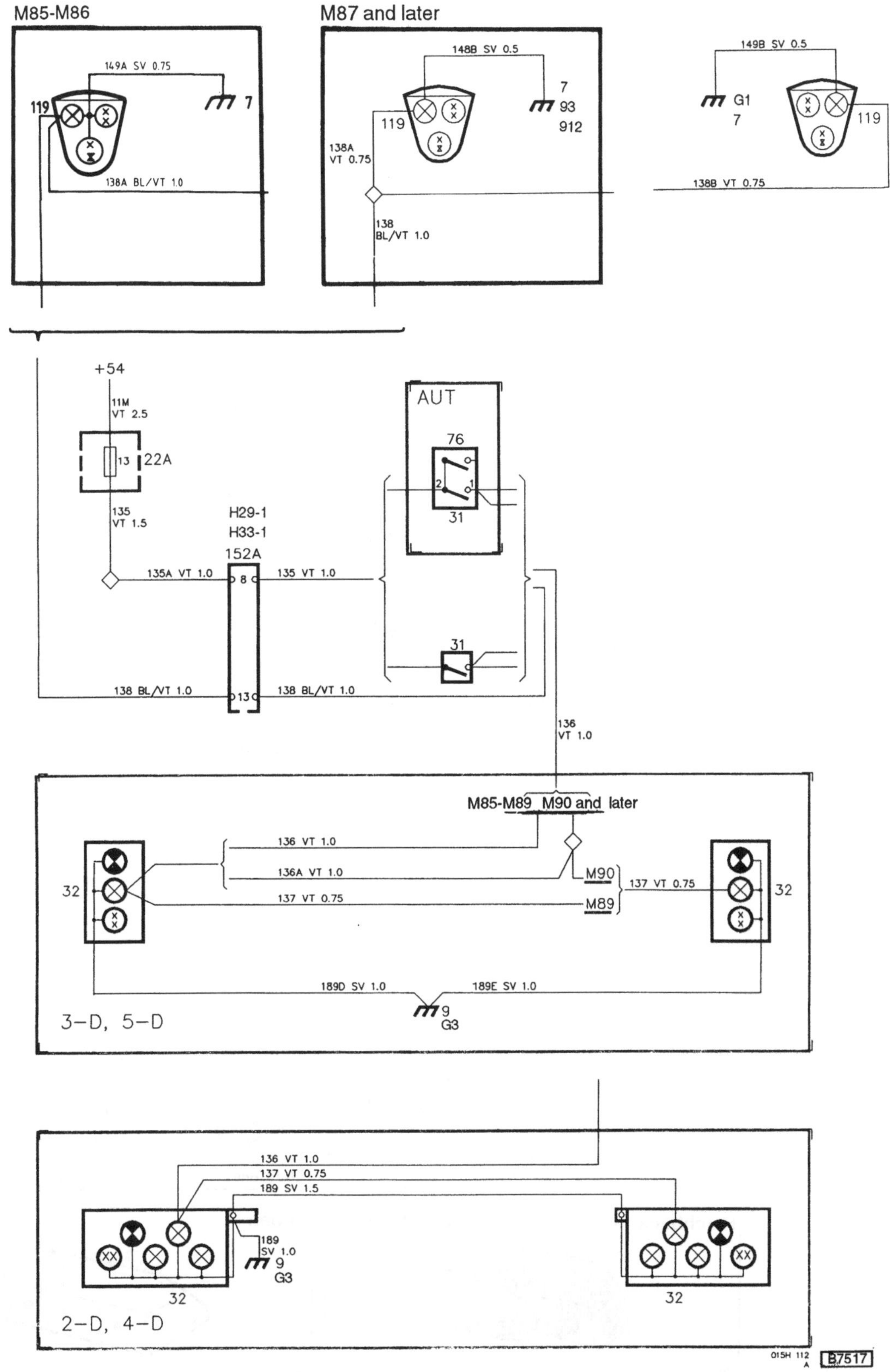

Brake Lights

Operation

The brake lights are supplied voltage through the brake light switch (29). When the brake pedal is depressed, the brake light switch closes and completes the circuit to the brake lights. The brake switch is mounted on the upper part of the brake pedal and is adjustable. On 1986 and later models, a center brake light is used. See **351 Lighting** for information on changing bulbs. For troubleshooting information on the ABS warning lights see the **Anti-lock Braking System (ABS)** wiring diagram.

Fault Tracing

1. Check fuse 31.
2. Check for voltage at the brake light switch. Check for voltage at the brake light sockets when the brake pedal is depressed.

NOTE

A wooden rod can be used to hold the brake pedal down during troubleshooting.

---

3. Check the adjustment of the brake light switch. The brake lights should come on when the brake pedal is depressed 10 mm (0.4 in.) or more. If necessary, adjust the switch as described in **364 Electric Controls and Switches.**

Component Locations

22 main fuse/relay panel
29 brake light switch at the brake pedal
30 brake lights in the rear clusters
109 third (center) brake light

Connector Locations

1985-1990

57 3-pin connector (hatchback only) in the luggage compartment
59 2-pin connector in the luggage compartment
60 1-pin connector in luggage compartment at the rear light
123 4-pin connector at hatchback, left side
152A 29-pin bulkhead connector at main fuse/relay panel

1991 and later

H2-17 2-pin connector at hatchback left side, or on sedan under parcel shelf
H2-18 2-pin connector at hatchback, left side
H2-42 2-pin connector at luggage compartment lid
H2-43 2-pin connector at left hinge, convertible
H3-13 3-pin connector in luggage compartment, left air outlet
H4-12 4-pin connector in trunk lid, left air outlet
H29-1 29-pin bulkhead connector behind main fuse/relay panel (1991-1991 1/2)
H33-1 33-pin bulkhead connector behind main fuse/relay panel (1991 1/2 and later)

Ground Locations

1985-1990

9 ground point in luggage compartment

1991and later

G3 ground point in luggage compartment
G5 ground point under the rear seat
G9 ground point in luggage compartment
G19 ground point in hatchback

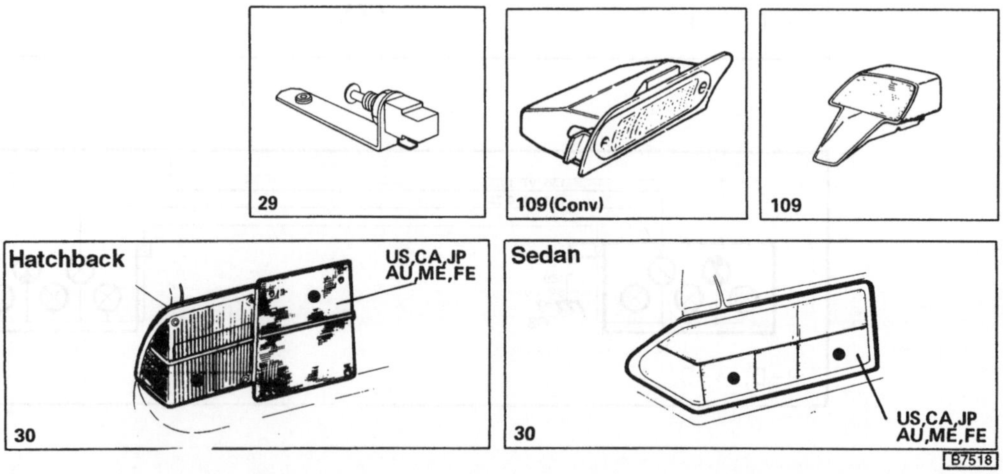

Brake Lights

1985-1989

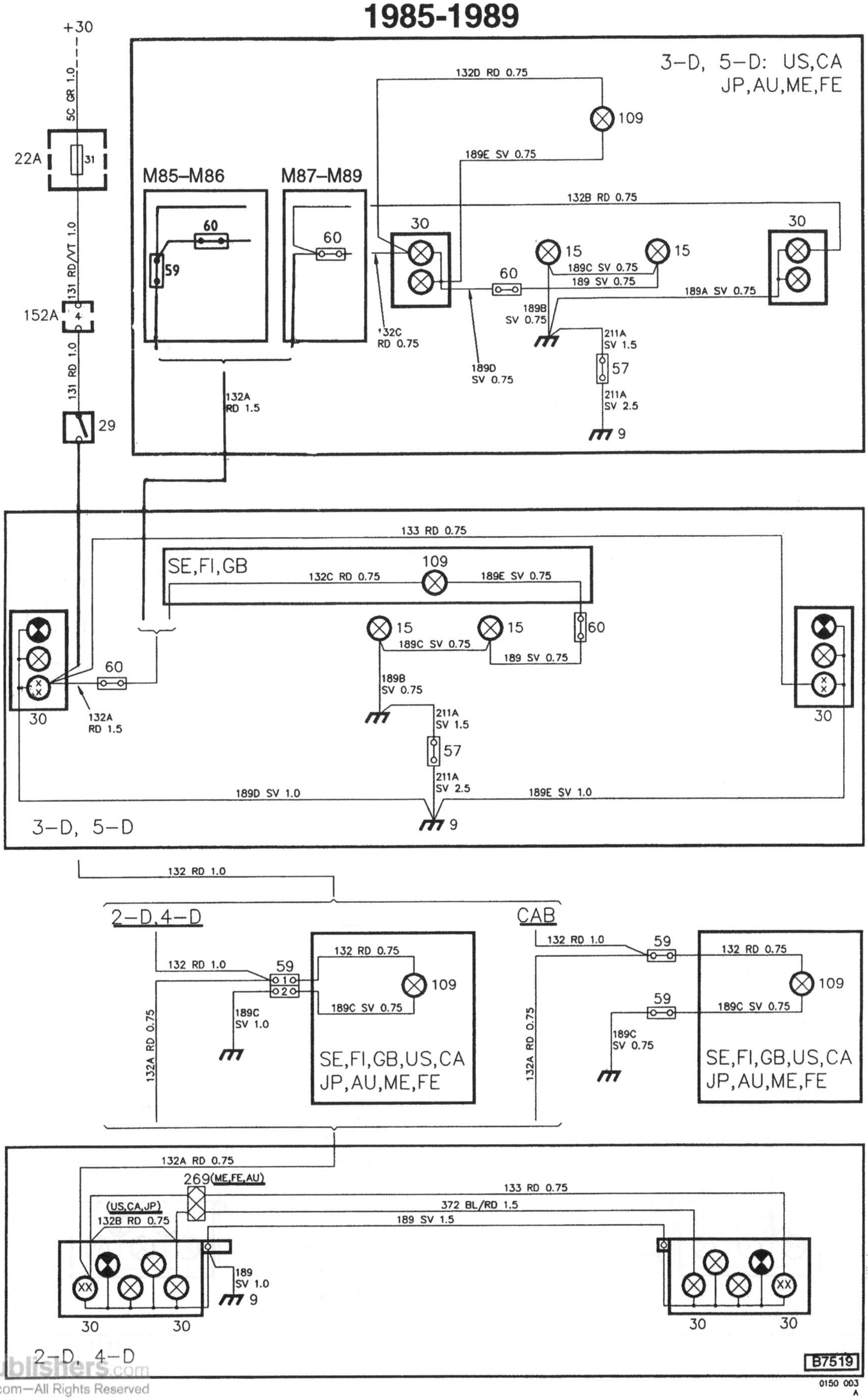

Brake Lights
1990-1993

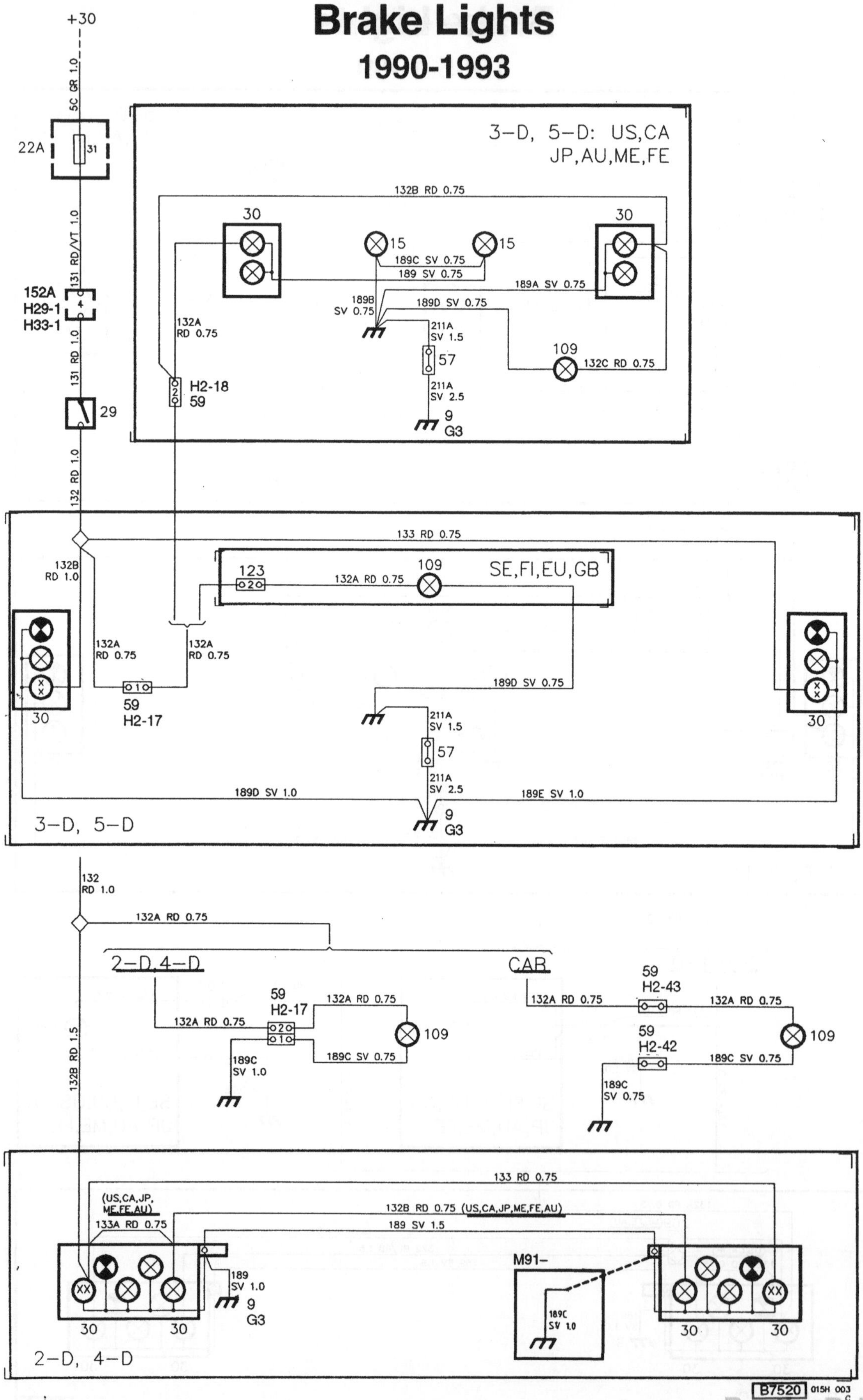

Burglar Alarm

Operation

The burglar alarm consists of a control unit, a main switch, a siren, switches in the doors, hood, and luggage compartment, and a remote control unit. On some later models, a motion detector, a glass breakage detector, a service-mode switch, and a rear seat switch are used. The alarm system is integrated with the central locking system. See the **Central Locking** wiring diagram found later in this repair group.

Fault Tracing

1. Check the fuses.
2. Check that the batteries in the remote control unit are good condition. Replace the batteries if in doubt.
3. Check the function of the LED on the left speaker grille.
4. Check all switches for proper operation.

Component Locations

1 battery at right front of engine compartment
4 starter motor under intake manifold
20 ignition switch
22A main fuse/relay panel in left wheel housing
27 left-hand direction indicator lamps
28 right-hand direction indicator lamps
53 interior light switch in center console
54 door switches on the door jamb
55 luggage compartment light in storage area
56 luggage compartment light switch in storage area
77 neutral/safety switch (automatics only) at gear selector lever
82 seatbelt warning relay under the rear seat
151 interior lighting time delay relay under the rear seat
274 fuse for burglar alarm under rear seat
275 alarm siren in engine compartment near right headlamp
276 alarm hood switch under the hood at front of wheel housing
288 alarm switch at center console
289 alarm control unit under the rear seat
304 alarm motion detector under the rear seat
305 alarm LED at left speaker grille
313 alarm relay under the rear seat
314 alarm seat switch in driver seat
315 alarm seat switch in passenger seat
316 alarm diode under the rear seat in the wiring harness
401 auxiliary fuse/relay panel under rear seat
412 alarm service mode switch under left side passenger seat
413 alarm glass breakage detector under passenger seat
423 alarm start interlock relay (convertible) under rear seat
424 alarm hazard warning light relay (convertible) under rear seat at the control unit
425 door lock switch in center console
426 switch for rear seat under the rear seat

Connector Locations

1985-1990

57A 3-pin white connector
57B 3-pin red connector
57D 3-pin green connector
57E 3-pin blue connector
58 12-pin connector behind instrument panel to the left of the steering column
59 2-pin connector in luggage compartment near left hinge
60 1-pin connector (3) one for the alarm siren in engine compartment, two under left speaker grille for the LEDs
122 8-pin connector for alarm under the rear seat
123 4-pin connector one in each B pillar
152A 29-pin bulkhead connector at main fuse/relay panel

1991 and later

H1-18 1-pin connector under center console
H1-21 1-pin connector for LED, under left speaker grille
H1-22 1-pin connector for LED, under left speaker grille
H1-28 1-pin connector under rear seat
H2-40 2-pin connector for the alarm siren, behind headlamp
H2-41 2-pin connector under passenger seat, right side
H2-42 2-pin connector in trunk lid
H2-43 2-pin connector at left hinge, convertible
H2-53 2-pin connector under the rear seat
H8-3 8-pin connector under the rear seat
H10-8 10-pin connector under rear seat at alarm control unit
H12-1 12-pin connector behind instrument panel, left of steering column
H33-1 33-pin bulkhead connector behind main fuse/relay panel (1991 1/2 and later

Ground Locations

1985-1990

117 ground point at the handbrake

1991 and later

G1 ground point at radiator cross member
G4 ground point between ignition switch and handbrake
G5 ground point under the rear seat
G5 ground point under the rear seat
G25 ground point at the gearbox

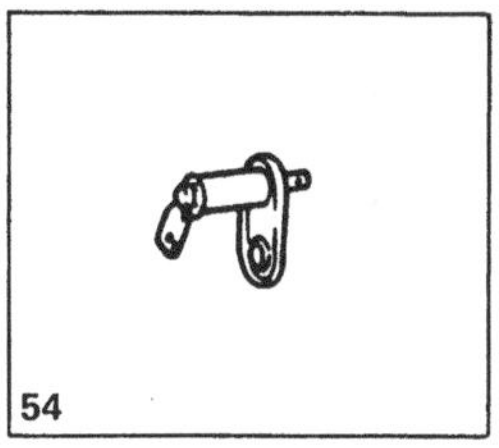

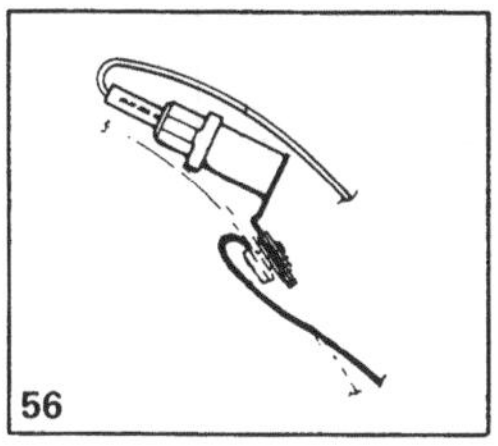

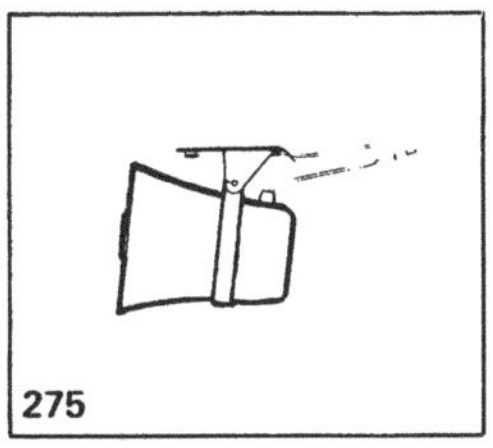

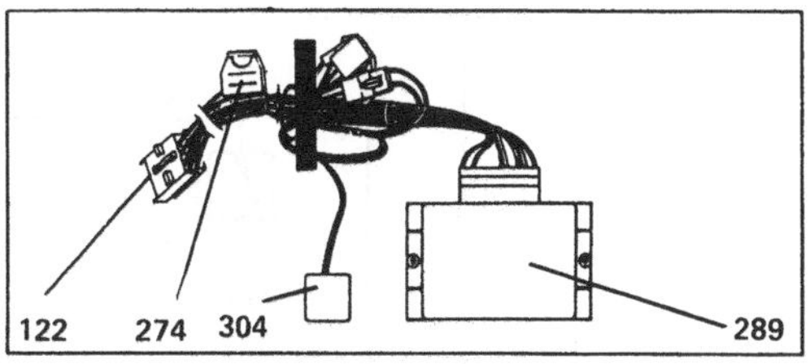

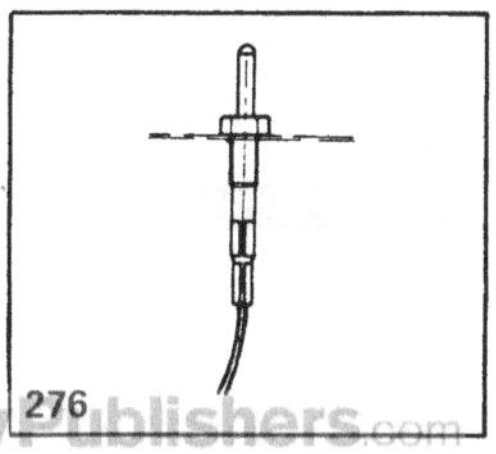

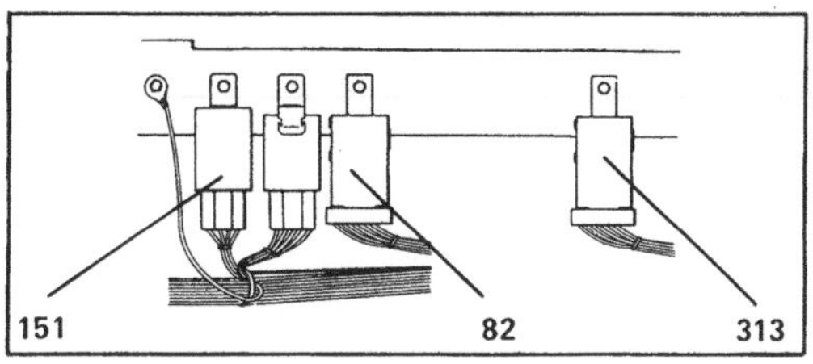

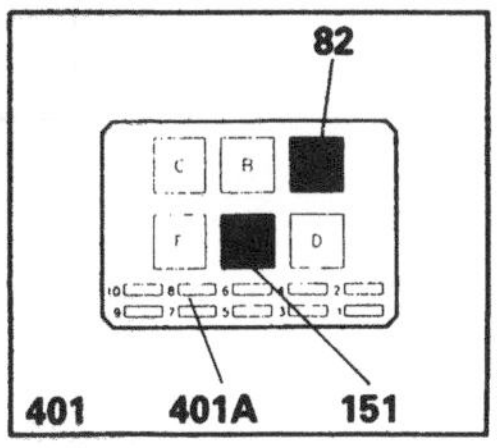

B7521

Burglar Alarm—ex. Convertible

1985-1986

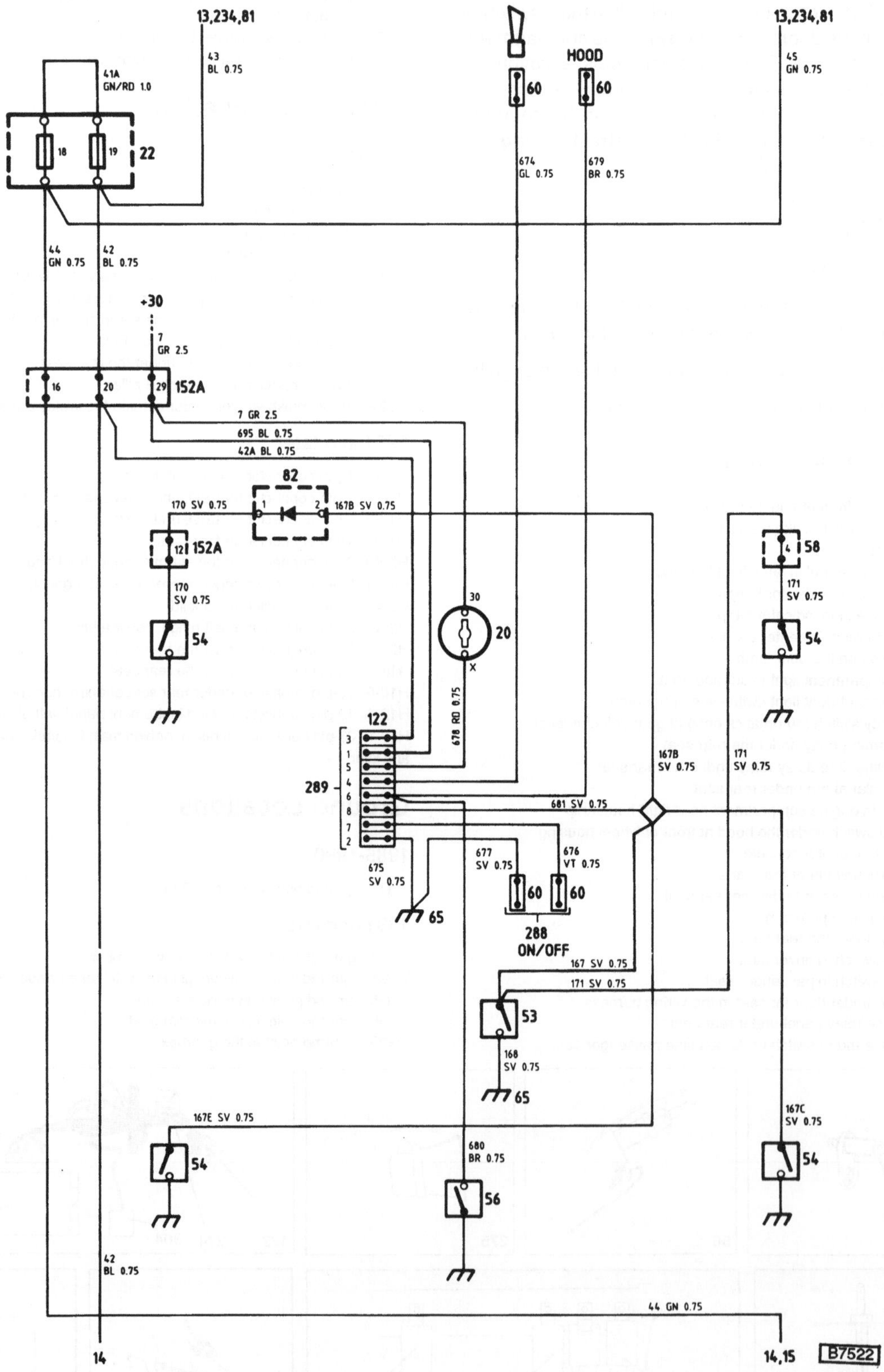

Burglar Alarm—ex. Convertible

1987-1990

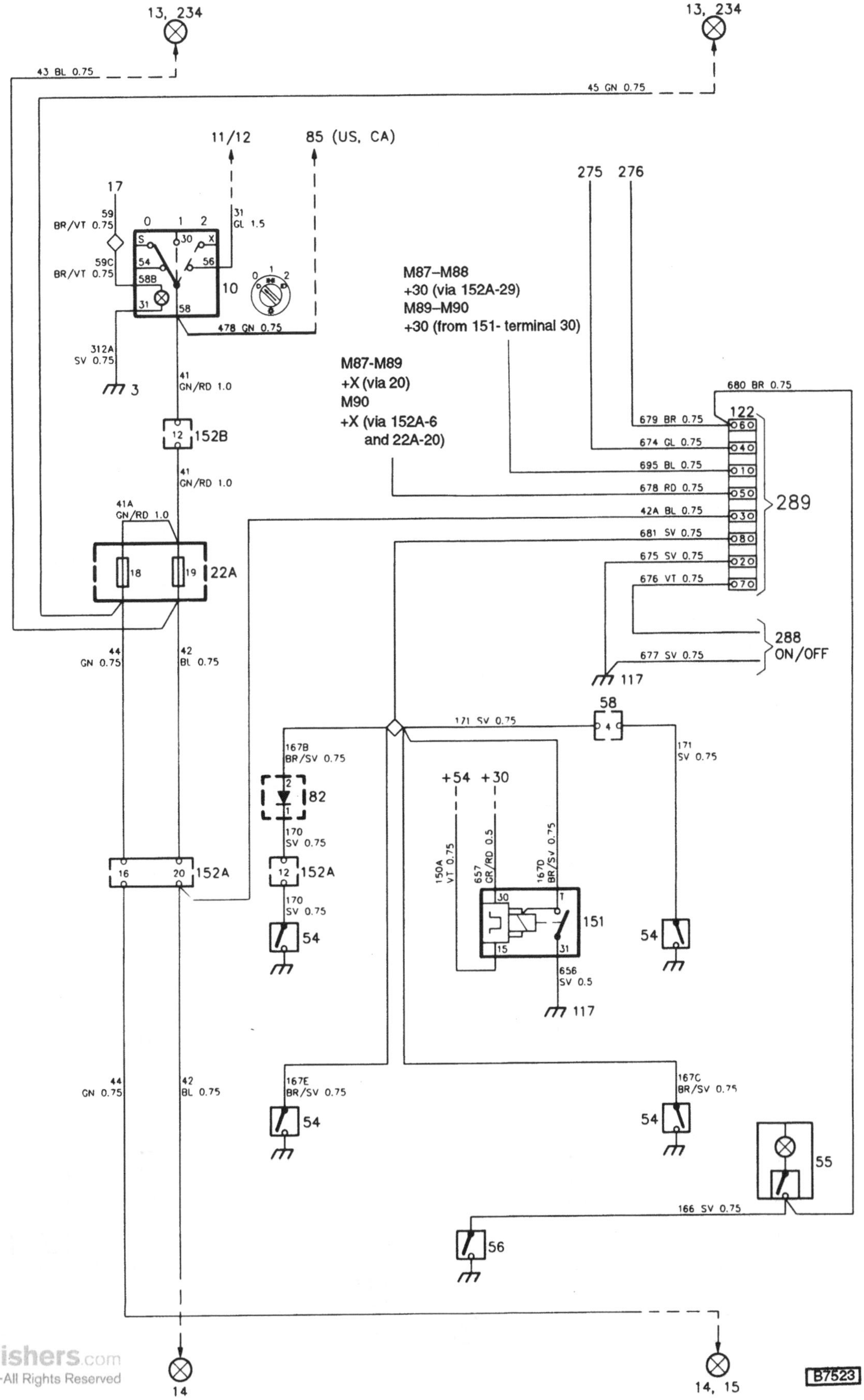

Burglar Alarm—ex. Convertible

1991-1993

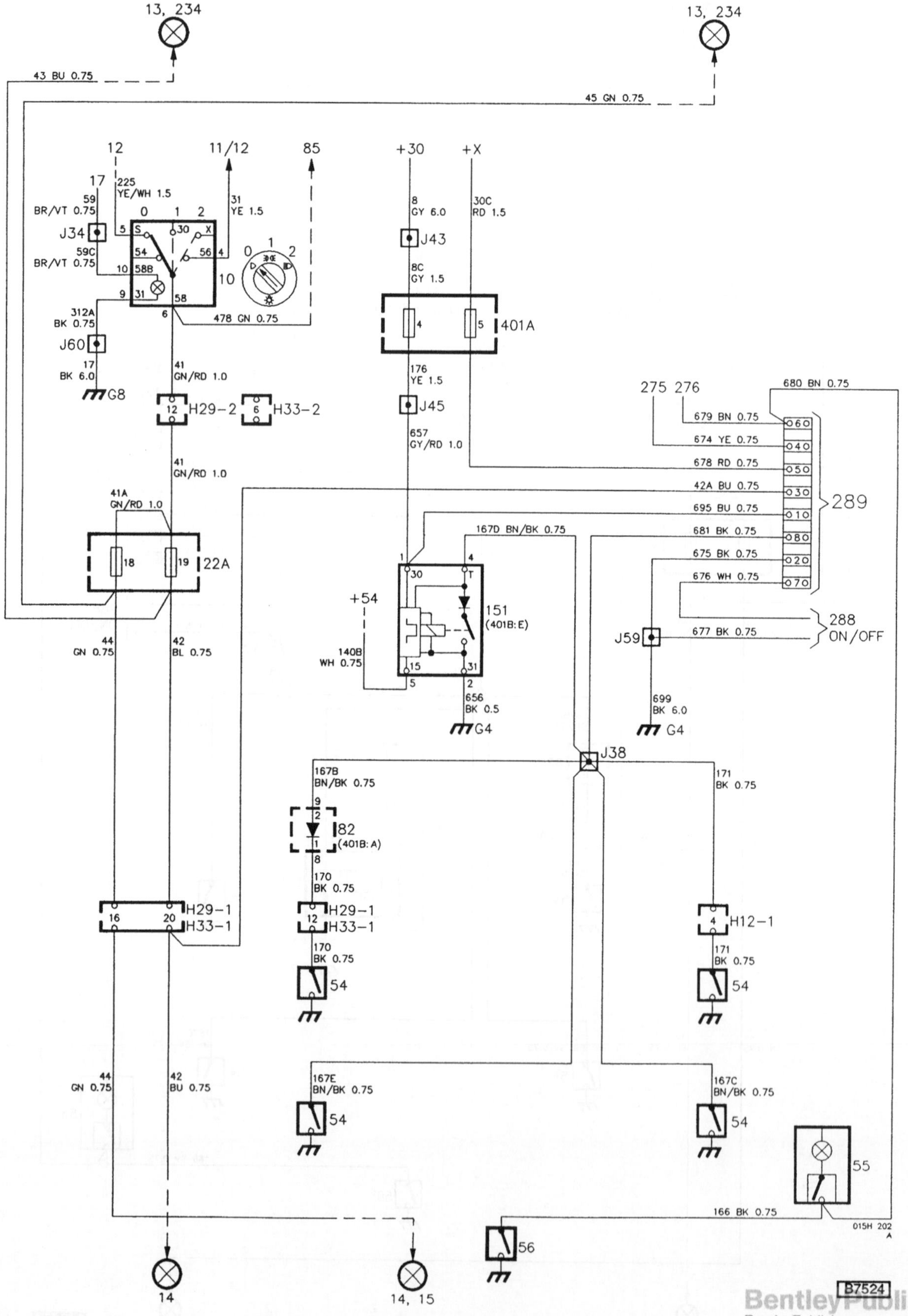

Burglar Alarm—Convertible

1987-1988

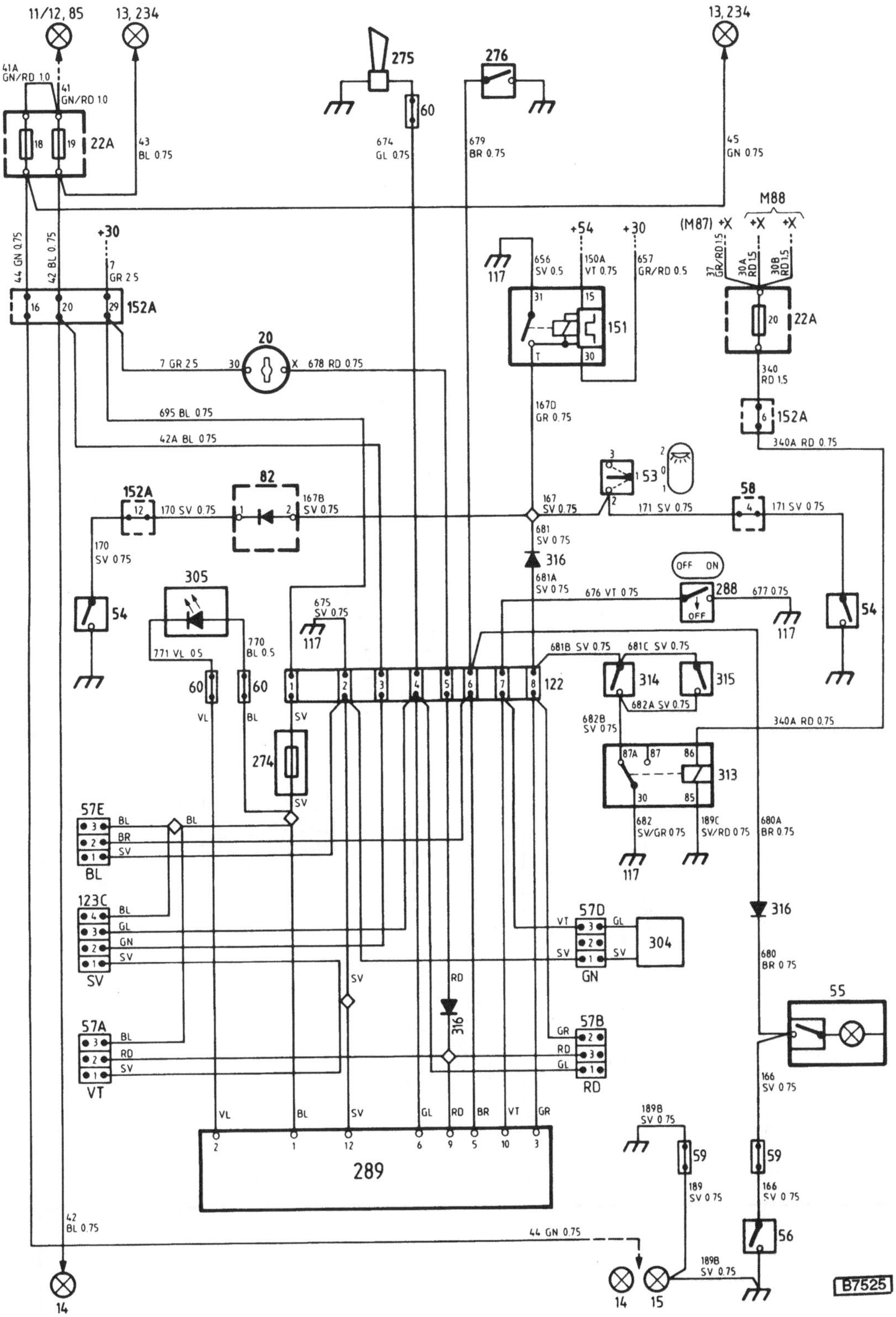

Burglar Alarm—Convertible

1989

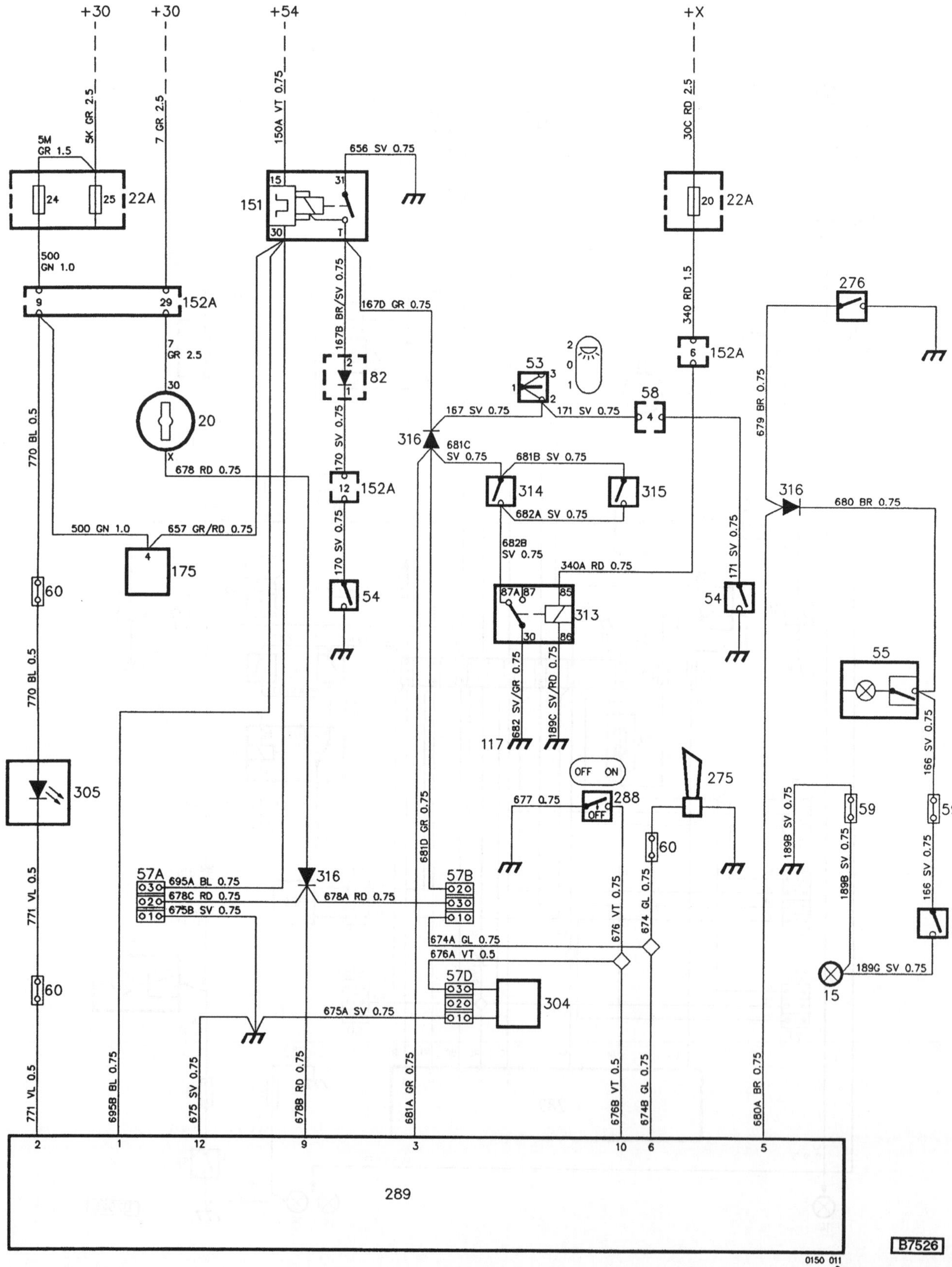

Burglar Alarm—Convertible

1990

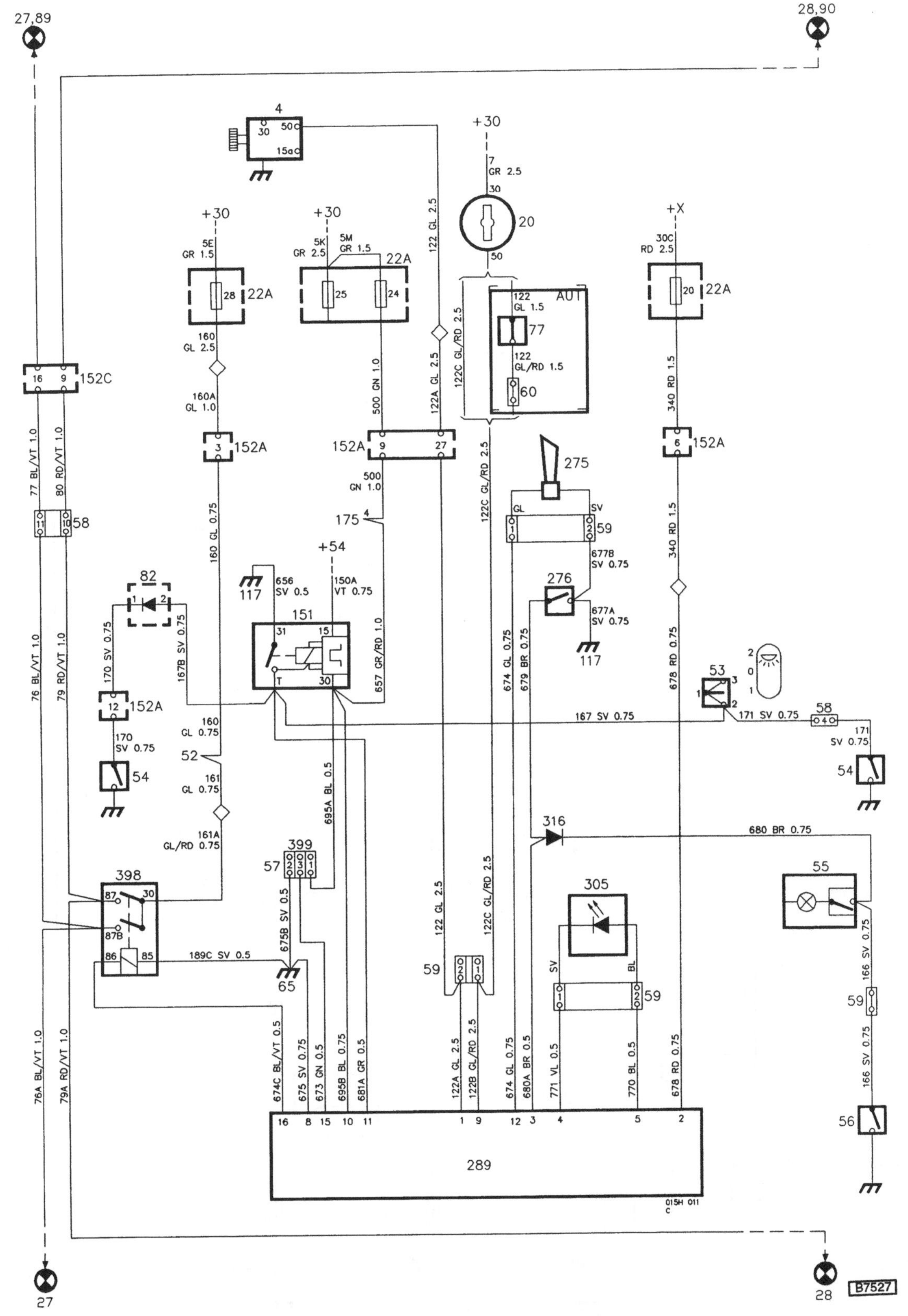

Burglar Alarm—Convertible

1991

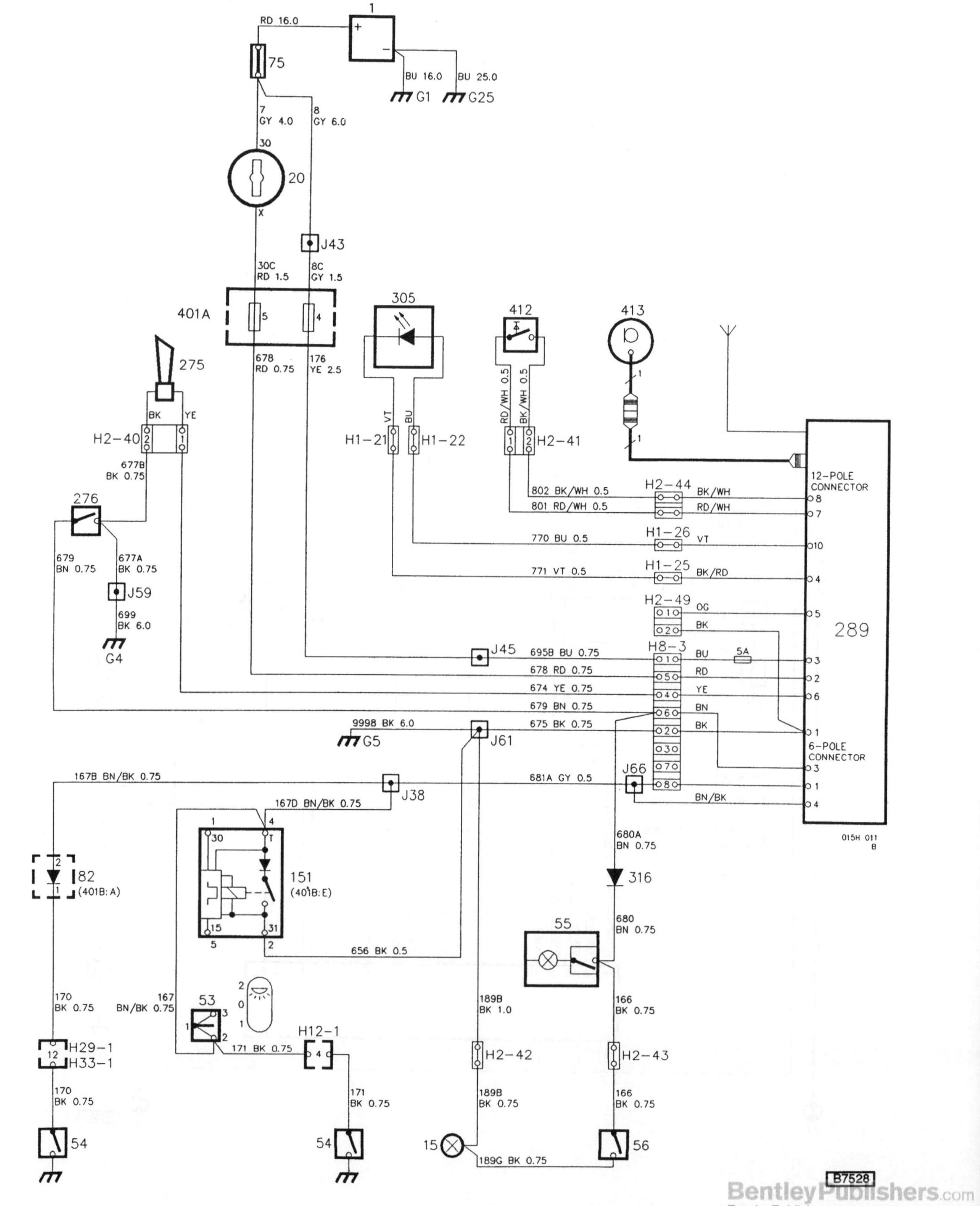

Burglar Alarm—Convertible

1992 and later

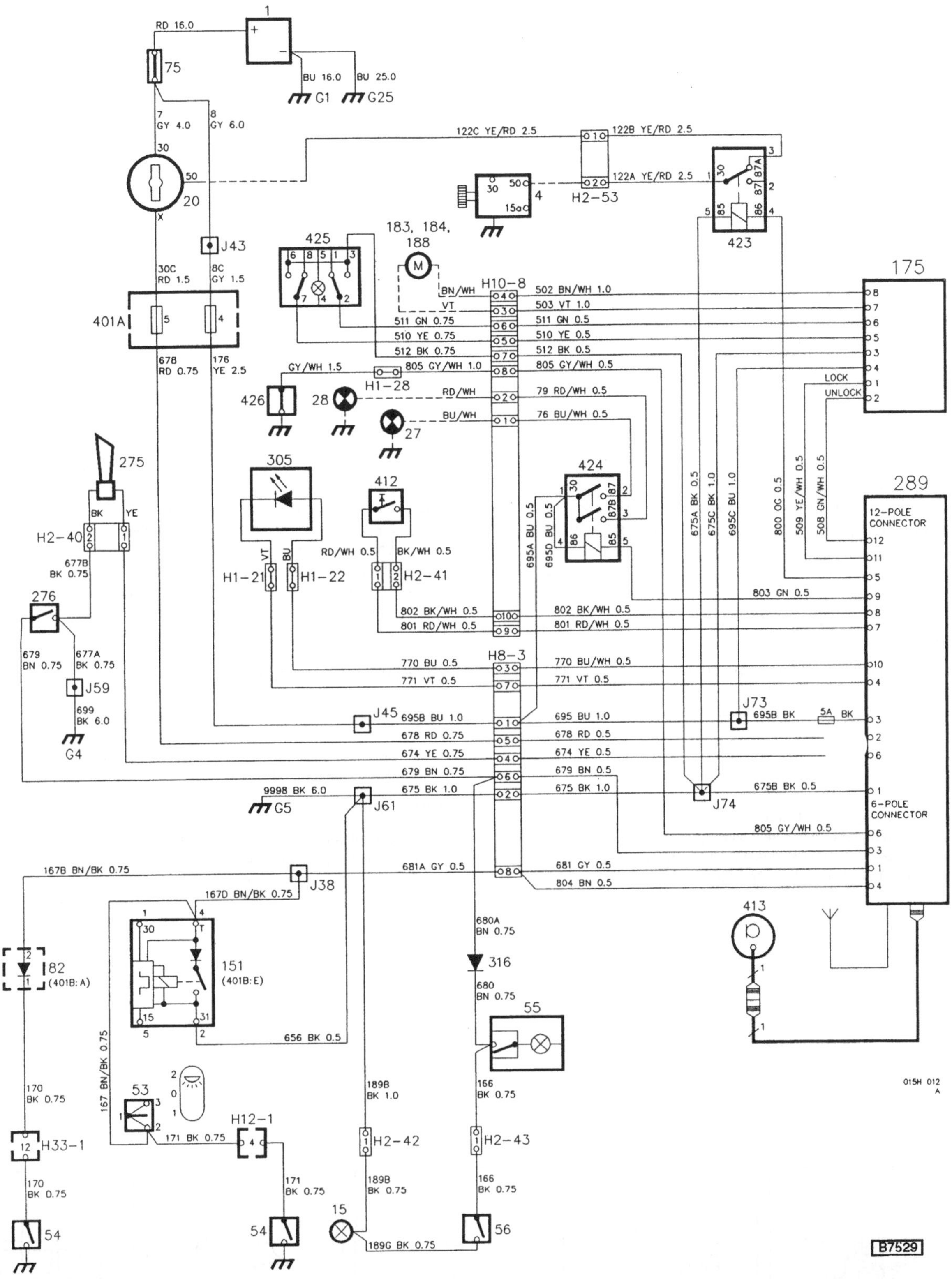

Central Locking

Operation

All of the door locks and the luggage compartment lid are locked or unlocked when the key is turned in the driver's door. The central locking control unit (175) actuates all of the lock motors via pin 7 and pin 8.

See **830 Doors, Windows, and Lids** for more information on removing the lock motors and other door components.

Fault Tracing

1. Check the fuses. Check the voltage supply to control unit (175) pin 4. This pin should always be live.
2. Check that pin 7 of the control unit (175) is live when the door key is in the unlock position. Check that pin 8 of the control unit (175) is live when the key is in the unlock position.

CAUTION
The lock motors should only be tested through pin 7 and pin 8 of the control unit (175). Do not 'bench test' the lock motors using any other power supply sources as this may overload and damage the motors.

NOTE
When testing pin 7 and pin 8, the supply will only be live for approximately 1 second.

Component Locations

22A main fuse/relay panel at left wheel housing
75 distribution block at right wheel housing
175 central locking control unit behind instrument panel right side
183 key switch for the driver's central door locking in driver door
184 central door lock motor in passenger door
185 central door lock motor in right rear door
186 central door lock motor in left rear door
188 central door lock motor in tailgate lock
401 auxiliary fuse/relay panel under rear seat

Connector Locations

1985-1990

57 3-pin connector (2) in driver door and in engine compartment near left door hinge
59 2-pin connector (5) right front door, engine compartment near right door hinge, left rear door, right rear door, luggage compartment near central lock motor
123 4-pin connector (2) one in each B pillar
152A 29-pin bulkhead connector at main fuse/relay panel

1991 and later

H2-4 2-pin connector at engine compartment near right door hinge
H2-8 2-pin connector at engine compartment near right door hinge
H2-20 2-pin connector in luggage compartment, left wheel housing
H2-30 2-pin connector in left rear door
H2-31 2-pin connector in right rear door
H2-32 2-pin connector in luggage compartment near the central lock motor
H3-7 3-pin connector in engine compartment near left door hinge
H3-15 3-pin connector in the driver door
H4-4 4-pin connector at right B pillar behind trim
H4-5 4-pin connector at left B pillar behind trim
H10-1 10-pin connector behind instrument panel, left of steering column
H12-1 12-pin connector behind instrument panel, left of steering column
H29-2 29-pin bulkhead connector behind main fuse/relay panel (1991-1991 1/2)
H33-2 33-pin bulkhead connector behind main fuse/relay panel (1991 1/2 and later)

Ground Locations

1985-1990

65 ground point at the handbrake
117 ground point at the handbrake

1991 and later

G1 ground point at radiator cross member
G8 ground point behind instrument panel
G25 ground point at gearbox

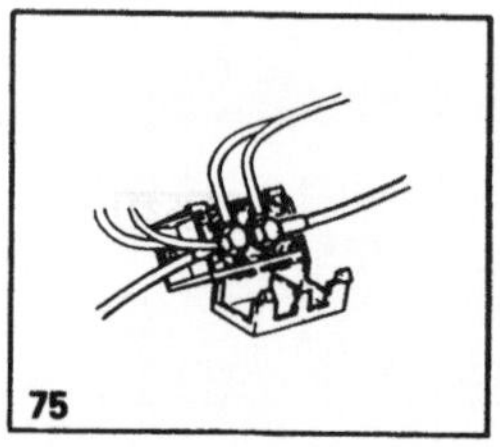

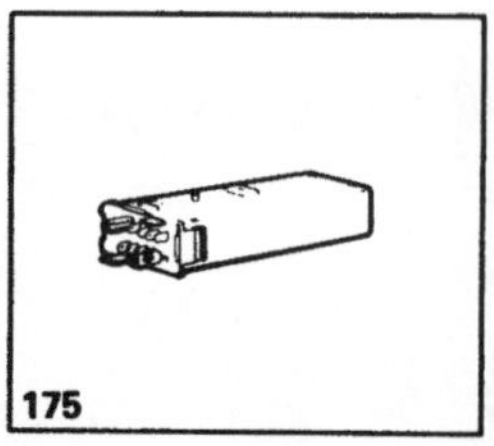

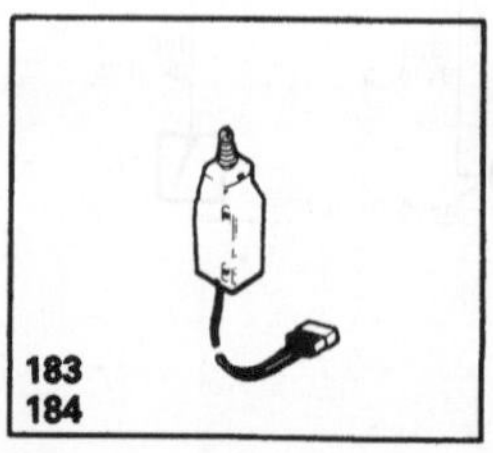

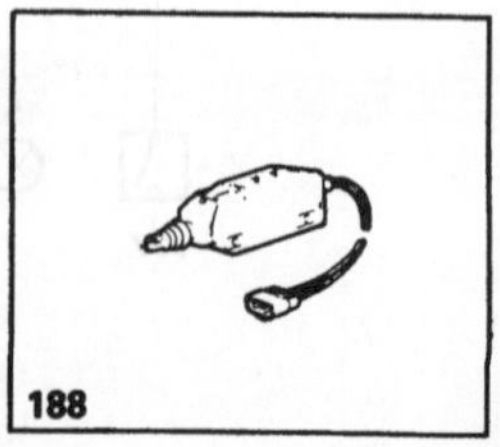

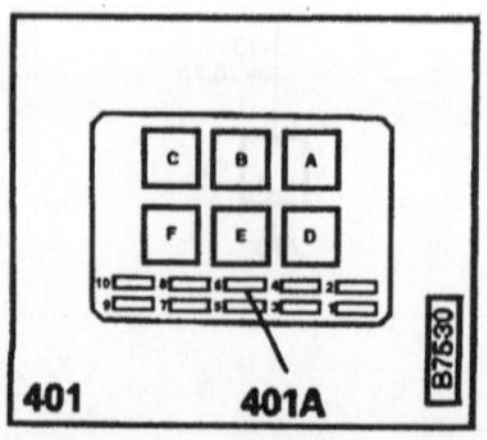

Central Locking

1985-1990

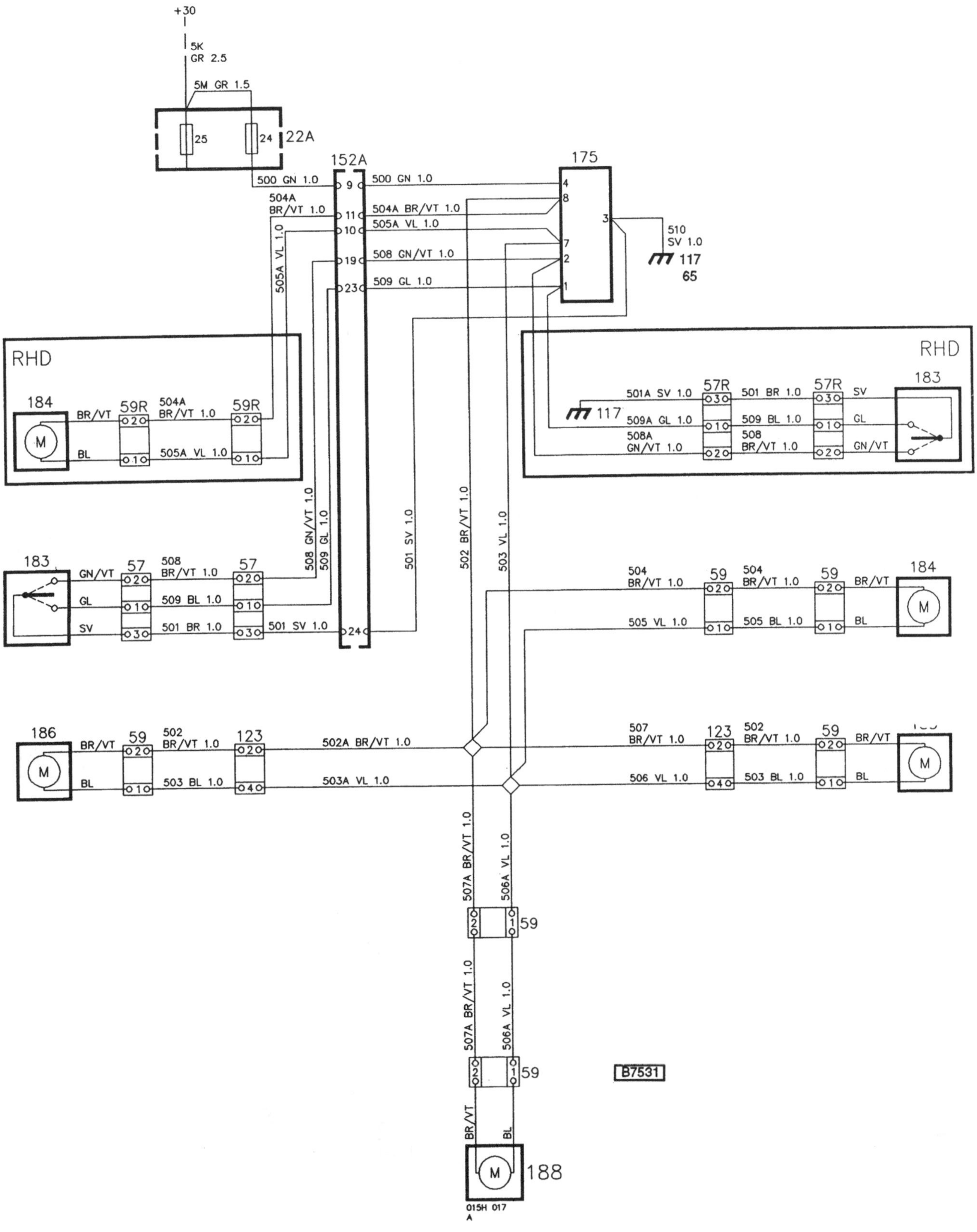

Central Locking

1991 and later (ex. 1992 and later convertible)

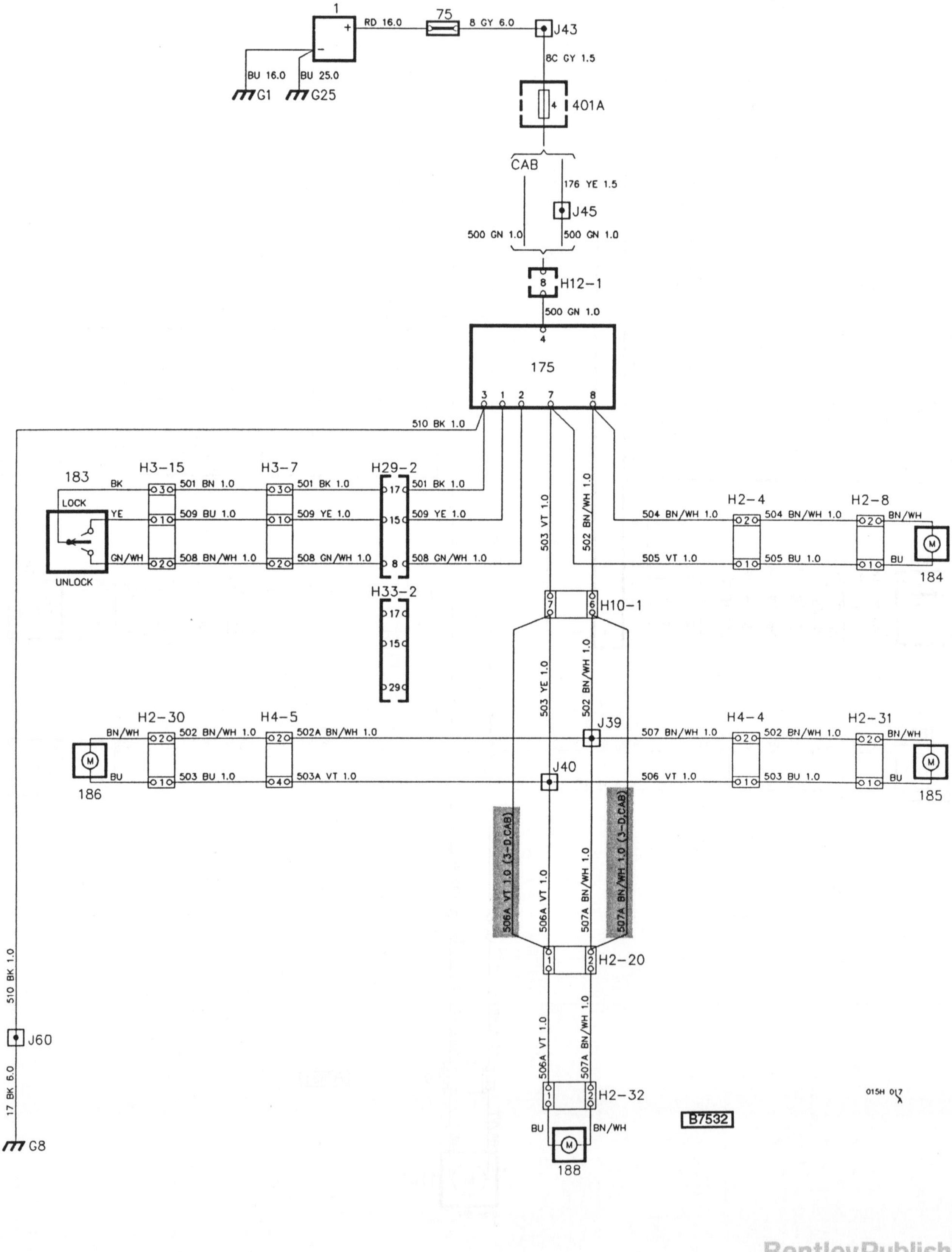

Central Locking

1992 and later Convertible

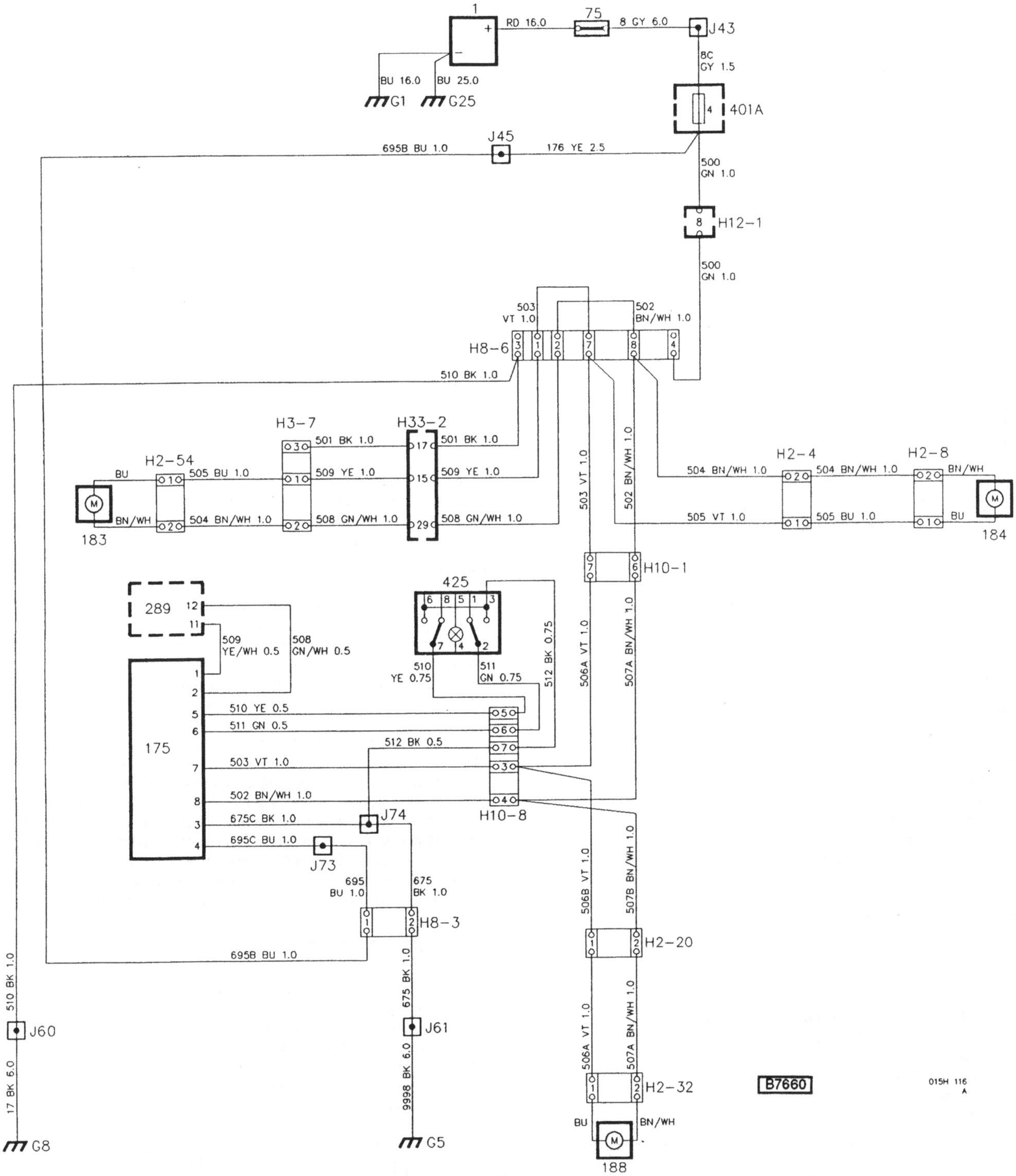

Combined Instruments

Operation

The combined instruments, or instrument cluster, monitors various operating functions. See **853 Dashboard and Consoles** for more information on removing and testing the instrument cluster and instrument panel. Additional test and repair information on sensors related to instrument warnings can be found in **364 Electric Controls and Switches** or in the section pertaining to the warning function such as **5 BRAKES** or **2 ENGINE**.

Fault Tracing

1. Check the fuses.
2. Check for power and ground to the instrument cluster.

Component Locations

2 alternator at back of engine compartment
8 lighting relay
10 lighting switch
16 rheostat for instrument cluster
18 instrument lighting
20 ignition switch
21 ignition switch relay in main fuse/relay panel
22A main fuse/relay panel at left wheel housing
25 hazard warning switch
42 brake fluid level switch in brake fluid reservoir
43 handbrake switch under the handbrake lever
44 oil pressure switch at left side of engine, above oil filter
45 coolant temperature transmitter at front of engine near thermostat housing
46 fuel level transmitter in the fuel tank, accessible in luggage compartment
47 instrument cluster
47A fuel level gauge
47B fuel warning light
47C coolant temperature gauge
47D oil pressure warning light
47E charging lamp
47F brake warning light
47G high beam warning light
47H left-hand direction indicator warning lamp
47I right-hand direction indicator warning lamp
47J rear window heater warning lamp
47K shift-up indicator
47M handbrake warning lamp
47N rear fog light warning lamp
47P CHECK ENGINE light
47Q ABS warning lamp
47S passive seatbelt warning lamp
47T SRS airbag warning lamp (see SRS warnings)
47U cruise control warning lamp
49 clock
75 distribution block at right wheel housing
110 tachometer
116 rear heated window switch
132 speed transmitter at the rear of the instrument cluster
141 selector for cruise control at direction indicator stalk
146 ignition amplifier in front of left wheel housing
148 ashtray lighting
153 cigarette lighter lamp
154 heater control lighting
158 negative distribution in main fuse/relay panel
159 +15 distribution in main fuse/relay panel
161 rear fog light switch
176 EZK control unit at inner left fender
200 LH control unit forward of the right front door, behind trim
291 ABS control unit (1989) in engine compartment, left wheel housing
291 ABS control unit (1990 and later) under rear seat, right side
306 logic box for passive seatbelts under rear seat, left side
331 SRS airbag control unit behind instrument panel, left grille.

Connector Locations

1985-1990

57 3-pin connector (2) Under spare tire cover, under rear seat, left side
58 12-pin connector behind instrument panel to left of steering column
59 2-pin connector at main fuse/relay panel, convertibles
60 1-pin connector under center console
98 10-pin connector behind instrument panel to left of steering column
123 4-pin connector at hatchback, left side
152A 29-pin bulkhead connector at main fuse/relay panel
152B 29-pin bulkhead connector at main fuse/relay panel
152C 29-pin bulkhead connector at main fuse/relay panel
267 radio connector in instrument panel

1991 and later

H2-3 2-pin connector behind the instrument cluster
H2-6 2-pin connector behind instrument cluster,left of steering column
H2-7 2-pin connector in instrument panel at light switch
H2-10 2-pin connector behind cigarette lighter
H2-27 2-pin connector at main fuse/relay panel
H3-4 3-pin connector at main fuse/relay panel
H4-6 4-pin connector in instrument panel left side
H4-7 4-pin connector at the LH control unit, forward of right front door
H4-15 4-pin connector under rear seat, next to fuse/relay panel 401
H10-2 10-pin connector at the SRS airbag control unit, left speaker grille
H29-1 29-pin bulkhead connector at main fuse/relay panel (1991–1991 1/2)
H29-2 29-pin bulkhead connector at main fuse/relay panel (1991–1991 1/2)
H29-3 29-pin bulkhead connector at main fuse/relay panel (1991–1991 1/2)
H33-1 33-pin bulkhead connector at main fuse/relay panel (1991 1/2 –on)
H33-2 33-pin bulkhead connector at main fuse/relay panel (1991 1/2 –on)
H33-3 33-pin bulkhead connector at main fuse/relay panel (1991 1/2 –on)

Ground Locations

3/G8 ground point behind instrument panel
7/G1 ground point at radiator cross member
9/G3 ground point in luggage compartment
G6 ground distribution at main fuse/relay panel
117 ground point under the handbrake
211/G25 ground point at the gearbox
257/G32 ground point at alternator bracket

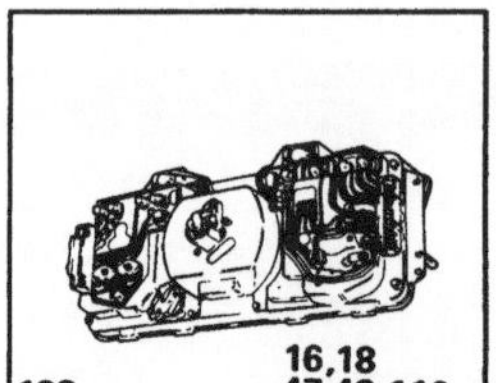
132 16,18 47,49,110

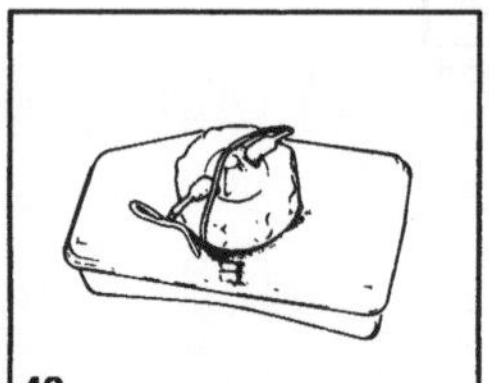
42

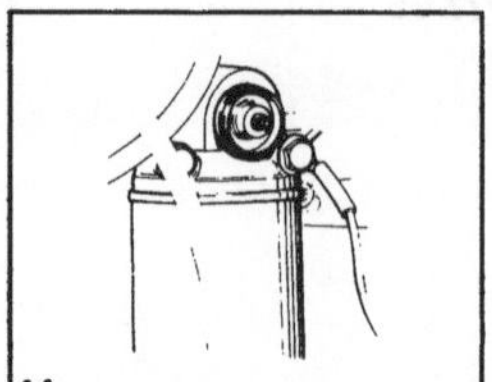
44

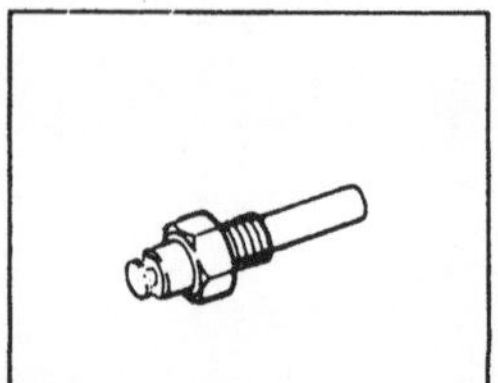
45

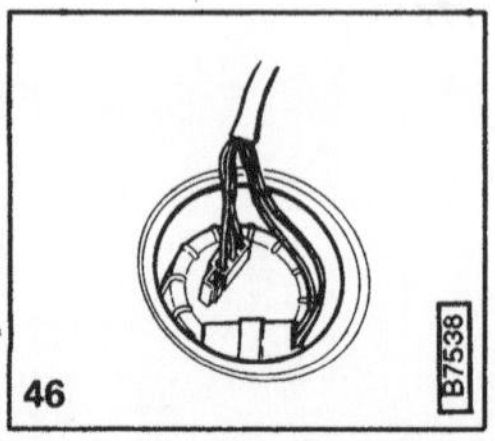

46

Combined Instruments

1985-1987

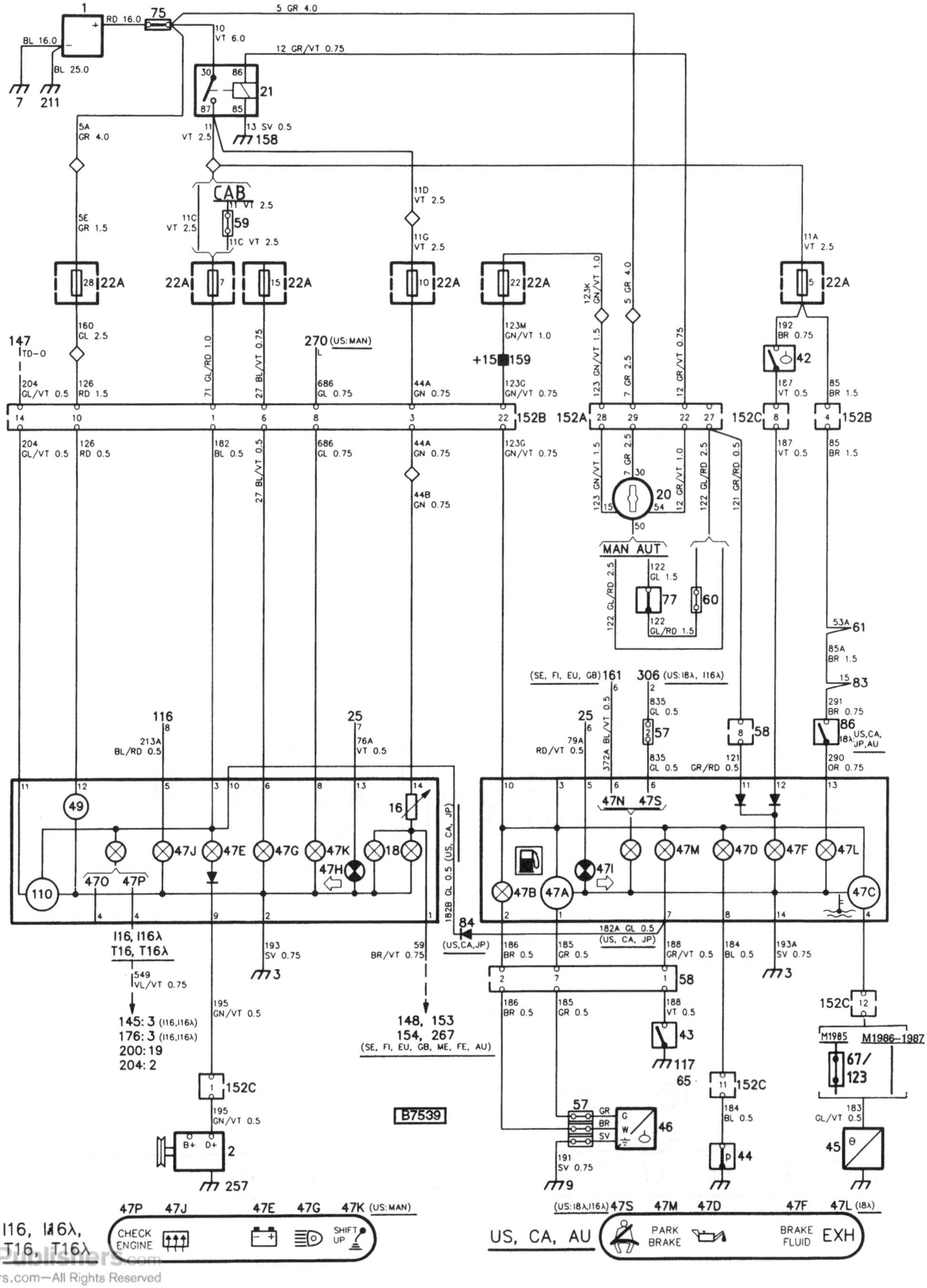

Combined Instruments

1988

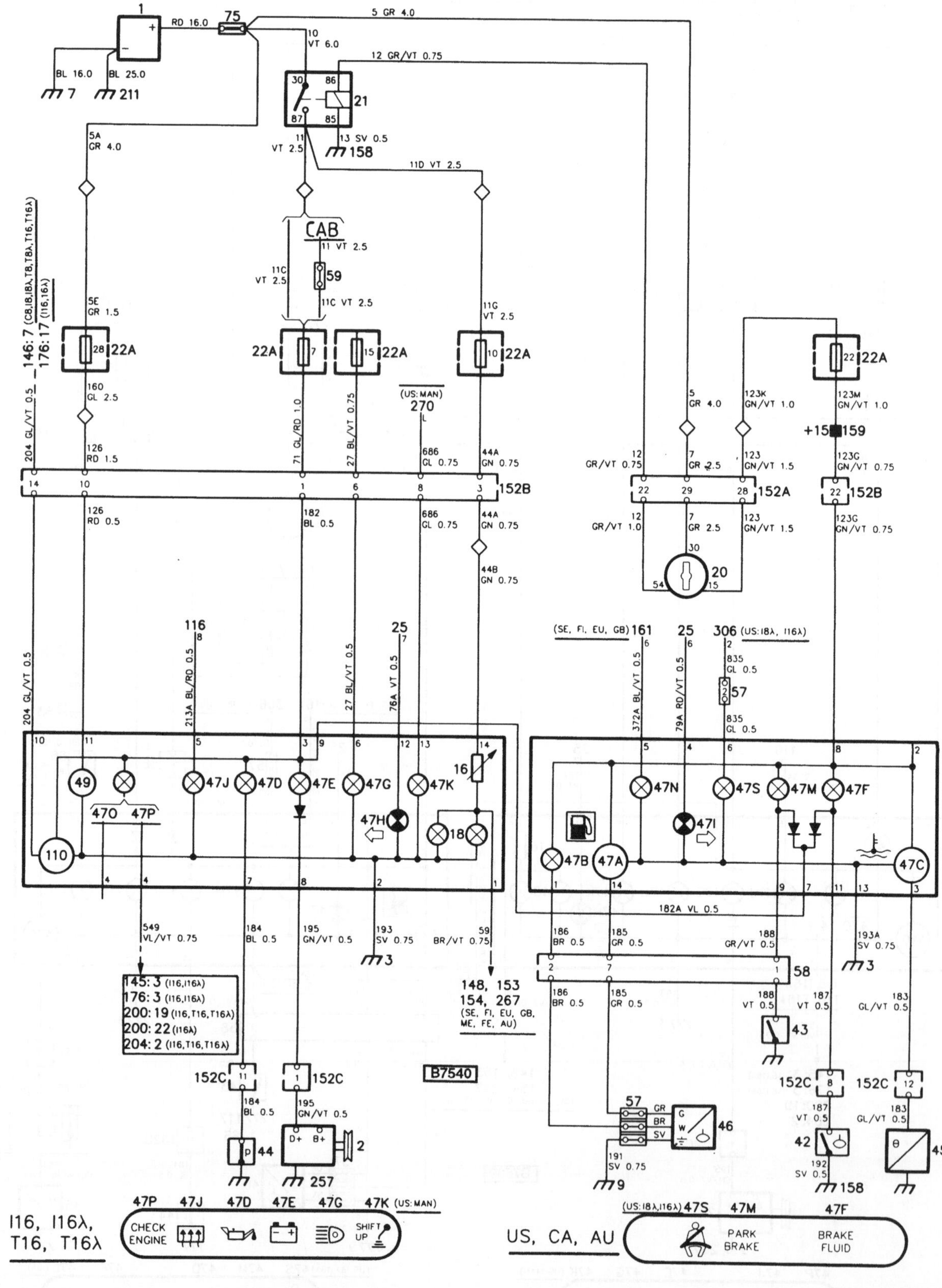

Combined Instruments

1989-1990

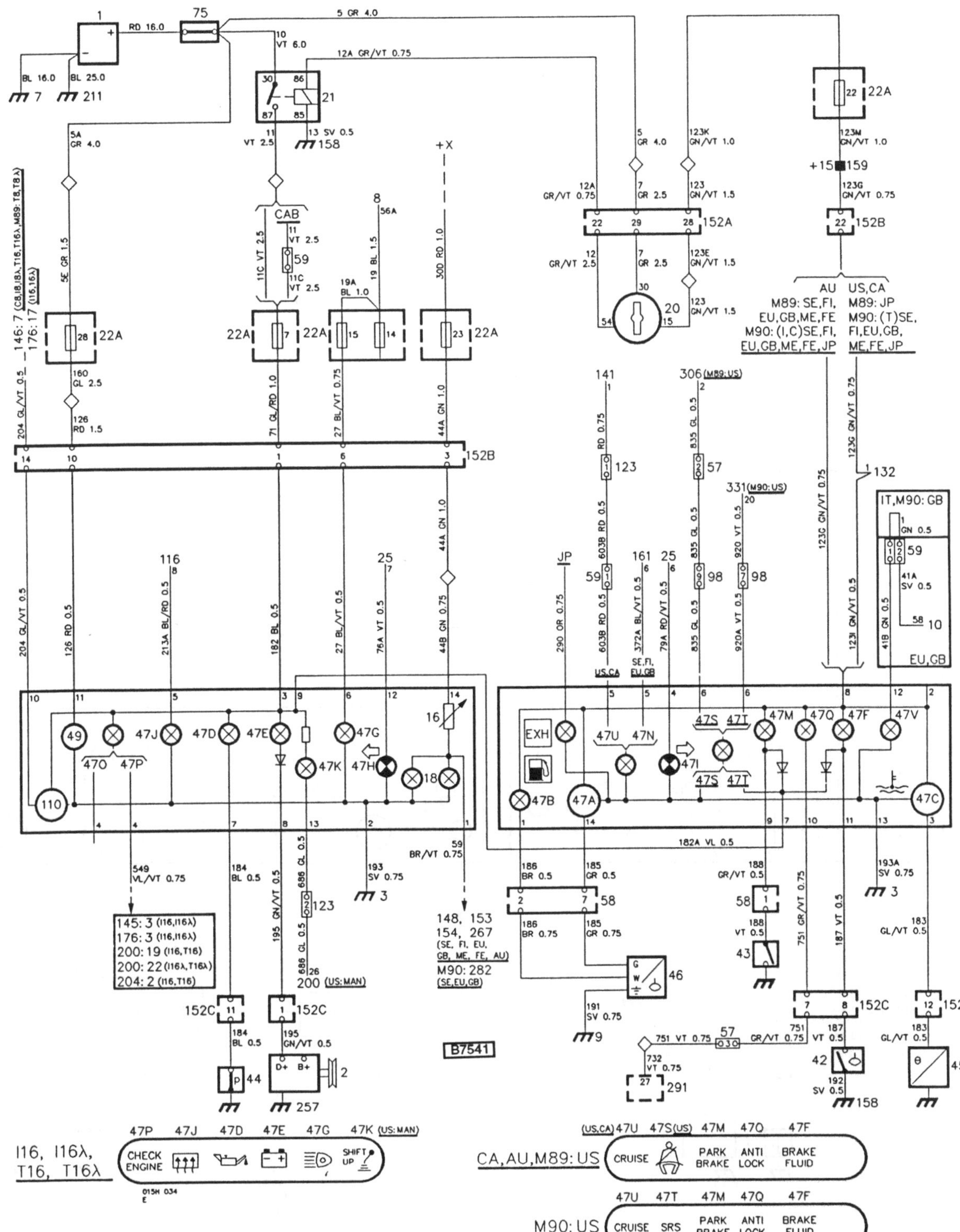

Combined Instruments

1991 and later

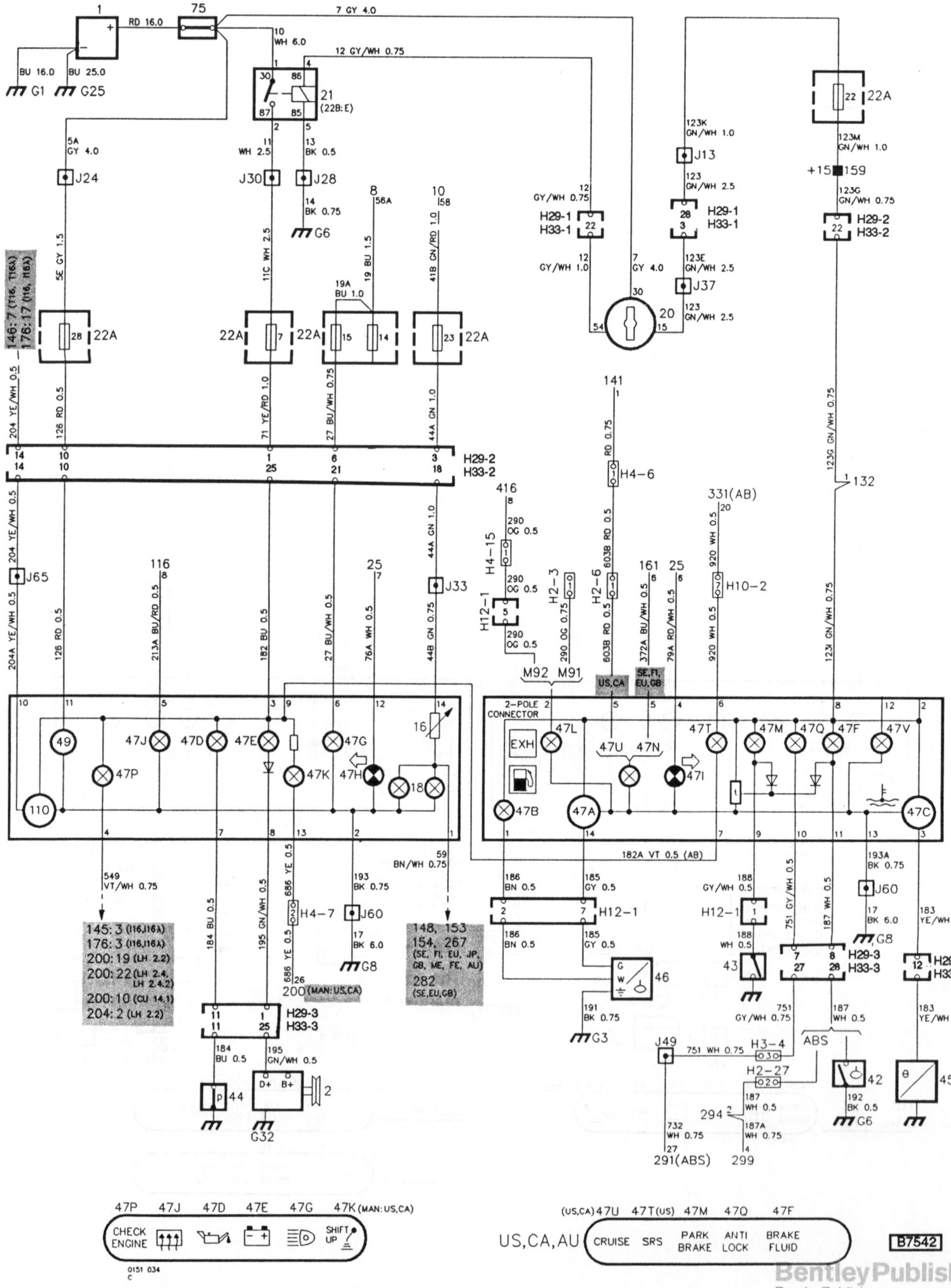

Convertible Top

Operation

The convertible top is raised and lowered through two hydraulic cylinders actuated by an electric pump. The handbrake must be applied before the top can be operated. When lowering the top, relay (278) is energized and current flows to the motor. When raising the top, relay (277) is energized and current flows in the opposite direction to the motor. See **812 Convertible Top** for repair information on the convertible top.

Fault Tracing

1. Check the fuses. Check the voltage supply to the main switch.
2. Check the operation of both relays.
3. Check for voltage at the electric pumps when the switch is actuated. Check that the voltage polarity changes when the switch is toggled the opposite way.

Component Locations

1 battery at right front of engine compartment
22A main fuse/relay panel in left wheel housing
43 handbrake switch under the handbrake
75 distribution block at right wheel housing
181 power sunroof or convertible switch in center console
273 motor for the convertible hydraulic pump under rear seat, right side
277 relay for raising the convertible top (1987-1988) under the rear seat on the bulkhead
277 relay for raising the convertible top (1989 on) under the rear seat in fuse/relay panel 401
278 relay for lowering the convertible top (1987-1988) under the rear seat on the bulkhead
278 relay for lowering the convertible top (1989 on) under the rear seat in fuse/relay panel 401
283 diodes for convertible top (1987-1988) under rear seat in the wiring harness
290 fuse for the convertible top in engine compartment, next to distribution block 75
290A fuse for the convertible top in engine compartment, next to distribution block 75
290B fuse for the convertible top in engine compartment, next to distribution block 75
401 auxiliary fuse/relay panel under rear seat

Connector Locations

1987-1990

58 12-pin connector behind instrument panel to left of steering column
60 1-pin connector (2) under the center console
98 10-pin connector behind instrument panel to left of steering column
152B 29-pin bulkhead connector at main fuse/relay panel

1991 and later

H1-17 1-pin connector under center console
H1-27 1-pin connector in engine compartment, next to distribution block 75
H1-27 1-pin connector in engine compartment, next to distribution block 75
H2-48 2-pin connector at the convertible top motor
H12-1 12-pin connector behind instrument panel, left of steering column
H29-2 29-pin bulkhead connector behind main fuse/relay panel (1991-1991 1/2)
H33-2 33-pin bulkhead connector behind main fuse/relay panel (1991 1/2 and later)

Ground Locations

1987-1990

117 ground point at the handbrake
211 ground point at the gearbox

1991 and later

G1 ground point at radiator cross member
G4 ground point between ignition switch and handbrake
G5 ground point under the rear seat
G25 ground point at the gearbox

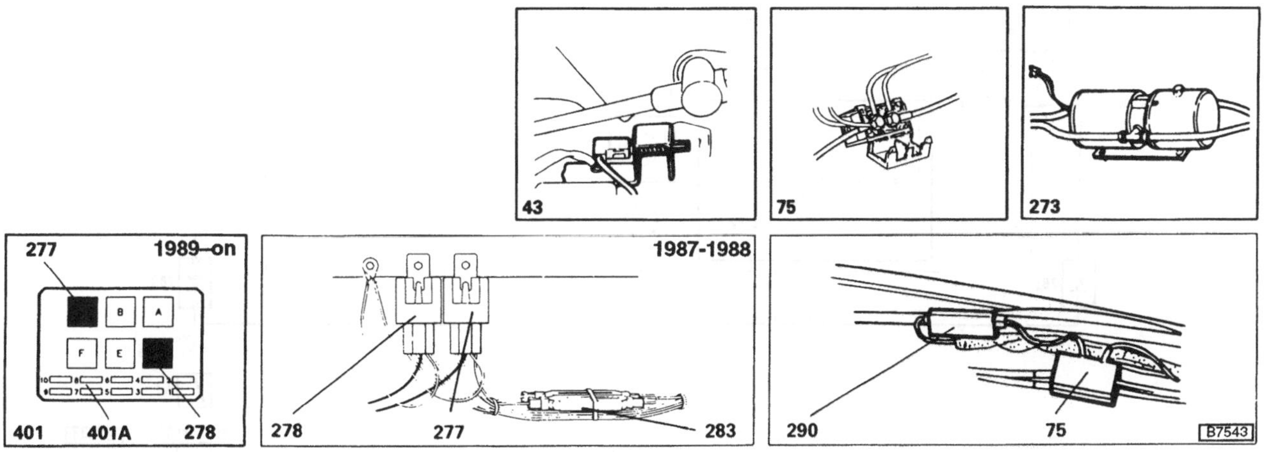

Convertible Top

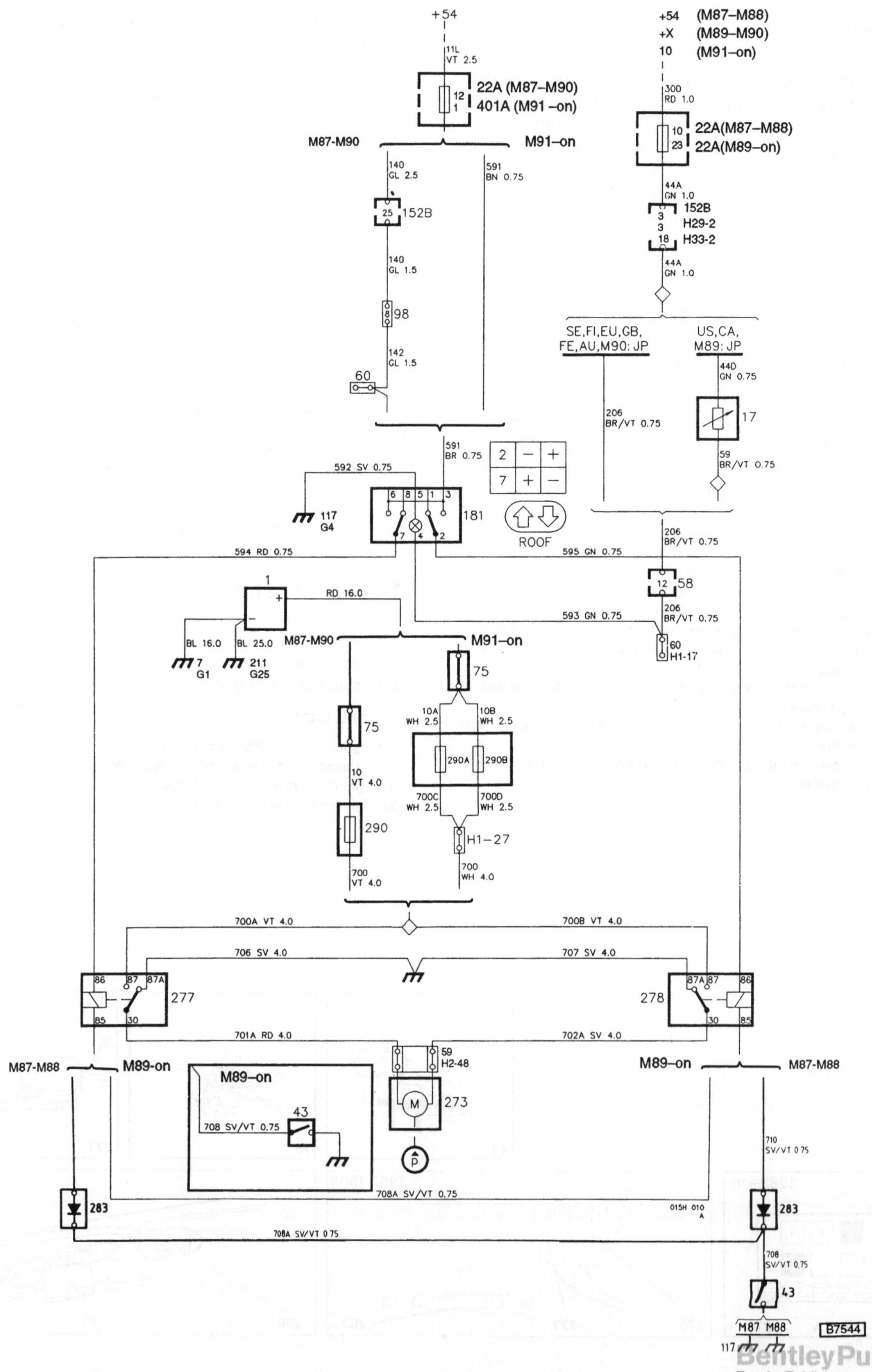

Cruise Control

Operation

Cruise control is an electro-mechanical system that uses a vacuum pump and vacuum actuator to control the accelerator linkage. The speed of the car is electronically sensed by a speedometer-mounted sensor. A clutch switch (cars with manual transmission) and a brake switch disable the system if either pedal is depressed. Disabling the system, via these two switches, is done electrically by opening the circuit to ground, and pneumatically by releasing the vacuum to the actuator. See **368 Cruise Control** for additional information, including repair information.

Fault Tracing

1. Check the fuses.
2. Check the voltage supply to selector switch (141) and cruise control unit (131) with the ignition key in the run position.
3. On 1989 and later cars, check for voltage to the speed transmitter (132) at the back of the instrument cluster.

WARNING
After completing cruise control electrical tests, the cruise control system should be road tested. Particular attention should be paid to the brake and clutch canceling functions. Be sure any faults found are repaired before driving the car.

Component Locations

22A main fuse/relay panel at left wheel housing
29 brake light switch at brake pedal cluster
31 backup light switch in gear selector housing
47U cruise control warning lamp
76 switch for raising idle speed (automatic transmission) under center console at shift lever
128 speed control relay behind instrument panel, left side
131 control unit for cruise control behind instrument panel, left side
132 speed transmitter at the rear of the instrument cluster
133 clutch switch for cruise control at the clutch pedal
134 brake switch for cruise control at the brake pedal
141 selector for cruise control at direction indicator stalk
159 +15 distribution in main fuse/relay panel
177 APC control unit at left wheel housing
179 APC solenoid valve above radiator fan on radiator
187 cruise control vacuum pump in engine compartment left wheel housing
233 cruise control vacuum switch in engine compartment left wheel housing

Connector Locations

1985-1990 models

59 2-pin connector behind instrument panel, left side
60 1-pin connector (2) left side near bulkhead connectors
98 10-pin connector behind instrument panel to the left of the steering column
123 4-pin connector behind instrument panel, left side
152A 29-pin bulkhead connector at main fuse/relay panel
152B 29-pin bulkhead connector at main fuse/relay panel
152C 29-pin bulkhead connector at main fuse/relay panel

1991 and later

H1-2 1-pin connector next to bulkhead connectors, left side behind instrument panel
H1-5 1-pin connector behind instrument panel to the left of the steering column
H1-8 1-pin connector behind instrument panel to the left of the steering column
H1-16 1-pin connector under center console next to shift selector (automatics)
H1-16 1-pin connector under center console next to shift selector (automatic)
H2-6 2-pin connector behind instrument cluster, left of steering column
H4-6 4-pin connector at instrument panel left side
H4-16 4-pin connector behind instrument panel to the left of the steering column
H10-1 10-pin connector behind instrument panel, left of steering column
H29-1 29-pin bulkhead connector behind main fuse/relay panel (1991-1991 1/2)
H29-2 29-pin bulkhead connector behind main fuse/relay panel (1991-1991 1/2)
H29-3 29-pin bulkhead connector behind main fuse/relay panel (1991-1991 1/2)
H33-1 33-pin bulkhead connector behind main fuse/relay panel (1991 1/2 –on)
H33-2 33-pin bulkhead connector behind main fuse/relay panel (1991 1/2 –on)
H33-3 33-pin bulkhead connector behind main fuse/relay panel (1991 1/2 –on)

Ground Locations

1985-1990 models

3 ground point behind instrument panel
7 ground point at radiator cross member

1991 and later models

G1 ground point at radiator cross member
G3 ground point in luggage compartment
G8 ground point behind instrument panell;

29

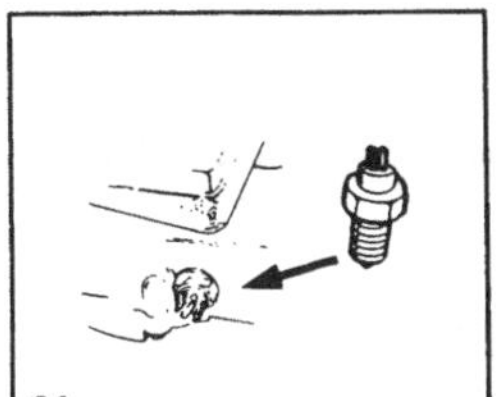
31

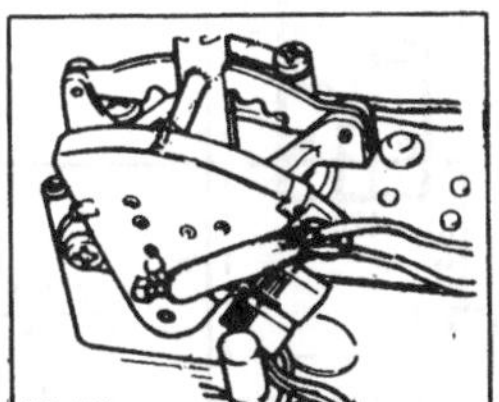
31,76

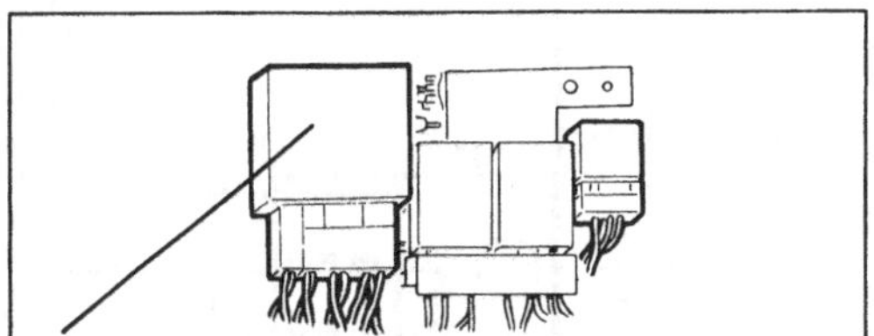
131

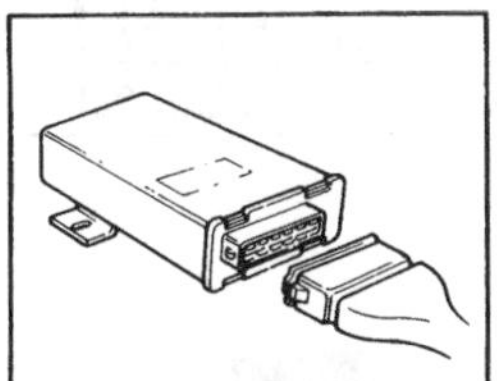
131 (M1992 and later)

133
134

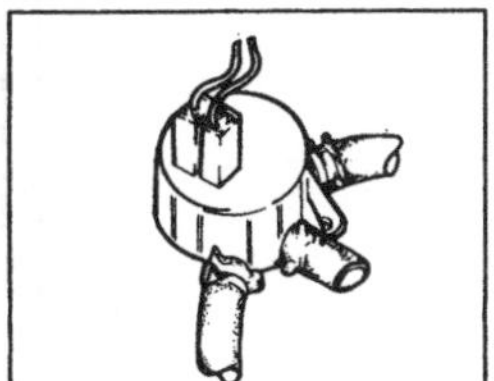
179

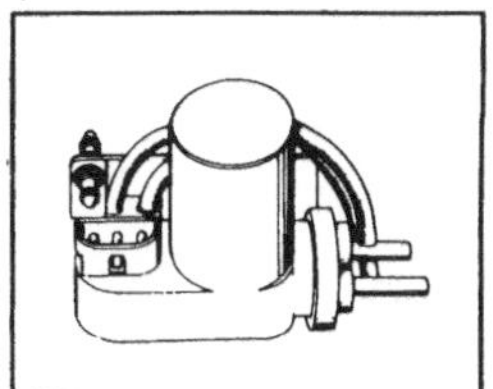
187

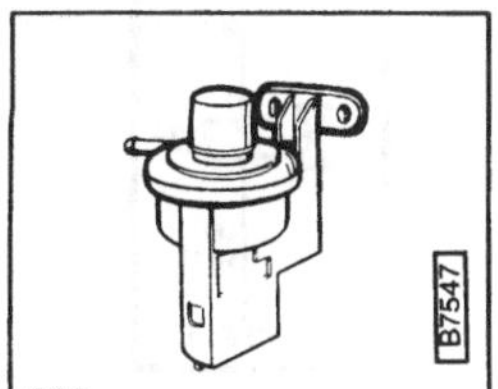

233

Cruise Control

1985-1988

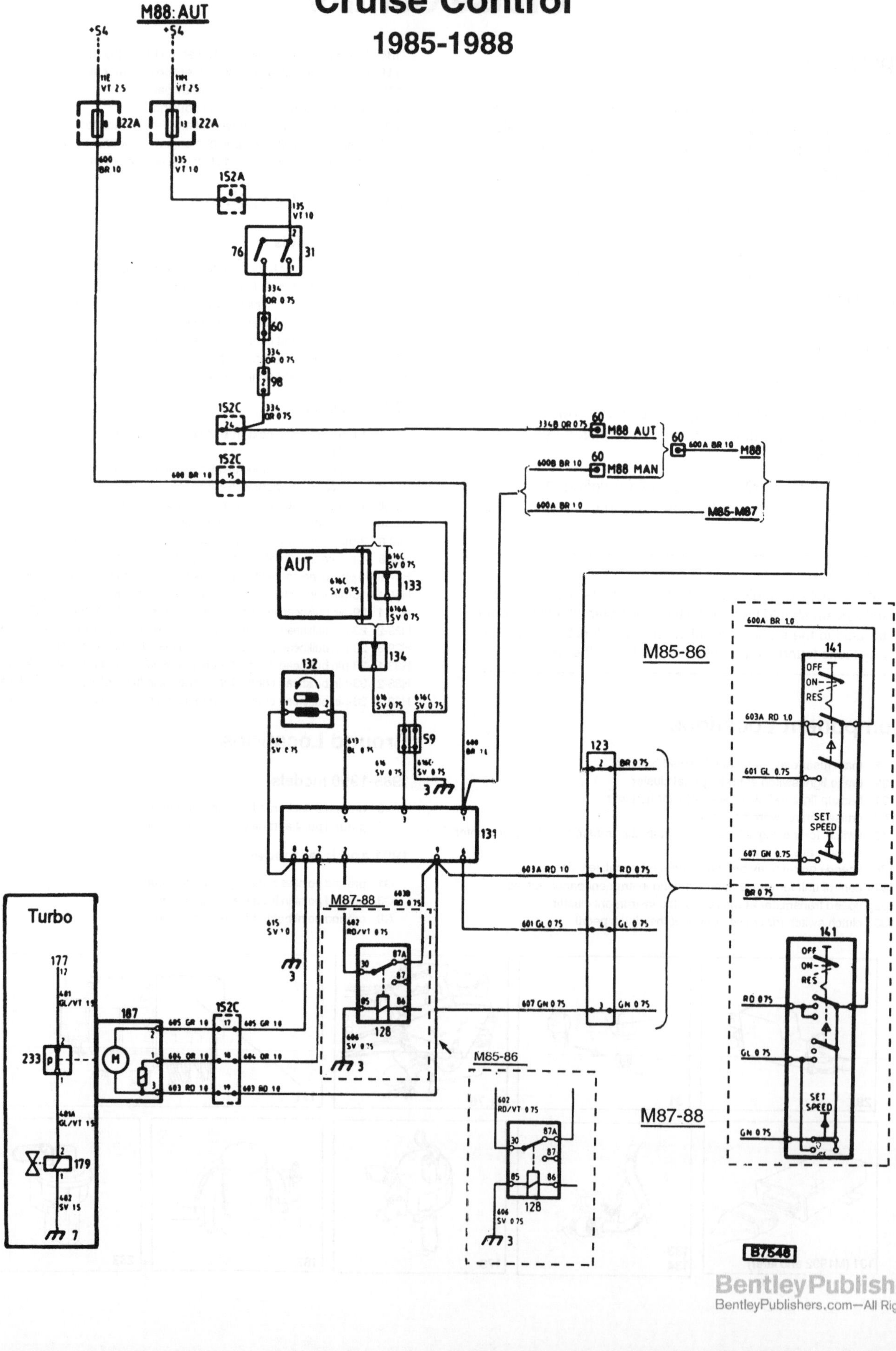

Cruise Control

1989-1991

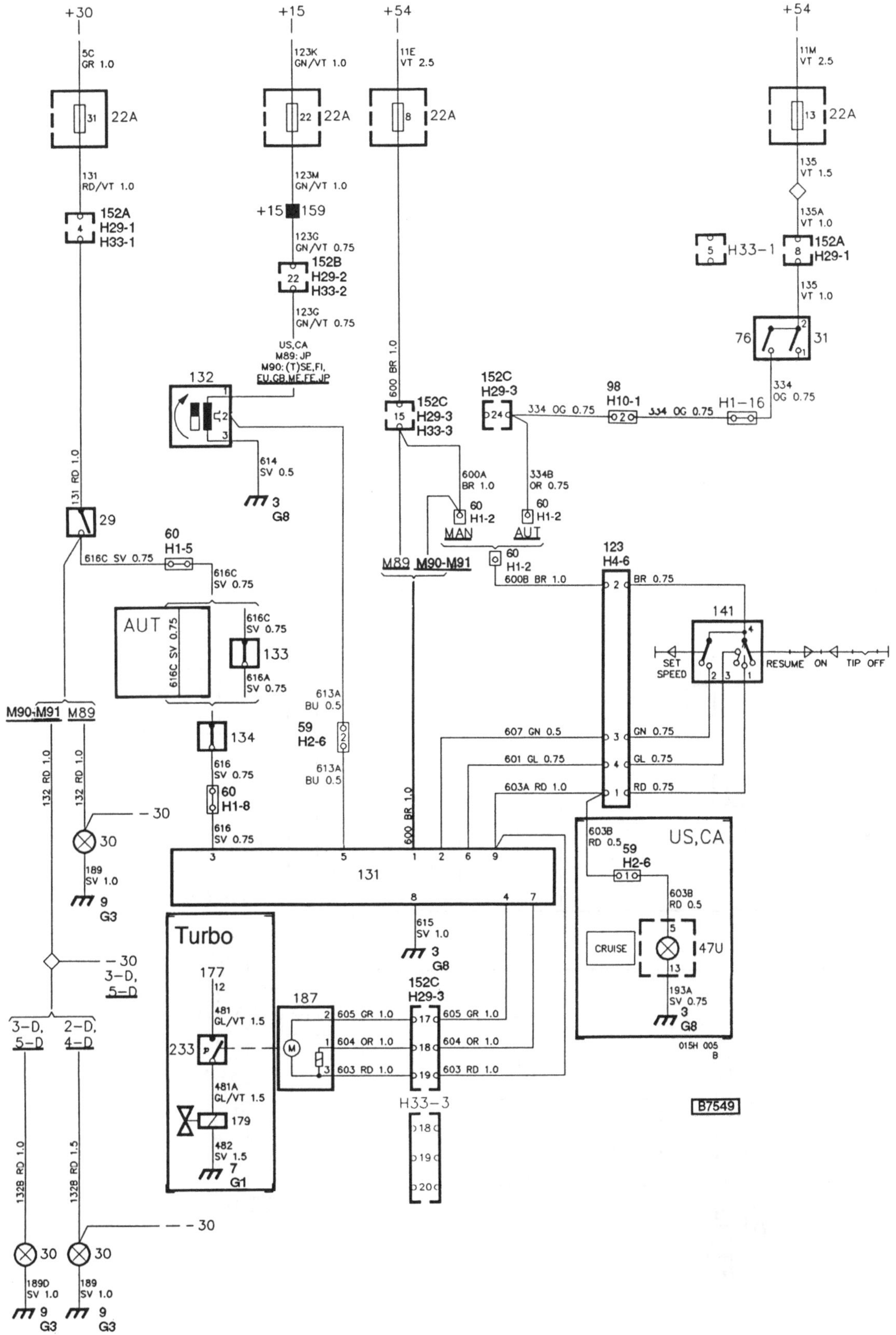

Cruise Control

1992 and later

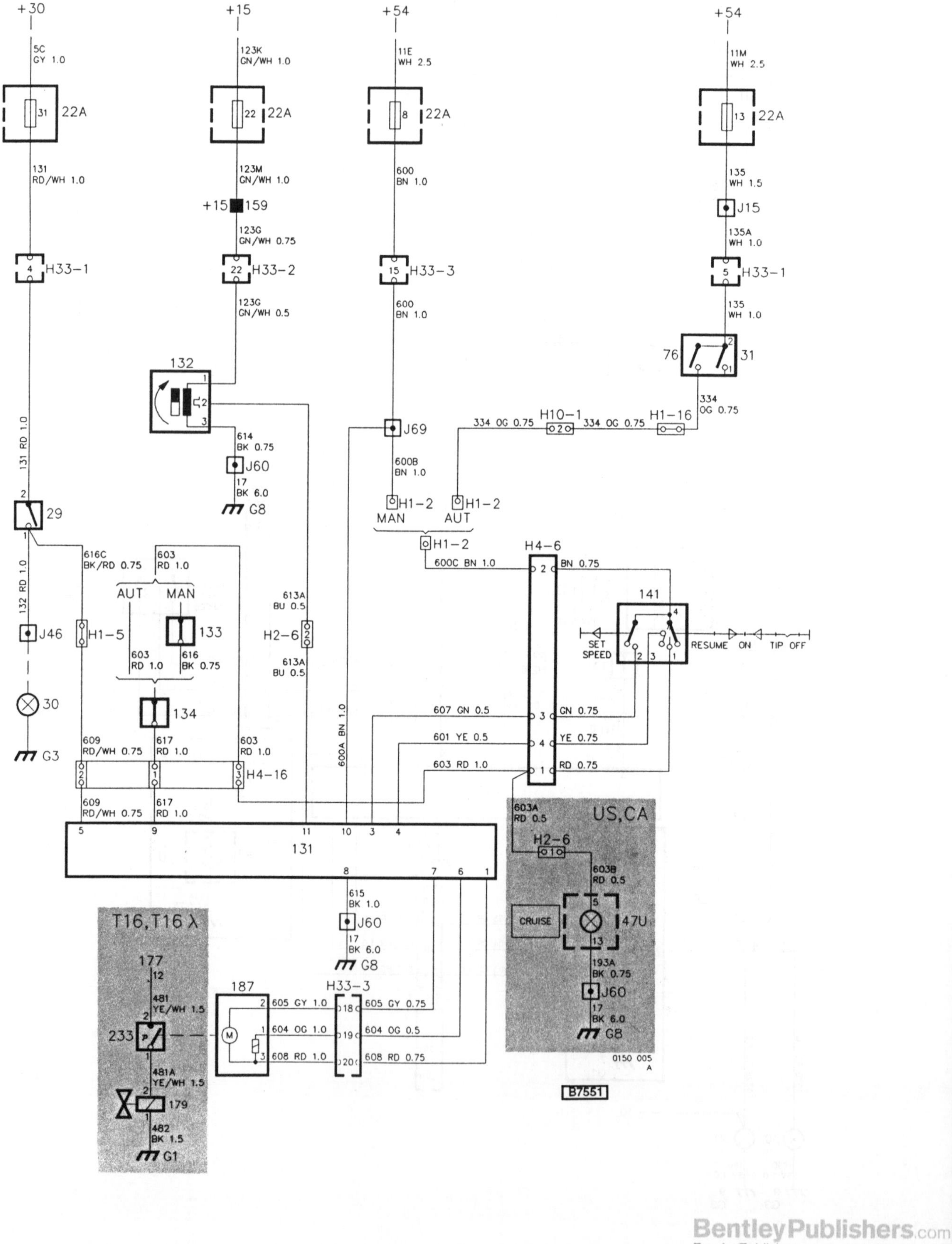

Direction Indicators and Hazard Warning Lights

Operation

The direction indicator stalk (24) switch connects either the left or right direction lights to the flasher relay (23). This creates a load at terminal 49a of the flasher relay (23). With the load connected, the flasher relay (23) begins cycling on and off. The direction indicator stalk (24) also supplies constant voltage to the cornering light (118) in the front cluster when signaling left or right. For information on changing bulbs see **351 Lighting**.

NOTE

- If the direction indicators on one side operate correctly, the flasher relay (23) is OK and can be ruled out as a problem.
- The direction indicators will operate erratically if a bulb is burned out or the wrong type of bulb is installed.

Fault Tracing

1. Check for voltage to the direction indicator switch terminal 1 and the flasher relay terminal 49 whenever the key is in the run position.
2. Check for ground at the flasher relay (23), terminal 31.

Component Locations

22A main fuse/relay panel
23 flasher relay under instrument panel, left side
24 direction indicator switch on steering column
25 hazard warning switch in instrument panel
27 direction indicator lights at left clusters
28 direction indicator lights at right clusters
47H left direction indicator in instrument cluster
47I right direction indicator in instrument cluster
89 side direction indicator in left fender (1986 and later)
90 side direction indicator in right fender (1986 and later)
118 cornering lights in the front clusters
158 negative distribution in main fuse/relay panel

Connector Locations

1985-1990

58 12-pin connector behind the instrument panel to left of steering column
152B 29-pin bulkhead connector at main fuse/relay panel
152C 29-pin bulkhead connector at main fuse/relay panel

1991 and later

H12-1 12-pin behind instrument panel, left of steering column
H29-2 29-pin bulkhead connector behind main fuse/relay panel (1991-1991 1/2)
H29-3 29-pin bulkhead connector behind main fuse/relay panel (1991-1991 1/2)
H33-2 33-pin bulkhead connector behind main fuse/relay panel (1991 1/2 and later)
H33-3 33-pin bulkhead connector behind main fuse/relay panel (1991 1/2 and later)

Ground Locations

1985-1990

3 ground point behind instrument panel
7 ground point at radiator cross member
9 ground point in luggage compartment
93 ground point at left wheel housing

1991 and later

G1 ground point at radiator cross member
G3 ground point in luggage compartment
G6 ground distribution at main fuse/relay panel
G8 ground point behind instrument panel
G12 ground point at left wheel housing

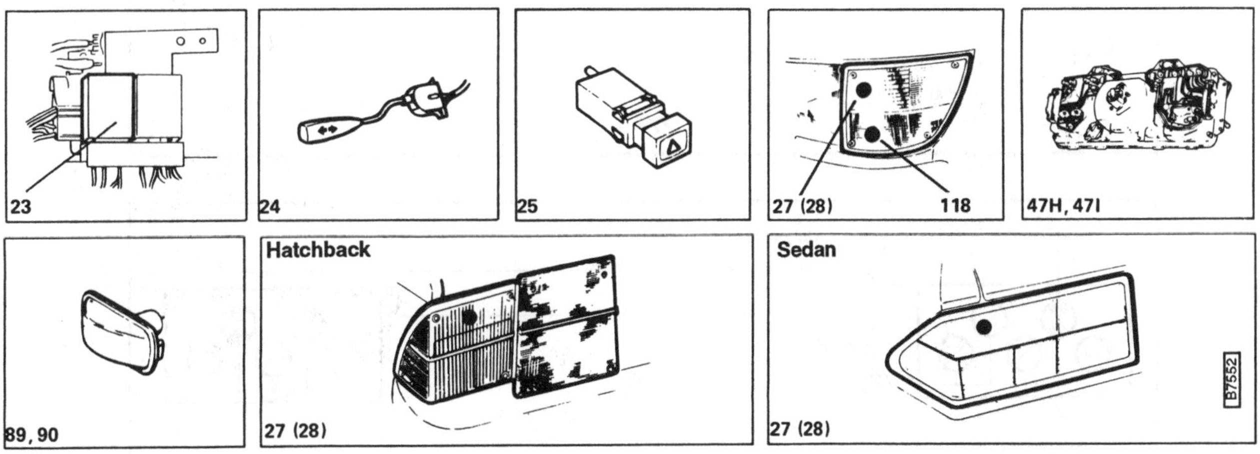

Direction Indicators and Hazard Warning Lights

1985-1990

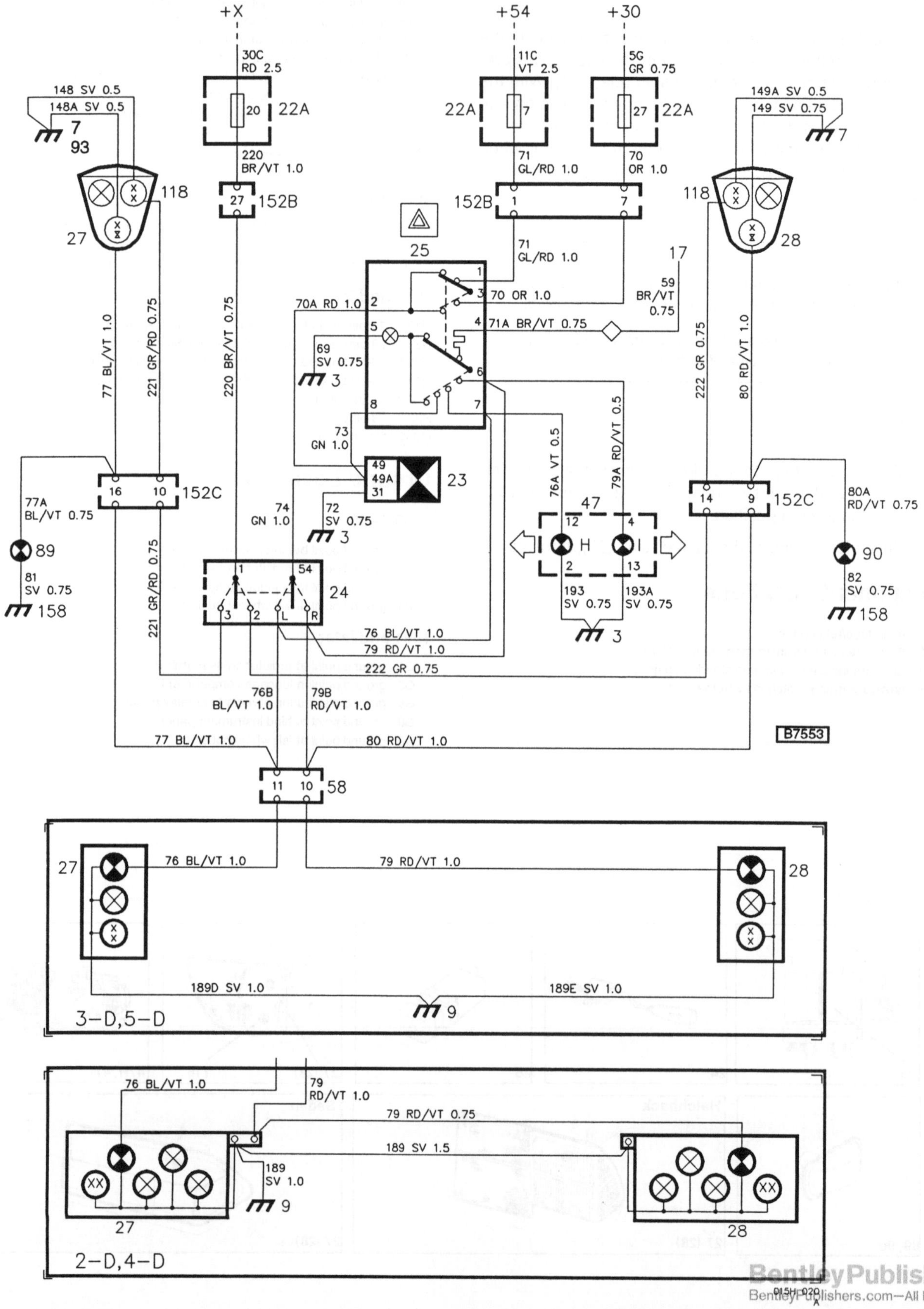

Direction Indicators and Hazard Warning Lights

1991 and later

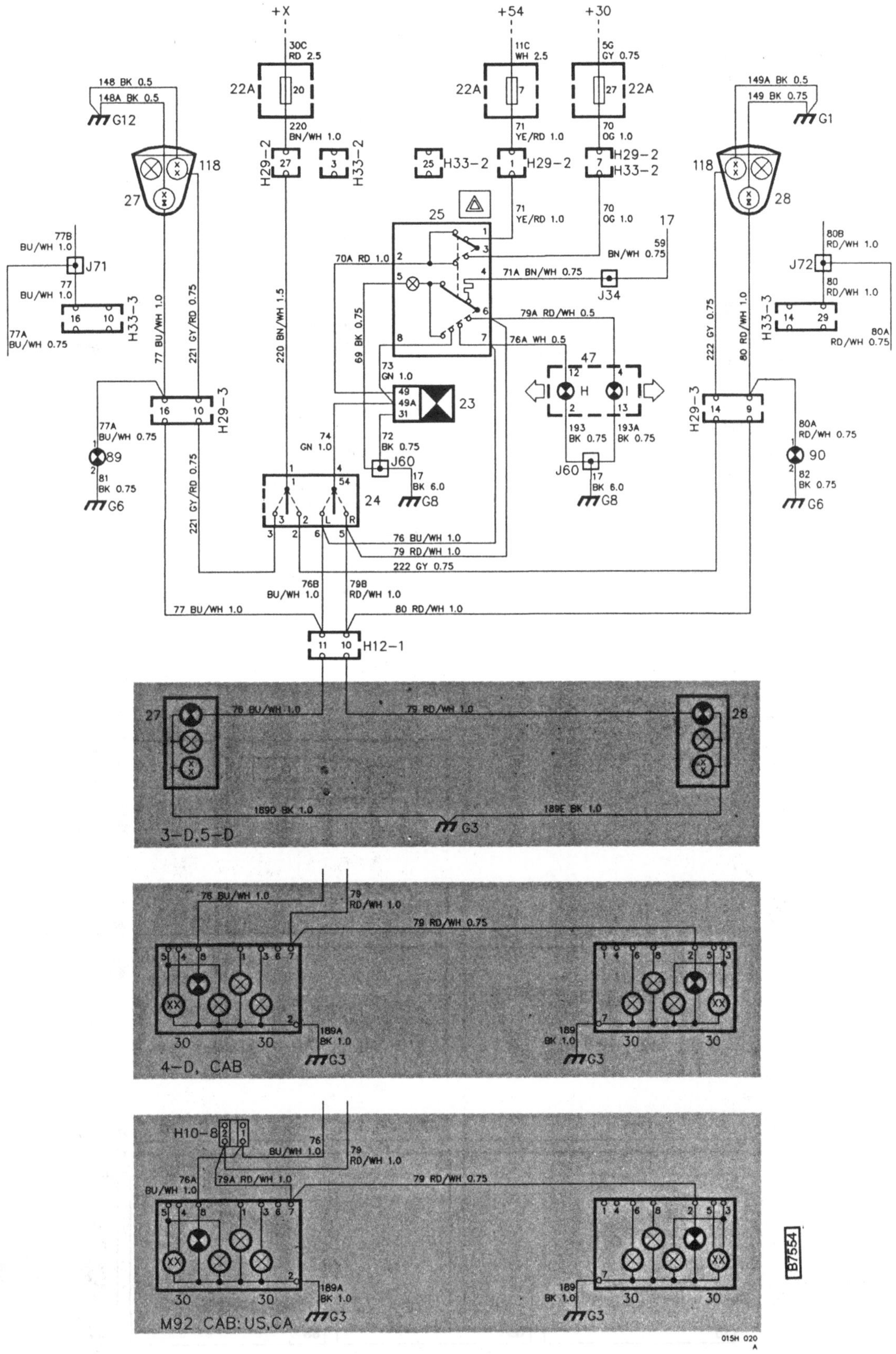

Foglamps

Operation

Foglamps come on only when the park or low beam headlights are on. Voltage to control the foglamp relay (107) is supplied by the foglamp switch (88), and ground is supplied by the high beam headlight filaments. When the high beams are switched on, the foglamp relay (107) is de-energized because there is voltage at both terminals 85 and 86.

Fault Tracing

1. Check fuse 21. Check that the highbeam headlights are not blown.
2. Turn the headlight switch (10) to the park position and turn the foglamp switch (88) on. Check for ground at terminal 85 of the fog lamp relay (107), and check for voltage at terminal 86.

Component Locations

10 lighting switch in instrument panel
22A main fuse/relay panel at left wheel housing
85 foglamps under front bumper
88 switch for foglamps in instrument panel
107 relay for foglamps in main fuse/relay panel
174 relay for daytime driving lights

Connector Locations

1985-1990

152B 29-pin bulkhead connector at main fuse/relay panel
152C 29-pin bulkhead connector at main fuse/relay panel

1991 and later

H29-2 29-pin bulkhead connector behind main fuse/relay panel (1991-1991 1/2)
H29-3 29-pin bulkhead connector behind main fuse/relay panel (1991-1991 1/2)
H33-2 33-pin bulkhead connector behind main fuse/relay panel (1991 1/2 and later)
H33-3 33-pin bulkhead connector behind main fuse/relay panel (1991 1/2 and later)

Ground Locations

1985-1990

3 ground point behind instrument panel
7 ground point at radiator cross member

1991 and later

G1 ground point at radiator cross member
G8 ground point behind instrument panel

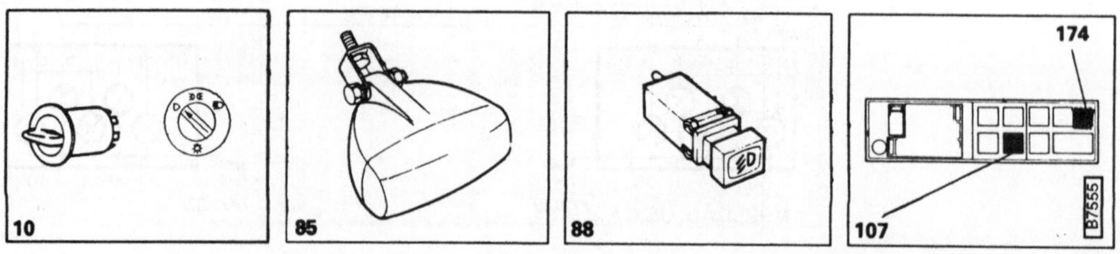

Foglamps

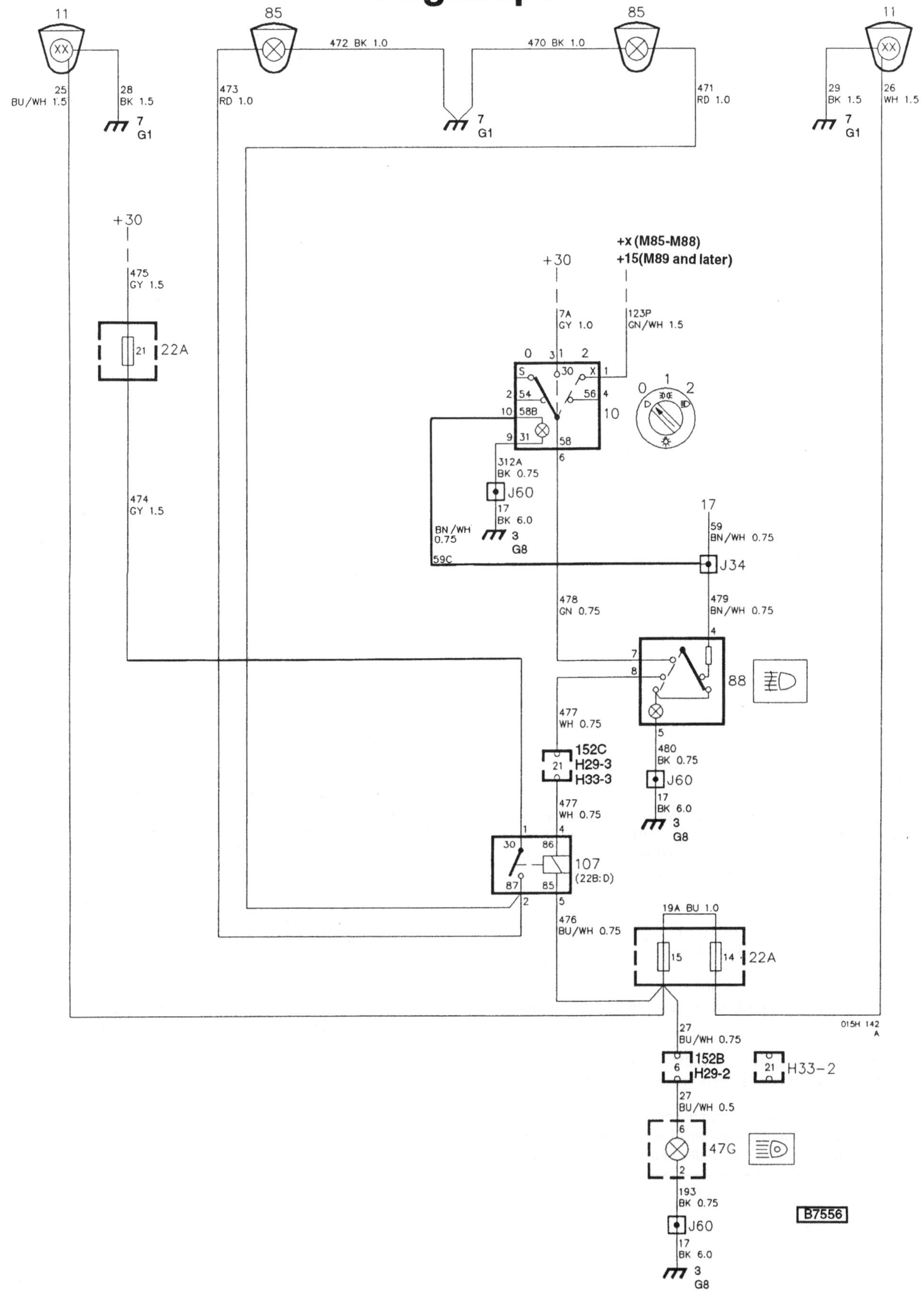
11
85
85
11
472 BK 1.0
470 BK 1.0
25
BU/WH 1.5
28
BK 1.5
7
G1
473
RD 1.0
7
G1
471
RD 1.0
29
BK 1.5
7
G1
26
WH 1.5
+30
475
GY 1.5
21
22A
+x (M85-M88)
+15(M89 and later)
+30
7A
GY 1.0
123P
GN/WH 1.5
0 3 1 2
S 30 X 1
2 54 56 4
10 58B
9 31 58
10
6
0 1 2
312A
BK 0.75
J60
17
BK 6.0
3
G8
BN/WH
0.75
59C
474
GY 1.5
17
59
BN/WH 0.75
J34
478
GN 0.75
479
BN/WH 0.75
4
7
8
88
5
477
WH 0.75
152C
H29-3
H33-3
480
BK 0.75
J60
17
BK 6.0
3
G8
477
WH 0.75
1 4
30 86
107
(22B:D)
87 85
2 5
476
BU/WH 0.75
19A BU 1.0
15
14
22A
015H 142
A
27
BU/WH 0.75
152B
H29-2
H33-2
6
21
27
BU/WH 0.5
6
47G
2
193
BK 0.75
J60
17
BK 6.0
3
G8
B7556

Fuel System

Operation

All the cars covered by this manual have a variant of the Bosch LH fuel injection system. **Table a** shows the applications for the LH system. This is a pulsed, electronic system.

This section covers only the electrical fault tracing, component location, and wiring. Information on troubleshooting the fuel injection system can be found in **240 Fuel Injection**.

Table a. LH Fuel Injection Variants

System	Year/Model	Identification
LH 2.2	1985-1988 Turbo 1986-1987 Non-Turbo	Three-wire AIC idle valve, metal air mass meter with a sealed mixture adjustment screw, and a throttle dashpot
LH 2.4	1989 and later Turbo 1988-1990 Non-Turbo	Two-wire AIC idle valve, plastic air mass meter without a mixture adjustment screw
LH 2.4.2	1991and later Non-Turbo	Three-wire AIC idle valve, similar plastic air mass meter without a mixture adjustment screw, and a throttle potentiometer

Fault Tracing

CAUTION

- Before making electrical tests on the fuel system, please read and understand the warnings given at the beginning of **300 Electrical—General**.

- Always wait at least 40 seconds before disconnecting the main connector at the LH control unit (200). The main fuel injection relay has a timing feature that maintains power for about 30 seconds. Damage to the control unit will result if the connector is removed too soon.

- On models with LH 2.4 and LH 2.4.2, activate and read any stored fault codes before disconnecting the LH control unit (200) connector. The LH diagnostic memory will be erased if the battery is disconnected or if the control unit connector is removed. See **240 Fuel Injection** for information on LH fault diagnosis.

- Use only a digital multimeter (DMM) or LED test light when making electrical checks/tests on the LH fuel system circuit.

- Where possible, probe the rear of connectors when making electrical checks. Otherwise the small contacts in the connectors may be damaged causing faulty or intermittent operation.

1. Check the fuses. Check the ground wires at the cylinder head lifting lug.
2. Using an LED test light, check for a pulsed signal between pin 1 of the LH control unit (200) and ground while cranking the engine. This is the tach signal from the ignition system.
3. Using an LED test light, check for a pulsed signal at the fuel injector connectors while cranking the engine. Positive (+) battery voltage is always present at the connectors and the LH control unit switches the ground side of the circuit to turn the injectors on and off.

Component Locations

22 main fuse/relay panel in engine compartment
47 combined instruments
47P CHECK ENGINE light
75 voltage distribution block in engine compartment right bulkhead
76 switch for raising idle speed (automatic transmission) at center console at shift lever
94 starting injector in throttle housing
101 supply fuel pump in the fuel tank
102 fuel pump relay at the LH control unit 200, forward of r.f. door
103 main fuel pump in the fuel tank
132 speed transmitter at the rear of the instrument cluster
136 lambda (oxygen) exhaust sensor in the exhaust manifold
144 boost pressure switch (turbo)
146 ignition amplifier/control unit in engine compartment, left bulkhead
147 ignition pulse amplifier at main fuse/relay panel, position D
156 A/C compressor relay at main fuse/relay panel, engine compartment
159 +15 distribution terminal in main fuse/relay panel, engine compartment
166 A/C pressure switch for cooling fan at receiver/dryer unit, engine compartment
176 EZK control unit in engine compartment, left bulkhead
200 LH control unit forward of the right front door, behind trim
202 engine temperature transmitter at intake manifold flange, between cylinders 2 and 3
203 throttle angle transmitter on throttle housing
204 LH diagnostic connector, in engeine compartment, behind right-hand wheel housing
205 LH air mass meter in the throttle intake hose
206 fuel injectors at intake manifold
229 fuel injection main relay at the LH control unit 200, forward of r.f. door
271 lambda preheater (integral with lambda sensor)
272 AIC valve (idle speed motor) at front left side of cylinder head
285 fuse for lambda sensor in engine compartment, next to fresh air intake
321 charcoal canister valve (ELCD) in engine compartment, ahead of left wheel housing
323 fuel pump in fuel tank
347 LH diagnostic connector, under rear seat, right side
389 NTC resistor under intake manifold (1990 and later)
390 EGR modulating valve ahead of left wheel housing (1990 and later)

Fuel System

Connector Locations

1985-1990

204 LH test connector in engine compartment, behind right wheel housing
322 coding connector at the LH control unit 200, forward of r.f. door
347 diagnostic test connector under rear seat, right side

1991 and later

H1-9 1-pin connector near lambda sensor, right wheel housing
H1-12 1-pin connector in engine compartment, right side near air intake
H1-16 1-pin connector under center console next to shift selector (automatic)
H1-19 1-pin connector near LH control unit, behind right A pillar trim
H2-13 2-pin connector at the fuel pump in the storage area
H2-33 2-pin connector near the lambda sensor in the engine compartment
H3-14 3-pin connector near LH control unit, behind right A pillar trim
H4-7 4-pin connector at the LH control unit, forward of right front door
H6-1 6-pin connector in engine compartment right wheel housing
H6-2 6-pin connector in engine compartment right wheel housing
H29-1 29-pin bulkhead connector behind main fuse/relay panel (1991-1991 1/2)
H29-2 29-pin bulkhead connector behind main fuse/relay panel (1991-1991 1/2)
H29-3 29-pin bulkhead connector behind main fuse/relay panel (1991-1991 1/2)
H33-1 33-pin bulkhead connector behind main fuse/relay panel (1991 1/2 and later)
H33-2 33-pin bulkhead connector behind main fuse/relay panel (1991 1/2 and later)
H33-3 33-pin bulkhead connector behind main fuse/relay panel (1991 1/2 and later)

Ground Locations

1985-1990

201 ground point for fuel injection at engine lifting lug, 2 points

1991 and later

G1 ground point at radiator cross member
G3 ground point in luggage compartment
G7 ground point at engine lifting lug
G25 ground point at gearbox

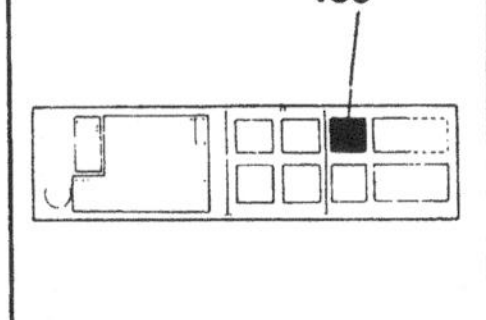

22

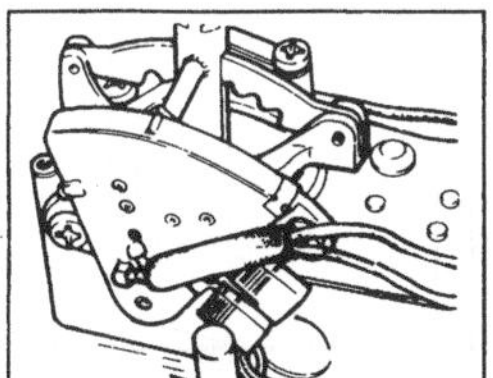
76

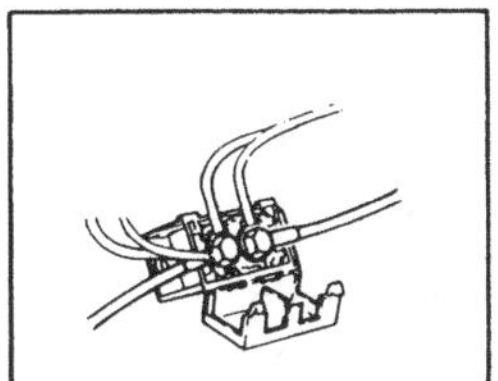
75

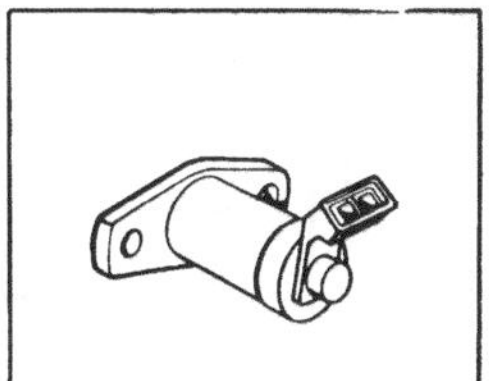
94

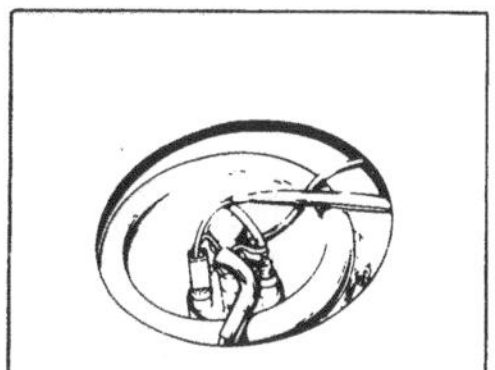
101,103

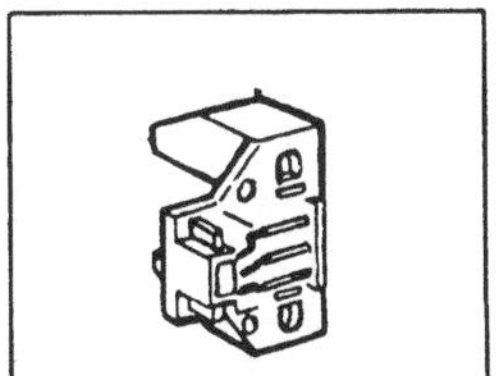
132

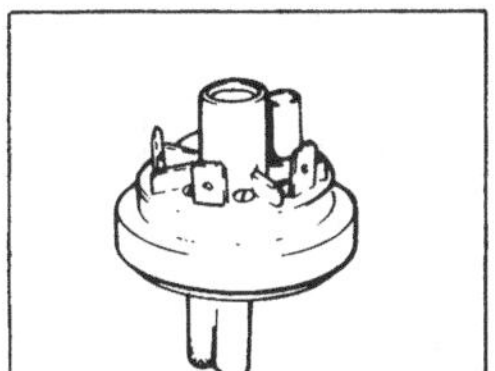
144

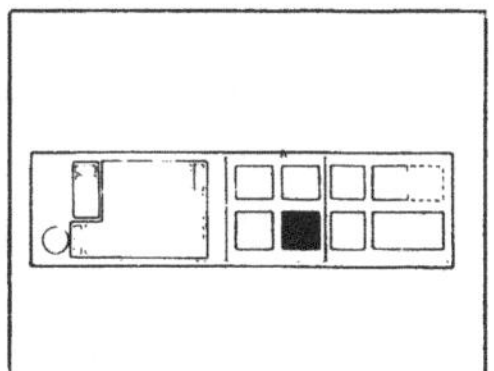
147

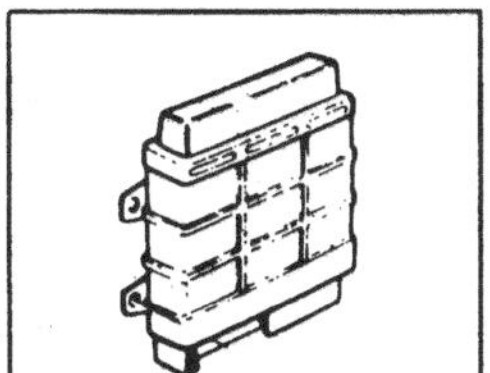
176

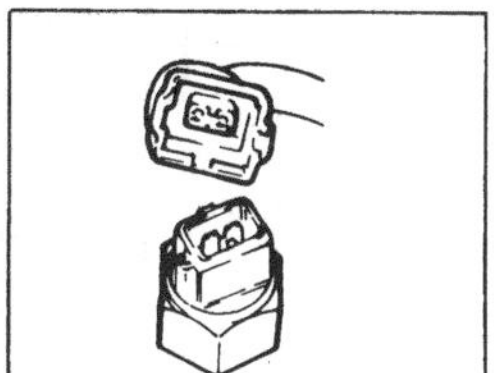
202

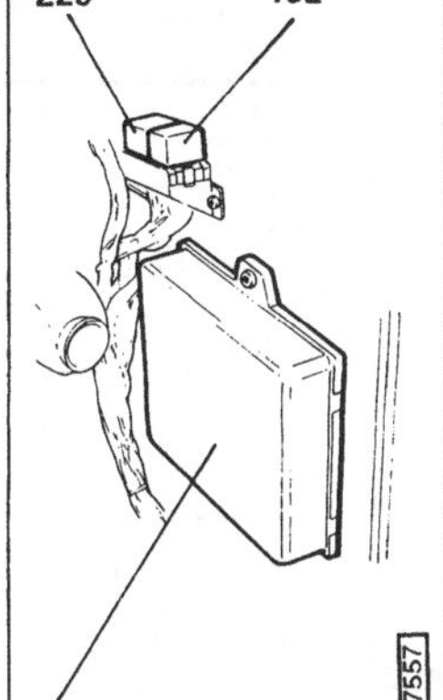

200

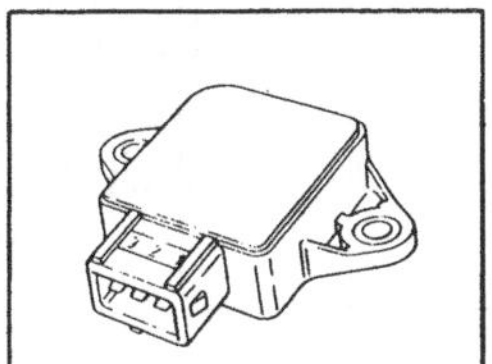
203

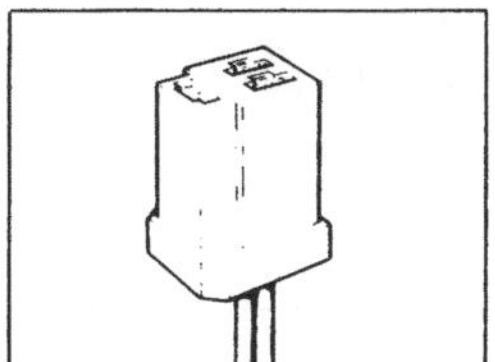
204

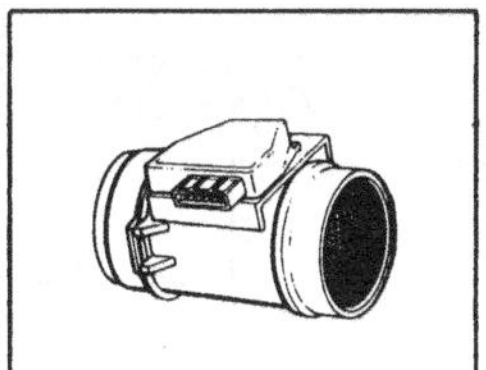
205

206

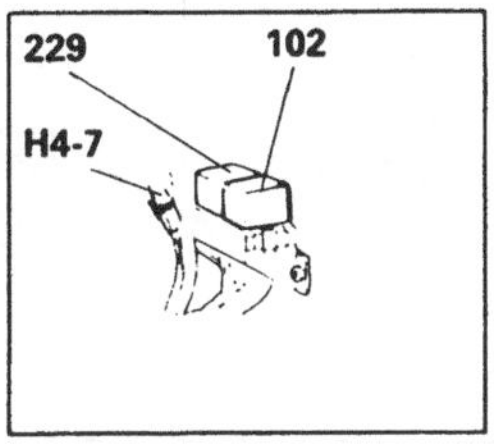

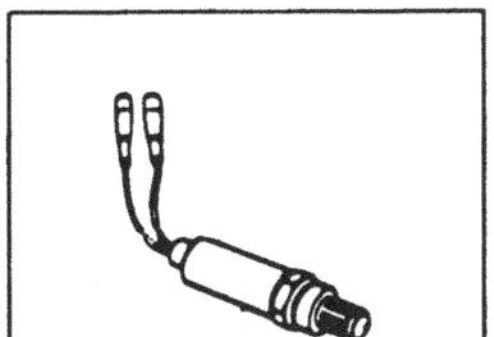
136
271

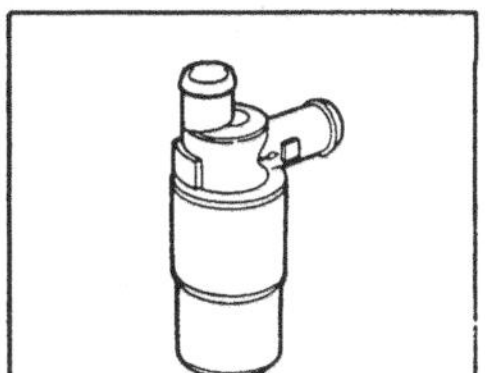
272 (LH)

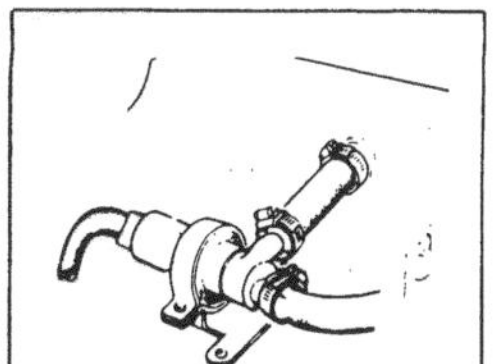
272

321

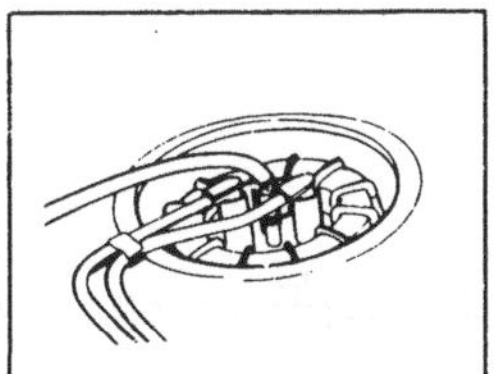
323

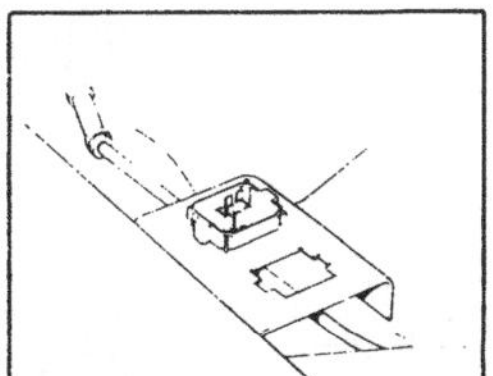
347

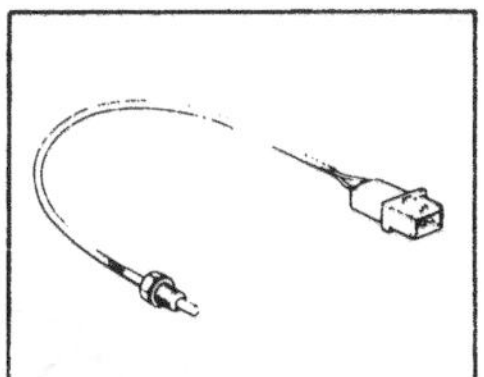
389

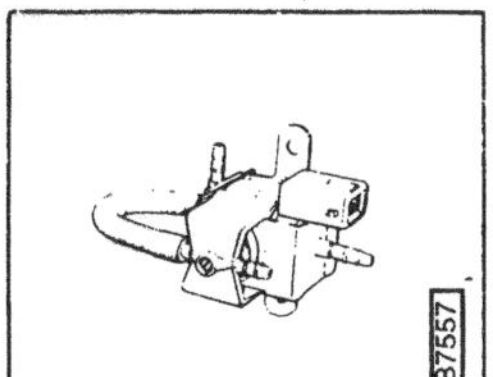

390

LH 2.2 Fuel System

1985-1986 (all)

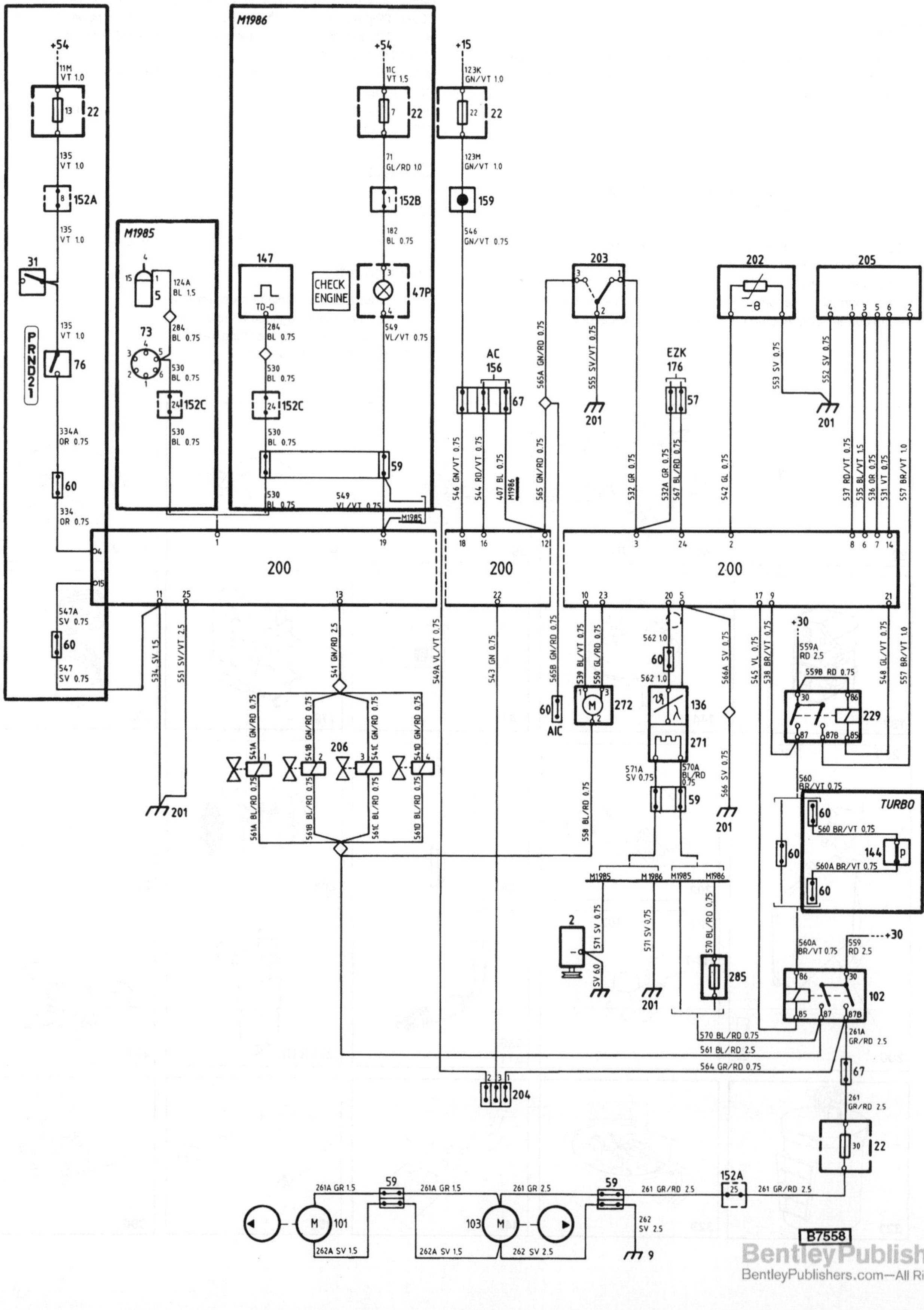

LH 2.2 Fuel System

1987-1988 turbo

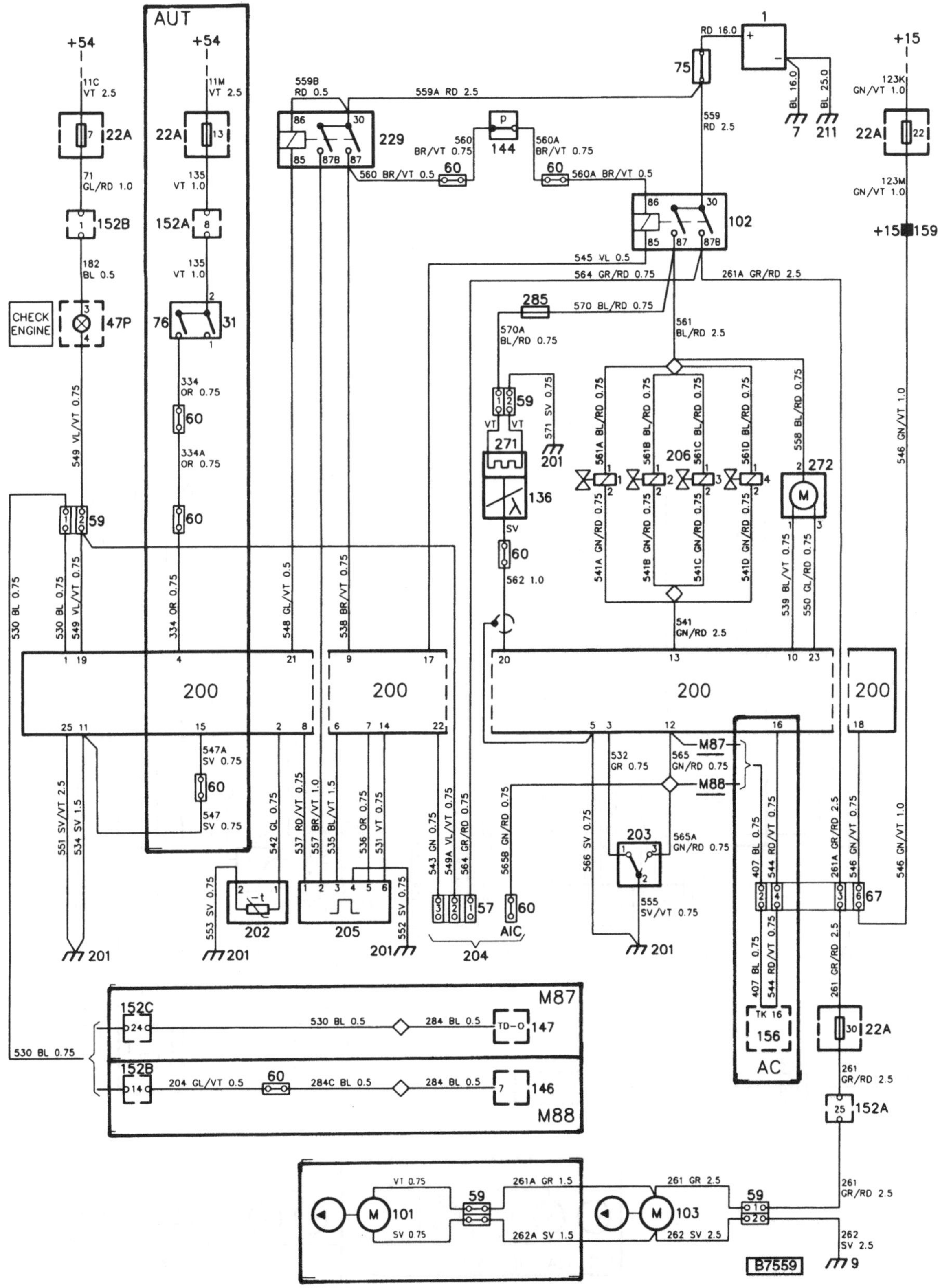

LH 2.2 Fuel System

1987 non-turbo

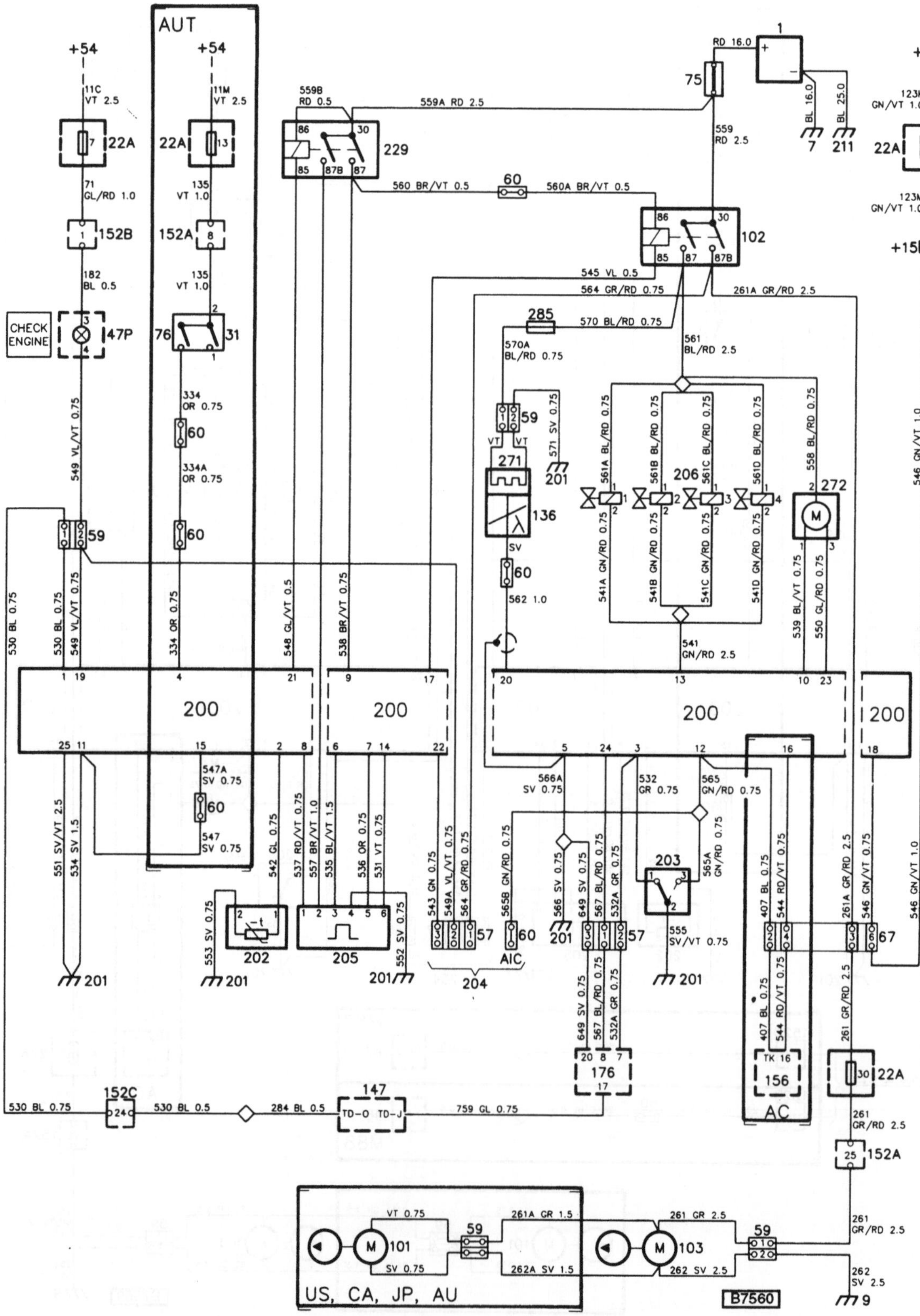

LH 2.4 Fuel System

1988 non-turbo

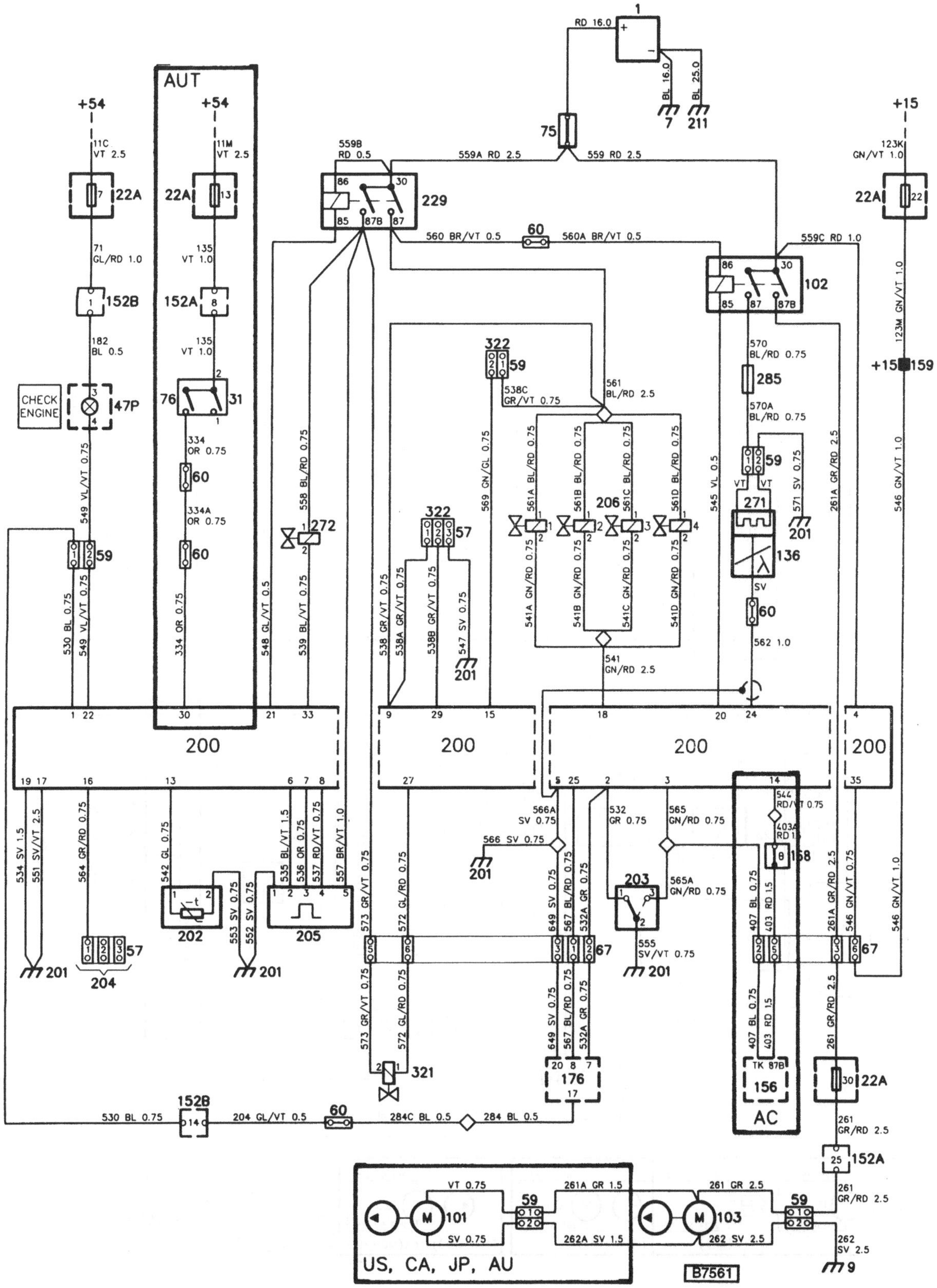

LH 2.4 Fuel System

1989-1990 (all)

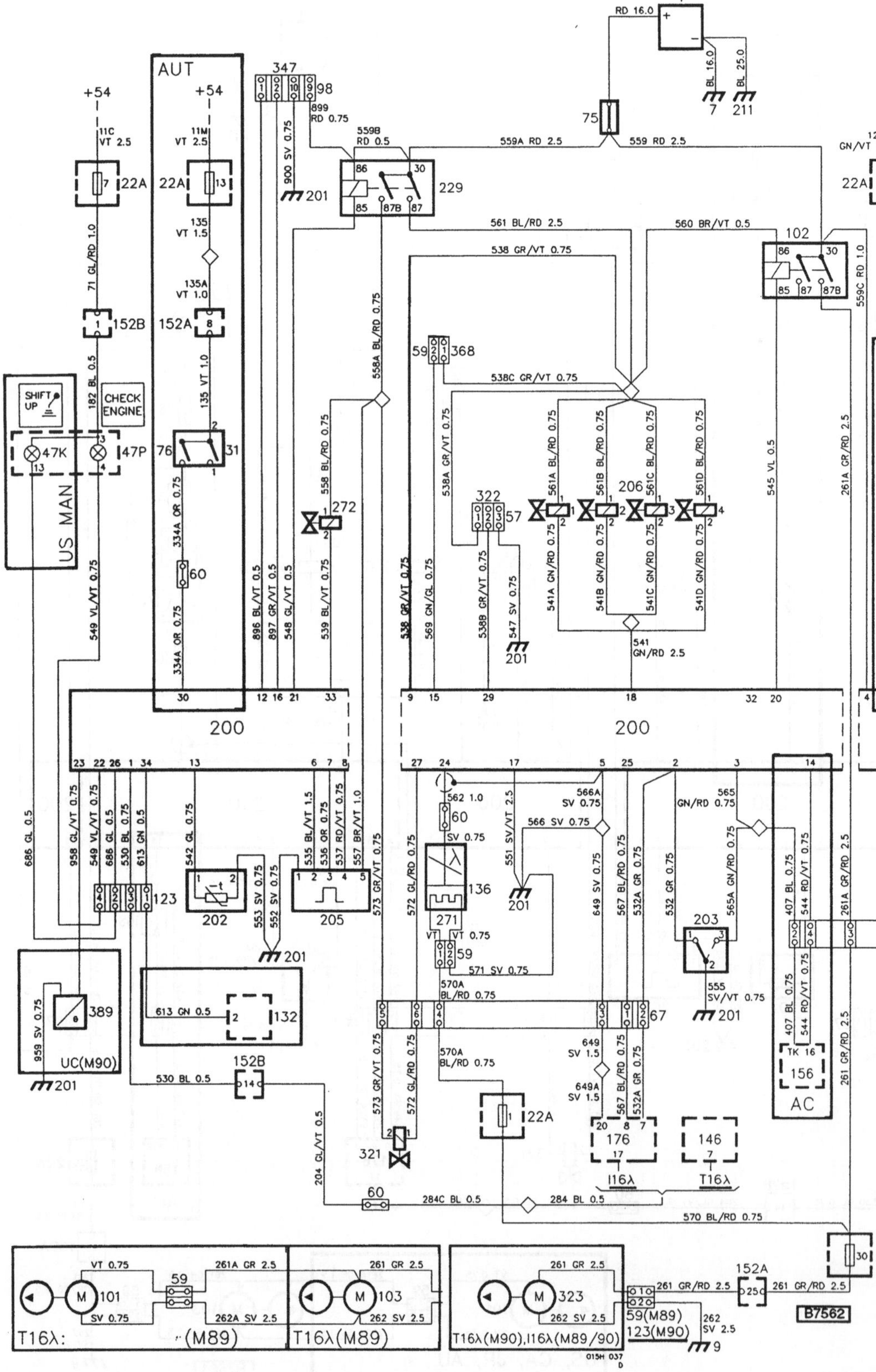

LH 2.4 Fuel System with 29-pin connector

1991-1992 turbo

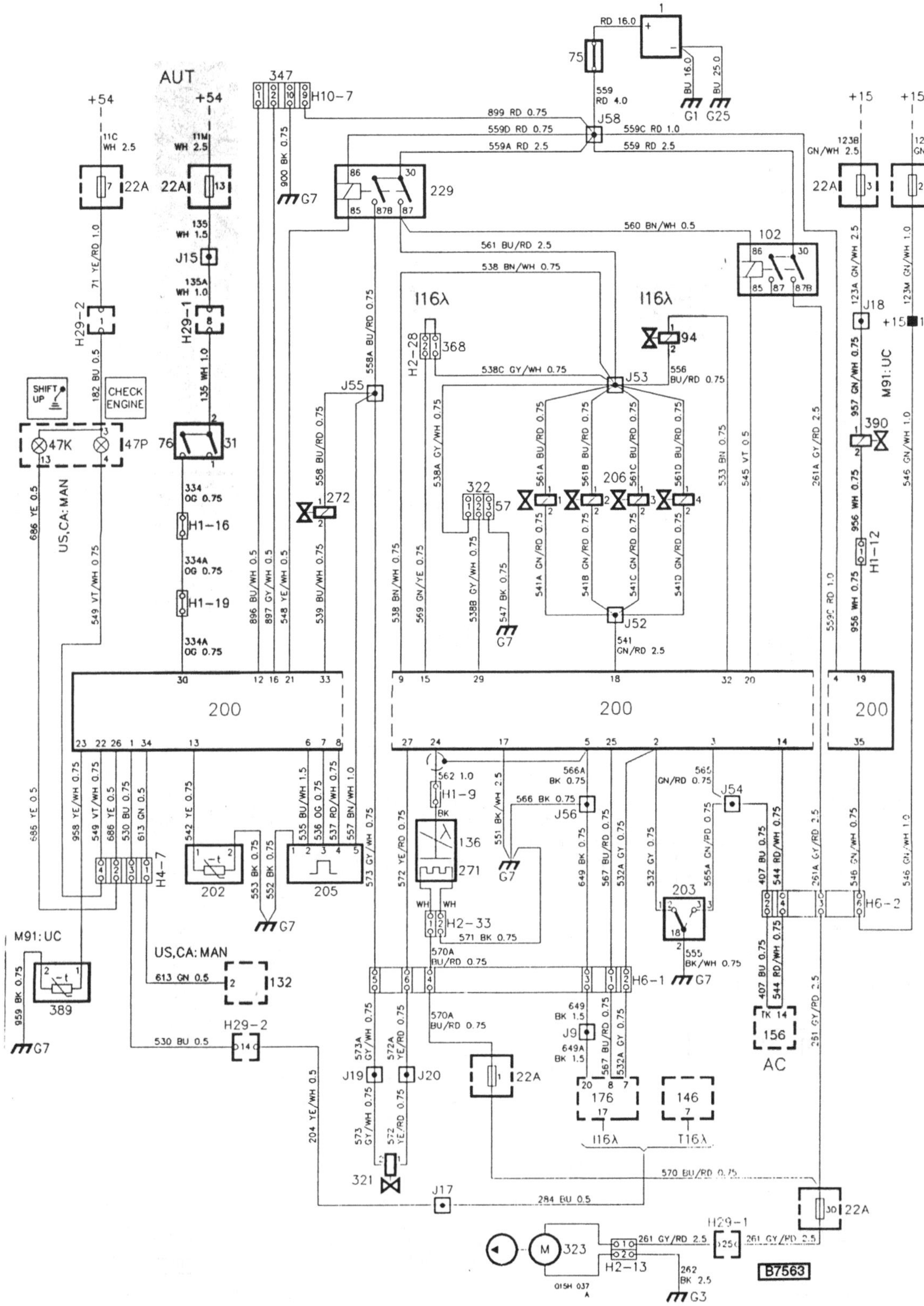

LH 2.4 Fuel System with 33-pin connector

1991 and later Turbo

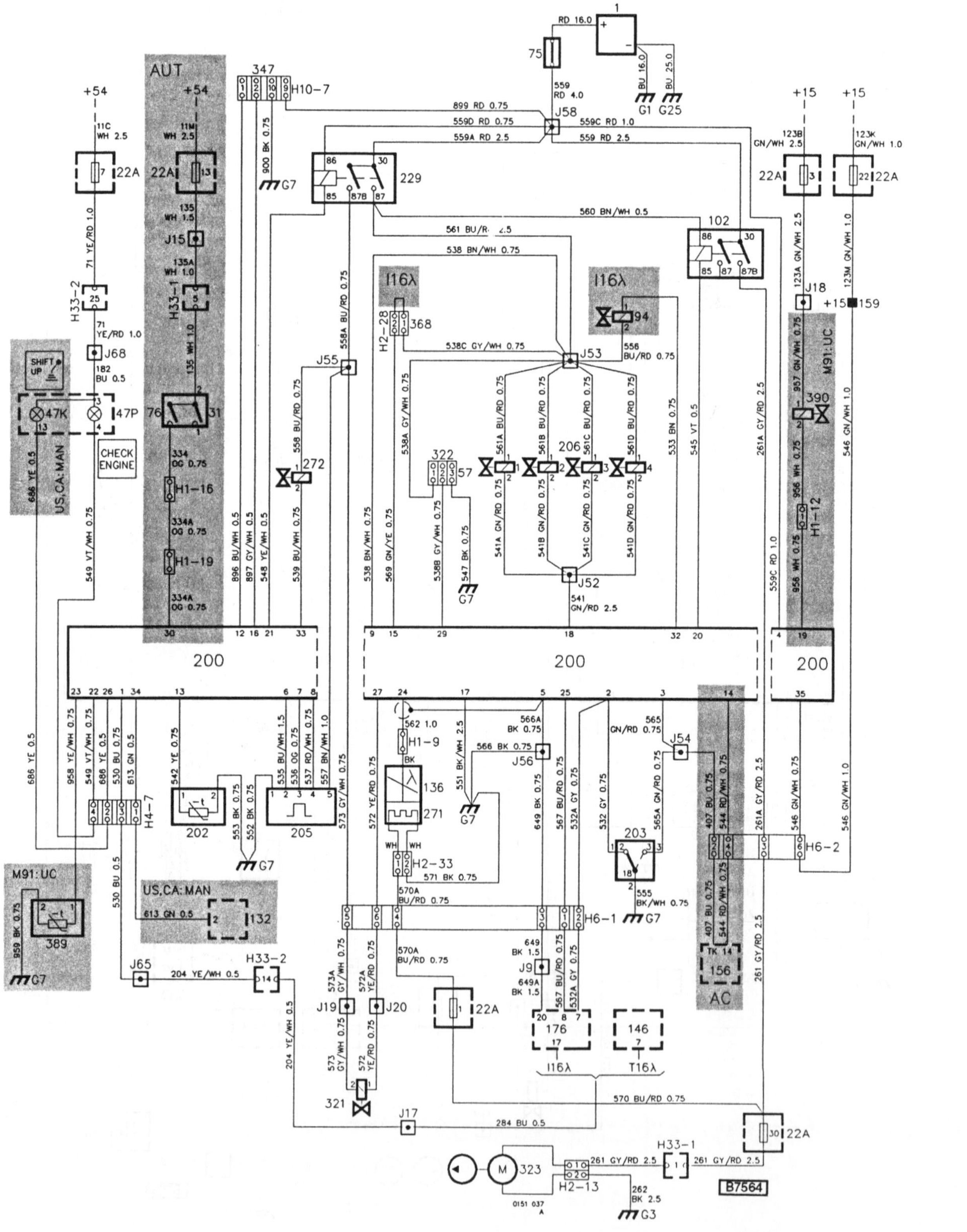

LH 2.4.2 Fuel System with 29-pin connector

1991-1992 non-turbo

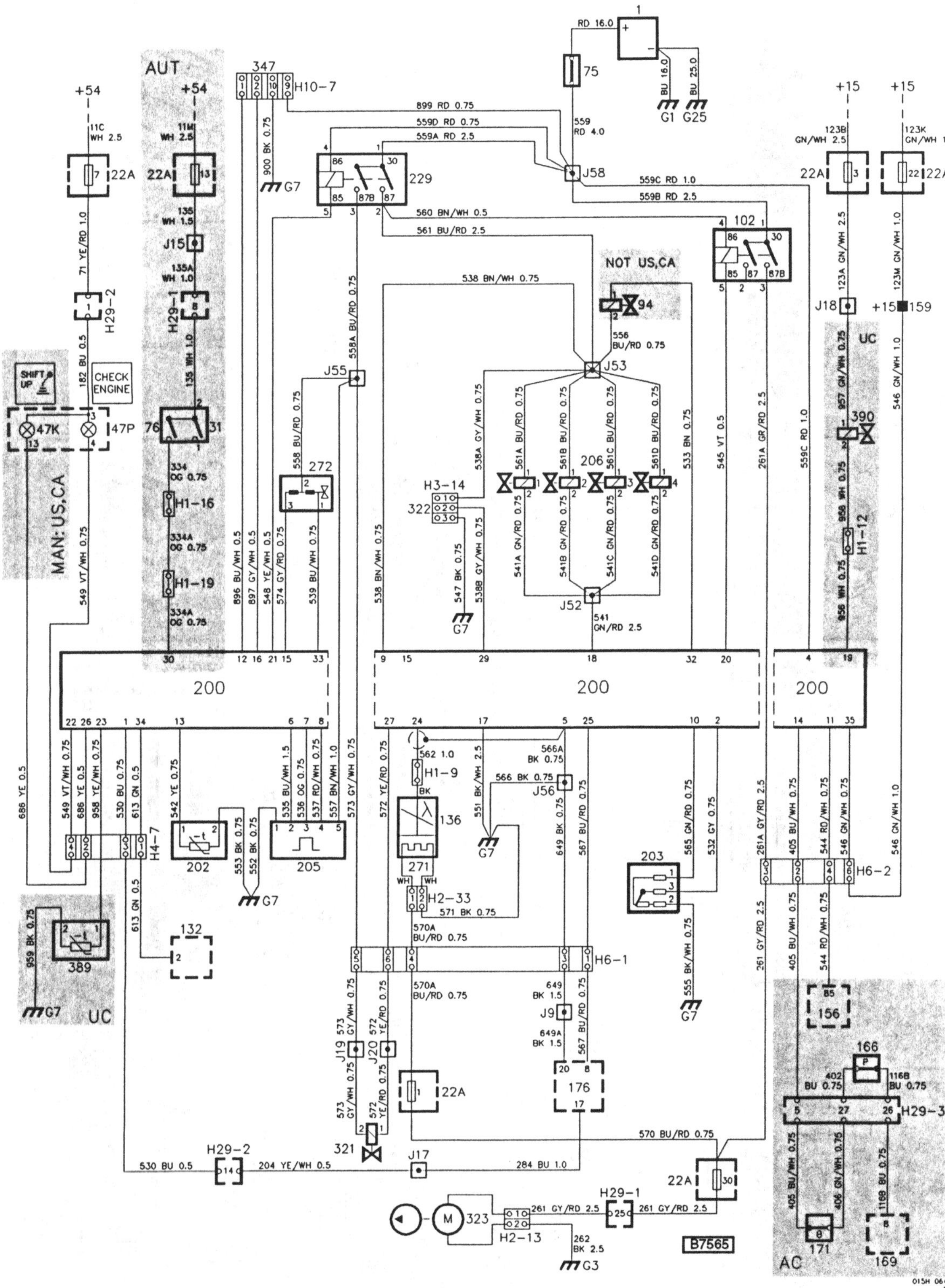

LH 2.4.2 Fuel System with 33-pin connector

1991 and later Non-turbo

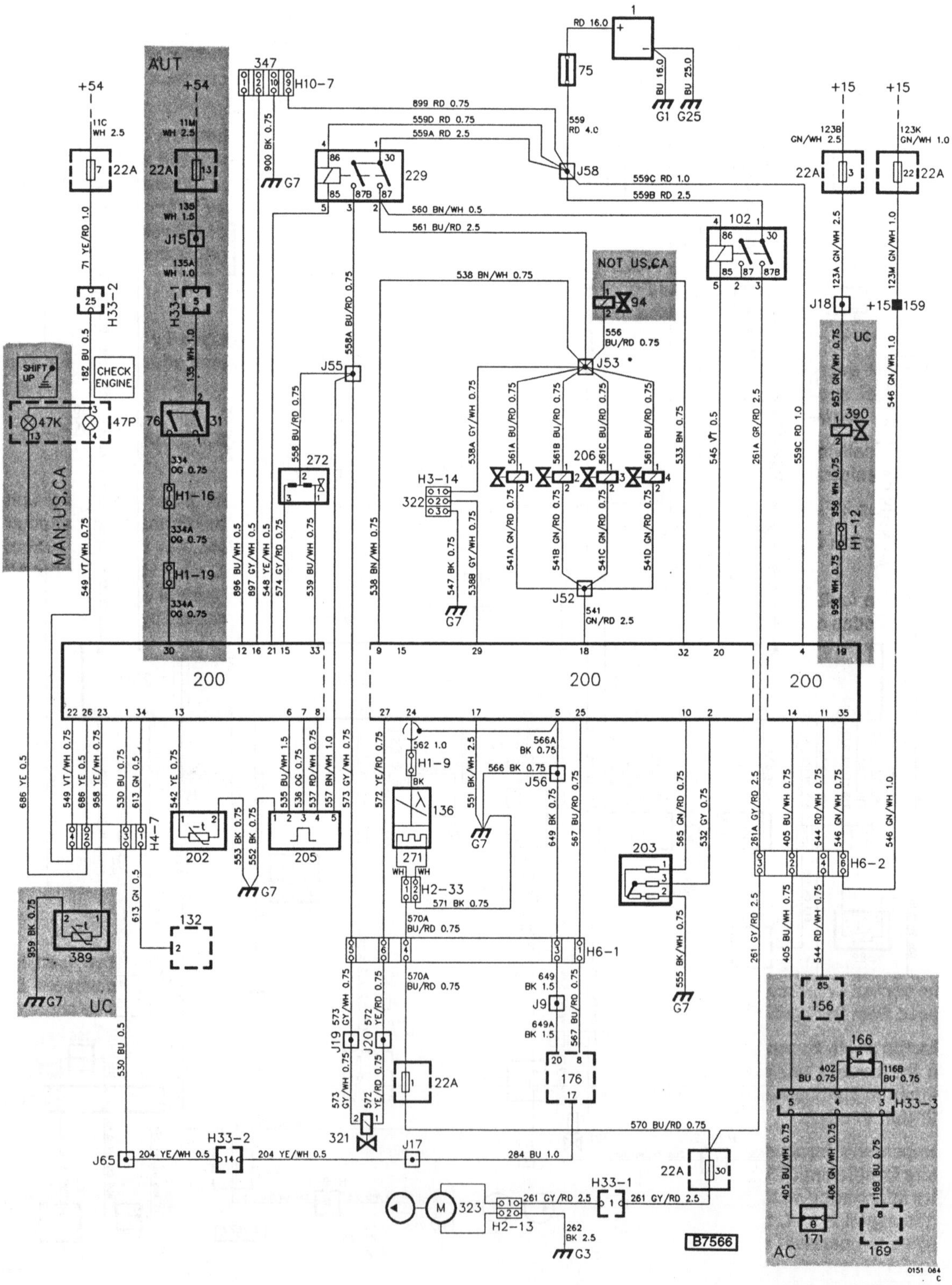

Glovebox Lighting

Operation

The glovebox light (19) is controlled by the glovebox switch (160). With the glovebox door open, the switch is closed lighting up the lamp. To change the bulb, pry off the lens.

Fault Tracing

1. Check the fuse. Check for voltage at the lamp assembly whenever the key is in the run position and the headlight switch is in the on position.
2. Check for continuity to ground to the left of the steering wheel behind the instrument panel.

Component Locations

10 light switch in instrument panel
19 glovebox light in glovebox
22 main fuse/relay panel
160 glovebox light switch in side of glovebox

Connector Locations

1985-1990

152B 29-pin bulkhead connector at main fuse/relay panel

1991 and later

H29-2 29-pin bulkhead connector behind main fuse/relay panel (1991-1991 1/2)
H33-2 33-pin bulkhead connector behind main fuse/relay panel (1991 1/2 and later)

Ground Locations

1985-1990

3 ground point behind instrument panel

1991 and later

G8 ground point behind instrument panel

19

160

B7567

Glovebox Lighting

1985-1986

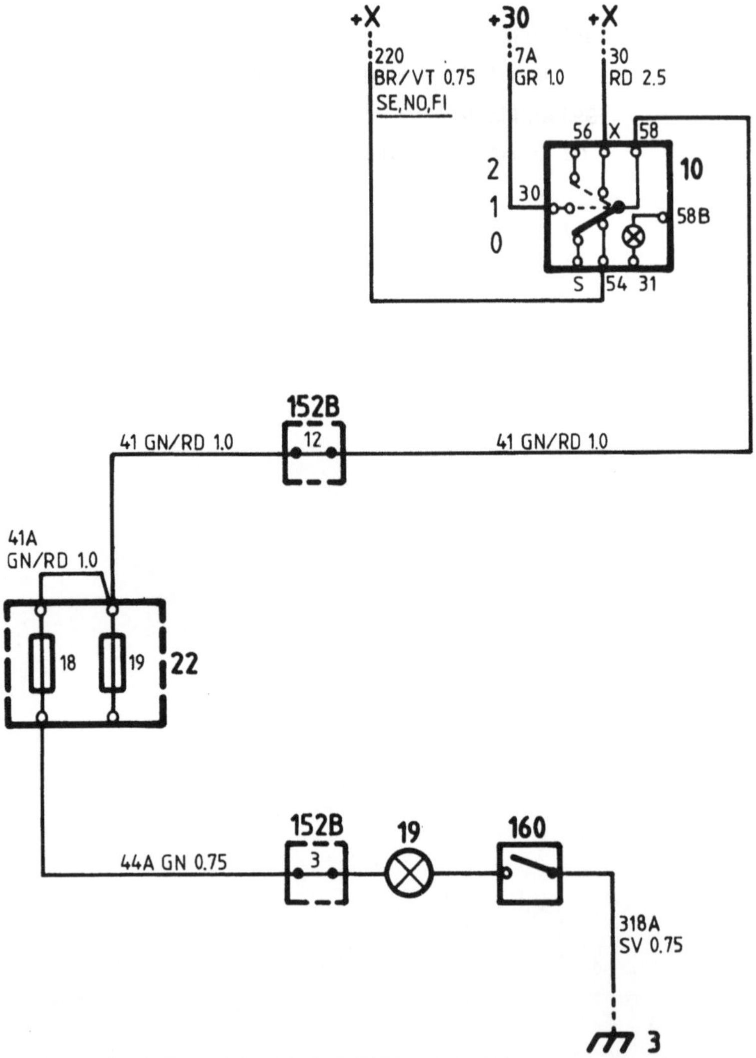

1987 and later

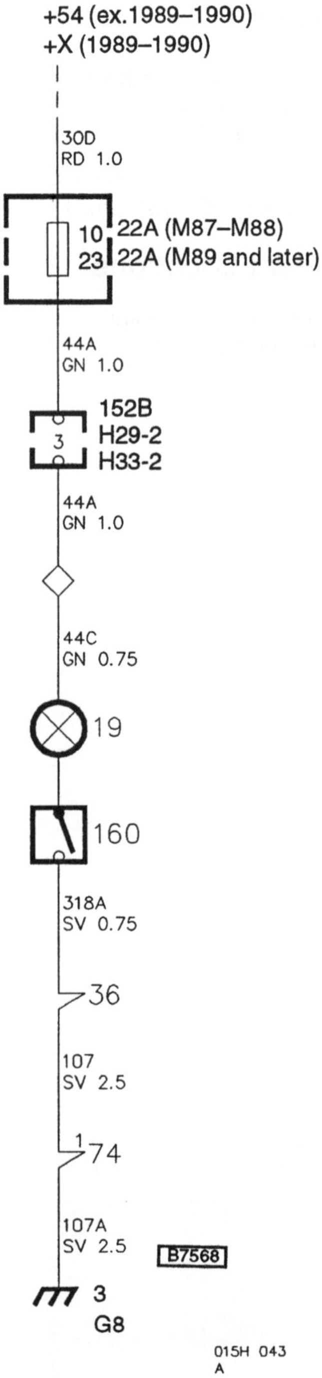

015H 043
A

Headlights

Operation

The headlights receive power from the headlight switch whenever the key is in the run position. All power to the headlights must pass through the headlight relay and four separate fuses. Each beam (i.e. left low beam) is protected by a separate fuse. Two types of headlights are used on the cars covered by this manual. On 1985 and 1986 models, a sealed beam is used. On 1987 and later cars, a replaceable headlight bulb is used.

Fault tracing

Always begin by checking fuses and connectors. Check and tighten all grounds. See **351 Lighting** for repair information, including changing bulbs.

Component Location

- 8 lighting relay in main fuse/relay panel
- 10 lighting switch in instrument panel
- 11 headlight high beam
- 12 headlight low beam
- 47G headlight high beam warning light in instrument cluster
- 75 terminal distribution in right bulkhead near exhaust manifold
- 158 negative distribution terminal in main fuse/relay panel
- 215 dip switch

Connector Locations

1985-1990

- 58 12-pin connector behind instrument panel to left of steering column
- 60 1-pin connector behind instrument panel to left of steering column
- 152A 29-pin bulkhead connector
- 152B 29-pin bulkhead connector

1991 and later

- H2-7 2-pin connector at the light switch
- H12-1 12-pin connector in left side of instrument panel
- H29-2 29-pin red bulkhead connector (1991–1991 1/2)
- H33-3 33-pin gray bulkhead connector (1991 and later)

Ground Locations

1985-1990

- 3 ground point behind instrument panel
- 7 ground point at radiator cross member

1991 and later

- G1 ground point at radiator cross member
- G6 ground distribution in main fuse/relay panel
- G8 ground point behind instrument panel
- G25 ground point on gearbox

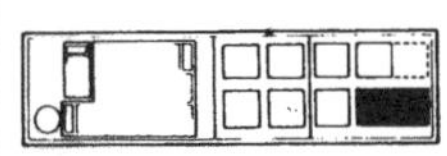
8

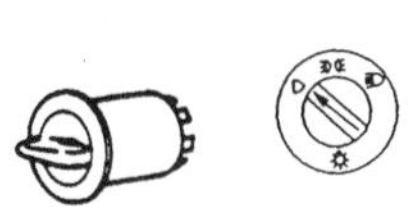
10

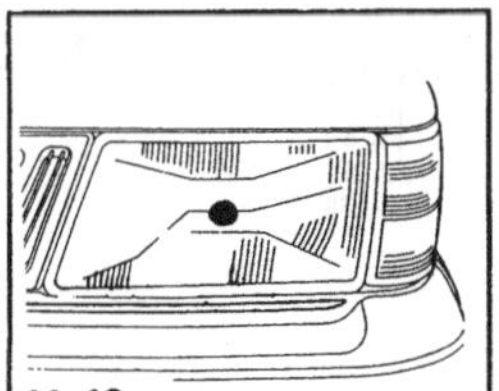
11, 12

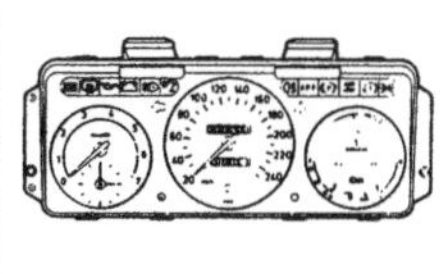
47G

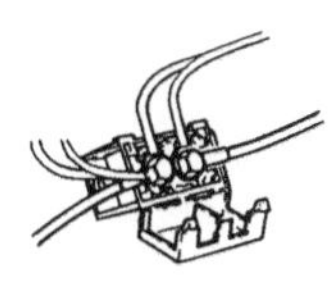
75

215

Headlights

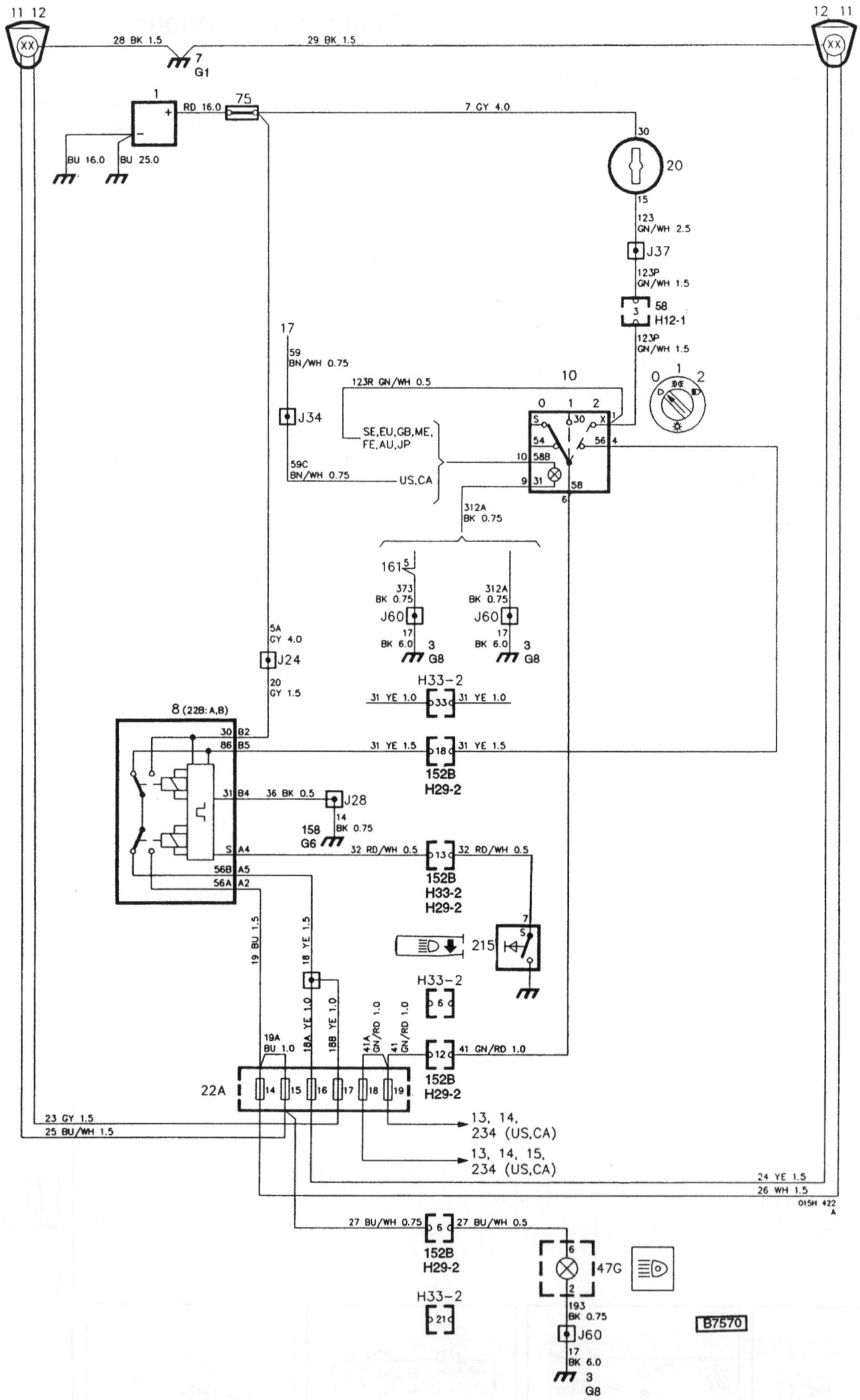
11 12
12 11
28 BK 1.5
29 BK 1.5
7
G1
1
RD 16.0
75
7 GY 4.0
BU 16.0
BU 25.0
30
20
15
123
GN/WH 2.5
J37
123P
GN/WH 1.5
58
H12-1
123P
GN/WH 1.5
17
59
BN/WH 0.75
123R GN/WH 0.5
10
0 1 2
J34
SE,EU,GB,ME,
FE,AU,JP
59C
BN/WH 0.75
US,CA
312A
BK 0.75
161
373
BK 0.75
J60
17
BK 6.0
3
G8
312A
BK 0.75
J60
17
BK 6.0
3
G8
5A
GY 4.0
J24
20
GY 1.5
H33-2
31 YE 1.0
31 YE 1.0
8 (22B: A,B)
31 YE 1.5
31 YE 1.5
152B
H29-2
36 BK 0.5
J28
14
BK 0.75
158
G6
32 RD/WH 0.5
32 RD/WH 0.5
152B
H33-2
H29-2
19 BU 1.5
18 YE 1.5
215
H33-2
19A
BU 1.0
18A YE 1.0
18B YE 1.0
41A
GN/RD 1.0
41
GN/RD 1.0
41 GN/RD 1.0
152B
H29-2
22A
23 GY 1.5
25 BU/WH 1.5
13, 14,
234 (US,CA)
13, 14, 15,
234 (US,CA)
24 YE 1.5
26 WH 1.5
27 BU/WH 0.75
27 BU/WH 0.5
152B
H29-2
47G
H33-2
193
BK 0.75
J60
17
BK 6.0
3
G8
B7570

Heated Rear Window and Cigarette Lighter

Heated Rear Window Operation

The heated rear window element (115) is activated by the switch (116) on the instrument panel. A rear window heater warning light (47J) will light up in the instrument cluster. The circuit has a time delay relay (113) that will turn off the supply to the heating element after about 10 minutes. Convertible models have a mercury switch (279) that deactivates the circuit if the convertible top is down.

Heated Rear Window Fault Tracing

1. Check the fuse.
2. Check for voltage at the heating element. Check the continuity of the heating element to find breaks in the rear window grid.
3. Check for voltage at time delay relay (113) terminal 30 and terminal 87.

Cigarette Lighter Operation

The cigarette lighter works after it is pushed in only with the key in the run position.

Cigarette Lighter Fault Tracing

1. Check the fuse.
2. Remove the cigarette lighter element and carefully check for voltage between the center contact and ground with the key on.
3. Inspect the heating element for breaks. Check the continuity of the heating element.

Component Locations

- 20 ignition switch
- 22A main fuse/relay panel at left wheel housing
- 47 instrument cluster
- 47J rear window heater warning lamp
- 48 cigarette lighter
- 113 time delay relay for heated rear window under rear seat, auxiliary fuse/relay panel
- 115 rear window heater element
- 116 rear window heater switch
- 158 negative distribution in main fuse/relay panel
- 159 +15 distribution in main fuse/relay panel
- 207 heated rear view mirrors on front doors
- 279 mercury switch for convertible heated rear window at left side top mounting
- 401 auxiliary fuse/relay panel under rear seat

Connector Locations

1985-1990

- 57 3-pin connector in luggage compartment, left air outlet
- 59 2-pin connector in luggage compartment, left air outlet
- 152A 29-pin bulkhead connector at main fuse/relay panel
- 152B 29-pin bulkhead connector at main fuse/relay panel

1991 and later

- H2-10 2-pin connector behind cigarette lighter
- H3-13 3-pin connector in luggage compartment, left air outlet
- H10-1 10-pin connector behind instrument panel, left of steering column
- H29-2 29-pin bulkhead connector behind main fuse/relay panel (1991-1991 1/2)
- H33-2 33-pin bulkhead connector behind main fuse/relay panel (1991 1/2 and later)

Ground Locations

1985-1990

- 3 ground point behind instrument panel

1991 and later

- G3 ground point in luggage compartment
- G5 ground point under the rear seat
- G8 ground point behind instrument panel
- G19 ground point at hatchback
- G26 ground point at right C pillar on sedans

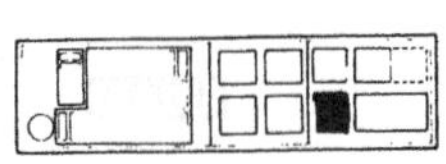

22A

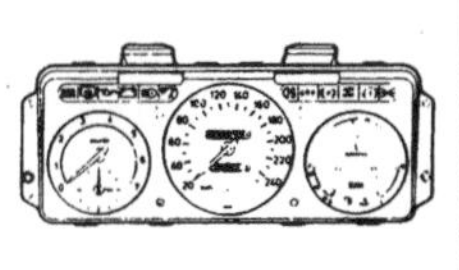

47, 47J

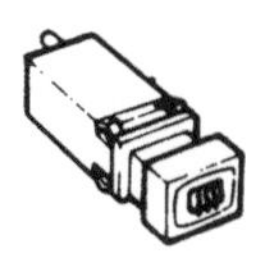

116

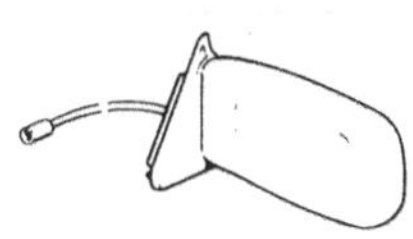

207

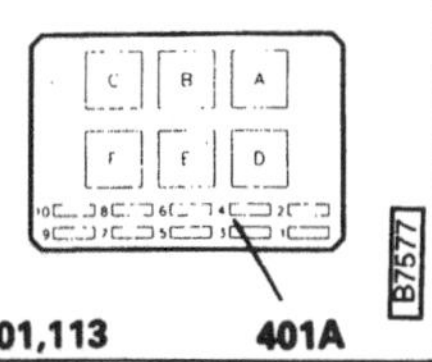

401,113 401A

Heated Rear Window and Cigarette Lighter

1985 on (ex. Convertible)

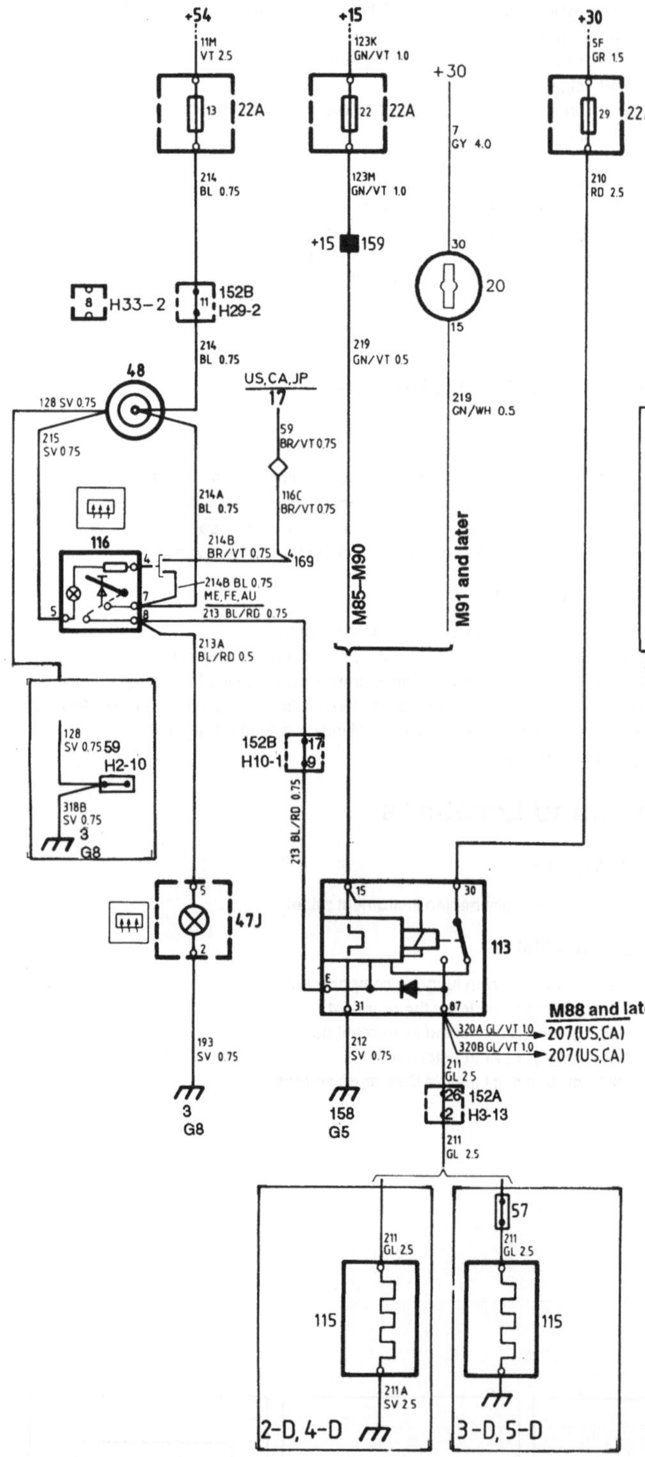

1987 on (Convertible)

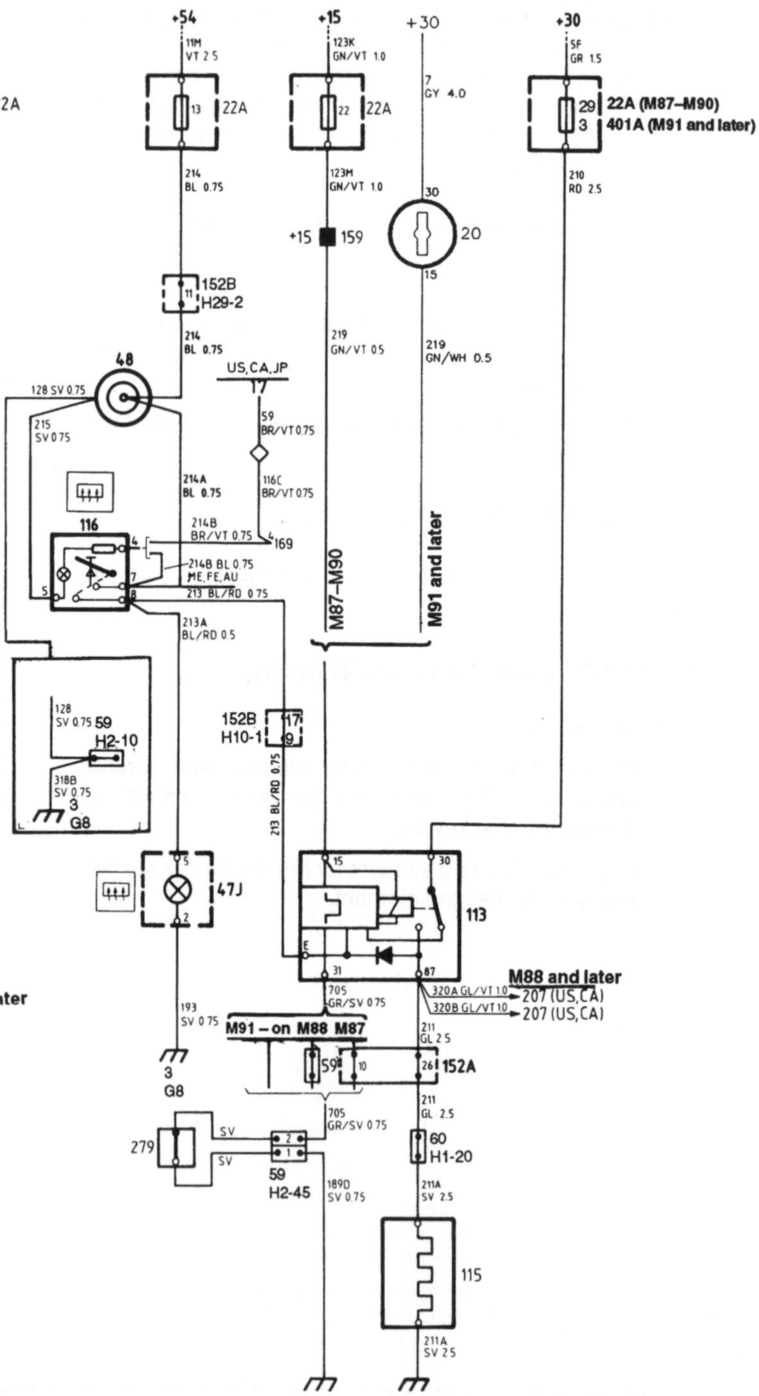

B7578

Heated Seats

Operation

The front heated seats are thermostatically controlled. Some models have a rheostat (252) on the instrument panel for temperature control. The heating element is located under the bottom seat cover and the seat back. A thermostatic button under the bottom cover turns the heating element on and off. The passenger seat switch (121) prevents the heater from turning on if the passenger seat is not occupied.

For information on removing the seats and seat covers, see **851 Interior Trim and Upholstery**.

Fault Tracing

1. Check the fuse. On 1991 and later models, the fuse is located in the fuse/relay panel (401) under the rear seat.
2. Check for voltage at the seat heating pad element (64) and check for continuity through the element.
3. Check the seat switch (121) under the passenger seat. This switch should be closed (continuity through switch) when a passenger is in the seat.
4. Check the thermostat button under the bottom seat cover. The thermostat should be closed (continuity) or open (no continuity) depending on the temperature of the switch.

Heated seat thermostat
- switch closed................ below 54°F (12°C)
- switch open.................. above 82°F (28°C)

NOTE
A commercial aerosol freeze spray can be sprayed on the thermostat button to simulate cold temperatures for testing. Always check compatibility of the spray with components and upholstery before using.

Component Locations

22A main fuse/relay panel at left wheel housing
64 heated seat pad with thermostat in the seat cushion and backrest
121 passenger seat switch for heated seats under the passenger seat
252 driver heated seat rheostat in instrument panel
254 driver seat temperature transmitter in driver seat
401 auxiliary fuse/relay panel under rear seat

Connector Locations

1985-1990

59 2-pin connector under each front seat
60 1-pin connector under the center console
98 10-pin connector to the left of the steering column
152A 29-pin bulkhead connector at main fuse/relay panel
152B 29-pin bulkhead connector at main fuse/relay panel

1991 and later

H8-1 8-pin connector under the passenger seat
H8-2 8-pin connector under the driver seat
H10-1 10-pin connector behind instrument panel, left of steering column
H10-6 10-pin connector at the auxiliary fuse/relay panel under rear seat

Ground Locations

1985-1990

65 ground point at the handbrake
117 ground point at the handbrake

1991 and later

G4 ground point between ignition switch and handbrake

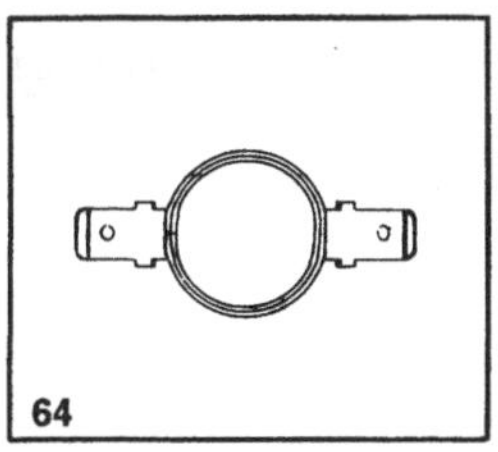

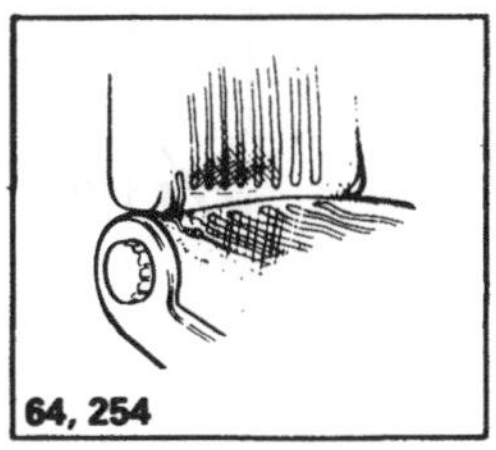

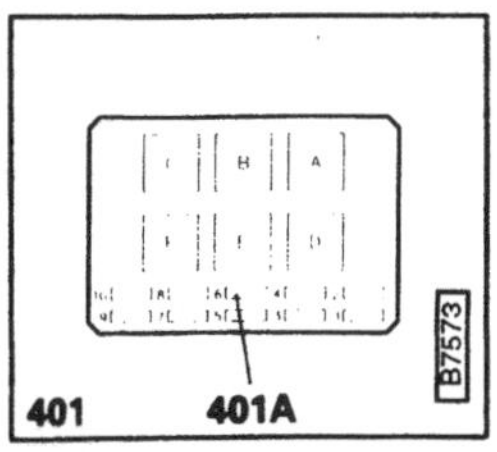

Heated Front Seats without Rheostat

1985-1988

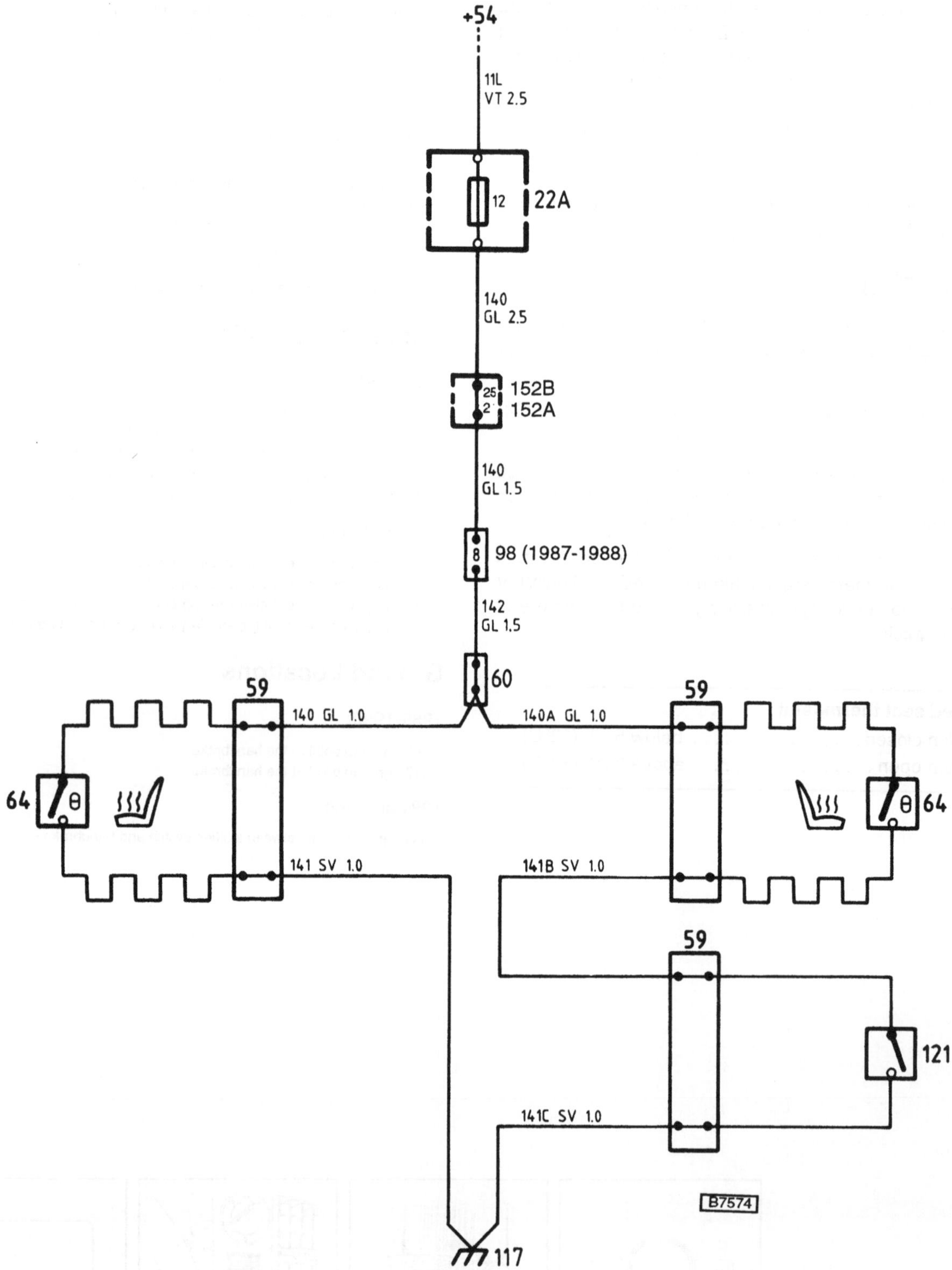

Heated Front Seats with Rheostat

1987-1990

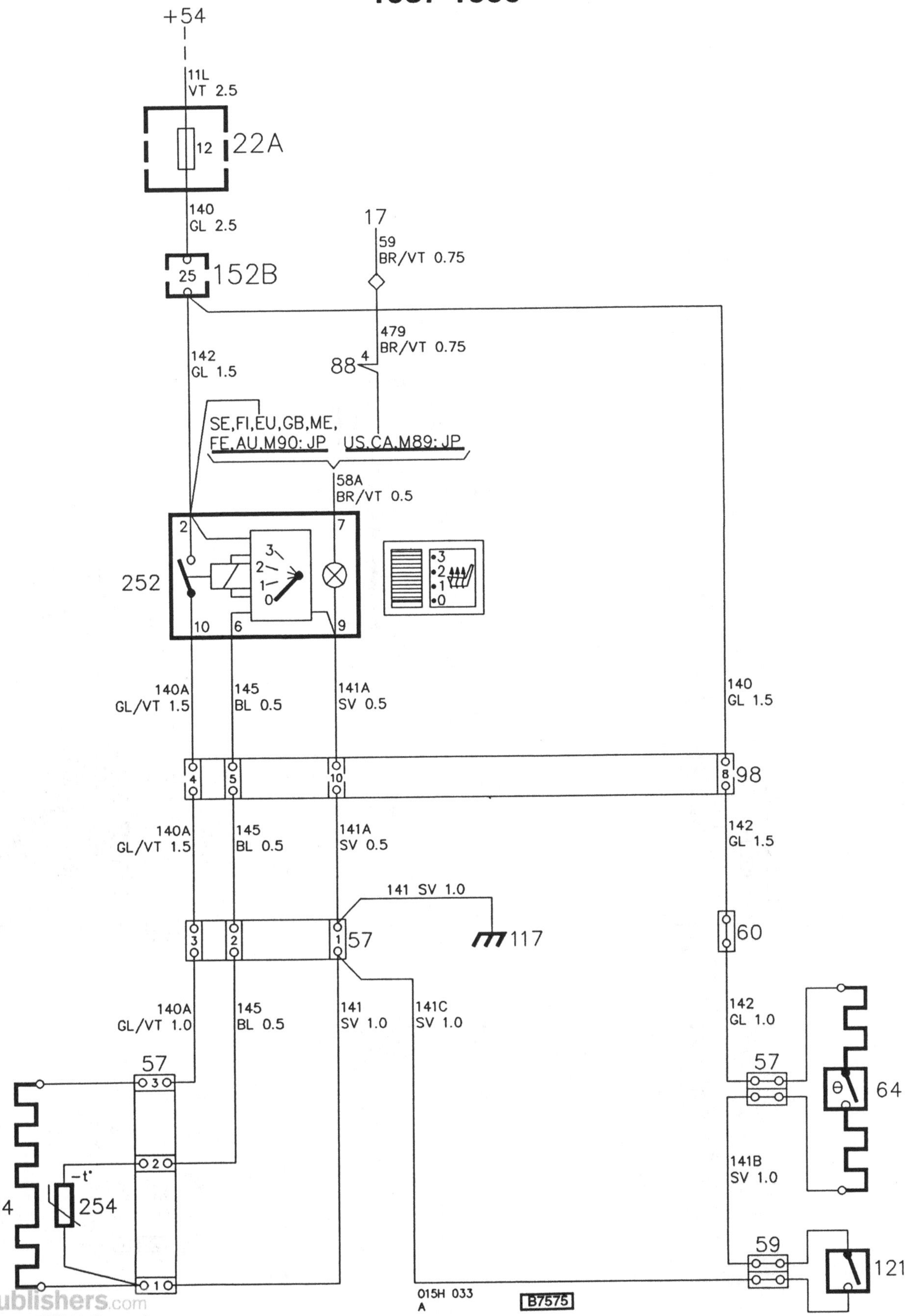

Heated Front Seats with Rheostat

1991 and later

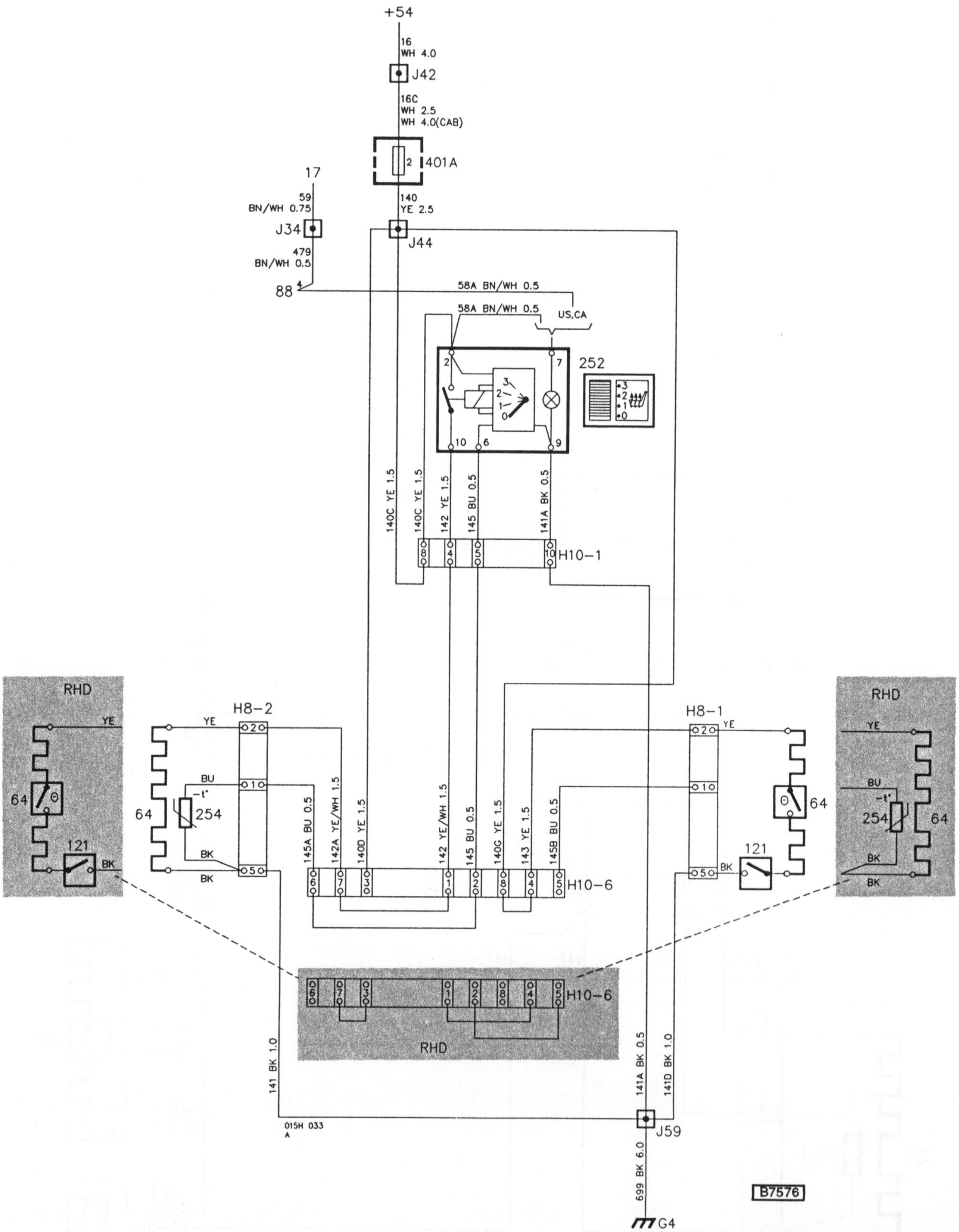

Horns

Operation

The horns can be sounded with the key in the run position. The buttons on the steering pad (41) complete the circuit to ground through the slip ring assembly (352/336) in the steering wheel and the horn relay (68).

WARNING
Some cars covered by this manual are equipped with an explosive SRS airbag device. Do not attempt to service the wiring in the steering wheel or steering column assembly. The airbag may accidentally ignite, causing personal injury. See an authorized Saab dealer for any repairs necessary to the SRS wiring or components.

For information on removing and repairing the steering wheel see **641 Steering Column.**

Fault Tracing

1. Check that the fuse is okay.
2. Remove the horn relay (68) and jumper pins 30 and 87, the horn should sound. If not, there is a problem between the fuse and the horns.
3. Check for ground at pin 85 of the horn relay when the horn button is depressed. If ground is not present, check the slip ring assembly.
4. Check for positive (+) battery voltage at pin 86 of the horn relay when the key is in the run position. If there is no voltage check the wiring from the ignition switch (20) through the ignition switch relay (21).

Component Locations

20 ignition switch in center console
21 ignition switch relay in main fuse/relay panel
22A main fuse/relay panel
40A horns located on right side behind headlight
41 horn switch in steering wheel pad
68 horn relay in main fuse/relay panel
75 distribution block at right wheel housing
158 negative distribution in main fuse/relay panel
289 burglar alarm control unit (convertible) under rear seat
336 slip ring contact behind steering wheel (cars with SRS Airbag)
352 slip ring switch behind steering wheel

Connector Locations

1985-1990

152A 29-pin bulkhead connector behind main fuse/relay panel
152B 29-pin bulkhead connector behind main fuse/relay panel

1991 and later

H29-1 29-pin bulkhead connector behind main fuse/relay panel (1991-1991 1/2)
H29-2 29-pin bulkhead connector behind main fuse/relay panel (1991-1991 1/2)
H33-1 33-pin bulkhead connector behind main fuse/relay panel (1991 1/2 and later)
H33-2 33-pin bulkhead connector behind main fuse/relay panel (1991 1/2 and later)

Ground Locations

1985-1990

7 ground point at radiator cross member
211 ground point at gearbox

1991 and later

G1 ground point at radiator cross member
G6 ground distribution at main fuse/relay panel
G25 ground point at gearbox

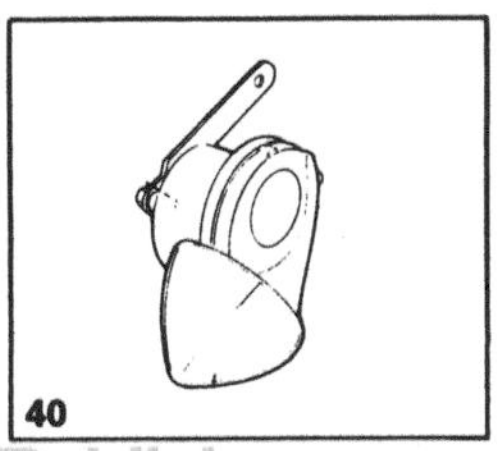

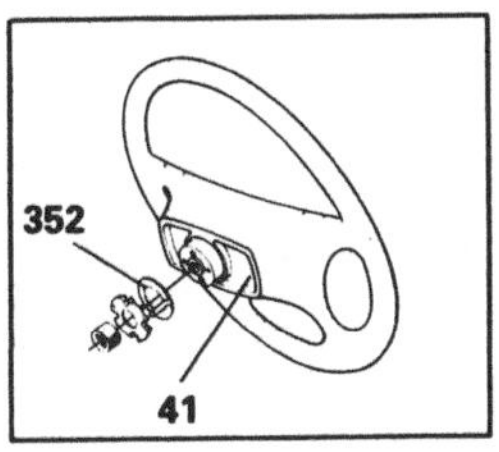

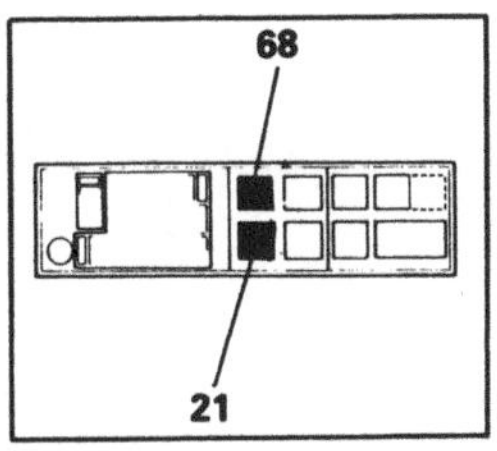

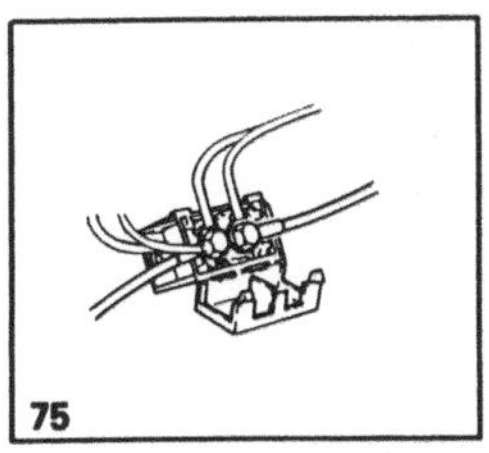

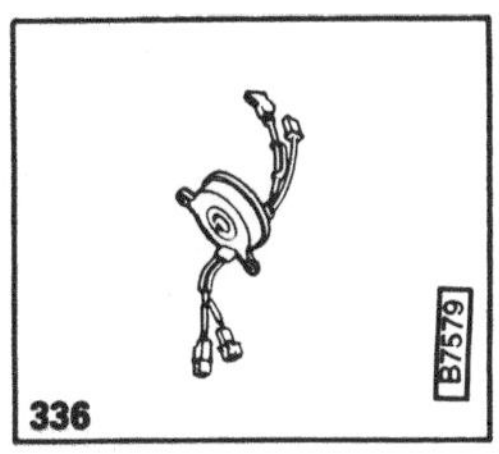

Horns

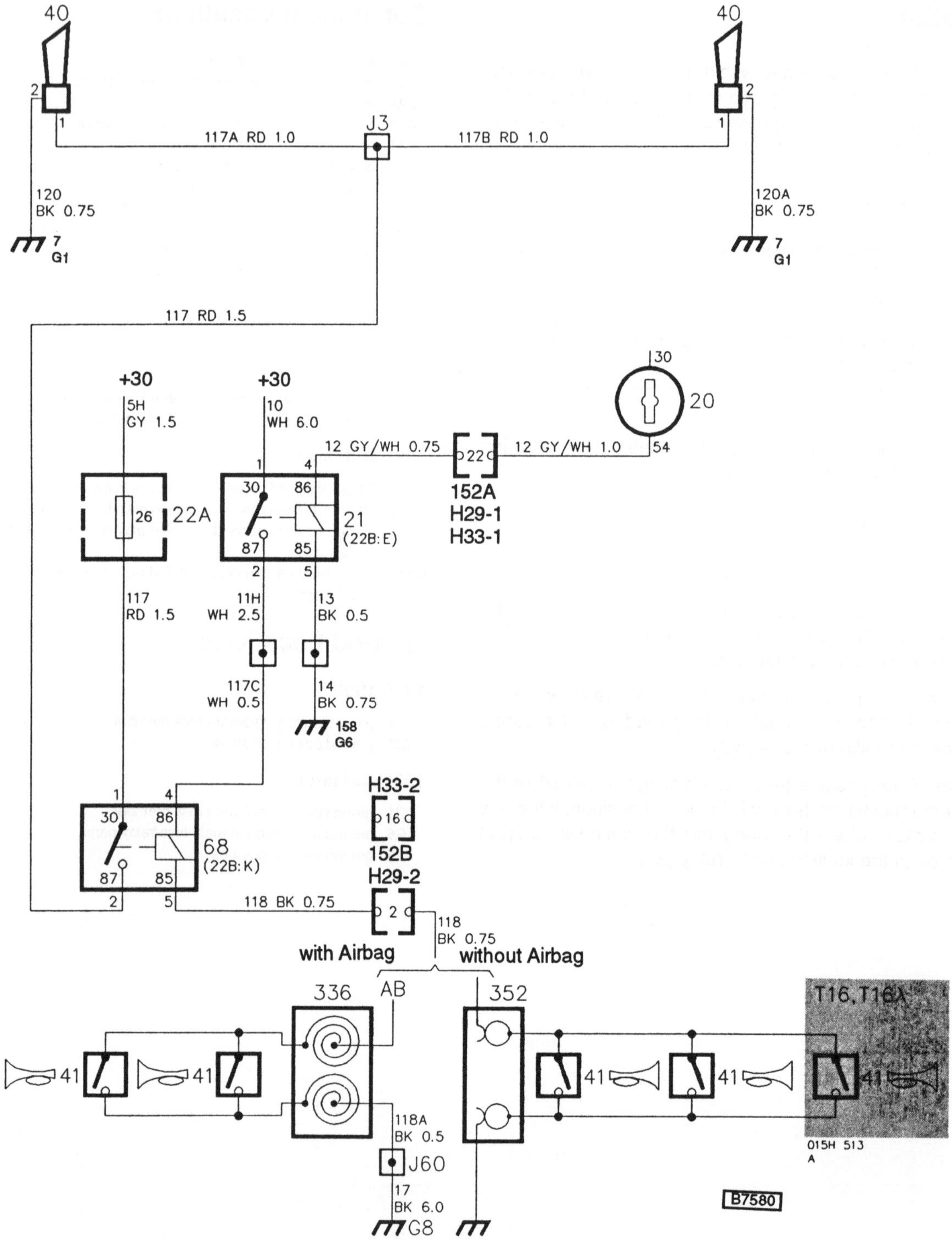
40
40
J3
117A RD 1.0
117B RD 1.0
120
BK 0.75
120A
BK 0.75
7
G1
117 RD 1.5
+30
5H
GY 1.5
+30
10
WH 6.0
30
20
12 GY/WH 0.75
22
12 GY/WH 1.0
54
152A
H29-1
H33-1
26
22A
21
(22B:E)
117
RD 1.5
11H
WH 2.5
13
BK 0.5
117C
WH 0.5
14
BK 0.75
158
G6
H33-2
16
152B
H29-2
68
(22B:K)
118 BK 0.75
118
BK 0.75
with Airbag
without Airbag
336
AB
352
T16, T16A
41
118A
BK 0.5
J60
17
BK 6.0
G8
015H 513
A
B7580

Ignition System

Operation

The ignition system creates and regulates the high voltage pulses that fire the spark plugs. Two versions of a Hall-effect electronic ignition system are used on the cars covered by this manual. Complete ignition system repair information is covered under **340 Ignition System.**

Ignition System Applications

- Turbo models — basic Hall ignition
- Non-turbo models — EZK ignition (with knock sensor)

WARNING

Any time work is being done on the ignition system it should first be disabled as described in **340 Ignition System**. The ignition system operates at lethal voltages. People with pacemakers or weak hearts should not expose themselves to the ignition system. Extra caution must be taken when working on or servicing the engine while it is running or if the key is on.

Fault Tracing

Read and understand the warnings and cautions given in **300 Electrical—General** before working on the ignition system. See **340 Ignition System** for fault tracing.

Component Locations

5 ignition coil on top of radiator
6 distributor at end of cylinder head
47P Check Engine light
73 TSI socket in main fuse/relay panel
110 Tachometer in instrument cluster
145 EZK test connector next to main fuse/relay panel
146 Ignition Control Unit (non-EZK ignition)
Ignition amplifier (EZK ignition)(in front of left wheel housing)
147 Ignition pulse amplifier main fuse/relay panel, position D (1986-1987)
158 ground (–) distribution in main fuse/relay panel
159 +15 distribution in main fuse/relay panel
176 EZK control unit on inner left fender
178 EZK knock sensor on engine block, below intake manifold
200 LH control unit behind passenger side A pillar trim
203 throttle angle transmitter on throttle housing
320 ignition coil with integrated amplifier on inner right fender, near battery (1988-1989)
345 EZK crankshaft sensor behind crank pulley (1989 and later)

Connector Locations

1985-1990

57 3-pin connector in engine compartment right side
59 2-pin connector in engine compartment right side
60 1-pin connector near main fuse/relay panel
67 6-pin connector in engine compartment right side

1991 and later

H1-18 1-pin connector under center console
H2-53 2-pin connector under the rear seat
H3-3 3-pin connector to left of starter motor
H3-5 3-pin connector in engine compartment, left of air intake
H4-1 4-pin connector at left wheel housing
H4-1 4-pin connector behind instrument panel right side behind trim
H6-1 6-pin connector in engine compartment right wheel housing
H6-2 6-pin connector in engine compartment right wheel housing
H29-1 29-pin bulkhead connector behind main fuse/relay panel (1991-1991 1/2)
H29-2 29-pin bulkhead connector behind main fuse/relay panel (1991-1991 1/2)
H33-1 33-pin bulkhead connector behind main fuse/relay panel (1991 1/2 and later)
H33-2 33-pin bulkhead connector behind main fuse/relay panel (1991 1/2 and later)

Ground Locations

1985-1990

7 ground point at radiator cross member
158 ground (–) distribution in main fuse/relay panel
211 ground point at gearbox

1991 and later

G1 ground point at radiator cross member
G6 ground distribution at main fuse/relay panel
G7 ground point at engine lifting lug
G8 ground point behind instrument panel
G12 ground point at left wheel housing
G25 ground point at gearbox

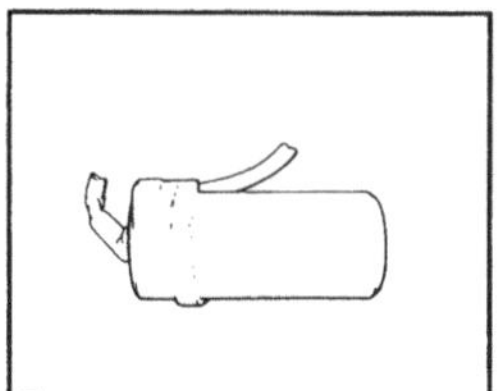
5

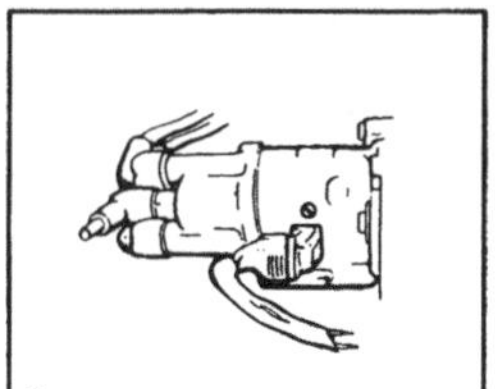
6

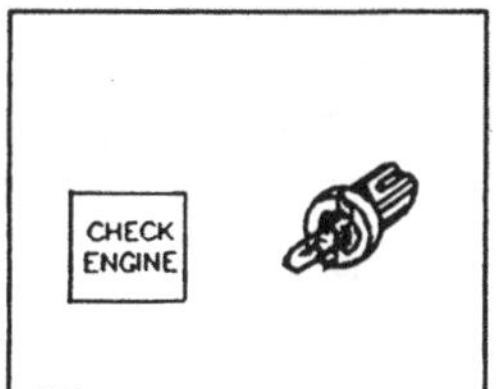

47P

145

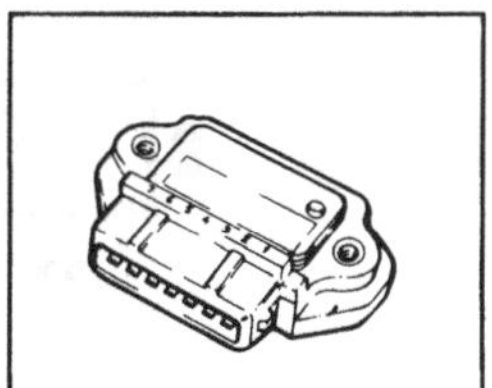
146

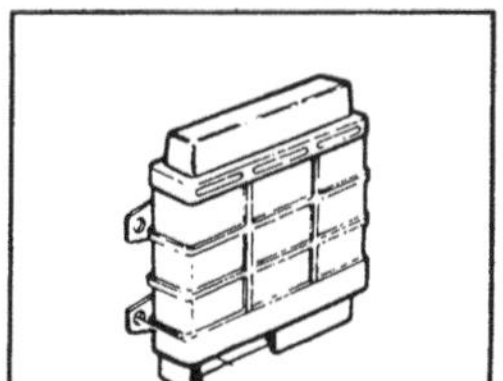
176

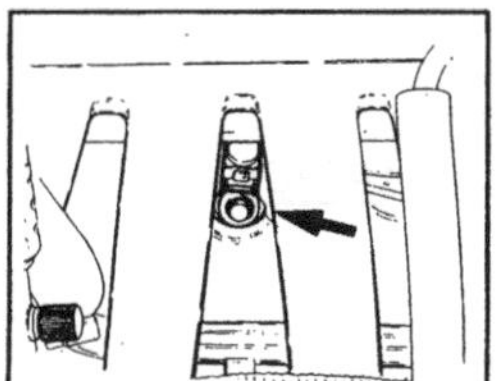
178

203

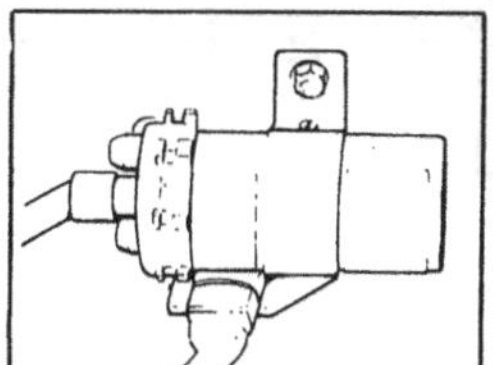
320

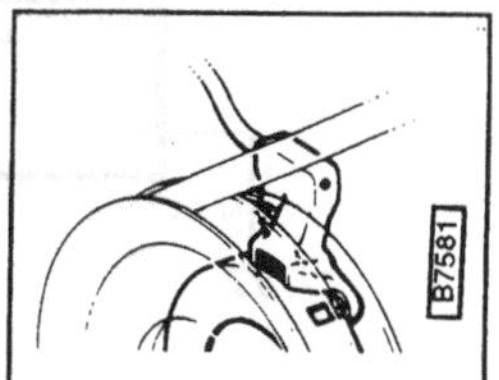

345

Ignition System—turbo

1985

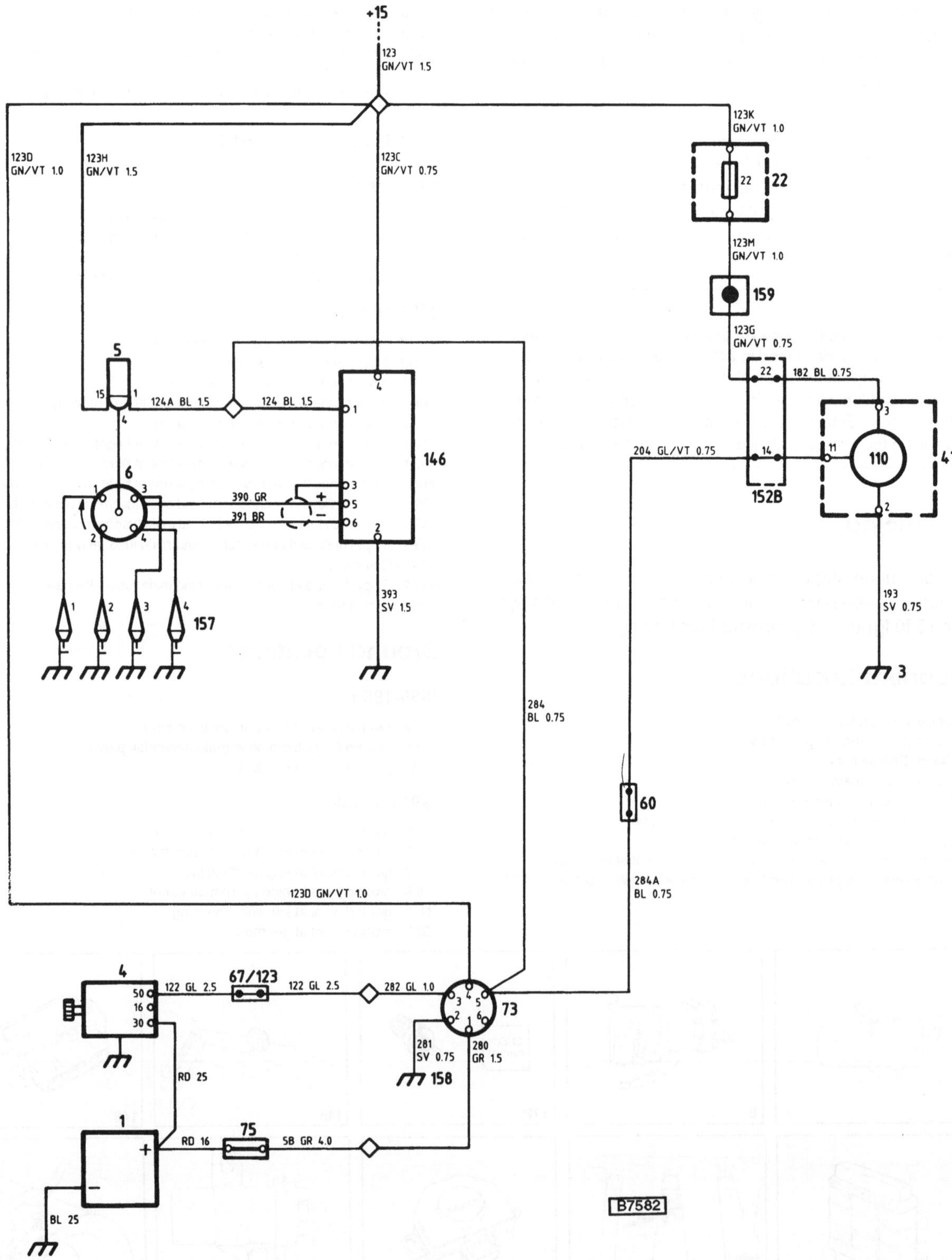

Ignition System—turbo

1986-1987

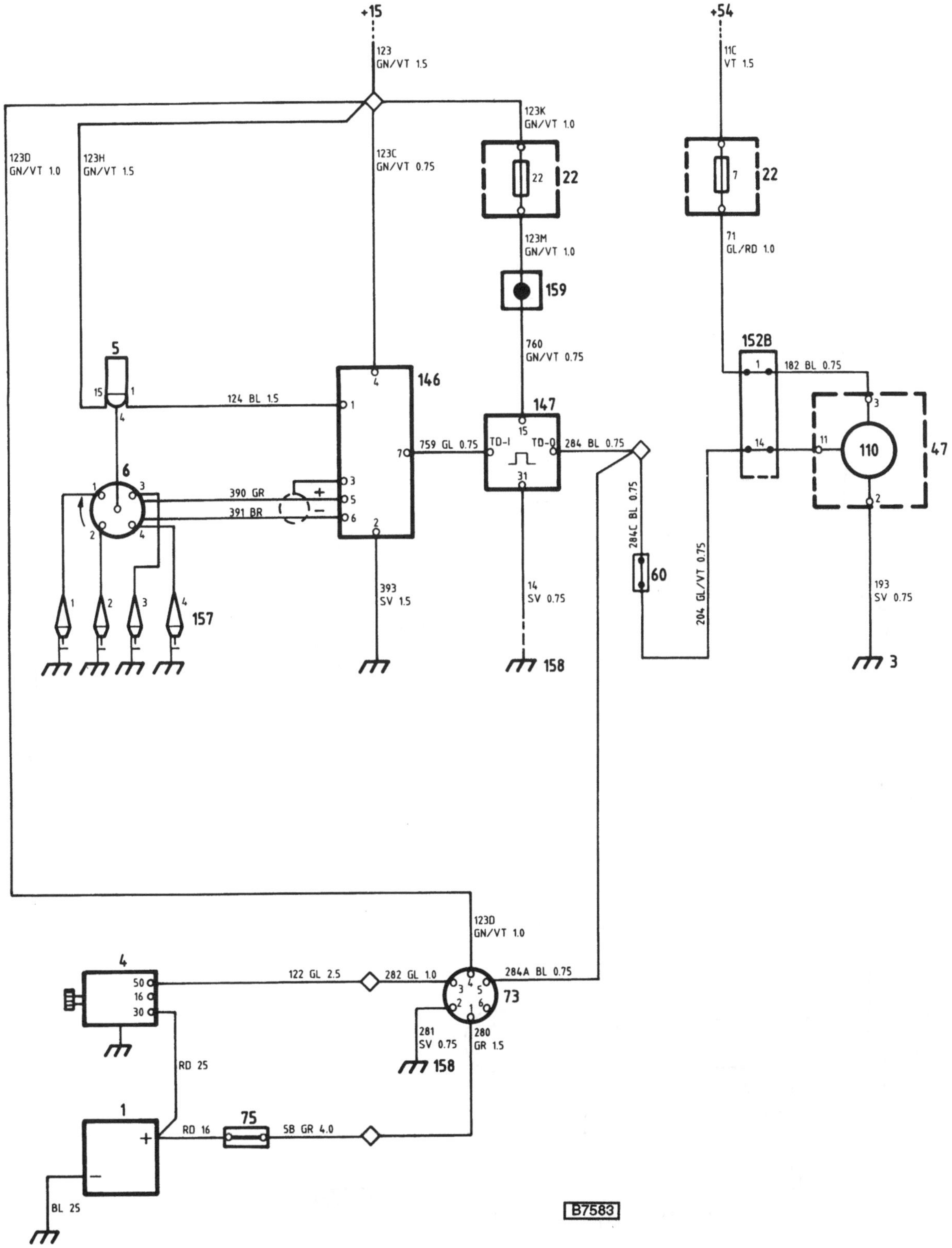

Ignition System—turbo

1988 and later

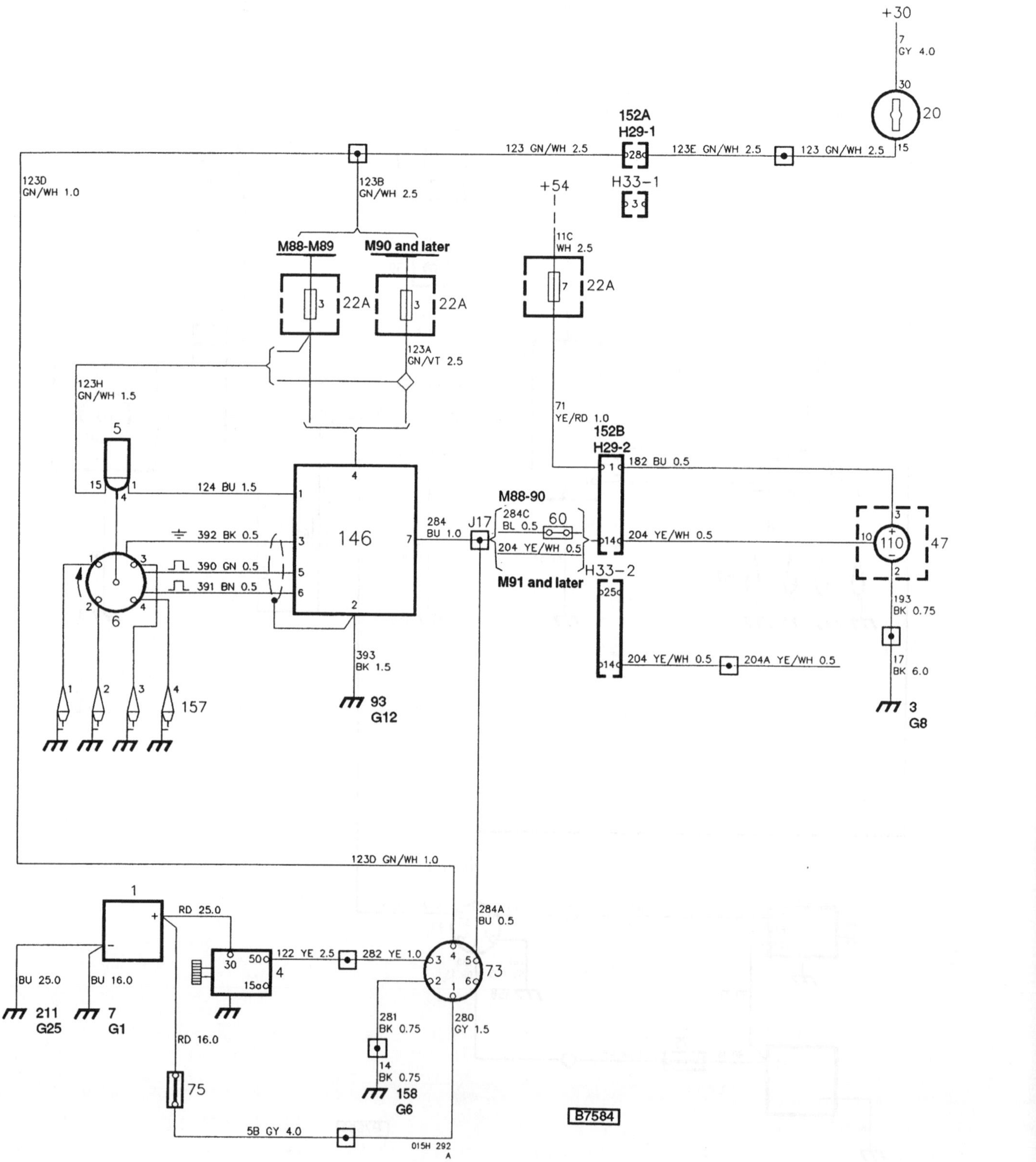

EZK Ignition System—non-turbo
1986-1987

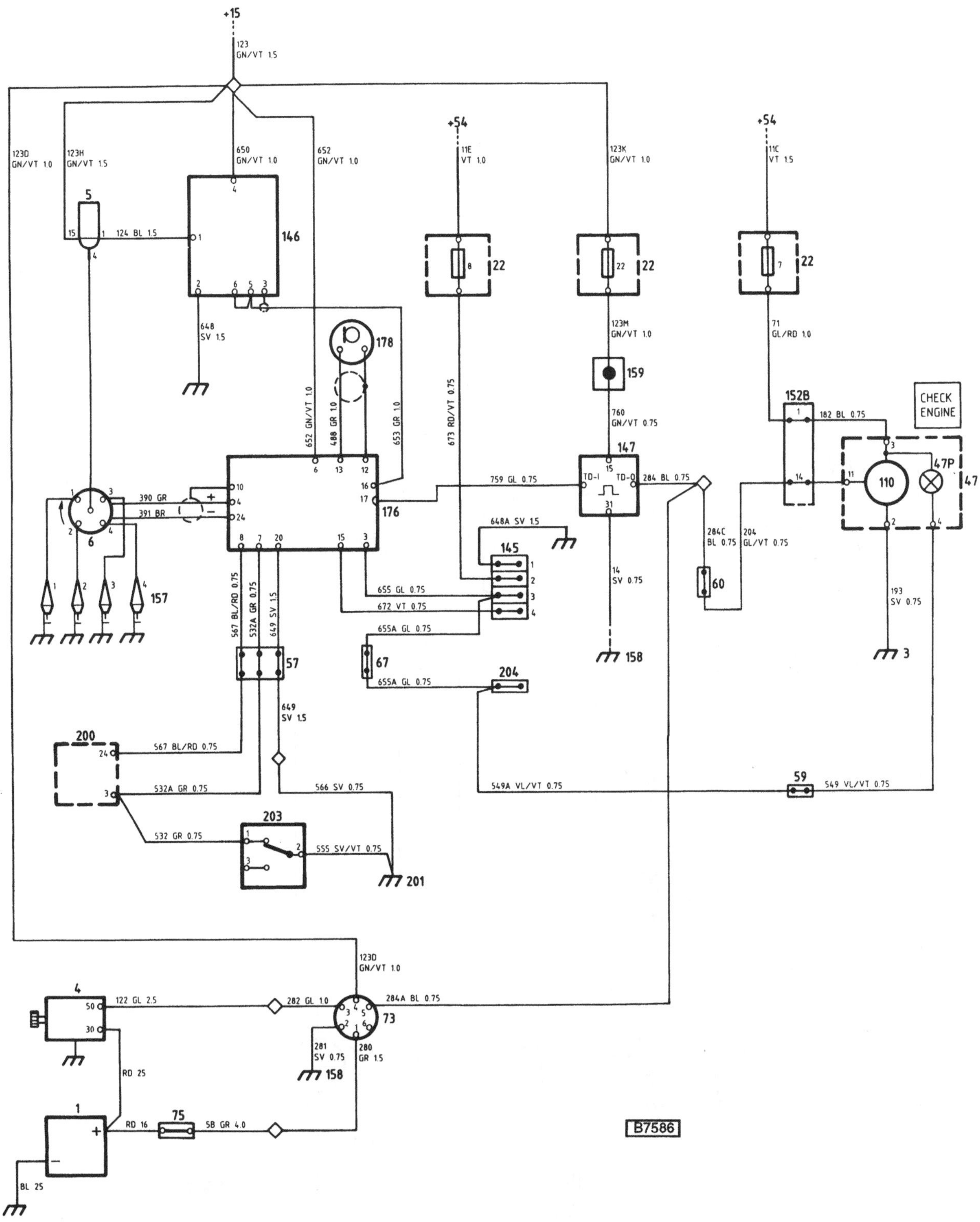

EZK Ignition System—non-turbo

1988

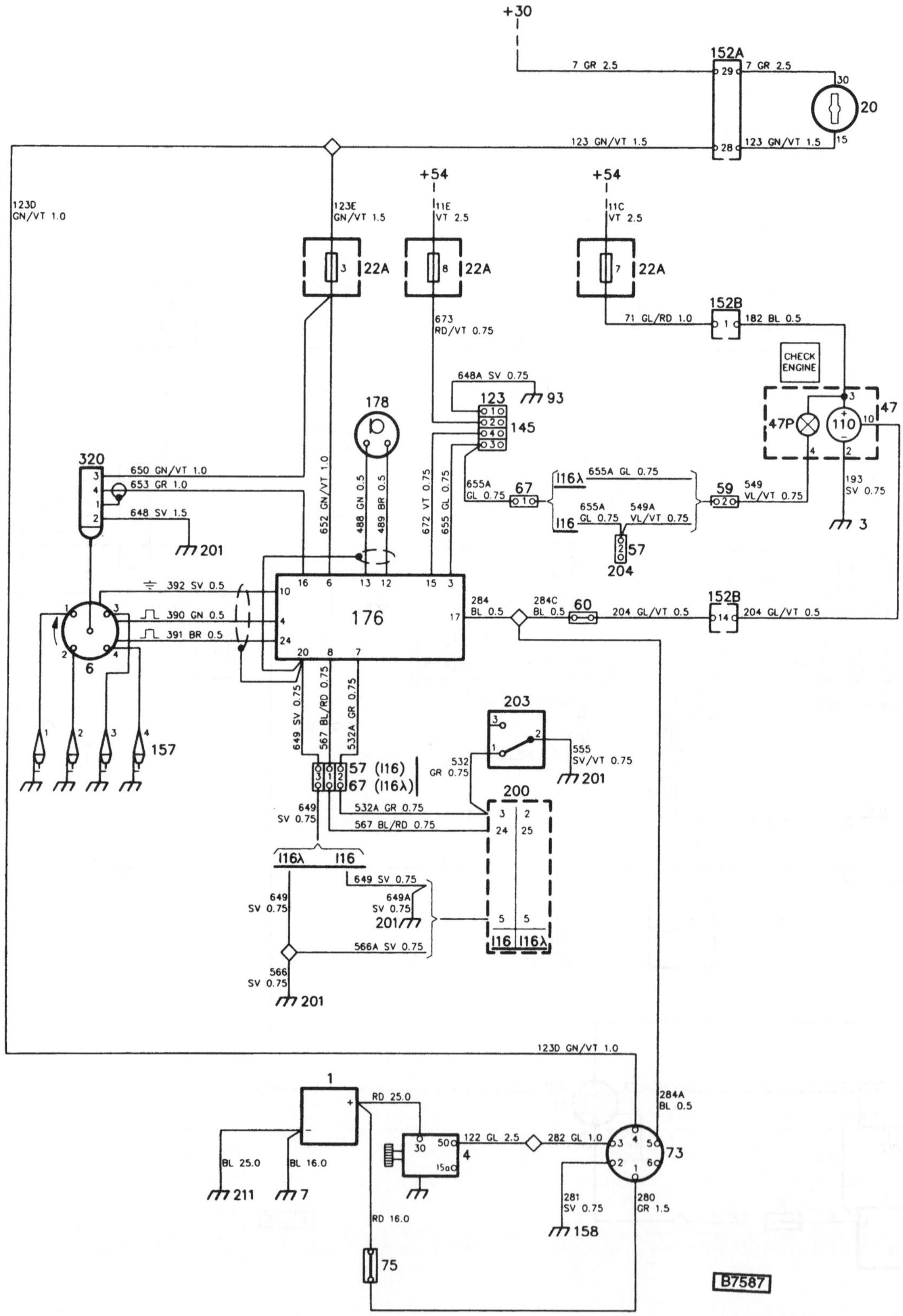

EZK Ignition System—non-turbo

1989

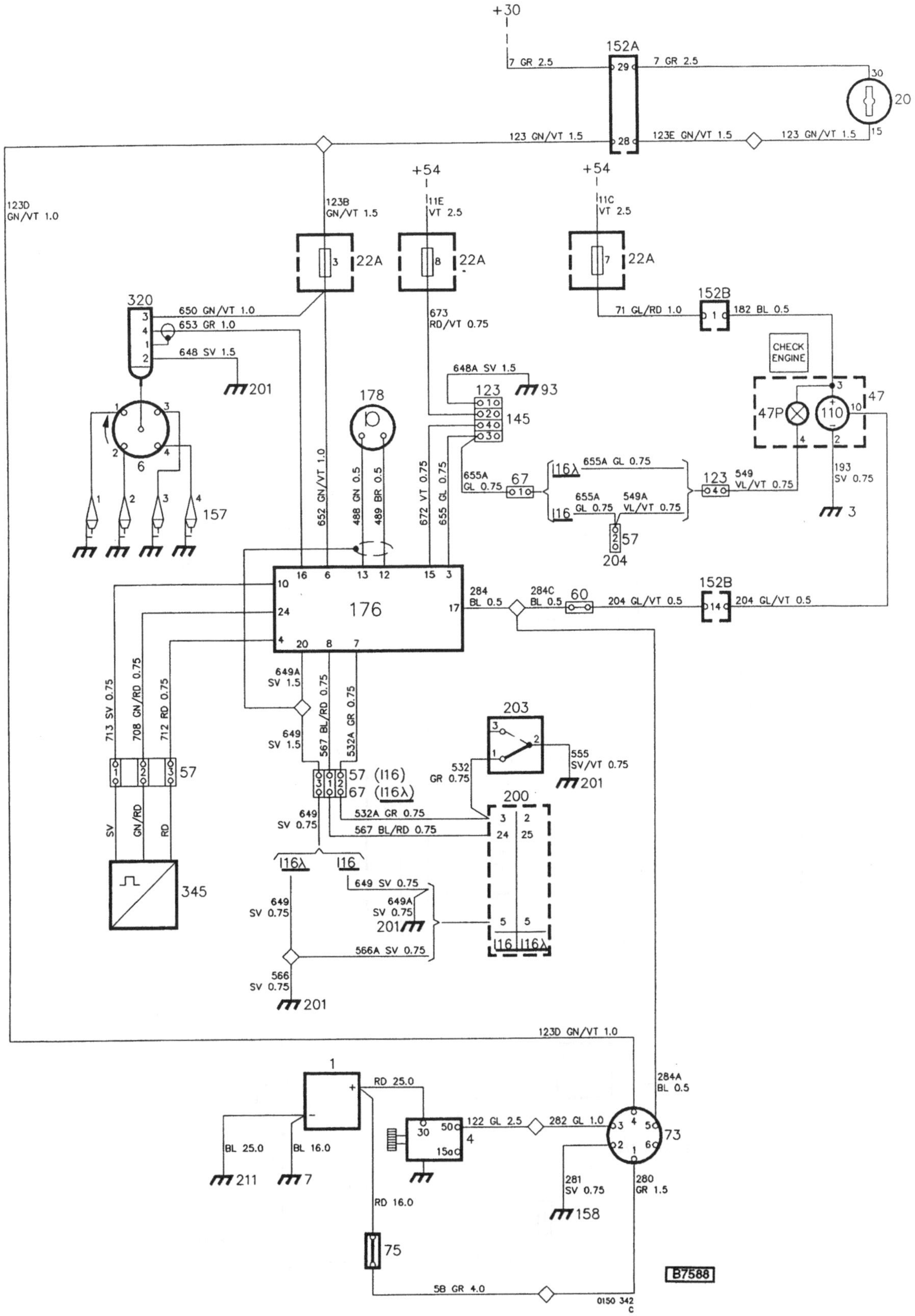

EZK Ignition System—non-turbo

1990

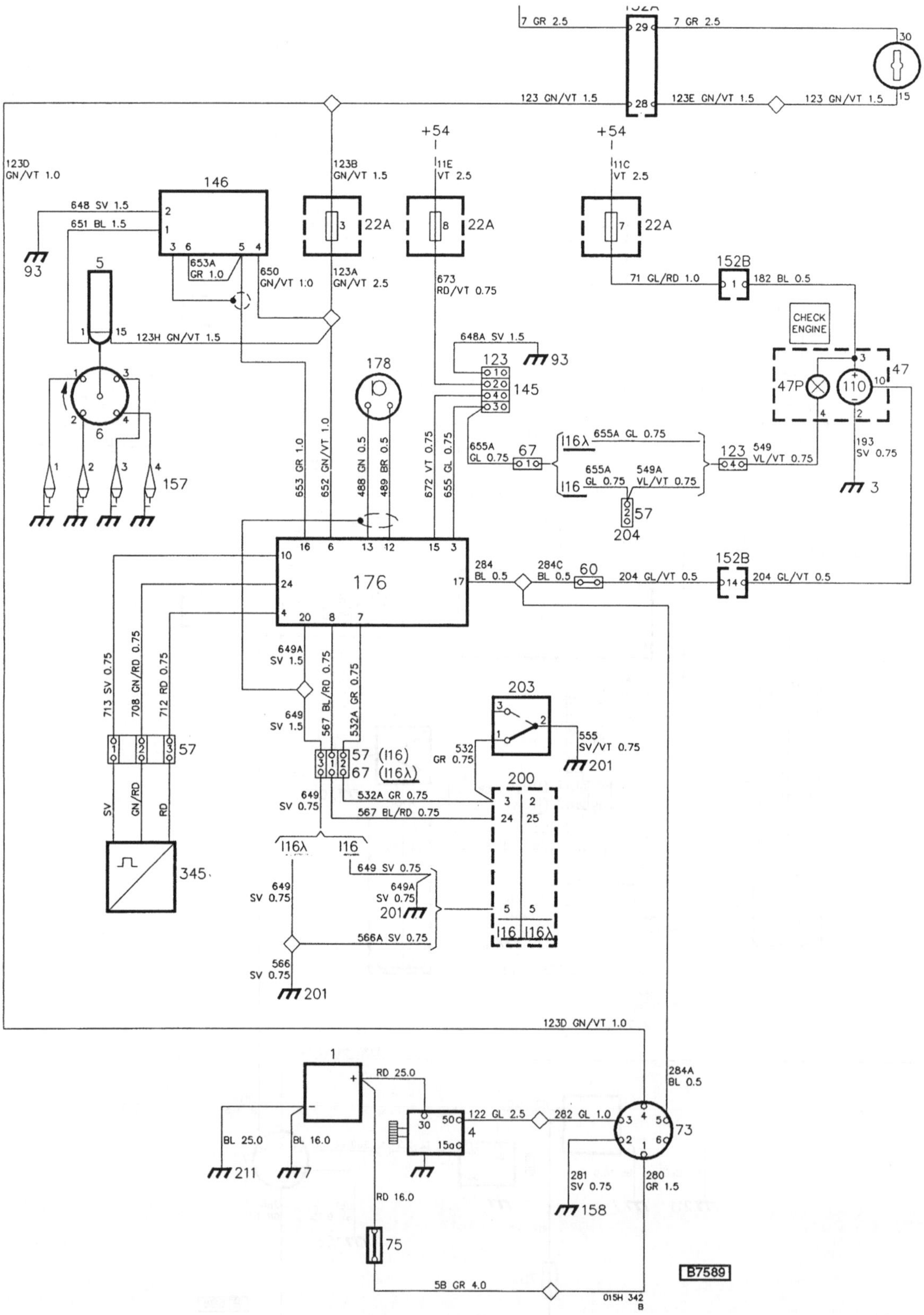

EZK Ignition System—non-turbo

1991 and later

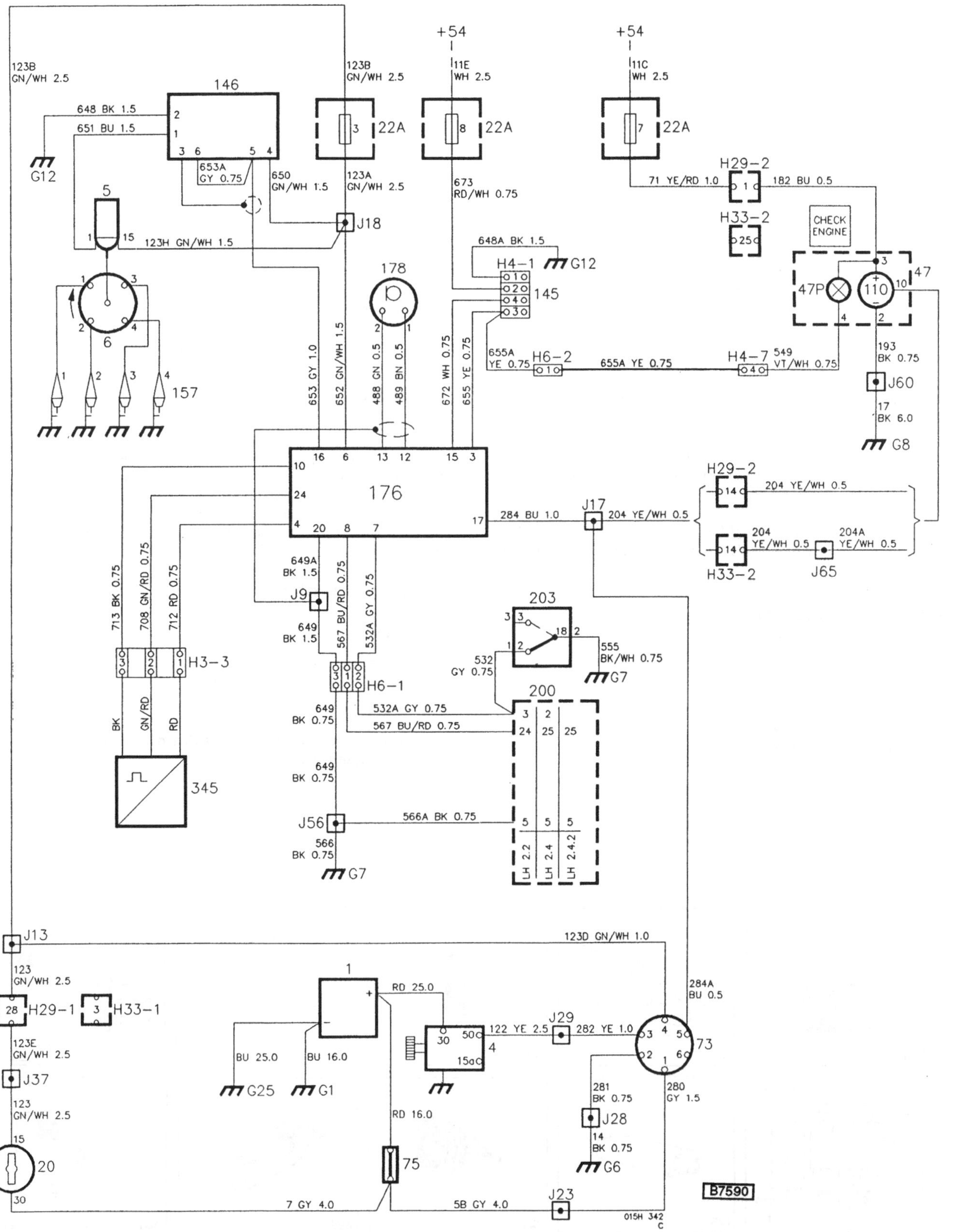

Interior Lighting

Operation

The interior lighting is controlled by the door switches on the door jambs. The door switches can be overridden by the center console switch or the combination roof light/switch (except convertible). A time delay relay is located under the back seat. On hatchback models, a cargo area light is controlled by a switch located below the lock striker plate .

The roof light bulbs or switches can be removed after removing the screws (where applicable) and carefully prying away the lens. The ignition switch light can be pulled straight down out of its holder after carefully peeling up the shift lever boot and reaching under the console. See **432 Manual Transmission Controls** for information on removing the center console.

Fault Tracing

1. Check the fuse. Check for battery voltage at the lamp assembly at all times.
2. Check for ground at the lamp assembly whenever a door is opened or when the center console switch is toggled.
3. Check door switches for a good ground through the chassis screw or switch assembly. Check the wiring to each door switch.

Component Locations

22 main fuse/relay panel
50 roof lamp/switch in the headliner
51 front roof lamp at rear view mirror
52 ignition switch light next to ignition switch, center console
53 interior light switch in center console
54 door switches on the door jamb
82 seat belt warning relay under the rear seat
151 time delay relay under the rear seat
175 central locking control unit behind instrument panel right side
225 reading lamp at rear seat
289 burglar alarm control unit under the rear seat
316 diode for the burglar alarm
401 auxiliary fuse/relay panel under rear seat

Connector Locations

1985-1990

57 3-pin connector under center console
58 12-pin connector behind instrument panel, left of steering column
59 2-pin connector in luggage compartment, left wheel housing
59 2-pin connector under roof lining at rear view mirror
59 2-pin connector at reading lamps (convertible)
60 1-pin connector under center console
98 10-pin connector to left of steering column
123 4-pin connector at hatchback, left side
152A 29-pin bulkhead connector at main fuse/relay panel
152B 29-pin bulkhead connector at main fuse/relay panel

1991 and later

H2-21 2-pin connector in luggage compartment, left wheel housing
H2-35 2-pin connector above roof lining at the rear view mirror
H2-46 2-pin connector at left reading lamp, convertible
H2-47 2-pin connector at right reading lamp, convertible
H12-1 12-pin connector behind instrument panel, left of steering column
H29-1 29-pin bulkhead connector behind main fuse/relay panel (1991-1991 1/2)
H33-1 33-pin bulkhead connector behind main fuse/relay panel (1991 1/2 and later)

Ground Locations

1985-1990

65 ground point under the handbrake
117 ground point under the handbrake

1991 and later

G4 ground point between ignition switch and handbrake
G5 ground point under the rear seat

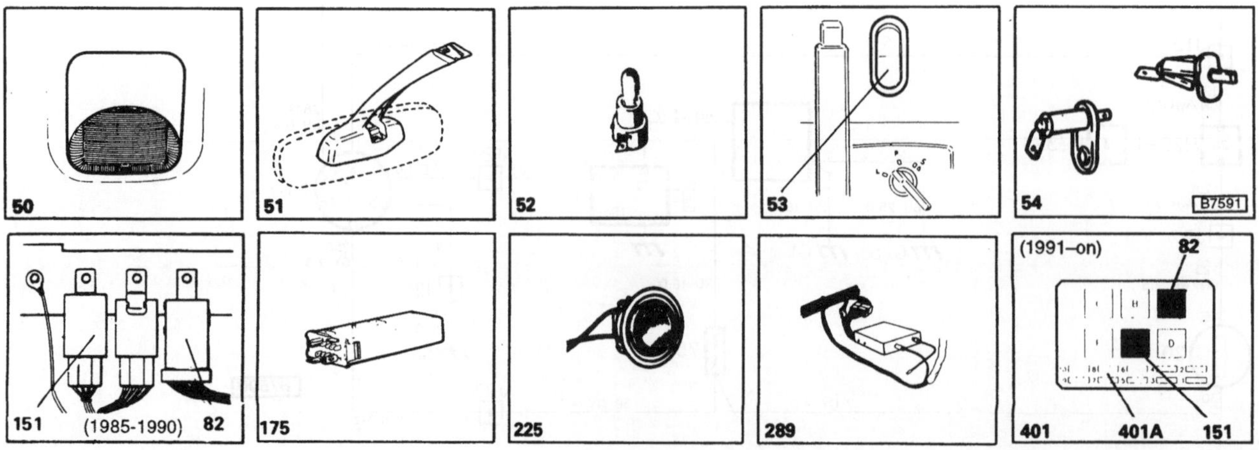

Interior Lighting—ex. Convertible

1985-1986

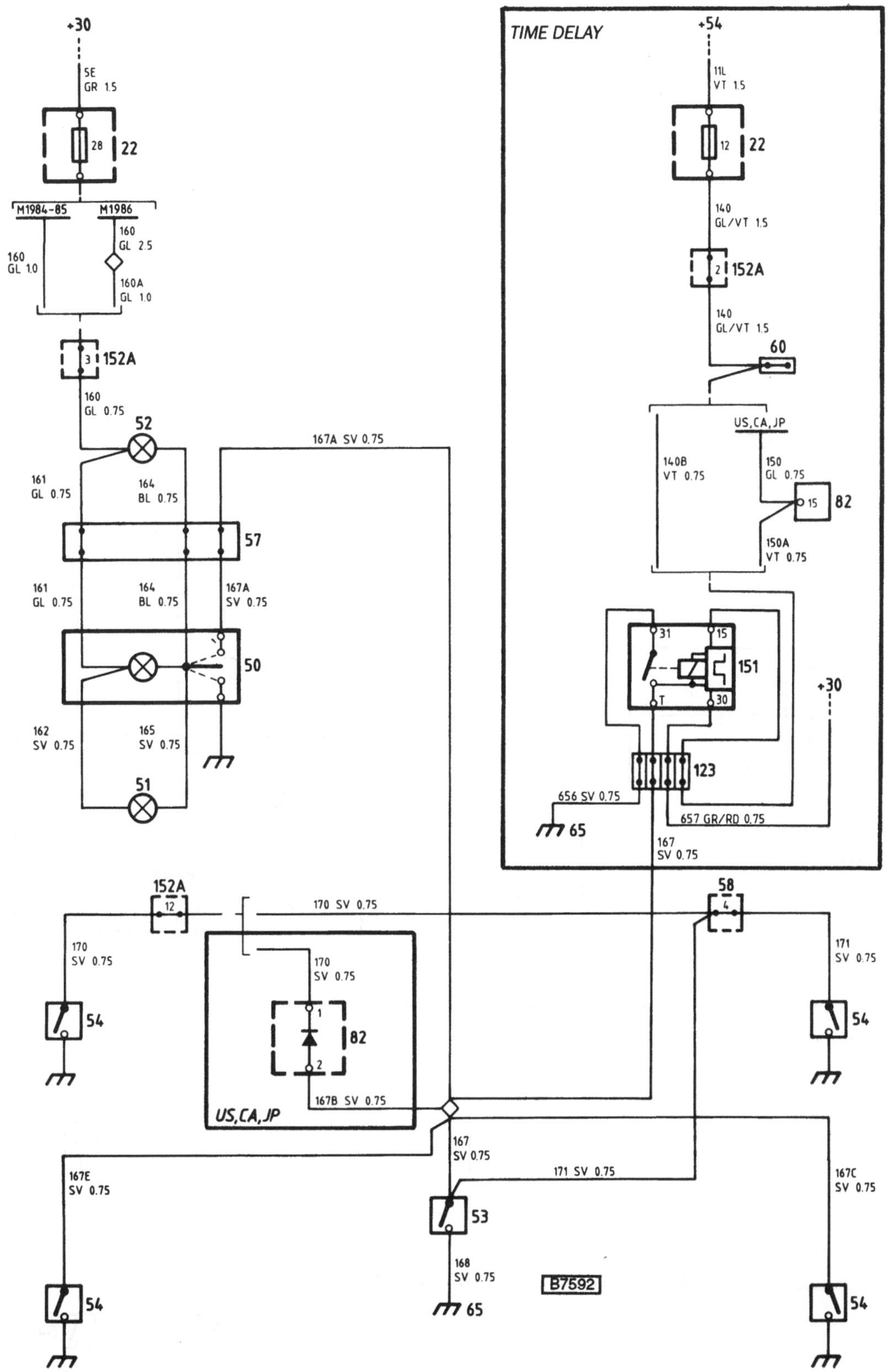

Interior Lighting—ex. Convertible
1987-1990

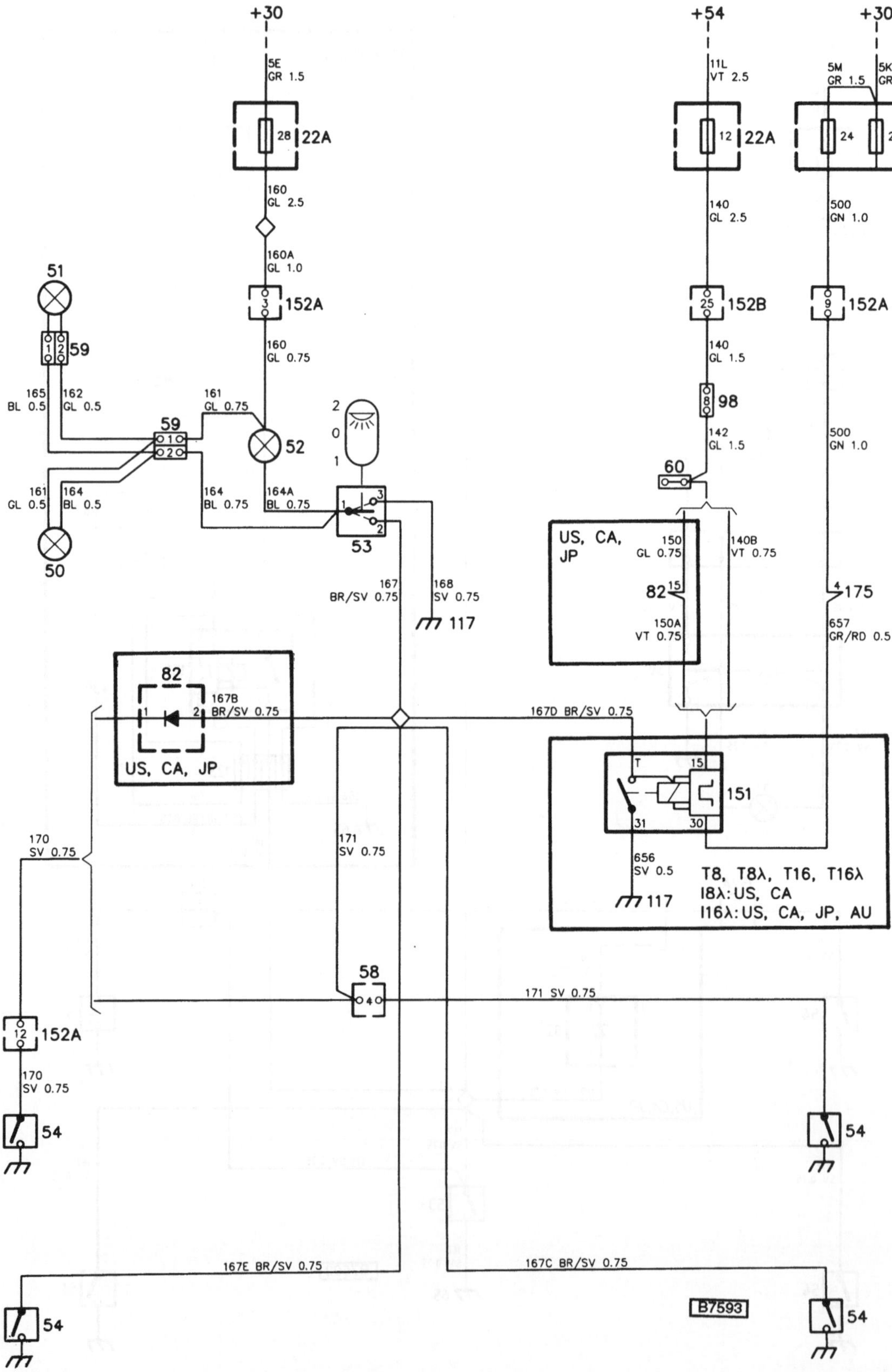

Interior Lighting—ex. Convertible

1991 and later

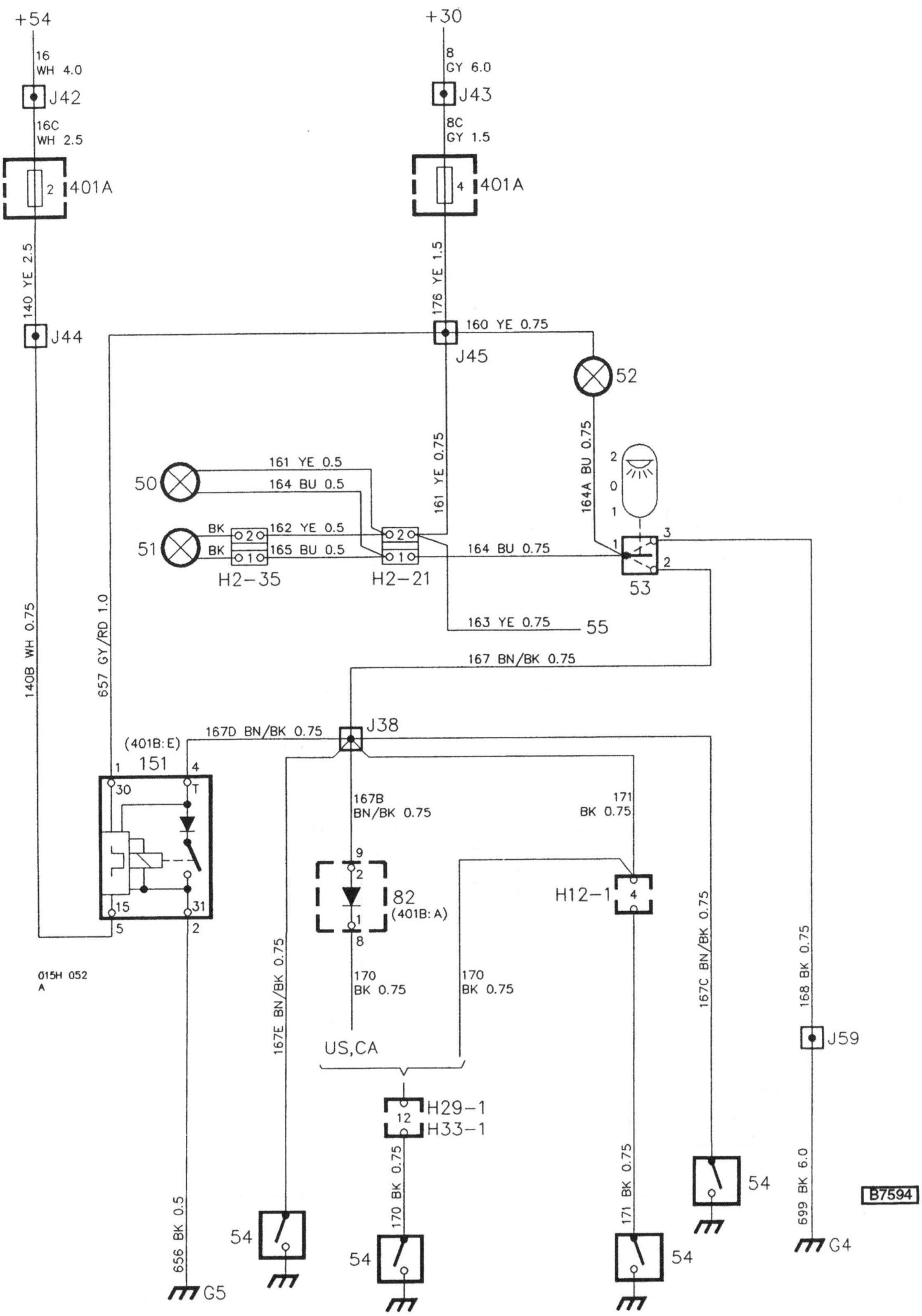

Interior Lighting — Convertible

1987-1990

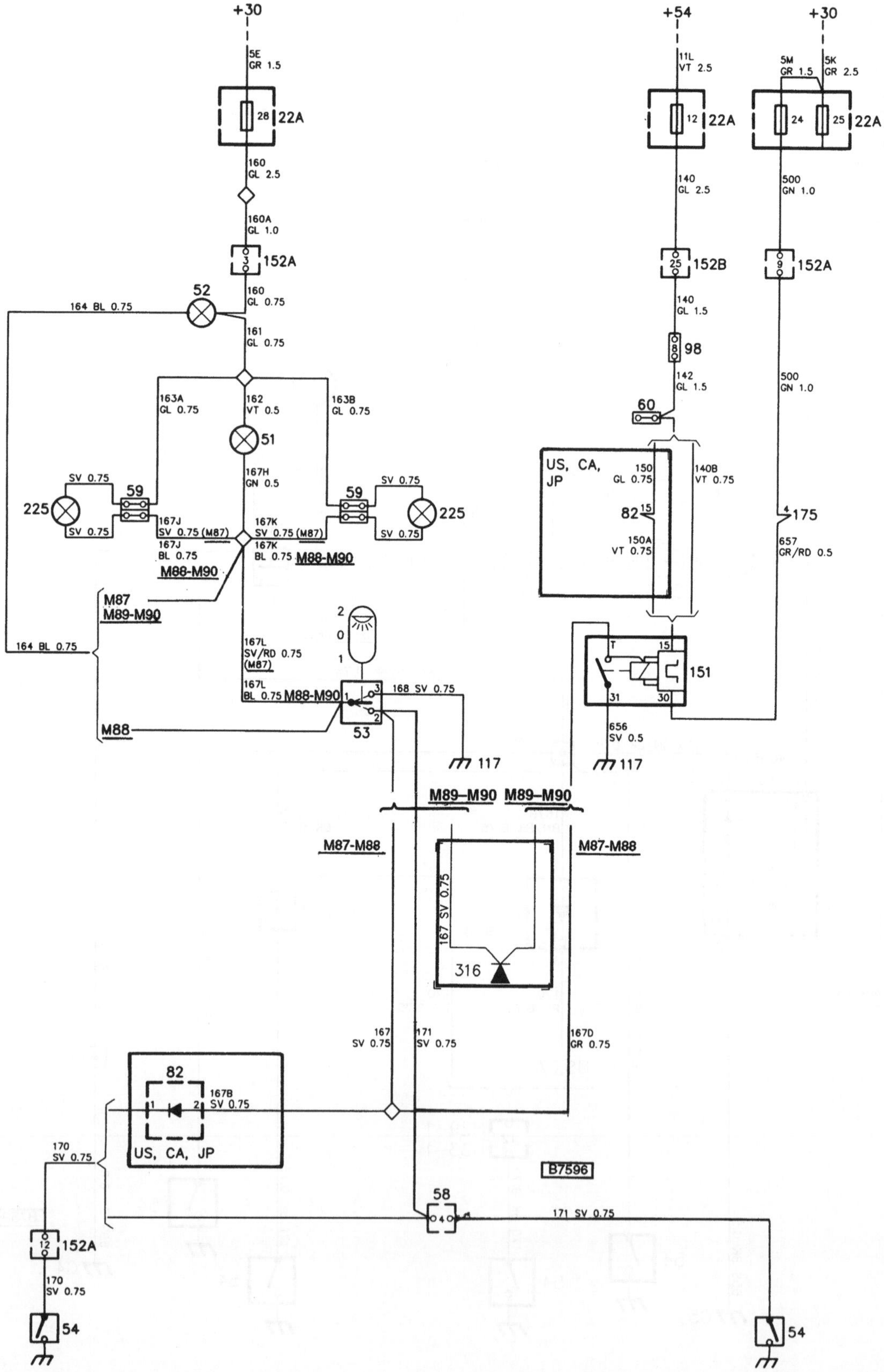

Interior Lighting — Convertible

1991 and later

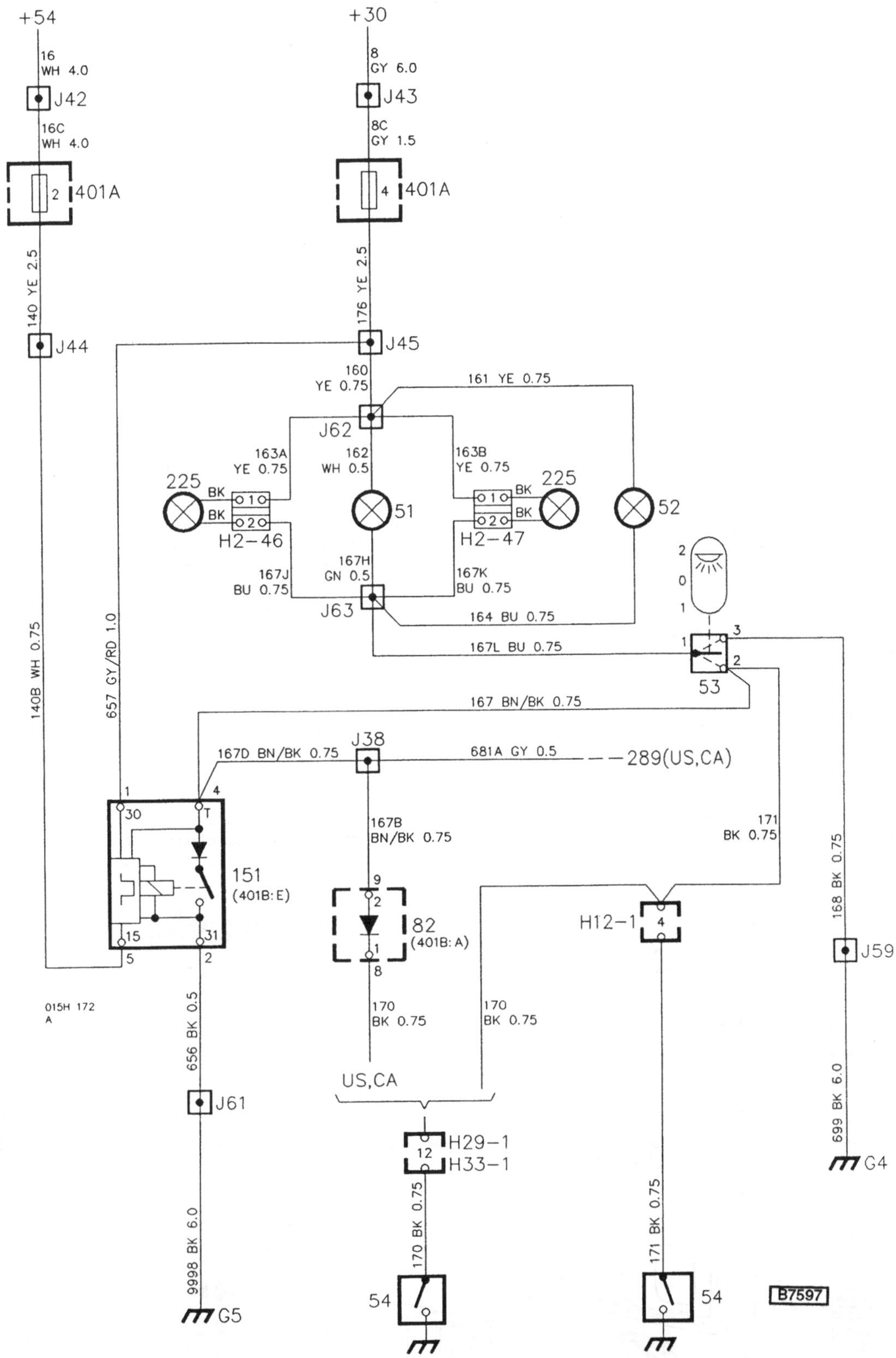

Lighting for Controls

Operation

The Lighting for Controls wiring diagrams cover instrument panel and switch illumination. Some of the switches have replaceable bulbs. See **364 Electric Controls and Switches** for information on replacing switch bulbs. See **853 Dashboard and Consoles** for information on removing the instrument cluster. See **432 Manual Transmission Controls** or **444 Automatic Transmission Controls** for details on removing the center console.

Fault Tracing

1. Check the fuses and check for voltage at rheostat (16) and rheostat (17) with the key in the run position.
2. Check that the rheostats are fully clockwise.
3. Check the ground point to the left of the steering column behind the instrument panel, and the ground point under the center console at the handbrake lever.

Component Locations

10 light switch in instrument panel
16 rheostat in instrument cluster
17 rheostat in instrument panel
18 combined instrument lighting
22 main fuse/relay panel, left wheel housing
25 hazard warning light switch in instrument panel
47 cigarette lighter
48 instrument cluster
88 fog lamp switch in instrument panel
91 gear position light (automatic) in center console
116 heated rear window switch in instrument panel
143 A/C recirculation switch in instrument panel
148 ashtray light in instrument panel
153 cigarette light in instrument panel
154 heater control lights in instrument panel
162 driver power window switch light in center console
163 passenger power window switch light in center console
169 A/C switch in instrument panel
181 sunroof switch light in center console
189 lockout switch for rear power windows in center console
190 left rear power window switch in center console
191 right rear power window switch in center console
252 driver seat rheostat light in instrument panel
386 switch for power windows and sunroof in center console
425 power door lock switch in center console

Connector Locations

1985-1990

57 3-pin connector (3) one behind left B pillar trim, one behind right B pillar trim, one under driver's seat
58 12-pin connector behind instrument panel, to left of steering column
59 2-pin connector in instrument panel, behind cigarette lighter
60 1-pin connector (3) one in each rear door, one under center console
98 10-pin connector behind instrument panel, for radio

1991 and later

H1-6 1-pin connector at right B pillar, behind trim
H1-7 1-pin connector at left B pillar, behind trim
H1-17 1-pin connector under center console
H2-10 2-pin connector behind cigarette lighter
H3-10 3-pin connector at right B pillar, behind trim
H3-11 3-pin connector at left B pillar, behind trim
H10-1 10-pin connector behind instrument panel, left of steering column
H12-1 12-pin connector behind instrument panel, left of steering column
H29-2 29-pin bulkhead connector behind main fuse/relay panel (1991-1991 1/2)
H33-2 33-pin bulkhead connector behind main fuse/relay panel (1991 1/2–on)

Ground Locations

1985-1990

3 ground point behind instrument panel
117 ground point under the handbrake

1991 and later

G4 ground point between ignition switch and handbrake
G8 ground point behind instrument panel

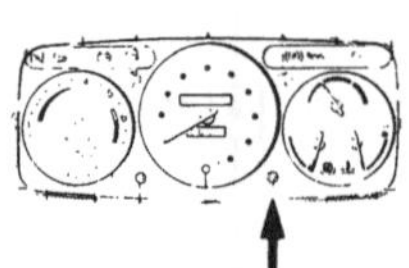
16

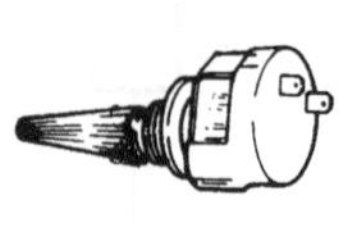
17

25

18
47

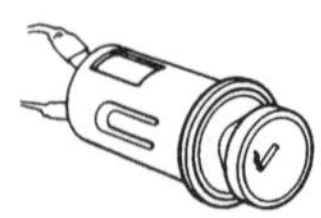
48

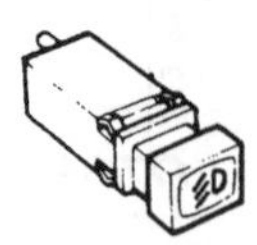
88

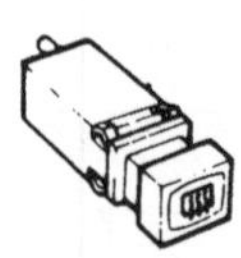
116

143

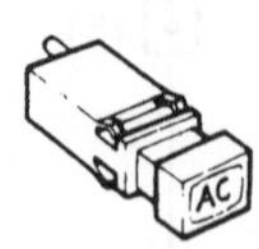
169

162
163
189
190
191

181

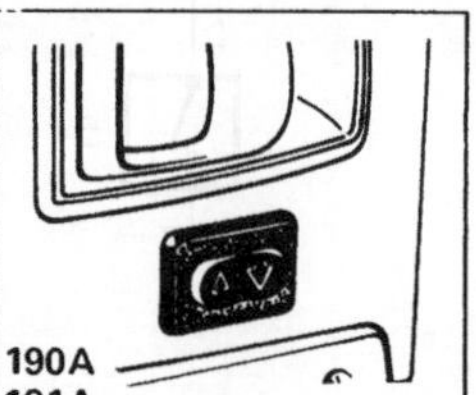
190A
191A

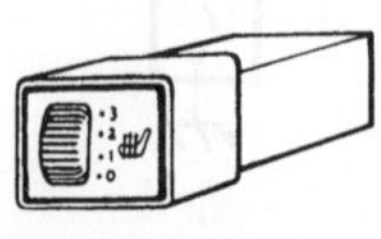
252

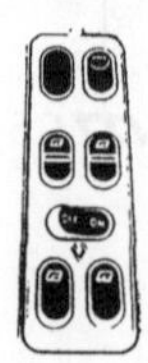
386

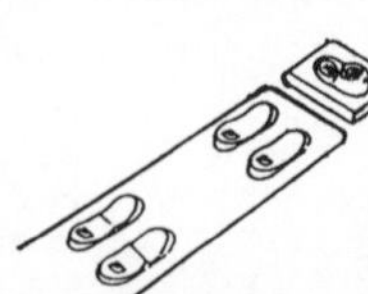
425 (M1992 - on)

B7598

Lighting for Controls

1985-1986

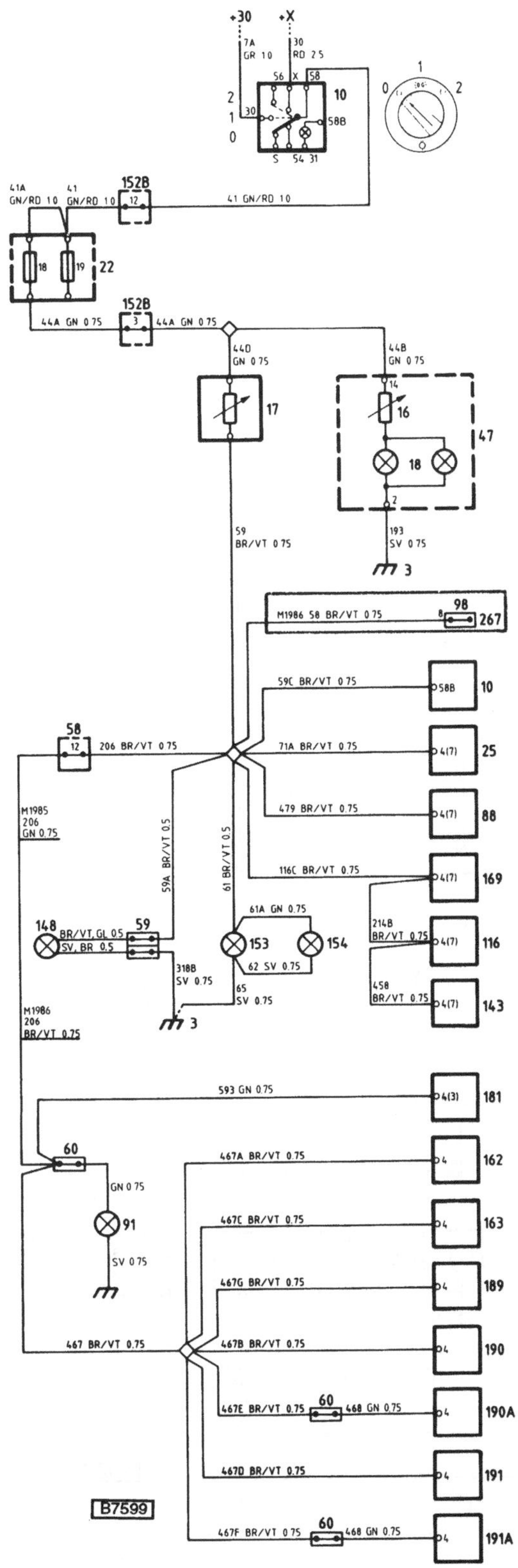

Lighting for Controls

1987-1990

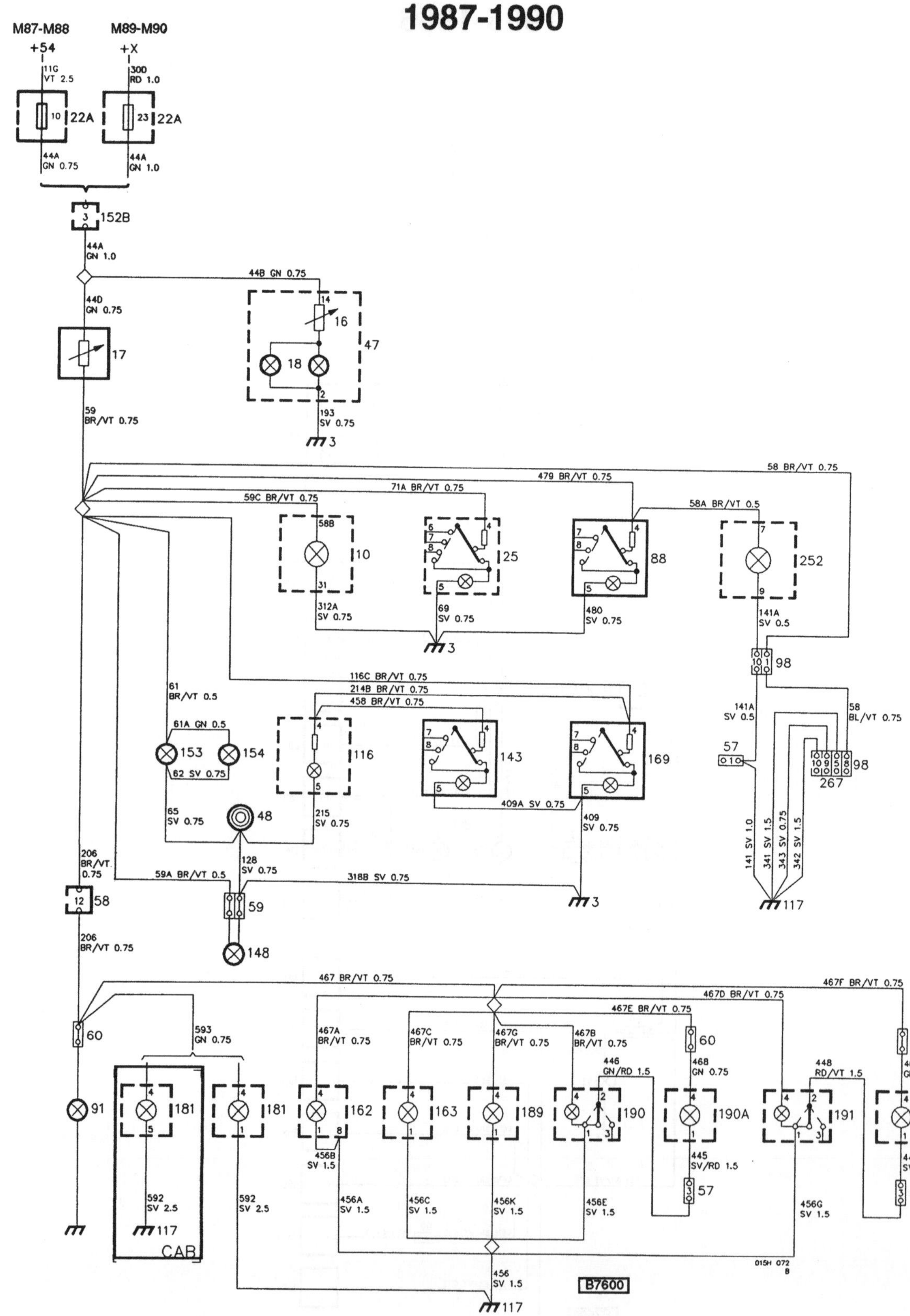

Lighting for Controls

1991 and later

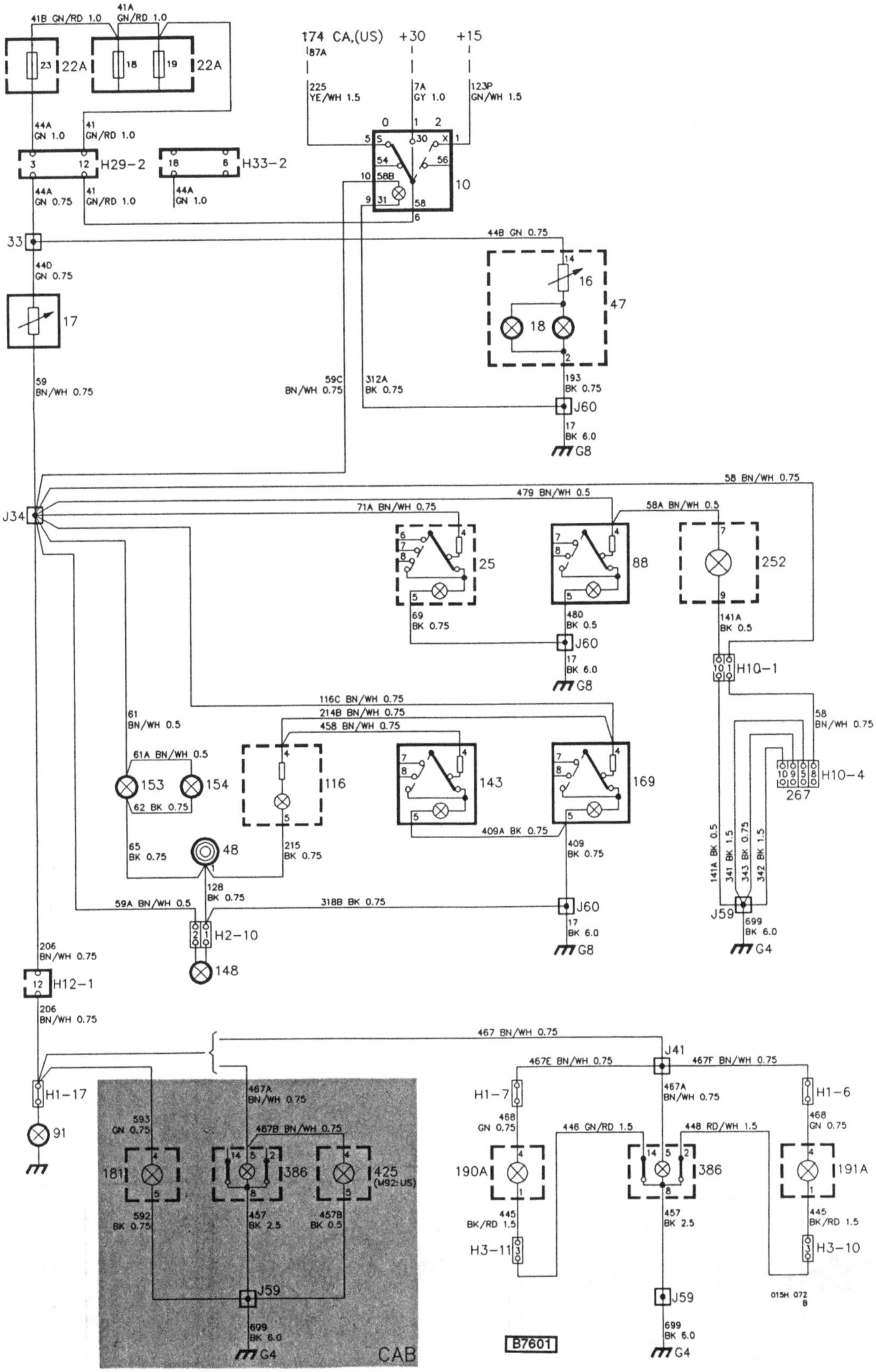

Luggage Compartment Lighting

Operation

The luggage compartment light (55) is controlled by a luggage compartment light switch located behind the left trunk hinge on sedans and beneath the striker plate on hatchbacks. With the trunk open, the switch is closed to light the lamp. The luggage compartment light (55) also has a switch to disable the light at all times. To change the bulb, pry off the lens. Convertible models use a mercury switch (position sensitive) rather than a mechanical switch.

Fault Tracing

1. Check the fuse.
2. Check for battery voltage at the lamp assembly.
3. Check for ground at the light switch.

Component Locations

1 battery
15 license plate light on the rear sill (convertible only)
22A main fuse/relay panel
52 ignition switch light in center console
55 luggage compartment light in storage area
56 luggage compartment light switch in storage area
75 terminal block at right engine bulkhead
401 auxiliary fuse/relay panel under rear seat

Connector Locations

1985-1990

57 3-pin connector at left side of luggage compartment (1985-1986)
59 2-pin connector at left side of luggage compartment (1987-1990)
152A 29-pin bulkhead connector at main fuse/relay panel

1991 and later

H2-21 2-pin connector in luggage compartment, left wheel housing
H2-43 2-pin connector at left hinge, convertible
H2-42 2-pin connector in trunk lid

Ground Locations

1991 and later

G1 ground point at radiator cross member
G5 ground point under the rear seat
G25 ground point at gearbox

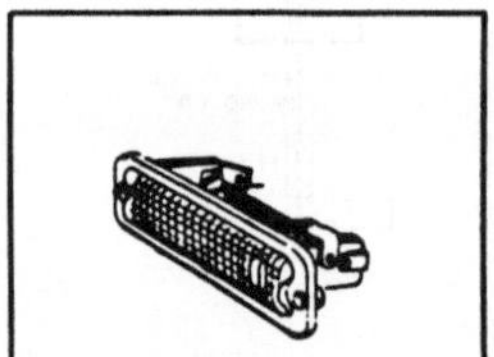
55

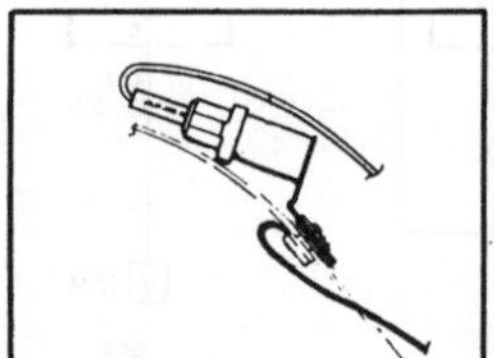
56 (conv)

56

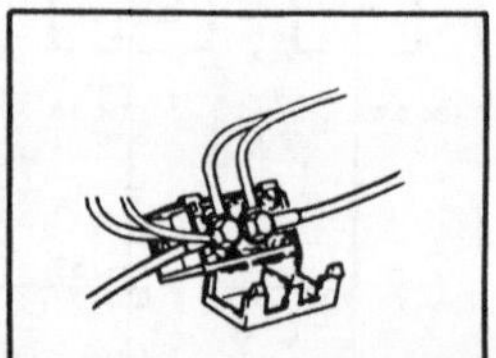
75

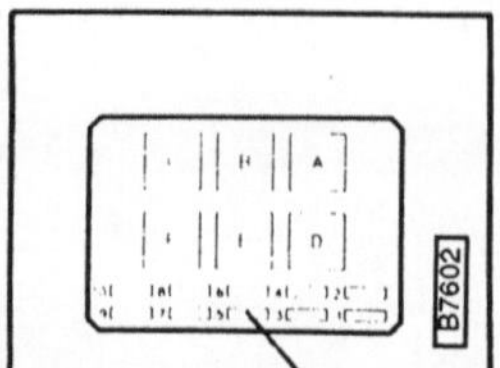

401 401A

Luggage Compartment Lighting

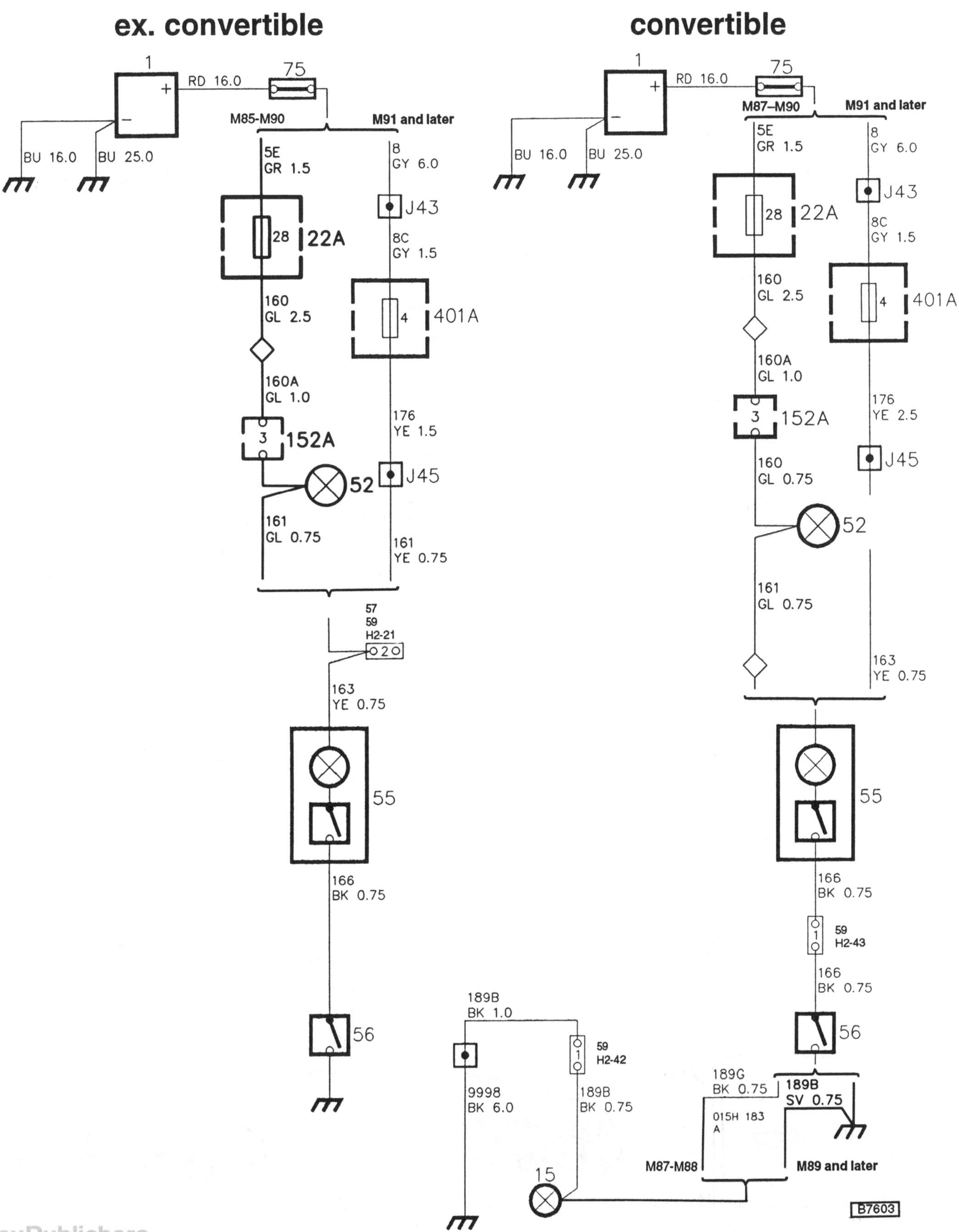

Parking Lights

Operation

When the light switch (10) is in position 1, battery voltage powers the parking lights and the side marker lights. See **351 Lighting** for repair information and information on replacing bulbs.

Fault Tracing

1. Check for battery voltage at terminal 3 (30) of the light switch (10). Check fuses 18 and 19.

Component Locations

10 light switch in instrument panel
13 park lights
14 rear lights
15 license plate lights
22 main fuse/relay panel in engine compartment
234 side marker lights in the front cluster

Connector Locations

1985-1990

57 3-pin connector in luggage compartment, near air outlet
59 2-pin connector in luggage compartment
81 driving lights, below front bumper
152A 29-pin bulkhead connector at main fuse/relay panel
152B 29-pin bulkhead connector at main fuse/relay panel

1991 and later

H2-7 2-pin connector behind instrument panel
H2-18 2-pin connector at hatchback, left side
H2-19 2-pin connector in trunk lid
H3-13 3-pin connector in luggage compartment, left air outlet
H4-12 4-pin connector in trunk lid, left air outlet
H29-1 29-pin bulkhead connector behind main fuse/relay panel (1991-1991 1/2)
H29-2 29-pin bulkhead connector behind main fuse/relay panel (1991-1991 1/2)
H33-1 33-pin bulkhead connector behind main fuse/relay panel (1991 1/2 and later)
H33-2 33-pin bulkhead connector behind main fuse/relay panel (1991 1/2 and later)

Ground Locations

1985-1990

3 ground point behind instrument panel
7 ground point at radiator cross member
9 ground point in luggage compartment
93 ground point at left wheel housing (1989-1990)

1991 and later

G1 ground point at radiator cross member
G3 ground point in luggage compartment
G8 ground point behind instrument panel
G9 ground point in luggage compartment
G12 ground point at left wheel housing
G19 ground point at hatchback

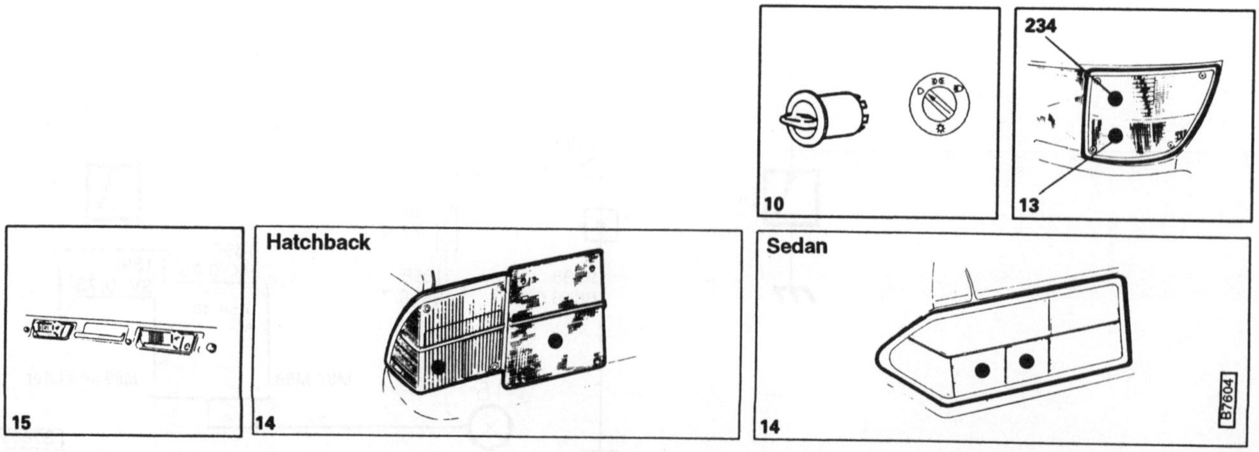

Parking Lights

1985-1986

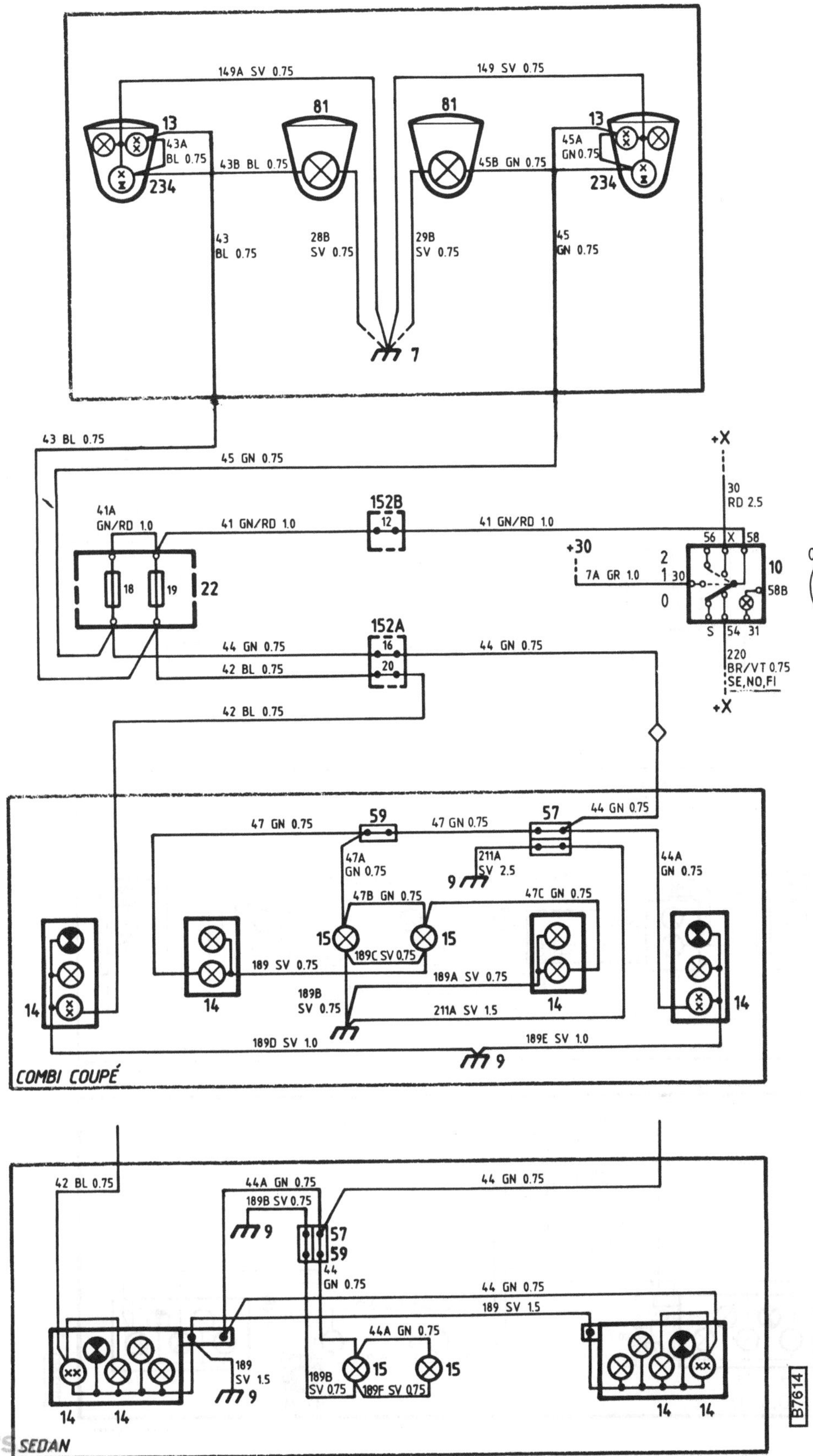

Parking Lights

1987 and later

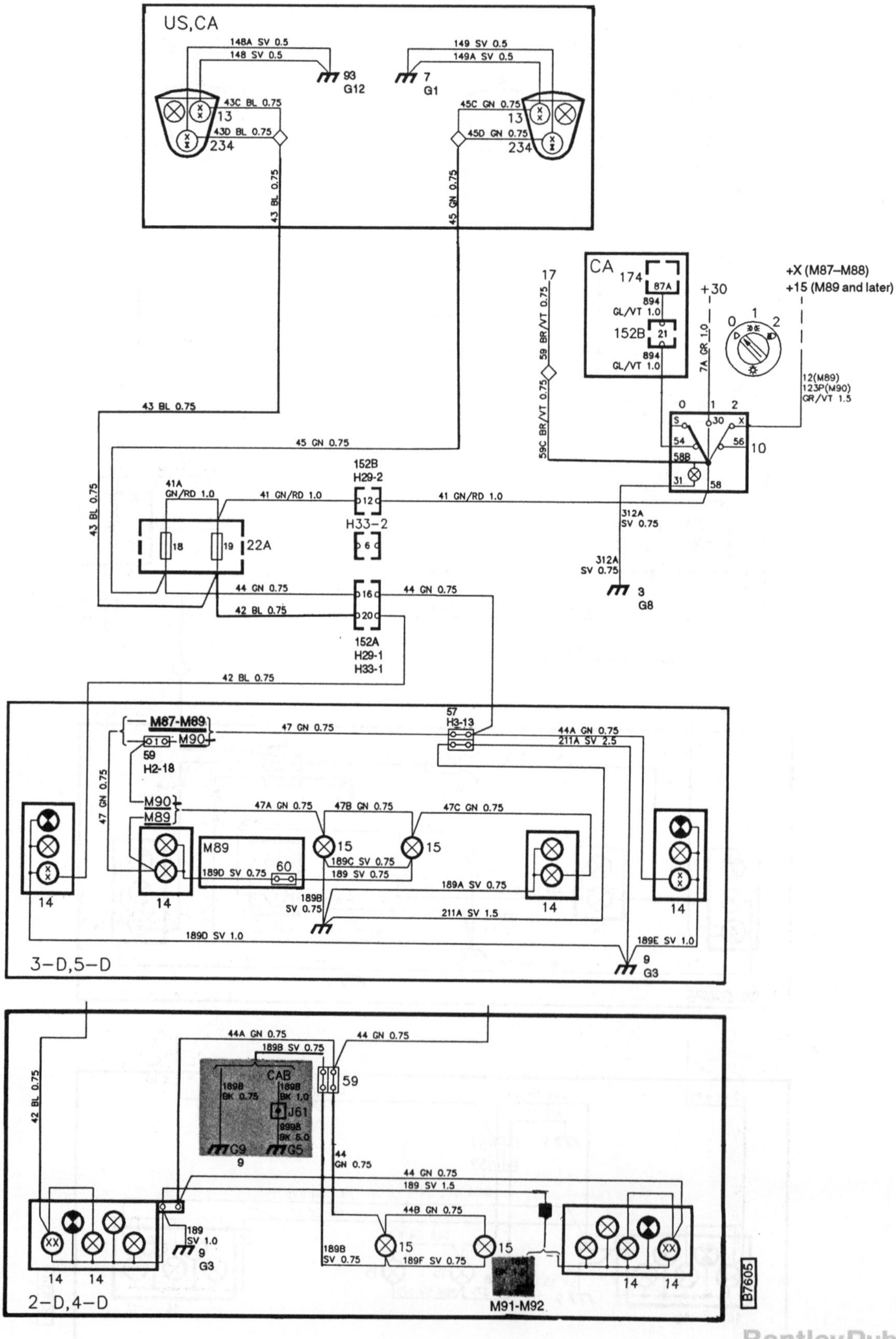

Passive Seat Belts

Operation

The passive seat belt consists of a motor-driven, two-point diagonal strap and manual lap belt. The diagonal strap is controlled and monitored by the logic box (306). The belt travel is controlled by two limit switches at the A pillar and at the B pillar. A warning lamp on the dash (47S) will flash or remain continuously lit if a fault is detected in the seat belt travel. The passive seat belt fuses, relays, and logic box are located under the rear seat on the driver's side.

WARNING
If the seat belt warning lamp on the instrument cluster lights up at any time, do not drive the car. Have an authorized Saab dealer inspect and correct any faults in the system before the car is driven.

Fault Tracing

CAUTION
Always disconnect the connector to the logic box (306) before making any measurements.

1. Check the fuses under the rear seat. Also check fuses 5, 12, and 13 in the main fuse relay panel (22).
2. Using a voltmeter, check for battery voltage between the disconnected logic box connector terminals and ground with the key in the run position. See **Table a.**

Table a. Passive Seat Belt Logic Box Connector Test Voltage

Logic box connector terminal	Comments
11	reverse gear engaged
12	ignition switch in run position
18	via the relay coil (312P)
19	via the relay coil (311P)
22	via the relay coil (312D)
23	via the relay coil (311D)
25	via the g sensor (309)

3. Using an ohmmeter, check for continuity between the specified logic box connector terminal and ground with the key off. See **Table b.**

Table b. Passive Seat Belt Continuity Checks

Logic box connector terminal	Resistance (ohms)	Comments
1	0	belt in the carriage
2	approx. 13	bulb resistance
4	approx. 22	coil resistance
5	open circuit	carriage at A pillar
6	0	—
7	0	passenger door open
8	approx. 22	coil resistance
9	open circuit	carriage at A pillar
10	0	—
13	0	—
14	0	belt in the carriage
17	open circuit	—
20	0	lap strap not fastened
21	open circuit	—
24	0	driver door open

Component Locations

20 ignition switch
22A main fuse/relay panel at left wheel housing
31 backup light switch
47S passive seat belt warning lamp
70 seat belt switch, driver side between driver and passenger seat
72 seat belt warning light in center of instrument panel
76 switch for raising idle speed (automatic transmission) in center console at shift lever
82 seat belt warning relay under the rear seat
208D passive seat belt door switch in driver door
208P passive seat belt door switch in passenger door
306 logic box for passive seat belts under rear seat, left side
307D passive seat belt belt reel at driver seat
307P passive seat belt belt reel at passenger seat
308D passive seat belt motor with limit switches at driver seat
308P passive seat belt motor with limit switches at passenger seat
309 passive seat belt g sensor under the rear seat, left side
310D passive seat belt fuse for driver seat under rear seat
310P passive seat belt fuse for passenger seat under rear seat
311D passive seat belt motor relay for driver seat under rear seat
311P passive seat belt motor relay for passenger seat under rear seat
312D passive seat belt motor relay for driver seat under rear seat
312P passive seat belt motor relay for passenger seat under rear seat

Passive Seat Belts

Connector Locations

57 3-pin connector behind instrument panel, to left of steering column
58 12-pin connector behind instrument panel, to left of steering column
59 2-pin connector (2) in engine compartment, one at top left hinge, one at top right hinge
60 1-pin connector (4) two in each door, behind door trim
98 10-pin connector behind instrument panel to left of steering column
122 8-pin connector for burglar alarm under the rear seat
152A 29-pin bulkhead connector in main fuse/relay panel
152B 29-pin bulkhead connector in main fuse/relay panel

Ground Locations

3 ground point behind instrument panel
65 ground point at handbrake
117 ground point at handbrake

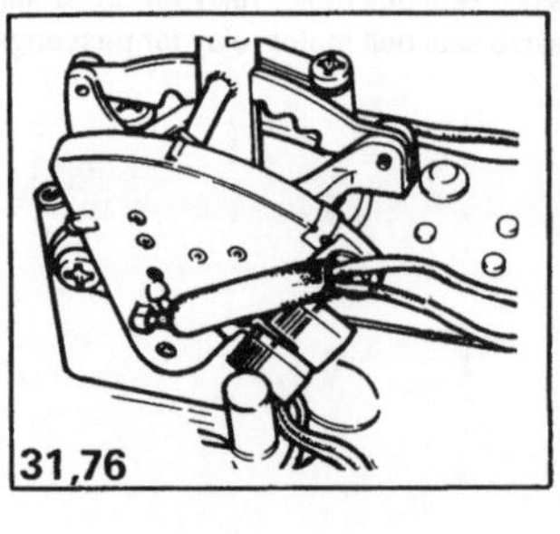

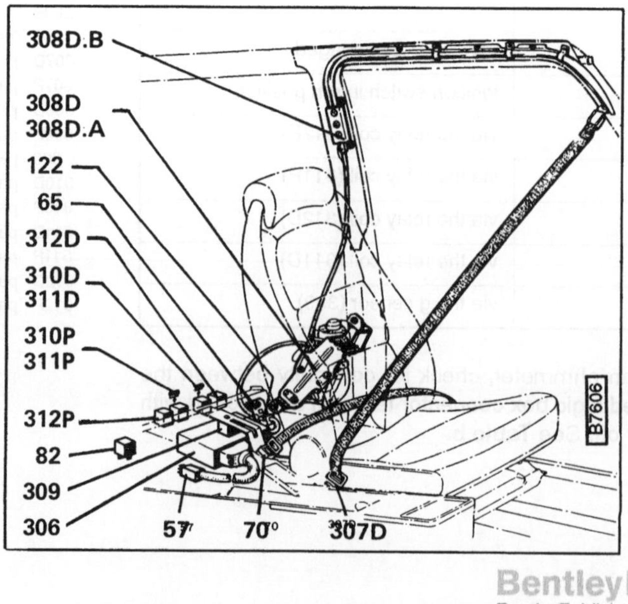

Passive Seat Belts

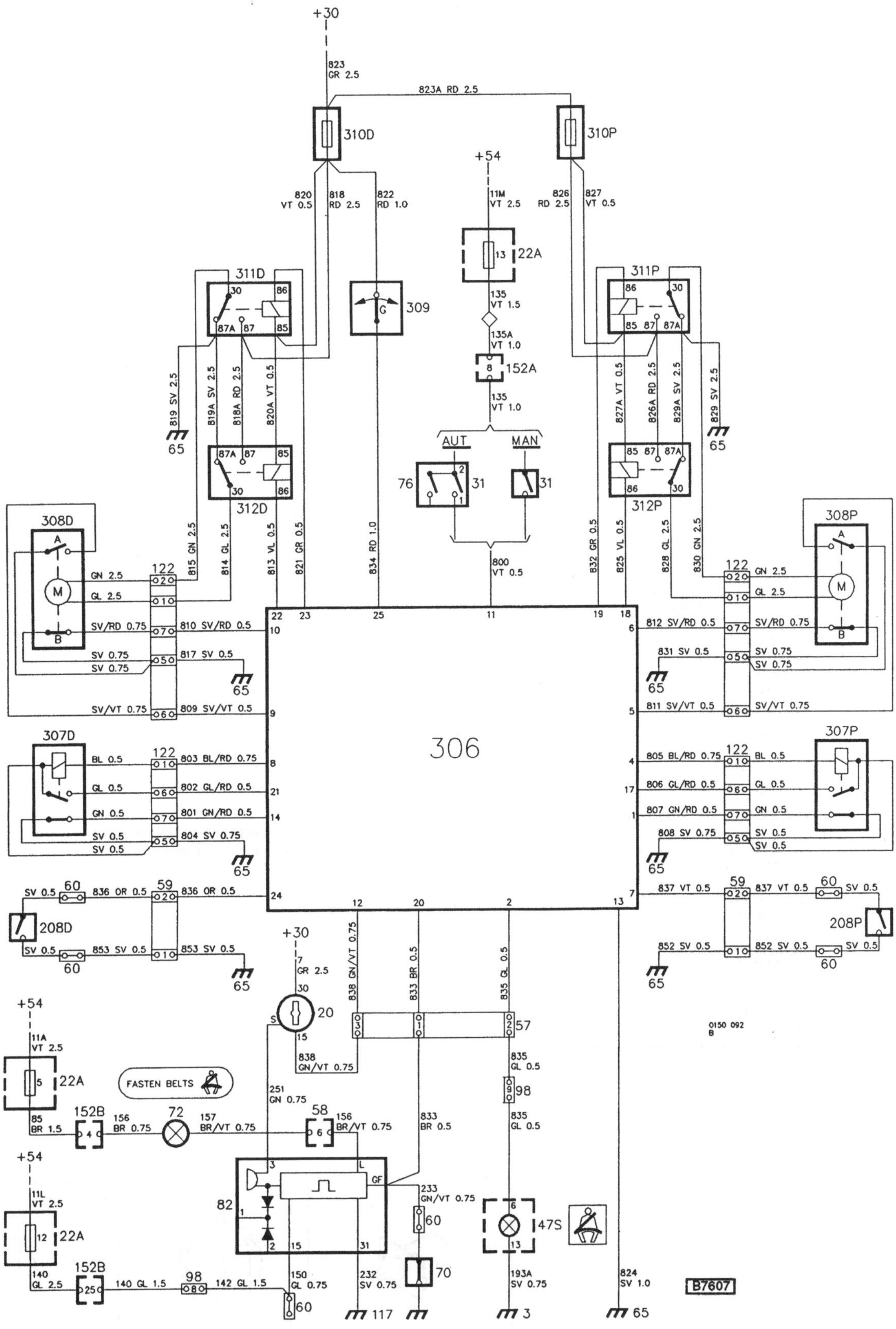

Power Mirrors

Operation

Each power mirror is controlled by a four-way switch. Horizontal mirror movement is controlled by a motor in the mirror assembly. Vertical movement is also controlled by the same motor and a solenoid. On 1988 and later models, the mirror glass is electrically heated and is controlled by the heated rear window circuit. See **Heated Rear Window** elsewhere in this section.

See **830 Doors, Windows, and Lids** for more information on removing the power mirrors and switches.

Fault Tracing

1. Check the fuses. Check the power supply to the switch with the key on.
2. Use the table below and the wiring diagram to test for ground and positive power at the mirror.

Left mirror			
Switch Position	**+ at motor**	**– at motor**	**+ at solenoid**
left	316 grey-red	317 green-white	
right	317 green-white	316 grey-red	
up	316 grey-red	317 green-white	315 white
down	317 green-white	316 grey-red	315 white
Right mirror			
Switch Position	**+ at motor**	**– at motor**	**+ at solenoid**
left	321 grey-red	320 green-white	
right	320 green-white	321 grey-red	
up	321 grey-red	320 green-white	319 white
down	320 green-white	321 grey-red	319 white

Component Locations

22A main fuse/relay panel at left wheel housing
113 time delay relay for heated rear window under rear seat, auxiliary fuse/relay panel
124 left power mirror switch in instrument panel
125 right power mirror switch in instrument panel
126 left power mirror motor and solenoid in the left mirror
127 right power mirror motor and solenoid in the right mirror
158 negative distribution in main fuse/relay panel
207 heated rear view mirrors on front doors

Connector Locations

1985-1990

152B 29-pin bulkhead connector at main fuse/relay panel

1991 and later

H1-1 1-pin connector in engine compartment, upper left door hinge
H1-3 1-pin connector in engine compartment, upper right door hinge
H1-14 1-pin connector in the left front door
H1-15 1-pin connector in the right front door
H4-2 4-pin connector in engine compartment, near top left door hinge
H4-3 4-pin connector in engine compartment, near top right door hinge
H4-8 4-pin connector in the left front door
H4-9 4-pin connector in the right front door
H29-1 29-pin bulkhead connector behind main fuse/relay panel (1991-1991 1/2)
H29-2 29-pin bulkhead connector behind main fuse/relay panel (1991-1991 1/2)
H33-1 33-pin bulkhead connector behind main fuse/relay panel (1991 1/2 and later)
H33-2 33-pin bulkhead connector behind main fuse/relay panel (1991 1/2 and later)

Ground Locations

1985-1990

3 ground point behind instrument panel

1991 and later

G6 ground distribution at main fuse/relay panel
G8 ground point behind instrument panel

124
125

126
127

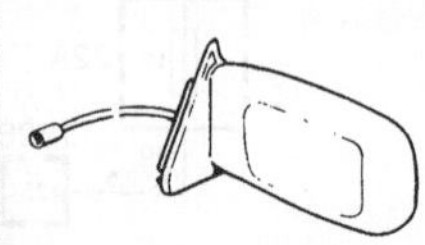

207

B7608

Power Mirrors

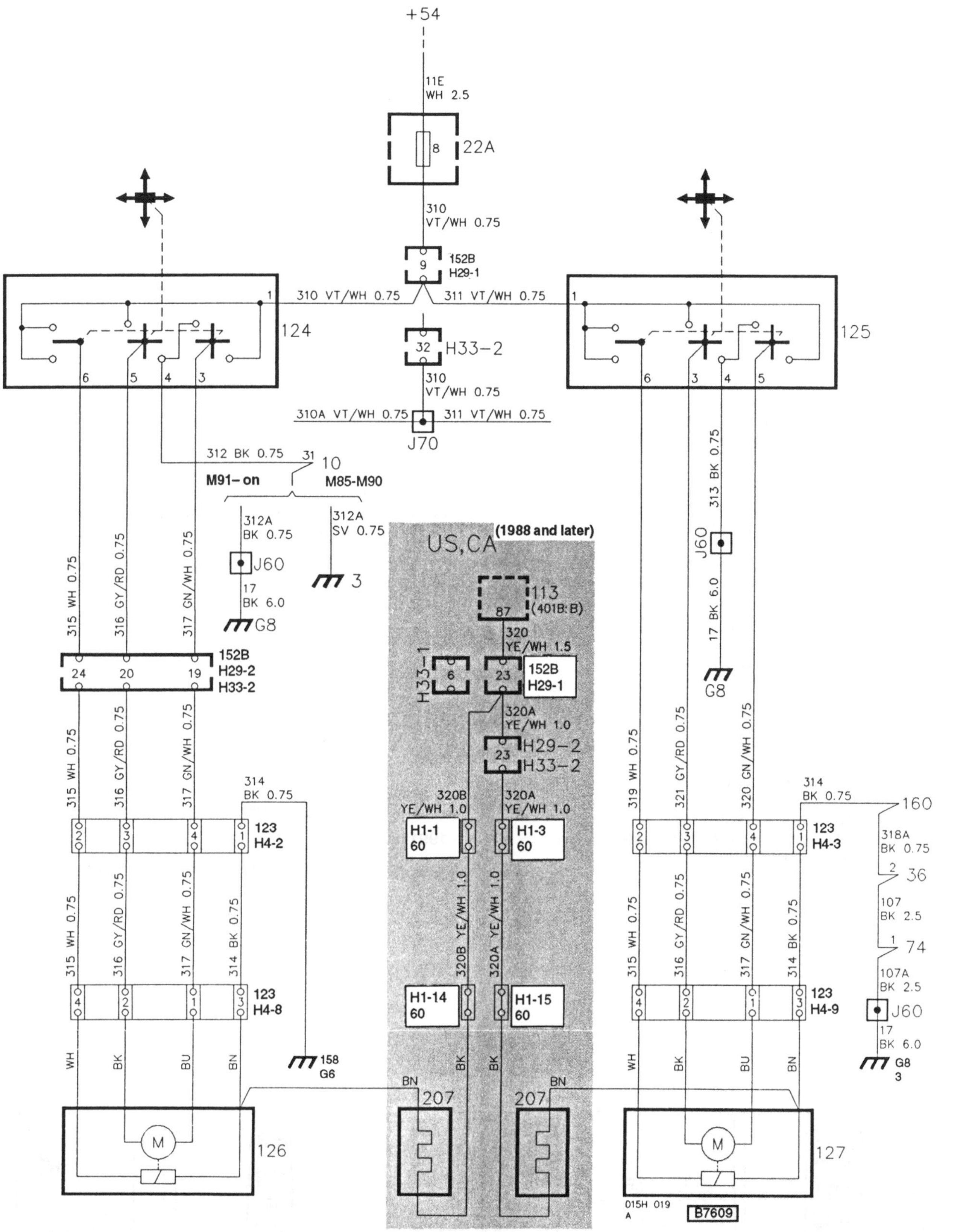

Power Windows

Operation

The power windows are controlled by console switches, and, if applicable, door-mounted switches. Electric motors in each door raise and lower the window. Motor direction is controlled by changing the voltage polarity at the motor. On 4-door cars, a lockout switch—labelled ON/OFF allows the driver to disable operation of the rear power windows.

NOTE
The console switch assembly (386) on 1991 and later models also contains the relays for the automatic window down feature. These relays cannot be replaced separately.

See **830 Doors, Windows, and Lids** for information on removing the window motors and glass. Additional information can be found in **364 Electric Controls and Switches** for replacing the switch and switch illumination bulbs.

Fault Tracing

1. Check the fuses. Check the power supply to the switch.
2. At the window motor, check for battery voltage with the switch pushed in one direction. Reverse the test leads and push the switch in the other direction. There should again be battery voltage.

Component Locations

22A main fuse/relay panel at left wheel housing
162 switch for driver power window in center console
163 switch for passenger power window in center console
164 power window motor in left front door
165 power window motor in right front door
189 lockout switch for rear power windows
190 switch for left rear power window in center console
190a switch for left rear power window in left rear door
191 switch for right rear power window in center console
191a switch for right rear power window in right rear door
193 power window motor in left rear
194 power window motor in right rear
287 power window relay for automatic control
1989-1990 Convertible under rear seat in fuse/relay panel. 401
1989-1990 ex. Convertible under rear seat, left-hand side
386 switch assembly for power windows and sunroof (where applicable) in center console
401 auxiliary fuse/relay panel under rear seat

Connector Locations

1985-1990

57 3-pin connector behind trim at left or right B pillar
58 12-pin connector behind instrument panel, left of steering column
59 2-pin connector at top left front door hinge or top right front door hinge, (2) in the main fuse/relay panel, (2) under center console
60 1-pin connector under center console
152A 29-pin bulkhead connector at main fuse/relay panel
152B 29-pin bulkhead connector at main fuse/relay panel

1991 and later

H1-6 1-pin connector at right B pillar, behind trim
H1-7 1-pin connector at left B pillar, behind trim
H1-17 1-pin connector under center console
H2-5 2-pin connector in engine compartment, near top left door hinge
H2-11 2-pin connector in engine compartment, near top right door hinge
H3-10 3-pin connector at right B pillar, behind trim
H3-11 3-pin connector at left B pillar, behind trim
H12-1 12-pin connector behind instrument panel, left of steering column
H29-1 29-pin bulkhead connector behind main fuse/relay panel (1991-1991 1/2)
H29-2 29-pin bulkhead connector behind main fuse/relay panel (1991-1991 1/2)
H33-1 33-pin bulkhead connector behind main fuse/relay panel (1991 1/2 and later)
H33-2 33-pin bulkhead connector behind main fuse/relay panel (1991 1/2 and later)

Ground Locations

1985-1990

65 ground point at handbrake
117 ground point at handbrake

1991 and later

G4 ground point between ignition switch and handbrake

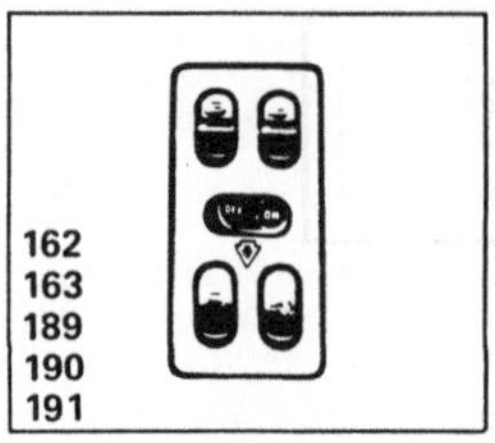

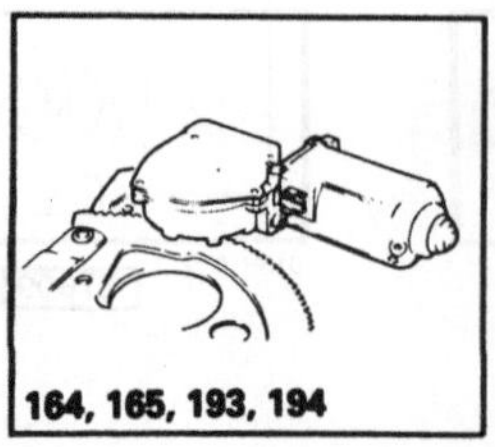

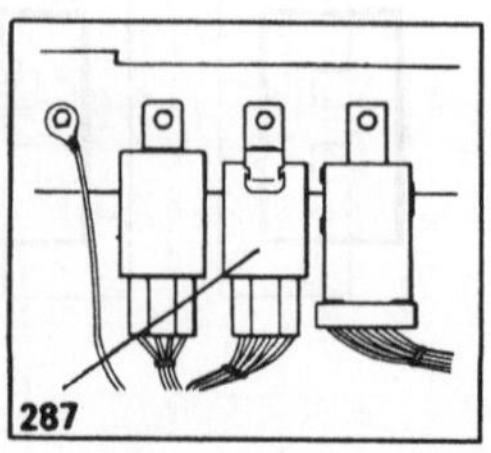

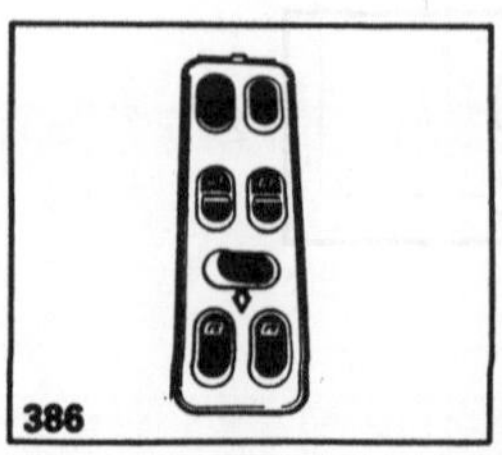

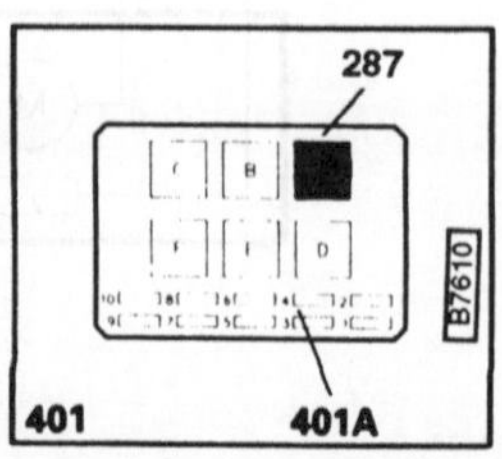

Power Windows

1985

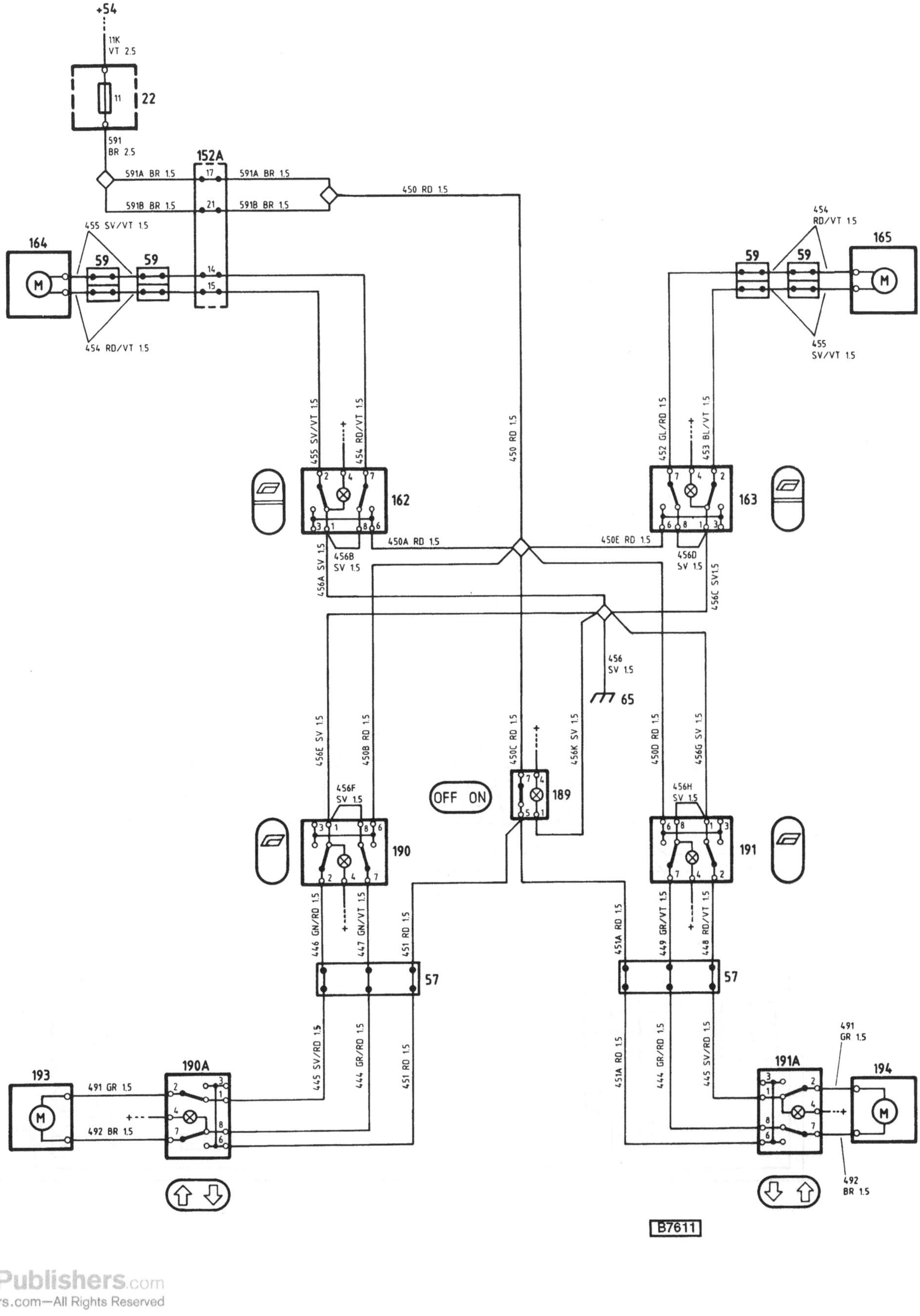

Power Windows

1986-1990

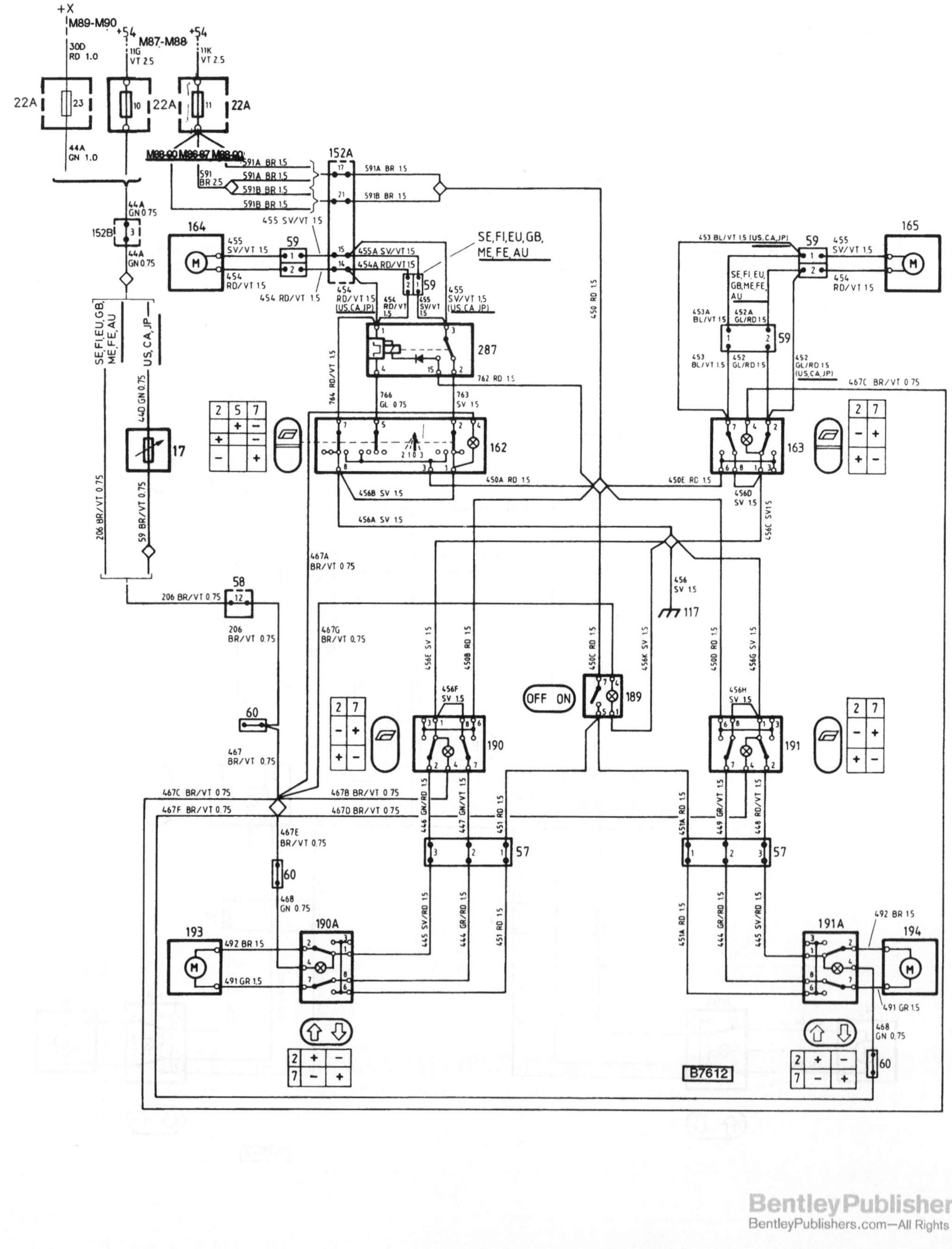

B7612

Power Windows

1991 and later

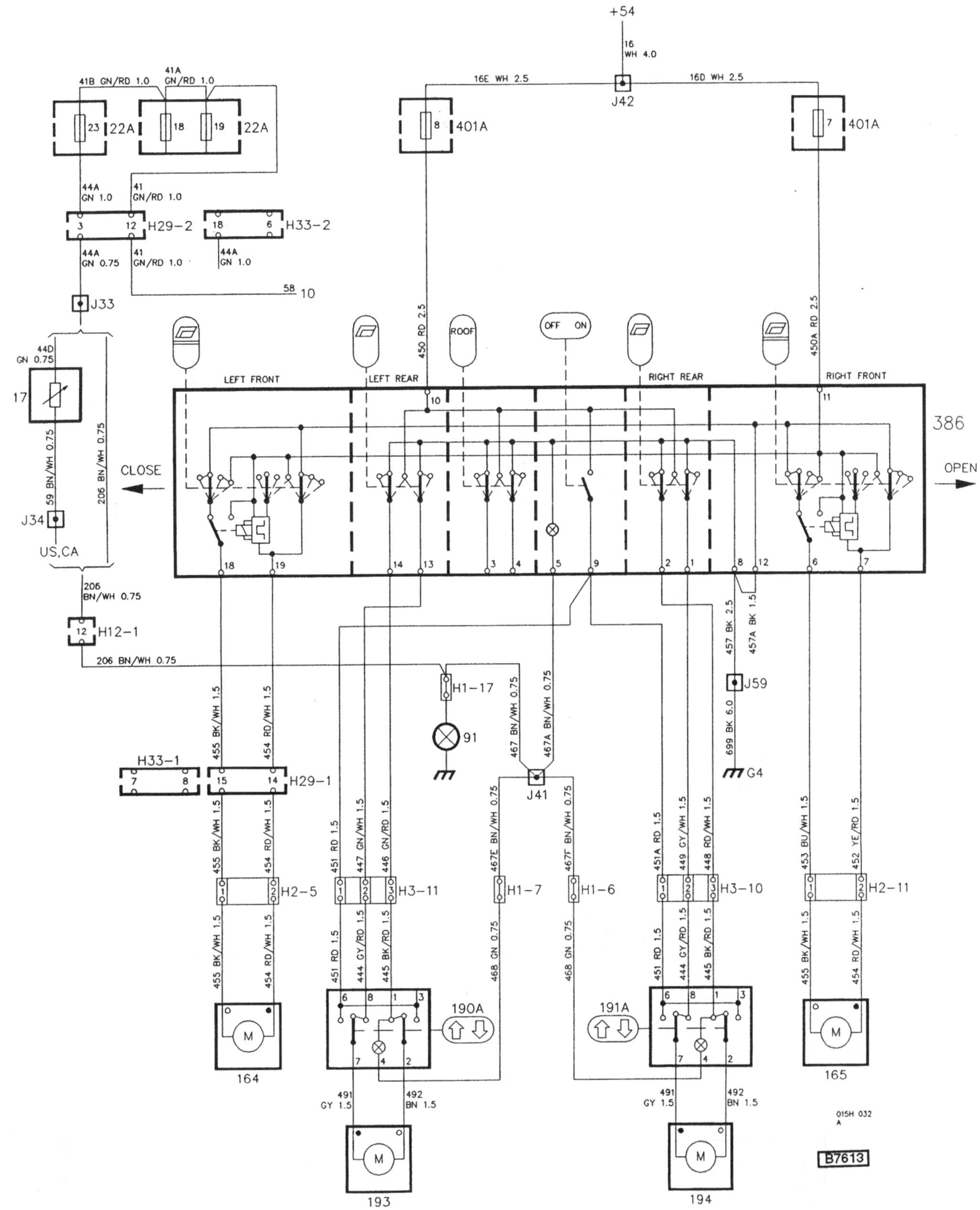

Power Windows—Convertible

1987-1990

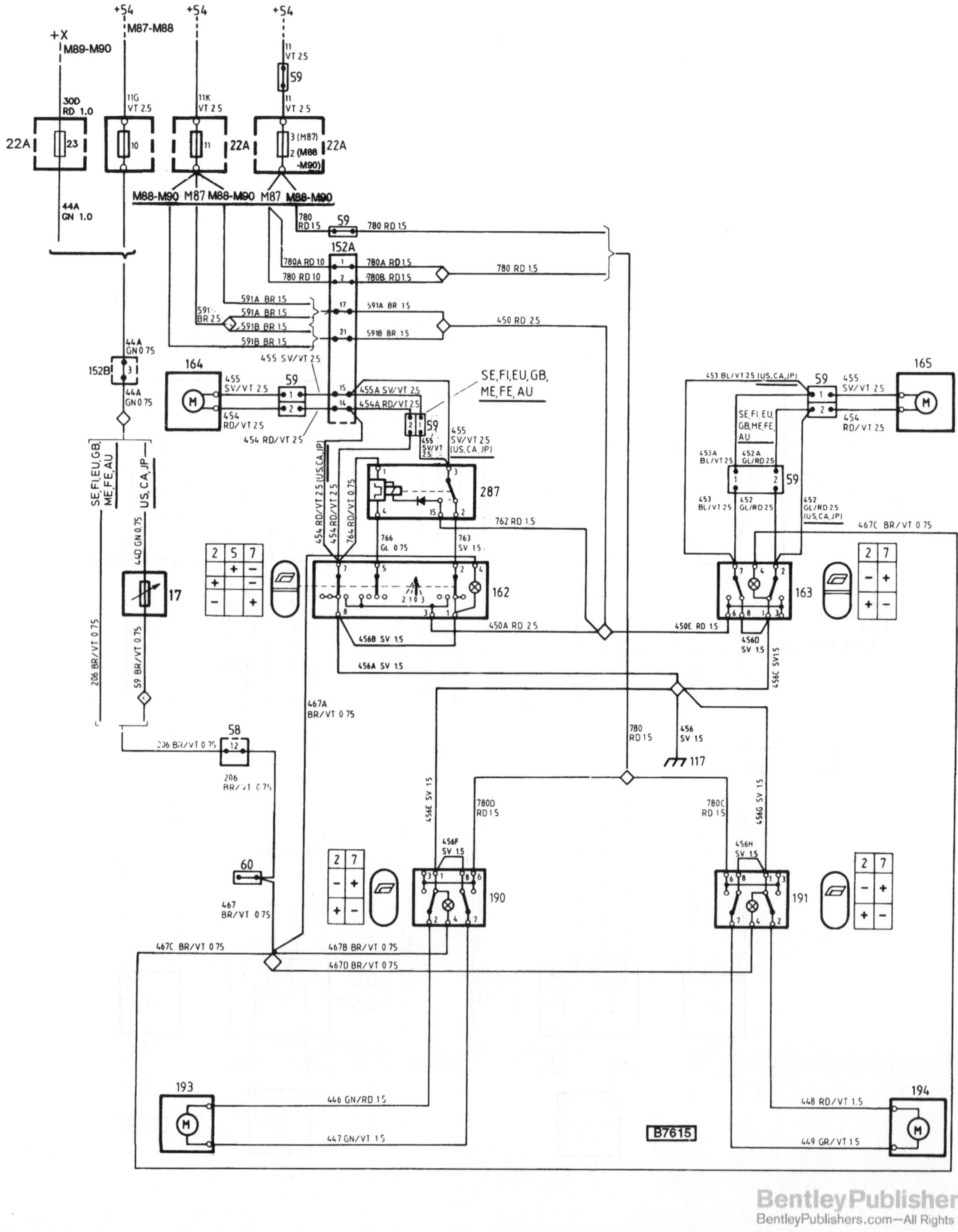

Power Windows—Convertible

1991 and later

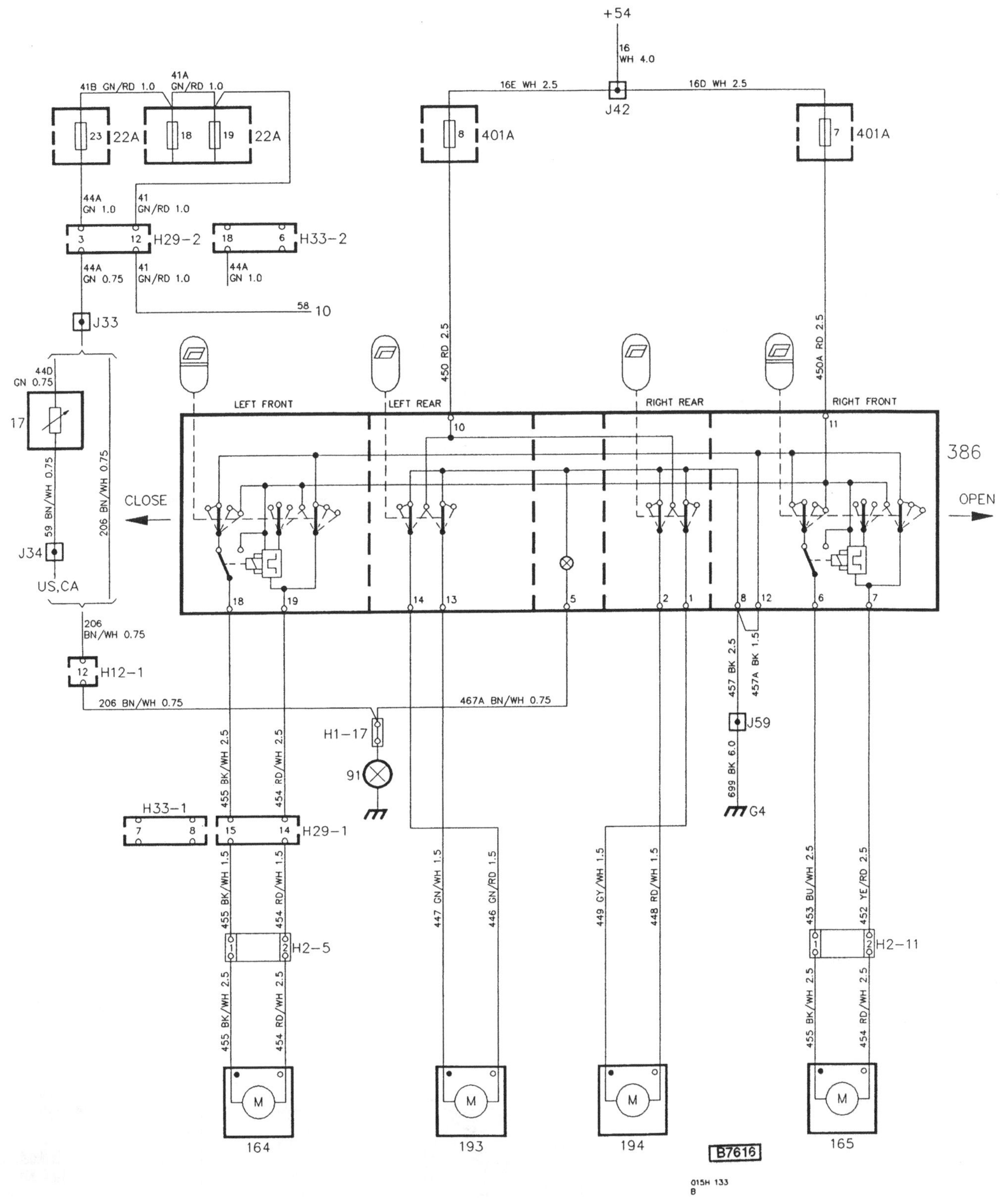

Radiator Cooling Fan

Operation

The cooling fan, located behind the radiator, is thermostatically controlled by the coolant temperature switch (39) located at the top left of the radiator. A coolant fan time delay relay (26) or radiator fan relay (396) is used depending on model year.

For information on the electric radiator cooling fan for the air conditioner, see **Air Conditioner** elsewhere in this section. For cooling system repair information, see **261 Radiator and Cooling System**.

Fault Tracing

1. Check the fuses.
2. Check the coolant fan time delay relay (26) by checking for voltage at the relay panel sockets (terminals 30 and 87).
3. Check the coolant temperature switch by jumpering the switch terminals with the key on. The fan should now run.

Component Locations

- 22A main fuse/relay panel at left wheel housing
- 26 coolant fan time delay relay at front left wheel housing
- 37 radiator cooling fan motor behind radiator
- 39 coolant fan temperature switch at upper left corner of radiator
- 156 A/C compressor relay at main fuse/relay panel, engine compartment
- 159 +15 distribution in main fuse/relay panel
- 396 radiator fan relay at main fuse/relay panel

Connector Locations

1985-1990

- 59 2-pin connector at radiator fan motor
- 60 1-pin connector behind radiator

1991 and later

- H2-1 2-pin connector at radiator fan motor

Ground Locations

1985-1990

- 7 ground point at radiator cross member

1991 and later

- G1 ground point at radiator cross member
- G6 ground distribution at main fuse/relay panel

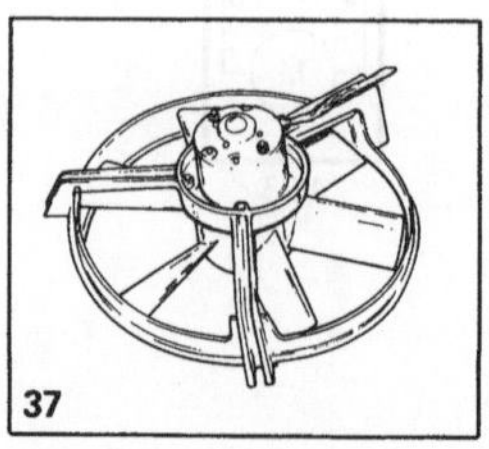

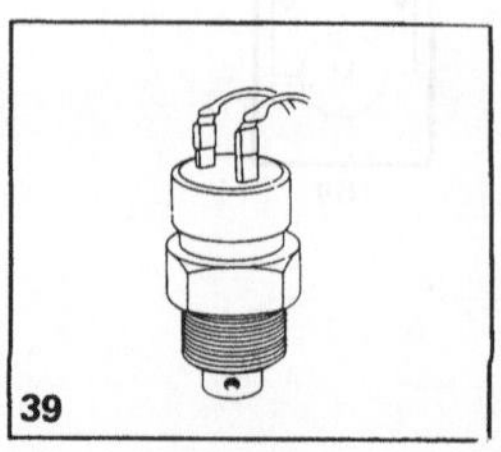

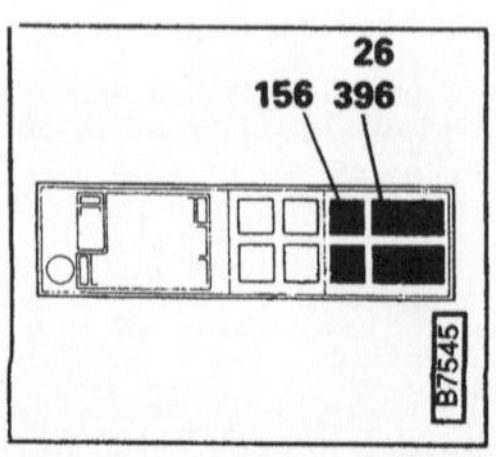

Radiator Cooling Fan

1985-1989 **1990** **1991 and later**

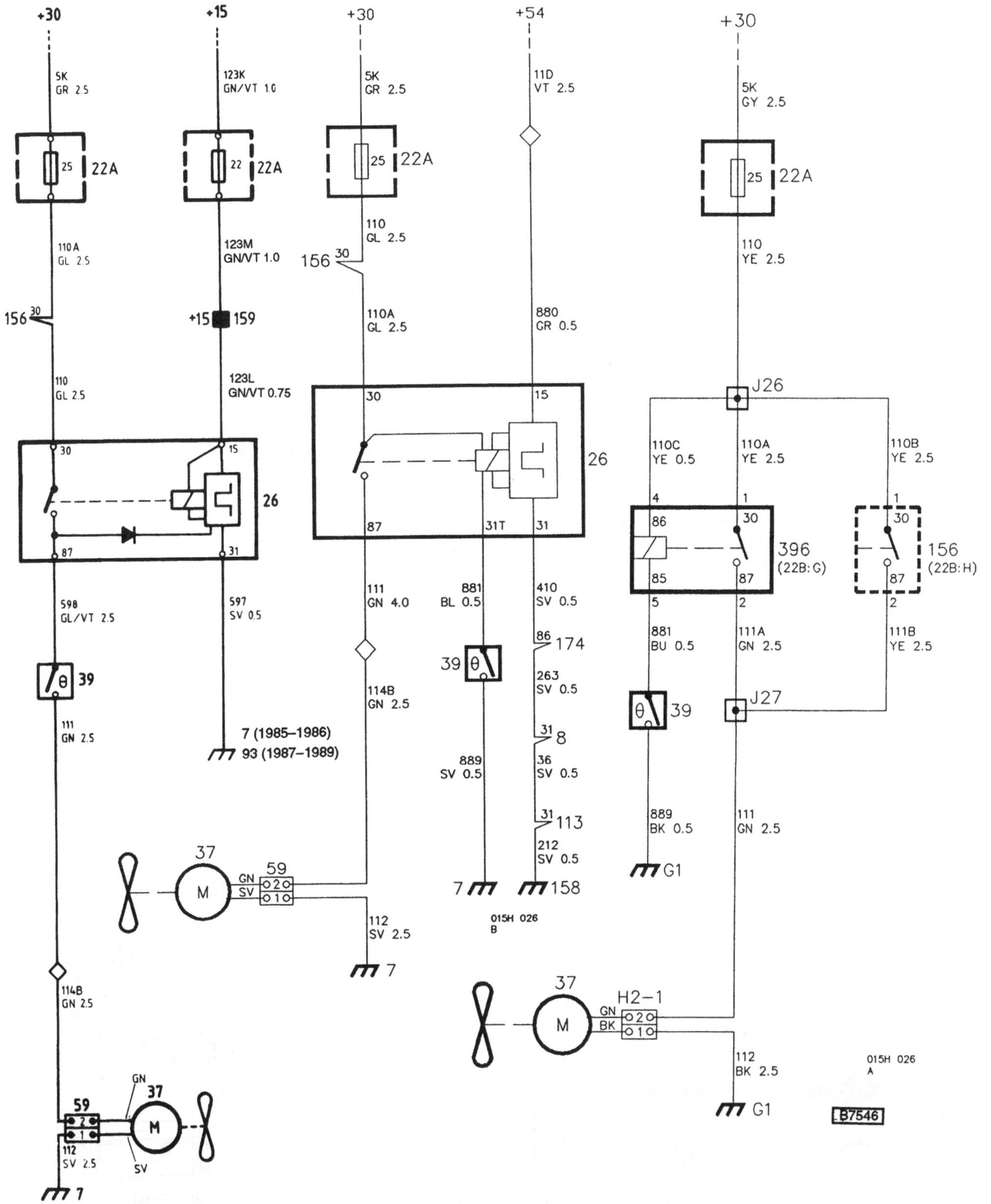

Radio

Operation

Radio wiring is provided to the opening in the instrument panel on all models. Some models have wiring for a graphic equalizer and compact disc player in the center console. An electric antenna is available on some models. A coaxial antenna cable is installed from the left rear up to the instrument panel.

See **853 Dashboard and Consoles** for information on removing the radio. See **851 Interior Trim and Upholstery** for information on removing the antenna.

Fault Tracing

1. Check the fuses, including the ones in the back of the instrument panel opening. See **Component Locations** below.
2. To check the coaxial cable with an ohmmeter, unplug the cable at both ends. There should be no continuity (open circuit) between the center and outer conductors. There should be no continuity (open circuit) between either conductor and chassis ground.
3. To check the coaxial cable for internal continuity, unplug the cable at both ends. Short the conductors together at one end of the cable with a piece of wire. Test for approximately 0 ohms (short circuit) at the other end of the cable between the conductors.

Component Locations

22A main fuse/relay panel at left wheel housing
265 electric radio antenna in left rear body panel
266 radio speakers (4) two in front, two in rear
349 radio contact box with amplifier in instrument panel
350 contact box for CD player or equalizer in center console top compartment
351 audio amplifier under rear seat below ABS control unit (291)
401 auxiliary fuse/relay panel under rear seat

Connector Locations

1985-1990

59 2-pin connector near electric antenna
60 1-pin connector near each speaker
98 10-pin connector in instrument panel, radio compartment
123 4-pin connector in instrument panel, (2) in radio compartment
152A 29-pin bulkhead connector in main fuse/relay panel
267 radio connector in instrument panel

1991 and later

H2-14 2-pin connector in left rear wheel housing in luggage compartment
H2-51 2-pin connector for left rear speaker
H2-52 2-pin connector for right rear speaker
H6-4 6-pin connector behind instrument panel
H8-7 8-pin connector behind instrument panel
H8-8 8-pin connector behind instrument panel
H10-1 10-pin connector behind instrument panel, left of steering column
H10-5 10-pin connector behind instrument panel
H12-3 12-pin connector behind instrument panel
H12-4 12-pin connector behind instrument panel

Ground Locations

1985-1990

3 ground point behind instrument panel

1991 and later

G4 ground point between ignition switch and handbrake
G5 ground point under rear seat
G27 ground point for electric antenna

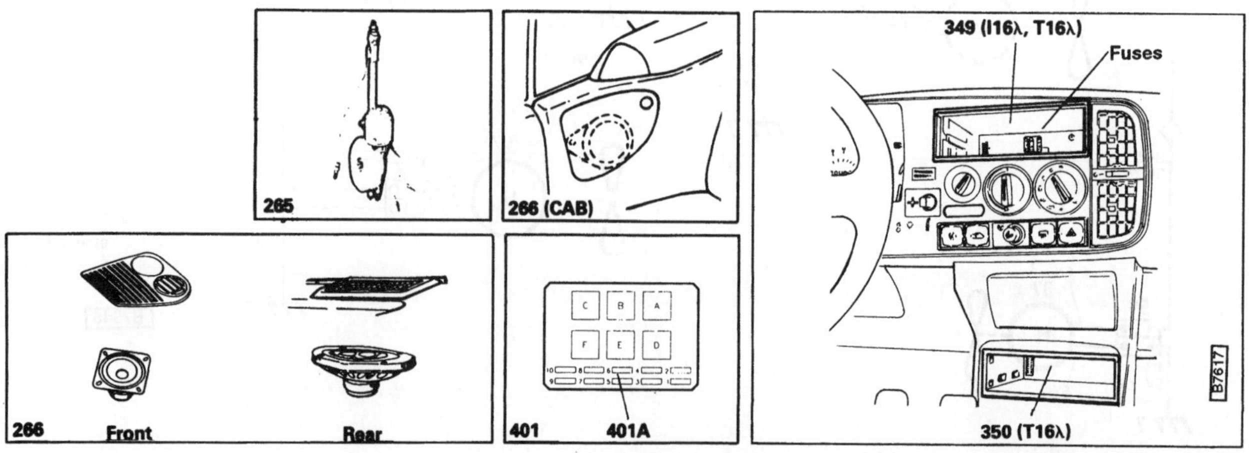

Radio

1985

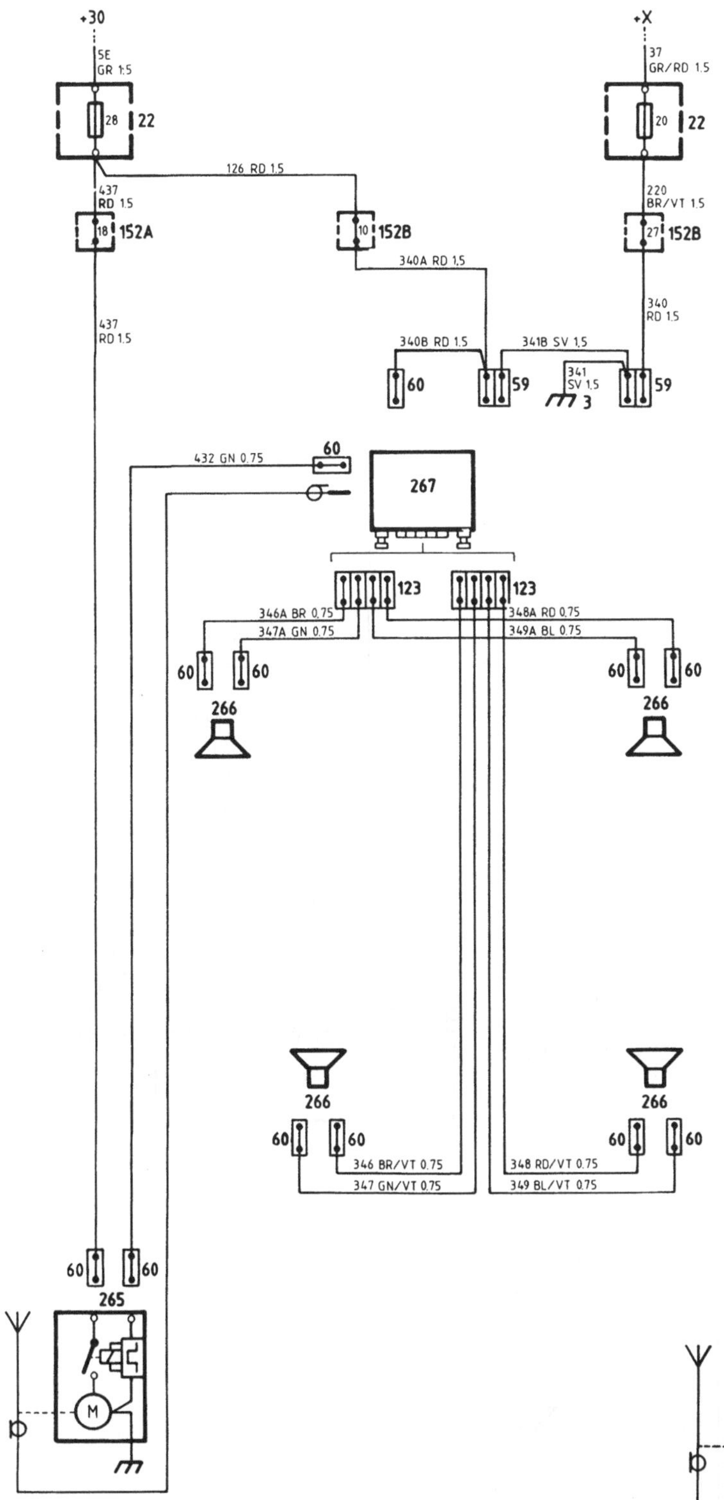

1986

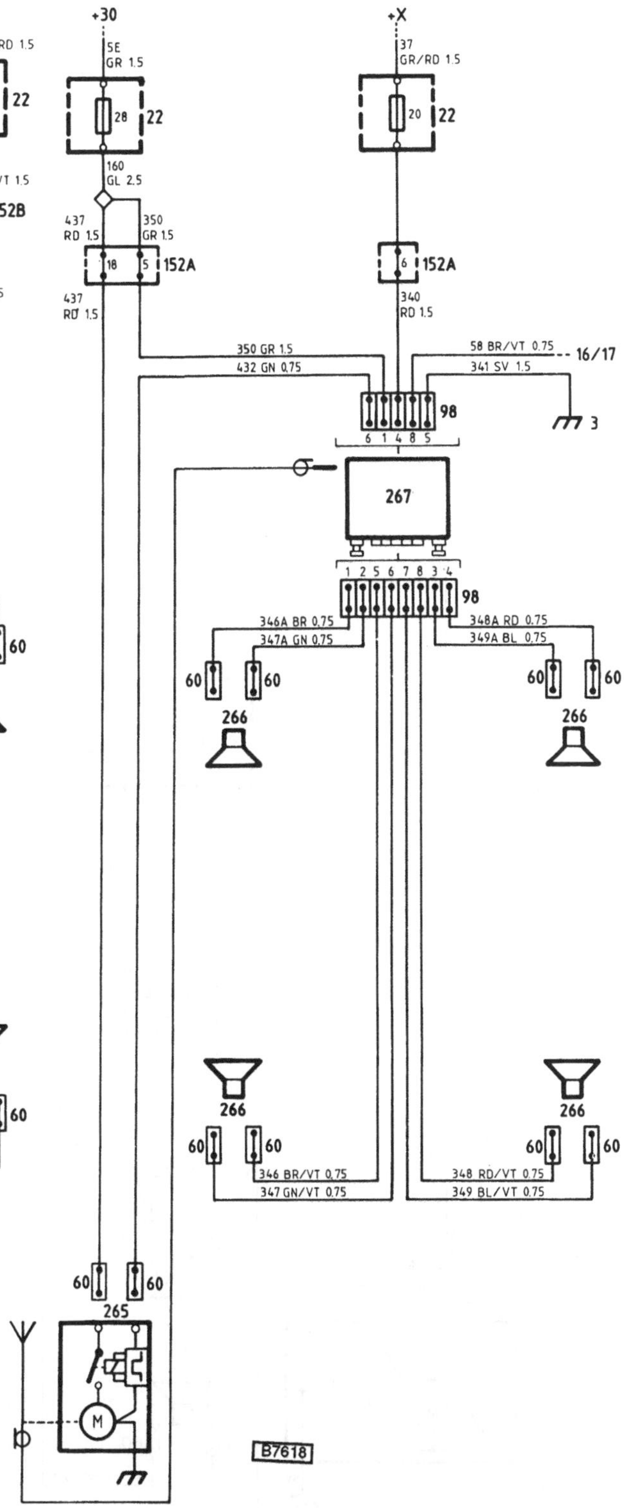

Radio

1987-1990

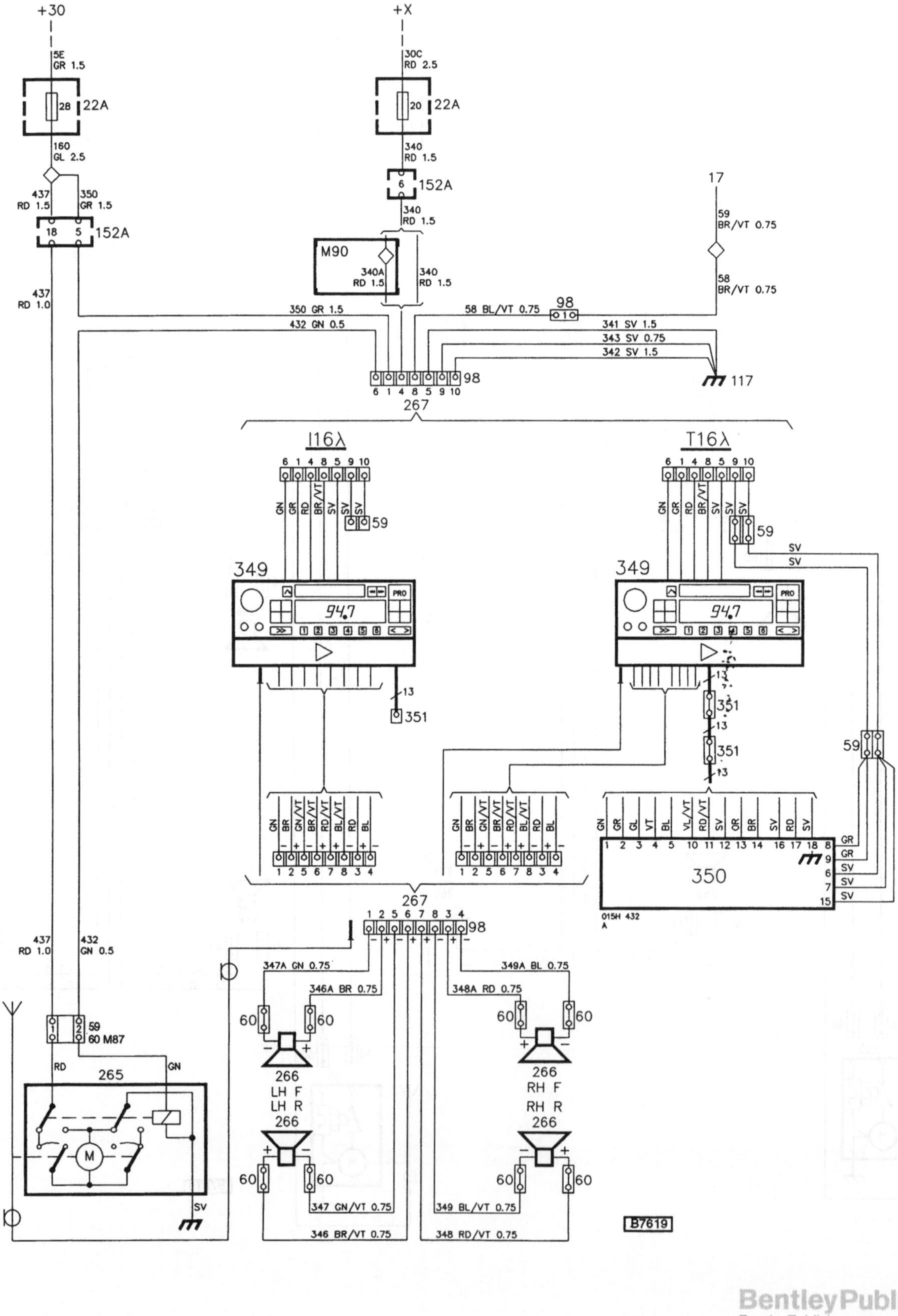

Radio

1991

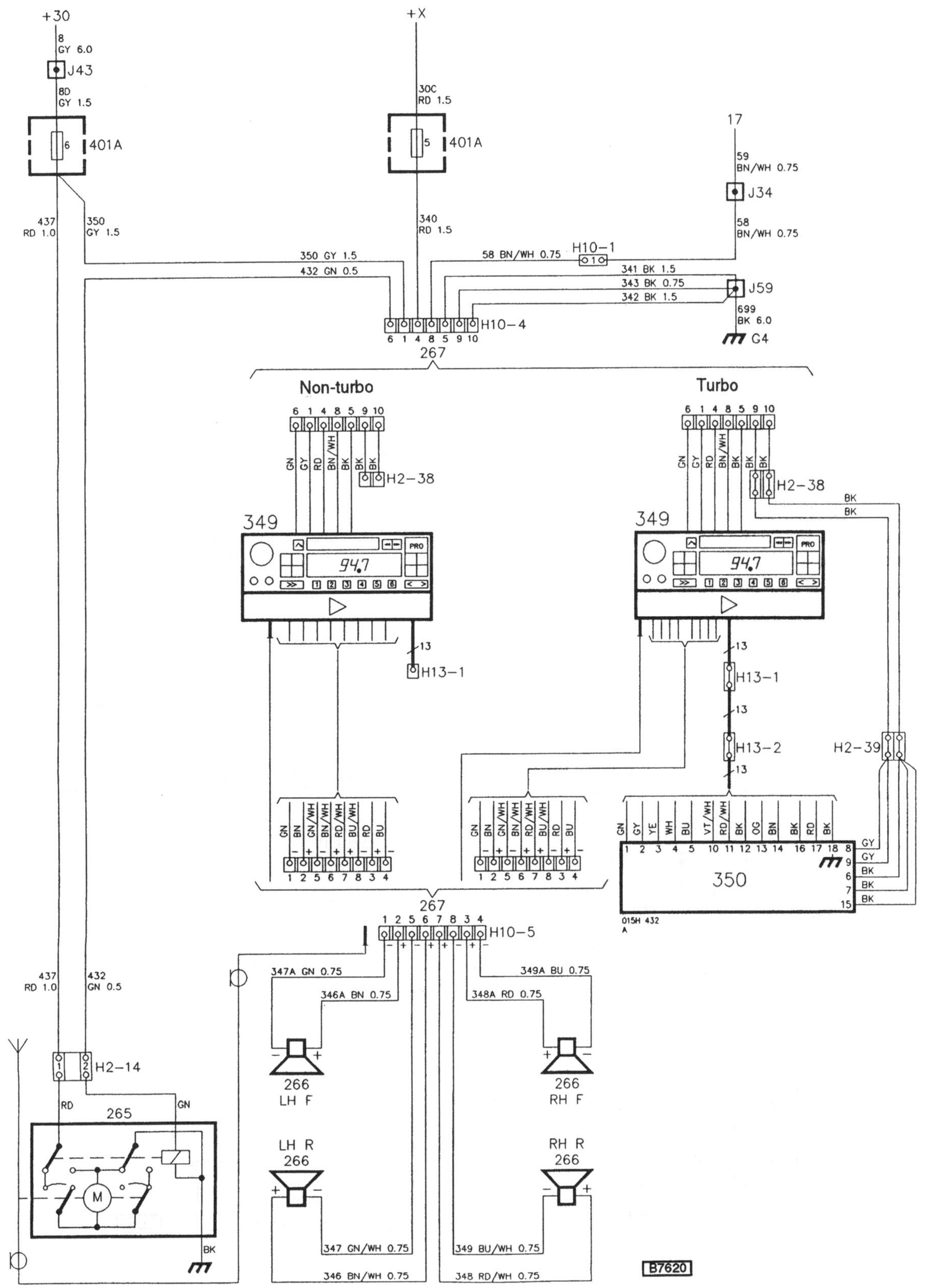
+30
8
GY 6.0
J43
8D
GY 1.5
401A
437
RD 1.0
350
GY 1.5
+X
30C
RD 1.5
340
RD 1.5
17
59
BN/WH 0.75
J34
58
BN/WH 0.75
350 GY 1.5
432 GN 0.5
58 BN/WH 0.75
H10-1
341 BK 1.5
343 BK 0.75
342 BK 1.5
J59
699
BK 6.0
G4
H10-4
267
Non-turbo
Turbo
H2-38
349
H13-1
H13-2
H2-39
350
015H 432
A
H10-5
347A GN 0.75
346A BN 0.75
349A BU 0.75
348A RD 0.75
266
LH F
RH F
LH R
RH R
432
GN 0.5
H2-14
RD
GN
265
BK
347 GN/WH 0.75
346 BN/WH 0.75
349 BU/WH 0.75
348 RD/WH 0.75
B7620

Radio

1992 and later

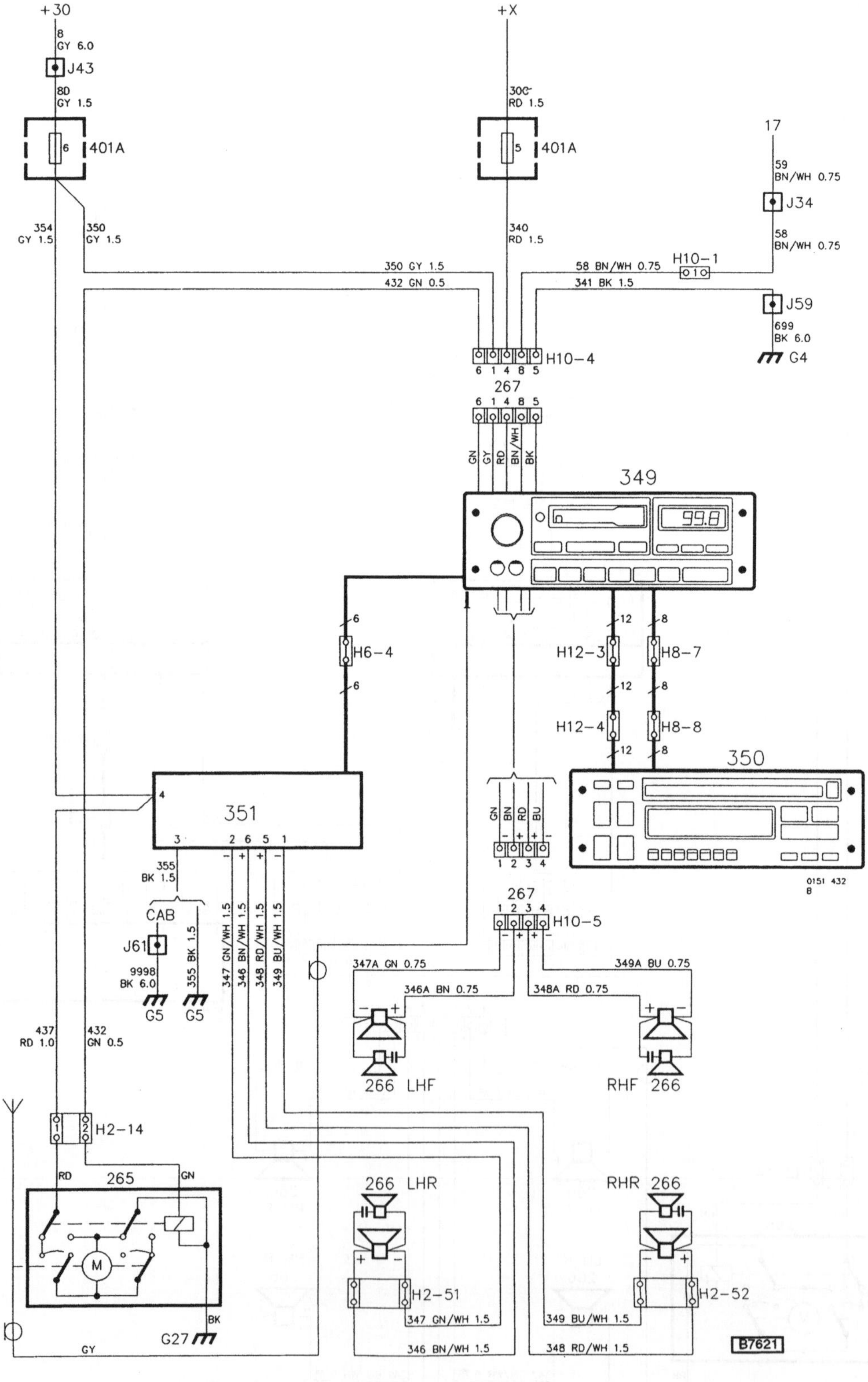

Seat Belt and Ignition Key Warning

Operation

The seat belt and ignition key warning system warns the driver to fasten his seat belt. The system also warns the driver whenever the key is left in the ignition and the door is open.

For information on removing and repairing the ignition lock and ignition switch see **432 Manual Transmission Controls.**

Fault Tracing

1. Check the fuses.
2. Check that the door switch wire has continuity to ground when the door is opened, and has no continuity to ground when the door is shut.
3. Check for voltage at terminals 3, 15, and L of the seat belt warning relay (82) with the key in the run position.

Component Locations

20 ignition switch in center console
22 main fuse/relay panel
54 door switches on the door jamb
69 passenger seat belt switch under seat cover
70 seat belt switch, driver side between driver and passenger seat
72 seat belt warning light at center of instrument panel
82 seat belt warning relay under the rear seat

Connector Locations

1985-1990

58 12-pin connector to left of steering column
60 1-pin connector under center console
98 10-pin connector to left of steering column
152A 29-pin bulkhead connector behind main fuse/relay panel
152B 29-pin bulkhead connector behind main fuse/relay panel

1991 and later

H3-18 3-pin connector under passenger seat
H8-1 8-pin connector under passenger seat
H8-2 8-pin connector under driver seat
H12-1 12-pin connector behind instrument panel, left of steering column
H29-2 29-pin bulkhead connector behind main fuse/relay panel (1991-1991 1/2)
H33-2 33-pin bulkhead connector behind main fuse/relay panel (1991 1/2 and later)

Ground Locations

1985-1990

117 ground point under handbrake

1991 and later

G4 ground point between ignition switch and handbrake

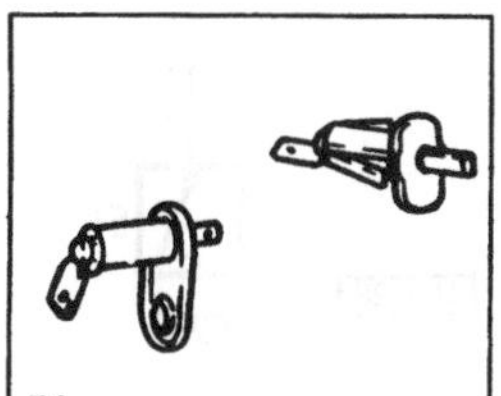
54

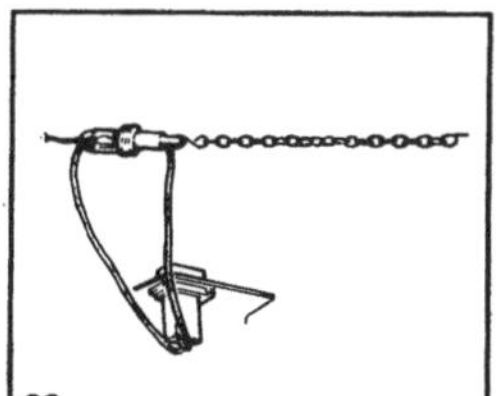
69

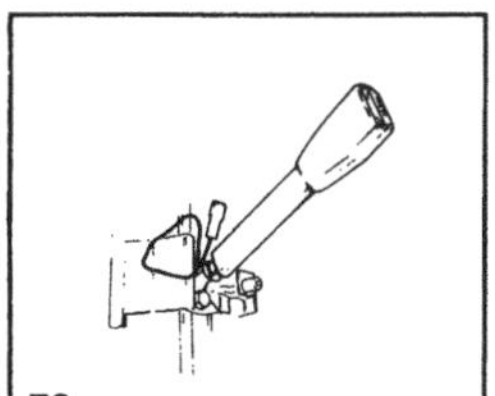
70

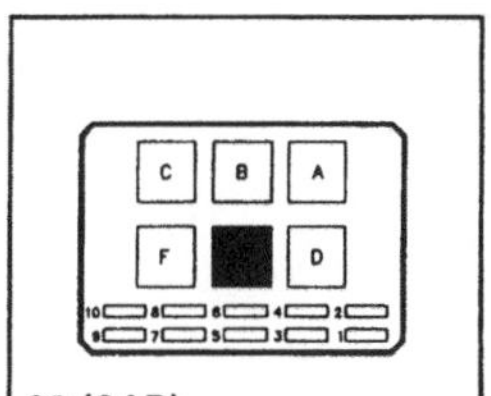

82 (CAB)

82

Seat Belt and Ignition Key Warning

1985-1986

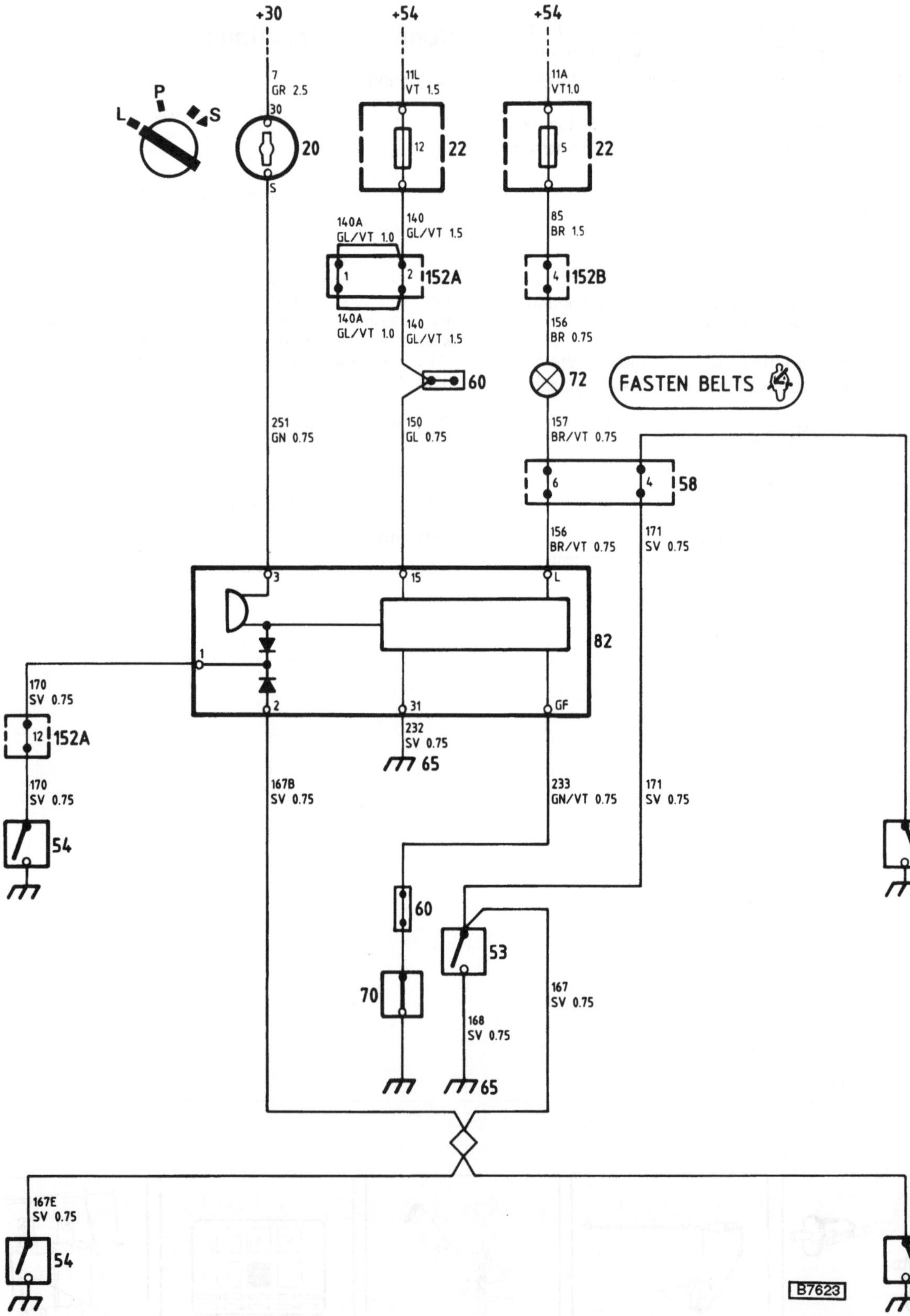

Seat Belt and Ignition Key Warning

1987-1990

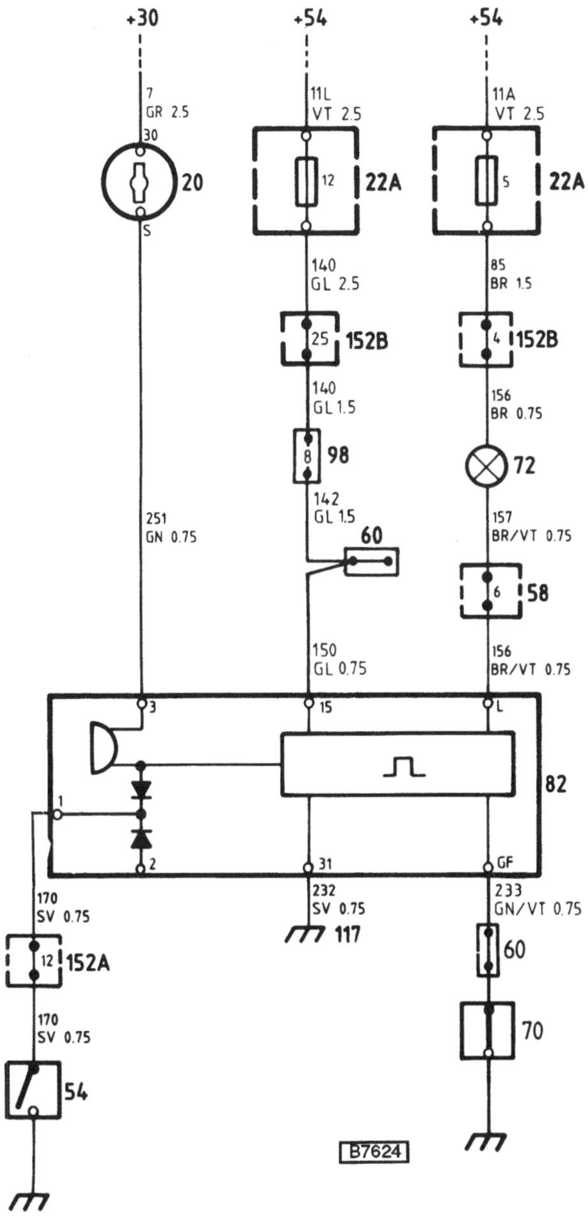

1991 and later

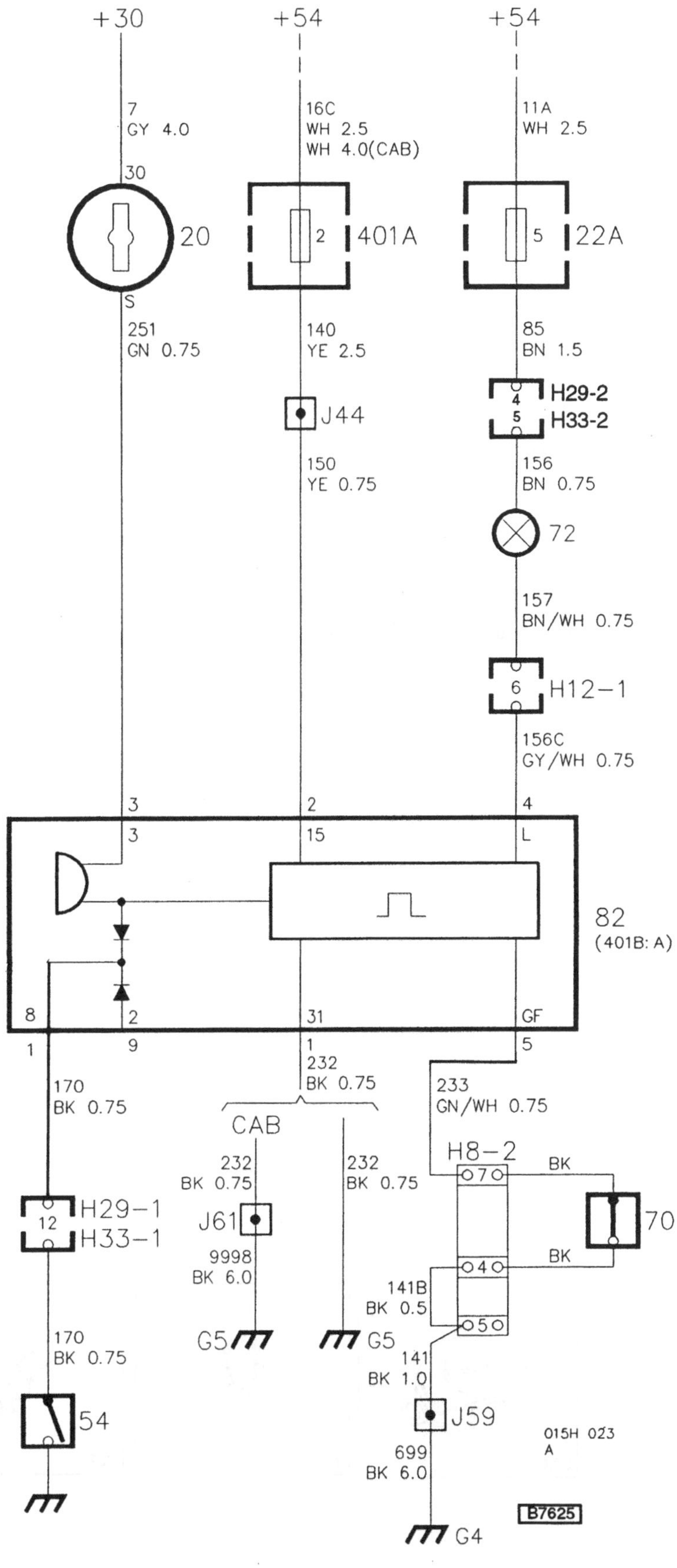

Shift-up Indicator

Operation

The shift-up indicator is installed on cars with manual transmission only. The light in the instrument cluster lights up indicating to the driver when to shift up. The light will stay off if the car is in fifth gear, if the engine temperature is below 95°F (35°C), or if the throttle is in the rest position. A time delay function of 1.7 seconds prevents the light from flashing and causing irritation. This time delay function is incorporated in the shift-up indicator relay (270). 1989 and later models have shift-up incorporated into the LH control unit.

NOTE

For shift-up wiring diagrams on 1989 and later models, see **Combined Instruments** elsewhere in this section.

Fault Tracing

1. Check the fuses with the key in the run position.
2. Check for voltage at the shift-up relay (270) terminal G and terminal 15. The car should be in first gear and the engine temperature cold (less than 95°F).
3. Check the shift-up lamp (47K) in the instrument cluster.

Component Locations

22A main fuse/relay panel
47K shift-up indicator lamp
146 ignition amplifier in front of left wheel housing
147 ignition pulse amplifier in main fuse/relay panel, position D
158 ground (–) distribution in main fuse/relay panel
176 EZK control unit at inner left fender
203 throttle angle transmitter on throttle housing
261 throttle contacts for shift-up indicator at throttle housing
262 temperature switch for shift-up indicator between center intake ports of intake manifold
263 vacuum switch for shift-up indicator at left wheel housing
264 5 gear switch for shift-up indicator at front side cover of gearbox
270 shift-up indicator relay in main fuse/relay panel

Connector Locations

59 2-pin connector behind the instrument panel
152B 29-pin bulkhead connector at main fuse/relay panel

Ground Locations

3 ground point behind instrument panel
201 ground point for fuel injection at engine lifting lug, 2 points

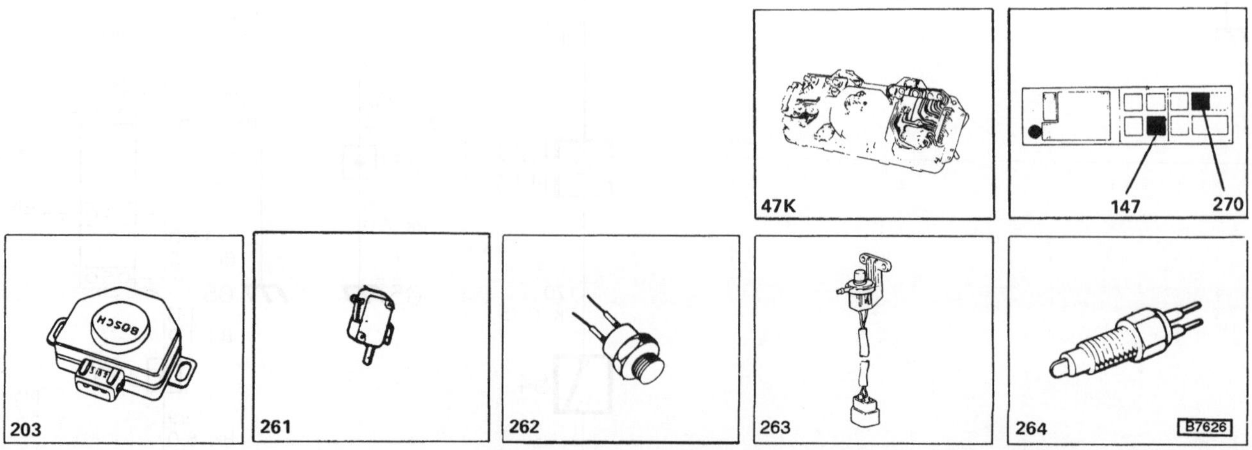

Shift-up Indicator

1985-1988

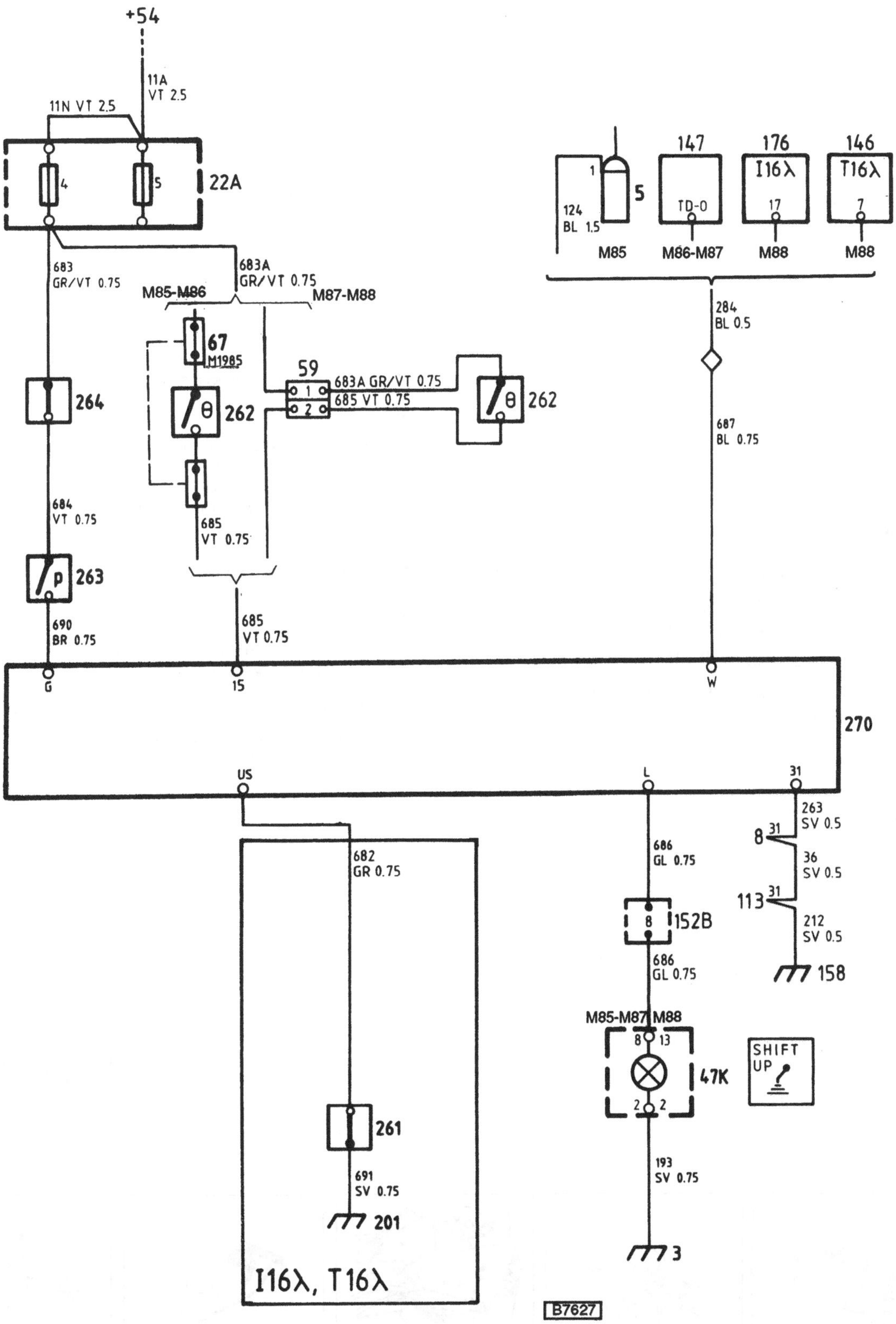

Starting System

Operation

For information on removing and repairing the starter see **331 Starter.** For more information on replacing and adjusting the neutral safety switch on automatics, see **444 Automatic Transmission Controls**.

WARNING
Batteries generate explosive gasses. Keep sparks and open flame away. Do not smoke.

CAUTION
- Do not disconnect the battery when the engine is running. The alternator could be destroyed.
- Always disconnect the negative (–) battery cable first when servicing the starter. Reconnect the negative (–) battery cable last.

Fault Tracing

1. Check for battery voltage at terminal 30 of the starter.
2. Check for battery voltage at terminal 50 of the starter when the key is in the start (crank) position. If no voltage is present, check the ignition switch and ignition relay.
3. Check for continuity between the engine block and battery negative (–) terminal.
4. On 1992 convertible models, check the start interlock relay 423 for the burglar alarm, located under the rear seat.

Component Locations

1 battery
4 starter motor under intake manifold
20 ignition switch at center console
75 distribution block at right wheel housing
77 neutral safety switch (automatics only) at gear selector lever
289 burglar alarm control unit (1992 and later Convertible) under rear seat
423 burglar alarm start interlock relay (1992 and later Convertible) under rear seat

Connector Locations

1985-1990

60 2-pin connector at main fuse/relay panel
152A 29-pin bulkhead connector behind main fuse/relay panel

1991 and later

H1-18 1-pin connector under center console
H2-53 2-pin connector under rear seat
H29-1 29-pin bulkhead connector behind main fuse/relay panel (1991-1991 1/2)
H33-1 33-pin bulkhead connector behind main fuse/relay panel (1991 1/2 and later)

Ground Locations

1985-1990

7 ground point at radiator cross member
211 ground point at gearbox

1991 and later

G1 ground point at radiator cross member
G25 ground point at gearbox

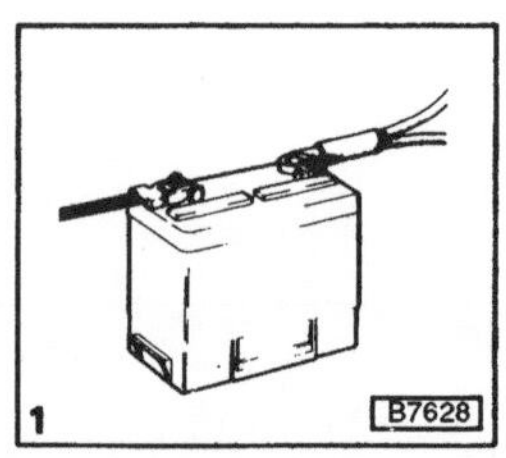

1

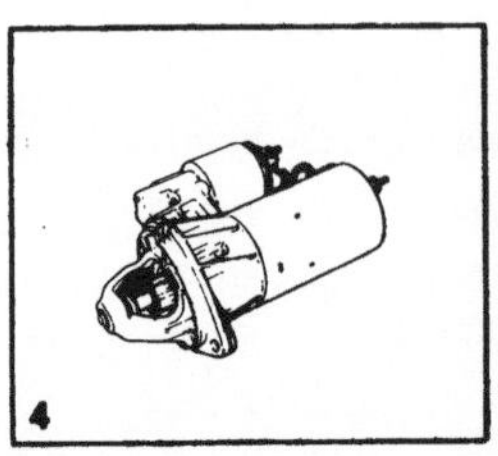
4

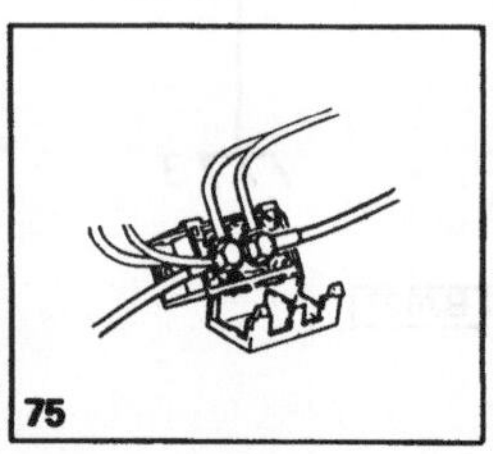
75

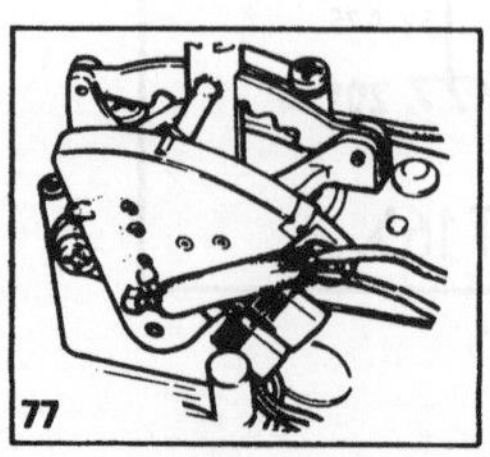
77

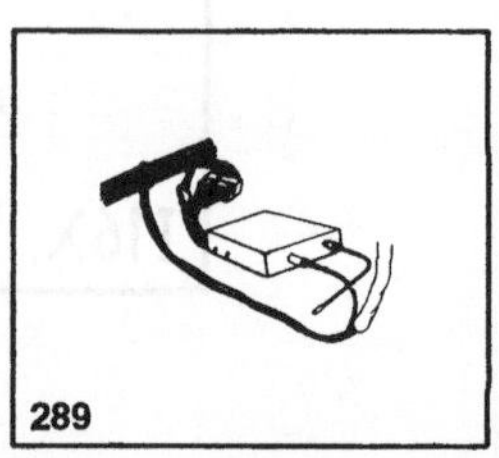
289

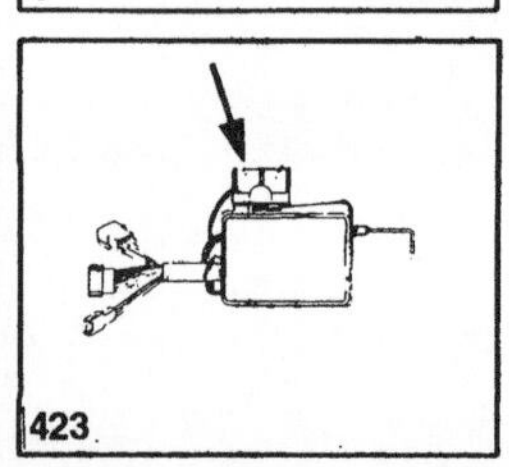
423

Starting System
1985-1990

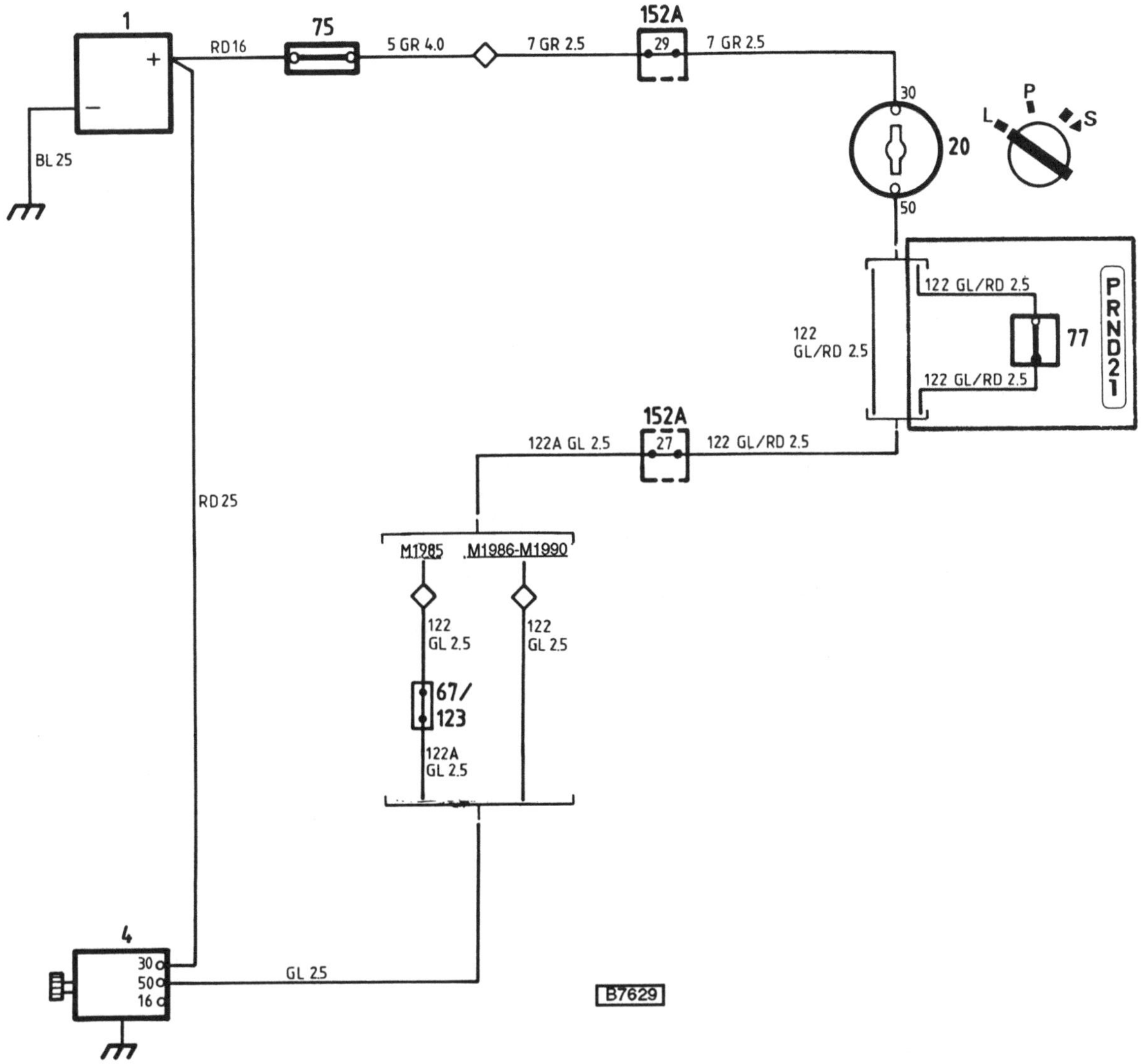

Starting System

1991 and later

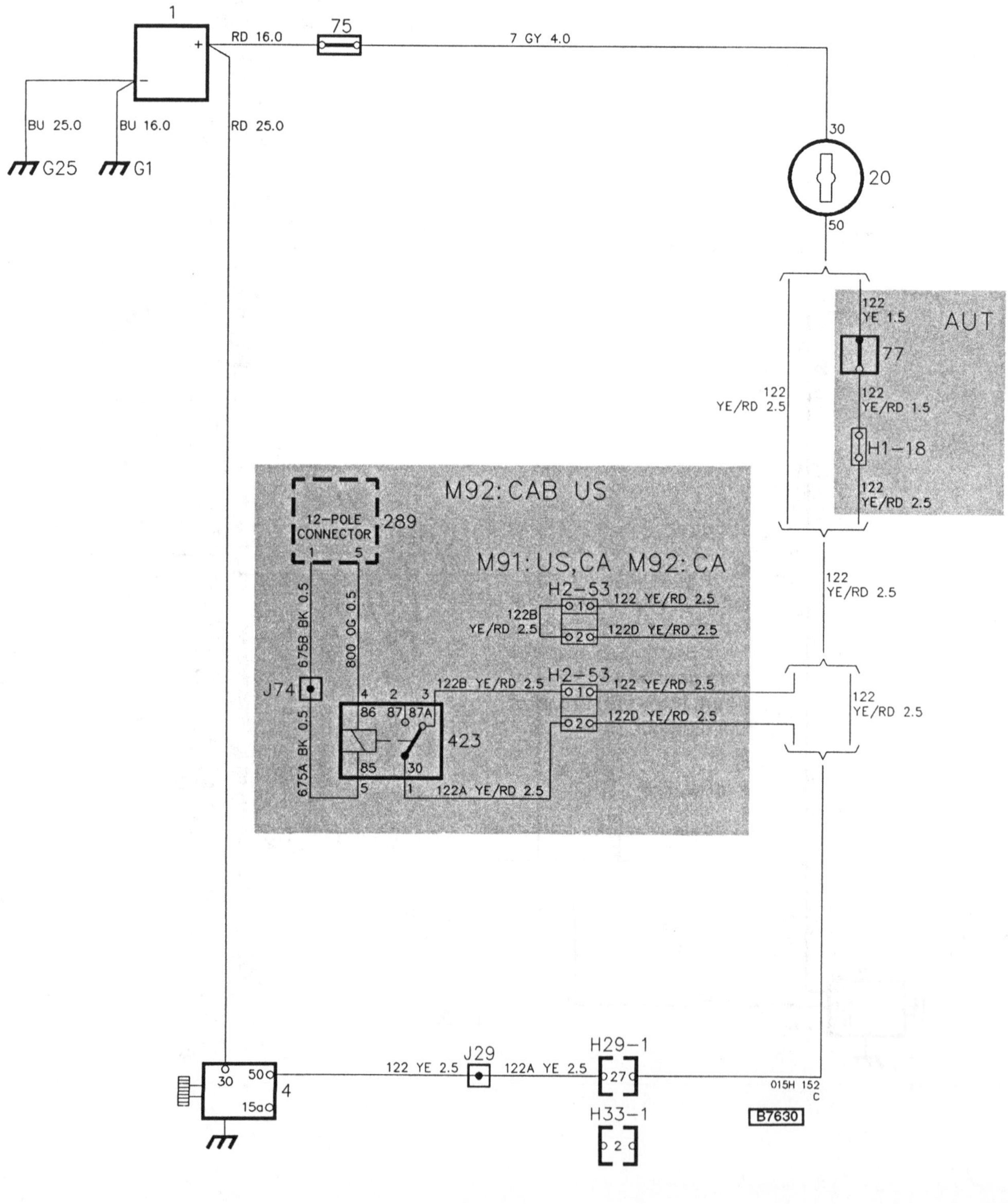

Supplemental Restraint System (SRS)—Airbag

Operation

The supplemental restraint system (SRS) consists of two front sensors, a control unit, a safety sensor, and the airbag with gas generator. If a fault should occur in the system, the SRS lamp in the instrument cluster will flash when the ignition switch is turned to the run position. Fault codes in the system's memory can be read by an authorized Saab dealer using special diagnostic equipment.

WARNING

- The airbag is an explosive device. Do not attempt to remove the steering wheel or service the wiring in the steering column assembly without first disarming the SRS system. You could accidentally ignite the airbag, causing serious personal injury. See an authorized Saab dealer for any repairs to the SRS system or to disarm the system.

- The components and wiring used in the system should not be altered or repaired as they are unique in design and construction. Additionally, do not connect ground wires from other systems to the mounting screws for the front sensors or the electronic unit since this may cause disturbances in the system, rendering it inoperative.

Fault Tracing

The following fault tracing should be used as a guide only. Comprehensive fault tracing should be carried out by an authorized Saab dealer who has the necessary diagnostic equipment and training.

1. Check the fuses.

Component Locations

1	battery at right front of engine compartment
20	ignition switch
21	ignition switch relay in main fuse/relay panel
22A	main fuse/relay panel at left wheel housing
47E	charging lamp in instrument cluster
47K	shift up indicator lamp
47T	SRS airbag warning lamp (see SRS warnings)
75	distribution block at right wheel housing
158	ground (–) distribution in main fuse/relay panel
331	SRS airbag control unit behind instrument panel, left grille. See SRS warning
332A	SRS airbag left sensor at left wheel housing in engine compartment. See SRS warning.
332B	SRS airbag right sensor at right wheel housing in engine compartment. See SRS warning.
333	SRS airbag in steering wheel. See SRS warning.
336	slip ring contact in steering wheel

Connector Locations

1990

98	10-pin connector behind instrument panel to left of steering column
98	10-pin connector next to SRS airbag electronic unit
123	4-pin connector at the SRS airbag electronic unit
123	4-pin connector at the left SRS airbag sensor
123	4-pin connector at the right SRS airbag sensor
152A	29-pin bulkhead connector at main fuse/relay panel
152B	29-pin bulkhead connector at main fuse/relay panel
330	SRS airbag 10-pin connector in center console under rubber bellows. See SRS warning
335	SRS airbag 2-pin orange connector below steering column. Pins are short circuited when connector is separated. See SRS warnings

1991 and later

H3-16	3-pin connector at the SRS airbag left front sensor
H3-17	3-pin connector at the SRS airbag right front sensor
H4-10	4-pin connector at the SRS airbag control unit
H4-10	4-pin connector at the SRS airbag at the control unit
H4-11	4-pin connector at the SRS airbag at the control unit
H10-1	10-pin connector behind instrument panel, left of steering column
H10-2	10-pin connector at the SRS airbag control unit, left speaker grille. See warning.
H12-1	12-pin connector behind instrument panel, left of steering column
H29-1	29-pin bulkhead connector behind main fuse/relay panel (1991-1991 1/2)
H29-2	29-pin bulkhead connector behind main fuse/relay panel (1991-1991 1/2)
H33-1	33-pin bulkhead connector behind main fuse/relay panel (1991 1/2 –on)
H33-2	33-pin bulkhead connector behind main fuse/relay panel (1991 1/2 –on)

Ground Locations

7	ground point at radiator cross member
211	ground point at gearbox
334	SRS airbag ground point for electronic unit adjacent to electronic unit. See SRS warning.
400	SRS airbag redundant ground point behind instrument panel, left side. See SRS warning.

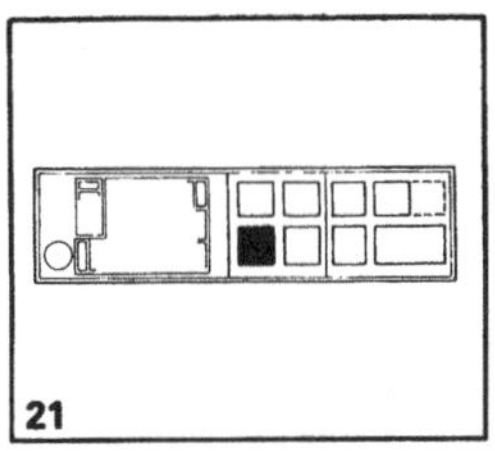
21

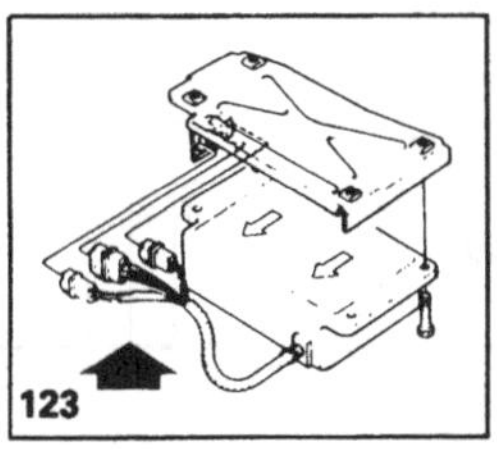
123

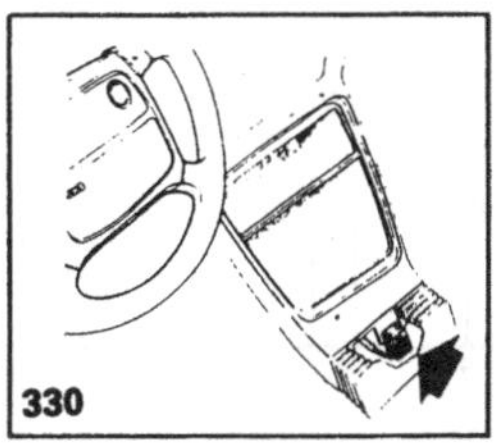
330

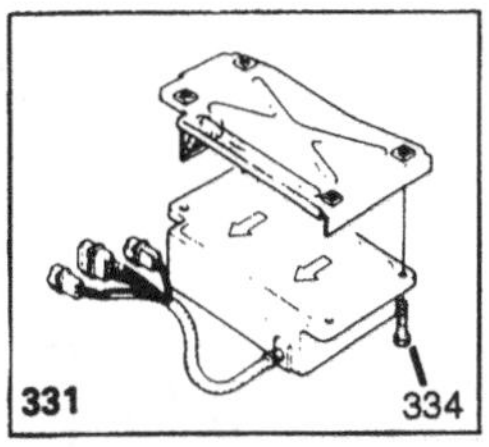

331

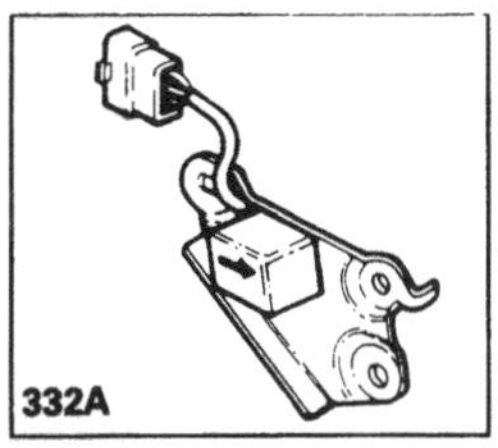
332A

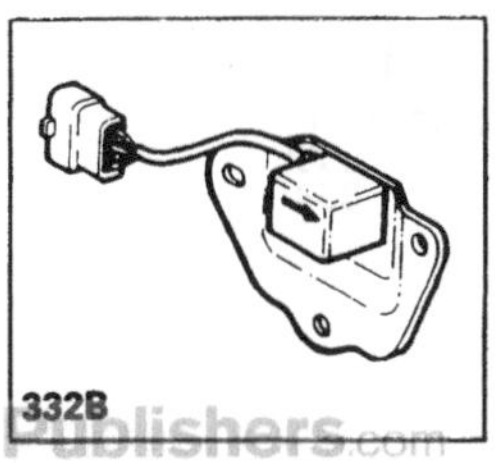
332B

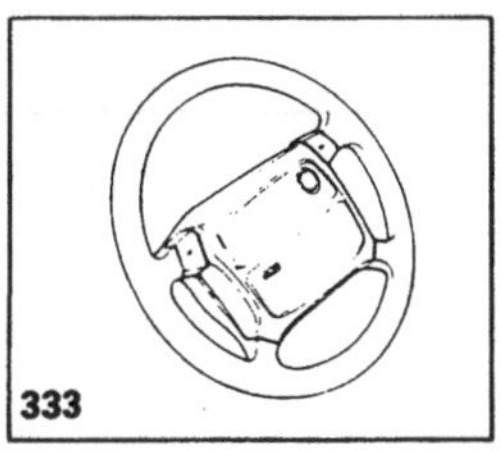
333

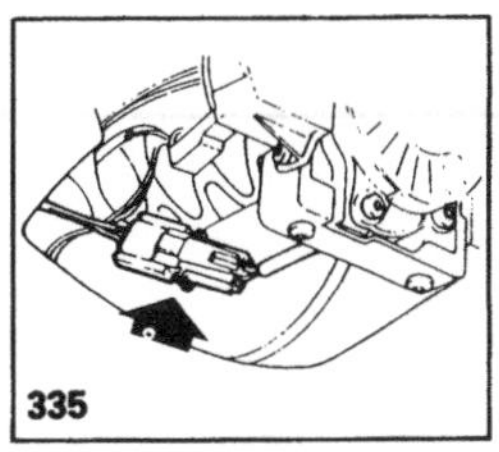
335

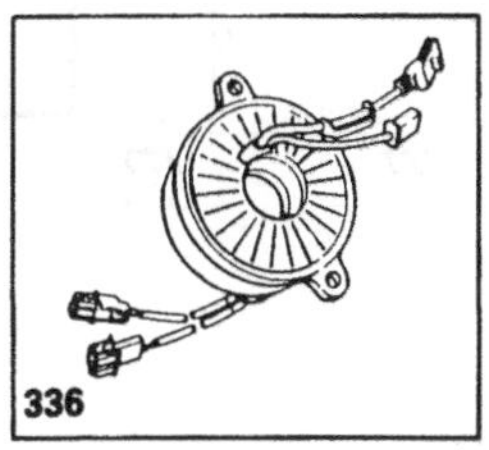
336

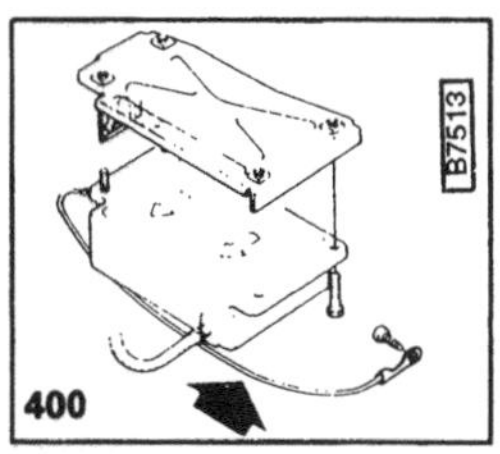

400

Supplemental Restraint System (SRS)—Airbag

1990

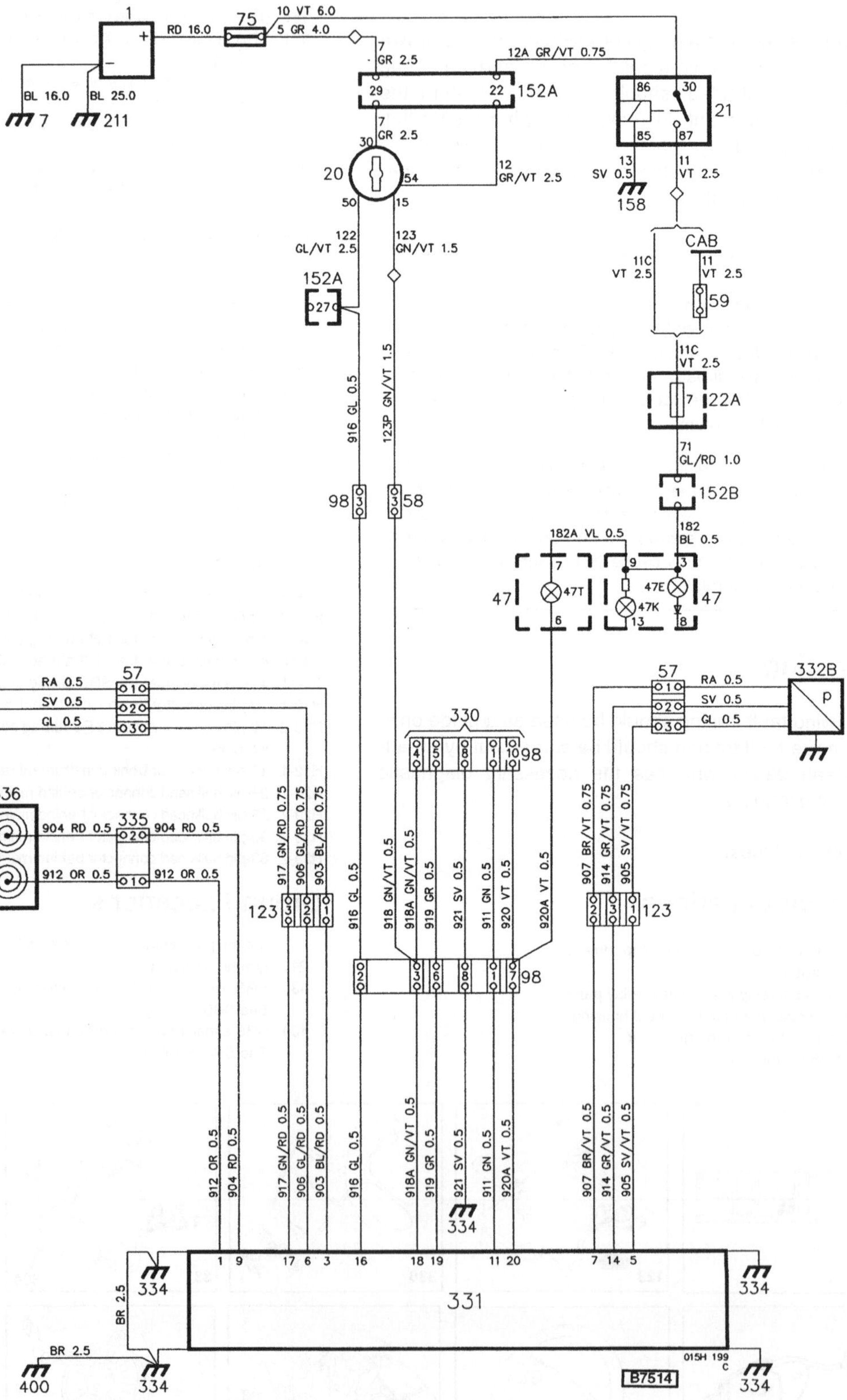

Supplemental Restraint System (SRS)—Airbag

1991 and later

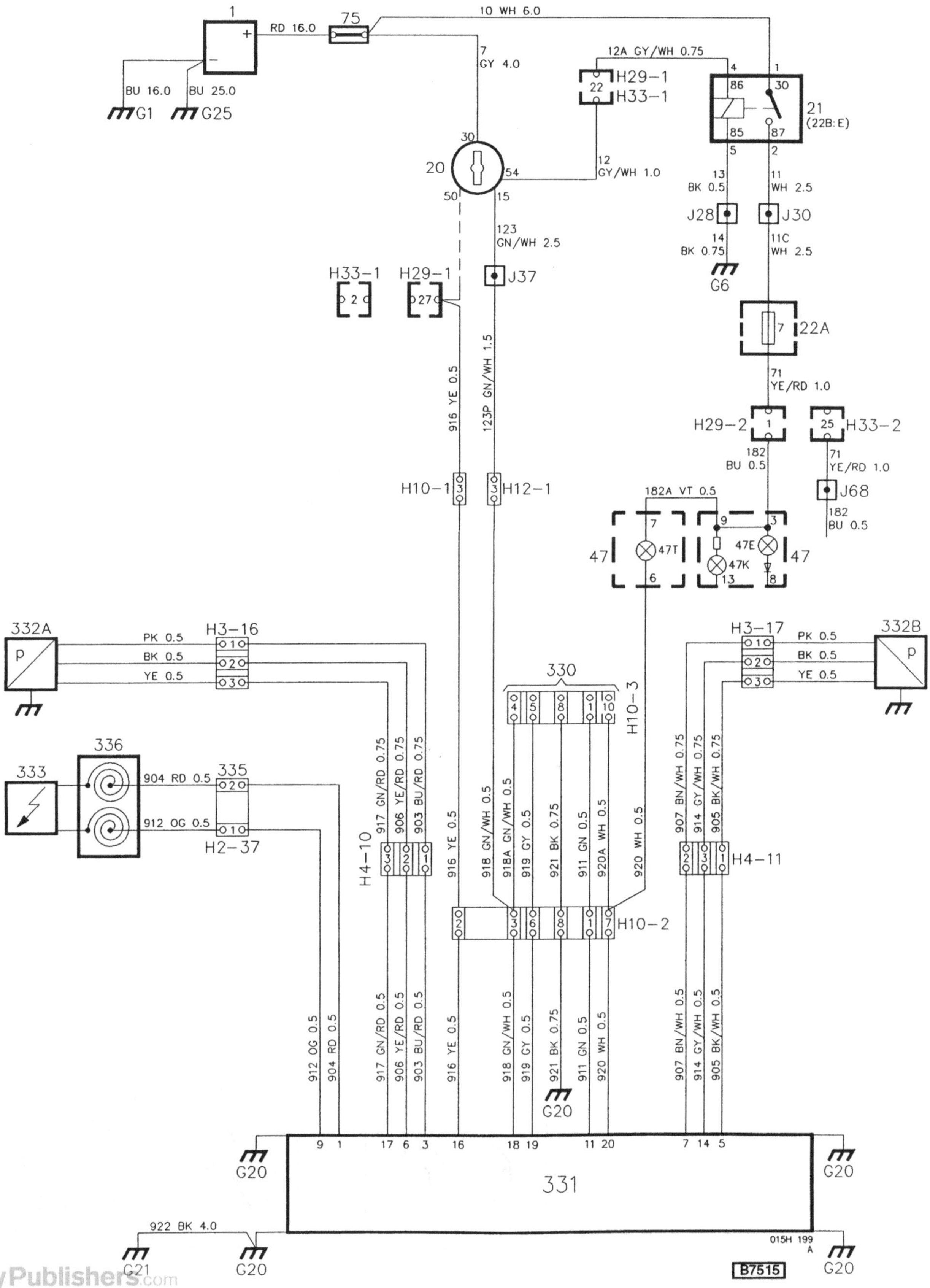

Ventilation Fan

Operation

The ventilation fan, located behind the instrument panel on the right side, is controlled by the fan and air distribution switches. Normally there are three fan speeds. If the air distribution switch (149) is switched to position 1, the fan switch (35) is overridden and the fan runs at high. The resistor pack provides various resistor values for different fan speeds. The resistor pack (74) is thermally protected internally and will blow the fuse if overheated.

For information on the related A/C systems, see the **Air Conditioner (A/C)** wiring diagram elsewhere in this section.

For information on repairing the heating and air conditioning system, see **854 Heating and Air Conditioning.**

Fault Tracing

1. Check the fuse. If the fuse blows repeatedly, check the resistor pack (74).

Component Locations

22A main fuse/relay panel at left wheel housing
35 ventilation fan switch in instrument panel
36 ventilation fan motor under right speaker grille
74 resistor pack for ventilation fan under left speaker grille
149 main switch for ventilation fan integrated in air distribution switch
150 air distribution switch in instrument panel

Connector Locations

1985-1990

152B 29-pin bulkhead connector at main fuse/relay panel

1991 and later

H29-2 29-pin bulkhead connector behind main fuse/relay panel (1991-1991 1/2)
H33-2 33-pin bulkhead connector behind main fuse/relay panel (1991 1/2 and later)

Ground Locations

1985-1990

3 ground point behind instrument panel

1991 and later

G8 ground point behind instrument panel

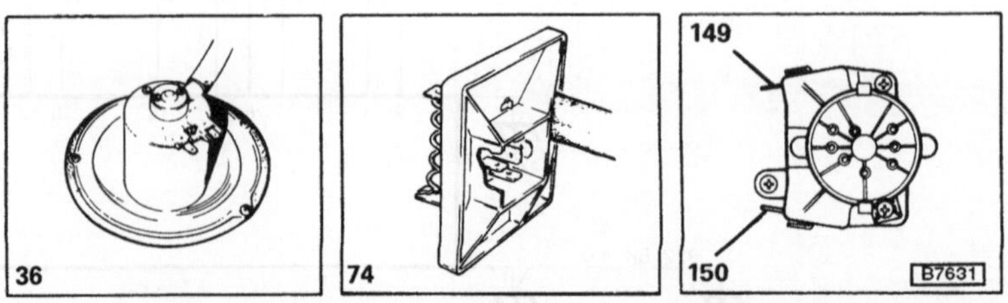

Ventilation Fan

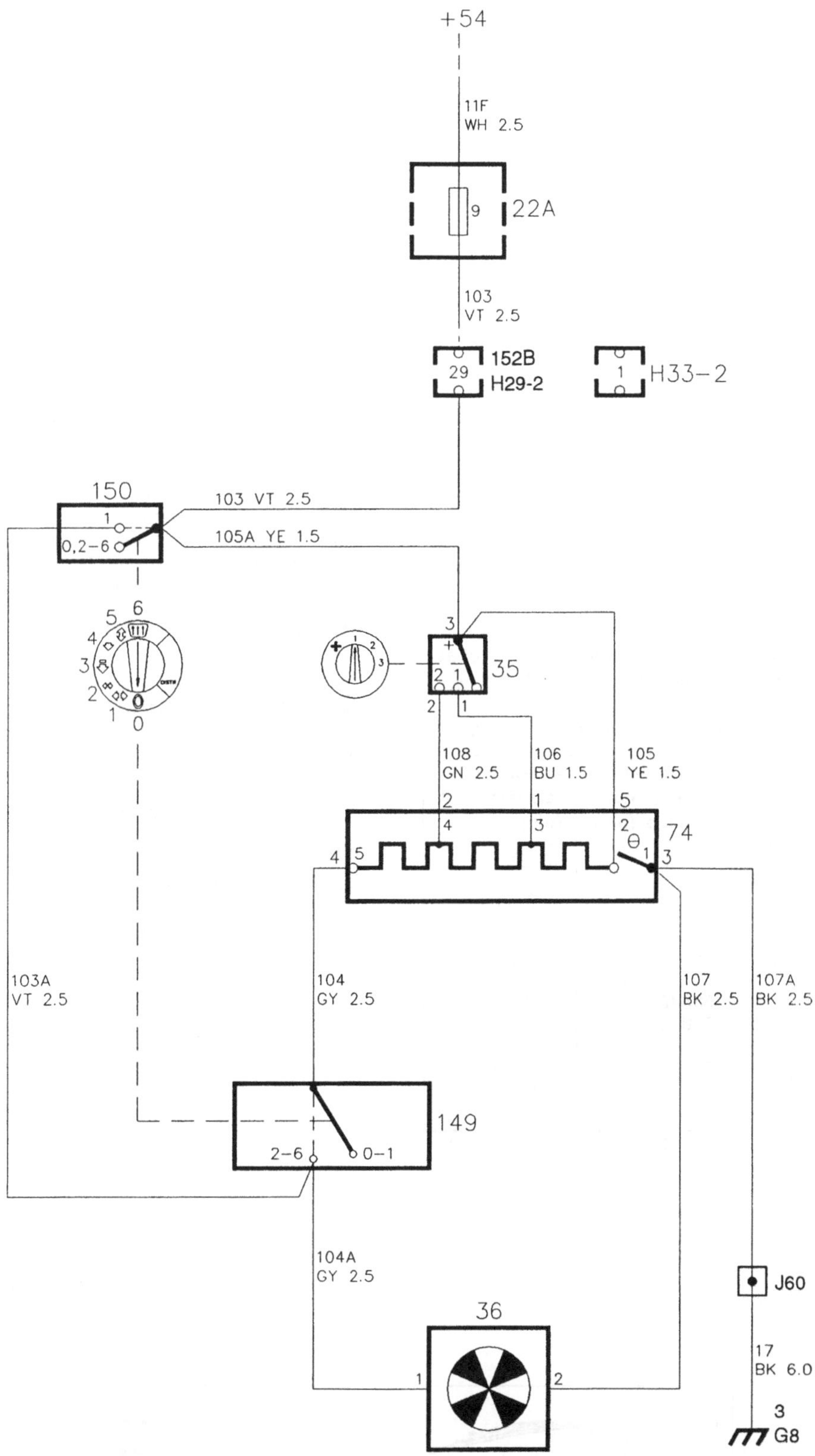

Wipers and Washers

Operation

The windshield wipers have three speeds controlled by the wiper stalk switch (61). The intermittent speed is controlled by the intermittent windshield wiper relay (83). The windshield wiper motor (62) has a park switch that interrupts voltage to correctly park the wiper blades at the base of the windshield after the stalk switch has been shut off.

The headlight wipers are controlled by the wiper stalk switch (61). When the wiper stalk switch is pulled toward the driver, the windshield washer pump (63) is energized and the wipers will run (via the intermittent relay 83) for 5 cycles. The headlight wiper motor (66), has a park switch that interrupts voltage to correctly park the wiper blades at the base of the headlight after the stalk switch has been released.

See **363 Wipers and Washers** for information on removing and installing the wiper motor mechanism, the headlight wipers, and the washer pump.

Fault Tracing

1. Check the fuses. Check for approximate battery voltage with the key in the run position using the table below.

Table a. Wiper and Washer Electrical Tests

Position	Switch terminal (pin)	Windshield wiper motor terminal (pin)	Intermittent relay terminal (pin)
0 off	53A (1)	53A (2)	15 (2)
I INT	53A (1) INT (5)	53A (2)	15 (2), I (8)
II slow	53A (1), 53 (2)	53A (2) 53 (5)	15 (2)
III fast	53A (1) 53B (4)	53A (2) 53B (4)	15 (2)
A* wash	53A (1) 54 (7), 31B (3), 53 (2)	53 (5)	T (5), 15 (2), 53M (1)

* check for voltage at the windshield washer pump (63), and headlight wiper motors

Component Locations

22A main fuse/relay panel
61 wiper stalk switch on steering column
62 windshield wiper motor at engine compartment firewall
63 windshield washer motor below fluid bottle in engine compartment
66 headlight wiper motor behind headlight assembly
83 intermittent windshield wiper relay to left of steering column, behind instrument panel
158 ground (–) distribution in main fuse/relay panel

Connector Locations

1985-1990

59 2-pin connector at the washer bottle
152B 29-pin bulkhead connector at main fuse/relay panel
152C 29-pin bulkhead connector at main fuse/relay panel

1991 and later

H3-1 3-pin connector at front of engine compartment, right wiper motor
H3-2 3-pin connector at front of engine compartment, left wiper motor
H29-2 29-pin bulkhead connector behind main fuse/relay panel (1991-1991 1/2)
H29-3 29-pin bulkhead connector behind main fuse/relay panel (1991-1991 1/2)
H33-2 33-pin bulkhead connector behind main fuse/relay panel (1991 1/2 and later)
H33-3 33-pin bulkhead connector behind main fuse/relay panel (1991 1/2 and later)

Ground Locations

1985-1990

3 ground point behind instrument panel
7 ground point at radiator cross member

1991 and later

G1 ground point at radiator cross member
G6 ground distribution in main fuse/relay panel
G8 ground point behind instrument panel

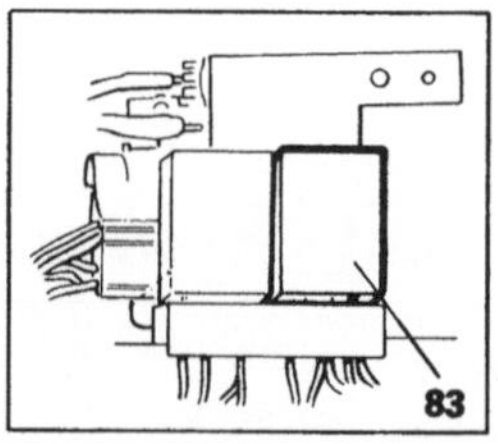
83

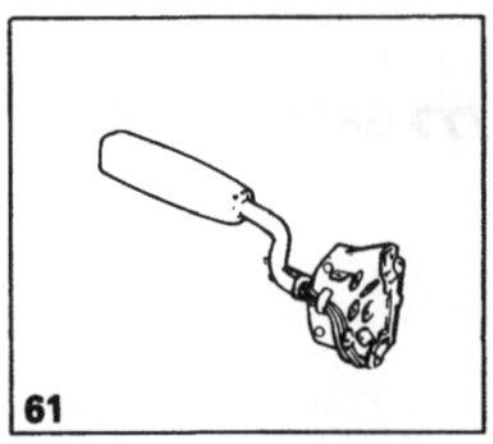
61

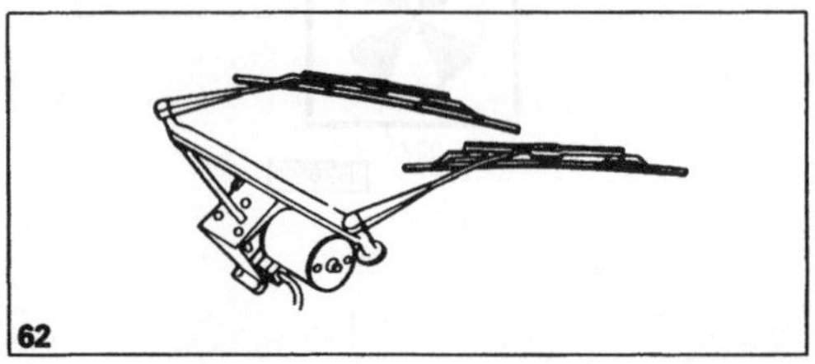
62

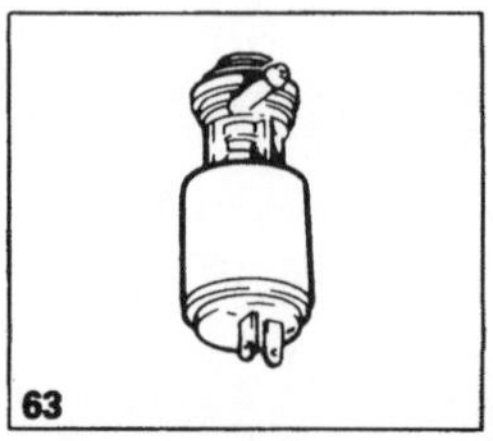
63

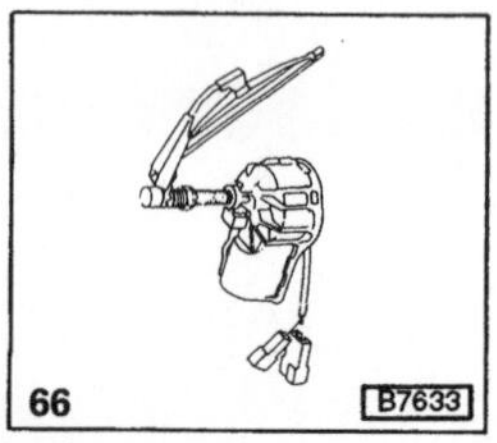
66

B7633

Wipers and Washers

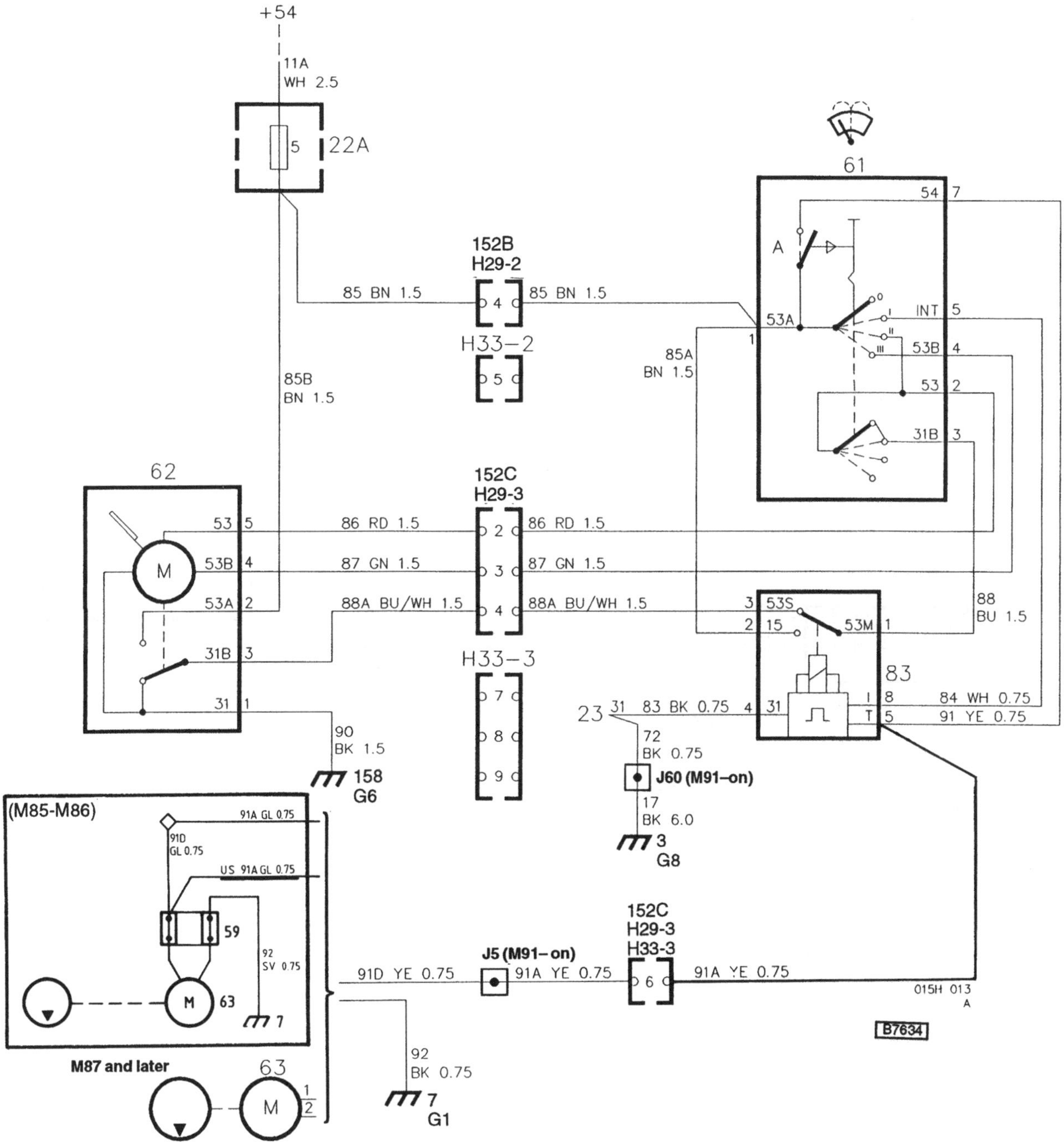

Headlight Wipers

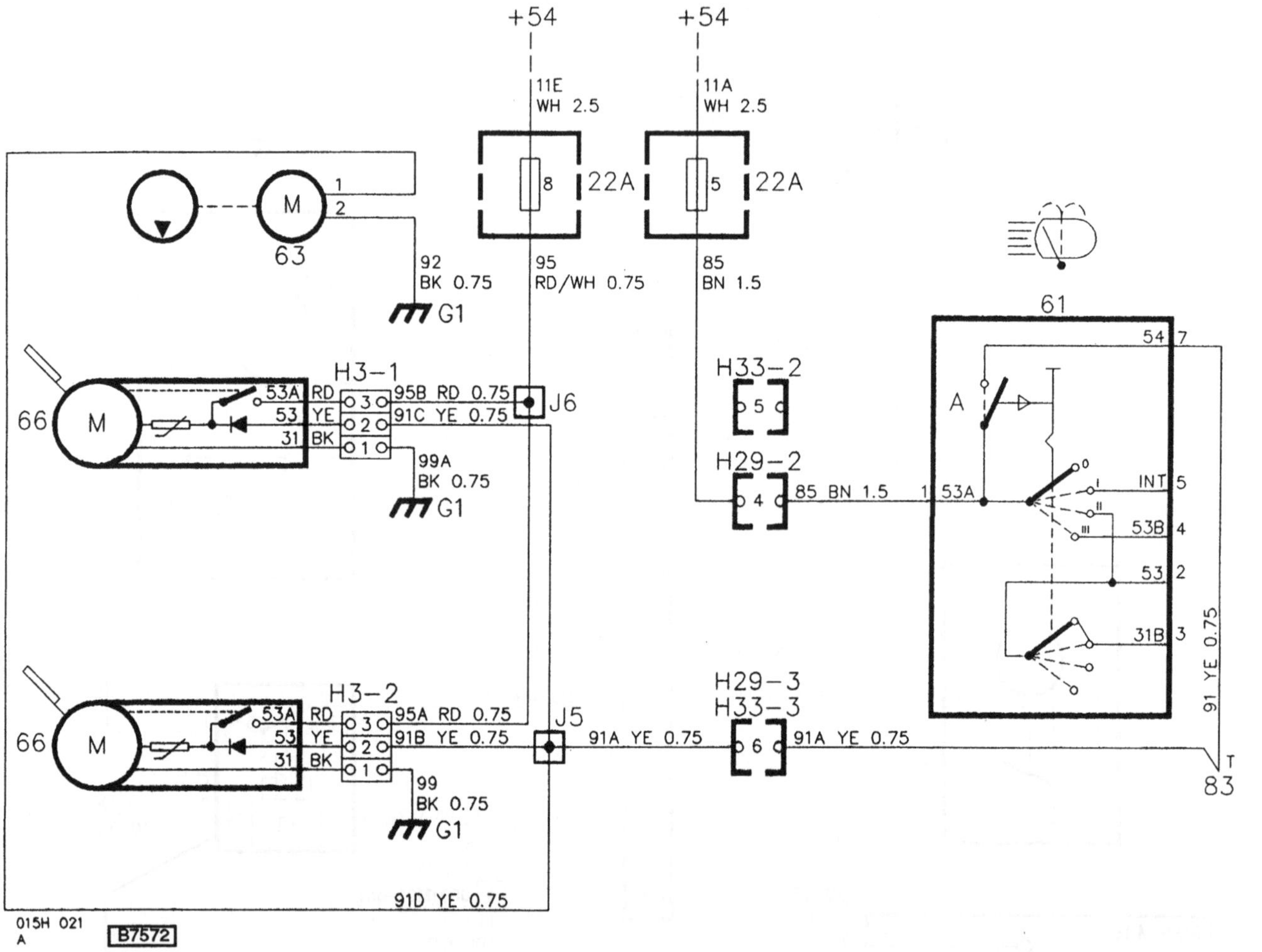

Component List—Numerical
(page 1 of 4)

1 battery—right front of engine compartment
2 alternator—back of engine compartment
3 ground point—behind instrument panel
4 starter motor—under intake manifold
5 ignition coil—on top of radiator
6 ignition distributor—end of cylinder head
7 ground point—radiator cross member
8 headlight relay—main fuse/relay panel
9 ground point—luggage compartment
10 lighting switch
11 high beam headlight
12 low beam headlight
13 parking lights
14 rear lights
15 license plate lights
16 rheostat for instrument cluster
17 rheostat for instrument panel lighting
18 instrument lighting
19 glove compartment light
20 ignition switch
21 ignition switch relay—main fuse/relay panel
21A ignition switch relay—in fuse/relay panel (401) under rear seat
22/22A main fuse/relay panel—left wheel housing
23 flasher relay—behind instrument panel, left side
24 direction indicator stalk switch
25 hazard warning switch
26 coolant fan time delay relay—front left wheel housing
27 left-hand direction indicator lamps
28 right hand direction indicator lamps
29 brake light switch—brake pedal cluster
30 brake lamps
31 backup light switch—gear selector housing
32 backup lights
33 rear fog lights
35 ventilation fan switch—instrument panel
36 ventilation fan motor—under right speaker grille
37 radiator cooling fan motor—behind radiator
38 recirculation valve—behind instrument panel, near right side A pillar
39 coolant fan temperature switch—upper left corner of radiator
40 horn—engine compartment
41 horn switch—in steering wheel
42 brake fluid level switch—brake fluid reservoir
43 handbrake switch—under handbrake
44 oil pressure switch—left side of engine, above oil filter
45 coolant temperature transmitter—front of cylinder head
46 fuel level transmitter—in fuel tank, accessible in luggage compartment
47 instrument cluster
47A fuel level gauge
47B fuel warning light
47C coolant temperature gauge
47D oil pressure warning light
47E charging lamp in instrument cluster
47F brake warning light
47G full—beam warning light
47H left-hand direction indicator warning lamp
47I right-hand direction indicator warning lamp
47J rear window heater warning lamp
47K shift-up indicator lamp
47M handbrake warning lamp
47N rear fog light warning lamp
47P CHECK ENGINE light
47Q ABS warning lamp
47S passive seat belt warning lamp
47T SRS airbag warning lamp (see warnings)
47U cruise control warning lamp
48 cigarette lighter
49 clock
50 roof lamp/switch—to left of driver
51 front roof lamp—center rear view mirror
52 ignition switch light—next to ignition switch, center console
53 interior light switch—center console
54 door switches—on door jamb
55 luggage compartment light—storage area
56 luggage compartment light switch—storage area
57 3-pin connector
58 12-pin connector
59 2-pin connector
60 1-pin connector
61 wiper stalk switch—steering column
62 windshield wiper motor—engine compartment firewall
63 windshield washer motor—below fluid bottle in engine compartment
64 heated seat pad with thermostat—in seat cushion and backrest
65 ground point—at handbrake
66 headlight wiper motor—behind headlight assembly
67 6-pin connector
68 horn relay—main fuse/relay panel
69 passenger seat belt switch—under seat cover
70 seat belt switch, driver side—between driver and passenger seat
71 seat belt switch, passenger side—between driver and passenger seat
72 seat belt warning light—center of instrument panel
73 TSI (test service instrument) socket—in main fuse/relay panel
74 resistor pack for ventilation fan—under left speaker grille
75 distribution block—right wheel housing
76 switch for raising idle speed (automatic transmission)—center console
77 neutral safety switch (automatics only)—gear selector lever
82 seat belt warning relay—under rear seat
83 intermittent windshield wiper relay—behind instrument panel, left side
85 fog lamps
88 switch for fog lamps
89 left fender direction indicator
90 right fender direction indicator
91 gear indicator light—at gear selector lever (automatics)
93 ground point—left wheel housing
94 starting injector—throttle housing
98 10-pin connector
101 supply fuel pump—in fuel tank
102 fuel pump relay—at LH control unit 200, forward of r.f. door
103 main fuel pump—in fuel tank
107 relay for extra lamps
109 third brake lamp (high mount brake light)
110 tachometer
113 time delay relay for heated rear window—under rear seat
115 rear window heater element
116 rear heated window switch
117 ground point—at handbrake
118 cornering lights—in front clusters
119 side backup lights in front light clusters
121 passenger seat switch for heated seats—under passenger seat
122 8-pin connector
123 4-pin connector
124 left power mirror switch—instrument panel
125 right power mirror switch—instrument panel
126 left power mirror motor—in left mirror
127 right power mirror motor—in right mirror
128 speed control relay—behind instrument panel, left side
131 control unit for cruise control—behind instrument panel, left side
132 speed transmitter—at rear of instrument cluster
133 clutch switch for cruise control—at clutch pedal
134 brake switch for cruise control—at brake pedal
136 lambda exhaust sensor—in exhaust manifold
141 selector for cruise control—direction indicator stalk
143 recirculation switch
144 boost pressure switch (turbo)
145 EZK test connector—next to main fuse/relay panel
146 ignition control unit (non-EZK ignition)
ignition amplifier (EZK ignition)—in front of left wheel housing

Component List—Numerical

(page 2 of 4)

147 ignition pulse amplifier—main fuse/relay panel, position D
148 ashtray lighting
149 over-ride switch for ventilation fan—integrated in air distribution switch
150 air distribution switch—instrument panel
151 interior lighting time delay relay—under rear seat
152A 29-pin bulkhead connector—main fuse/relay panel (white)
152B 29-pin bulkhead connector—main fuse/relay panel (red)
152C 29-pin bulkhead connector—main fuse/relay panel (black)
153 cigarette lighter lamp
154 heater control lighting
155 relay for A/C cooling fan—main fuse/relay panel
156 A/C compressor relay—main fuse/relay panel
157 sparkplug
158 negative (ground) distribution in main fuse/relay panel
159 +15 distribution in main fuse/relay panel
160 glovebox light switch—side of glovebox
161 rear fog light switch
162 switch for driver power window—center console
163 switch for passenger power window—center console
164 power window motor—left front door
165 power window motor—right front door
166 A/C pressure switch for cooling fan—on receiver/dryer unit
167 A/C throttle switch—throttle housing
168 A/C coolant temperature switch—radiator hose near distributor
169 A/C switch—instrument panel
170 A/C compressor—top of engine near firewall
171 A/C anti freeze-up thermostat—behind instrument panel, right side
172 A/C cooling fan—behind radiator
173 A/C compressor diode—engine wiring harness, rear of valve cover
174 relay for daytime driving lights
175 central locking control unit—behind instrument panel right side
176 EZK control unit—inner left fender
177 APC control unit–under back seat (1985), ahead of left wheel housing (1986 and later)
178 knock sensor—engine block, below intake manifold
179 APC boost pressure solenoid valve—above radiator fan on radiator
180 boost pressure sensor behind instrument panel to left of steering column
181 power sunroof or convertible switch
182 power sunroof motor—luggage compartment at spare tire compartment
183 key switch for driver's central door locking—driver door
184 central door lock motor—passenger door
185 central door lock motor—right rear door
186 central door lock motor—left rear door
187 cruise control vacuum pump—engine compartment left wheel housing
188 central door lock motor—tailgate lock
189 lockout switch for rear power windows
190 switch for left rear power window—center console
190a switch for left rear power window—left rear door
191 switch for right rear power window—center console
191a switch for right rear power window—right rear door
193 power window motor—left rear door
194 power window motor—right rear door
200 LH fuel injection control unit—forward of right front door, behind trim
201 LH ground point for fuel injection—engine lifting lug, 2 points
202 LH engine temperature transmitter—intake manifold flange, between cylinders 2 and 3
203 LH throttle angle transmitter—on throttle housing
204 LH test connector—engine compartment, behind right wheel housing
205 LH air mass meter—in throttle intake hose
206 fuel injectors—intake manifold
207 heated rear view mirrors—front doors
208D passive seat belt door switch—driver door
208P passive seat belt door switch—passenger door
211 ground point—gearbox
215 headlight dip switch
225 reading lights, convertible—back seat
229 LH fuel injection main relay—at LH control unit 200, behind A pillar trim at passenger door
233 cruise control vacuum switch—engine compartment left wheel housing
234 side marker lights in front light clusters
252 driver heated seat rheostat—instrument panel
254 driver seat temperature thermostat—in driver seat
257 ground point—alternator bracket
261 throttle contacts for shift-up indicator—throttle housing
262 temperature switch for shift-up indicator—between center intake ports of intake manifold
263 vacuum switch for shift-up indicator—left wheel housing
264 5 gear switch for shift-up indicator—front side cover of gearbox
265 electric radio antenna—left rear body panel
266 radio speakers—(4) two in front, two in rear
267 radio connector—instrument panel
270 shift-up indicator relay—main fuse/relay panel
271 lambda preheater (integral with lambda sensor)
272 AIC valve (LH idle speed motor)—front left side of cylinder head
273 motor for convertible hydraulic pump—under rear seat, right side
274 fuse for burglar alarm under rear seat
275 burglar alarm siren—engine compartment near right headlamp
276 burglar alarm hood switch—under hood at front of wheel housing
277 relay for raising convertible top (1987-1988)—under rear seat on bulkhead
277 relay for raising convertible top (1989 and later)—under rear seat in fuse/relay panel 401
278 relay for lowering convertible top (1987-1988)—under rear seat on bulkhead
278 relay for lowering convertible top (1989 and later)—under rear seat in fuse/relay panel 401
279 mercury switch for convertible heated rear window—left side top mounting
283 diodes for convertible top (1987-1988)—under rear seat in wiring harness
285 fuse for lambda sensor—engine compartment, next to fresh air intake
287 power window relay for automatic control—under rear seat
288 burglar alarm switch—center console
289 burglar alarm control unit—under rear seat
290A fuse for convertible top—engine compartment, next to distribution block 75
290B fuse for convertible top—engine compartment, next to distribution block 75
291 ABS control unit (1989)—engine compartment, left wheel housing
291 ABS control unit (1990 and later)—under rear seat, right side
292 ABS system relay (1989)—ABS distribution box, engine compartment, left wheel housing
292 ABS system relay (1990 and later)—ABS distribution box, behind right wheel housing
293 ABS pump relay (1989)—ABS distribution box, left wheel housing
293 ABS pump relay (1990 and later)—ABS distribution box, behind right wheel housing
294 ABS pressure switch—on 302 ABS electrical distribution box (1990 and later)—engine compartment, behind right wheel housing
295 ABS master valve—on hydraulic unit, engine compartment
296 ABS valve block—forward of left wheel housing
297 ABS hydraulic pump motor—engine compartment at hydraulic unit
298A ABS wheel sensor—left front wheel
298B ABS wheel sensor—right front wheel
298C ABS wheel sensor—left rear wheel
298D ABS wheel sensor—right rear wheel
299 ABS brake fluid level sensor—brake fluid reservoir
300 ABS ground point—at front of hydraulic unit, engine compartment
302 ABS electrical distribution box (1989)—engine compartment, forward of left wheel housing
302 ABS electrical distribution box (1990 and later)—engine compartment, behind right wheel housing

Component List—Numerical

(page 3 of 4)

302A ABS fuse holder (1989)—engine compartment, forward of left wheel housing
302A ABS fuse holder (1990 and later)—engine compartment, behind right wheel housing
303B ABS diode (1989)—engine compartment, forward of left wheel housing
303B ABS diode (1990 and later)—engine compartment, behind right wheel housing
304 burglar alarm motion detector—under rear seat
305 burglar alarm LED—left speaker grille
306 logic box for passive seat belts—under rear seat, left side
307D passive seat belt belt reel—driver seat
307P passive seat belt belt reel—passenger seat
308D passive seat belt motor with limit switches—driver seat
308P passive seat belt motor with limit switches—passenger seat
309 passive seat belt g sensor—under rear seat, left side
310D passive seat belt fuse for driver seat—under rear seat
310P passive seat belt fuse for passenger seat—under rear seat
311D passive seat belt motor relay for driver seat—under rear seat
311P passive seat belt motor relay for passenger seat—under rear seat
312D passive seat belt motor relay for driver seat—under rear seat
312P passive seat belt motor relay for passenger seat—under rear seat
313 burglar alarm relay—under rear seat
314 burglar alarm seat switch—driver seat
315 burglar alarm seat switch—passenger seat
316 burglar alarm diode—under rear seat in wiring harness
320 ignition coil with integrated amplifier—inner right fender, near battery
321 charcoal canister valve (ELCD)—engine compartment ahead of left wheel housing
322 LH coding connector—at LH control unit 200, forward of r.f. door
323 fuel pump—in fuel tank
330 SRS airbag 10-pin connector—center console under rubber bellows. See SRS warning.
331 SRS airbag control unit—behind instrument panel, left grille. See SRS warning.
332A SRS airbag left sensor—left wheel housing in engine compartment. See SRS warning.
332B SRS airbag right sensor—right wheel housing in engine compartment. See SRS warning.
333 SRS airbag—in steering wheel. See SRS warning.
334 SRS airbag ground point for electronic unit—adjacent to electronic unit. See SRS warning.
335 SRS airbag 2-pin orange connector—pins are short circuited when connector is separated. See SRS warning.
336 SRS airbag slip ring contact—in steering wheel. See SRS warning.
345 EZK crankshaft sensor—behind crankshaft pulley
347 LH diagnostic test connector—under rear seat, right side
349 radio contact box with amplifier—instrument panel
350 contact box for CD player or equalizer—center console top compartment
351 audio amplifier—under rear seat below ABS control unit (291)
352 slip ring switch—in steering wheel
386 switch assembly for power windows and sunroof—center console
389 LH NTC resistor—under intake manifold
390 LH EGR modulating valve—ahead of left wheel housing
396 radiator fan relay—main fuse/relay panel
397 ABS diagnostic test socket—near hydraulic unit in engine compartment, wiring harness
400 SRS airbag redundant ground point—behind instrument panel, left side. See SRS warning.
401 auxiliary fuse/relay panel—under rear seat
412 burglar alarm service mode switch—under passenger seat
413 burglar alarm glass breakage detector—under passenger seat
423 burglar alarm start interlock relay (convertible)—under rear seat
424 burglar alarm hazard warning light relay (convertible)—under rear seat at control unit
425 door lock switch—center console
426 switch for rear seat—under rear seat

G1 ground point—radiator cross member
G3 ground point—luggage compartment
G4 ground point—between ignition switch and handbrake
G5 ground point—under rear seat
G6 ground distribution—main fuse/relay panel
G7 ground point—engine lifting lug
G8 ground point—behind instrument panel
G9 ground point—luggage compartment
G12 ground point—left wheel housing
G16 ground point—front of ABS hydraulic unit
G19 ground point—hatchback
G20 ground point—SRS airbag ground at electronic unit. See SRS warning.
G21 ground point—SRS airbag redundant ground. See SRS warning.
G25 ground point—gearbox
G26 ground point—right C pillar on sedans
G27 ground point—electric antenna
G32 ground point—alternator bracket
H1-1 1-pin connector—engine compartment, upper left door hinge
H1-2 1-pin connector—next to bulkhead connectors, left side behind instrument panel
H1-3 1-pin connector—engine compartment, upper right door hinge
H1-5 1-pin connector—behind instrument panel to left of steering column
H1-6 1-pin connector—right B pillar, behind trim
H1-7 1-pin connector—left B pillar, behind trim
H1-8 1-pin connector—behind instrument panel to left of steering column
H1-9 1-pin connector—near lambda sensor, right wheel housing
H1-10 1-pin connector—near A/C compressor in engine compartment
H1-12 1-pin connector—engine compartment, right side near air intake
H1-14 1-pin connector—in left front door
H1-15 1-pin connector—in right front door
H1-16 1-pin connector—under center console next to shift selector (automatic)
H1-17 1-pin connector—under center console
H1-18 1-pin connector—under center console
H1-19 1-pin connector—near LH control unit, behind right A pillar trim
H1-21 1-pin connector—for LED, under left speaker grille
H1-22 1-pin connector—for LED, under left speaker grille
H1-27 1-pin connector—engine compartment, next to distribution block 75
H1-28 1-pin connector—under rear seat
H2-1 2-pin connector—at radiator fan motor
H2-2 2-pin connector—next to A/C cooling fan (172)
H2-3 2-pin connector—behind instrument cluster
H2-4 2-pin connector—engine compartment near right door hinge
H2-5 2-pin connector—engine compartment near top left door hinge
H2-6 2-pin connector—behind instrument cluster, left of steering column
H2-7 2-pin connector—instrument panel at light switch
H2-8 2-pin connector—engine compartment near right door hinge
H2-10 2-pin connector—behind cigarette lighter
H2-11 2-pin connector—engine compartment, near top right door hinge
H2-12 2-pin connector—under rear seat, left side
H2-13 2-pin connector—at fuel pump in storage area
H2-14 2-pin connector—left rear wheel housing in luggage compartment
H2-17 2-pin connector—hatchback, left side or sedan, under parcel shelf
H2-18 2-pin connector—hatchback, left side
H2-19 2-pin connector—trunk lid
H2-20 2-pin connector—luggage compartment, left wheel housing
H2-21 2-pin connector—luggage compartment, left wheel housing
H2-22 2-pin connector—engine compartment near ABS hydraulic unit
H2-23 2-pin connector—engine compartment right side partition
H2-24 2-pin connector—under rear seat, left side
H2-25 2-pin connector—under rear seat, right side
H2-27 2-pin connector—main fuse/relay panel
H2-30 2-pin connector—left rear door
H2-31 2-pin connector—right rear door
H2-32 2-pin connector—luggage compartment near central lock motor
H2-33 2-pin connector—near lambda sensor in engine compartment
H2-34 2-pin connector—left wheel housing
H2-35 2-pin connector—above roof lining at rear view mirror

Component List—Numerical
(page 4 of 4)

H2-40 2-pin connector—for burglar alarm siren, behind headlamp
H2-41 2-pin connector—under passenger seat, right side
H2-42 2-pin connector—trunk lid
H2-43 2-pin connector—left hinge, convertible
H2-46 2-pin connector—left reading lamp, convertible
H2-47 2-pin connector—right reading lamp, convertible
H2-48 2-pin connector—at convertible top motor
H2-51 2-pin connector—left rear speaker
H2-52 2-pin connector—right rear speaker
H2-53 2-pin connector—under rear seat
H3-1 3-pin connector—front of engine compartment, right wiper motor
H3-2 3-pin connector—front of engine compartment, left wiper motor
H3-3 3-pin connector—left of starter motor
H3-4 3-pin connector—main fuse/relay panel
H3-5 3-pin connector—engine compartment, left of air intake
H3-7 3-pin connector—engine compartment near left door hinge
H3-10 3-pin connector—right B pillar, behind trim
H3-11 3-pin connector—left B pillar, behind trim
H3-13 3-pin connector—luggage compartment, left air outlet
H3-14 3-pin connector—near LH control unit, behind right A pillar trim
H3-15 3-pin connector—in driver door
H3-16 3-pin connector—SRS airbag left front sensor
H3-17 3-pin connector—SRS airbag right front sensor
H3-18 3-pin connector—under passenger seat
H4-1 4-pin connector—left wheel housing
H4-1 4-pin connector—instrument panel right side behind trim
H4-2 4-pin connector—engine compartment, near top left door hinge
H4-3 4-pin connector—engine compartment, near top right door hinge
H4-4 4-pin connector—right B pillar behind trim
H4-5 4-pin connector—left B pillar behind trim
H4-6 4-pin connector—instrument panel left side
H4-7 4-pin connector—at LH control unit, forward of right front door
H4-8 4-pin connector—in left front door
H4-9 4-pin connector—in right front door
H4-10 4-pin connector—SRS airbag at control unit
H4-11 4-pin connector—SRS airbag at control unit
H4-12 4-pin connector—trunk lid, left air outlet
H4-15 4-pin connector—under rear seat, next to fuse/relay panel 401
H4-16 4-pin connector—behind instrument panel to left of steering column
H6-1 6-pin connector—engine compartment right wheel housing
H6-2 6-pin connector—engine compartment right wheel housing
H6-4 6-pin connector—behind instrument panel
H8-1 8-pin connector—under passenger seat
H8-2 8-pin connector—under driver seat
H8-3 8-pin connector—under rear seat
H8-7 8-pin connector—behind instrument panel
H8-8 8-pin connector—behind instrument panel
H10-1 10-pin connector—behind instrument panel, left of steering column
H10-2 10-pin connector—at SRS airbag control unit, left speaker grille
H10-5 10-pin connector—behind instrument panel
H10-6 10-pin connector—at auxiliary fuse/relay panel under rear seat
H10-8 10-pin connector—under rear seat at burglar alarm control unit
H12-1 12-pin connector—behind instrument panel, left of steering column
H12-3 12-pin connector—behind instrument panel
H12-4 12-pin connector—behind instrument panel
H29-1 29-pin bulkhead connector—behind main fuse/relay panel
H29-2 29-pin bulkhead connector—behind main fuse/relay panel
H29-3 29-pin bulkhead connector—behind main fuse/relay panel
H33-1 33-pin bulkhead connector—behind main fuse/relay panel
H33-2 33-pin bulkhead connector—behind main fuse/relay panel
H33-3 33-pin bulkhead connector—behind main fuse/relay panel

Component List—Alphabetical

(page 1 of 4)

Component	No.
5th gear switch for shift-up indicator—front side cover of gearbox	264
+15 distribution terminal—main fuse/relay panel, engine compartment	159
ABS brake fluid level sensor—brake fluid reservoir	299
ABS control unit (1989)—engine compartment, left wheel housing	291
ABS control unit (1990 and later)—under rear seat, right side	291
ABS diagnostic test socket—near hydraulic unit in engine compartment, wiring harness	397
ABS diode (1989)—engine compartment, forward of left wheel housing	303B
ABS diode (1990 and later)—engine compartment, behind right wheel housing	303B
ABS electrical distribution box (1989)—engine compartment, forward of left wheel housing	302
ABS electrical distribution box (1990 and later)—engine compartment, behind right wheel housing	302
ABS fuse holder (1989)—engine compartment, forward of left wheel housing	302A
ABS fuse holder (1990 and later)—engine compartment, behind right wheel housing	302A
ABS ground point—at front of hydraulic unit, engine compartment	300
ABS hydraulic pump motor—engine compartment at hydraulic unit	297
ABS master valve—on hydraulic unit, engine compartment	295
ABS pressure switch—on ABS hydraulic unit	294
ABS pump relay (1989)—ABS distribution box, left wheel housing	293
ABS pump relay (1990 and later)—ABS distribution box, behind right wheel housing	293
ABS system relay (1989)—ABS distribution box, engine compartment, left wheel housing	292
ABS system relay (1990 and later)—ABS distribution box, behind right wheel housing	292
ABS valve block—forward of left wheel housing	296
ABS warning lamp	47Q
ABS wheel sensor—front wheel, left	298A
ABS wheel sensor—front wheel, right	298B
ABS wheel sensor—rear wheel, left	298C
ABS wheel sensor—rear wheel, right	298D
A/C anti freeze-up thermostat—instrument panel, right side	171
A/C compressor—top of engine near firewall	170
A/C compressor diode—engine wiring harness, rear of valve cover	173
A/C compressor relay—main fuse/relay panel, engine compartment	156
A/C coolant temperature switch—radiator hose near distributor	168
A/C cooling fan—behind radiator	172
A/C pressure switch for cooling fan—on receiver/dryer unit	166
A/C switch—instrument panel	169
A/C throttle switch—throttle housing	167
AIC valve (LH idle speed motor)—front left side of cylinder head	272
air distribution switch—instrument panel	150
alternator—back of engine compartment	2
APC boost pressure solenoid valve—above radiator fan on radiator	179
APC control unit—under back seat (1985), ahead of left wheel housing (1986 and later)	177
ashtray lighting	148
audio amplifier—under rear seat below ABS control unit (291)	351
auxiliary fuse/relay panel—under rear seat	401
backup light switch—gear selector housing	31
battery—right front of engine compartment	1
battery—right front of engine compartment	1
boost pressure sensor—behind instrument panel to left of steering column	180
boost pressure switch (turbo)	144
brake fluid level switch—brake fluid reservoir	42
brake lamps	30
brake light switch—brake pedal cluster	29
brake switch for cruise control—at brake pedal	134
brake warning light	47F
burglar alarm control unit—under rear seat	289
burglar alarm diode—under rear seat in wiring harness	316
burglar alarm glass breakage detector—under passenger seat	413
burglar alarm hazard warning light relay (convertible)—under rear seat at control unit	424
burglar alarm hood switch—under hood at front of wheel housing	276
burglar alarm LED—left speaker grille	305
burglar alarm motion detector—under rear seat	304
burglar alarm relay—under rear seat	313
burglar alarm seat switch—driver seat	314
burglar alarm seat switch—passenger seat	315
burglar alarm service mode switch—under passenger seat	412
burglar alarm siren—engine compartment near right headlamp	275
burglar alarm start interlock relay (convertible)—under rear seat	423
burglar alarm switch—center console rear seat at control unit	288
central door lock motor—left rear door	186
central door lock motor—passenger door	184
central door lock motor—right rear door	185
central door lock motor—tailgate lock	188
central locking control unit—behind instrument panel right side	175
charcoal canister valve(ELCD)—engine compartment ahead of left wheel housing	321
charging lamp—in instrument cluster	47E
CHECK ENGINE light	47P
cigarette lighter	48
cigarette lighter lamp	153
clock	49
clutch switch for cruise control—at clutch pedal	133
coding connector—at LH control unit 200, behind A pillar trim	322
combined instruments	47
connector, 1-pin	60
connector, 2-pin	59
connector, 3-pin	57
connector, 4-pin	123
connector, 6-pin	67
connector, 8-pin for burglar alarm—under rear seat	122
connector, 10-pin	98
connector, 12-pin	58
connector, 29-pin—main fuse/relay panel	152A
connector, 29-pin—main fuse/relay panel	152B
connector, 29-pin—main fuse/relay panel	152C
connector, 1-pin—engine compartment, upper left door hinge	H1-1
connector, 1-pin—next to bulkhead connectors, left side behind instrument panel	H1-2
connector, 1-pin—engine compartment, upper right door hinge	H1-3
connector, 1-pin—behind instrument panel to left of steering column	H1-5
connector, 1-pin—right B pillar, behind trim	H1-6
connector, 1-pin—left B pillar, behind trim	H1-7
connector, 1-pin—behind instrument panel to left of steering column	H1-8
connector, 1-pin—near lambda sensor, right wheel housing	H1-9
connector, 1-pin—near A/C compressor in engine compartment	H1-10
connector, 1-pin—engine compartment, right side near air intake	H1-12
connector, 1-pin—in left front door	H1-14
connector, 1-pin—in right front door	H1-15
connector, 1-pin—under center console next to shift selector (automatic)	H1-16
connector, 1-pin—under center console	H1-17
connector, 1-pin—under center console	H1-18
connector, 1-pin—near LH control unit, behind right A pillar trim	H1-19
connector, 1-pin—for LED, under left speaker grille	H1-21
connector, 1-pin—for LED, under left speaker grille	H1-22
connector, 1-pin—engine compartment, next to distribution block 75	H1-27
connector, 1-pin—under rear seat	H1-28
connector, 2-pin—at radiator fan motor	H2-1
connector, 2-pin—next to A/C cooling fan (172)	H2-2
connector, 2-pin—behind instrument cluster	H2-3
connector, 2-pin—engine compartment near right door hinge	H2-4
connector, 2-pin—engine compartment near top left door hinge	H2-5
connector, 2-pin—behind instrument cluster, left of steering column	H2-6
connector, 2-pin—instrument panel at light switch	H2-7

Component List—Alphabetical

(page 2 of 4)

Component	Ref.
connector, 2-pin—engine compartment near right door hinge	H2-8
connector, 2-pin—behind cigarette lighter	H2-10
connector, 2-pin—engine compartment, near top right door hinge	H2-11
connector, 2-pin—under rear seat, left side	H2-12
connector, 2-pin—at fuel pump in storage area	H2-13
connector, 2-pin—left rear wheel housing in luggage compartment	H2-14
connector, 2-pin—hatchback, left side or sedan, under parcel shelf	H2-17
connector, 2-pin—hatchback, left side	H2-18
connector, 2-pin—trunk lid	H2-19
connector, 2-pin—luggage compartment, left wheel housing	H2-20
connector, 2-pin—luggage compartment, left wheel housing	H2-21
connector, 2-pin—engine compartment near ABS hydraulic unit	H2-22
connector, 2-pin—engine compartment right side partition	H2-23
connector, 2-pin—under rear seat, left side	H2-24
connector, 2-pin—under rear seat, right side	H2-25
connector, 2-pin—main fuse/relay panel	H2-27
connector, 2-pin—left rear door	H2-30
connector, 2-pin—right rear door	H2-31
connector, 2-pin—luggage compartment near central lock motor	H2-32
connector, 2-pin—near lambda sensor in engine compartment	H2-33
connector, 2-pin—left wheel housing	H2-34
connector, 2-pin—above roof lining at rear view mirror	H2-35
connector, 2-pin—for burglar alarm siren, behind headlamp	H2-40
connector, 2-pin—under passenger seat, right side	H2-41
connector, 2-pin—trunk lid	H2-42
connector, 2-pin—left hinge, convertible	H2-43
connector, 2-pin—left reading lamp, convertible	H2-46
connector, 2-pin—right reading lamp, convertible	H2-47
connector, 2-pin—at convertible top motor	H2-48
connector, 2-pin—left rear speaker	H2-51
connector, 2-pin—right rear speaker	H2-52
connector, 2-pin—under rear seat	H2-53
connector, 3-pin—front of engine compartment, right wiper motor	H3-1
connector, 3-pin—front of engine compartment, left wiper motor	H3-2
connector, 3-pin—left of starter motor	H3-3
connector, 3-pin—main fuse/relay panel	H3-4
connector, 3-pin—engine compartment, left of air intake	H3-5
connector, 3-pin—engine compartment near left door hinge	H3-7
connector, 3-pin—right B pillar, behind trim	H3-10
connector, 3-pin—left B pillar, behind trim	H3-11
connector, 3-pin—luggage compartment, left air outlet	H3-13
connector, 3-pin—near LH control unit, behind right A pillar trim	H3-14
connector, 3-pin—in driver door	H3-15
connector, 3-pin—under passenger seat	H3-18
connector, 4-pin—left wheel housing	H4-1
connector, 4-pin—instrument panel right side behind trim	H4-1
connector, 4-pin—engine compartment, near top left door hinge	H4-2
connector, 4-pin—engine compartment, near top right door hinge	H4-3
connector, 4-pin—right B pillar behind trim	H4-4
connector, 4-pin—left B pillar behind trim	H4-5
connector, 4-pin—instrument panel left side	H4-6
connector, 4-pin—at LH control unit, forward of right front door	H4-7
connector, 4-pin—in left front door	H4-8
connector, 4-pin—in right front door	H4-9
connector, 4-pin—trunk lid, left air outlet	H4-12
connector, 4-pin—under rear seat, next to fuse/relay panel 401	H4-15
connector, 4-pin—behind instrument panel to left of steering column	H4-16
connector, 6-pin—engine compartment right wheel housing	H6-1
connector, 6-pin—engine compartment right wheel housing	H6-2
connector, 6-pin—behind instrument panel	H6-4
connector, 8-pin—under passenger seat	H8-1
connector, 8-pin—under driver seat	H8-2
connector, 8-pin—under rear seat	H8-3
connector, 8-pin—behind instrument panel	H8-7
connector, 8-pin—behind instrument panel	H8-8
connector, 10-pin—behind instrument panel, left of steering column	H10-1
connector, 10-pin—at SRS airbag control unit, left speaker grille	H10-2
connector, 10-pin—behind instrument panel	H10-5
connector, 10-pin—at auxiliary fuse/relay panel under rear seat	H10-6
connector, 10-pin—under rear seat at burglar alarm control unit	H10-8
connector, 12-pin—behind instrument panel, left of steering column	H12-1
connector, 12-pin—behind instrument panel	H12-3
connector, 12-pin—behind instrument panel	H12-4
connector, 29-pin—at bulkhead, behind main fuse/relay panel	H29-1
connector, 29-pin—at bulkhead, behind main fuse/relay panel	H29-2
connector, 29-pin—at bulkhead, behind main fuse/relay panel	H29-3
connector, 33-pin—at bulkhead, behind main fuse/relay panel	H33-1
connector, 33-pin—at bulkhead, behind main fuse/relay panel	H33-2
connector, 33-pin—at bulkhead, behind main fuse/relay panel	H33-3
contact box for CD player or equalizer—center console, top compartment	350
control unit for cruise control—behind instrument panel, left side	131
coolant fan temperature switch—upper left corner of radiator	39
coolant fan time delay relay—front left wheel housing	26
coolant temperature gauge	47C
coolant temperature transmitter—front of cylinder head	45
crankshaft sensor—behind crankshaft pulley	345
cruise control vacuum pump—engine compartment left wheel housing	187
cruise control vacuum switch—engine compartment left wheel housing	233
cruise control warning lamp—instrument cluster	47U
diodes for convertible top (1987-1988)—under rear seat in wiring harness	283
direction indicator lamps, left-hand	27
direction indicator lamps, right-hand	28
direction indicator stalk switch	24
direction indicator warning lamp, left-hand	47H
direction indicator warning lamp, right-hand	47I
distribution block—right wheel housing	75
door lock switch—center console	425
door switches—on door jamb	54
driver heated seat rheostat—instrument panel	252
driver seat temperature thermostat—in driver seat	254
EGR modulating valve—ahead of left wheel housing	390
electric radio antenna—left rear body panel	265
EZK control unit—engine compartment, left bulkhead	176
EZK knock sensor—engine block, below intake manifold	178
EZK test connector—next to main fuse/relay panel	145
fender direction indicator, left	89
fender direction indicator, right	90
flasher relay—behind instrument panel, left side	23
fog lamps	85
front roof lamp—center rear view mirror	51
fuel injectors—intake manifold	206
fuel level gauge	47A
fuel level transmitter—in fuel tank, accessible in luggage compartment	46
fuel pump—in fuel tank	323
fuel pump relay—at LH control unit 200, forward of r.f. door behind A pillar trim	102
fuel warning light	47B
full beam warning light	47G
fuse for burglar alarm under rear seat	274
fuse for convertible top—engine compartment, next to distribution block 75	290A
fuse for convertible top—engine compartment, next to distribution block 75	290B
fuse for lambda sensor—engine compartment, next to fresh air intake	285
gear indicator light—at gear selector lever (automatics)	91
glovebox light	19
glovebox light switch—side of glovebox	160
ground distribution—main fuse/relay panel	G6
ground point—alternator bracket	257
ground point—alternator bracket	G32
ground point—behind instrument panel	3
ground point—behind instrument panel	G8

Component List—Alphabetical

(page 3 of 4)

Component	No.
ground point—between ignition switch and handbrake	G4
ground point—C pillar, right on sedans	G26
ground point—electric antenna	G27
ground point—engine lifting lug	G7
ground point—front of ABS hydraulic unit	G16
ground point for fuel injection—engine lifting lug, 2 points	201
ground point—gearbox	211
ground point—gearbox	G25
ground point—at handbrake	117
ground point—at handbrake	65
ground point—hatchback	G19
ground point—radiator cross member	7
ground point—radiator cross member	G1
ground point—luggage compartment	G3
ground point—luggage compartment	9
ground point—luggage compartment	G9
ground point—under rear seat	G5
ground point—wheel housing, left	93
ground point—wheel housing, left	G12
handbrake switch—under handbrake	43
handbrake warning lamp	47M
hazard warning switch	25
headlight dip switch	215
headlight switch	10
headlight wiper motor—behind headlight assembly	66
heated rear view mirrors—front doors	207
heated seat pad with thermostat—in seat cushion and backrest	64
heater control lighting in instrument cluster	154
high beam headlight	11
horn—engine compartment	40
horn relay—main fuse/relay panel	68
horn switch—in steering wheel	41
ignition amplifier—in front of left wheel housing	146
ignition coil—on top of radiator	5
ignition coil with integrated amplifier—inner right fender, near battery	
ignition control unit—in front of left wheel housing	320
ignition distributor—end of cylinder head	6
ignition pulse amplifier—main fuse/relay panel, position D	147
ignition switch	20
ignition switch light—next to ignition switch, center console	52
ignition switch relay—main fuse/relay panel	21
ignition switch relay—in fuse/relay panel (401) under rear seat	21A
instrument cluster	47
instrument lighting	18
interior light switch—center console	53
interior lighting time delay relay—under rear seat	151
intermittent windshield wiper relay—behind instrument panel, left side	83
key switch for driver's central door locking—driver door	183
knock sensor—engine block, below intake manifold	178
lambda exhaust sensor—in exhaust manifold	136
lambda preheater (integral with lambda sensor)	271
LH air mass meter—in throttle intake hose	205
LH coding connector—at LH control unit 200, forward of r.f. door	322
LH diagnostic test connector (sedan, hatchback)—under rear seat, right side	347
LH diagnostic test connector (convertible)—under center console bellows	347
LH diagnostic test connector—under rear seat, right side	347
LH engine temperature transmitter—intake manifold flange, between cylinders 2 and 3	202
LH fuel injection control unit—forward of right front door, behind trim	200
LH fuel injection main relay—at LH control unit 200, behind A pillar trim, passenger door	229
LH NTC resistor—under intake manifold	389
LH test connector—engine compartment, behind right wheel housing	204
license plate lights	15
lighting relay—main fuse relay panel	8
lockout switch for rear power windows	189
logic box for passive seat belt s—under rear seat, left side	306
low beam headlight	12
luggage compartment light—storage area	55
luggage compartment light switch—storage area	56
main fuel pump—in fuel tank	103
main fuse/relay panel—in engine compartment	22A
mercury switch for convertible heated rear window—left side top mounting	279
motor for convertible hydraulic pump—under rear seat, right side	273
negative (ground) distribution in main fuse/relay panel	158
neutral safety switch (automatics only)—gear selector lever	77
oil pressure switch—left side of engine, above oil filter	44
oil pressure warning light	47D
over-ride switch for ventilation fan—integrated in air distribution switch (150)	149
parking lights	13
passenger seat switch for heated seats—under passenger seat	121
passenger seat belt switch—under seat cover	69
passive seat belt belt reel—driver seat	307D
passive seat belt belt reel—passenger seat	307P
passive seat belt door switch—driver door	208D
passive seat belt door switch—passenger door	208P
passive seat belt fuse for driver seat—under rear seat	310D
passive seat belt fuse for passenger seat—under rear seat	310P
passive seat belt g sensor—under rear seat, left side	309
passive seat belt motor relay for driver seat—under rear seat	311D
passive seat belt motor relay for passenger seat—under rear seat	311P
passive seat belt motor relay for driver seat—under rear seat	312D
passive seat belt motor relay for passenger seat—under rear seat	312P
passive seat belt motor with limit switches—driver seat	308D
passive seat belt motor with limit switches—passenger seat	308P
passive seat belt warning lamp	47S
power mirror motor, left—in left mirror	126
power mirror motor, right—in right mirror	127
power mirror switch, left—instrument panel	124
power mirror switch, right—instrument panel	125
power sunroof or convertible switch	181
power window motor—left front door	164
power window motor—left rear door	193
power window motor—right front door	165
power window motor—right rear door	194
power window relay for automatic control—under rear seat	287
radiator cooling fan motor—behind radiator	37
radiator fan relay—main fuse/relay panel	396
radio connector—instrument panel	267
radio contact box with amplifier—instrument panel	349
radio speakers—(4) two in front, two in rear	266
reading lights, convertible—back seat	225
rear lights	14
rear fog lights	33
rear fog light switch	161
rear fog light warning lamp	47N
rear window heater element	115
rear window heater warning lamp	47J
recirculation switch—instrument cluster	143
recirculation valve—behind instrument panel, near right side A pillar	38
relay for A/C cooling fan—main fuse/relay panel	155
relay for daytime driving lights	174
relay for extra foglamps	107
rear heated window switch	116
relay for raising convertible top (1987-1988)—under rear seat on bulkhead	277
relay for raising convertible top (1989 and later)—under rear seat in fuse/relay panel 401	277
relay for lowering convertible top (1987-1988)—under rear seat on bulkhead	278

Component List—Alphabetical

(page 4 of 4)

Component	No.
relay for lowering convertible top (1989 and later)—under rear seat in fuse/relay panel 401	278
resistor pack for ventilation fan—under left speaker grille	74
rheostat for instrument cluster	16
rheostat for instrument panel lighting	17
roof lamp/switch—to left of driver	50
seat belt switch, driver side—between driver and passenger seat	70
seat belt warning light—center of instrument panel	72
seat belt warning relay—under rear seat	82
selector for cruise control—direction indicator stalk	141
shift-up indicator lamp	47K
shift-up indicator relay—main fuse/relay panel	270
side backup lights in front light clusters	119
side marker lights in front light clusters	234
slip ring switch—in steering wheel	352
sparkplug—center of valve cover	157
speed control relay—behind instrument panel, left side	128
speed transmitter—at rear of instrument cluster	132
SRS airbag 2-pin orange connector—pins are short circuited when connector is separated. See SRS warning.	335
SRS airbag 10-pin connector—center console under rubber bellows. See SRS warning.	330
SRS airbag—in steering wheel. See SRS warning.	333
SRS airbag control unit—behind instrument panel, left grille. See SRS warning.	331
SRS airbag ground point for electronic unit—adjacent to electronic unit. See SRS warning.	334
SRS airbag left sensor—left wheel housing in engine compartment. See SRS warning.	332A
SRS airbag redundant ground point—behind instrument panel, left side. See SRS warning.	400
SRS airbag right sensor—right wheel housing in engine compartment. See SRS warning.	332B
SRS airbag slip ring contact—in steering wheel. See SRS warning.	336
SRS airbag warning lamp. See SRS warning.	47T
starter motor—under intake manifold	4
starting injector—throttle housing	94
supply fuel pump—in fuel tank	101
switch assembly for power windows and sunroof—center console	386
switch for driver power window—center console	162
switch for fog lamps	88
switch for left rear power window—center console 190	
switch for left rear power window—left rear door	190a
switch for passenger power window—center console	163
switch for raising idle speed (automatic transmission)—center console	76
switch for rear seat—under rear seat 426	
switch for right rear power window—center console	191
switch for right rear power window—right rear door	191a
tachometer	110
temperature switch for shift-up indicator—between center intake ports of intake manifold	262
third brake lamp (high mount brake light)	109
throttle angle transmitter—on throttle housing	203
throttle contacts for shift-up indicator—throttle housing	261
time delay relay for heated rear window—under rear seat	113
TSI (test service instrument) socket—in main fuse/relay panel	73
vacuum switch for shift-up indicator—left wheel housing	263
ventilation fan switch—instrument panel	35
ventilation fan motor—under right speaker grille	36
voltage distribution block—engine compartment right bulkhead	75
windshield washer motor—below fluid bottle in engine compartment	63
windshield wiper motor—engine compartment firewall	62
wiper stalk switch—steering column	61

400 Transmission—General

TABLES

GENERAL

The repair groups found within the **4 TRANSMISSION** section contain manual and automatic transmission service and repair information. Some of the information contained in these repair groups is intended to be used as a guide for the professional or experienced automotive technician. In addition, some of the operations require special equipment. Transmission and final drive rebuilding are specialized repairs requiring many special Saab tools. Therefore, these operations are not within the scope of this manual and have not been included.

Procedures for changing the manual transmission lubricant, automatic transmission fluid, or final drive fluid (automatics only) are covered in **1 LUBRICATION AND MAINTENANCE**. Removal of the transmission is covered in **2 ENGINE**.

Transmission Assembly

The Saab transmission is a unique design in that it is bolted directly to the bottom of the engine. The top section of the transmission housing serves as the engine oil sump and the engine oil is completely sealed off from the manual transmission lubricant.

Power is transferred from the engine down to the transmission via a chain at the front of the engine/transmission assembly. Power then flows through the gearset to the final drive and out to the drive axles to drive the front wheels.

MANUAL TRANSMISSION

The 5-speed transmission is shown in Fig. 1. The transmission case contains the primary chain, the mainshaft, and the layshaft. At the end of the transmission is the integral final drive housing containing the differential, and the ring and pinion gears. The transmission and final drive housing has one common bath of lubricating oil.

The clutch, pressure plate, and release bearing can all be serviced without removing the engine or transmission. In addition, the clutch slave cylinder and engine flywheel can be serviced in the car as well. See **2 ENGINE** for information on removing the flywheel.

WARNING

The clutch disc contains asbestos fibers. Asbestos materials can cause asbestosis. Always wear an OSHA approved respirator and protective clothing when handling components containing asbestos. Do not use compressed air. Do not grind, heat, weld, or sand on or near any asbestos materials. Only approved cleaning equipment should be used to service the clutch disc or areas containing asbestos dust or asbestos fibers.

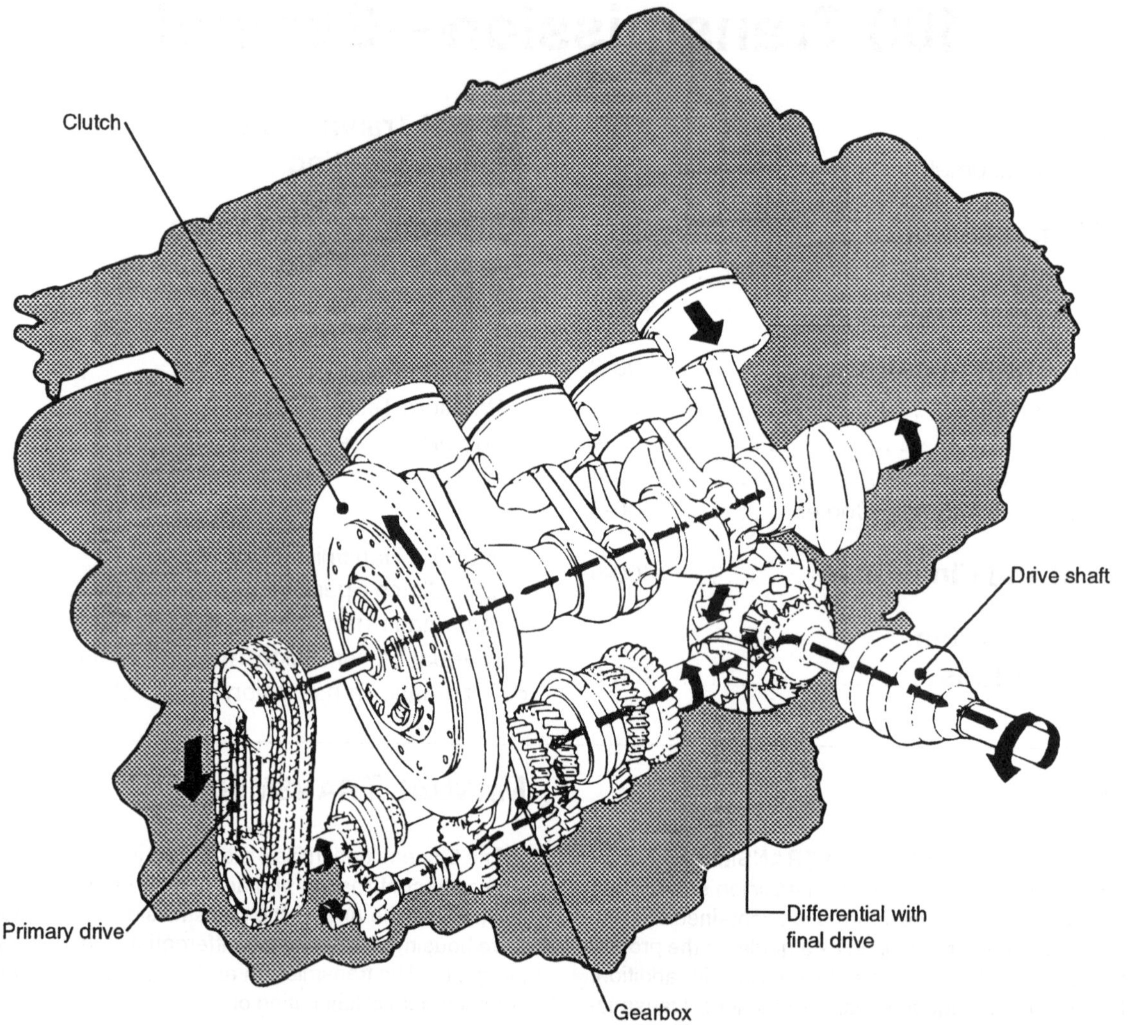

Fig. 1. Manual transmission with integral final drive.

Fig. 2. shows the internal components of the transmission. A description of several of the main components is given. Remember that power flows from the engine to the transmission via the clutch and primary chain. From the chain, power flows down into the gearbox, and to the final drive where the drive axles attach.

Clutch, Pressure Plate, Release Bearing

The clutch disc, pressure plate, and release bearing are located at the front of the car behind a plastic cover. They are removed and installed together. The clutch disc transmits power to the primary chain via a splined input shaft. See **411 Clutch**.

Clutch Master and Slave Cylinders

The clutch is hydraulically actuated. The master cylinder supplies pressure to the slave cylinder. The brake fluid reservoir supplies the clutch master cylinder with hydraulic fluid. The slave cylinder moves the release bearing, which presses on the fingers of the pressure plate allowing the clutch disc to rotate independent of engine speed. See Fig. 3. See **412 Clutch Control**.

AUTOMATIC TRANSMISSION

The automatic transmission is shown in Fig. 4. At one end of the transmission is the integral final drive housing. Inside the final drive housing are the differential and the ring and pinion gears. The final drive housing, containing oil for lubricating the final drive gears, is completely sealed off from the transmission case. The transmission case contains automatic transmission fluid (ATF) for the ATF pump, the hydraulic controls, and the planetary gear system. See Fig. 5. ATF does not circulate unless the engine is running, so the transmission parts are only partially lubricated when the engine is turned off.

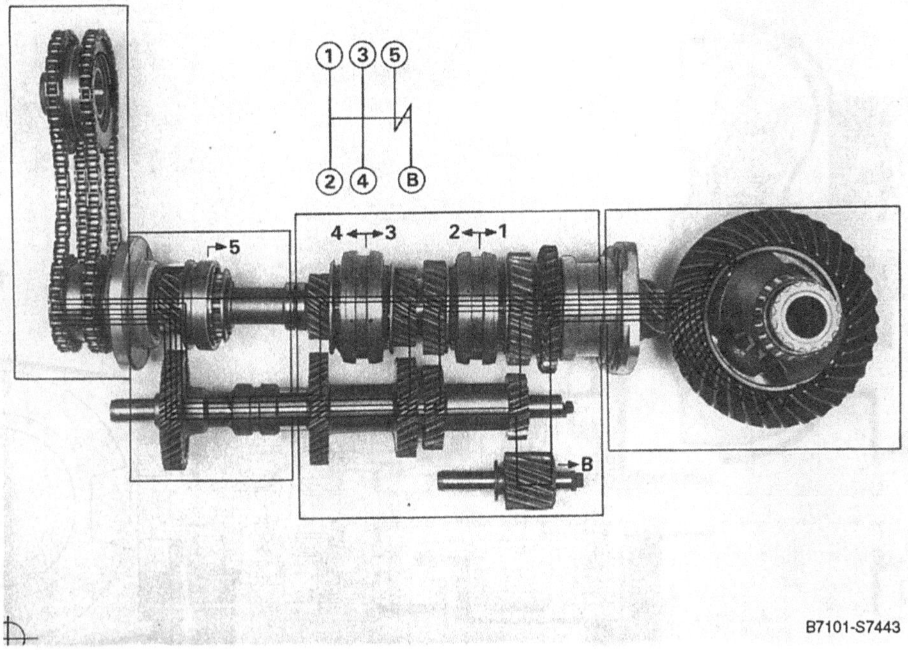

Fig. 2. Schematic view of manual transmission with primary chain (left) and final drive (right).

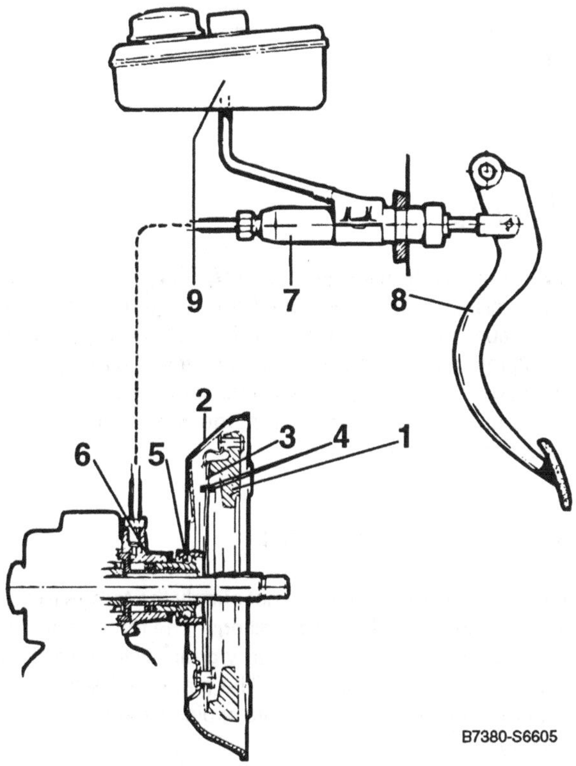

1. Pressure plate
2. Housing
3. Diaphragm spring
4. Pivot rings
5. Release bearing
6. Slave cylinder
7. Master cylinder
8. Clutch pedal
9. Clutch fluid reservoir

Fig. 3. Clutch components.

1. ATF dipstick filler tube
2. Torque convertor housing
3. Front sump
4. Rear sump
5. Front axle driver
6. Engine oil sump
7. Transmission throttle cable

Fig. 4. Automatic transmission with integral final drive.

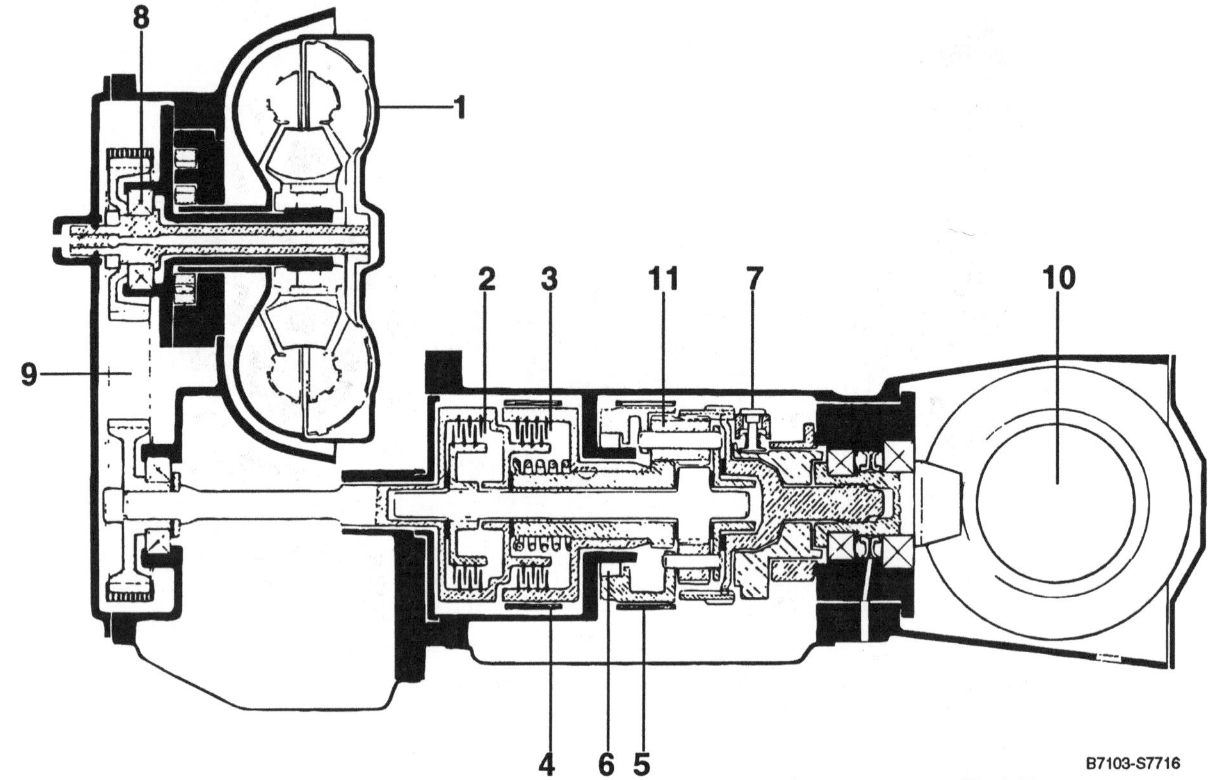

1. Torque converter
2. Front clutch
3. Rear clutch
4. Front band
5. Rear band
6. One-way clutch
7. Governor
8. ATF pump
9. Transmission chain
10. Final drive
11. Planetary gearset

Fig. 5. Cutaway view of automatic transmission.

CAUTION

- The following rules must be observed if a car with automatic transmission has to be towed:
- The selector lever must be in position **N.**
- The ATF level must be topped up an additional 2.0 l (2.1 qt.). After towing, drain the fluid so the transmission has the correct level.
- The car must not be towed a distance greater than 40 to 50 km. (25 to 30 miles).
- The towing speed must not exceed 25 mph (40 km/h)

Torque Converter

The torque converter is a doughnut-shaped assembly located between the engine and the transmission. The torque converter is a fluid coupling. The torque converter transmits engine output to the transmission, and also multiplies engine torque at low vehicle speeds. The engine and transmission must be removed and separated to service the torque converter and drive plate.

ATF Pump

The automatic transmission fluid (ATF) circulates through the transmission under pressure. The ATF pump that creates this pressure is located at the top front of the transmission. The splined pump driveshaft is connected directly to the torque converter housing, so ATF circulates whenever the engine is running. ATF is drawn from the ATF sump through a strainer into the pump.

Planetary Gear System

A torque converter alone cannot supply the torque multiplication needed for all driving conditions. The torque converter therefore drives through a planetary gear system which can operate at different drive ratios. The planetary gear system consists of two sun gears, two sets of pinions, a pinion carrier and a ring gear.

The two hydraulically-operated multiple-disc clutches control the routing of power to the planetary gearsets. The front clutch connects the converter to the forward sun gear. In reverse, the rear clutch connects the converter to the reverse sun gear.

Hydraulic System and ATF Cooler

The ATF pump supplies the system with fluid. A governor provides automatic control of the upshift and downshift points by routing fluid to the valve body shift valves. The shift valves provide hydraulic control of transmission functions by opening and closing ports. The transmission throttle cable controls a shift valve for downshifting during hard acceleration.

All models have an external ATF cooler that is one of two types: an engine coolant heat exchanger, or an air heat exchanger. See Fig. 6. Additionally, some models have an inline thermostat. For more information see **472 Automatic Transmission Seals and Gaskets**.

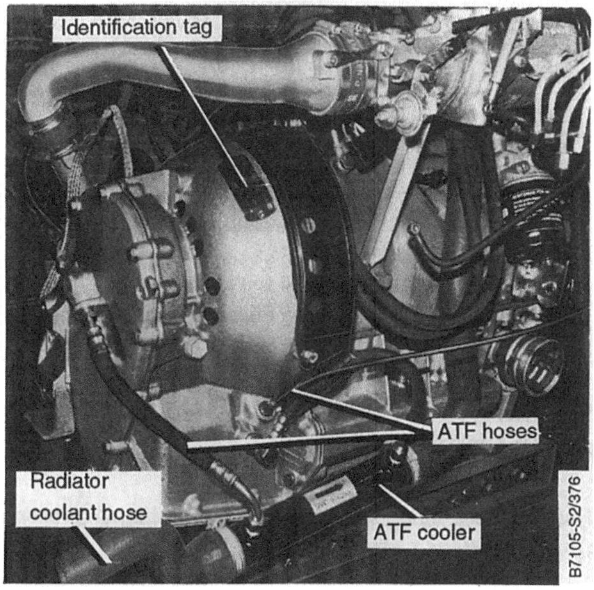

Fig. 6. Automatic with ATF cooler. Cooler shown is engine coolant type.

DIFFERENTIAL AND FINAL DRIVE

The final drive, including differential, is at the rear of the transmission case. On manual transmissions the final drive is lubricated by the transmission lubricant. On automatics the final drive has its own lubricant, separate from the ATF.

IDENTIFICATION CODES AND SPECIFICATIONS

Due to different power characteristics and performance requirements, there are minor variations of the basic transmission. On manual transmissions, the different versions are identified by code letters found stamped in the front case. See Fig. 7. On automatics, an identification tag is located on the torque converter housing. See Fig. 6 above. **Table a** gives specifications for the manual transmission. **Table b** gives specifications for the automatic transmission. For more information on checking or changing the fluids, see **1 LUBRICATION AND MAINTENANCE**.

Fig. 7. Identification stamp (arrow) on manual transmission.

Table a. Manual Transmission and Clutch Specifications

Clutch type	Fichtel & Sachs
Clutch diameter*	
non-turbo	204 mm (8 in.)
turbo	216 mm (8.5 in.)
Transmission oil capacity	3.0 l (3.5 qt.)
Transmission oil type	Engine oil SAE 10 W 30 or SAE 10 W 40
Clutch/brake master cylinder fluid	DOT 4 brake fluid

* approximately

Table b. Automatic Transmission Specifications

Type	Borg Warner 37
ATF volume	8.0 l (8.5 U.S. qts.)
ATF grade	Ford spec M2C-33F/G
Final drive fluid volume	1.25 l (1.3 U.S. qts.)
Final drive fluid type	SAE 10W 30

MANUAL TRANSMISSION TROUBLESHOOTING

This troubleshooting section applies to problems affecting the manual transmission.

The source of most problems is apparent from the symptoms. For example, difficulty in engaging a gear or imprecise shifting may be caused by air trapped in the clutch hydraulic system or it may be a more serious internal transmission problem. Other symptoms, such as clutch slippage, vibration or shuddering when releasing the clutch may be solved by repairing or replacing the clutch components only.

Noises may be the result of a failure in the gear train, or of the final drive. What appears to be a transmission oil leak may be engine oil leaking from a faulty rear crankshaft oil seal, especially if the leak is near the bottom of the transmission housing. See **2 ENGINE** for information on crankshaft oil seals.

NOTE

To pinpoint the location of leaks, have the undercarriage and transmission degreased. Remember that air turbulence from driving can cause fluids to appear to be leaking from other locations.

Transmission problems fall into two categories: those that can be fixed by external service, and those that require disassembly of the transmission. Problems that at first appear to be caused by internal faults, such as gear shifting difficulty or noisy operation, may be corrected by external adjustments such as gearshift lever adjustment, or bleeding the clutch hydraulic system.

Begin with a thorough visual inspection, both in the engine compartment and from beneath. Check all parts of the gearshift mechanism for wear that might cause misalignment and shifting difficulty. Look for wet spots that may indicate oil leaks. High or low oil level, or the wrong type of oil may be the cause of hard shifting or noise. Accurate pinpointing of leaks may require that the suspected area be cleaned and reinspected.

Correct leaks and oil level before acting on suspected internal problems. The gearshift mechanism is covered in **432 Manual Transmission Controls.** Checking and correcting oil level is covered in **1 LUBRICATION AND MAINTENANCE.**

Observation of the symptoms is the key to isolating and identifying transmission problems. A road test is an important step. Determining whether the problem is present in all gears, only during acceleration, when the clutch is engaged, or in some other special condition may help isolate the source of the problem.

Table c lists manual transmission and clutch symptoms, their probable causes, and recommended corrective actions. The numbers in bold type in the corrective action column refer to the repair group in this section where the suggested repairs are described.

Table c. Manual Transmission Troubleshooting

Symptom	Corrective action
1. Difficult or noisy shifting	a. Insufficient oil. Correct oil level. See **1 LUBRICATION AND MAINTENANCE**
	b. Adjust incorrect idle speed. See **2 ENGINE**
	c. Bleed or repair clutch hydraulic system. See **412**
1. Difficult or noisy shifting (cont'd)	d. Lubricate or replace binding clutch disc on input shaft. See **411**
	e. Repair worn gearshift linkage. See **432**
	f. Replace seized input shaft bearing in flywheel. See **2 ENGINE**
	g. Replace worn or damaged internal gear train components.
2. Transmission fails to engage a gear or jumps out of gear	a. Repair worn gearshift linkage or loose shift console. See **432**
	b. Repair worn or damaged internal gear train components.
3. Poor acceleration or slipping	a. Replace worn or burnt clutch friction surfaces. See **411**
4. Clutch not fully engaging	a. Inspect clutch disc for binding on input shaft. See **411**
	b. Replace release bearing, flywheel bearing for input shaft, clutch disc, or pressure plate. See **411**
	c. Replace oil soaked clutch disc and faulty oil seals. See **411**
	d. Check for faulty engine/transmission mounting. See **2 ENGINE**

continued

Table c. Manual Transmission Troubleshooting

AUTOMATIC TRANSMISSION TROUBLESHOOTING

This troubleshooting section applies to problems affecting the automatic transmission.

External adjustments to the transmission are covered in this manual, but some of the troubleshooting information in this section describes problems which can only be remedied by disassembly and internal repair. Internal transmission repairs require specialized knowledge and equipment and are, therefore, beyond the scope of this manual. The publisher recommends that such repairs be left to an authorized Saab dealer or other qualified automatic transmission repair shop.

NOTE

- To pinpoint the location of leaks, have the undercarriage and transmission degreased. Remember that air turbulence from driving can cause fluids to appear to be leaking from other locations.

- If the car has been in a collision, it is possible that the magnet located in the ATF sump has been dislodged to block the oil pickup tube, or that the oil pickup tube has been knocked loose. Remove the oil pan and check for the proper location of the magnet and tube. See **1 LUBRICATION AND MAINTENANCE.**

Basic Principles

In order for automatic transmission troubleshooting to provide meaningful results, the engine must be in good mechanical condition and properly tuned. See **1 LUBRICATION AND MAINTENANCE** for engine tuning.

Inspect the transmission for external damage, loose or missing fasteners, and for any obvious leaks. Many automatic transmission problems can be traced to an incorrect ATF level, incorrect ATF type, contaminated ATF, or to misadjusted transmission controls. Also check the final drive fluid level and type. See **1 LUBRICATION AND MAINTENANCE** for information on checking and changing the ATF and final drive fluid.

Check to see if the ATF is dirty or has a burned odor. A burned odor indicates overheated fluid. This may be accompanied by burned clutches, as well as friction material which may be clogging the valve body passages.

Minor automatic transmission problems can often be corrected merely by performing the following tasks. In any event, do these things before beginning any internal tear down or repair.

Preliminary service
- drain and refill the ATF
- drain and refill the final drive fluid
- clean or replace the ATF filter screen
- check the operation of the shift lever and cable
- check the operation of the transmission throttle cable

Table d is given as a guide for experienced transmission repair specialists. Special tools and techniques are required to repair major faults. The engine–transmission assembly must be removed from the car to disassemble the transmission.

Table d. Automatic Transmission Troubleshooting

Symptom	Corrective Action
1. No drive (car will not move)	a. Check ATF level. See **1 LUBRICATION AND MAINTENANCE** b. Check shift cable adjustment. See **444** c. Check for clogged fluid screen. See **1 LUBRICATION AND MAINTENANCE**
2. Does not shift properly	a. Check transmission throttle cable. See **444** b. Check for binding governor. See **444** c. Check pipes between governor and valve body. See **444**
3. Slips when shifting	a. Check throttle cable. See **444**
4. Starts in wrong gear	a. Check governor for binding. See **444**
5. Incorrect upshift speed	a. Check throttle cable adjustment. See **444** b. Check fluid pressure at idle. See **444** c. Check for binding governor. See **444**
6. Parking pawl does not hold	a. Check shift cable adjustment. See **444**
7. Noise	a. Check ATF level. See **1 LUBRICATION AND MAINTENANCE** b. Check final drive fluid level. See **1 LUBRICATION AND MAINTENANCE** c. Check shift cable adjustment. See **444** d. Check for broken engine drive plate. See **2 ENGINE**
8. Fluid leaks	a. Check all sealing surfaces. See **472** b. Check ATF pump seal. See **472** c. Check pinion bearing seal.

continued

Table d. Automatic Transmission Troubleshooting

Stall Speed Test

This test is used to check for faults in the torque converter and planetary gear system when there is no other apparent cause for poor performance and acceleration. The test results are meaningless if the engine is not running properly. A precise tachometer must be used for the rpm measurements, as dashboard instruments are not sufficiently accurate.

CAUTION
The stall speed test should be as short as possible and should never extend beyond 10 seconds maximum. Prolonging the test may overheat the transmission and damage the seals or internal components.

To perform stall speed test

1. Drive the car to normal operating temperature. On turbo engines, connect a pressure boost gauge and unplug the APC solenoid.

NOTE
For more information on the APC solenoid and its location, see **2 ENGINE.**

---

2. Connect a tachometer according to the instrument manufacturer's instructions, so that it can be read from the driver's seat, then start the engine.

3. Set the parking brake and depress the foot brake firmly to hold the vehicle stationary.

4. Place the selector lever in **D**.

WARNING
Be certain that no one is in front of or behind the car during the test.

---

5. While holding the car stationary with the brakes, floor the accelerator for no more than 10 seconds. Note the tachometer readings. The rpm should increase, and then hold steady. Maximum rpm achieved under these conditions is the stall speed.

Stall speed specifications depend on the transmission code numbers found on the torque converter housing. Shown below is a general range for stall speeds and can be used as a guide. See an authorized Saab dealer for exact stall speed specifications.

General stall speed specifications
- 2250 to 2700 rpm
- charging pressure* (turbo models only) . . . . 0.40 bar

---

*APC solenoid valve unplugged

A stall speed that is 300 hundred rpm below the specified range is probably due to reduced engine performance. A stall speed that is 800 rpm or more below the specified range indicates a faulty torque converter.

If the stall speed is substantially higher than normal, there is slippage in the forward clutch or lack of fluid to the torque converter. All of these faults require that the transmission be removed and disassembled for repair.

Pressure Test

A main pressure test will reveal internal leaks, sticking control valves, or other troubles in the hydraulic controls. Engine idle speed must be correctly adjusted.

Tool required
- ATF pressure gauge . . . . . . .Saab tool no. 87 91 592 or equivalent

---

To perform pressure test

1. Connect the pressure gauge to the main pressure tap on the transmission, as shown in Fig. 8.

2. Make sure the engine and transmission are at normal operating temperature, let the engine idle.

3. Set the parking brake.

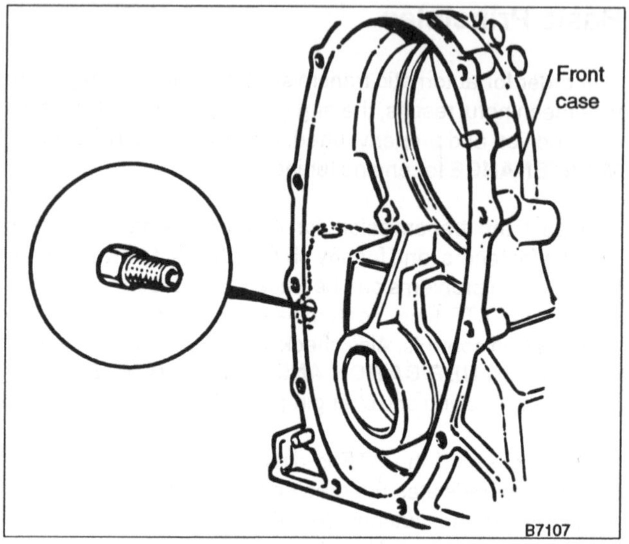

Fig. 8. Main pressure tap on transmission case, located on exhaust side of engine.

4. Depress the foot brake firmly to hold the vehicle stationary.

5. With the gear selector in **D**, check the main pressure. Correct test values are given below.

Automatic transmission line pressure
- 4.2 to 4.9 bar (61 to 71 psi) at idle, gear selector in D

---

6. Put the gear selector in **P** and shut off the engine. Disconnect the pressure gauge hose and reinstall the plug for the pressure tap. Torque the plug. Top up the ATF if necessary.

Pressure tap tightening torque
- 5 to 7 Nm (44 to 62 in-lb)

Any pressure that is higher or lower than specified usually indicates a malfunctioning valve body and valves, probably due to contamination. The valve body can be removed, cleaned, and reinstalled with the transmission in place by removing the ATF pan and filter screen. See **1 LUBRICATION AND MAINTENANCE**. However, the valve body contains many precision parts which must be reassembled in their exact locations. Because of the complexity of the valve body assembly, we recommend that these repairs be left to an authorized Saab dealer or other qualified repair shop.

Low pressure may also indicate a worn ATF pump or internal ATF pump leaks past seals, gaskets, and metal mating surfaces. These repairs require that the transmission be removed and disassembled.

411 Clutch

GENERAL

The clutch, pressure plate, and release bearing can all be serviced without removing the engine or transmission. In addition, the clutch slave cylinder and engine flywheel can be serviced after removing the clutch and pressure plate. See **2 ENGINE** for information on removing the flywheel.

The clutch, pressure plate, slave cylinder, and release bearing are located behind a plastic cover at the front of the engine. A splined input shaft transmits engine power from the clutch disc to the primary chainwheel. The clutch is non-adjustable and is actuated by the hydraulic slave cylinder. Brake fluid from the brake fluid reservoir supplies the clutch master cylinder.

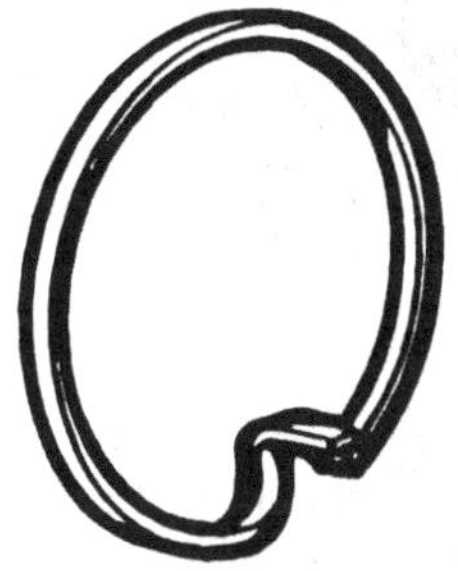

Fig. 1. Spacer ring (Saab tool no. 87 91 618) to disengage pressure plate.

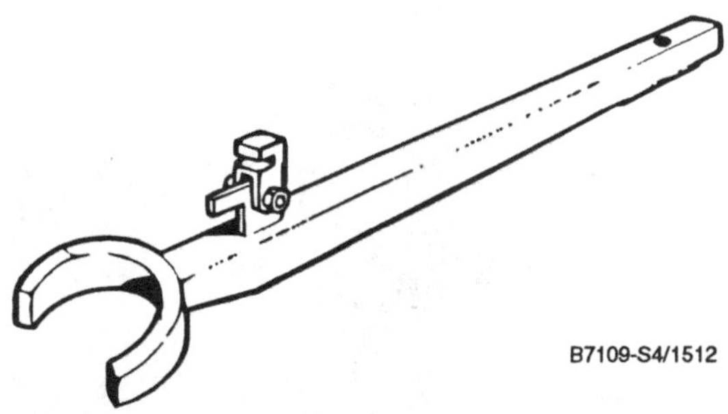

Fig. 2. Lever arm (Saab tool no. 83 94 033) to hold pressure plate fingers if clutch hydraulic system is not working.

CLUTCH, PRESSURE PLATE, RELEASE BEARING AND SLAVE CYLINDER

The clutch, pressure plate, release bearing and slave cylinder are removed and installed as a unit. This is because there is little clearance in the flywheel housing.

To remove clutch, pressure plate, release bearing and slave cylinder

Tools required
- spacer ring . . . . Saab tool no. 87 91 618 and /or 83 90 023
- lever arm (optional) . . . . Saab tool no. 83 94 033

NOTE
Two different clutch assemblies were installed on the cars covered by this manual. Most early cars (approximately 1988 and earlier models) were equipped with a smaller clutch that required a different spacer ring (Saab tool no. 83 90 023) for its removal. All newer cars and all clutch replacement parts use a larger spacer ring (Saab tool no. 87 91 618). Therefore, early cars still equipped with the original smaller clutch will need both special tools to replace the clutch. See an authorized Saab dealer for more information.

1. Disconnect the negative (–) battery cable. Remove the hood. Remove the air intake rubber bellows. On turbo models, remove the turbo pipes above the clutch cover.

2. Remove the left radiator fan as described in **261 Radiator and Cooling System.**

3. Remove the plastic clutch cover. See Fig. 3.

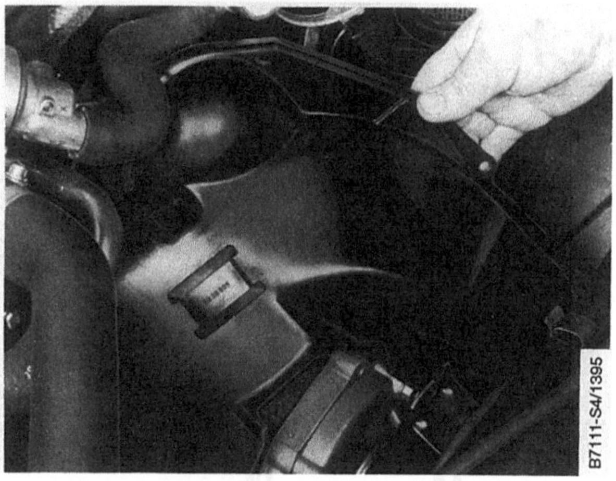

Fig. 3. Plastic cover being removed.

4. Have a helper depress and hold the clutch pedal. Insert the special spacer ring under the fingers of the pressure plate. See Fig. 4. Release the pedal.

NOTE

If the Saab tool no. 87 91 618 or tool no. 83 90 023 is not available, a length of sparkplug wire can be used under the fingers of the pressure plate.

Fig. 4. Spacer being inserted between fingers and housing of pressure plate.

NOTE

• Inserting the spacer creates the necessary clearance to remove the pressure plate, clutch, release bearing, and slave cylinder from the flywheel housing.

• If a helper is not available, wedge a long stick against the clutch pedal and the seat.

• If the clutch hydraulic system doesn't work, the special Saab tool no. 83 94 033 will be required to compress the pressure plate fingers in order to install the spacer ring.

---

5. Remove the spring clip and cover as shown in Fig. 5. Unscrew the plastic propeller from the input shaft.

CAUTION

The plastic propeller circulates oil to the bearing. Failure to reinstall the propeller after completing repairs may cause bearing seizure.

Fig. 5. Spring clip and cover to be removed to withdraw input shaft.

6. Thread an 8 mm bolt with flat washer into the front of the clutch input shaft. Using a pry bar, slide the clutch input shaft out by levering against the flat washer. See Fig. 6.

CAUTION

Be careful not to hit the radiator or fan with the pry bar or input shaft.

Fig. 6. Clutch input shaft being pried out using special tool (Saab tool no. 83 94 033).

NOTE

The shaft does not need to be fully removed.

---

7. Remove the three internal hex-head screws for the slave cylinder. See Fig. 7.

Fig. 7. Slave cylinder screws being removed.

NOTE

• The three internal hex-head screws can be difficult to loosen. Clean the head of the screw so the wrench will grip tightly.

• Do not disconnect the hydraulic line to the slave cylinder.

---

8. Remove the bolts holding the pressure plate assembly to the flywheel. Remove as a unit, the pressure plate, clutch disc, release bearing, and slave cylinder.

CAUTION

Do not step on the clutch pedal while the slave cylinder is removed.

---

9. Inspect the flywheel and pressure plate assembly for burn marks or scoring. Inspect the input shaft bearing in the flywheel and replace if necessary. See **2 ENGINE** for more information.

Note which side of the clutch disc faces the flywheel as a guide to reassembly. Carefully inspect the clutch input shaft seal located at the transmission case surface. If leaking, replace the seal using an appropriate seal driver. Apply grease to the seal lip before installing the shaft.

NOTE

Inspect for oil leakage at the rear of the transmission chain housing, directly below the clutch release bearing. If oil is present on the housing, it may be leaking past the chain tensioner bolts. Because oil will quickly ruin a new clutch disc, it is recommended that the oil leak be corrected by removing the tensioner bolts and reinstalling them using a thread sealer. Accessing the bolts requires supporting the engine/transmission assembly and removing the front engine mount and the front transmission chain cover. Be sure to use a new gasket at the front chain cover.

To install clutch, pressure plate, release bearing and slave cylinder

1. Working at the bench, use a press and cylinder of the correct diameter to depress the pressure plate fingers uniformly so the spacer ring can be installed.

NOTE

The diameter of the cylinder used to compress the pressure plate fingers should be equivalent to the diameter of the release bearing contact area.

---

2. Lubricate the input shaft splines with molybdenum disulfide grease. Assemble the pressure plate, clutch, release bearing, and slave cylinder in the proper order.

3. Install the pressure plate, clutch disc, release bearing, and slave cylinder into the flywheel housing. See Fig. 8.

Fig. 8. Pressure plate, clutch disc, and release bearing with slave cylinder being installed into flywheel housing. Also shown is spacer ring (arrow).

4. Hand tighten two of the pressure plate bolts to hold the assembly in place. Slide the input shaft into engagement with the clutch splines, then tap the shaft into the case with a plastic hammer until it bottoms out. See Fig. 9.

Fig. 9. Input shaft being tapped into case. Snap ring will lock shaft in the case.

5. Install the three slave cylinder screws using thread sealing compound.

Tightening torque

- slave cylinder mounting screws ..... 6 to 14 Nm (53 to 124 in-lb)

---

6. Install the plastic propeller, input shaft cover and spring clip.

7. Install and tighten the remaining pressure plate bolts.

8. Remove the spacer ring from the pressure plate while a helper depresses the clutch pedal. If the pedal is spongy the system needs to be bled, see **412 Clutch Control.**

CAUTION

Only press the pedal enough to remove the spacer. Otherwise the slave cylinder seal may be damaged.

---

9. Push the plastic sleeve toward the flywheel while someone holds the clutch pedal down. See Fig. 10. Finally, check the clutch pedal to make sure it doesn't bind.

NOTE

The plastic shield will automatically position itself when the clutch pedal is released.

Fig. 10. Plastic sleeve being pushed toward flywheel while clutch pedal is held down.

10. Reinstall the plastic cover, the hood, the air intake rubber bellows, and the turbo pipes (where applicable).

412 Clutch Control

GENERAL

The clutch master cylinder and slave cylinder can be serviced without removing the engine or transmission. However, the clutch slave cylinder can only be removed together with the pressure plate assembly, clutch disc, and release bearing. See **411 Clutch** for information on removing these components.

The clutch is non-adjustable. Typical problems with the clutch hydraulic system are fluid leaks, and air trapped in the system that creates a spongy feel to the clutch pedal. Fluid leaks are usually apparent around the component. Internal leakage past a seal can occur which results in a loss of pressure when the pedal is pushed.

To replace clutch master cylinder

1. Working in the engine compartment, pinch off the fluid supply hose from the brake fluid reservoir to the clutch master cylinder. See Fig. 1.

Fig. 1. Fluid reservoir **B** for clutch master cylinder and master cylinder **A**. Arrow shows fluid supply hose.

2. Remove the hydraulic pipe from the master cylinder. See Fig. 2. Remove the clip that holds the pipe to the body.

CAUTION
Brake fluid can damage paint and bodywork. Immediately rinse any spills with water.

NOTE
On cars equipped with ABS, it may be necessary to remove the alternator belt tensioner and move the alternator to one side to reach the master cylinder.

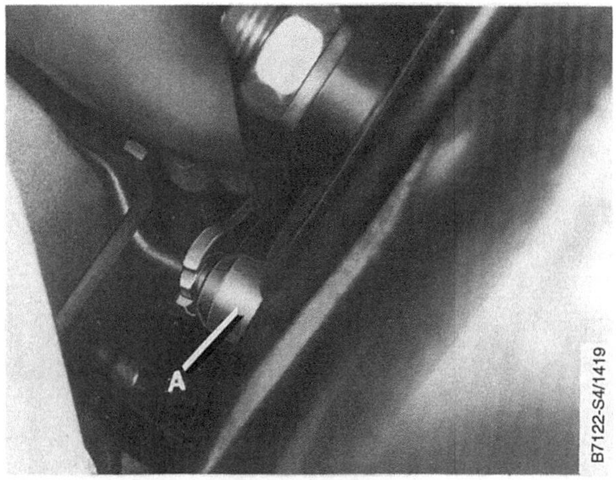

Fig. 2. Clutch master cylinder hydraulic pipe being removed. Master cylinder is located at **A**.

3. Inside the car, remove the left hand footwell air duct. See Fig. 3. Remove the clevis pin and the master cylinder self-locking mounting nuts. See Fig. 4.

WARNING
Always replace self-locking nuts with new ones.

---

4. Installation is the reverse of removal. Bleed the system as described below. Replenish the fluid in the brake master cylinder reservoir as necessary.

NOTE
Do not allow the reservoir to empty, otherwise the entire system will need to be bled after the reservoir is refilled.

412

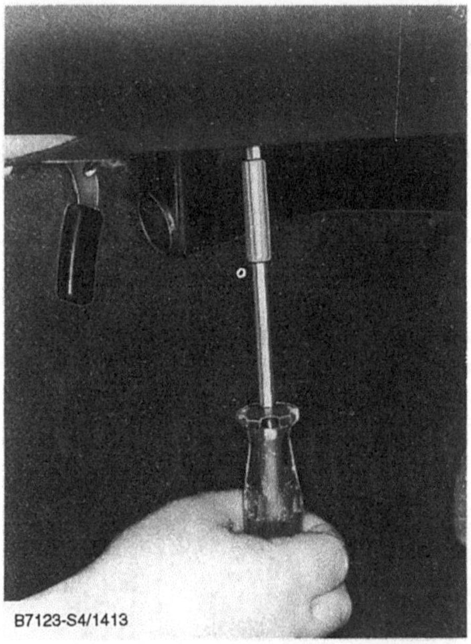

Fig. 3. Left hand footwell air duct being removed.

Fig. 4. Clevis pin (arrow) and clutch master cylinder self-locking nuts **A** to be removed. Master cylinder is removed from engine compartment.

To bleed clutch master and slave cylinders

Tool required

- cooling system pressure tester with cap or equivalent

Bleeding the clutch master cylinder requires that the brake fluid reservoir be pressurized using special equipment. Saab recommends using a coolant system pressure tester and an adaptor that screws onto the brake fluid reservoir. A brake system pressure bleeding kit can also be used for this procedure. These tools and kits are available from automotive tool manufacturers and supply stores.

1. Place a 6 mm. (1/4 in.) inside diameter hose on the slave cylinder bleeder nipple. See Fig. 5. Put the other end of the hose in a container.

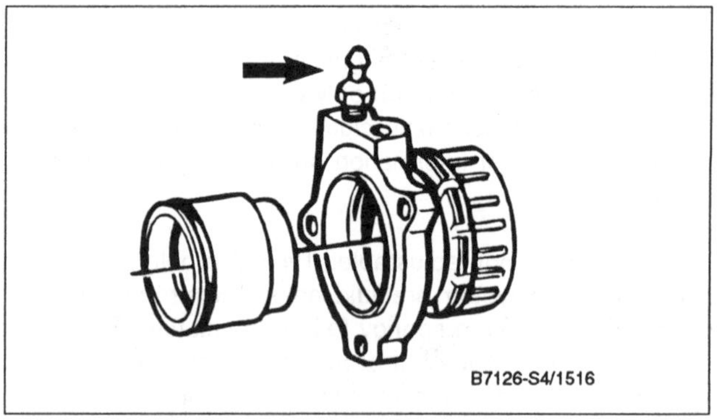

Fig. 5. Slave cylinder bleeder nipple (arrow).

2. Fill the master cylinder reservoir with brake fluid. Place the cooling system pressure tester on the master cylinder.

3. Open the bleeder nipple 1/2 of a turn.

4. Slowly pump up the tester only once or twice until fluid flows out of hose. Allow the fluid to flow until it is free of bubbles.

CAUTION
Too much pressure applied to the system will damage the seals.

---

5. Shut off the bleeder nipple, recheck the fluid level and check the clutch pedal operation.

NOTE
Do not allow the reservoir to empty, otherwise the entire system will need to be bled after the reservoir is refilled.

---

432 Manual Transmission Control

MANUAL TRANSMISSION CONTROL

This section covers repair and adjustment of the gear lever and gear selector rod. Also covered is replacement of the ignition lock cylinder and switch. Many of the fasteners used in the console area are Torx® T25 fasteners. Plan to have this size wrench on hand before performing any work. Also, before doing any adjustments, check the all engine mounts to be sure they are tight and not broken.

To remove gear lever housing

Tool required

- gear lever housing nut wrench .... Saab tool no. 87 90 370

1. Disconnect the negative (–) battery cable and remove the driver's seat. Pull up the handbrake, engage reverse and remove the key.

2. Remove the gear shift bellows, and the two front console screws. See Fig. 1.

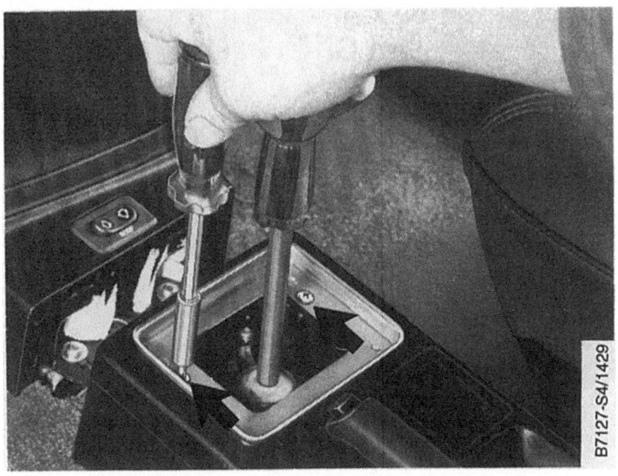

Fig. 1. Console front screws (arrows) being removed.

3. Remove the screws under the rear ashtray. See Fig. 2. Disconnect the wires to the interior light switch and remove the console.

NOTE

After removing all console screws, insert the key and disengage reverse gear to help in removing the console.

Fig. 2. Console rear screws (arrows) being removed.

4. Using the special Saab wrench, remove the gear lever housing nuts. See Fig. 3. Label and remove the wires to the ignition switch and reverse switch.

NOTE

- If replacing just the ignition switch, remove the cover plate screws shown in Fig. 4 then remove the two ignition switch screws shown in Fig. 5.

- An impact driver will help remove stuck screws.

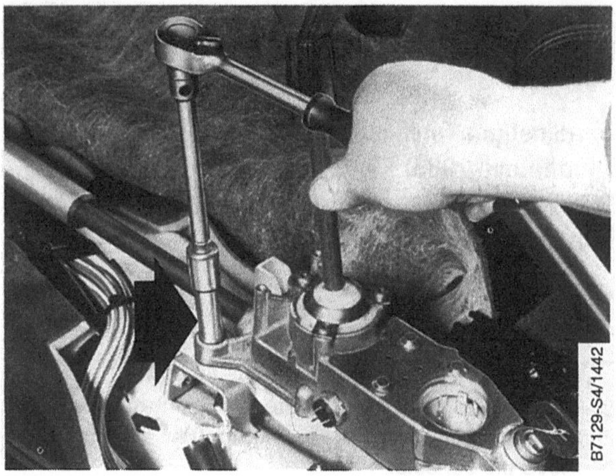

Fig. 3. Gear lever housing nuts being removed with special wrench, Saab tool no. 87 90 370.

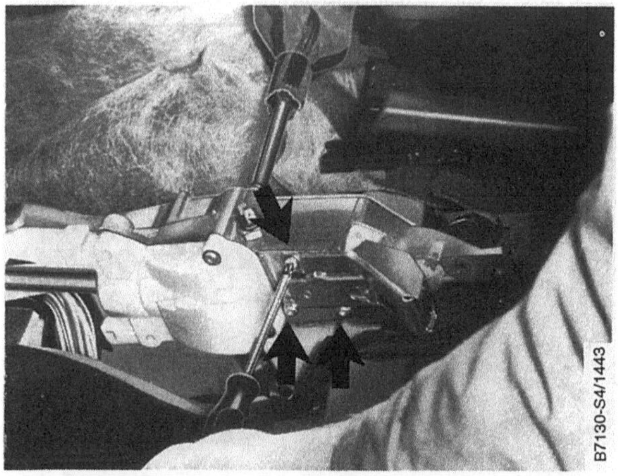

Fig. 4. Cover plate screws (arrows) being removed to replace ignition switch.

Fig. 5. Ignition switch being removed.

NOTE

- Engage the slot in the ignition switch with the tab in the gear lever housing when reinstalling the switch.

- For ignition switch replacement, no further disassembly is required. Adjust the gear shift control as described below.

---

5. Raise and rotate the gear lever housing on its side. Remove the screws securing the plastic bottom cover, remove the housing. See Fig. 6.

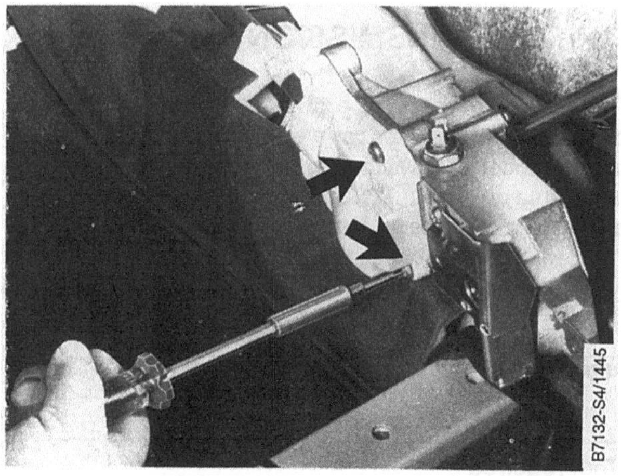

Fig. 6. Plastic bottom cover screws (arrows) being removed.

To remove ignition lock barrel

1. Remove the gear lever housing as described above.

2. Mount the housing in a vise and turn the key to the position shown in Fig. 7.

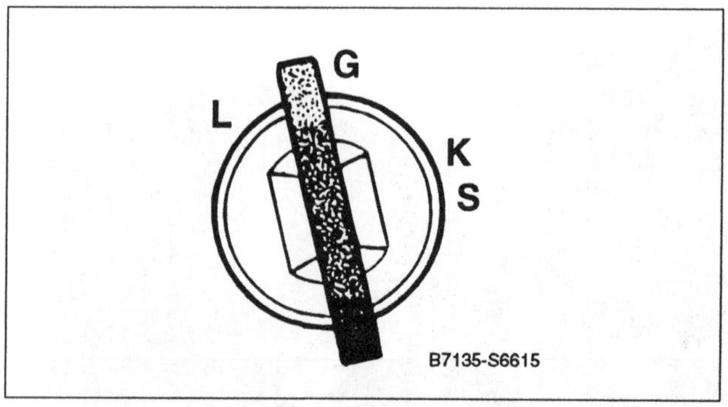

Fig. 7. Position of ignition key to remove lock barrel. View is from above key.

3. Using a stiff wire tool inserted in the hole in the housing, push in the spring loaded pin in the housing while pulling the lock barrel out with the key. See Fig. 8.

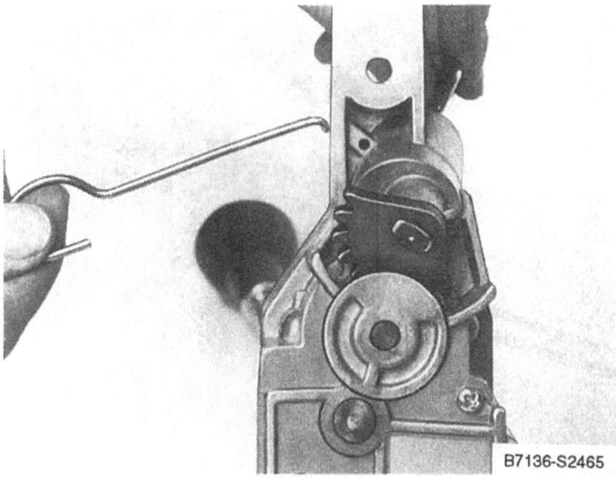

Fig. 8. Ignition lock barrel pin being pushed in. Pull on the key while pushing the pin.

NOTE
If the ignition key is missing, drill a hole in the side of the housing and drive in the pin as shown in Fig. 9.

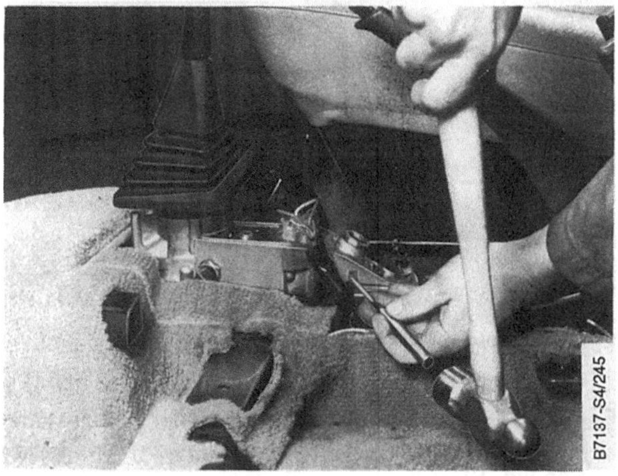

Fig. 9. Ignition lock barrel being removed when key is missing. Drill hole where the punch is located.

To install ignition lock barrel

1. Turn key in the new lock barrel to the position shown in Fig. 7 above.

2. Check that the tooth segment driver pins in the gear lever housing are in the correct position to accept the lock barrel.

3. Push in the spring loaded pin on the lock barrel and insert the barrel in the gear lever housing until it clicks.

To remove gear selector rod

1. Remove the gear lever housing as described above.

2. Loosen the clamp on the selector rod where the rod comes through the firewall. See Fig. 10.

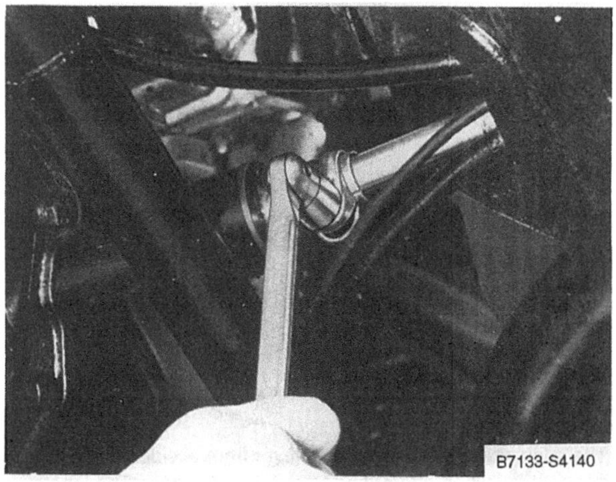

Fig. 10. Selector rod clamp being loosened.

3. Remove the rod from inside the passenger compartment.

4. After installing the rod, adjust the gear shift control as described below. Torque the clamp bolt.

Shift rod clamp bolt
• tightening torque . . . . . . .30 to 35 Nm (22 to 25 ft-lb)

NOTE
Inspect and replace any worn rubber bushings. Worn bushings may cause hard shifting or jumping out of gear.

To disassemble gear shift lever

1. Remove the gear lever housing as described above.

2. Remove the three screws and remove the gear shift lever. See Fig. 11.

3. Measure and record the length of the latch stud as shown in Fig. 12.

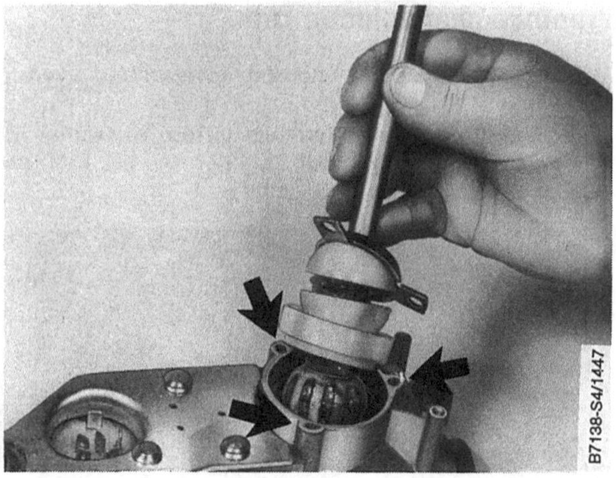

Fig. 11. Gear shift lever being removed after removing screws located at arrows.

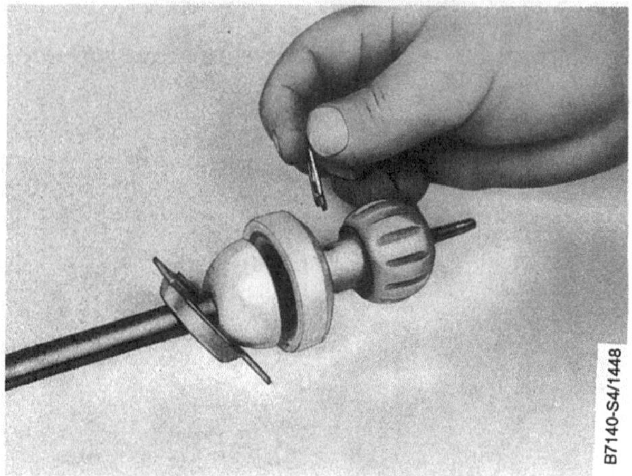

Fig. 13. Gear shift lever being disassembled.

CAUTION

The latch stud prevents the shift lever from accidentally going into reverse when attempting to shift into 5th gear. The length of the latch stud must not be altered or the lockout feature may not work.

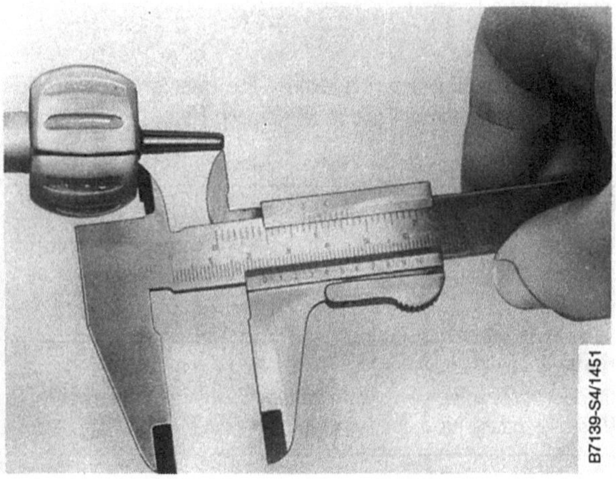

Fig. 12. Gear shift lever latch stud being measured. The length is adjustable.

Gear shift lever latch stud

- projecting length
  22±0.5 mm (55/64 in.)

---

4. Remove the wire clip, sleeve, spring, and bearings. See Fig. 13.

5. Unscrew the latch stud from the rod. See Fig. 14. Remove the rubber bellows and shift knob.

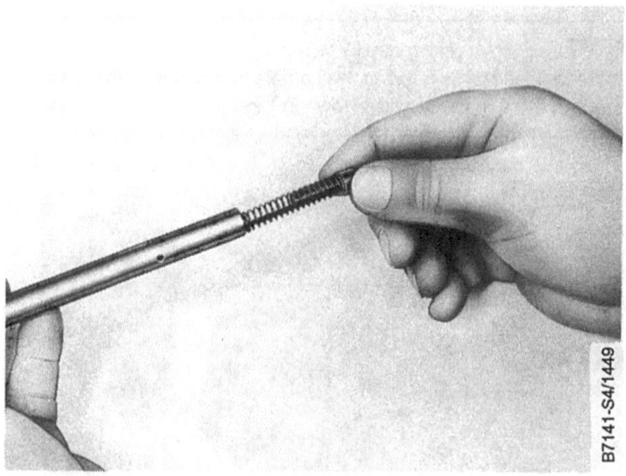

Fig. 14. Gear shift lever latch stud being unscrewed.

6. Inspect and replace any worn components. Adjust the latch stud projecting length to the value given above. See Fig. 12.

7. Reassemble the lever and housing. Grease all sliding components with soft EPI grease. See Fig. 15 for an exploded view of the assembly.

To adjust gear shift control (up to mid 1986)

1. Remove just the console as described earlier. Engage reverse gear, and release the clamp on the shift rod. See Fig. 10 above.

2. Lock the gear shift lever in position by inserting a 6 mm (1/4 in.) pilot tool in the hole as shown in Fig. 16.

NOTE

A drill bit or drift works well as a pilot tool.

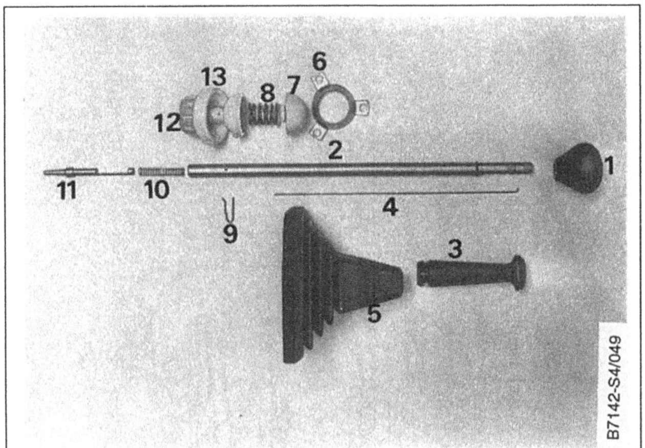

1. Shift knob
2. Gear lever
3. Sleeve
4. Latch rod
5. Rubber bellows
6. Plate
7. Bearing
8. Spring
9. Wire clip
10. Spring
11. Latch stud
12. Carrier
13. Bearing

Fig. 15. Gear shift lever assembly exploded view.

Fig. 16. Gear shift lever being locked into reverse prior to tightening shift rod clamp. Turbo model shown with console bellows (arrow) removed. Other models are similar.

3. Tighten the clamp on the shift rod to the torque given below. Remove the pilot tool and reinstall the console.

Shift rod clamp bolt

• tightening torque . . . . 30 to 35 Nm (22 to 25 ft-lb)

To adjust gear shift control (from mid 1986)

Tool required

• shift rod alignment tool . . . . Saab tool no. 87 91 576

---

1. Remove only the console as described earlier under removing the gear selector housing. Engage third gear and insert the special tool in the differential cover as shown in Fig. 17.

Fig. 17. Shift rod alignment tool (arrow) installed in back of differential cover. Shift lever must be in third gear.

2. Loosen the shift rod clamp bolt. See Fig. 10 above.

3. Lock the gear shift lever in third gear by inserting a 6 mm (1/4 in.) pilot tool in the hole as shown in Fig. 18.

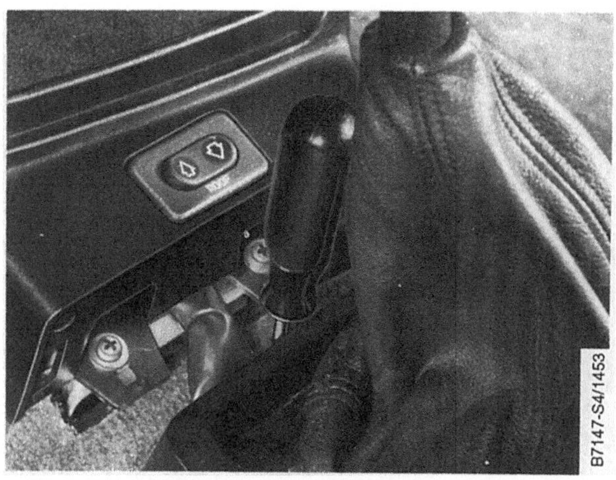

Fig. 18. Gear shift lever being locked in 3rd gear prior to tightening shift rod clamp.

NOTE

A drill bit or drift works well as a pilot tool.

4. Tighten the shift rod clamp. Remove the special tool from the differential cover and remove the pilot tool. Reinstall the console.

Tightening torque

- Shift rod clamp bolt. . . . . 30 to 35 Nm (22 to 25 ft-lb)

To adjust centering of gear shift lever in neutral gate (up to mid 1986 only)

The gear shift lever is in its proper neutral position if it can be shifted straight forward into third gear without having to "fish" around for the gear. No adjustment can be done on models from mid 1986 on.

1. To adjust models up to mid 1986, pull the carpeting away from the front of the shift lever, then engage third gear.

2. Loosen the two retainer screws. Move the bracket so the rollers drop into the plastic block grooves. See Fig. 19.

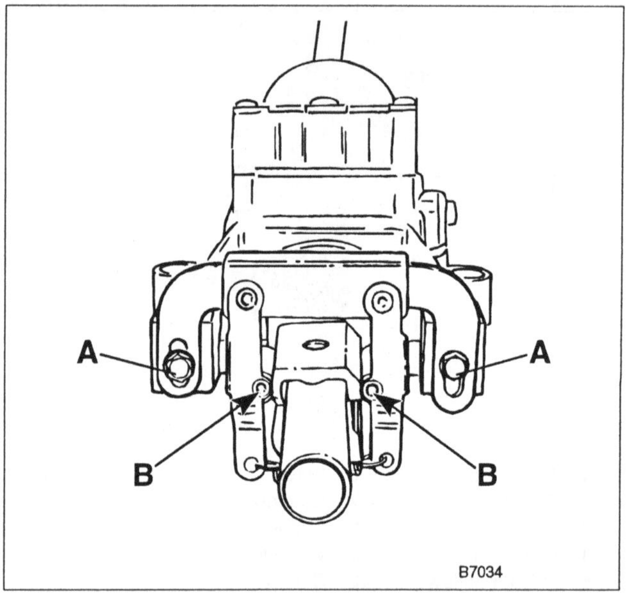

Fig. 19. Gear shift centering mechanism retainer screws **A** and rollers **B**. Adjustment is done with gearshift in 3rd gear.

3. Tighten the retaining screws and recheck the adjustment as described above. Reinstall the console pieces.

444 Automatic Transmission Controls

TRANSMISSION CONTROLS

To check shift lever and linkage

1. Move the selector lever to the **N** position. Press the pawl button and move the lever slightly back and forth.
2. Release the pawl button and check that the lever is positioned exactly at position **N**. If not, then adjust the gear selector cable as described below.

CAUTION
Correct adjustment of the gear selector cable is vital to the functioning and life of the transmission.

To remove gear selector cable

Tool required

- gear selector housing wrench — Saab tool no. 87 90 370

---

1. Disconnect the negative (–) battery cable. Move the front seats to their fully rear position. Apply the handbrake.
2. Move the gear selector to **P** and remove the ignition key.
3. Remove the indicator plate and slide it toward the shift knob. See Fig. 1.
4. Remove the cover screws and slide the plastic console up enough to disconnect the wires to the interior light switch. Remove the console. See Fig. 2.
5. Working at the transmission, remove the cable retaining bolt. See Fig. 3.
6. Slide back the spring loaded cable sleeve and unhook it from the transmission shift rod. See Fig. 4.
7. Remove the gear selector housing nuts using special Saab wrench no. 87 90 370. Rotate the housing onto its side then remove the circlip and pin from the end of the cable.

Fig. 1. Shift indicator plate being moved to reveal cover screw (arrow) below. Screw on opposite side (not shown) must also be removed.

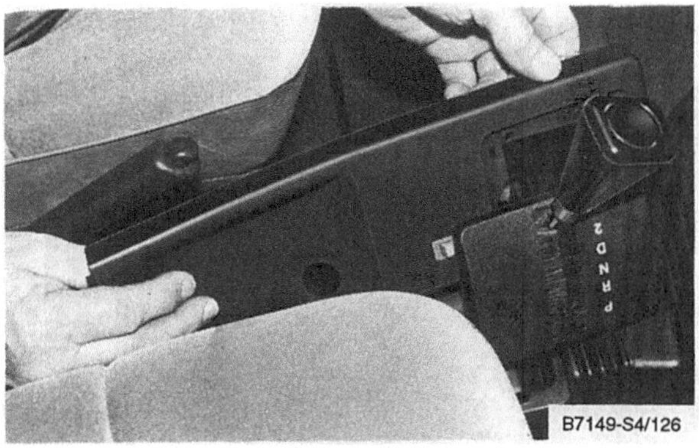

Fig. 2. Gear selector housing console being removed.

Fig. 3. Transmission cable retaining bolt (arrow).

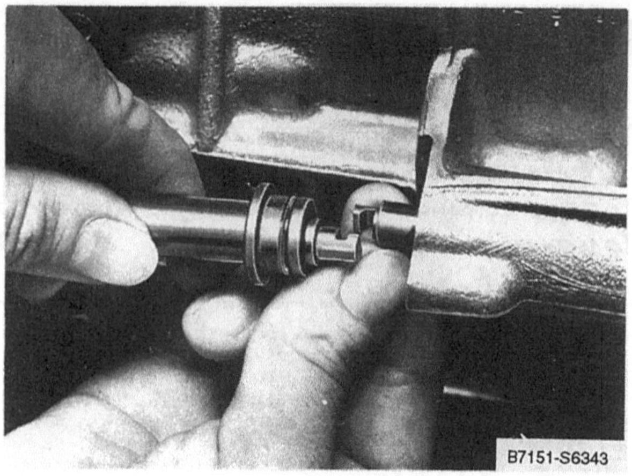

Fig. 4. Transmission cable being unhooked from transmission shift rod. Shift lever must be in **P**.

8. Remove the setscrew that secures the cable to the lever. See Fig. 5. Loosen the cable jacket nuts and remove the cable.

NOTE
The cable jacket nuts hold the cable to the bottom of the shift lever housing.

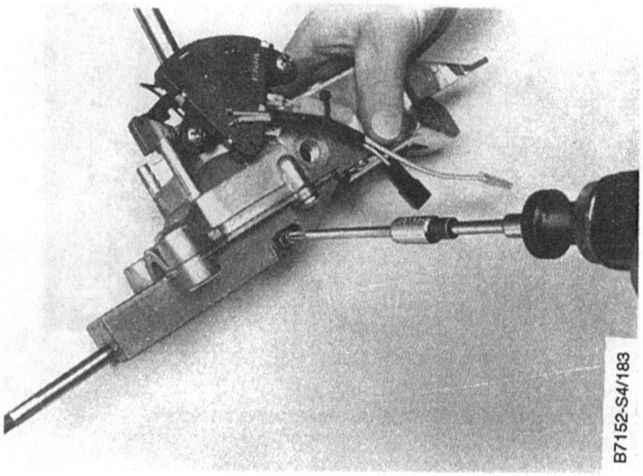

Fig. 5. Transmission cable setscrew being removed using internal-hex head wrench. Housing removed from car for clarity.

9. Remove the carpeting, heater duct, and the old cable.

To install and adjust gear selector cable

NOTE
Proper transmission operation and service life depends on the correct adjustment of the selector cable.

---

1. Install the new cable and check for a proper seal where the cable passes through the firewall.

2. Hook the cable to the transmission shift rod in the transmission, then insert the cable into the transmission case and install the retaining bolt to the case.

3. In the car, install the end of the cable into the selector lever hole and secure with the pin and circlip. Tighten the cable jacket nuts.

4. Withdraw the wire in the cable as far as possible, corresponding to position **P** as if the cable were attached to the shift lever. Move the cable in two notches corresponding to position **N**.

5. Move the gear selector to position **N** without disturbing the cable position. Install and tighten the cable setscrew. See Fig. 5 above.

Tightening torque
• Cable setscrew. . . . . 2.5 Nm (23 in-lb)

---

6. Install the gear selector housing nuts using the special Saab wrench no. 87 90 370.

CAUTION
The gear selector housing must be securely fastened to the floor console.

---

7. Perform a final check by moving the selector lever to position **D**. Release the pawl button and move the lever to position **2**. Increasing resistance should be felt in both directions. Slight movement should be felt before the lever is locked by the selector lever housing detent. If not, then the cable is not adjusted correctly.

The remaining is the reverse of removal. Be sure to check the operation of the neutral safety switch, interior lighting switch and the parking pawl.

To remove and install transmission throttle cable

The transmission throttle cable enters the transmission below the torque converter on the driver's side. It moves a cam in the transmission valve body. After replacment, the cable must be adjusted and the ATF must be topped up.

1. Disconnect the cable from the throttle lever and from its attachment to the retaining plate on the throttle housing.

2. Drain the ATF from the transmission and remove the forward sump cover. See **472 Automatic Transmission Seals and Gaskets**. Remove the ATF strainer.

CAUTION
- Do not dislodge any of the pipes in the valve body. The transmission will have to be removed from the car to reinstall these pipes.
- Do not allow any dirt to enter the transmission, or damage may result.

3. Detach the throttle cable from the cam on the control valve at the valve body. See Fig. 6. Unscrew the cable from the transmission housing. See Fig. 7.

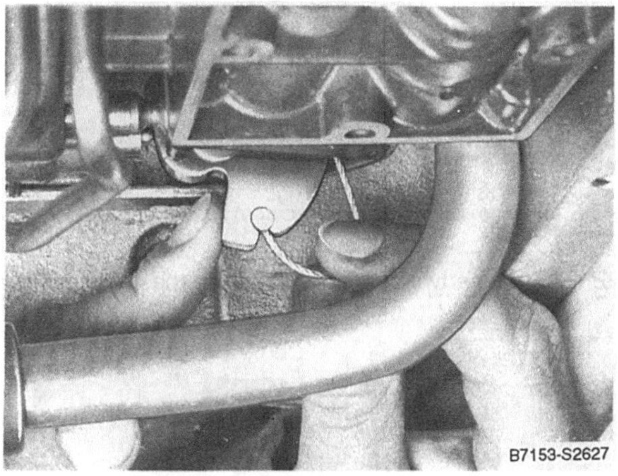

Fig. 6. Transmission throttle cable being unhooked from cam at valve body.

Fig. 7. Transmission throttle cable being unscrewed from transmission housing.

4. Connect a new cable to the transmission housing. Hook the cable end onto the cam at the control valve at the valve body.Make sure that the cable wire is seated in the cam guide.

5. Working at the other end of the cable, straighten the cable and pull the wire until slight tension of the cam is felt. Using pliers, clamp the stop clip next to the threaded cable sheath. See Fig. 8.

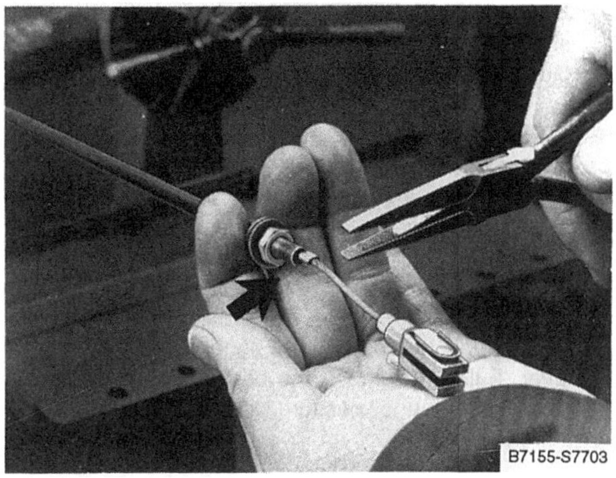

Fig. 8. Stop clip (arrow) being clamped on a new cable. The cable must be straight and slightly tensioned against cam in valve body.

6. Connect the cable to the throttle housing. Floor the accelerator pedal and adjust the nuts until the highest point of the cam on the control valve is aligned as shown in Fig. 9. This is the kickdown position.

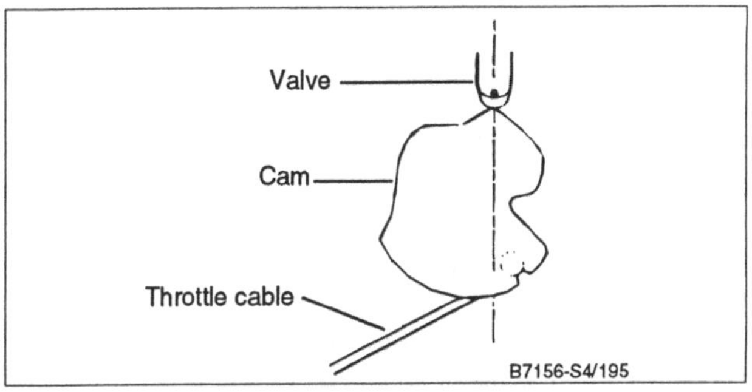

Fig. 9. Cam in valve body in kickdown position for initial throttle cable adjustment at throttle housing.

The remaining is the reverse of removal. Use a new ATF strainer if clogged or damaged. Install a new sump gasket and refill the transmission with new ATF. Check and correct any binding in the accelerator pedal linkage. Finally, do the precision adjustment described below.

To adjust transmission throttle cable

Precise adjustment of the cable should be done after installation of a new cable or if the kickdown function is inoperative. A helper is needed to complete the adjustment procedure.

Tool required
- ATF pressure gauge — Saab tool no. 87 91 592 or equivalent

1. Remove the pressure tap on the transmission and install the pressure gauge. See **400 Transmission–General** for information on performing a pressure test.

2. Block the wheels and apply the handbrake. Start the engine and check for correct idle speed with the gear selector in position **P**.

3. Disconnect the transmission throttle cable from the throttle lever.

4. Firmly apply the footbrake. Engage **D** with the selector lever. Pull and release the cable wire several times. Record the gauge pressure with the wire released.

NOTE
If pressure does not return to the same value, the control valve in the valve body may be sticking and needs to be cleaned.

---

5. Reconnect the cable to the throttle lever. Line pressure should increase by 0.1 bar (1.5 psi). If pressure does not increase by 0.1 bar (1.5 psi), adjust the cable at the throttle lever while watching the pressure gauge. The correct pressure values are given below.

Line pressure
- 4.2 to 4.9 bar (61 to 71 psi)

To remove governor

The governor controls when the transmission shifts. Most problems with the governor occur from dirty ATF. See **400 Transmission–General** for more information on troubleshooting the governor.

The governor can be removed without disassembling the transmission. Extreme cleanliness must be observed when installing the sump cover. Refill the transmission with ATF as described in **1 LUBRICATION AND MAINTENANCE**.

1. Apply the handbrake and block the wheels. Set the selector lever to **N**. Drain the automatic transmission fluid.

NOTE
Do not drain the final drive fluid.

---

2. Remove the support member under the sump.

3. Clean the bottom part of the transmission around the rear sump pan. Remove the sump bolts.

4. Remove the pan by rotating it 1/8 of a turn to gain clearance around the cross member. See Fig. 10.

Fig. 10. ATF sump pan being removed from car.

5. Remove the two screws and remove the governor. See Fig. 11.

Fig. 11. Governor being removed from transmission.

NOTE
To gain access to the governor screws, rotate one of the wheels while holding the other stationary. This will change the position of the governor.

Installation is the reverse of removal. Be sure the cover plate for the governor faces toward the pinion bearing housing. Reinstall the ATF sump pan with a new gasket. See **472 Automatic Transmission Seals and Gaskets**. Fill the transmission with new fluid, and check for leaks.

Tightening torques
- ATF sump pan bolts . . . . 8 to 12 Nm (71 to 106 in-lb)
- ATF sump pan drain plug . . 5 to 8 Nm (44 to 71 in-lb)

---

472 Automatic Transmission Seals and Gaskets

GENERAL

This repair group covers replacement of the automatic transmission pan (sump) gasket and replacement of the automatic transmission fluid cooler only. Other automatic transmission gaskets and seals require disassembly of the transmission, which is not covered in this manual.

Automatic Transmission Pan Gasket

The gaskets used on sump pans are easily replaced in the car by removing the bolts around the pan. See Fig. 1. Refill the sump with new ATF, then start the car and select **P** for 15 seconds, select **D** for 15 seconds and select **R** for 15 seconds to circulate the ATF. Reselect **P** and check the fluid level at idle. Shut off the engine and check for leaks.

WARNING
Always firmly apply the handbrake before starting the engine.

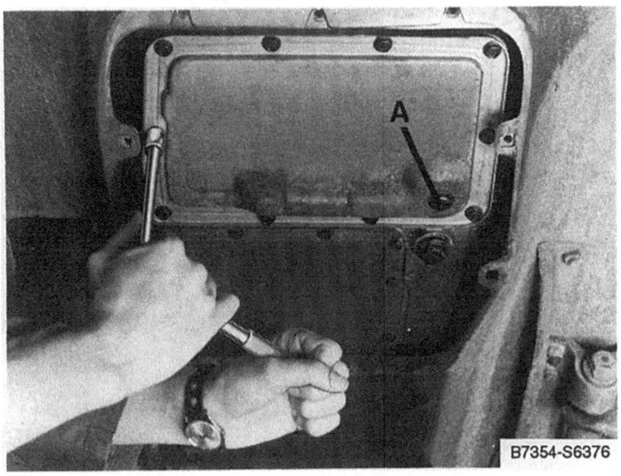

Fig. 1. ATF front cover sump bolts being removed. Rear sump cover can be seen in background. The ATF drain plug is located at **A**.

Tightening torques
- ATF sump pan bolts . . . 8 to 12 Nm (71 to 106 in-lb)
- ATF sump pan drain plug. . . . . 5 to 8 Nm (44 to 71 in-lb)

Automatic Transmission Specifications
- Grade of ATF—Ford Specification M2C-33F (alternate: Ford Specification G)
- ATF capacity 8.5 quarts (8.0 liters)

CAUTION
- Use only Ford spec M2C-33F/G ATF. Do not use other types of fluid.
- Do not allow any dirt to enter the transmission or damage may result.

472

Other areas where leakage can occur are at the torque converter-to-ATF pump seal and at the front cover gasket. The pump seal is integral with the ATF pump housing. Replacing this seal requires that the front cover, primary chain and sprocket, and the ATF pump be removed from the transmission housing.

The pinion bearing housing seals can leak allowing ATF and final drive fluid to mix. The only remedy is to disassemble the transmission and replace the seals. Special tools and techniques are required for the above procedures and are best left to an authorized Saab dealer.

Automatic Transmission Fluid Cooler

The cooler should be replaced if it is visibly leaking. The cooler can also leak internally, causing the engine coolant to mix with the ATF. When this happens, the ATF usually has a milky appearance.

To replace ATF cooler

1. Drain the engine coolant as described in **261 Radiator and Cooling System.**

2. Disconnect the coolant hose clamps and the ATF hoses. Then remove the cooler. See Fig. 2. Some models are equipped with an ATF thermostat in-line with the ATF cooler.

Fig. 2. ATF cooler. Arrows show coolant hose clamps and ATF hoses to be removed.

ATF cooler connection nut
- 13 to 16 Nm (10 to 12 ft-lb)

Installation is the reverse of removal. Always replace O-rings and gaskets. Be sure to top up the ATF.

CAUTION
- Always replace the ATF and flush out all ATF hoses if coolant contamination has occurred.
- If a reconditioned transmission is being installed, flush out the ATF cooler and hoses with clean ATF.

---

500 Brakes—General

TABLES

GENERAL

All cars covered by this manual are equipped with disc brakes at all four wheels. See Fig. 1. The brake system is hydraulically actuated by the master cylinder and has power assist. The brakes are self-adjusting. The parking brake system is mechanical using cables connected to the calipers. Some later models are equipped with an anti-lock braking system (ABS).

Models without ABS use a vacuum booster for power assist. Models with ABS provide power assist through the ABS unit.

The brake disc, also known as the rotor, is mounted to each wheel hub. Two brake pads, one on either side of the disc are located within the caliper. The caliper piston applies force to the brake pads. When the brakes are applied, the friction-surfaced brake pads slow the car. See Fig. 2. A combination of flexible and rigid brake lines connect the calipers to the hydraulic system and master cylinder.

The brake pedal directly actuates the master cylinder. The master cylinder creates pressure in two hydraulic circuits. The two hydraulic circuits form the dual-diagonal braking system where diagonally opposite wheels are braked by one circuit. The system assures a level of braking control if one circuit should fail.

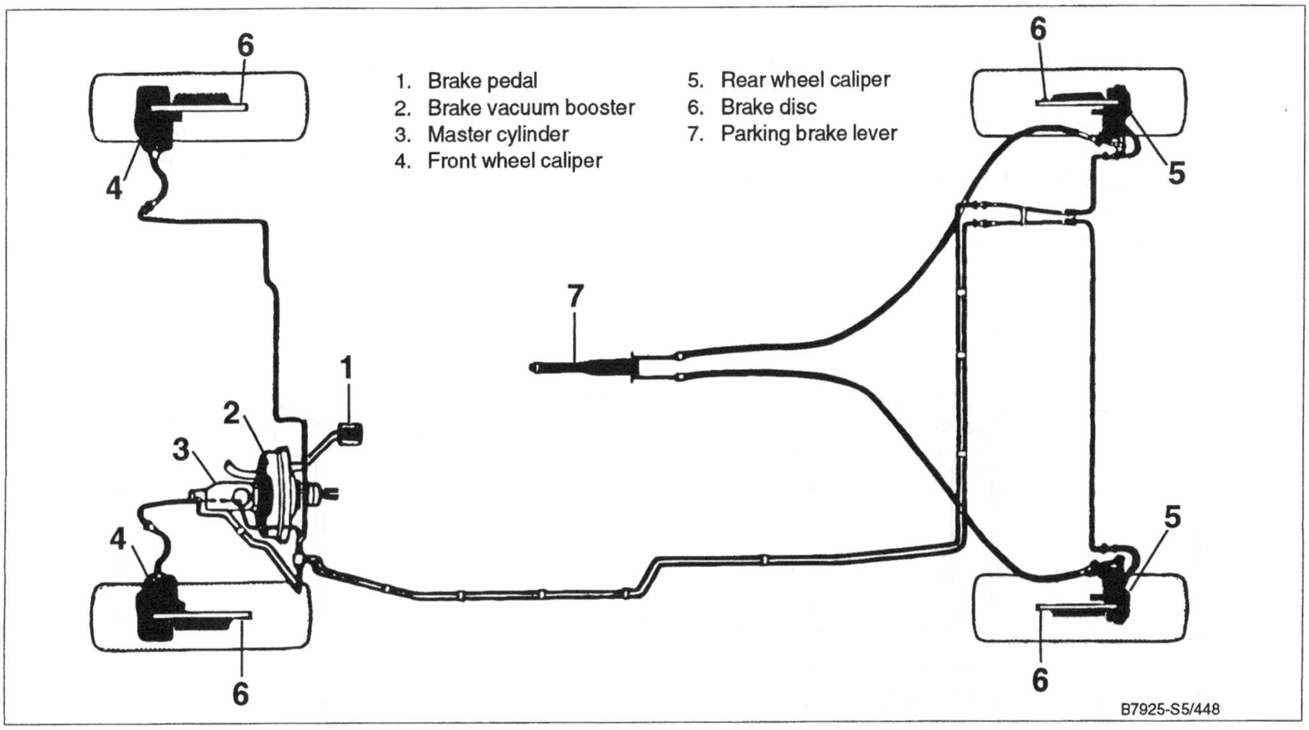

Fig. 1. Four wheel disc brake system

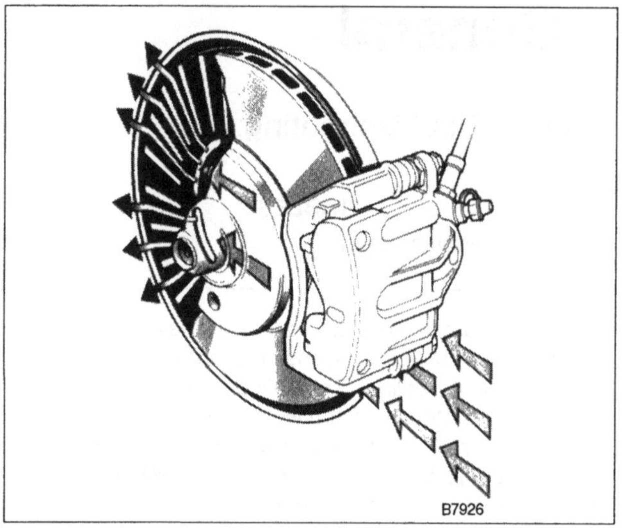

Fig. 2. Brake disc and caliper. Airflow over the brakes (arrows) helps cool the caliper and disc.

Anti-Lock Braking System (ABS)

The ABS system, used on some late models, was developed to ensure optimum braking ability under widely varying road and weather conditions. See Fig. 3. The system monitors and automatically controls braking at the front wheels individually and the rear wheels as a pair—preventing the wheels from locking during emergency braking situations. Steering control is maintained during braking and the shortest possible stopping distance is achieved. The ABS system only operates during emergency stops or hard braking. During most normal braking situations, the system is not in operation.

The ABS system can be simplified into four basic components that operate in a closed loop. See Fig. 4. The wheel speed sensors at all four wheels measure wheel speed. The control unit constantly compares these signals to determine when one wheel is not rotating at the same speed as the others—indicating a potential lock-up condition. Using this information, the control unit regulates the opening and closing of the solenoid valves to adjust the fluid (hydraulic) pressure to the calipers.

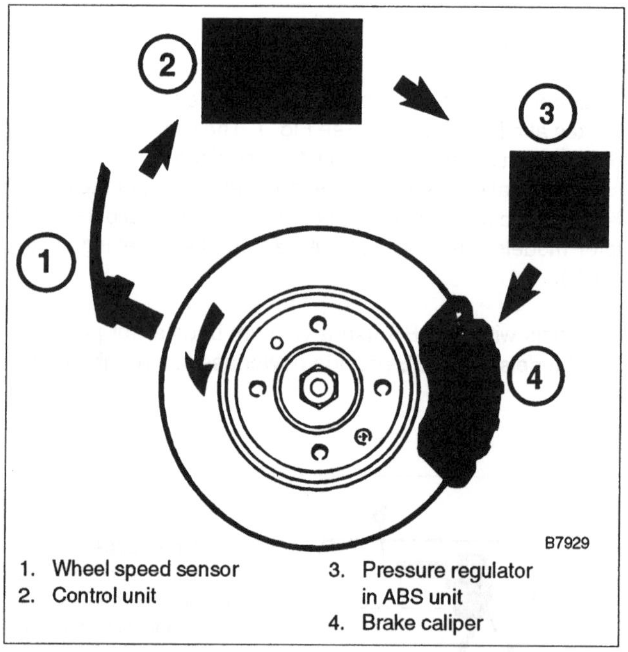

Fig. 4. Basic representation ABS system closed-loop operation.

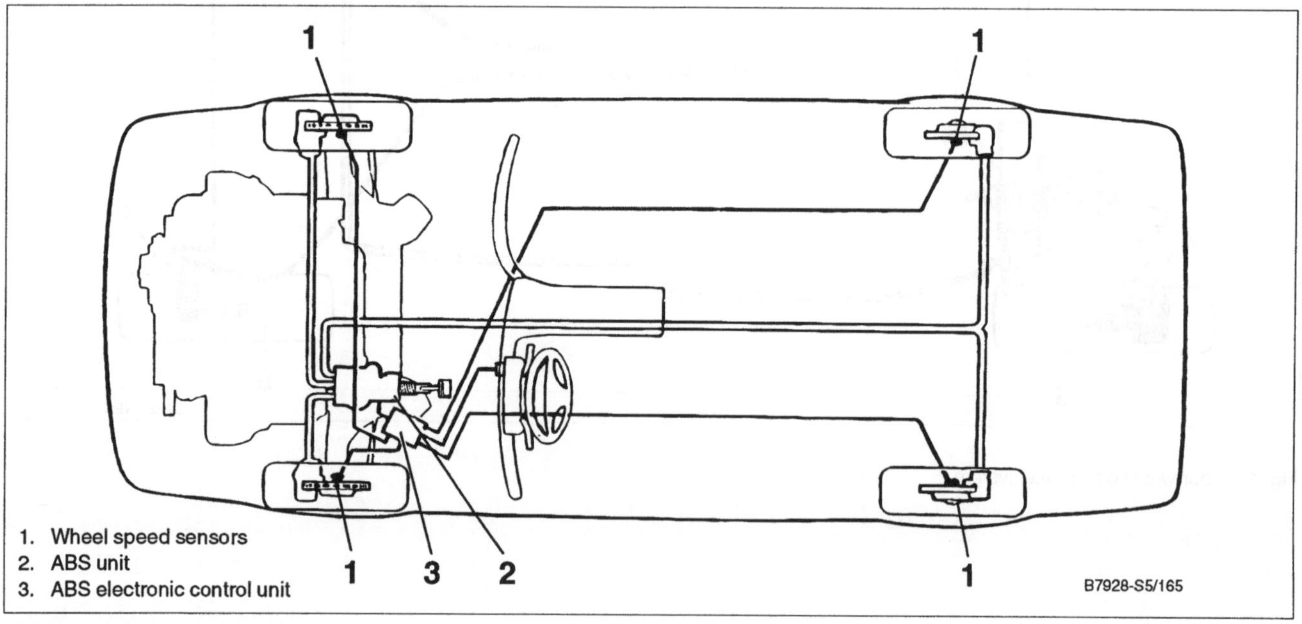

Fig. 3. Schematic of ABS system.

The ABS unit is shown in Fig. 5. The pump pressurizes the accumulator to provide power assist normally provided by the vacuum booster. The accumulator holds enough pressure in the system to provide power assist for approximately twenty stops if the pump should fail. The ABS system is able to monitor itself and if there is a sensor or control system failure it will revert to the non-ABS mode.

NOTE
For information on replacing the ABS unit, see **521 Brake Master Cylinder**. For ABS wiring and ABS wheel speed sensor testing, see **371 Wiring Diagrams, Fuses and Relays**

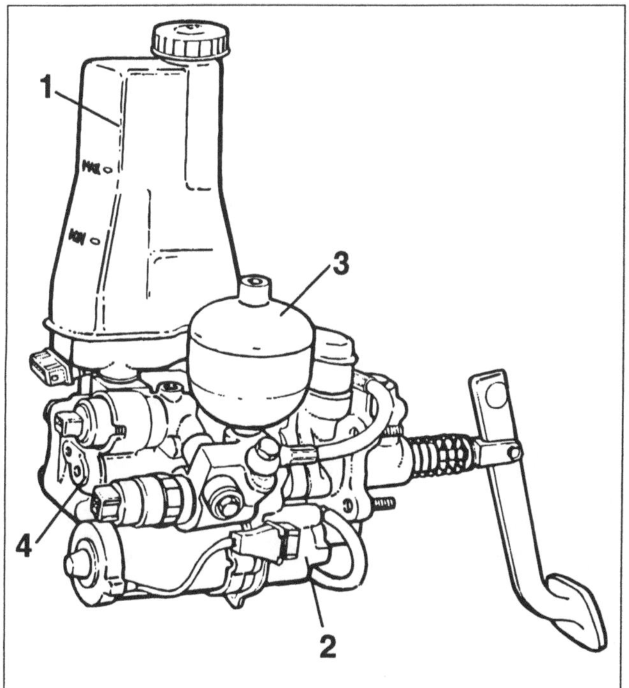

1. Brake fluid reservoir
2. Pump
3. Accumulator
4. Master cylinder

Fig. 5. ABS unit.

The ANTI-LOCK and the BRAKE FLUID warning lamps in the instrument cluster monitor the system. The ANTI-LOCK warning lamp will light up if a problem or fault occurs in the ABS system. The BRAKE FLUID warning lamp will light up if the brake fluid level in the reservoir falls too low. See an authorized Saab dealer if either warning lamps come on.

WARNING
If the ANTI-LOCK and the BRAKE FLUID warning lamp should come on, the car must not be driven. Brake failure may occur. Have the problem corrected promptly by an authorized Saab dealer.

WARNING
On cars with ABS, dangerously high pressure may be retained in the system, even though the car is not running or has not been driven recently. Before loosening any components, depressurize the system by disconnecting the battery negative cable and pressing the brake pedal 25 to 30 times, or until there is a noticeable increase in pedal effort. Never attempt to open the pressure accumulator.

NOTE
The control unit has the capability to detect and store fault codes, although retrieving these faults codes requires special test equipment. See an authorized Saab dealer for additional information on ABS fault codes.

To replace ABS wheel speed sensor

Replacement of wheel speed sensors is similar for either the front or the rear. The sensor is a sealed unit and cannot be repaired. Be sure to have a new fiber washer on hand when removing or replacing the sensor. The washer is used to precisely set the air gap between the sensor and the toothed wheel.

1. Working in the engine compartment, disconnect the negative battery cable and the wheel speed sensor connectors. See Fig. 6.

NOTE
The wheel speed sensor connectors for the rear wheels are located under the rear seat.

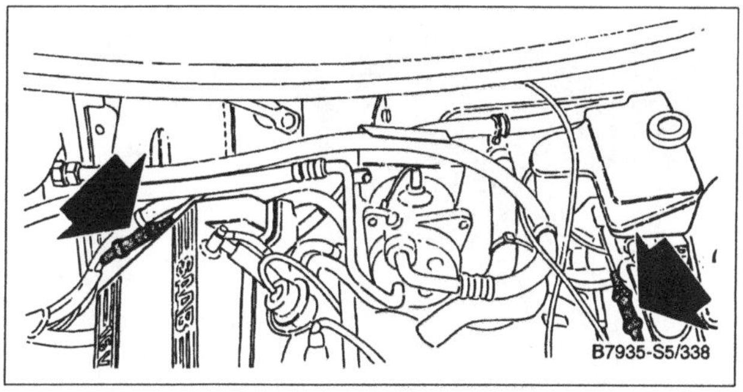

Fig. 6. Location of front wheel speed sensor connectors (arrows).

2. Raise the car and support it with jackstands. Remove the road wheels.

WARNING
Make sure the car is well supported with jackstands. Never use just a jack to support a car.

---

3. Remove the sensor harness from the steering member. Carefully pull the sensor lead through the rubber grommet in the wheel arch. See Fig. 7.

NOTE

The rubber grommets for the rear wheel speed sensors are located in the floor pan and can be found by tracing the harness from the wheel.

Fig. 7. Front ABS wheel speed sensor harness guide **A** and rubber grommet **B**.

4. Remove the wheel speed sensor. See Fig. 8 and Fig. 9.

Fig. 8. Front wheel speed sensor retaining screw being removed.

5. Remove the plastic sleeve from the sensor housing. See Fig. 10.

Fig. 9. Rear wheel speed sensor retaining screw being removed.

NOTE

When removing rear wheel speed sensors, remove the parking brake cable guide from the suspension spring link then remove the wheel speed sensor harness from the clips. See Fig. 11.

Fig. 10. Plastic sleeve being removed from sensor housing.

6. Clean out the sensor housing in the steering member or axle. Spin the wheel hub slowly and clean out any debris in the toothed wheel.

7. Working at the bench, clamp the sensor adjusting sleeve in a vice and remove the setscrew. See Fig. 12. Remove the sensor from the adjusting sleeve.

CAUTION

Clamp only the sensor sleeve. Do not clamp the metal part of the sensor, as this will damage it.

Fig. 11. Parking brake cable guide (**A**) and wire harness clip (**B**) holding wheel speed sensor harness to rear suspension spring link.

Fig. 12. Setscrew being removed from wheel speed sensor sleeve.

8. Clean any debris from the end of the wheel speed sensor and sensor sleeve using a wire brush. Lightly lubricate the sensor body.

9. Glue a new fiber spacer to the tip of the wheel speed sensor. Install the new sensor in the sensor sleeve.

CAUTION

The fiber spacer has a thickness of 0.65 mm to provide the sensor with the proper clearance to the toothed wheel. Do not spin the toothed wheel until the sensor has been completely installed. If the fiber washer is damaged, the sensor may be incorrectly installed.

Specification

- ABS wheel speed sensor air gap .... 0.65 mm (0.026 in.)

---

10. Install the wheel speed sensor and tighten the retaining screw. See Fig. 8 or 9 above.

11. Gently press the sensor body against the sensor wheel then tighten the setscrew.

The remainder of installation is the reverse of removal. Check that the ABS warning light comes on for several seconds when starting the engine. It should then go out.

BRAKES TROUBLESHOOTING

This troubleshooting section applies mainly to the mechanical and hydraulic components regardless of whether or not the car is equipped with ABS. It is important to realize that ABS only operates when one or more wheels are about to lock during emergency or otherwise hard braking. It is very unlikely that ABS components are the cause of more conventional brake noise or grabbing symptoms.

NOTE

This troubleshooting section does not specifically include troubleshooting for ABS. If an ABS system fault exists, see an authorized Saab dealer who has the necessary training and diagnostic equipment to properly service the system.

The most common brake problems and repairs involve the brake calipers, pads, and brake discs since these components are exposed to heat, water, chemicals, and corrosion.

Table a. Brake Troubleshooting

Symptom	Probable cause	Corrective action
1. Brake squeak or squeal	a. Normal condition b. Brake pads installed incorrectly c. Calipers or brake pads binding, brake caliper guide pins or caliper dirty or corroded d. Brake pad lining worn e. Brake pad linings heat-glazed or oil-soaked	a. See **516 Brake Discs** b. Check installation. **517 Brake Pads** c. Clean calipers and/or guide pins. **523 Brake Calipers**, **517 Brake Pads** d. Install new pads. **517 Brake Pads** e. Replace brake pads. Clean discs. Replace leaking calipers as required. **517 Brake Pads, 516 Brake Discs, 523 Brake Calipers**
2. No brakes, pedal goes to floor	a. Brake fluid level too low b. Master cylinder faulty c. Failed line or hose, loose union, or other major fluid leak	a. Check fluid level. Check for leaks. Fill and bleed system. **520 Brake Fluid and Brake Bleeding** b. Replace master cylinder. **521 Brake Master Cylinder** c. Check for leaks. Fill and bleed system. **522 Brake Lines**
3. Low pedal, (brakes properly bled, no signs of leakage)	a. Master cylinder faulty	a. Replace master cylinder. **521 Brake Master Cylinder**
4. Pedal feels spongy or brakes work only after pedal is pumped	a. Insufficient brake fluid b. Air in brake fluid c. Master cylinder faulty d. Leaking line and hose unions	a. Check fluid level. Check for leaks. Fill and bleed system. See **520 Brake Fluid and Brake Bleeding** b. Bleed system. **520 Brake Fluid and Brake Bleeding** c. Replace master cylinder. **521 Brake Master Cylinder** d. Check for leaks. Fill and bleed system. **522 Brake Lines**
5. High pedal effort required (braking performance poor)	a. Brake pads wet b. Brake pads heat-glazed or oil-soaked c. Vacuum booster or vacuum connections faulty (non-ABS cars only) d. Faulty master cylinder	a. Use light pedal pressure to dry brakes while driving. b. Replace brake pads. Clean discs. Replace leaking calipers as required. **517 Brake Pads, 516 Brake Discs, 523 Brake Calipers** c. Inspect vacuum lines and check valve. Test booster and replace as required. **541 Brake Vacuum Booster** d. Replace master cylinder. **521 Brake Master Cylinder**
6. Brakes pulsate, chatter, or grab	a. Warped brake discs b. Brake pads worn c. Brake pads heat-glazed or oil-soaked	a. Recondition or replace discs. **516 Brake Discs** b. Recondition brakes. **517 Brake Pads** c. Replace brake pads. Clean discs. Replace leaking calipers as required. **517 Brake Pads, 516 Brake Discs, 523 Brake Calipers**
7. Uneven braking, car pulls to one side	a. Incorrect tire pressures b. Brake pads heat-glazed or oil-soaked c. Brake caliper or brake pads binding d. Brake fluid contaminated e. Worn tires or suspension components	a. Check and correct tire pressures. See **1 LUBRICATION AND MAINTENANCE** b. Replace brake pads. Clean discs. Replace leaking calipers as required. **517 Brake Pads, 516 Brake Discs, 523 Brake Calipers** c. Clean and recondition brakes. Check parking brake adjustment. **517 Brake Pads, 516 Brake Discs, 523 Brake Calipers, 551 Parking Brake** d. Drain, flush, refill, and bleed brake system. **520 Brake Fluid and Brake Bleeding** e. See **6 STEERING AND WHEEL ALIGNMENT, 7 SUSPENSION**
8. Brakes drag, bind, or overheat	a. Parking brake dragging b. Brake caliper or brake pad binding c. Master cylinder faulty	a. Adjust parking brake. Replace caliper or caliper return spring. **551 Parking Brake** b. Clean and recondition brakes. **517 Brake Pads, 516 Brake Discs, 523 Brake Calipers** c. Replace master cylinder. **521 Brake Master Cylinder**
9. Brakes make loud clunk when backing up	a. Caliper yoke worn b. Caliper guide pins worn or pre-set bend in pin no longer exists	a. Replace caliper assembly. **523 Brake Calipers** b. Replace guide pins. **523 Brake Calipers**

516 Brake Discs

TABLES

GENERAL

Fig. 1 is an exploded view of the front disc brake. Two types of front discs—ventilated and solid—are used on cars covered by this manual, depending on models and model year. See **Table a.** The rear discs are solid on all models. The brake disc is bolted to the wheel hub and can be removed once the caliper is removed.

Table a. Front Brake Disc Applications

Model year and Model	Disc type
1985 through 1987 900	Solid
1985 and 1986 900s	Solid
1985 Turbo	Solid
1987 900s	Ventilated
1986 through 1987 Turbo, including convertible	Ventilated
1988 and later, All	Ventilated

Discs and brake pads are subject to the greatest wear and are the components most often requiring service. While it possible to restore the brakes by replacing only the brake pads, the discs should be resurfaced or replaced at the same time to achieve full braking performance and to maximize pad life.

INSPECTING AND REPLACING BRAKE DISCS

Inspecting the brake discs should be done during regular service or when the brakes are reconditioned. The discs should be checked for cracks, scoring, glazing and warpage.

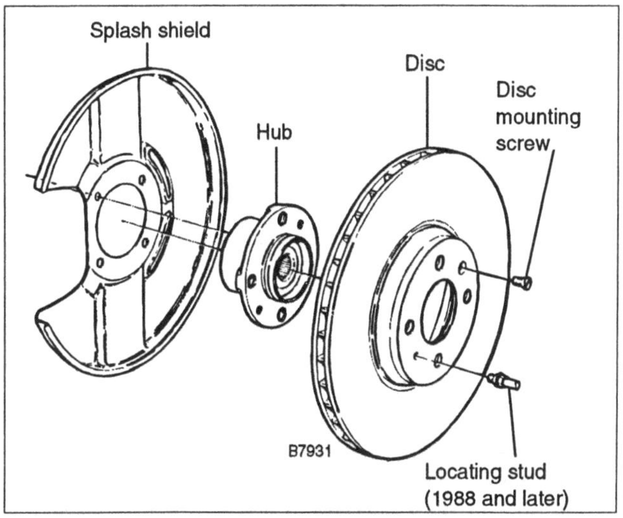

Fig. 1. Front disc assembly (ventilated front disc shown).

Brake discs must be replaced when they cannot be resurfaced without exceeding the specified minimum thickness given below.

Brake discs that are glazed or lightly scored may be resurfaced. Cracked discs should always be replaced. Discs should be replaced or resurfaced in pairs only. If the disc is being resurfaced, both faces should be resurfaced by equal amounts. See **Table b** and **Table c** for disc specifications.

Table b. Rear Disc Specifications

thickness (new)	9.0 mm (0.354 in.)
thickness (wear limit)	7.0 mm (0.276 in.)
brake surface lateral runout (maximum)	0.1 mm (0.004 in.)

Table c. Front Disc Specifications

	Solid disc	Ventilated disc (1986-1987)	Ventilated disc (1988 and later)
thickness (new)	12.7 mm (0.50 in.)	20 mm (0.784 in.)	23.5 mm (0.93 in.)
thickness (wear limit)	11.2 mm (0.441 in.)	18 mm (0.709 in.)	21.5 mm (0.85 in.)
brake surface lateral runout (maximum)	0.1 mm (0.004 in.)	0.1 mm (0.004 in.)	0.08 mm (0.003 in.)

NOTE
If the brake pads are soaked with oil, grease, or brake fluid, the cause of the contamination must be found and corrected before new brake pads or discs are installed.

To remove brake disc

New replacement brake discs should be cleaned with a grease-free solvent, such as brake cleaner, before installing the caliper and brake pads. The procedure below applies to front and rear brake discs.

1. Raise the car and support it with jackstands. Remove the wheel.

WARNING
Make sure the car is well supported with jackstands. Never use just a jack to support a car.

2. Remove the brake pads as described in **517 Brake Pads**.

WARNING
Friction materials in the brake pads may contain asbestos fibers. Avoid breathing any asbestos fibers or dust. Always use a respirator. Do not create dust by grinding or sanding, or by cleaning the pads or disc with compressed air.

3. Without removing the flexible brake hose, remove the caliper as described in **523 Brake Calipers.** Suspend the caliper from the body using stiff wire. See Fig. 2.

CAUTION
Do not allow the removed caliper to hang by the flexible brake hose.

NOTE
If the brake hose remains connected, it will not be necessary to bleed the brakes.

4. Remove the disc mounting screws or road wheel locating stud (where applicable). See Fig. 3.

Fig. 2. Caliper removed from disc and suspended using stiff wire.

Fig. 3. Arrows show locating stud and disc mounting screw.

NOTE
• On 1985 through 1987 models, the disc is mounted with two screws. On 1988 and later cars, the disc is mounted with a screw and a locating stud.

• Use a penetrating oil in advance to help loosen corroded screws. Keep penetrating oil away from rubber parts.

5. Remove the brake disc from the wheel hub. See Fig. 4.

NOTE
If the disc is frozen to the hub, tap it loose with a soft faced mallet.

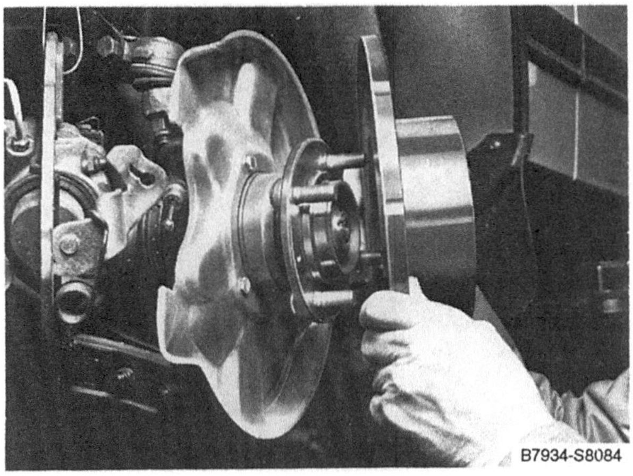

Fig. 4. Disc being removed from hub.

Installation is the reverse of removal. Use a new caliper mounting bolt locking plate when installing the caliper. Reinstall the brake pads and gently pump the brake pedal several times to seat the pads on the disc. Install the wheels and tighten the wheel lugs. Test the footbrake and parking brake functions.

Tightening torques

- front brake caliper to steering member (1985 to 1987) . . . . . . 110 to 130 Nm (81 to 96 ft-lb)
- front brake pad carrier to steering member (1988 models and later) . . . . . 70 to 110 Nm (52 to 81 ft-lb)
- rear brake caliper to axle (1985 to 1987) . . . . 70 to 90 Nm (52 to 66 ft-lb)
- rear brake pad carrier to axle (1988 models and later) . . . . 40 to 54 Nm (30 to 49 ft-lb)
- road wheel locating stud to wheel hub . . . . 25 to 30 Nm (18 to 22 ft-lb)
- wheel to wheel hub
  - lug nuts . . . . 90 to 110 Nm (66 to 88 ft-lb)
  - lug bolts . . . . 105 to 125 Nm (77 to 92 ft-lb)
- all other fasteners
  - M5 bolt . . . . 5 Nm (44 in-lb)
  - M6 bolt . . . . 10 Nm (89 in-lb)
  - M8 bolt . . . . 20 Nm (15 ft-lb)
  - M10 bolt . . . . 40 Nm (30 ft-lb)

517 Brake Pads

GENERAL

Since the disc brake assemblies and the parking brake are self-adjusting it is not possible to judge the condition of the brake pads from brake pedal travel or from the parking brake lever. Pads should be inspected during regular service or when any abnormal brake noise is noticed. The brake discs should also be inspected during pad replacement as described in **516 Brake Discs**.

INSPECTING BRAKE PADS

Inspection of the brake pads can be done after removing the road wheel. The pads are made up of friction material and are designed to wear away. Pad thickness is measured by how much friction material is left. See Fig. 1.

Fig. 1. Disc brake pads as viewed through opening in caliper (road wheel removed). Minimum brake pad thickness shown by dimension **A**.

WARNING
- Do not reuse any self-locking nuts or bolts. These fasteners are designed to be used only once and should be replaced anytime they are loosened or removed.
- Friction materials in the brake pads may contain asbestos fibers. Avoid breathing any asbestos fibers or dust. Always use a respirator. Do not create dust by grinding or sanding, or by cleaning the pads or disc with compressed air.

NOTE
The front brake pads will usually wear out before the rear pads owing to the fact that the front brakes do most of the stopping.

Always replace brake pads in complete sets. If the old pads are to be reinstalled, make sure they are refitted in their original locations. Inspect other parts such as springs, shims, hoses, and cables and replace as necessary.

Brake Pad Lining

Model	Minimum thickness (dimension A)
1985 to 1987 models	1.0 mm (0.04 in.)
1988 and later models	4.0 mm (0.16 in.)

REPLACING BRAKE PADS

The brake pads can be easily replaced without disconnecting the flexible brake hose or the handbrake cable. This eliminates the need to bleed the brake system. In some cases, special tools may be required to replace the pads. Read the procedures completely before starting the job.

> **CAUTION**
> Different brake pads are used on different models. In addition, some models require different inner and outer front brake pads. Check with an authorized Saab dealer for the latest in parts information.

To replace front brake pads (1985-1987 models)

Fig. 2 shows the front brake assembly used on 1985 through 1987 models. Owing to the self-adjusting caliper assemblies, the caliper piston ratchets out as the brake pads are used up. In order to fit the thicker new pads to the caliper, the piston needs to be screwed back into the caliper housing and this requires a special tool. See Fig. 3.

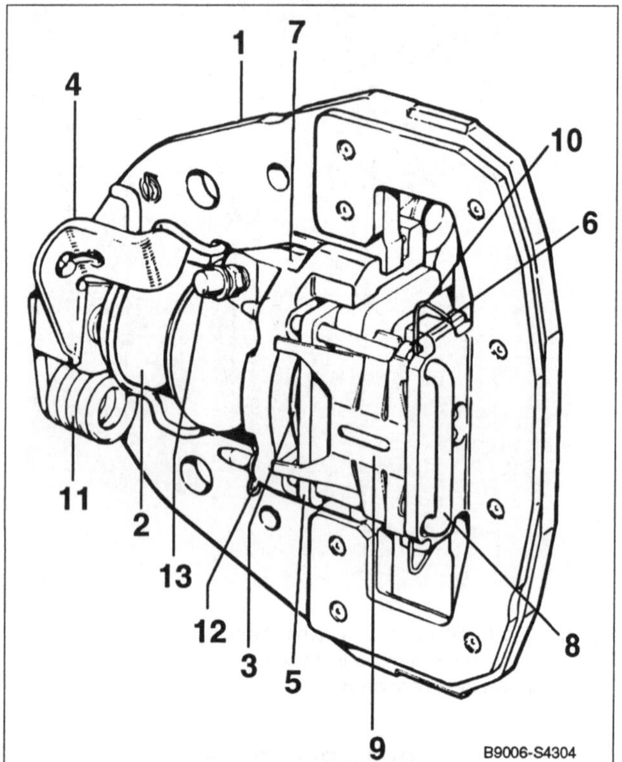

1. Sliding yoke
2. Indirect piston
3. Direct piston
4. Parking brake lever
5. Inner brake pad
6. Outer brake pad
7. Piston housing
8. U-clip
9. Damper (anti-rattle clip)
10. Anti-rattle spring
11. Parking brake lever return spring
12. U-clip retaining pin
13. Bleeder

Fig. 2. Front brake assembly used on 1985 through 1987 models.

1. Raise the car and support it with jackstands. Remove the front wheels.

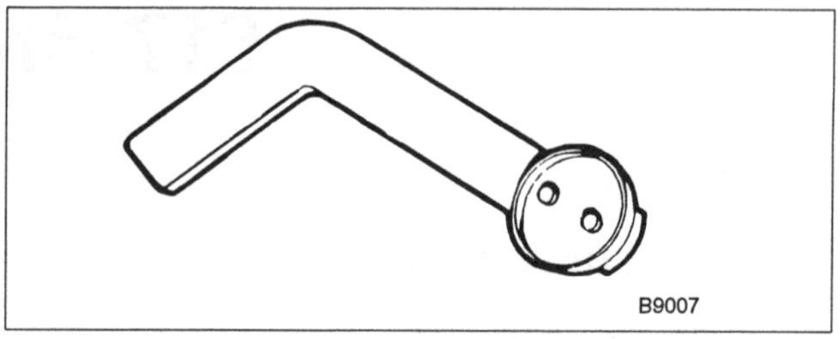

Fig. 3. Special brake piston wrench (Saab tool no. 89 96 043) needed for pad replacement.

> **WARNING**
> Make sure the car is well supported with jackstands. Never use just a jack to support a car.

2. Release the parking brake. Remove any rust and debris from the brake disc.

3. Working at the caliper, push the parking brake lever towards the sliding yoke and unhook the parking brake cable from the lever.

> **CAUTION**
> The parking brake lever must operate correctly. If the lever sticks, the pads will not fully release from the brake disc. This can lead to premature brake pad wear or even brake system failure due to the constant heat.

> **NOTE**
> Check the operation of the parking brake lever by pushing it fully towards the yoke and then releasing it. The return spring should cause the lever to snap back and contact the yoke. If the parking brake lever does not fully release, the lever is sticking in the caliper or the lever return spring is damaged. If the lever cannot be made to move freely, the caliper must be replaced as described under **523 Brake Calipers.**

4. Reconnect the parking brake cable to the lever. Check that the parking brake cables are correctly adjusted by checking for a small gap between the parking brake lever and yoke. Measure the gap as shown in Fig. 4. If the clearance is incorrect, adjust the parking brake cables as described in **551 Parking Brake.**

5. Remove the small retaining pin at the bottom of the U-clip, then pull the U-clip from the caliper. If necessary, gently pry the U-clip out evenly using a large screwdriver.

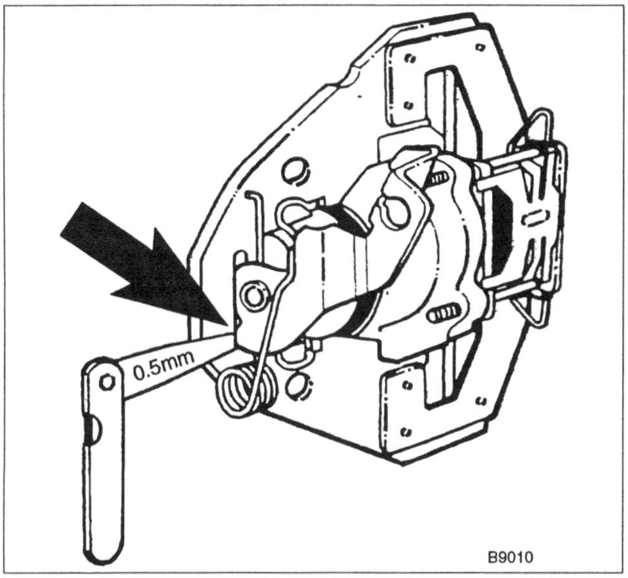

Fig. 4. Parking brake lever gap being measured. Cable must be connected to lever when making measurement.

Parking brake adjustment	
parking brake lever to caliper clearance	0.5±0.1 mm (0.019±0.003 in.)

NOTE
Observe the position of the damper (anti-rattle clip) behind the U-clip. During installation, the damper should be reinstalled in the same position.

6. Remove the inner and outer pads from the caliper. The pads can be removed from the caliper by grabbing the pad using large pliers. As an alternative, a special brake pad puller is available to help remove the pads. See Fig. 5.

Fig. 5. Brake pads being removed from caliper. Special brake pad puller (Saab tool no. 89 95 771) shown.

CAUTION
Do not step on the brake pedal or apply the parking brake with the brake pads removed. The piston will be forced out of the caliper making it impossible to reinstall the pads without removing the caliper.

7. Clean the caliper assembly using brake cleaner and a small wire brush. Be careful not to damage rubber seals with the wire brush. Check that the yoke slides freely in its groove in the caliper.

8. Lubricate the sliding surfaces between the caliper and yoke using Gleitmo 540 lubricant, which is available from an authorized Saab dealer. See Fig. 6.

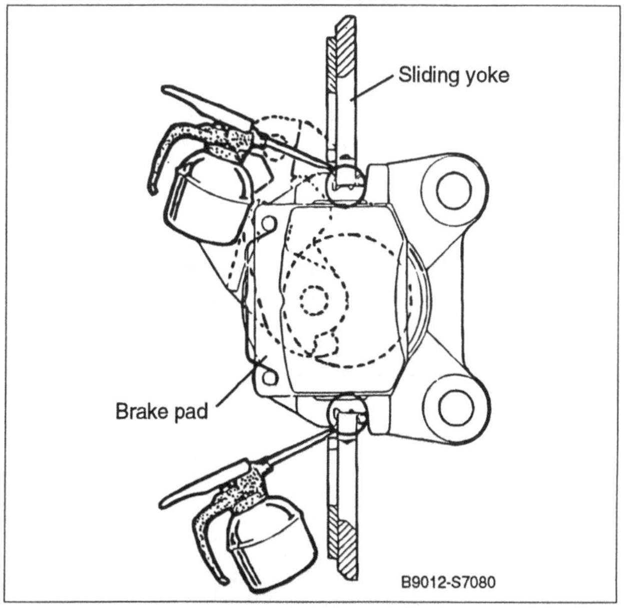

Fig. 6. Sliding surfaces of yoke to be lubricated using special grease (Gleitmo 540).

9. Using the special brake piston wrench, engage the pins of the wrench in the piston face and rotate the piston clockwise. See Fig. 7. Remove the wrench and push the piston into the caliper. If necessary, gently pry the piston in using a screwdriver. Repeat the process until the pads can be fitted.

CAUTION
Do not thread the piston in beyond the face of the caliper. See Fig. 8. The piston seal and dust cover can be damaged. If the piston is threaded all the way in and there still is not enough clearance to install the pads, install the wrench to the piston and carefully slide the yoke away from the caliper. Continue to thread the piston in until the pads can be fitted.

NOTE

The piston can only be rotated a small amount at one time with the wrench. After rotating the piston about 1/4 turn, remove the wrench and push the piston in.

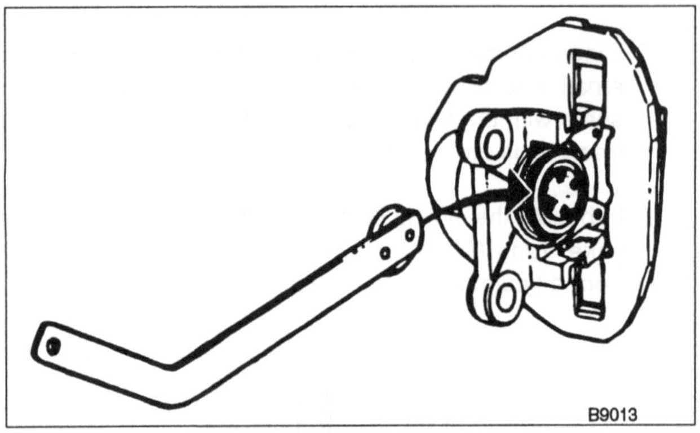

Fig. 7. Special wrench (Saab part no. 89 96 043) used to screw caliper piston in. Rotate piston clockwise only.

NOTE

Pushing the caliper piston in may cause brake fluid to overflow the master cylinder reservoir. To prevent this, use a clean syringe or some equivalent to remove some fluid from the reservoir.

Fig. 8. Piston shown flush to face of caliper (arrow). Piston must not be threaded beyond this point. If necessary, carefully slide yoke away from caliper to rotate piston in further.

10. Install the new brake pads, damper and U-clip. Install a new retaining pin in the U-clip.

NOTE

Saab recommends that the U-clips and dampers (anti-rattle clips) be replaced whenever replacing brake pads. The U-clips are pre-bent to help prevent brake noise (clunking noise when backing up). During normal use, the bends in the U-clips tend to straighten out and the damper spring tension tends to weaken.

11. Pull up the handbrake lever in the passenger compartment 7 notches (clicks). Start the engine and gently pump the brake pedal several times. Release the lever and then raise it 9 to 10 notches. The parking brake should now be fully applied.

NOTE

Pulling up the hand brake lever 7 notches blocks the self adjusting mechanism of the caliper and prevents over adjustment. If the caliper is over adjusted, the brake pads will drag.

12. Install the road wheels and lower the car. Check the brake fluid level and top up as necessary.

WARNING

New brake pads require some break-in. Allow for slightly longer stopping distances for the first 100 to 150 miles of city driving, and avoid hard stops.

Tightening torque

- wheel to wheel hub (lug nuts) . . . . 90 to 110 Nm (66 to 88 ft-lb)

To replace rear brake pads (1985-1987 models)

Fig. 9 shows the rear brake assembly on 1985 through 1987 models.

1. Raise the car and support it with jackstands. Remove the rear wheels.

WARNING

Make sure the car is well supported with jackstands. Never use just a jack to support a car.

2. Using a 2.5 mm drift, tap out the brake pad retaining pins. See Fig. 10.

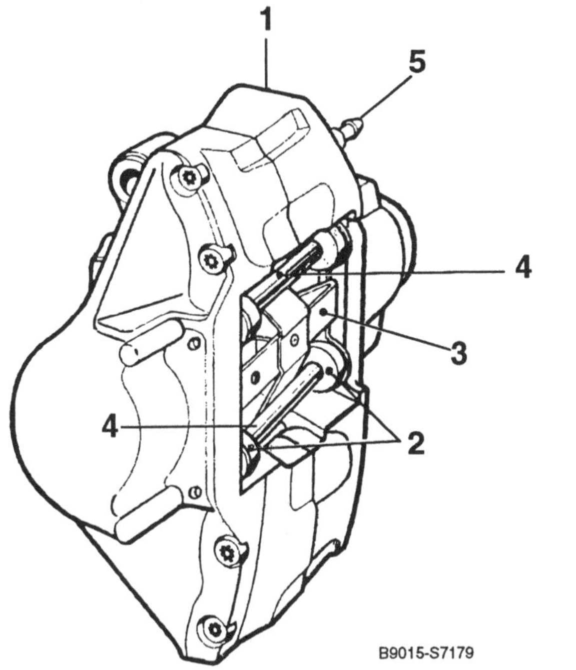

1. Caliper housing
2. Brake pads
3. Damper (anti-rattle clip)
4. Brake pad retaining pins
5. Bleeder

Fig. 9. Rear brake assembly used on 1985 through 1987 models.

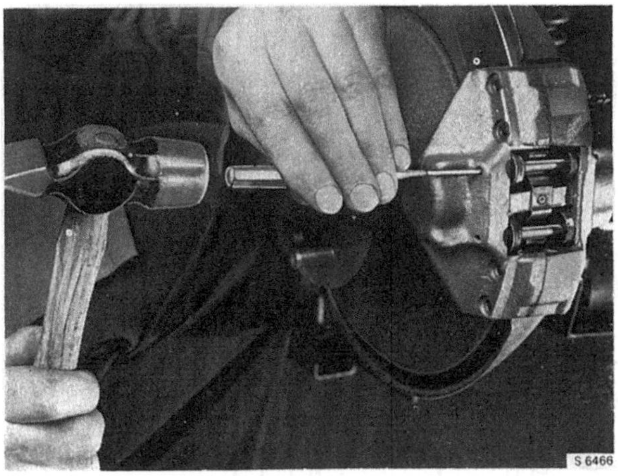

Fig. 10. Brake pad retaining pins being tapped out of caliper.

NOTE

Observe the position of the damper (anti-rattle clip) behind the retaining pins. During installation, the damper should be reinstalled in the same position.

---

3. Remove the inner and outer pads together with pad shims.

NOTE

- The brake pad shims are located on the rear of the pads. Label the shims during removal, as they are different between the left and right wheels.

- The pads can be removed from the caliper by grabbing them with a large pliers. As an alternative, a special brake pad puller is available to help remove the pads. See Fig. 11.

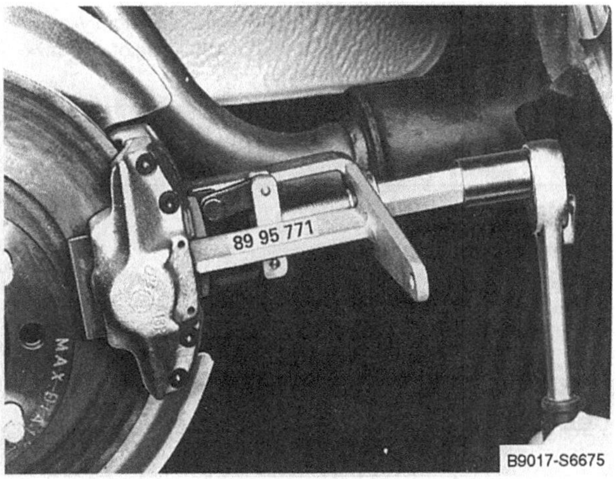

Fig. 11. Brake pads being removed from caliper. Special brake pad puller (Saab tool no. 89 95 771) shown.

CAUTION

Do not step on the brake pedal with the brake pads removed. The piston will be forced out of the caliper.

---

4. Clean and inspect the caliper assembly, the pad retaining pins, and the shims using brake cleaner and a small wire brush. Replace any worn components. Be careful not to damage rubber seals with the wire brush. Inspect the piston rubber dust cover and replace if necessary.

5. Using a screwdriver, gently push the piston back into the caliper. Install the new brake pads with the shims.

CAUTION

Push the piston in just far enough to install the new brake pads. Pushing the piston in too far may damage the piston seals.

NOTE

Pushing the caliper piston in may cause brake fluid to overflow the master cylinder reservoir. To prevent this, use a clean syringe or some equivalent to first remove some fluid from the reservoir.

---

6. Install the damper (anti-rattle clip) and the pad retaining pins. Tap the retaining pins in using a drift until they are fully seated.

7. Install the road wheels and lower the car. Gently pump the brake pedal several times to seat the pads. Check the brake fluid level and top up as necessary.

WARNING
New brake pads require some break-in. Allow for slightly longer stopping distances for the first 100 to 150 miles of city driving, and avoid hard stops.

Tightening torque
- wheel to wheel hub (lug nuts) . . . . 90 to 110 Nm (66 to 88 ft-lb)

To replace front brake pads (1988 and later models)

Fig. 12 shows the front brake assembly used on 1988 and later models.

1. Raise the car and support it with jackstands. Remove the front wheels.

WARNING
Make sure the car is well supported with jackstands. Never use just a jack to support a car.

---

2. Working through the opening in the caliper, press the brake piston back into the caliper housing using large pliers. See Fig. 13.

3. While holding the lower guide pin stationary, remove the brake caliper mounting bolt from the end of the guide pin. See Fig. 14.

4. Swing the caliper up out of the way and remove the brake pads. See Fig. 15.

CAUTION
Do not step on the brake pedal with the brake pads removed. The piston will be forced out of the caliper damaging the seals.

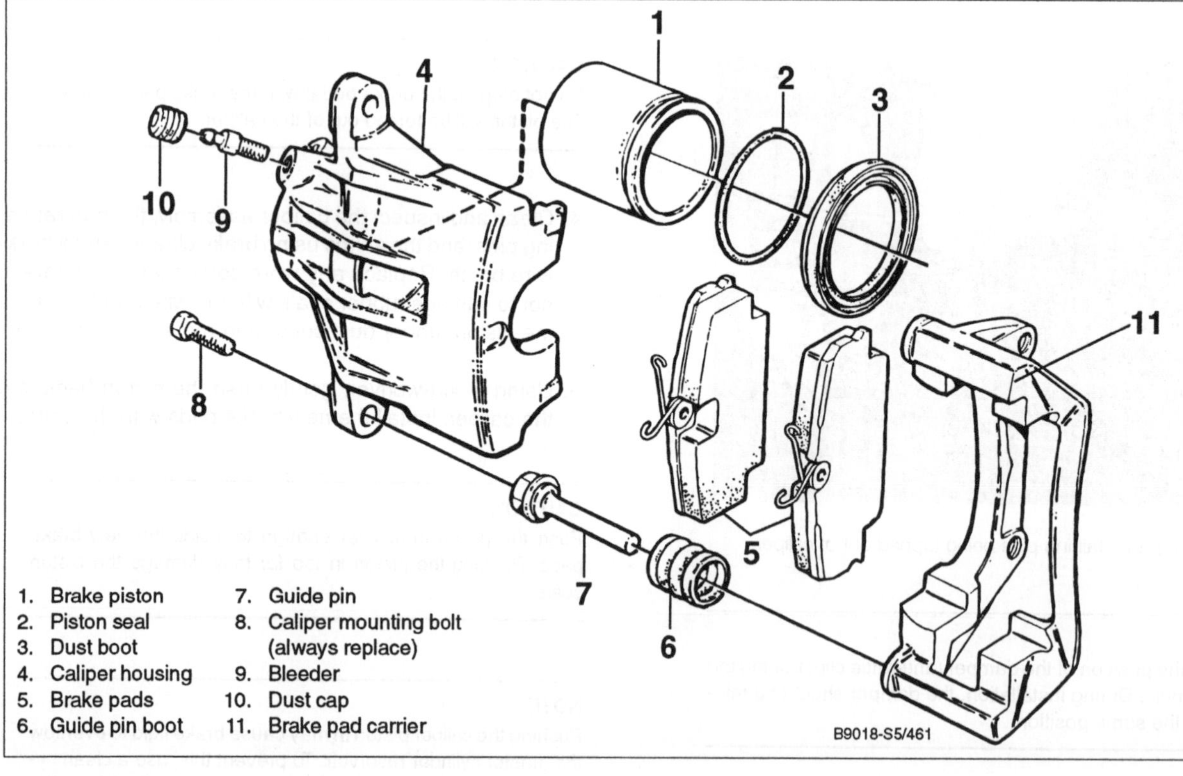

1. Brake piston
2. Piston seal
3. Dust boot
4. Caliper housing
5. Brake pads
6. Guide pin boot
7. Guide pin
8. Caliper mounting bolt (always replace)
9. Bleeder
10. Dust cap
11. Brake pad carrier

Fig. 12. Exploded view of front brake assembly (1988 and later models).

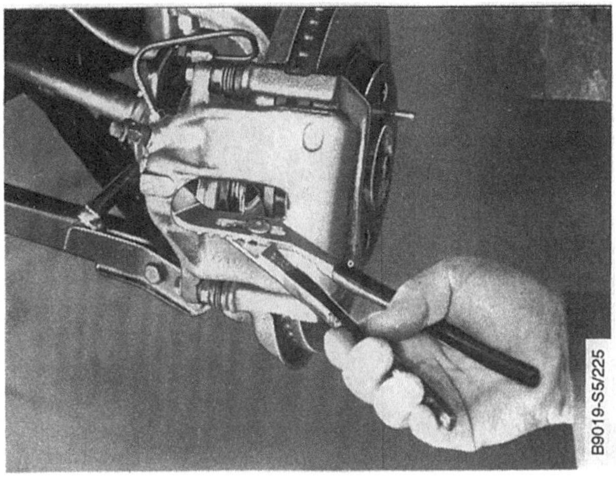

Fig. 13. Piston being pressed into caliper housing.

Fig. 14. Caliper mounting bolt being removed from lower guide pin. Open-end wrench (arrow) being used to hold guide pin stationary.

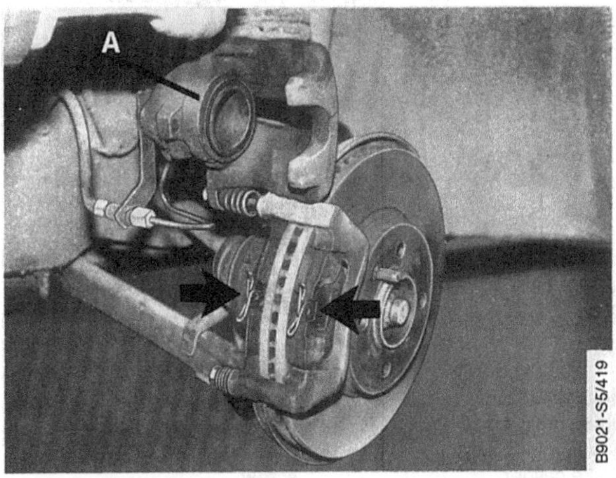

Fig. 15. Caliper (**A**) pivoted up to remove brake pads. Arrows show brake pads.

NOTE

To prevent the piston from creeping out of the caliper, place a strong rubber band around the piston and caliper to hold the piston in.

---

5. Clean and inspect the caliper assembly, the guide pin, and the pad carrier using brake cleaner and a small wire brush. Replace any worn parts. Be careful not to damage rubber seals with the wire brush. Inspect the rubber dust covers for the piston and the guide pins. Replace if necessary. Lightly lubricate the guide pins.

6. Using large pliers and a protective cloth covering the piston, press the piston back into the caliper just enough to install the new pads.

7. Install the new pads into the pad carrier and pivot the caliper down into position. If necessary, push the lower guide pin in as the caliper is lowered into position.

8. Install a new caliper mounting bolt in the lower guide pin and tighten it while holding the guide pin stationary.

WARNING

Do not reuse the caliper self-locking mounting bolts. These are designed to be used only once and should be always be replaced.

Tightening torque

- caliper mounting bolt to guide pin .... 30 to 35 Nm (22 to 26 ft-lb)

---

9. Install the road wheels and lower the car. Gently pump the brake pedal several times to seat the pads. Check the brake fluid level and top up as necessary.

Tightening torque

- wheel lug bolts...... 105 to 125 Nm (77 to 92 ft-lb)

WARNING

New brake pads require some break-in. Allow for slightly longer stopping distances for the first 100 to 150 miles of city driving, and avoid hard stops.

To replace rear brake pads (1988 and later models)

Fig. 16 shows the rear brake assembly used on 1988 and later models.

1. Raise the car and support it with jackstands. Remove the rear wheels.

WARNING
Make sure the car is well supported with jackstands. Never use just a jack to support a car.

---

2. Disconnect the parking brake cable from the lever at the caliper. See Fig. 17.

3. Remove the plug covering the piston adjusting screw. See Fig. 18.

4. Loosen the piston adjusting screw using a 4 mm internal-hex head wrench until the piston moves against its stop. See Fig. 19.

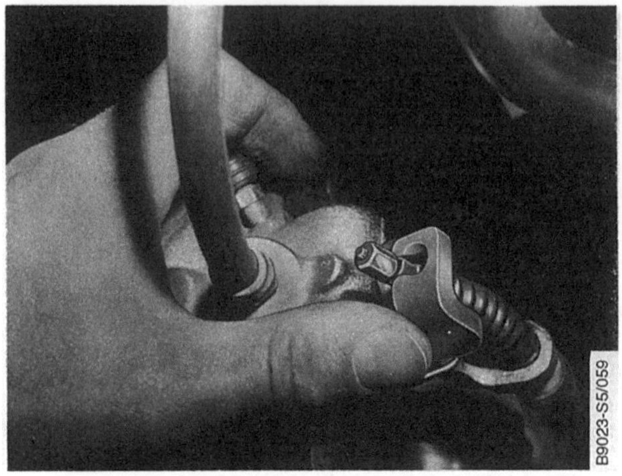

Fig. 17. Parking brake cable being disconnected from caliper.

5. Remove the dust caps from the guide pins then unscrew the upper and lower guide pins using a 7 mm internal-hex head wrench. See Fig. 20.

6. Remove the pad retaining clip. See Fig. 21.

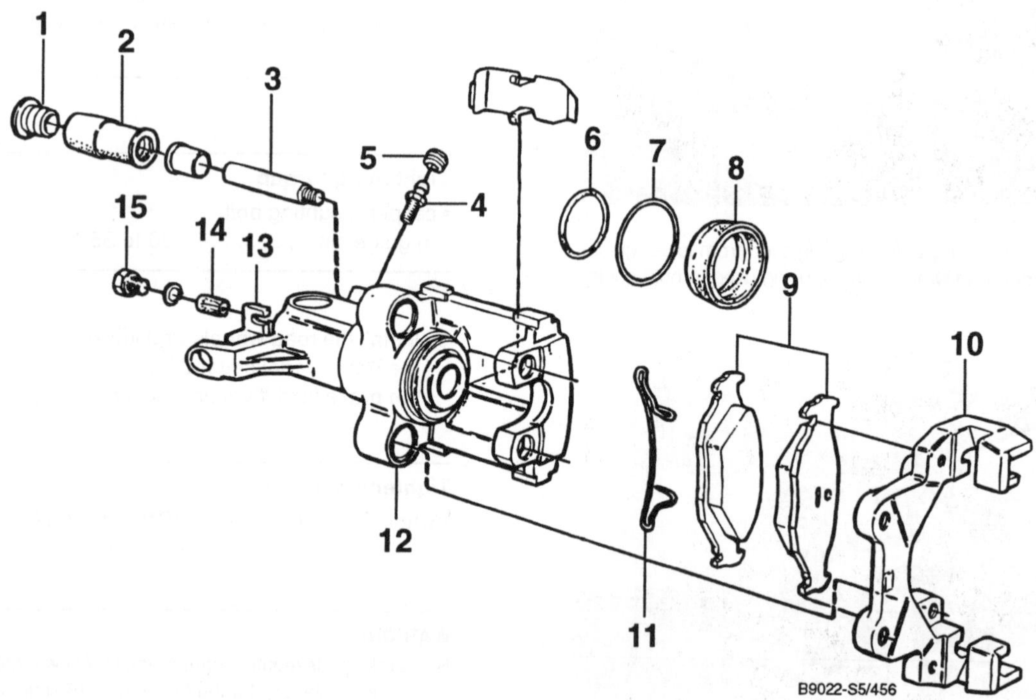

1. Guide pin plug
2. Boot
3. Guide pin (1 of 2 shown)
4. Bleeder
5. Dust cap
6. Piston seal
7. Piston seal
8. Dust boot
9. Brake pads
10. Brake pad carrier
11. Pad retaining clip
12. Brake caliper
13. Parking brake lever
14. Piston adjusting screw
15. Plug for piston adjusting screw

Fig. 16. Exploded view of rear brake assembly used on 1988 and later models.

Fig. 18. Piston adjusting screw plug being removed using wrench (arrow).

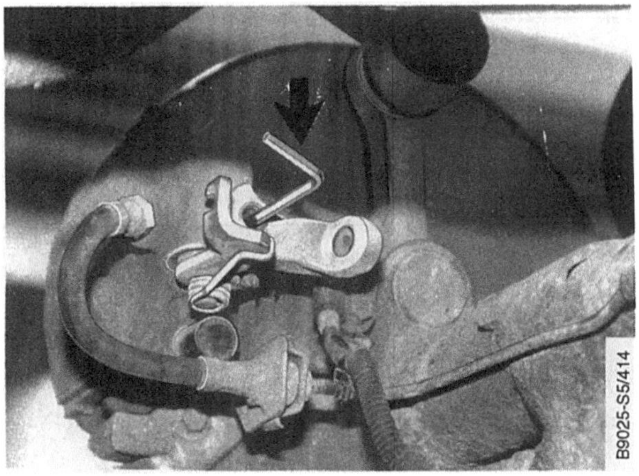

Fig. 19. Internal-hex head wrench (arrow) to loosen piston adjusting screw.

Fig. 20. Lower guide pin being removed from caliper. Upper guide pin shown at arrow.

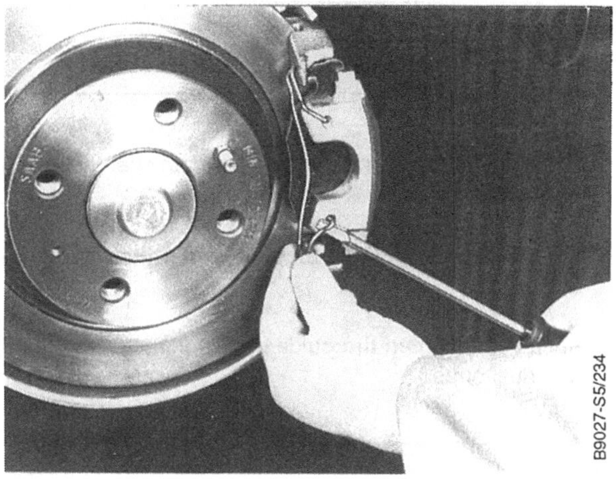

Fig. 21. Pad retaining clip being removed.

7. Remove the caliper from the pad carrier. Remove the brake pads from the carrier. See Fig. 22.

Fig. 22. Caliper being removed from pad carrier. Brake pad with spring clip (arrow) is on piston side of caliper.

CAUTION

Do not step on the brake pedal with the brake pads removed. The piston will be forced out of the caliper.

NOTE

To prevent the piston from creeping out of the caliper, place a strong rubber band around the piston and caliper to hold the piston in.

---

8. Clean and inspect the caliper assembly, the guide pins, and the pad carrier using brake cleaner and a small wire brush. Replace any worn parts. Inspect the rubber dust covers for the piston and the guide pins. Replace if necessary.

CAUTION
Be careful not to damage rubber seals with the wire brush.

9. Using large pliers and a protective cloth over the piston, press the piston back into the caliper.

10. Install the new pads in the pad carrier then install the caliper.

11. Install and tighten the guide pins then install the guide pin dust caps.

NOTE
No lubrication is specified for the rear guide pins.

Tightening torque
- rear caliper to pad carrier (guide pins) .... 25 to 30 Nm (18 to 22 ft-lb)

12. Install the pad retaining clip as shown above in Fig. 21.

13. Turn the piston adjusting screw all the way. See Fig. 19 above. Then back the screw out 1/4 to 1/2 turn. Check that the brake disc rotates freely.

14. Install the plug for the adjusting screw then reconnect the parking brake cable.

15. Check the adjustment of the parking brake cable as described in **551 Parking Brake.** Adjust the cable if necessary.

16. Install the road wheels and lower the car. Gently pump the brake pedal several times to seat the pads. Check the brake fluid level and top up as necessary.

WARNING
New brake pads require some break-in. Allow for slightly longer stopping distances for the first 100 to 150 miles of city driving, and avoid hard stops.

Tightening torque
- wheel to wheel hub (lug bolts) .... 105 to 125 Nm (77 to 92 ft-lb)

520 Brake Fluid and Brake Bleeding

TABLES

GENERAL

Brake fluid has the job of transmitting the force applied at the pedal to the brake pads. In order for the fluid to properly transmit the force, it must be free of air and moisture. Because of the moisture-absorbing properties of brake fluid, it is essential that the fluid be replaced on a regular basis to prevent corrosion of brake parts and brake system failure. See **110 Maintenance Program** for the specified intervals for changing brake fluid.

BLEEDING AND CHANGING BRAKE FLUID

Bleeding is the process used to purge air bubbles or flush contaminated fluid from the system. Bleeding brakes periodically is good preventative maintenance, but it is also essential any time the system has been opened, such as when disconnecting a brake line. A spongy feeling brake pedal is a sign that brakes may need to be bled.

Bleeding the brake system can be done using pressure bleeding equipment, vacuum bleeding equipment, or by conventional means using two people.

WARNING
- Brake fluid is poisonous. Wear safety glasses when working with brake fluid. Wear rubber gloves to prevent brake fluid from entering the bloodstream through cuts of scratches. Do not siphon brake fluid with your mouth.
- Brake fluid is considered a hazardous waste. Catch brake fluid in a suitable container. Dispose of brake fluid in a manner that is mindful of the environment and meets ordinances governing disposal of wastes.
- On cars equipped with ABS, the ABS system is self-pressurizing and maintains very high hydraulic pressure, as much as 210 bar (3045 psi). Before working on any part of the ABS system, switch off the ignition and disconnect the cable from the battery negative (-) terminal.

NOTE
It is good practice to flush the clutch hydraulic system whenever the brake fluid is being changed. See **412 Clutch Control** to bleed the clutch hydraulic system.

Bleeding the Brake System

The procedure below is for bleeding the brakes using a manual method and requires a helper. When using other methods—such as vacuum or pressure bleeding equipment, follow the instructions given by the equipment's manufacturer.

CAUTION
Use only DOT 4 brake fluid. Do not use silicone based (DOT 5) brake fluid.

To bleed brake system (cars without ABS)

Bleeding the brakes or replacing the brake fluid must be done in a particular order as shown in **Table a.**

CAUTION
Brake fluid absorbs moisture from the air. Add only unused fluid to the system and only from well-sealed or previously unopened containers. Do not use fluid expelled from the system, even if it is fresh fluid.

Table a. Order for Bleeding Brakes (Cars without ABS)

1985-1987 models	1988-1989 models
1. Right front wheel	1. Left rear wheel
2. Left rear wheel	2. Right front wheel
3. Left front wheel	3. Right rear wheel
4. Right rear wheel	4. Left front wheel

1. Check the level of the brake fluid in the master cylinder reservoir. If only bleeding the brakes, top up the reservoir using fresh brake fluid from a sealed container.

520

• Brake fluid — DOT 4

NOTE

• When replacing the brake fluid, remove as much of the old fluid from the reservoir as possible, then fill the reservoir with new brake fluid. Plan to have approximately 1.2 liters of new fluid on hand to flush the system.

• Brake fluid is very harmful, especially to paint and rubber parts. Spilled brake fluid should be cleaned up immediately.

2. Starting at the first caliper listed in **Table a,** place a hex wrench over the caliper bleeder. Then connect a length of clear tubing to the bleeder. Place the other end of the tube in a container. See Fig. 1.

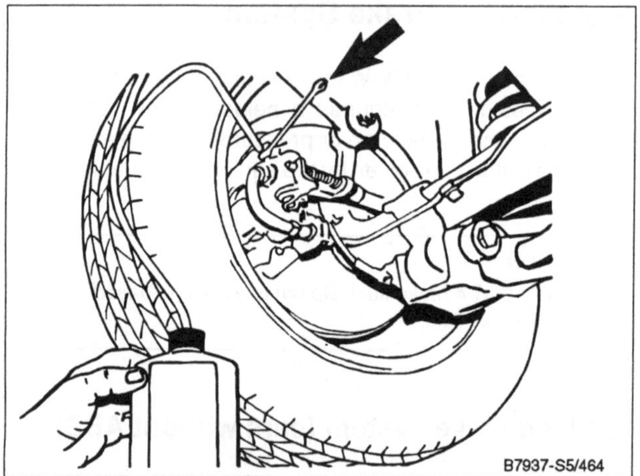

Fig. 1. Tube and container to bleed brake caliper. Arrow shows wrench on bleeder.

NOTE

The bleeder is a small hex-head fitting with a hollow center and usually has a rubber dust boot over it. See Fig. 2.

3. Have a helper pump the brake pedal slowly three times and hold it down.

4. Open (unscrew) the bleeder about 1/2 turn. As the pedal goes to the floor and the fluid stops flowing, close the bleeder. Release the pedal and repeat step 3 until the pedal feels hard or all of the air has been expelled from the system.

NOTE

When replacing the brake fluid, remove at least 1 pint (500 cc) of brake fluid from each caliper to completely expel the old fluid.

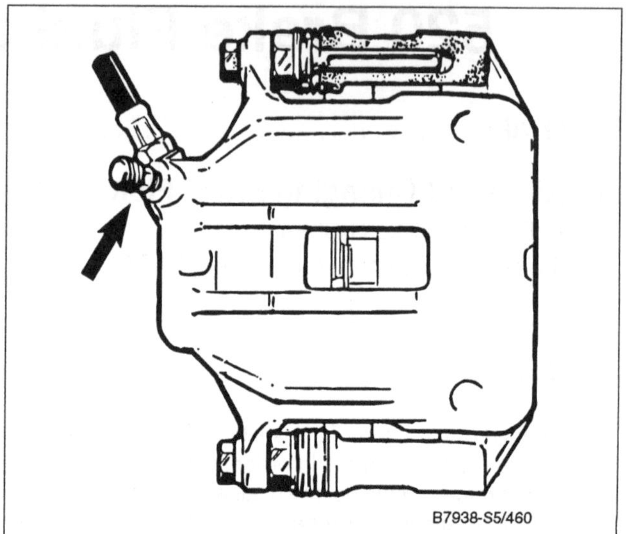

Fig. 2. Front caliper bleeder (arrow).

5. Repeat the above from procedure (step 1 through step 4) bleeding the next caliper in the order given above in **Table a.**

To bleed brake system (cars with ABS)

On cars with ABS the front brakes must be bled first. The front brakes can be bled using pressure bleeding equipment or they can be bled manually. The rear brakes are bled using the ABS pump as described below.

1. Bleed the front brakes using the technique described above. Bleed the right front caliper then the left front caliper.

NOTE

• When replacing the brake fluid, remove as much of the old fluid from the reservoir as possible, then fill the reservoir with new brake fluid.

• When replacing the brake fluid, remove at least 1 pint (500 cc) of brake fluid from each caliper to completely expel the old fluid.

2. Bleed the rear calipers by connecting a length of clear tubing to the bleeder and placing the other end in a container. See Fig. 3.

3. Have a helper turn the ignition key to run (engine off) then press the brake pedal. The ABS pump will come on.

4. Working at the caliper, loosen (unscrew) the bleeder about 1/2 turn and let the fluid flow until it is free of bubbles. Close the bleeder.

Fig. 3. Rear caliper being bled on model with ABS.

5. Check the level of the fluid in the reservoir. If necessary, repeat steps 2 through 4 at the other rear caliper.

CAUTION

- Do not allow the ABS pump to run more than 2 minutes or it may overheat. After running 2 minutes, allow the pump to cool for 10 minutes.

- Do not allow the ABS pump to run without fluid. Damage to the pump may result.

Top up the brake fluid in the reservoir as necessary and replace any dust caps.

521 Brake Master Cylinder

MASTER CYLINDER AND FLUID RESERVOIR

The master cylinder, on cars without ABS, has two chambers that create pressure in two hydraulic circuits. One chamber controls the right front and left rear brakes. The second chamber controls the left front and right rear brakes. This design—known as dual diagonal braking—assures a high level of braking control if one circuit fails. Fig. 1 shows an exploded view of the master cylinder used on cars without ABS.

The master cylinder has an integral brake fluid reservoir. The fluid reservoir has chambers for the brake master cylinder and for the clutch master cylinder (where applicable). The reservoir cap has a float switch that signals the driver when the fluid level is low.

CAUTION
Whenever the brake system is open, plug all openings to prevent dirt from entering the system. Even a minute particle of dirt can cause problems.

On cars with ABS, the master cylinder has three brake circuits: Individual circuits for each front wheel and one circuit for the rear wheels. The master cylinder is integrated with other ABS components to form the ABS unit. See Fig. 2.

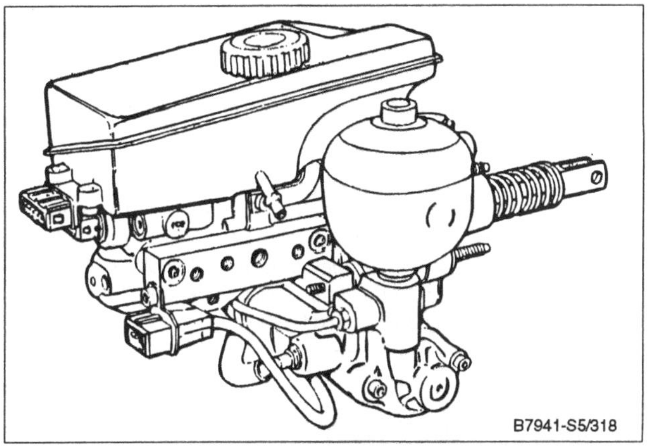

Fig. 2. ABS braking unit with integral master cylinder.

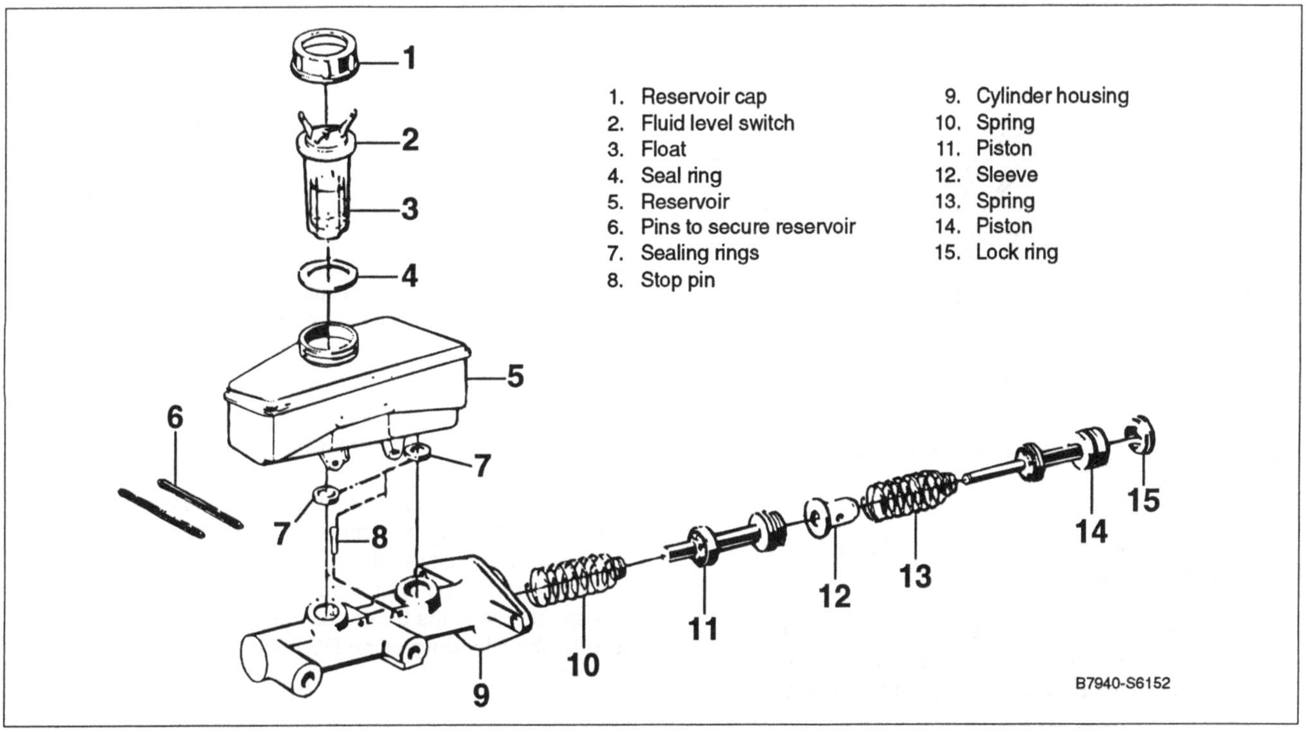

Fig. 1. Exploded view of brake master cylinder.

The master cylinder displaces brake fluid when the brake pedal is depressed. The displaced fluid moves a piston in each caliper, which in turn presses the brake pads against the rotor. The master cylinder must displace fluid without losing pressure or leaking. When it cannot do this, it must be replaced. If there is any leakage from around the master cylinder the master cylinder is probably faulty and should be replaced. For troubleshooting information see **500 Brakes—General**.

NOTE

- Unexplained fluid loss from the reservoir could be due to brake fluid leaking from the master cylinder into the vacuum booster. The fluid is drawn into the engine through the vacuum booster hose. Look for seepage at the master cylinder to vacuum booster flange. Replace both the master cylinder and the vacuum booster as the diaphragm in the booster may be damaged.

- Check the master cylinder by holding the pedal down hard with the car stopped and the engine running. The pedal should feel solid and stay solid. If the pedal slowly falls to the floor, either the master cylinder is leaking internally, or fluid is escaping from the system. If no leaks can be found, the master cylinder is faulty and should be replaced.

To remove master cylinder

1. Disconnect the electrical wires from the fluid reservoir cap.

2. Remove the vacuum hose with check valve from the vacuum booster. See Fig. 3.

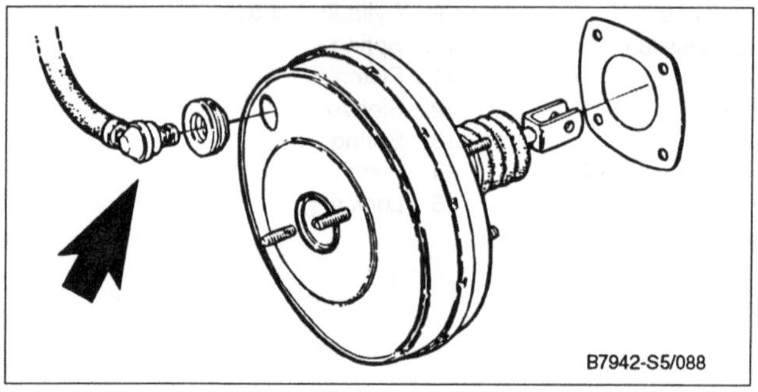

Fig. 3. Vacuum hose and check valve (arrow) removed from vacuum booster.

3. Using a syringe, remove as much brake fluid from the reservoir as possible.

4. On cars with manual transmission, pinch off the fluid supply hose to the clutch master cylinder. See Fig. 4. Remove the supply hose and plug the fittings.

Fig. 4. Special clamping pliers (arrow) used to pinch off supply line to clutch master cylinder.

WARNING

- Brake fluid is poisonous. Wear eye and hand protection when working with brake fluid.

- Brake fluid is considered a hazardous waste. Recycle or dispose of brake fluid in a manner that is mindful of the environment and meets ordinances governing disposal of wastes.

CAUTION

Brake fluid is very harmful, especially to paint and rubber parts. Spilled brake fluid should be cleaned up immediately.

---

5. Using a brake line (flare) wrench, loosen the brake line fittings and remove the two brake lines from the master cylinder. See Fig. 5.

NOTE

To make access to the brake lines easier, remove the mounting bolt holding the coolant expansion tank and move the tank out of the way. It is not necessary to drain the coolant.

---

6. Remove the two nuts holding the master cylinder to the vacuum booster. Remove the master cylinder.

Installation is the reverse of removal. Inspect the seal between the master cylinder and the vacuum booster and replace if necessary. A faulty seal can be the source of a vacuum leak, causing a reduction in braking performance and/or an erratic idle speed. Bleed the brakes as described in **520 Brake Fluid and Brake Bleeding** using DOT 4 brake fluid from a sealed container.

Fig. 5. Brake line being removed. Coolant tank (arrow) has been moved out of way.

CAUTION
Always start the brake line fittings by hand to prevent stripping the threads.

NOTE
It is good practice to flush the clutch hydraulic system whenever the brake fluid is being changed or the master cylinder is being replaced. See **412 Clutch Control** for details on bleeding the clutch hydraulic system.

To remove ABS unit

The ABS unit is removed from the car as a complete assembly. Certain replacement parts for the ABS unit are available from an authorized Saab dealer. See your dealer for the latest parts information.

WARNING
- Dangerously high pressure may be retained in the ABS system, even though the car is not running or has not been driven recently. Never attempt to open or disassemble the pressure accumulator, which contains a nitrogen gas charge and is under high pressure.
- If the ANTI-LOCK and the BRAKE FLUID warning lamps should come on, the car must not be driven. Brake failure may occur. Have the problem corrected promptly by an authorized Saab dealer.

---

1. Depressurize the ABS system before loosening any components by disconnecting the battery negative cable and pressing the brake pedal 25 to 30 times, or until there is a noticeable increase in pedal effort.

2. Remove the lower dashboard pad and footwell heating duct as described in **851 Interior Trim and Upholstery.**

3. Working at the brake pedal assembly, remove the padded insulating material. Then remove the defroster hose from the heater box.

4. Remove the brake pedal pushrod pin locking clip and the pushrod pin from the ABS unit. See Fig. 6.

CAUTION
Inspect the pushrod locking clip and clevis pin and replace them with new ones if they are in anyway damaged or deformed.

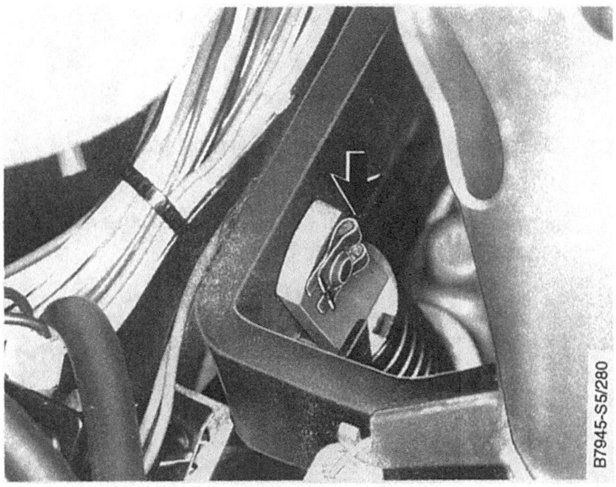

Fig. 6. Brake pedal pushrod locking clip (arrow). Note the orientation of the pin and locking clip.

5. Working at the front of the engine compartment, remove the air intake duct. See Fig. 7.

Fig. 7. Air intake duct being removed.

6. Remove the coolant expansion tank to body mounting bolt and move the tank aside. See Fig. 8. Do not disconnect any cooling hoses or remove the cap from the tank.

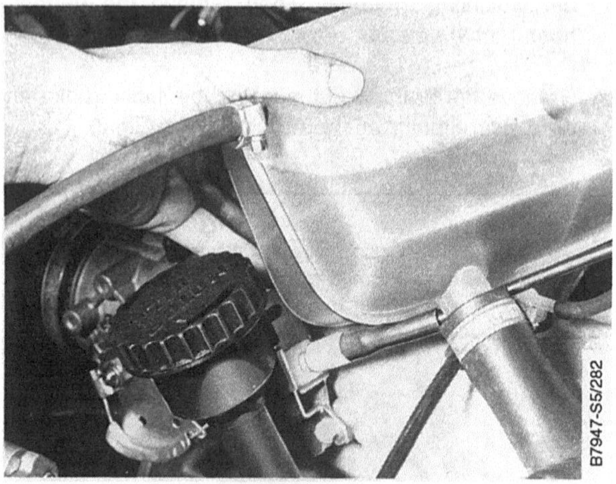

Fig. 8. Coolant expansion tank being moved out of way.

7. Disconnect all electrical connectors from the ABS unit. Remove the ground lead. See Fig. 9.

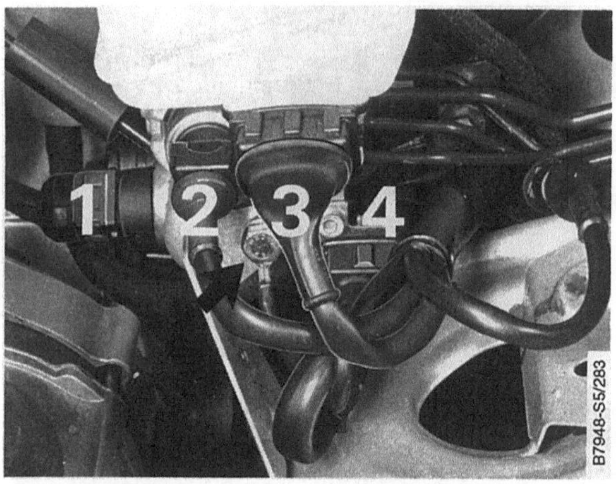

1. Pressure switch
2. Main valve
3. Brake fluid level indicator switch
4. ABS pump motor

Fig. 9. ABS unit electrical connectors. Arrow shows ground lead.

8. Remove the ABS unit mounting bracket. See Fig. 10.

9. Using a syringe, siphon off the brake fluid from the reservoir.

Fig. 10. ABS unit mounting bracket being removed.

WARNING

- Brake fluid is poisonous. Wear eye and hand protection when working with brake fluid.
- Brake fluid is considered a hazardous waste. Recycle or dispose of brake fluid in a manner that is mindful of the environment and meets ordinances governing disposal of wastes.

CAUTION

Brake fluid is very harmful, especially to paint and rubber parts. Spilled brake fluid should be cleaned up immediately.

---

10. On cars with manual transmission pinch off the supply hose to the clutch master cylinder. Remove the hose from the reservoir. See Fig. 11.

Fig. 11. Supply hose to clutch master cylinder being removed. Brake lines can be seen below hose fitting.

11. Label and remove the brake lines from the ABS unit using a brake line (flare) wrench.

NOTE
Plug or cover the brake lines and the openings in the ABS unit to prevent dirt from entering system.

12. Working in the passenger compartment above the pedals, remove the four nuts that hold the ABS unit to the bulkhead. See Fig. 12.

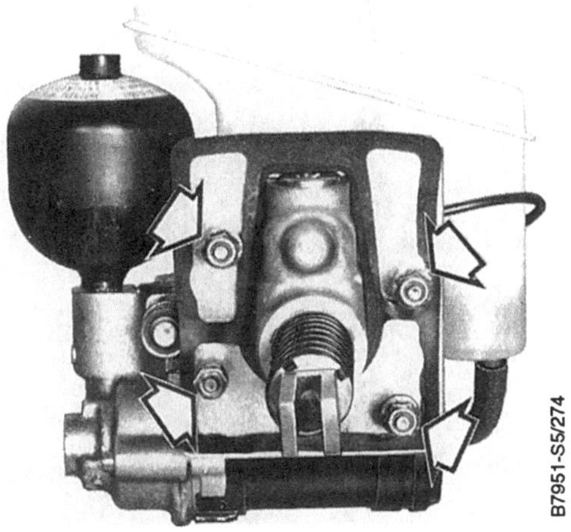

Fig. 12. ABS unit mounting nuts (arrows). Nuts are accessible in passenger compartment above pedal assembly.

To install ABS unit

1. Carefully install the ABS unit in the opening in the bulkhead. Position the ABS unit pushrod so it correctly aligns with the brake pedal.

CAUTION
Use care when installing the ABS unit through the bulkhead. The brake light switch and cruise control switches on the pedal assembly can be easily damaged.

2. Working in the passenger compartment, install the pedal pushrod pin and locking clip through the ABS unit pushrod. See Fig. 6 above.

3. Install and tighten the four new nuts that hold the ABS unit to the bulkhead. See Fig. 12 above.

WARNING
Always use new self-locking nuts to mount the ABS unit to the bulkhead.

Tightening torque
- ABS unit to bulkhead . . . . 26±4 Nm (19±3 ft-lb)

4. Connect the brake lines to the ABS unit. See Fig. 11 above.

CAUTION
Always start the brake line fittings by hand to prevent stripping the threads.

The remainder of installation is the reverse of removal. Fill the reservoir with DOT 4 brake fluid from a sealed container. Switch on the ignition and check that the ABS pump comes on. Bleed the system, including the clutch, as described in **520 Brake Fluid and Brake Bleeding** and **412 Clutch Control**.

Check that there are no leaks and that the warning lights go out after the car is started. Finally, test drive the car to check all brake functions. Recheck the brake fluid level.

522 Brake Lines

Brake Lines and Hoses

The brake lines carry brake fluid and transmit hydraulic pressure to the calipers at each wheel. They are rigid up to their connecting points with the flexible hoses near the wheels. The flexible hoses are used to accommodate suspension and steering movement. Normally, no maintenance is required other than inspecting the lines and hoses for leaks or deterioration. Fig. 1 shows the typical brake line and hose routing.

Brake lines are available from Saab pre-formed to fit without any bending or modification. See an authorized Saab dealer for the latest parts information.

WARNING
- On cars equipped with ABS, dangerously high pressure may be retained in the ABS system even though the car is not running or has not been driven recently. Before loosening any brake line fitting, depressurize the system by disconnecting the battery negative cable and pressing the brake pedal 25 to 30 times, or until there is a noticeable increase in pedal effort.
- Brake fluid is poisonous. Wear hand and eye protection.

WARNING
Brake fluid is considered a hazardous waste. Catch brake fluid in a suitable container and label it. Recycle or dispose of brake fluid in a manner that is mindful of the environment and meets ordinances governing disposal of wastes.

To replace brake lines or brake hoses

1. Clean the connections of the brake line at both ends.

2. Remove any brake line clamps or tie-downs. Loosen the fittings using a brake line (flare) wrench.

NOTE
Brake lines to the rear brakes travel under the carpeting in the passenger compartment on the driver side. See **8 BODY** for information on removing the seats, sill plates, and carpeting.

3. Remove the brake line or hose.

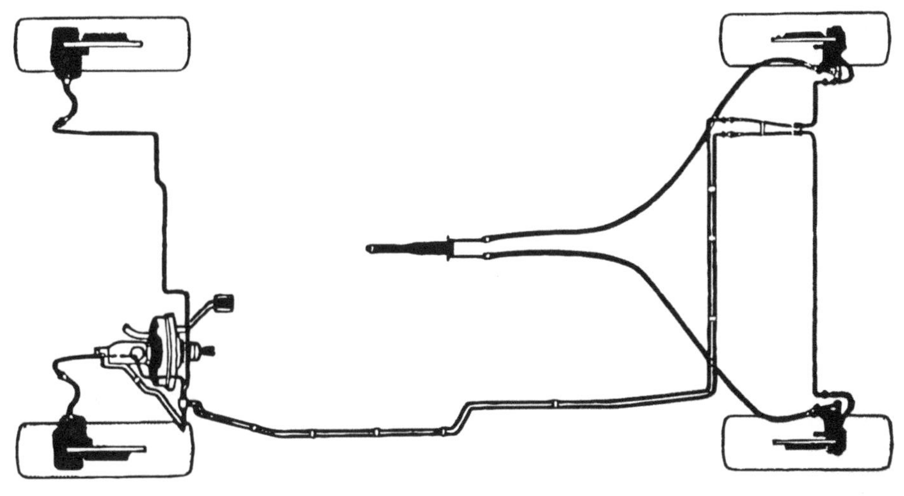

Fig. 1. Brake lines and hoses. Also shown is parking brake with cables.

CAUTION

- Brake fluid is harmful to paint and rubber parts. Spilled brake fluid should be cleaned up immediately.
- Plug the openings in the hydraulic system to prevent dirt from entering the system. Even a minute particle of dirt can cause problems.

4. Clean the inside of the new brake line or hose with compressed air.

5. Place the line or hose in position and hand tighten the fittings. When connecting lines to hoses, counter hold the fittings using two wrenches during tightening.

CAUTION

- Always start the brake line fittings by hand to prevent stripping the threads.
- When fitting flexible hoses, be sure the hose is not twisted and is properly positioned. Check that the hose does not contact any suspension or steering parts.

NOTE

When replacing front flexible brake hoses, jack the car up and support it so the road wheels are freely suspended and pointing straight ahead. This will ensure that there is the proper amount of slack in the hose when the car is jacked up.

6. Tighten the brake fittings using a brake line wrench, and, if necessary, reinstall any clamps or tie-downs.

CAUTION

Never attempt to bend or twist brake lines after the fittings have been tightened. If any lines are at all kinked they must be replaced.

The rest of installation is the reverse of removal. Fill the reservoir with DOT 4 brake fluid from a sealed container. Bleed the system as described in **520 Brake Fluid and Brake Bleeding**. Check that there are no leaks. Finally test drive the car to check all brake functions. Recheck the brake fluid level.

523 Brake Calipers

GENERAL

Two basic brake caliper configurations are used on cars covered by this manual. On 1985 through 1987 cars, the front calipers contain the parking brake assembly. On all 1988 and later cars, the rear calipers contain the parking brake assembly. On all cars, the front calipers are bolted to the steering swivel member while the rear calipers are bolted to the axle.

Brake fluid seeping or leaking from around the caliper piston is the result of a failed or damaged piston seal. The seal can be replaced. However, corrosion, scoring, or pitting on the piston will quickly destroy a new seal. It is recommended that the caliper be replaced to remedy a leaking piston seal and to prevent future problems.

WARNING

- Friction materials in the brake pads may contain asbestos fibers. Avoid breathing any asbestos fibers or dust. Do not create dust by grinding or sanding, or by cleaning the pads or disc with compressed air. Always use a respirator.

- Brake fluid is poisonous. Always wear hand and eye protection.

NOTE

- Saab recommends that the brake pad U-clips and dampers (anti-rattle clips) be replaced whenever replacing brake pads or calipers on 1985 through 1987 cars. The U-clips are prebent to help prevent brake noise (clunking noise when backing up). During normal use, the bends in the clip tend to straighten out and the damper spring tension weakens.

- If the parking brake lever does not fully release, the lever is sticking in the caliper or the lever return spring is damaged. This can lead to brake system failure and premature brake pad wear. If the lever cannot be made to move freely, the caliper must be replaced as described below.

NOTE

Always plug or cover the brake line or hose fittings whenever the system has been opened to prevent dirt from getting into the system. Even a minute particle of dirt can cause problems.

REPLACING BRAKE CALIPERS (1985-1987 MODELS)

To remove front calipers on 1985 through 1987 models

1. Raise the car and support it with jackstands. Remove the front wheels.

WARNING

Make sure the car is well supported on the jackstands. Never use just a jack to support a car.

---

2. Release the parking brake. Remove any rust and debris from the brake disc and caliper. Clean the hose fittings.

3. Working at the caliper, push the parking brake lever towards the sliding yoke and unhook the parking brake cable from the lever. Remove the small circlip then pull the cable from the sliding yoke.

4. Remove the brake pads as described in **517 Brake Pads.**

5. Pinch off the flexible hose using special brake hose clamping pliers. Unscrew the rigid brake line fitting at the brake hose. See Fig. 1.

523

CAUTION
If the special hose clamping pliers are not available, remove the brake fluid from the fluid reservoir using a syringe. Do not clamp the hose using any other type of pliers. Internal damage to the hose may result.

NOTE
Plug the hose fittings to prevent dirt from entering the system.

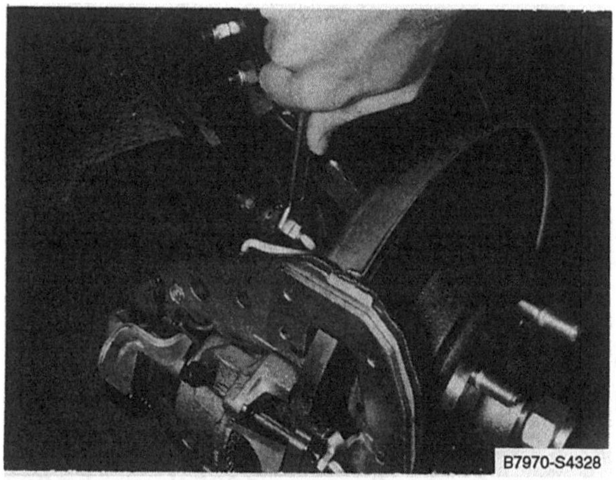

Fig. 1. Brake line fitting being unscrewed.

6. Working from the rear of the caliper, bend the locking plate tabs away from the caliper mounting bolt heads. Remove the two mounting bolts. See Fig. 2.

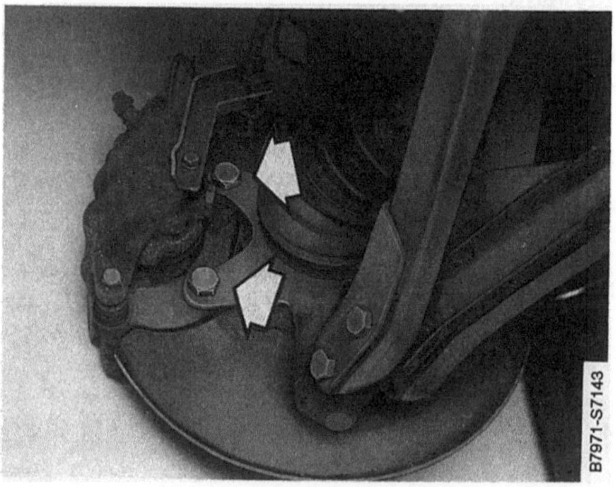

Fig. 2. Front caliper mounting bolts (arrows). Caliper shown is slightly different than those covered here.

Installation is the reverse of removal. Tighten the caliper mounting bolts. Then, using a new locking plate, lock the bolts in position by bending the tabs over the bolt heads. Do not twist or bend the flexible brake hose when connecting the brake line. Install the brake pads and reconnect the parking brake cable as described in **517 Brake Pads**. Bleed the system as described in **520 Brake Fluid and Brake Bleeding.** Adjust the parking brake cables as described in **551 Parking Brake.** Install the wheels and tighten the wheel lug nuts or bolts. Test all brake functions for correct operation.

NOTE
The front calipers have a sliding yoke that requires cleaning and lubrication as part of caliper replacement. Lubricate the sliding surfaces between the caliper and yoke using Gleitmo 540 lubricant, which is available from an authorized Saab dealer.

Tightening torques
- front caliper to steering member .... 110 to 130 Nm (81 to 96 ft-lb)
- wheel lug nuts .... 90 to 110 Nm (66 to 81 ft-lb)

To remove rear calipers on 1985 through 1987 models

Fig. 3. shows an exploded view of rear caliper used on 1985 to 1987 cars.

1. Raise the car and support it with jackstands. Remove the rear wheels.

WARNING
Make sure the car is well supported on the jackstands. Never use just a jack to support a car.

CAUTION
Do not place a jack under the rear axle to raise up the car. The axle may be damaged.

2. Remove any rust and debris from the brake disc and caliper. Clean the hose fittings.

3. Remove the brake pads as described in **517 Brake Pads.**

4. Remove the rigid brake line fitting at the brake caliper.

NOTE
Plug the hose fittings to prevent dirt from entering the system.

5. Bend the locking plate tabs away from the caliper mounting bolt heads. Remove the two mounting bolts holding the caliper to the axle.

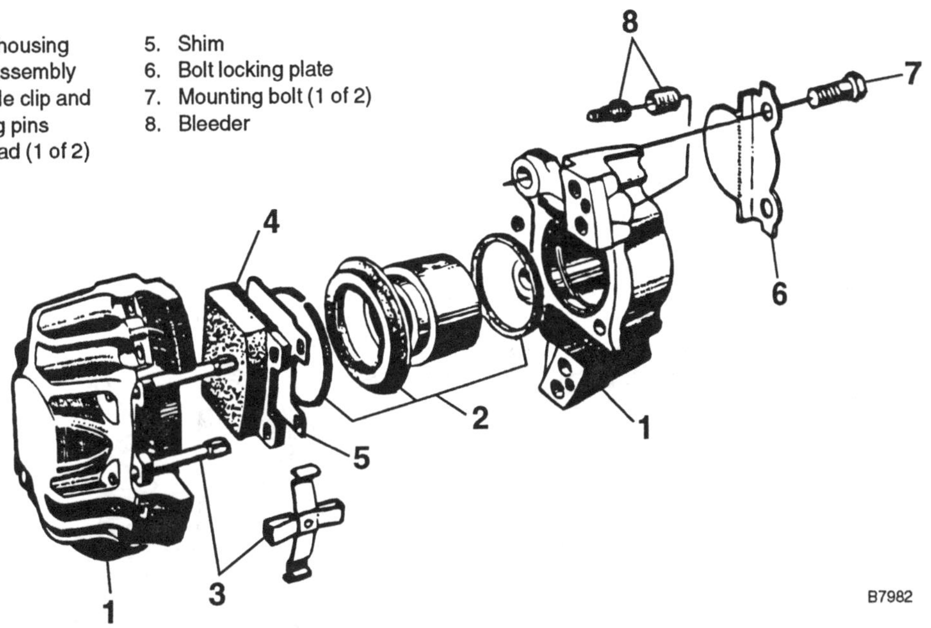

Fig. 3. Exploded view of rear calipers on 1985 to 1987 models.

Installation is the reverse of removal. Tighten the caliper mounting bolts. Then, using a new locking plate, lock the bolts in position by bending the tabs down over the bolt heads. Install the brake pads as described in **517 Brake Pads**. Tighten the brake hose and bleed the system as described in **520 Brake Fluid and Brake Bleeding**. Install the wheels and tighten the wheel lug nuts. Test all brake functions for correct operation.

Tightening torques
- rear caliper to axle . . . . . .70 to 90 Nm (52 to 66 ft-lb)
- wheel lug nuts . . . . .90 to 110 Nm (66 to 81 ft-lb)

REPLACING BRAKE CALIPERS (1988 AND LATER MODELS)

To remove front calipers on 1988 and later models

Fig. 4 is an exploded view of the front caliper used on 1988 and later models.

1. Raise the car and support it with jackstands. Remove the front wheels.

WARNING
Make sure the car is well supported on the jackstands. Never use just a jack to support a car.

---

2. Remove any rust and debris from the brake disc and caliper. Clean the hose fittings.

3. Using special brake hose clamping pliers, pinch off the flexible brake hose. Remove the brake line fitting to the caliper using a brake line (flare) wrench. See Fig. 5.

WARNING
If the special hose clamping pliers are not available, remove the brake fluid from the fluid reservoir using a syringe. Do not clamp the hose using any other type of pliers. Internal damage to the hose may result.

---

4. Remove the bracket that holds the brake hose to the caliper. See Fig. 6.

NOTE
Note how the bracket dimples fit into the caliper.

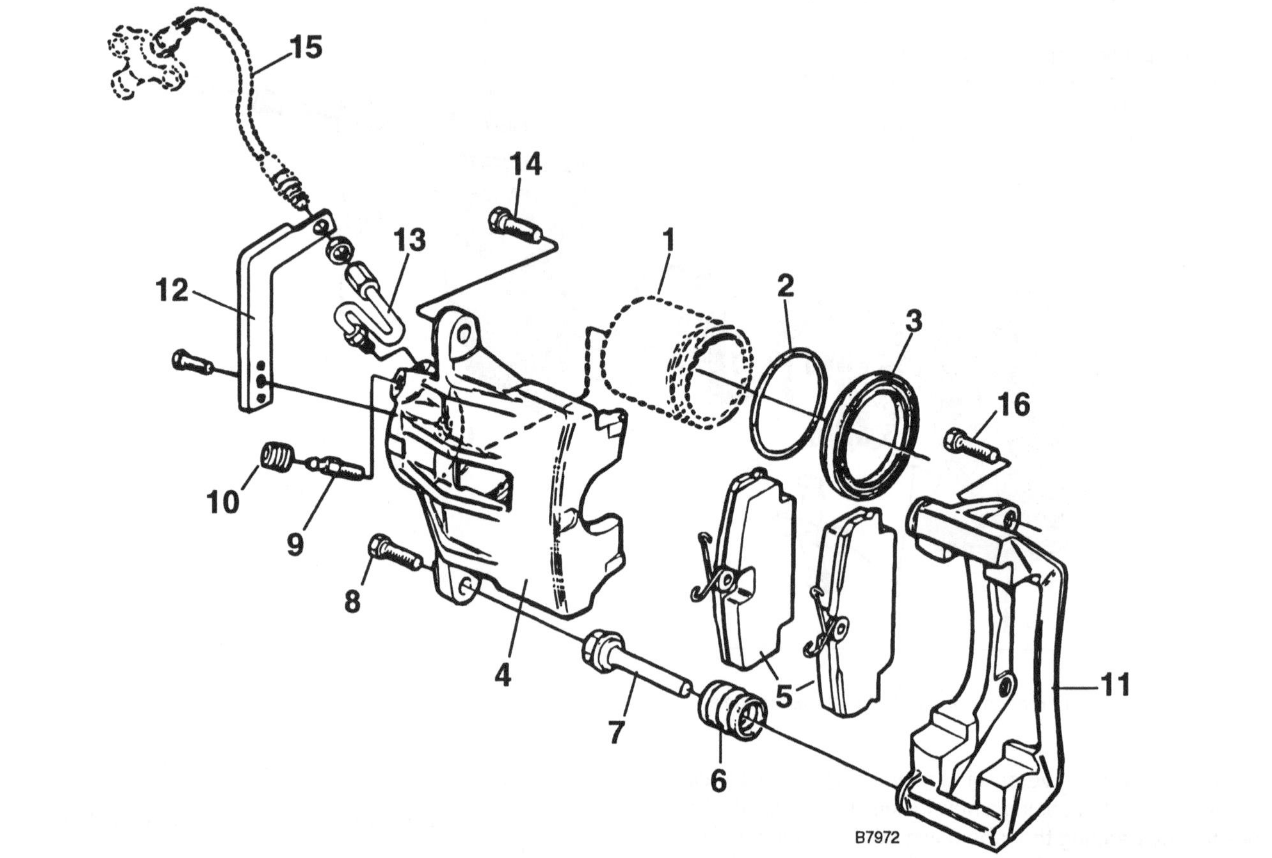

1. Brake piston
2. Piston seal
3. Dust boot
4. Caliper housing
5. Brake pads
6. Guide pin boot
7. Guide pin (1 of 2 shown)
8. Caliper mounting bolt (always replace)
9. Bleeder
10. Dust cap
11. Pad carrier
12. Brake hose bracket
13. Brake line
14. Caliper mounting bolt (always replace)
15. Brake hose
16. Brake pad carrier mounting bolt

Fig. 4. Exploded view of front caliper assembly used on 1988 and later models.

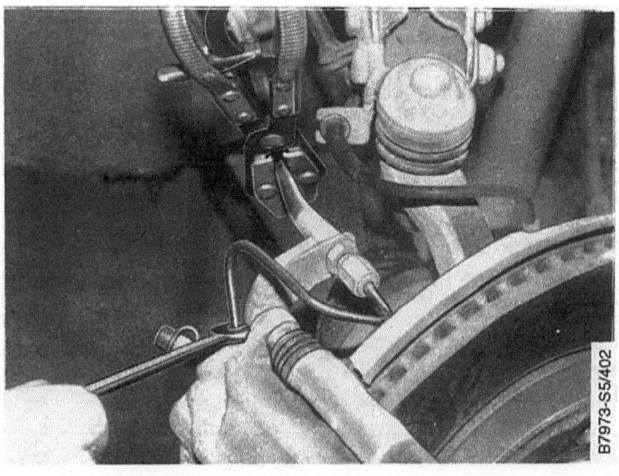

Fig. 5. Brake line fitting being removed.

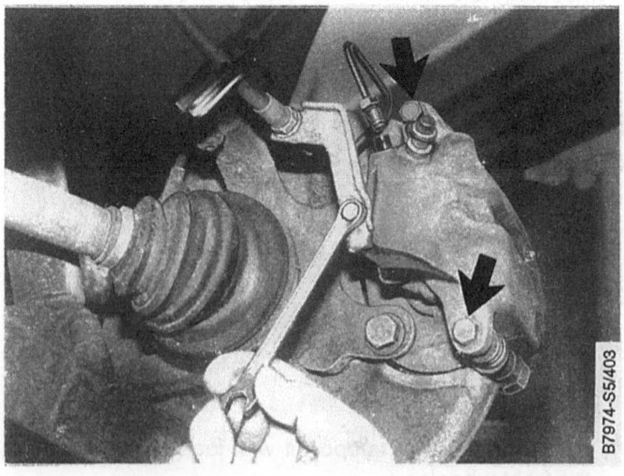

Fig. 6. Brake hose bracket being removed. Arrows show upper and lower caliper mounting bolts.

5. While holding the guide pins stationary, remove the upper and lower caliper mounting bolts. See Fig. 7. Remove the caliper.

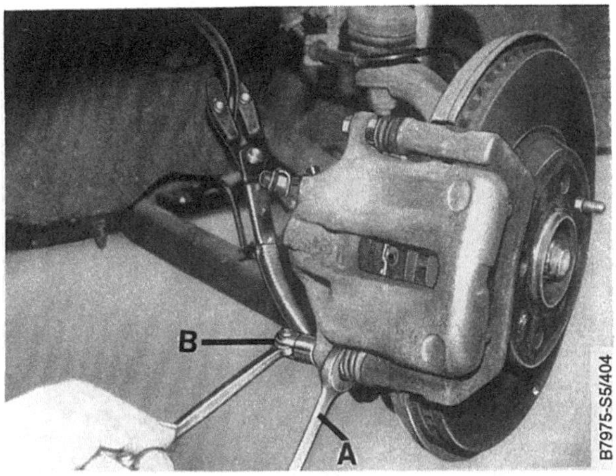

Fig. 7. Hold guide pins stationary using open-end wrench (**A**) when loosening caliper mounting bolts (**B**).

6. Remove the brake pad carrier by removing the two bolts holding it to the steering member.

Installation is the reverse of removal. Do not twist or bend the flexible brake hose when connecting the brake line. Inspect the rubber boots for the guide pins and replace as necessary. Use new self-locking bolts when installing the caliper to the pad carrier. Install the brake pads as described in **517 Brake Pads**. Bleed the system as described in **520 Brake Fluid and Brake Bleeding**. Install the wheels and tighten the wheel lug bolts. Test all brake functions for correct operation.

CAUTION

Always replace the front caliper mounting bolts with new ones. These bolts are designed to be used only once and may fail when used a second time.

Tightening torques

- caliper mounting bolt to guide pin .... 30 to 35 Nm (22 to 26 ft-lb)
- front pad carrier to steering member (1988 and later models)..... 70 to 110 Nm (52 to 81 ft-lb)
- wheel lug bolts...... 105 to 125 Nm (77 to 92 ft-lb)

To remove rear calipers on 1988 and later models

Fig. 8. shows the rear caliper used on 1988 and later models.

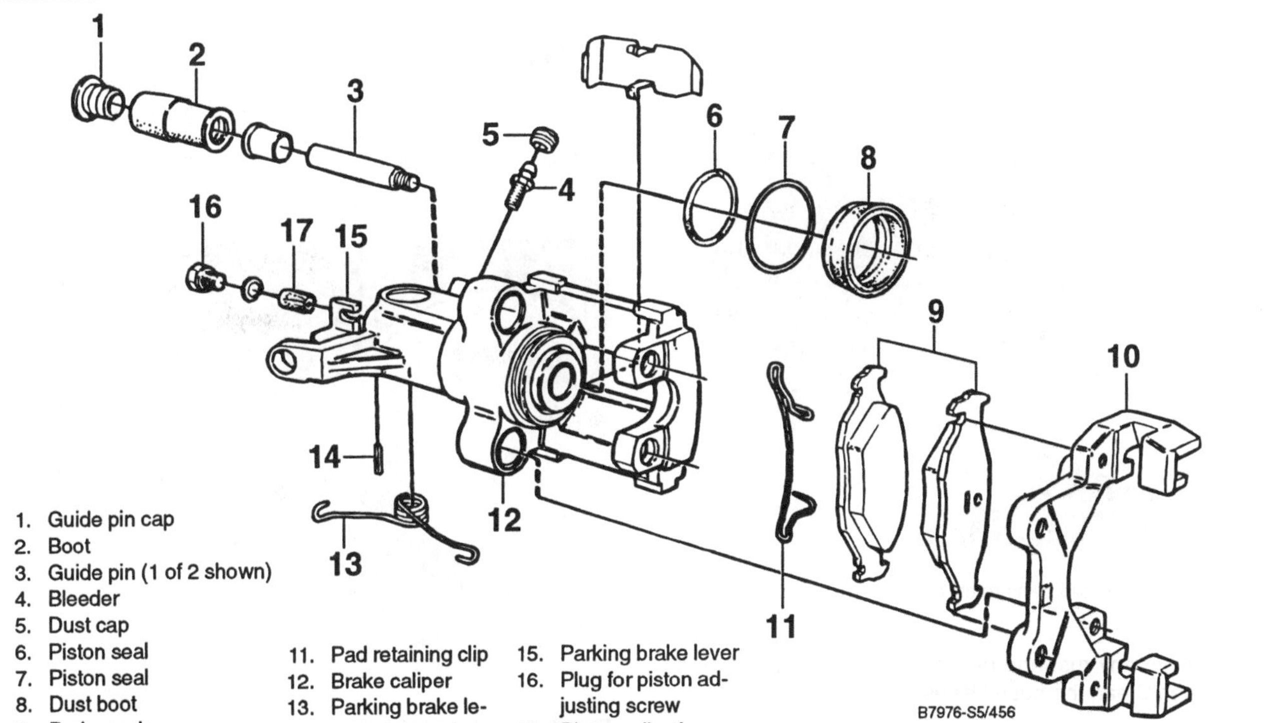

1. Guide pin cap
2. Boot
3. Guide pin (1 of 2 shown)
4. Bleeder
5. Dust cap
6. Piston seal
7. Piston seal
8. Dust boot
9. Brake pads
10. Brake pad carrier
11. Pad retaining clip
12. Brake caliper
13. Parking brake lever return spring
14. Pin
15. Parking brake lever
16. Plug for piston adjusting screw
17. Piston adjusting screw

Fig. 8. Exploded view of rear caliper assembly used on 1988 and later models.

1. Raise the car and support it with jackstands. Remove the rear wheel.

WARNING
Make sure the car is well supported on the jackstands. Never use just a jack to support the car.

2. Unhook the parking brake cable from the caliper lever. See Fig. 9. Pull the cable out of the caliper housing.

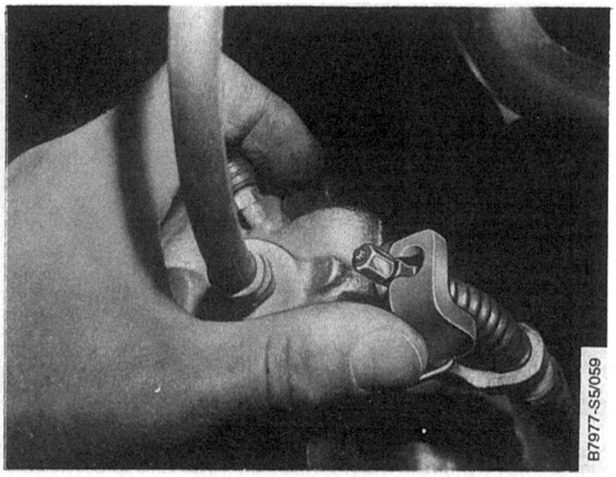

Fig. 9. Parking brake cable being unhooked from caliper.

3. Remove the pad retaining clip. See Fig. 10.

Fig. 10. Pad retaining clip being removed.

4. Remove any rust and debris from the brake disc and caliper. Clean the hose fittings.

5. Remove the threaded screw plug covering the piston adjusting screw. Then unscrew the piston adjusting screw using a 4 mm internal-hex head wrench. See Fig. 11.

NOTE
The piston caliper adjusting screw positions the piston to the pads. The screw needs to be backed out of the caliper so that the piston can be pushed away from the pads. More information on the piston adjusting screw is given in **517 Brake Pads**.

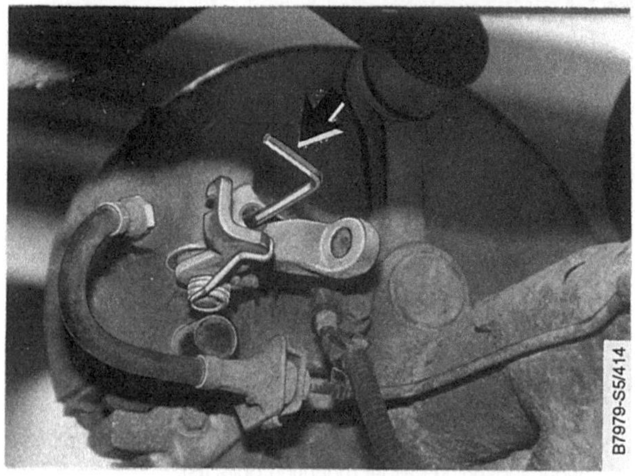

Fig. 11. Internal-hex head wrench (arrow) used to loosen caliper piston adjusting screw.

6. Using special brake hose clamping pliers, pinch off the flexible brake hose. Loosen slightly, but do not remove, the brake hose. See Fig. 12.

WARNING
If the special hose clamping pliers are not available, remove the brake fluid from the fluid reservoir using a syringe. Do not clamp the hose using any other type of pliers. Internal damage to the hose may result.

Fig. 12. Brake hose being loosened at rear caliper.

7. Remove the guide pin dust caps and then remove the two guide pins using a 7 mm internal-hex head wrench. See Fig. 13.

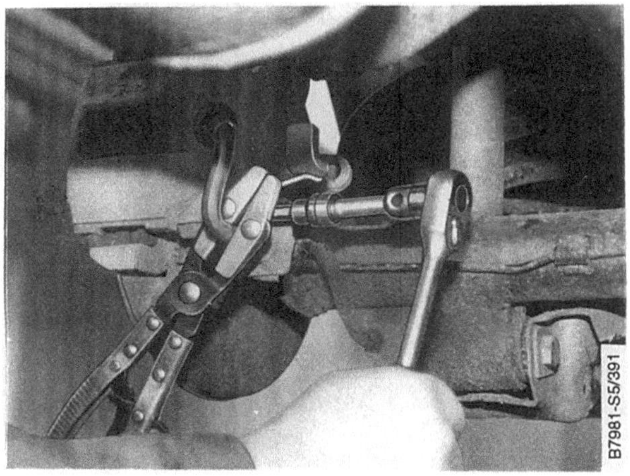

Fig. 13. Lower caliper guide pin being removed. Remove upper guide pin also.

8. Lift the caliper housing off the pad carrier and remove the brake pads.

9. While holding the flexible brake hose stationary, unscrew the caliper. Plug the hose and caliper openings to keep out dirt.

10. Remove the two bolts holding the pad carrier to the axle then remove the two bolts holding the dust shield. See Fig. 14.

Installation is the reverse of removal. Reinstall the brake pads and adjust the piston as described in **517 Brake Pads.** Tighten the hose and bleed the system as described in **520 Brake Fluid and Brake Bleeding.** Reconnect the parking brake cable and check the cable adjustment as described in **551 Parking Brake.** Install the wheels and tighten the wheel lug bolts. Test all brake functions for correct operation.

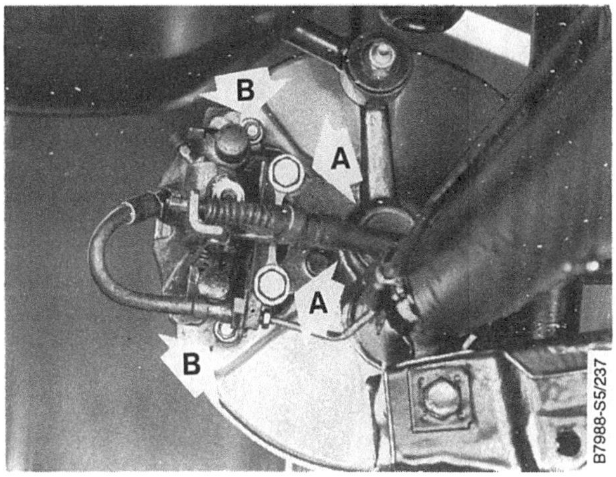

Fig. 14. Pad carrier bolts (**A**) and dust shield bolts (**B**).

Tightening torques

- rear caliper to pad carrier (pins). . . . . 25 to 30 Nm (18 to 22 ft-lb)
- brake pad carrier to rear axle (1988 and later models). . . . . . 40 to 54 Nm (30 to 40 ft-lb)
- wheel lug bolts. . . . . . 105 to 125 Nm (77 to 92 ft-lb)
- all other fasteners
  - M5 bolt . . . . 5 Nm (44 in-lb)
  - M6 bolt . . . . 10 Nm (89 in-lb)
  - M8 bolt . . . . 20 Nm (15 ft-lb)
  - M10 bolt . . . . 40 Nm (30 ft-lb)

524 Brake Pedal Mechanism

GENERAL

The brake pedal mechanism consists of the brake pedal and connecting linkage that transmits the mechanical force from the driver's foot to the master cylinder. The mechanism is located next to the steering column in the passenger compartment. See Fig. 1.

WARNING
On cars equipped with an SRS airbag system, do not attempt to service the wiring in the steering wheel, steering column assembly, or airbag control unit without first disarming the SRS system. The airbag may ignite causing injury. See an authorized Saab dealer for any repairs necessary to the SRS wiring or components.

To remove brake pedal

1. Remove the lower dashboard pad and footwell heating duct as described in **851 Interior Trim and Upholstery**.
2. Working at the brake pedal assembly, remove the padded insulating material then remove the defroster hose from the heater box.
3. Unhook the brake pedal return spring. See Fig. 2.

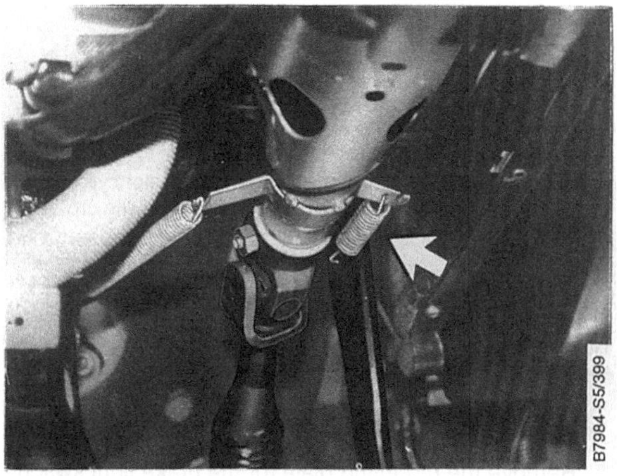

Fig. 2. Brake pedal return spring (arrow).

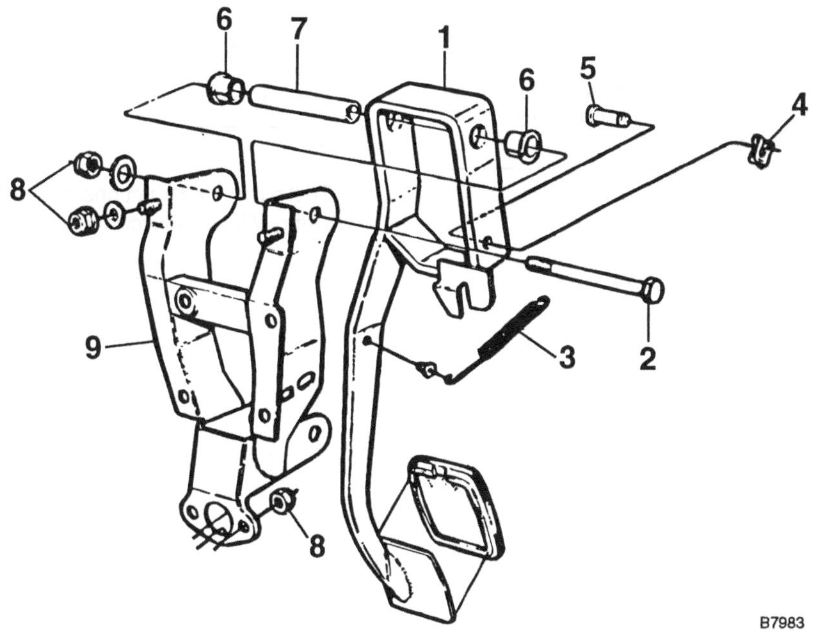

1. Brake pedal
2. Pedal spindle bolt
3. Return spring
4. Locking clip (cotter pin used on 1985 models, not shown)
5. Pushrod clevis pin
6. Bushing
7. Spacer tube
8. Self-locking nuts (always replace)
9. Bracket

Fig. 1. Exploded view of brake pedal assembly.

4. Remove the locking clip (or cotter pin) and the pushrod pin for the master cylinder pushrod. See Fig. 3.

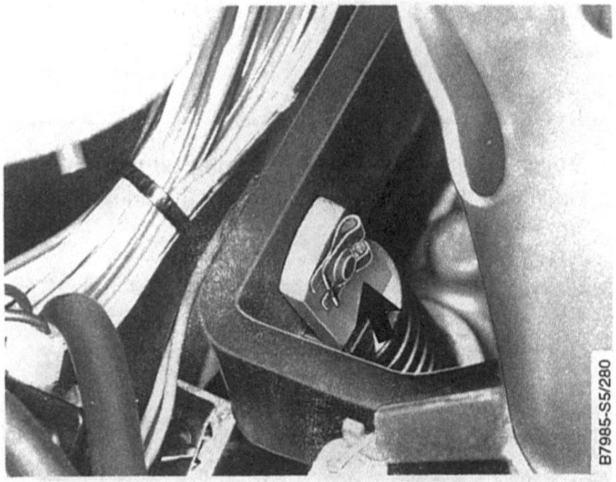

Fig. 3. Locking clip (arrow) for pushrod pin on master cylinder. 1985 models use a cotter pin instead of a locking clip.

NOTE

Note the orientation of the pushrod pin and locking clip (or cotter pin).

---

5. Remove the mounting plate for the brake light switch.

6. Remove the self-locking nut for the pedal spindle bolt. Remove the bolt.

WARNING

Always replace the self-locking nut with a new one. Self-locking nuts are designed to be used only once and may loosen and fall off when used a second time.

---

7. Remove the brake pedal. It may be necessary to rotate the pedal as it is being removed. Be sure to save the bushings and spacer tube. See Fig. 4.

NOTE

Inspect the bushings and spacer tube for any signs of wear and replace as necessary.

Fig. 4. Brake pedal being removed.

Installation is the reverse of removal. Always replace self-locking nuts with new ones. On 1985 models, use a new cotter pin. Adjust the brake light switch as described in **364 Electric Controls and Switches**.

CAUTION

Inspect the pushrod locking clip and replace it with a new one if it is in anyway damaged or deformed.

Test all braking functions and check that the brake lights are working. Where applicable, test cruise control functions.

541 Brake Vacuum Booster

GENERAL

The vacuum booster is used to assist braking by reducing the effort required of the driver to operate the master cylinder. Engine manifold vacuum, ported to the booster, acts on a diaphragm to boost the driver's own force on the master cylinder. See Fig. 1. The vacuum booster is a sealed unit and cannot be repaired.

NOTE
A vacuum booster is not used on cars with ABS. Instead, the brake assist is handled through the ABS unit. See **500 Brakes—General** for more information on the ABS system.

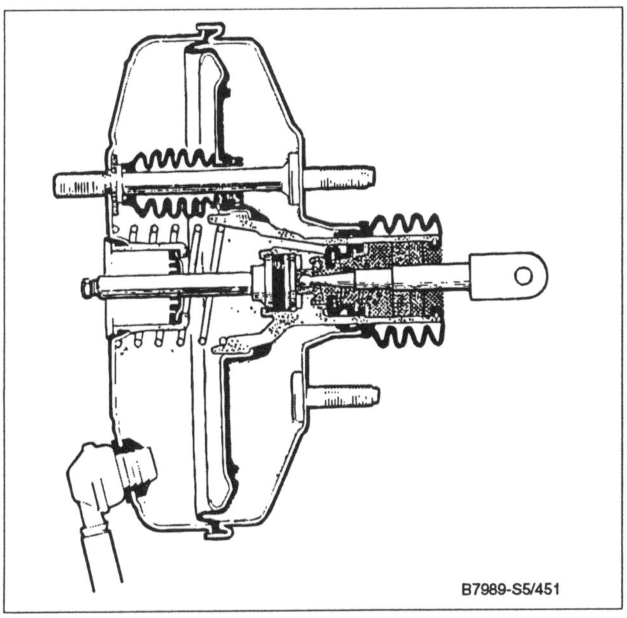

Fig. 1. Vacuum booster unit.

REPLACING VACUUM BOOSTER

The vacuum booster should be replaced if braking pedal effort is too high from a fault with the vacuum booster. If the booster is unable to hold vacuum after the engine is stopped, the vacuum check valve and seal may be defective or there may be a leak in the seal between the vacuum booster and the master cylinder.

A quick-check of the vacuum booster can be made by pumping the brake pedal approximately 10 times with the engine off. Then while holding the pedal down, start the engine. the pedal should fall slightly. If not, the booster may be faulty.

NOTE
For a more accurate check of the vacuum booster and check valve, create a vacuum in the booster using a hand-held vacuum pump. Apply vacuum at the vacuum hose from the intake manifold. The system should not leak. See Fig. 2.

Fig. 2. Vacuum booster unit and check valve (**A**). Apply vacuum at hose (arrow) to test unit.

Replacing the vacuum booster involves disconnecting the pushrod linkage at the brake pedal and removing the brake master cylinder first.

To remove vacuum booster unit

1. Disconnect the battery negative (–) cable.

2. Remove the lower dashboard pad and footwell heating duct as described in **851 Interior Trim and Upholstery**.

WARNING
On cars equipped with an SRS airbag system, do not attempt to service the wiring in the steering wheel, steering column assembly, or airbag control unit without first disarming the SRS system. The airbag may ignite causing injury. See an authorized Saab dealer for any repairs necessary to the SRS wiring or components.

---

3. Working at the brake pedal assembly, remove the padded insulating material then remove the defroster hose from the heater box.

4. Remove the locking clip and the pushrod pin for the vacuum booster pushrod. See Fig. 3.

NOTE

- 1985 models use a split (cotter) pin rather than a locking clip.
- Inspect the pushrod pin for wear and replace if necessary.

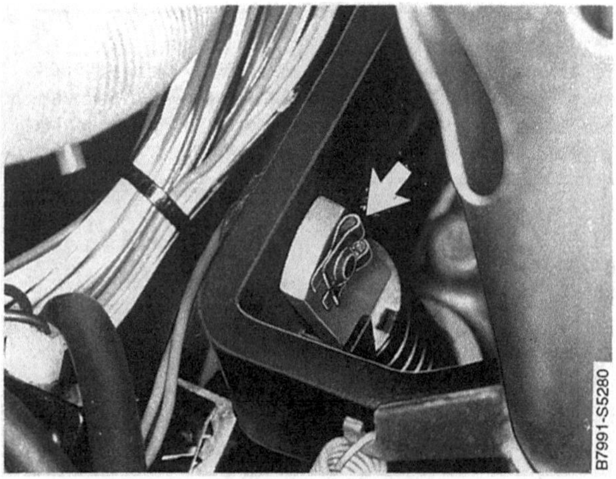

Fig. 3. Locking clip (arrow) for pushrod pin on master cylinder.

NOTE

Note the orientation of the pushrod pin and locking clip so they can be reinstalled correctly.

---

5. Working in the engine compartment, disconnect the vacuum supply hose from the brake booster. See Fig. 2 above.

6. Remove the master cylinder as described in **521 Brake Master Cylinder.**

7. Working in the passenger compartment, remove the four nuts holding the vacuum booster to the bulkhead. Remove the vacuum booster from the engine compartment.

Installation is the reverse of removal. Inspect the seal between the vacuum booster and the master cylinder and replace as necessary. Replace any self-locking nuts with new ones. Bleed the master cylinder as described in **521 Brake Master Cylinder.** Test all braking functions.

551 Parking Brake

GENERAL

The parking brake system, working through cables, moves a lever at the calipers to force the brake pads against the brake disc. On 1985 through 1987 models, the parking brake mechanism operates through the front calipers. On 1988 and later models, the parking brake mechanism operates through the rear calipers. The parking brake system is self-adjusting, so that as the pads wear the play between the pads and the brake disc is taken up automatically and the clearance is kept constant. The system should be periodically checked for proper operation.

PARKING BRAKE ASSEMBLY

The parking brake mechanically actuates the brake calipers and is independent of the hydraulic brake system. The ratcheting parking brake lever operates two separate, replaceable cables that connect to the calipers. If the parking brake mechanism at the caliper is damaged or seized, the caliper must be replaced.

NOTE
For additional information on checking the parking brake lever on the caliper, see **517 Brake Pads**. Information on replacing the calipers can be found in **523 Brake Calipers**.

To remove parking brake cables (1985-1987 models)

Fig. 1 shows the front parking brake system.

1. Raise the car and support it with jackstands.

WARNING
Make sure the car is well supported on the jackstands. Never use just a jack to support a car.

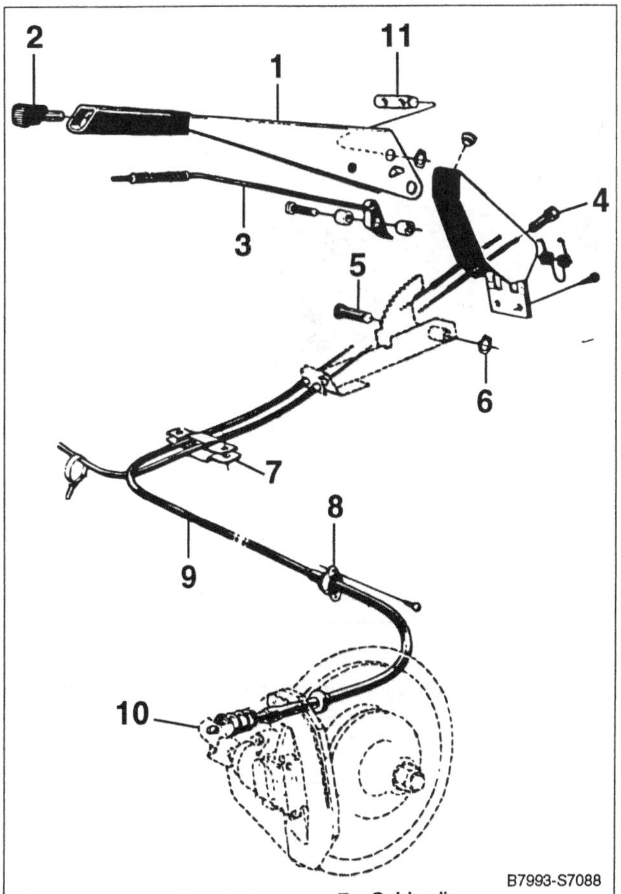

1. Parking (hand) brake lever
2. Release button
3. Release link
4. Cable adjusting nut (1 of 2)
5. Pin
6. Circlip
7. Cable clip
8. Wheel housing grommet
9. Cable
10. Caliper lever
11. Pivot pin

Fig. 1. Parking brake system used on 1985 through 1987 models.

2. Release the parking brake.

3. Remove the driver's seat and sill plates, then pull back the front carpeting to expose the heating duct. See **851 Interior Trim and Upholstery**.

4. Remove the center console beneath the handbrake lever as described in **432 Manual Transmission Controls**. Where applicable, remove the center storage console as described in **853 Dashboard and Consoles.**

5. Remove the center heating duct.

6. Remove the adjusting nuts from the cable ends at the parking brake lever. See Fig. 2.

Fig. 2. Parking brake cable adjusting nuts (arrow). Center console removed.

NOTE
Individual cables are available as replacement parts.

---

7. Remove the cable hold-down clip that anchors the cables to the floor. See Fig. 1 above.

NOTE
The cables cross over each other just ahead of the cable clip.

---

8. Working at the wheel housing, remove the rubber grommet that holds the cable to the body. Unhook the cable end from the caliper lever. Remove the small circlip the holds the cable to the caliper and slide the cable out of the caliper. See Fig. 3.

9. Working below the engine compartment, cut any cable ties from the parking brake cable. Then pull the cables through the bulkhead grommet and out of the passenger compartment. Pull the other end of the cable through the opening in the wheel housing.

Fig. 3. Parking brake cable (arrow) being disconnected at caliper lever.

NOTE
Note how the cable runs through the engine compartment so the new cable can be installed in the same location.

To install parking brake cables (1985-1987 models)

1. Working below the engine, inspect the rubber grommet in the bulkhead and replace it if necessary.

2. Install the cables through the bulkhead grommet into the passenger compartment.

NOTE
It may help to spray the rubber grommet and cable jacket with silicone spray as an aid to installation.

---

3. Working in the passenger compartment, install the cable ends into the pivot pin in the handbrake lever and thread the adjusting nuts on to the cable. See Fig. 2 above.

NOTE
Check that the cables cross over each other just inside the passenger compartment.

---

4. Install the cable hold-down clip over the cables.

5. Working under the car, feed the other end of the cable through the opening in the wheel housing and connect the cable end to the caliper lever. See Fig. 3 above.

The rest of the installation is the reverse of removal. Replace any cable ties as necessary. Apply the parking brake lever several times to stretch the cables then adjust the cables as described below. Check that the parking brake warning lamp operates correctly.

NOTE

Adjust the parking brake as described below before reinstalling the center console. The adjusting nuts cannot be reached as easily after the console has been fitted.

To adjust parking brake (1985-1987 models)

1. With the parking brake fully released, back off the adjusting nuts at the hand brake lever until no tension is on the actuating lever at the caliper. Then insert a 0.5 mm feeler gauge between the caliper lever and yoke as shown in Fig. 4.

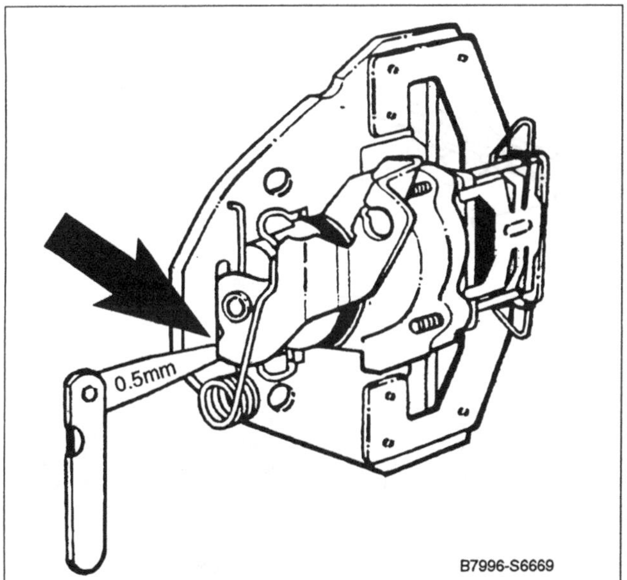

Fig. 4. Parking brake lever gap being measured at caliper. Cable must be connected to lever.

parking brake lever to yoke clearance	0.5±0.1 mm (0.019±0.003 in.)

2. Then working at the hand brake lever, adjust the parking brake cable adjusting nut(s) in the passenger compartment just until the feeler gauge falls out of the caliper. Recheck the gap with the feeler gauge.

NOTE

While a helper pulls up and then releases the handbrake lever in the passenger compartment, check that the caliper lever fully releases. If not, the lever is sticking in the caliper or the lever return spring is damaged. If the lever cannot be made to move freely, the caliper should be replaced. See **523 Brake Calipers.**

Check that the parking brake is fully applied when the hand brake lever is pulled up 9 to 10 notches (clicks). Check that the parking brake warning lamp works correctly.

To remove parking brake cables (1988 and later models)

Fig. 5 is an exploded view of the parking brake system used on 1988 and later models.

1. Remove the center console beneath the hand brake lever as described in **432 Manual Transmission Controls**. Release the parking brake after removing the center console.

2. Remove the rear seat cushion and rear sill plates as described in **851 Interior Trim and Upholstery**. Pull back the rear carpeting.

3. On 1991 and later models, remove the sound absorbing padding and the metal cover over the handbrake cables. See Fig. 6.

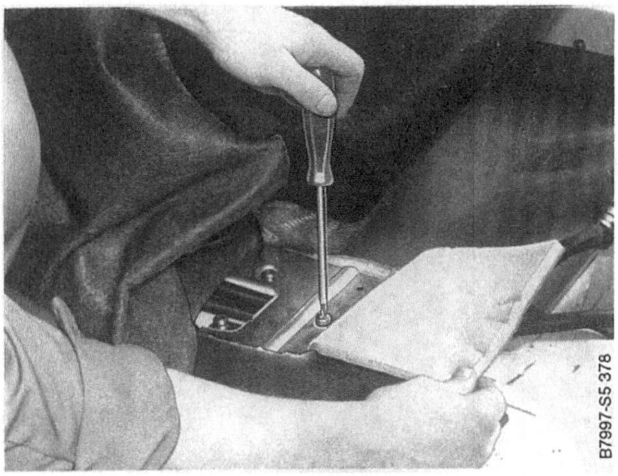

Fig. 6. Handbrake cable metal cover mounting screws being removed on 1991 and later models. Cover is mounted with 3 screws.

4. Remove the adjusting nut from the end of the cable. See Fig. 7.

5. Remove the cable locking plate and slide the cable out of the parking brake console. See Fig. 8.

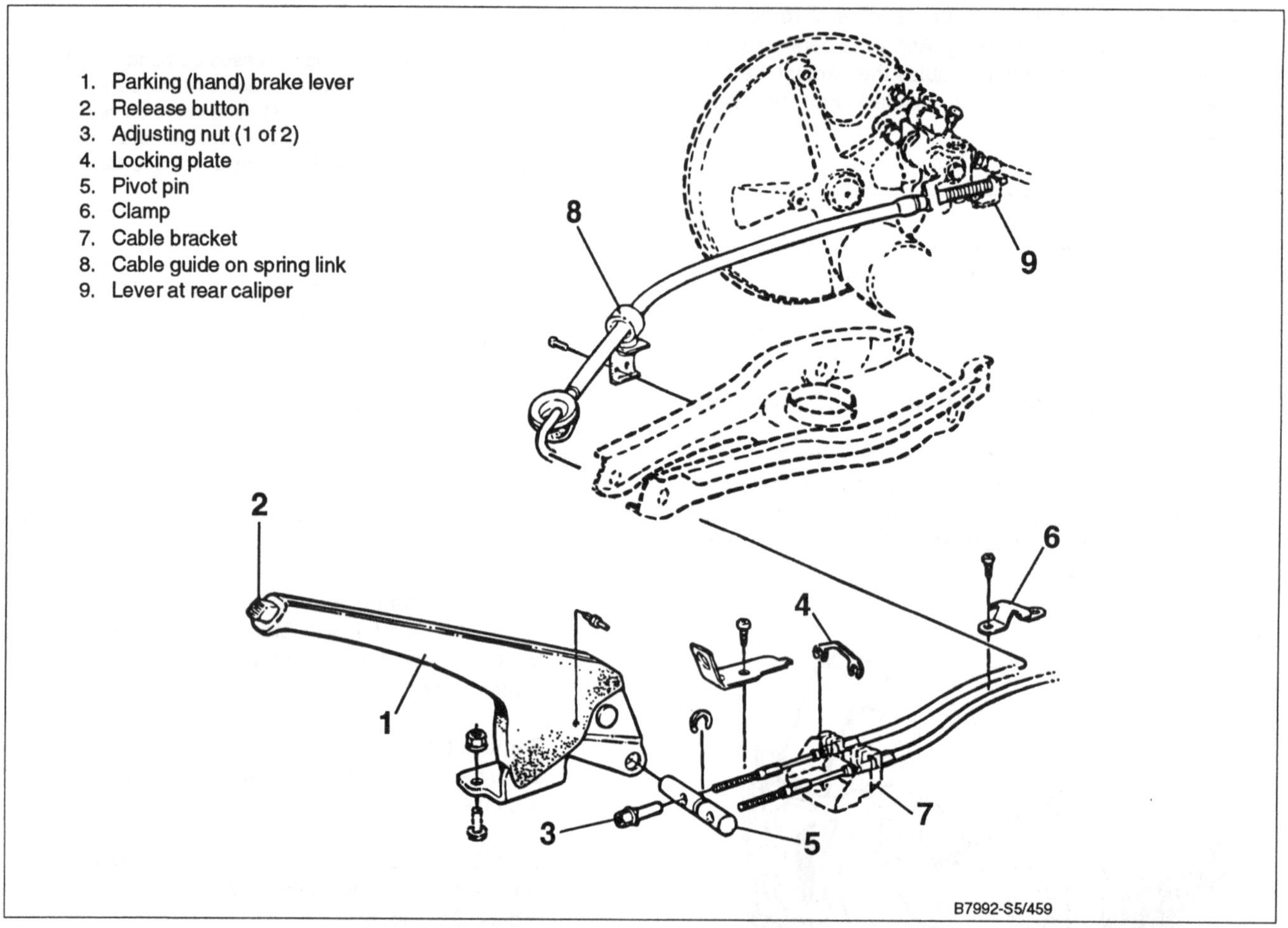

Fig. 5. Parking brake mechanism on 1988 and later models.

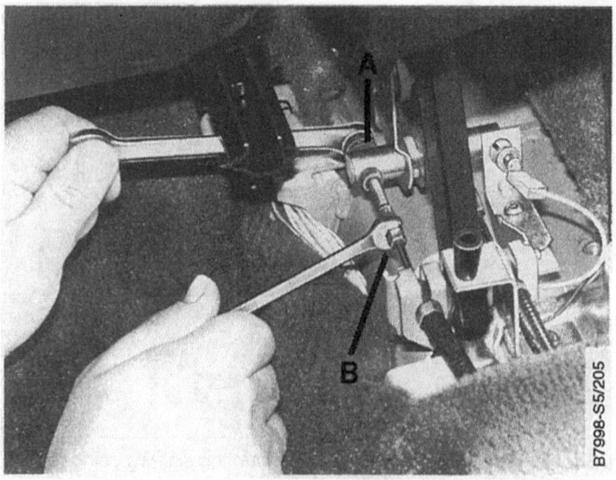

Fig. 7. Cable adjusting nut (**A**) being removed while cable (**B**) is held stationary.

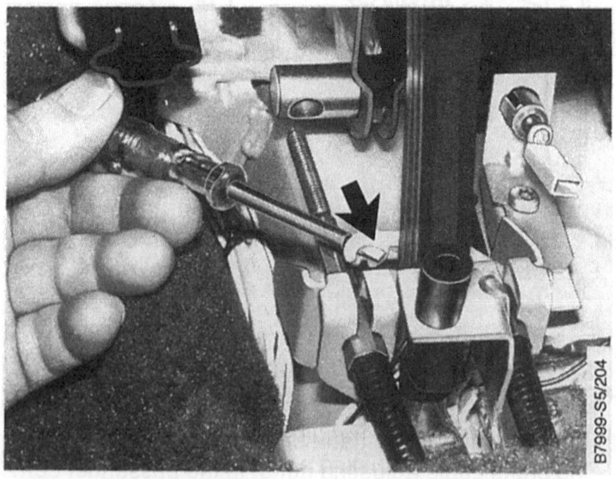

Fig. 8. Cable locking plate (arrow) being removed.

6. Remove the cable hold-down clamp under the carpeting in the center of the floor. See Fig. 9.

NOTE
On models with ABS, cut the tie wraps holding the wheel speed sensor to the parking brake cable.

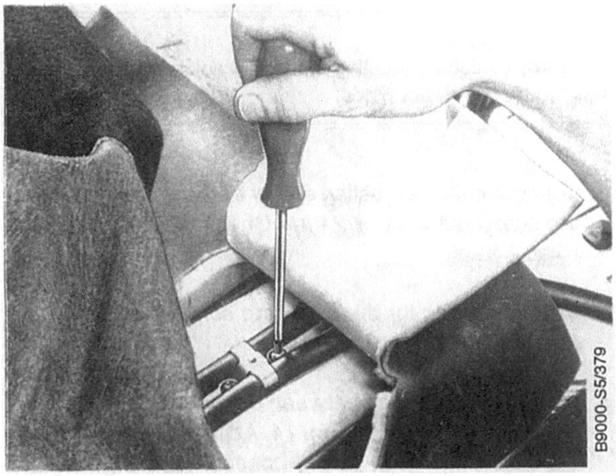

Fig. 9. Cable hold-down clamp being removed. Clamp is on center of floor.

7. Raise the rear of the car and support it with jackstands. Remove the rear wheels.

WARNING
Make sure the car is well supported on the jackstands. Never use just a jack to support a car.

---

8. Remove the cable guide from the rear suspension spring link. See Fig. 10.

Fig. 10. Cable guide on rear suspension link.

9. Remove the cable end from the brake caliper lever. See Fig. 11.

Fig. 11. Cable end being removed from caliper lever.

10. Working inside the passenger compartment, remove the cables. See Fig. 12.

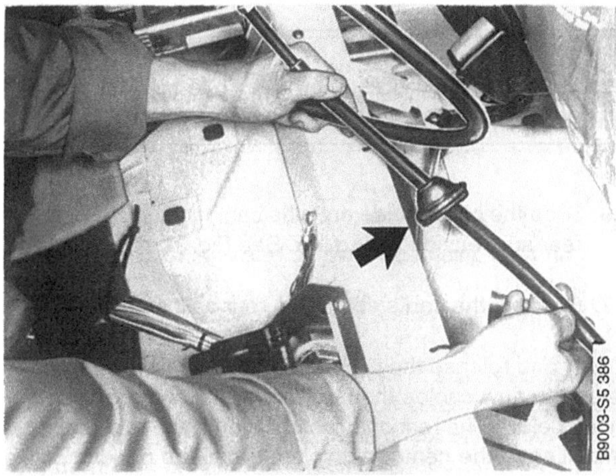

Fig. 12. Cables being removed. Rubber grommet (arrow) seals body opening.

To install parking brake cables (1988 and later models)

1. On 1991 and later cars, unscrew the adjusting sleeve locknut until the cable bottoms out in the adjusting sleeve. See Fig. 13. Then screw the locknut clockwise until it contacts the sleeve, plus an additional three turns. This will preset the cable length.

2. Insert the new cable through the hole under the rear seat and into the parking brake console. Install the cable adjusting nut(s) hand tight and then install the locking plate. See Fig. 8 above.

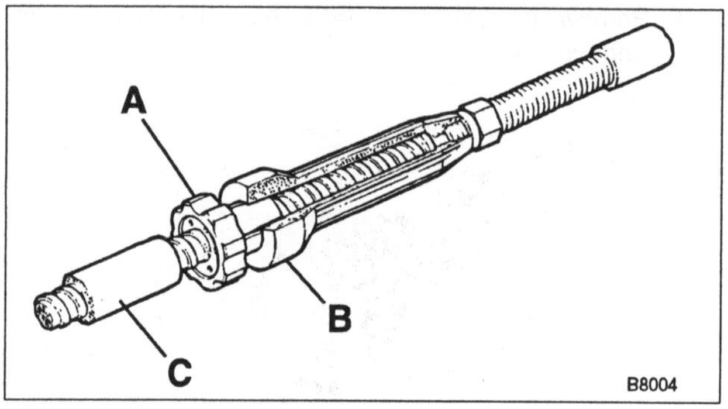

Fig. 13. Cable locknut (**A**) and adjusting sleeve (**B**). Loosen locknut then bottom out cable (**C**) in sleeve.

3. Install the cable hold-down clamp as shown in Fig. 9 above.

4. On 1991 and later cars, install the metal cover and sound absorbing padding.

5. Feed the cable through the body opening towards the caliper. Install the rubber grommet to the body.

NOTE

On cars with ABS, install a new wire tie to hold the wheel speed sensor harness to the parking brake cable.

---

6. Slide the cable guide onto the cable then install it to the rear suspension spring link. See Fig. 10 above.

7. Connect the cable end to the rear caliper lever.

Apply and release the parking brake lever several times to stretch the new cables then adjust the parking brake as described below. The rest of the installation is the reverse of removal. Leave the center console off until the parking brake has been adjusted. Check for correct operation of the parking brake warning light.

To adjust parking brake (1988-1990 models)

Two versions of parking brake cables are used on the cars covered by this manual, depending on model year. Adjusting procedures vary depending on cable installed.

On 1988 through 1990, the cables are adjusted at the parking brake console. On 1991 and later models the cables are adjusted at the adjusting sleeves under the rear seat cushion.

NOTE

On 1988 and later convertible models and on all 1991 and later models, the right hand cable is slightly longer than the left hand cable.

---

1. Remove the center console as described in **432 Manual Transmission Controls.** Release the parking brake after removing the center console.

2. Working at the caliper, remove the plug covering the piston adjusting screw then loosen the adjusting screw using a 4 mm internal-hex head wrench until the piston moves out against its stop.

NOTE

More information on locating the piston adjusting screw can be found in **517 Brake Pads.**

---

3. Turn the piston adjusting screw all the way in then back the screw out 1/4 to 1/2 turn. Check that the brake disc rotates freely.

4. Install the plug for the adjusting screw then reconnect the parking brake cable.

5. Insert a feeler gauge between the parking brake caliper lever and its stop. See Fig. 14. Adjust the cable through the adjusting nut(s) at the console (see Fig. 7 given earlier) until the feeler gauge falls out.

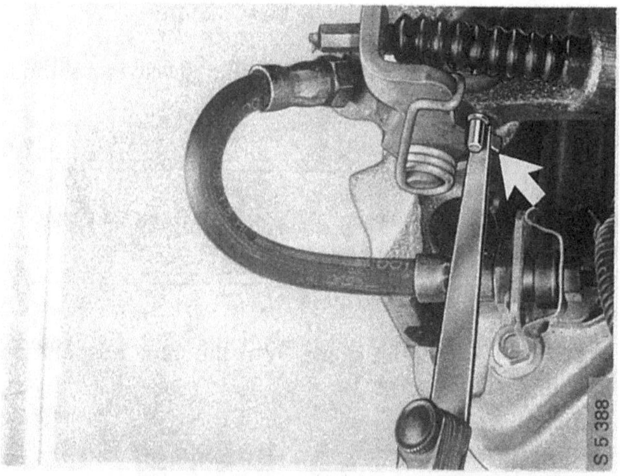

Fig. 14. Parking brake lever gap being measured with feeler gauge (arrow). Cable must be connected to lever.

parking brake lever to caliper clearance	1.0±0.5 mm (0.04±0.02 in.)

The rest of installation is the reverse of removal.

To adjust parking brake (1991 and later models)

1. Remove the rear seat cushion as described in **851 Interior Trim and Upholstery.** Release the parking brake before making adjustments to the parking brake.

2. Working at the caliper, remove the plug covering the piston adjusting screw, then loosen the adjusting screw using a 4 mm internal-hex head wrench until the piston moves out against its stop.

NOTE
More information on locating the piston adjusting screw can be found in **517 Brake Pads**.

---

3. Turn the piston adjusting screw all the way in, then back the screw out 1/4 to 1/2 turn. Check that the brake disc rotates freely.

4. Install the plug for the adjusting screw then reconnect the parking brake cable.

5. Working at the caliper, insert a feeler gauge (between the parking brake caliper lever and its stop. See Fig. 14 above.

parking brake lever to caliper clearance	0.5 to 2.0 mm (0.02 to 0.08 in.)

6. Locate the cable adjusting sleeve under the rear seat, then carefully pry the sleeve away from the sleeve's locknut to remove the slack in the cable. See Fig. 15. Then screw in the locknut (approximately 3 turns) until the feeler gauge falls out.

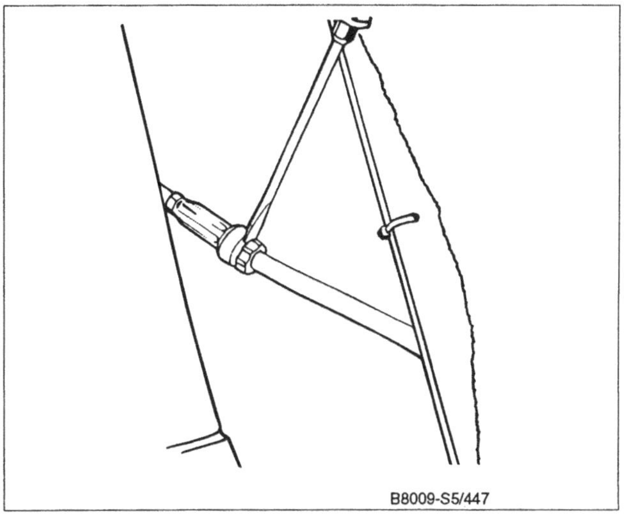
B8009-S5/447

Fig. 15. Locknut being pried away from adjusting sleeve to remove slack in cable.

NOTE
- Apply and release the parking brake several times and check the adjustment again. Make sure the lever moves freely.

- While a helper pulls up and then releases the handbrake lever in the passenger compartment, check that the caliper lever fully releases. If not, the lever is sticking in the caliper or the lever return spring is damaged. If the lever cannot be made to move freely, the caliper should be replaced. See **523 Brake Calipers.**

---

600 Steering and Wheel Alignment—General

TABLES

GENERAL

Steering

All of the cars covered by this manual have power steering. The steering rack is located at the rear of the engine compartment. The belt-driven power steering pump is mounted on a bracket at the back of the engine, on the passenger side. Fig. 1 shows the power steering system.

NOTE
For drive axle repair information, see **774 Wheel Bearings and Drive Axles.**

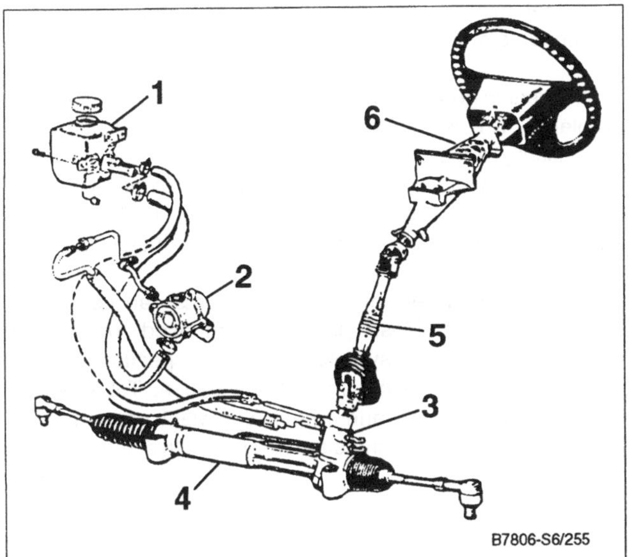

1. Power steering fluid reservoir
2. Power steering pump
3. Control valve
4. Steering rack
5. Intermediate shaft
6. Steering column

Fig. 1. Power-assisted steering system.

The steering column connects to the steering rack via an intermediate shaft an a universal joint. The column is collapsible in the event of a serious accident. Some models are equipped with an SRS airbag system. The airbag itself is located in the steering wheel pad. Tie rods with replaceable ends connect the steering rack to the steering swivel member.

WARNING
On cars equipped with an SRS airbag system, do not attempt to service the wiring in the steering wheel, steering column assembly, or airbag control unit without first disarming the SRS system. The airbag may ignite causing injury. See an authorized Saab dealer for any repairs necessary to the SRS wiring or components.

Wheel Alignment

Wheel alignment is the precise adjustment of the steering and suspension components to ensure that all wheels are orientated correctly, compared to the other wheels, the chassis and the direction of travel. Small changes can have a big effect on the how the car drives. Proper alignment provides the best compromise between responsiveness, stability, and tire wear.

Wheel alignment is adjustable only on the front wheels. Toe-in, Caster, and Camber are the three wheel alignment adjustments that are made using special front-end alignment equipment.

Toe-in is the difference between dimension A and dimension B shown in Fig. 2. If the wheels are exactly parallel, with the two dimensions equal, the toe-in will be zero. For proper steering and road control, the toe-in value must always be positive. In other words, dimension B must be greater than dimension A. This small amount of toe-in helps to offset the forces that tend to spread the tires outward as the car is travelling forward at speed.

Camber refers to the angle formed between the wheel and the vertical. See Fig. 3. If the wheel tilts outwards, the camber angle is said to be positive; if it tilts inwards, the angle is said to be negative. Proper wheel geometry on the Saab 900 requires positive camber. The camber angle influences cornering and directional stability.

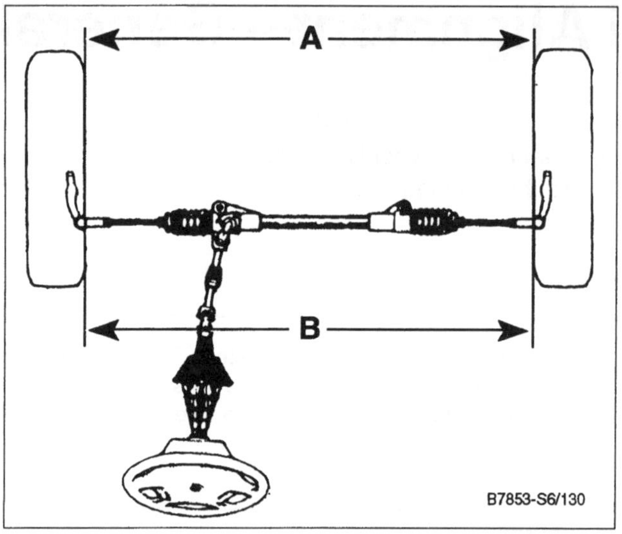

Fig. 2. Toe-in is the difference between dimension **A** and dimension **B**.

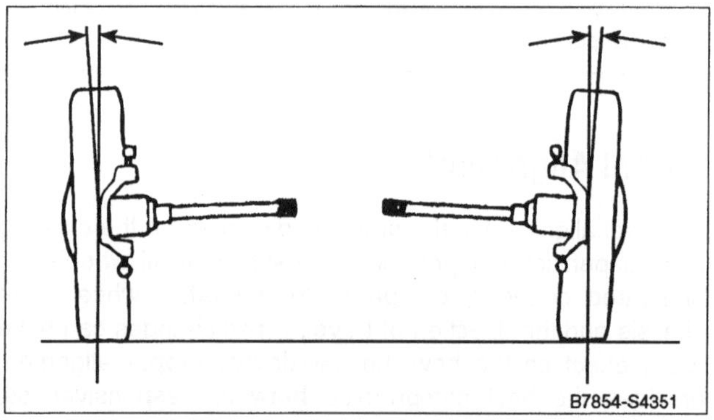

Fig. 3. Camber as viewed from front or rear.

Caster is the angle at which the swivel member (king pin) deviates from the vertical when viewed from the side. See Fig. 4. When the swivel member axis is inclined backwards, the caster is said top be positive; when the axis is inclined forward, the caster is said to be negative. The 900 is designed with positive caster. Positive camber helps to provide self-centering steering and directional stability.

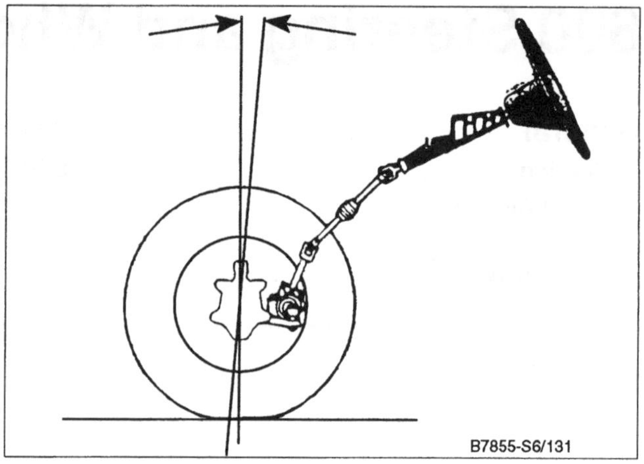

Fig. 4. Caster as viewed from the side.

TROUBLESHOOTING

This troubleshooting section applies to problems affecting steering and wheel alignment. Such problems are usually caused by faults in the steering system, but a problem such as consistently pulling to one side may also be caused by faulty brakes or a damaged drive-axle joint.

The steering system must precisely position the wheels so that the car is stable and controllable. Instability or imprecise road feel may be caused by worn or damaged steering components or incorrectly aligned wheels.

Tire wear is good indicator of steering and wheel alignment problems. Tire wear and incorrect inflation pressures can dramatically affect handling. Subtle irregularities in wheel alignment angles also affect stability. Mixing different types or sizes of tires, particularly on the same end of the car, can affect alignment and may unbalance a car's handling.

Steering and wheel alignment problems can usually be isolated and at least partially diagnosed by careful observation of the symptoms and inspection of the components that are the most likely cause. **Table a** lists symptoms of steering and wheel alignment problems and their probable causes, and suggests corrective actions.

Table a. Steering and Wheel Alignment Troubleshooting

Symptom	Probable cause	Corrective action
1. Pull to one side, wandering	**a.** Incorrect tire pressure	**a.** Check and correct tire pressures
	b. Defective/unevenly worn tire	**b.** Inspect tires and replace as needed
	c. Incorrect wheel alignment	**c.** Check and adjust wheel alignment. **601 Wheel Alignment**
	d. Faulty brakes (pulls only under braking)	**d.** See **5 BRAKES**
2. Little or no power assist when car/engine is cold	**a.** Worn control valve and seals	**a.** Replace steering rack. **644 Power Steering**
3. Steering heavy, noisy, poor return-to-center	**a.** Incorrect tire pressure	**a.** Check and correct tire pressures
	b. Low power steering fluid	**b.** Check power steering fluid and add as required. See **1 LUBRICATION AND MAINTENANCE**
	c. Loose, worn or broken power steering pump V-belt	**c.** Inspect V-belt. Tighten or replace as necessary. See **1 LUBRICATION AND MAINTENANCE**
	d. Worn front suspension parts, such as rubber bushings.	**d.** Inspect and replace any worn parts found. See **7 SUSPENSION**
	e. Worn or damaged steering column universal joint	**e.** Replace universal joint. **641 Steering Column**
	f. Faulty power steering pump	**f.** Test and, if necessary, replace pump. **644 Power Steering**
	g. Air in power steering fluid	**g.** Repair air leak and, if necessary, add fluid. **644 Power Steering**
4. Front-end vibration or shimmy	**a.** Unbalanced wheels/tires or front wheels out of alignment	**a.** Balance tires. Align front wheels. Check tires for uneven wear patterns
	b. Loose wheel lug bolts	**b.** Tighten lug bolts to proper torque
	c. Bent wheel rim (radial or lateral runout)	**c.** Inspect wheels and replace as necessary
	d. Tire(s) out-of-round (radial runout)	**d.** Measure tire radial runout and remount or replace as necessary
	e. Severely worn shock absorbers	**e.** Replace worn shocks and shock bushings. See **7 SUSPENSION**
	f. Worn control arm bushings	**f.** Replace bushings. **632 Control Arms**
5. Knocking, thumping or clicking noises from front end	**a.** Worn shock absorber bushings	**a.** Check and, if necessary, replace bushings. **731 Front Suspension**
	b. Worn front suspension bushings	**b.** Check and, if necessary, replace bushings. **731 Front Suspension**
	c. Loose suspension mounting bolts	**c.** Retorque all front suspension fasteners. **731 Front Suspension**
	d. Worn CV joints or wheel bearings	**d.** Replace worn CV joints or wheel bearings. **774 Wheel Bearings and Drive Axles**
6. Steering loose, imprecise	**a.** Incorrect tire pressure	**a.** Check and correct tire pressures
	b. Worn steering rack bushings	**b.** Replace bushings. **644 Power Steering**
	c. Worn tie rod end(s)	**c.** Replace tie rod end(s) and align wheels. **643 Tie Rod Ends, 601 Wheel Alignment**
	d. Faulty front wheel bearings	**d.** Replace wheel bearing. **774 Wheel Bearings and Drive Axles**
	e. Worn or damaged steering column universal joint	**e.** Replace universal joint. **641 Steering Column**
	f. Worn or damaged steering rack	**f.** Replace steering rack. **644 Power Steering**

601 Wheel Alignment

TABLES

WHEEL ALIGNMENT SPECIFICATIONS

The specifications and procedures given in this repair group are designed to be used in conjunction with special wheel alignment equipment. Only the alignment of the front wheels is adjustable, although rear wheel alignment can be checked. Checking wheel alignment is normally necessary whenever new tires are installed, if the tires are wearing abnormally, or if the car wanders to one side or handles irregularly.

Checking tire balance is recommended when doing a front wheel alignment.

There are two chassis used on the cars covered by this manual; a standard chassis and a sport chassis. To identify the type of chassis, measure the distance between the top of the wheel rim and the seam line between the fender and the hood. See Fig. 1.

Fig. 1. Chassis ride height measurement (dimension **B**).

Chassis Ride Height (dimension **B**)
- standard chassis.............. 243 mm (9.56 in.)
- sport chassis
  - with 15 in. wheel .... 230 mm (9.06 in.)
  - with 16 in. wheel .... 217 mm (8.54 in.)

WARNING
On cars equipped with an SRS airbag system, do not attempt to remove the steering wheel without first disarming the SRS system. The airbag may ignite causing injury. See an authorized Saab dealer for any repairs necessary to the SRS system.

When checking front wheel alignment, the car should be on a level surface with a full tank of gas. There should be no passengers or additional weight in the car. **Table a** lists wheel alignment specifications for cars covered by this manual. Always make sure that tire pressures are correct and that no suspension or steering components are worn.

Table a. Wheel Alignment Specifications

Measurement	Model	
	except sport chassis	**sport chassis**
Front		
Caster	2°±0.5	2°±0.25
Camber	0.5°±0.5	0.25°±0.25
Toe-in	2±1 mm (0.08±0.04 in.)	1.5±0.5 mm (0.06±0.02 in)
Swivel pin inclination	11.5°±1	11.5°±1
Steering angle—outer wheel	20°	20°
Steering angle—inner wheel	20.75°±0.50	20.75°±0.50
Rear		
Toe-in	4±1 mm (0.16±0.04 in.)	4±1 mm (0.16±0.04 in.)
Camber	–0.5°±0.25	–0.5°±0.25

CAUTION
The front wheel toe-in specifications listed in **Table a** are only applicable when making measurements directly from the rim, such as when using a toe bar. If optical alignment equipment is used, the specifications need to be corrected based on type of equipment being used. Most modern alignment equipment is based on a universal 28.64 in. circle.

NOTE
Swivel pin inclination, also known as king pin inclination, is not adjustable. If the swivel pin inclination is incorrect, the swivel member must be replaced as described under **774 Wheel Bearings and Drive Axles.**

Front Wheel Toe-in

Toe-in is adjusted by varying the length of the threaded tie rod in the tie rod end. Screwing the rod into the tie rod end will decrease toe-in and screwing the rod out will increase toe-in.

To adjust toe-in

1. Working from below, loosen the tie rod lock nut. See Fig. 2.

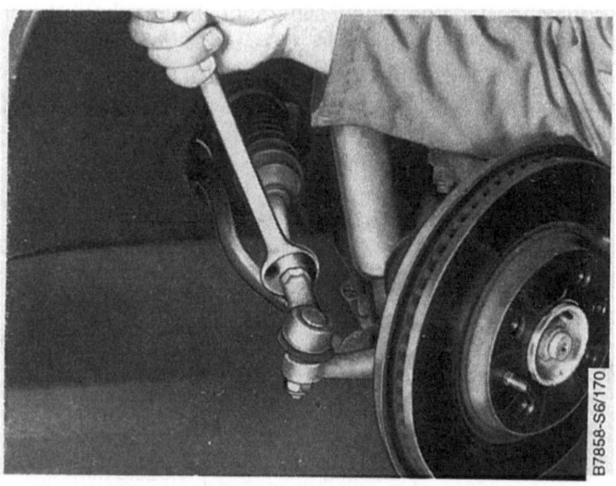

Fig. 2. Tie rod locknut being loosened. (Wheel shown removed for clarity.)

2. Using a locking pliers, turn the tie rod in or out to obtain the correct setting.

CAUTION
When the toe-in setting is correct, tighten the tie rod lock nut.- The length of the tie rod from the locknut to the stepped part of the rod must never exceed 125 mm (4.92 in). See Fig. 2. Also, the difference in length between the two rods must not exceed 2 mm (0.08 in.). For example, if the right rod measured 124 mm, the left rod must measure between 122 mm and 125 mm. See Fig. 3.

Tightening torque
- steering tie rod to tie rod end (locknut) . . . . 60 to 80 Nm (44 to 59 ft-lb)

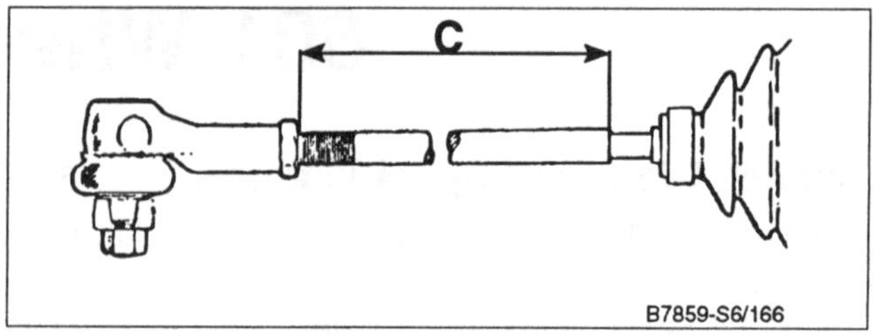

Fig. 3. Maximum permissible length (**C**) of tie rod must not exceed 125 mm (4.92 in.).

Front Wheel Camber and Caster

Camber and caster are adjusted by varying the number of shims under the upper control arm bearing brackets. See Fig. 4. The shims are available from an authorized Saab dealer in thicknesses of 0.5 mm, 1.0 mm and 2.0 mm.

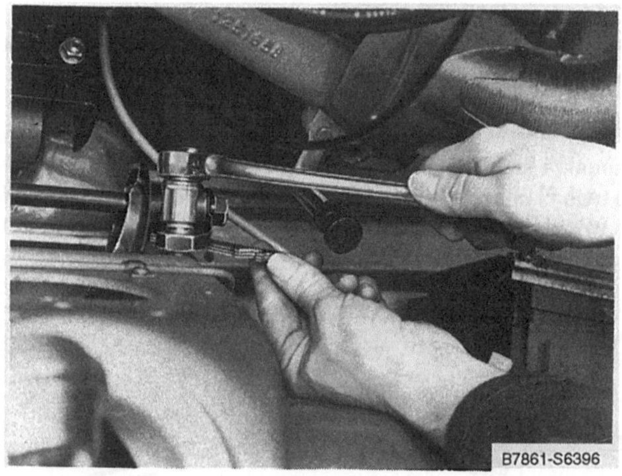

Fig. 4. Camber/caster shims being removed from upper control arm bearing bracket.

Camber is adjusted by changing the number of shims equally under both the front and rear brackets. Removing an equal number of shims from beneath both brackets will increase positive camber. Adding shims will reduce positive camber.

Caster is adjusted by changing the number of shims from either the left or the right bearing brackets. Moving shims from the front bearing bracket to the rear bearing bracket will increase the caster. Moving shims from the rear bracket to the front bracket will reduce caster. Once the final adjustments have been made, torque the bracket bolts.

Tightening torque
- upper control arm bearing bracket to body . . . . 40 to 55 Nm (30 to 40 ft-lb)

631 Ball Joints

CHECKING AND REPLACING BALL JOINTS

The ball joints locate the steering swivel member to the upper and lower control arms. See Fig. 1. The ball joints are permanently lubricated and do not require any service. If found to be worn, they must be replaced. See **632 Control Arms** for illustrations of ball joint mountings to the control arms.

WARNING
Do not reuse self-locking nuts. These parts are designed to be used only once and must always be replaced once loosened or removed.

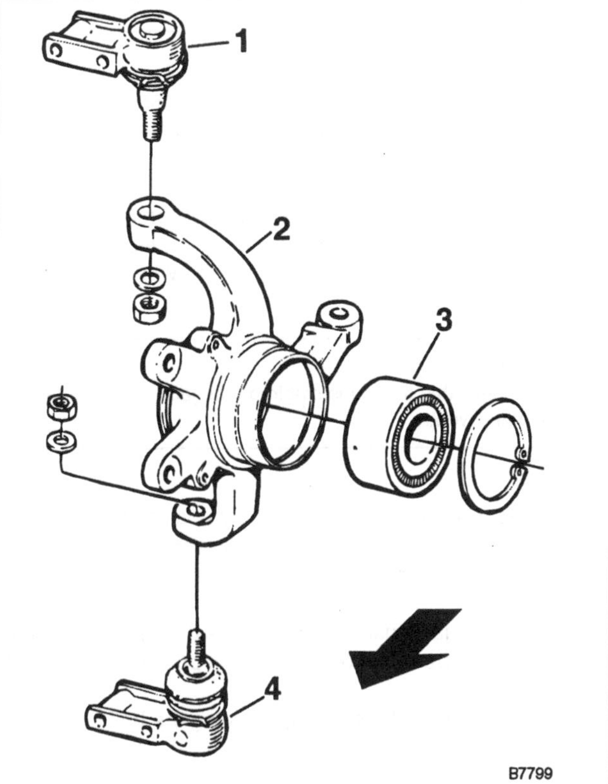

1. Upper ball joint
2. Swivel member
3. Wheel bearing and circlip
4. Lower ball joint

Fig. 1. Exploded view of steering swivel member.

To check ball joint play

To accurately check ball joints for wear, the coil spring tension on the control arm must be relieved. A special tool, as shown below, is available for this purpose.

1. With the car on the ground, insert spacer tool (Saab part no. 83 93 209) between the upper control arm and the body. See Fig. 2.

Fig. 2. Spacer tool (arrow) inserted between upper control arm and body. The spacer tool relieves suspension loading on the ball joints.

2. Loosen the wheel lug bolts or nuts, then raise the front of the car and support it securely on jack stands designed for the purpose. Remove the wheel.

3. Carefully inspect the ball joint rubber boots. If any boot is torn or cracked, the complete ball joint should be replaced.

4. Using water pump pliers, check the axial (up-down) play of the upper and lower ball joints. See Fig. 3. If the play is excessive, the ball joint must be replaced.

631

Fig. 3. Lower ball joint being checked for axial (up-down) play using water pump pliers.

Ball joint wear limit
- maximum axial play . . . . 2 mm (0.078 in.)

---

5. Using a pry bar, check the radial (side-to-side) play of the upper and lower ball joint. Gently pry the joint back and forth to check for play. See Fig. 4. If the play is excessive, the ball joint must be replaced.

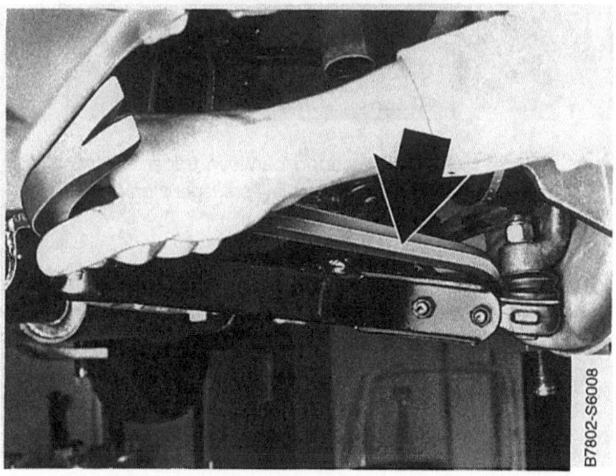

Fig. 4. Lower ball joint radial (side-to-side) play being checked using pry-bar (arrow).

Ball joint wear limit
- maximum radial play . . . . 1 mm (0.040 in.)

CAUTION
Use care not to damage the ball joint rubber boot with the pry bar.

If the ball joints are in good shape, loosely install the wheel. Lower the car and tighten the wheel bolts or nuts. Remove the spacer tool from the opening in the body.

Tightening torque
- road wheel to wheel hub
  lug nuts . . . . 90 to 110 Nm (66 to 81 ft-lb)
  lug bolts. . . . . 105 to 125 Nm (77 to 92 ft-lb)

To replace ball joint

The procedure below applies to both upper and lower ball joints.

WARNING
Do not reuse self-locking nuts. These parts are designed to be used only once and must always be replaced once loosened or removed.

---

1. With the car on the ground, insert spacer tool (Saab no. 83 93 209) between upper control arm and body. See Fig. 2 above.

2. Loosen the wheel lug bolts or nuts, then raise the front of the car and support it securely on jack stands designed for the purpose. Remove the wheel.

WARNING
Always check that the car is stable on the jack stands before starting any work.

---

3. Using a jack under the control arm, raise the lower control arm slightly. Remove the lower shock absorber mounting nut and pull the shock off its mount. See Fig. 5.

4. Lower the jack just until the drive axle is just contacting the body opening. Lock the jack in position for support.

5. Remove the self-locking nut holding the ball joint to the steering member.

6. Separate the ball joint using a press tool. See Fig. 6.

Fig. 5. Shock absorber being removed from control arm mounting.

CAUTION
Do not use a ball joint separator fork to remove the ball joint from the steering member. The steering member may be damaged as a result.

Fig. 6. Upper ball joint being pressed out of steering swivel member. Special press tool (Saab tool no. 89 95 409) shown.

7. Remove the two through-bolts holding the ball joint to the control arm. See Fig. 7. Remove the ball joint by sliding it out of the control arm.

Fig. 7. Ball joint-to-control arm mounting bolts shown at arrows. Upper ball joint shown, lower ball joint is similar.

CAUTION
If the upper ball joint is being removed, do not remove the bolt for the spring saddle as the spring is under compression.

Installation is the reverse of removal. Replace the ball joint bolts and all self-locking nuts. Torque the bolts and nuts to the values given below. Be sure to remove the spacer tool when the job is complete.

Tightening torques

- balljoint to steering swivel member
  (self-locking nut) ...... 35 to 50 Nm (26 to 37 ft-lb)
- balljoint to control arm bolts
  (self-locking nuts)...... 40 to 55 Nm (30 to 40 ft-lb)
- road wheel to wheel hub
  lug nuts............ 90 to 110 Nm (66 to 81 ft-lb)
  lug bolts .... 105 to 125 Nm (77 to 92 ft-lb)

632 Control Arms

GENERAL

The independent front suspension for each front wheel consists of two control arms, each of which is pivot-mounted to the unibody by bearing brackets fitted with rubber bushings. The steering swivel member is mounted to the control arms via ball joints. See Fig. 1.

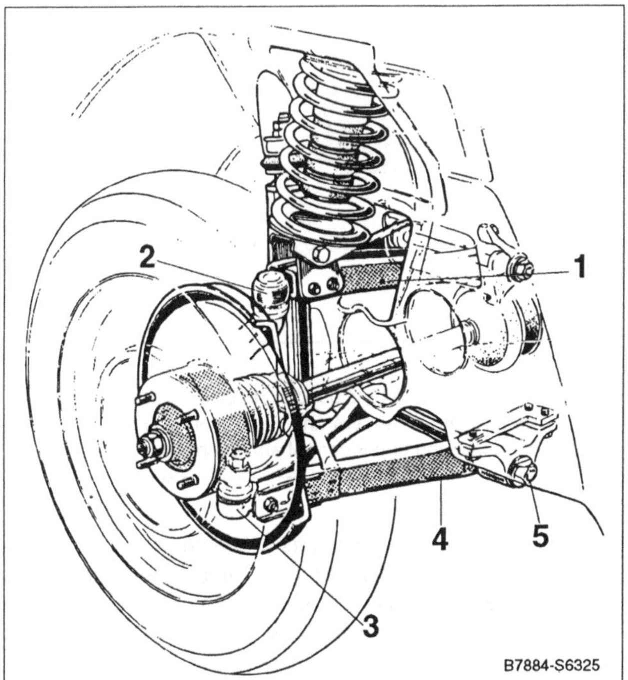

1. Upper control arm
2. Upper ball joint
3. Lower ball joint
4. Lower control arm
5. Bearing bracket with rubber bushing

Fig. 1. Cut-away view of control arms and front suspension.

CONTROL ARMS AND CONTROL ARM BUSHINGS

The control arms are not normally replaced unless damaged. The control arm bushings, on the other hand, are prone to wear under normal use and high mileage. Inspecting the control arms includes checking for distortion or damage due to collisions or road hazards. The bushings should be inspected for wear or deterioration. Fig. 2 shows an exploded view of the lower control arm and it bushings. The upper control arm is similar.

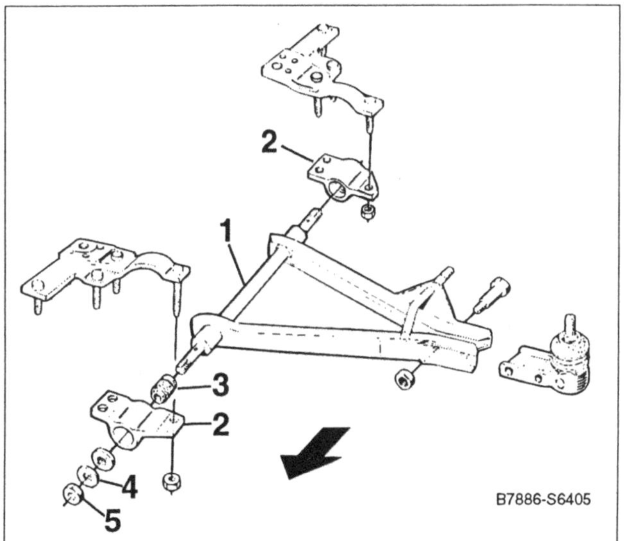

1. Lower control arm
2. Bearing brackets
3. Bushings
4. Lock washer
5. Locknut

Fig. 2. Exploded view of lower control arm.

WARNING

If the controls arms have been subjected to violent stresses as a result of a collision or a road hazard, they must be carefully inspected for bending or distortion. Be sure to also inspect the bearing brackets and their mountings to the body.

NOTE

Ball joint inspection is covered in **631 Ball Joints.** Steering tie rod end inspection is covered in **643 Tie Rod Ends**.

Worn control arm bushings are often noticed through knocking or clunking sounds when traveling over rough road surfaces. Before inspecting for damaged parts or worn bushings, first check the tightness of all fasteners using a torque wrench. See Fig. 3.

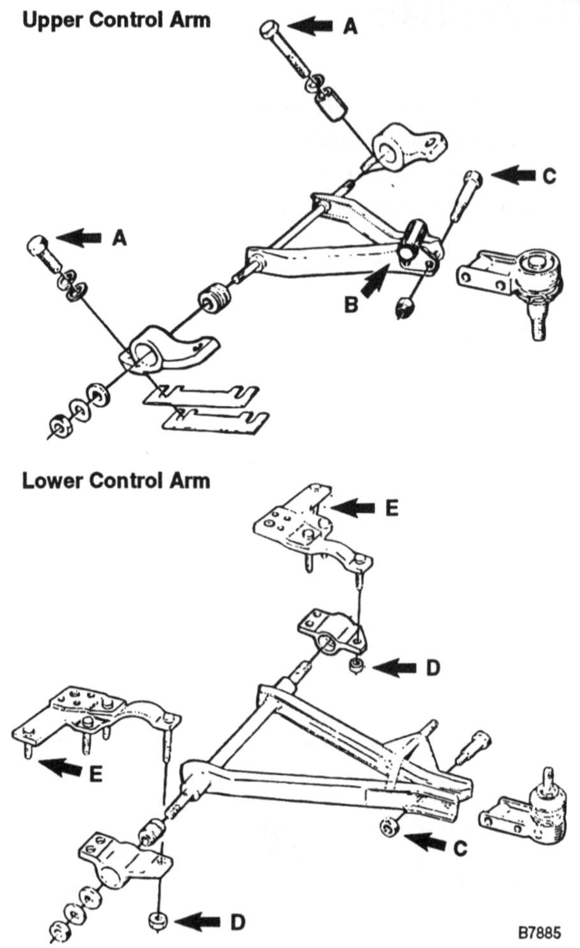

A Upper control arm bearing bracket
(front or rear) to body . . . . 40 to 55 Nm (30 to 40 ft-lb)

B Coil spring seat to
control arm . . . . 55 to 110 Nm (40 to 81 ft-lb)

C Ball joint to control arm
(upper or lower). . . . . 40 to 55 Nm (30 to 40 ft-lb)

D Lower control arm bearing bracket (front or rear) to body
M12 dry joint . . . . 70 to 95 Nm (52 to 70 ft-lb)
M12 lubricated joint
(waxed nut). . . . . 45 to 55 Nm (33 to 40 ft-lb)

E Lower control arm
bracket to body. . . . . 25 to 35 Nm (18 to 26 ft-lb)

Fig. 3. Exploded view of upper and lower control arm fasteners.

To replace control arm bushings

1. Remove the control arm as described below. Remove the bearing brackets from the control arm.

2. Press the old bushing out of the bracket. Thoroughly clean the bearing brackets. Apply soapy water to the new bushing and press it into the bracket. See Fig. 4.

CAUTION
Oil or grease must not be allowed to come into contact with the bushings. Use only soapy water.

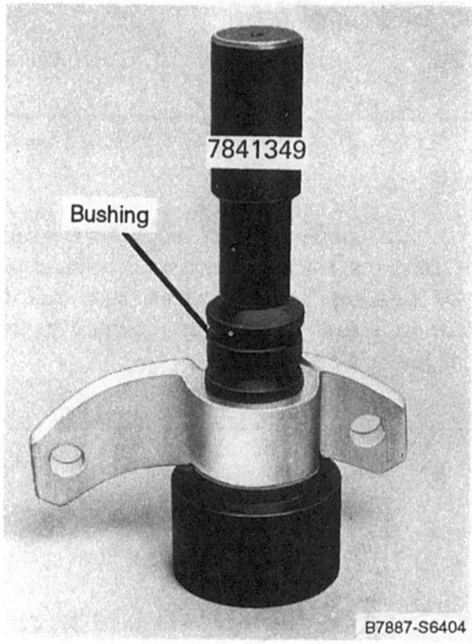

Fig. 4. Special press tool used to install control arm bushing.

NOTE
Saab tool no. 78 41 349 is used to replace lower control arm bushings (as shown above) and Saab tool no. 78 41 331 is used to replace the upper control arm bushings.

3. Install the bearing brackets to the control arm and position them as shown in Fig. 5. Tighten the locknut while holding the brackets in position. Recheck the angle of the bracket. Bend over the lock washer to lock the nut in place.

Bearing bracket to control arm (locknut and lock washer)
- upper control arm bracket to control arm . . . . 55 to 70 Nm (40 to 52 ft-lb)
- lower control arm bracket to control arm . . . . 75 to 90 Nm (55 to 66 ft-lb)

4. Install the control arm as described below.

Removing and Installing Control Arms

Removing and installing the driver's side upper control arm will require removal of the engine. With the engine installed, there is not enough clearance to pull the control arm from the

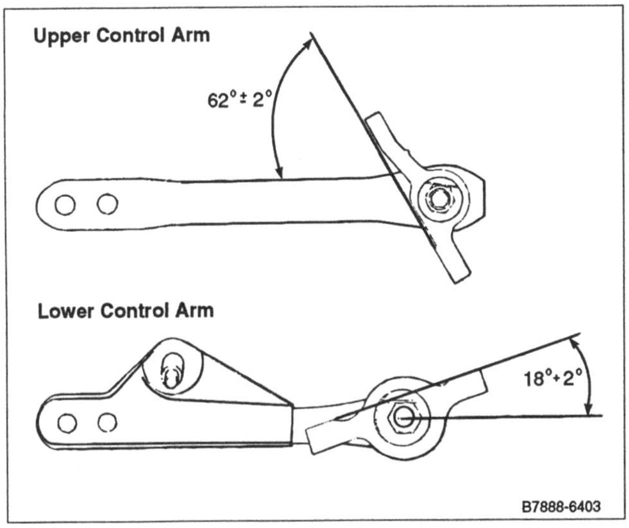

Fig. 5. Control arm bearing bracket installation angles.

body. Removing the passenger's side upper control arm requires that the engine be suspended with a hoist so that an engine mount can be removed. Special tools are required to remove the upper control arms. Read the procedures completely before starting any work.

WARNING

- Special spring compressing tools (Saab tool no. 88 18 791 or equivalent) are required to remove and install the upper control arms. Do not attempt to remove the upper control arms without the proper tools, as serious injury could result.

- Do not reuse self-locking nuts or locking washers. These parts are designed to be used only once and must be replaced once removed.

To remove upper control arm

1. If removing the driver's side control arm, remove the engine as described in **201 Engine Removal and Installation.**

2. Loosen the wheel lug bolts or nuts, then raise the front of the car and support it securely on jack stands designed for the purpose. Remove the front wheel.

WARNING

Always check that the car is stable on the jack stands before starting any work.

---

3. Using a jack under the control arm, raise the lower control arm slightly. Remove the lower shock absorber mounting nut and pull the shock off its mount. See Fig. 6. Remove the jack.

Fig. 6. Shock absorber being removed from control arm mounting.

4. Compress the suspension coil spring using a spring compressor (Saab no. 88 18 791 and 88 18 809 or equivalent). Remove the spring from the car. See Fig. 7.

WARNING

Work carefully as the spring is compressed. Check to make sure that the spring compressor is always securely seated on the spring.

Fig. 7. Compressed coil spring being removed from car.

5. Remove the spring seat and upper ball joint mounting bolts. See Fig 8. Remove the spring seat and ball joint from the control arm.

Fig. 8. Upper ball joint bolts (**1**) and spring seat bolt (**2**).

6. If removing the passenger's side control arm, use a suitable hoist to support the engine. Then remove the right side engine mount. See Fig. 9. Remove the engine mounting bracket from the engine block and engine support bracket.

CAUTION
Be careful not to damage the inner drive axle rubber boot when removing the engine mount bracket.

Fig. 9. Right side engine mount being removed. Mount is bolted in place from above and below. (Hydraulic engine mount shown, rubber engine mount is similar.)

7. Remove the four bolts holding the control arm to the body. Note the number of shims under each bearing bracket. See Fig. 10. Remove the control arm from the car.

CAUTION
The shims should be re-installed in their exact positions. The shims are used to align the front wheels. If the shims are incorrectly installed, a front wheel alignment will be necessary. See **601 Wheel Alignment.**

NOTE
The control arms are not symmetrical. If removing more than one control arm, mark their locations during removal.

Fig. 10. Upper control arm mounting bolts (arrows) as viewed from above.

Installation is the reverse of removal. If the bearing brackets were removed from the control arms, be sure to reinstall them as shown above in Fig. 5. Be sure to replace all self-locking nuts and locking washers. Torque all control arm fasteners to the values given in above in Fig. 2. Tighten the engine mounts to their final torques with the car on the ground and engine hoist removed. Check and adjust wheel alignment if necessary. See **601 Wheel Alignment.**

To remove lower control arm

1. Loosen the wheel lug bolts, then raise the front of the car and support it securely on jack stands designed for the purpose. Remove the front wheel.

2. Raise the lower control arm slightly using a jack at the outside of the control arm. Remove the shock absorber from its lower mount and remove the ball joint from the control arm. See Fig. 11. Remove the jack.

NOTE
Where applicable, move the anti-roll bar out of the way.

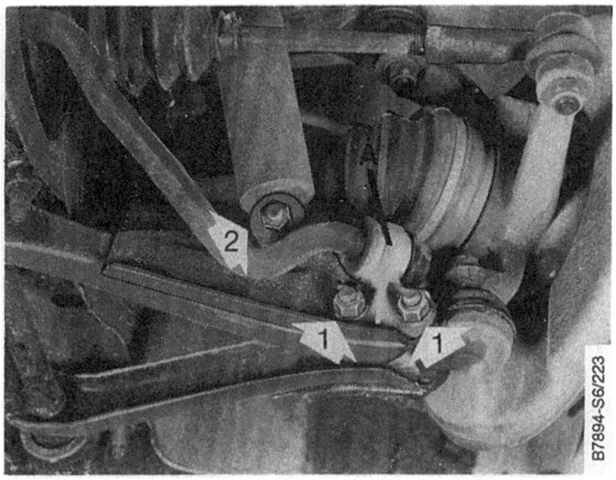

Fig. 11. Lower ball joint mounting bolts (**1**) and shock absorber mounting nut (**2**). Anti-roll bar mount shown at **A**.

3. Remove the six control arm bracket-to-body mounting nuts. See Fig. 12. Remove the control arm from the car.

Installation is the reverse of removal. If the bearing brackets were removed from the control arms, be sure to reinstall them as shown above in Fig. 5. Replace all self-locking nuts and torque all fasteners to the values given in Fig. 2. Check and adjust wheel alignment if necessary. See **601 Wheel Alignment.**

Fig. 12. Lower control arm mounting nuts (arrows).

641 Steering Column

GENERAL

The steering column assembly consists of three sections, connected by universal joints. See Fig. 1. The top perforated cage-like section is deformable and the steering wheel shaft is telescopic. The intermediate shaft is made of a flexible sheet steel, which will allow the steering column to collapse when subjected to high loads.

On 1990 and later models, an SRS airbag system is used, with the airbag itself located in the steering wheel pad.

WARNING
- The SRS (Supplemental Restraint System) uses an explosive device to inflate the airbag. Do not attempt to service the wiring in the steering wheel, steering column assembly, or airbag unit. The airbag may ignite causing serious injury. See an authorized Saab dealer for any repairs necessary to the SRS wiring or components, including removal of the steering wheel and steering column.
- Do not reuse self-locking nuts. These parts are designed to be used only once and must be replaced whenever loosened or removed.

CAUTION
- When spinning the steering wheel via the front road wheels (front wheels off the ground), move the wheels slowly from lock to lock to avoid damaging the collapsible steering column.
- Do not push or pull on the steering wheel (for example when trying to roll the car). Doing so can strain the steering column, causing it to collapse.

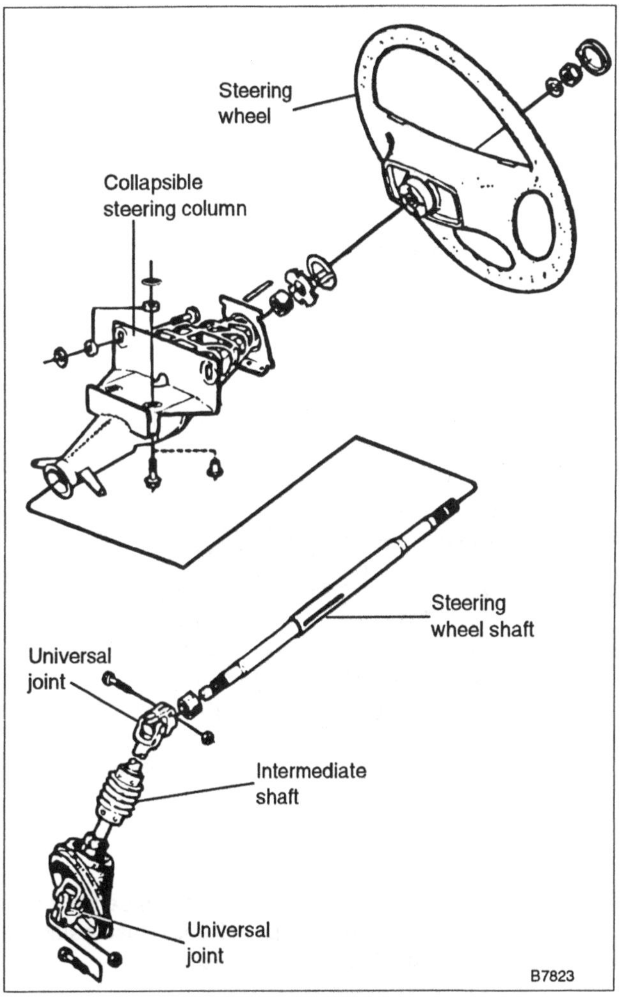

Fig. 1. Exploded view of steering column assembly.

To remove steering wheel

The procedure below does not cover removing the steering wheel on SRS (airbag) equipped cars. Cars with SRS can be identified by the three letter markings on the steering wheel pad.

WARNING
Removing the steering wheel on cars equipped with SRS airbag requires that the airbag unit be disarmed, removed, and stored in a safe place. Due to the inherent danger involved in this operation, it should be left to an authorized Saab dealer.

---

1. Position the steering wheel so that the road wheels are straight ahead.
2. On three-spoke steering wheels, carefully pry off the steering wheel pad and disconnect the electrical connectors for the horn. See Fig. 2.
3. On four-spoke steering wheels, pry out the badge in the center of the wheel.
4. Place matching marks on the steering wheel and the steering wheel shaft. Remove the center steering wheel nut and washer. See Fig. 3.

Fig. 2. Steering wheel pad being removed on cars with three-spoke wheels.

Fig. 4. Puller (Saab tool no. 89 96 258) being used to remove steering wheel. The two side bolts on the puller thread into the steering wheel itself and the center bolt is threaded down onto the steering wheel shaft.

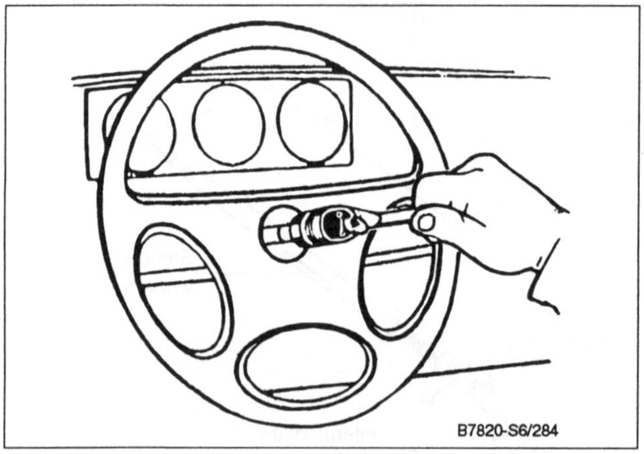

Fig. 3. Steering wheel mounting nut being removed.

5. Pull the steering wheel straight off the steering wheel shaft. If the wheel does not pull off easily, use a suitable puller to assist in removal. See Fig. 4.

CAUTION

Do not hammer on the steering wheel or steering wheel shaft in an attempt to loosen the steering wheel. This may damage the collapsible steering column.

Installation is the reverse of removal. Always replace the self-locking nut.

Tightening torque

- steering wheel to steering column (self-locking nut). . . . . 25-28 Nm (18 to 21 ft-lb)

To remove steering column assembly

The procedure below does not cover removing the steering column assembly on SRS (airbag) equipped cars. Cars with SRS can be identified by the three letter markings on the steering wheel pad.

WARNING

Removing the steering column assembly on cars equipped with SRS airbag requires that the airbag unit be disarmed, removed, and stored in a safe place. Due to the inherent danger involved in this operation, it should be left to an authorized Saab dealer.

---

1. Working from the engine compartment, remove the clamping bolt that holds the steering column universal joint to the steering rack. See Fig. 5. Push the steering column protective boot off the firewall bulkhead and into the passenger compartment.

2. Remove the steering wheel as described above.

3. Remove the center storage console and the lower instrument padding as described in **853 Dashboard and Consoles.**

4. Remove the steering column lower trim piece.

5. Remove the directional indicator and wiper/washer stalk switches from the steering column. Label the electrical connectors before unplugging them.

6. Unplug the connectors for the horn and cruise control (where applicable). Cut any cable ties holding the wiring harness to the steering column.

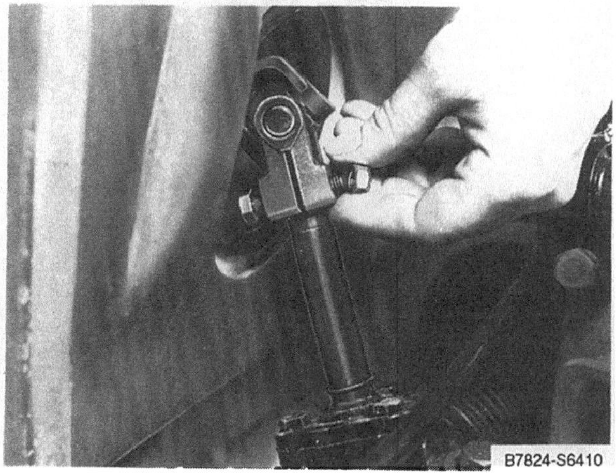

Fig. 5. Steering column universal joint clamping bolt being removed.

7. If applicable, remove the anti-chafing pad from the right side of the steering column.

NOTE
If an anti-chafing material is used, note its position before removing it.

---

8. Remove the four steering column-to-console bolts and spacers. See Fig. 6.

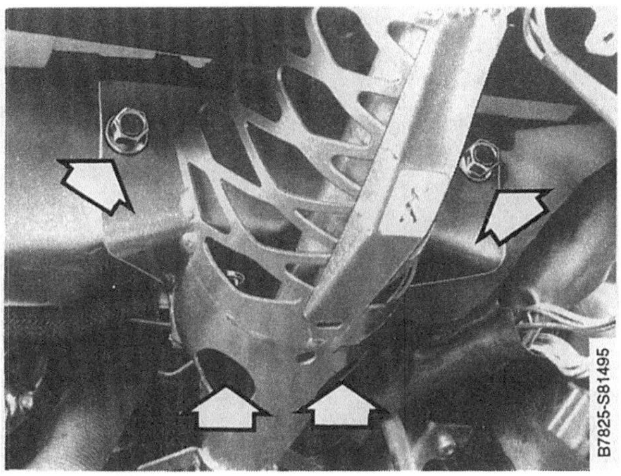

Fig. 6. Steering column mounting bolts (arrows).

9. Lower the column slightly and pull the entire assembly out of the car. See Fig. 7.

WARNING
Use care when handling the removed steering column assembly. Do not alter the position of the telescopic steering wheel shaft or the collapsible function of the column assembly may be impaired.

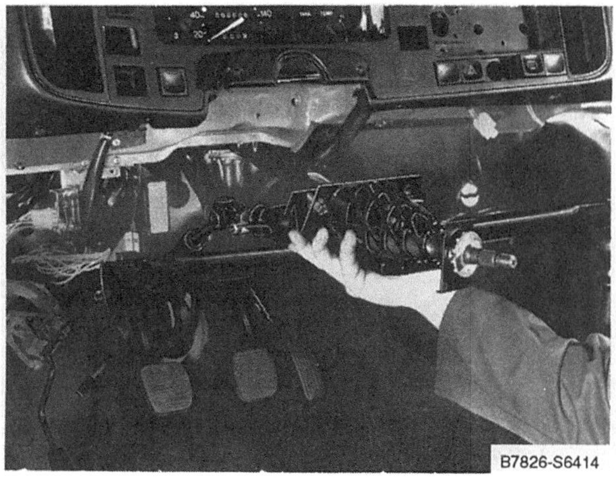

Fig. 7. Steering column assembly being removed from car.

Installation is the reverse of removal. Make sure that the steering column universal joint is fully seated on the steering rack pinion shaft so that the clamping bolt properly engages the groove in the shaft. See Fig. 8.

Fig. 8. Steering column-to-steering rack pinion shaft clamping bolt being installed. Bolt must engage groove in shaft.

Apply Loctite to the four steering column-to-body mounting bolts during installation. Replace all self-locking nuts. Check the universal joints for free movement. Universal joints that are seized or do not swivel freely should be replaced. Tighten all bolts to their recommended torques.

Tightening torques
- steering column to steering rack (clamping bolt, self-locking nut) . . . . 20 to 27 Nm (15 to 20 ft-lb)
- steering column to body (bolts, install with Loctite). . . . . 20 to 27 Nm (15 to 20 ft-lb)
- steering wheel to steering column (self-locking nut) . . . . 25-28 Nm (18 to 21 ft-lb)

---

643 Tie Rod Ends

GENERAL

The tie rod ends connect the steering rack tie rods to the steering swivel member. They thread onto the end of the tie rod and are securely fastened to the swivel member with a self-locking nut. No maintenance or lubrication is needed for the life of the joint. See Fig. 1.

CAUTION
Do not reuse self-locking nuts. These parts are designed to be used only once and must be replaced if loosened or removed.

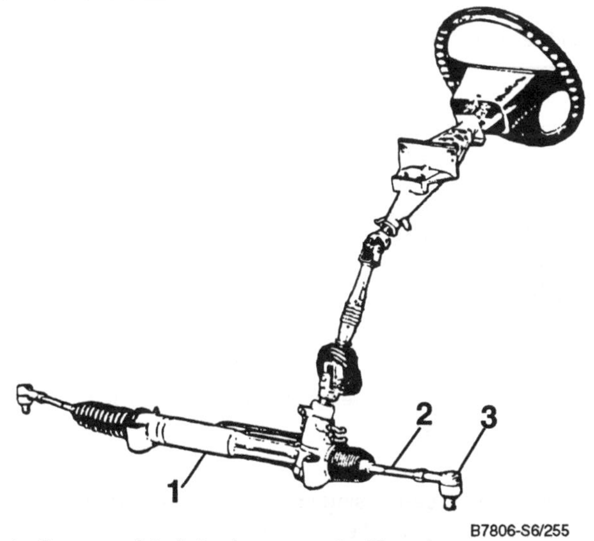

1. Power-assisted steering rack
2. Tie rod
3. Tie rod end

Fig. 1. Steering system.

Tie rod ends should be checked for wear at regular intervals. The rubber boots around the joints should also be inspected for tears, cracks or general deterioration. Replace the tie rod end if any faults are found.

Tie rod ends are best checked for wear with the front wheels off the ground. Grasp both sides of the tire and gently rock the wheel back and forth. If any noticeable play is present, the tie rod end is probably worn and should be replaced.

CAUTION
A new style tie rod end was introduced during the 1989 model year on the 900 models. This new tie rod end can be identified by a slot cut in the end of the threaded stud. See Fig. 2. The two tie rod ends must never be interchanged between the early cars and later cars, as steering geometry will be affected.

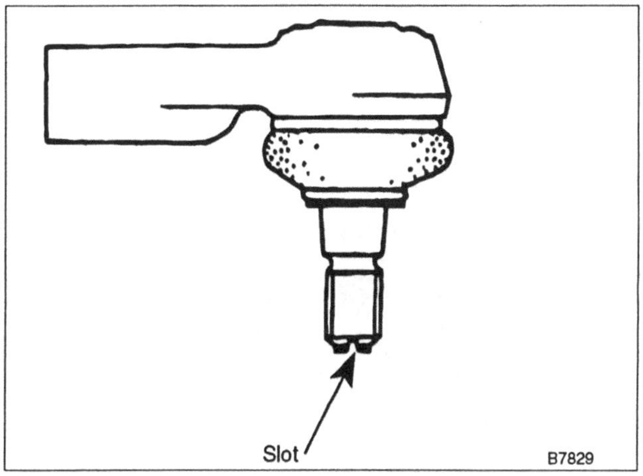

Fig. 2. New style tie rod end used on early 1989 and later models. This new tie rod end can be identified by slot cut into threaded end of stud.

To remove tie rod end

Removing the tie rod end will alter the alignment (toe-in) of the front wheels. It is highly recommended that the front wheels be aligned after replacing a tie rod end.

1. Loosen the wheel lug nuts or bolts, then raise the front of the car and support it securely on jack stands designed for the purpose. Remove the wheel.

WARNING
Always check that the car is stable on the jack stands before starting any work.

2. Loosen the locknut that secures the tie rod end to the tie rod. Remove the self-locking nut securing the tie rod end to the steering swivel member. See Fig. 3.

3. Using a press tool, remove the joint from the steering swivel member. See Fig. 4.

643

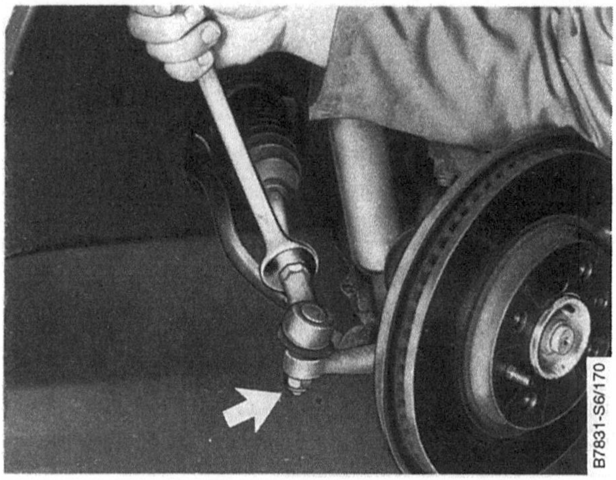

Fig. 3. Tie rod end locknut being loosened. Tie rod end self-locking nut shown at arrow.

B7896-S6420

Fig. 4. Tie rod end being removed from steering swivel member using press. Saab tool no. 89 95 409 shown.

WARNING

A tie rod end removal fork should not be used to remove the tie rod end. Do not hammer on the tie rod end in an attempt to remove it from the steering member. Doing this may damage the steering rack, the collapsible steering column, or the steering swivel member.

---

4. Mark the position of the tie rod end to the tie rod. Thread the tie rod end off the tie rod, counting the number of turns it takes to get it off.

NOTE

If the position of the tie rod end on the tie rod is carefully marked, the new tie rod end can be positioned where the old tie rod was prior to removal so the car can be safely driven.

---

5. Install the new tie rod end to the tie rod, threading it on the same number of turns counted above in step 4. Insert the tie rod end into the steering member and install and tighten a new self-locking nut.

Tightening torque

- tie rod end to steering member (self-locking nut). . . . . . . 50 to 60 Nm (37 to 44 ft-lb)

WARNING

The length of the tie rod from the locknut to the stepped part of the rod must never exceed 125 mm (4.92 in). See Fig. 2. Also, the difference in length between the two tie rods must not exceed 2 mm (0.08 in.). For example, if the right rod measured 124 mm, the left rod must measure between 122 mm and 125 mm. See Fig. 5.

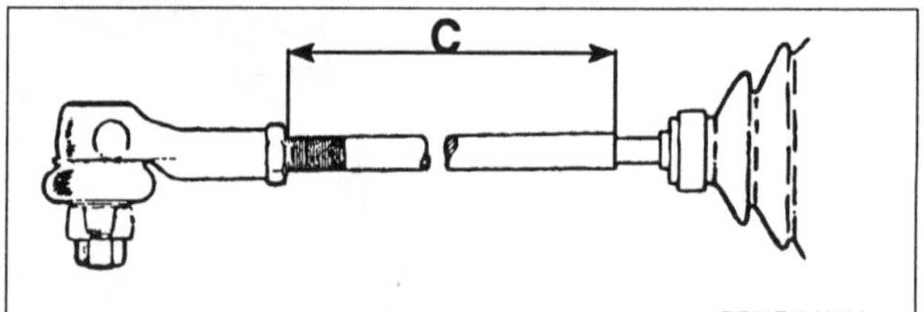

Fig. 5. Maximum permissible length (**C**) of tie rod must not exceed 125 mm (4.92 in.).

6. Adjust the toe-in using special wheel alignment equipment. Tighten the tie rod lock nut.

Tightening torque

- steering tie rod to tie rod end (locking nut). . . . 60 to 80 Nm (44 to 59 ft-lb)

---

644 Power Steering

POWER STEERING SYSTEM

The power-assisted steering rack is mounted at the lower rear of the engine compartment. A belt driven hydraulic pump supplies fluid under pressure to the control valve in the steering rack to help reduce steering effort. See Fig. 1.

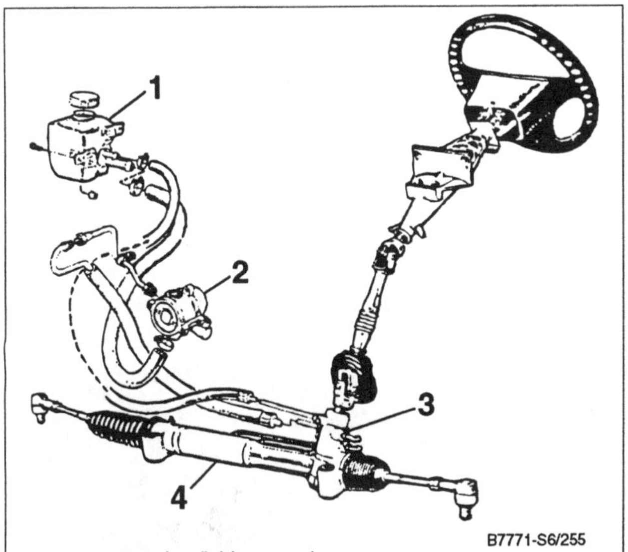

1. Power steering fluid reservoir
2. Power steering pump
3. Control valve (part of steering rack)
4. Steering rack

Fig. 1. Power steering system.

There is no routine maintenance for the power steering system other than checking the fluid level, see **1 LUBRICATION AND MAINTENANCE.** Checking the steering tie rod ends is described in **643 Tie Rod Ends.**

CAUTION

• Do not reuse self-locking nuts. These parts are designed to be used only once and must always be replaced whenever loosened or removed.

• Do not hold the steering wheel at full lock for extended periods with the engine running. This may overheat the pump and damage it.

To remove power steering rack

WARNING

Removing or replacing the steering rack will affect front wheel alignment. Always have the front end aligned after doing any work to the steering rack or related steering components. See **601 Wheel Alignment** for more information on aligning the wheels.

---

1. Loosen the wheel lug bolts, then raise the front of the car and support it securely on jack stands designed for the purpose. Remove the wheels.

WARNING

Always check that the car is stable on the jack stands before starting any work.

---

2. Drain the power steering fluid by removing the return hose from the fluid reservoir. See Fig. 2. Place the disconnected hose in a quart-sized container. Plug the open hose fitting at the reservoir.

NOTE

If necessary, connect an extension hose to the return hose.

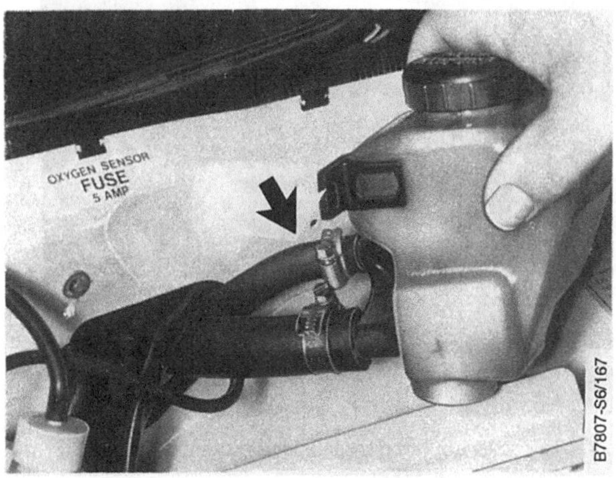

Fig. 2. Power steering pump return hose (arrow) at fluid reservoir.

644

3. Start the engine and allow the fluid to pump out of the system and into the container. Rotate the steering wheel slowly from lock to lock. Shut off the engine as soon as the system is empty. Reconnect the return hose to the reservoir.

4. Working at the wheels, loosen the tie rod end locknut at both tie rods. See Fig. 3.

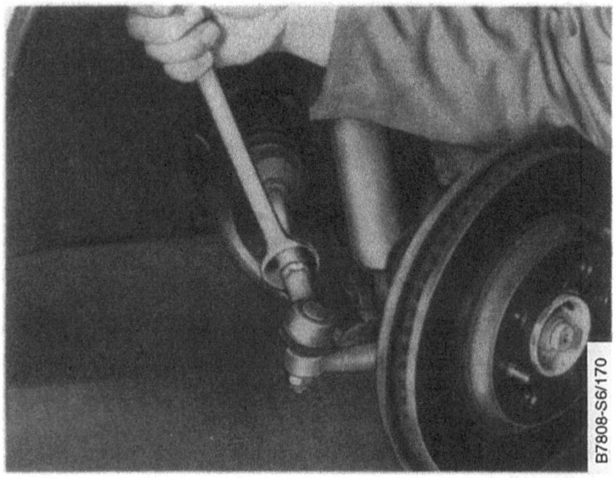

Fig. 3. Tie rod end locknut being loosened.

5. Using a vernier caliper, measure and record the tie rod length from the tie rod end to the stepped part tie rod. See Fig. 4.

NOTE

It may be necessary to push the bellows back slightly to access the stepped part of the tie rod.

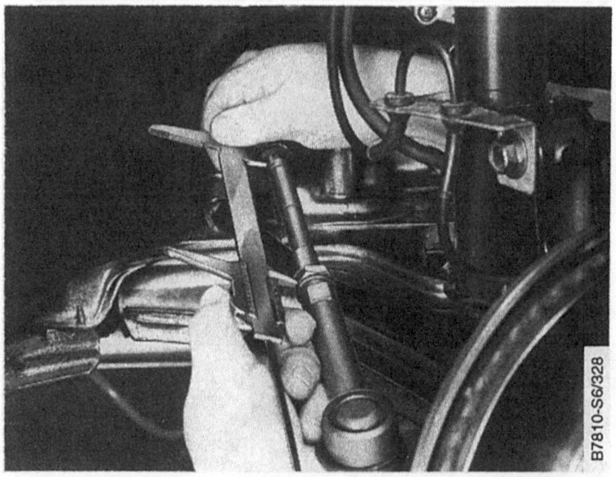

Fig. 4. Tie rod length being measured using vernier caliper. Car shown is Saab 9000.

6. Unscrew the tie rods from the tie rod ends. Hold the inner rubber bellows stationary while unscrewing the rods. See Fig. 5.

Fig. 5. Tie rod being unscrewed from tie rod end.

7. Thoroughly clean the area around the two hydraulic pipe fittings on the steering rack. See Fig. 6. Disconnect the pipes from the rack and plug the pipe openings. Remove the clamps holding the pipes to the rack.

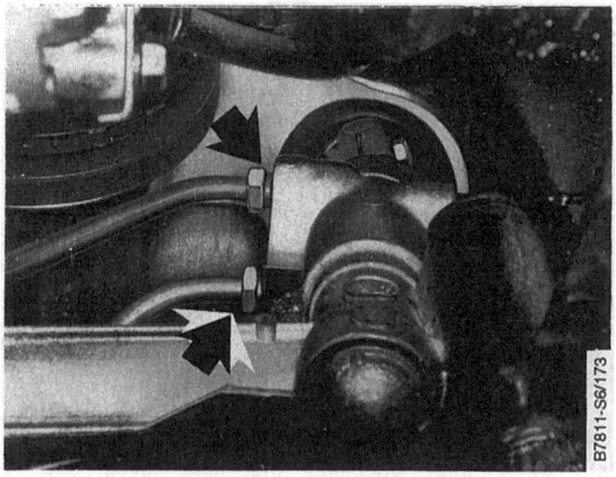

Fig. 6. Hydraulic pipe fittings on steering rack (arrows).

8. Working from the rear of the engine compartment, remove the clamping bolt holding the steering column universal joint to the steering rack. See Fig. 7.

9. Remove the bolts holding the steering rack to the body. See Fig. 8.

10. Separate the steering rack from the steering column universal joint.

11. Rotate the steering rack until its pinion shaft is pointing downward. See Fig. 9.

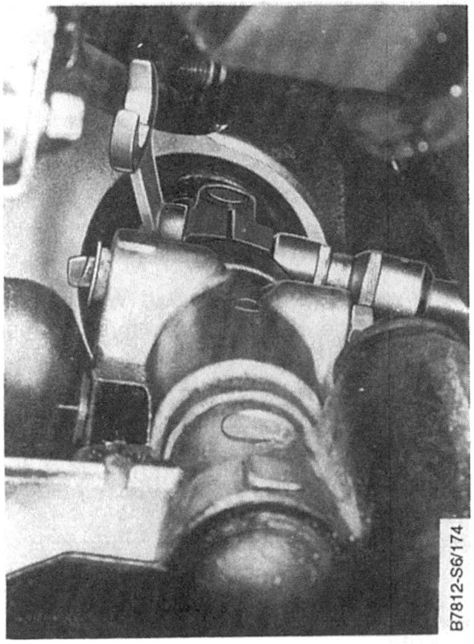

Fig. 7. Steering column clamping bolt being removed.

Fig. 8. Steering rack mounting bolts (arrows).

12. Slide the steering rack toward the passenger side of the car.

13. Turn the steering rack pinion shaft until the right-hand (passenger's) side tie rod is fully extended. Lift the left-hand (driver's) side tie rod over the subframe. See Fig. 10.

CAUTION
Turn the shaft using pliers. Place a rag over the pinion shaft to protect the splines.

---

14. Slide the entire rack toward the driver's side of the car and remove the rack assembly.

Fig. 9. Steering rack pinion shaft positioned correctly for removal of rack.

Fig. 10. Left-hand (driver's) side tie rod shown lowered from car body.

To install power steering rack

Inspect the rubber bellows and the rubber mounting bushings. Replace any worn or damaged parts.

1. Turn the pinion shaft until the driver's side tie rod is fully extended. Place the rack up and into the body with the pinion shaft pointing down. See Fig. 11.

2. Turn the pinion shaft until the passenger's side tie rod is fully extended. Lift the left-hand (driver's) side tie rod up and over the subframe member. Center the steering rack.

NOTE
When the steering rack is centered, the slot in the pinion shaft for the clamping bolt should be toward the rear of the car (shaft pointing downward).

Fig. 11. Steering rack being installed into body.

3. Rotate the entire rack up and connect the pinion shaft to the steering column universal joint. Loosely install the clamping bolt and new a self-locking nut.

CAUTION
Make sure the pinion shaft fully engages the universal joint. The clamping bolt must go through the slot in the pinion shaft.

---

4. Install and tighten the steering rack mounting bolts and the steering column clamping bolt.

Tightening torques
- power steering rack to body .... 60 to 80 Nm (44 to 59 ft-lb)
- universal joint to pinion shaft (clamping bolt and self-locking nut) .... 25 to 34 Nm (18 to 25 ft-lb)

---

5. Install and tighten the hydraulic pipes to the control valve. Replace the O-rings at the pipe connections where necessary.

Tightening torque
- power steering hose fittings .... 20 to 34 Nm (15 to 25 ft-lb)

---

6. Install new clamps to hold the pipes to the steering rack housing.

CAUTION
Use care not to pinch or damage the pressure-equalizing tube between the rubber bellows.

---

7. Thread the tie rods into the tie rod ends. Adjust the tie rod to the dimensions recorded above in Fig. 4. Tighten the locknuts.

Tightening torque
- tie rod end to tie rod ... 60 to 80 Nm (44 to 59 ft-lb)

---

8. Fill the reservoir with power steering fluid. With the engine off, slowly turn the steering wheel three or four times from lock to lock.

CAUTION
Do not use ATF in the power steering system. Use only GM (General Motors) Specification Power Steering Fluid.

Power steering fluid
- GM power steering fluid (GM 9985010), Texaco TL4634 or equivalent .... 75 cl (0.8 qt)

---

9. Install the wheels, lower the car and start the engine. Turn the wheels from lock to lock two more times.

Tightening torques
- wheel lug bolts...... 105 to 125 Nm (78 to 92 ft-lb)
- wheel lug nuts .... 90 to 110 Nm (66 to 81 ft-lb)

---

10. Check the front wheel alignment and adjust if necessary.

To remove power steering pump and lines

1. Drain the power steering fluid by removing the return hose from the fluid reservoir. See Fig. 2 above. Place the disconnected hose in a quart-sized container. If necessary, connect an extension hose to the return hose. Plug the open hose fitting at the reservoir.

2. Start the engine and allow the fluid to pump out of the system and into the container. Rotate the steering wheel slowly from lock to lock. Shut off the engine as soon as the system is empty. Reconnect the return hose to the reservoir.

3. Working in the engine compartment, cut the cable ties holding the main engine harness to the rear of the cylinder head. Rest the harness on top of the valve cover.

4. Slacken the pump bracket nut and bolt. Fully loosen the bracket adjusting screw. See Fig. 12. Remove the pump V-belt.

Fig. 12. Pump bracket nut (**1**), bracket bolt (**2**), adjusting screw (**3**), and outlet pipe (**4**).

5. Disconnect the fluid outlet pipe at the union shown in the above figure.

6. Working from below the car, remove the bracket's lower mounting bolt. Disconnect the inlet hose. See Fig. 13.

NOTE

Note the order of the rubber bushings, spacers, and washers for the lower mounting bolt.

---

7. Working from the engine compartment, remove the pump bracket top nut and bolt. Swivel the pump toward the wheel arch and remove it from the car. See Fig. 14.

Installation is the reverse of removal. Tighten the hose fittings to the specified torque. Adjust the belt tension as described in **1 LUBRICATION AND MAINTENANCE**. Fill and bleed the system as described above under the steering rack installation procedure.

Fig. 13. Pump bracket lower mounting bolt (**1**) and inlet hose (**2**) as viewed from below car.

Fig. 14. Power steering pump being removed.

Tightening torque

- power steering pump hose fittings .... 20 to 34 Nm (15 to 25 ft-lb)

---

700 Suspension—General

Front Suspension

The Saab 900 is equipped with an independent front suspension. See Fig. 1. The upper and lower control arms are pivot-mounted to the body. A double-acting shock absorber and coil spring controls suspension travel, which is limited by a rubber buffer. The steering swivel member is mounted to the control arms through ball joints.

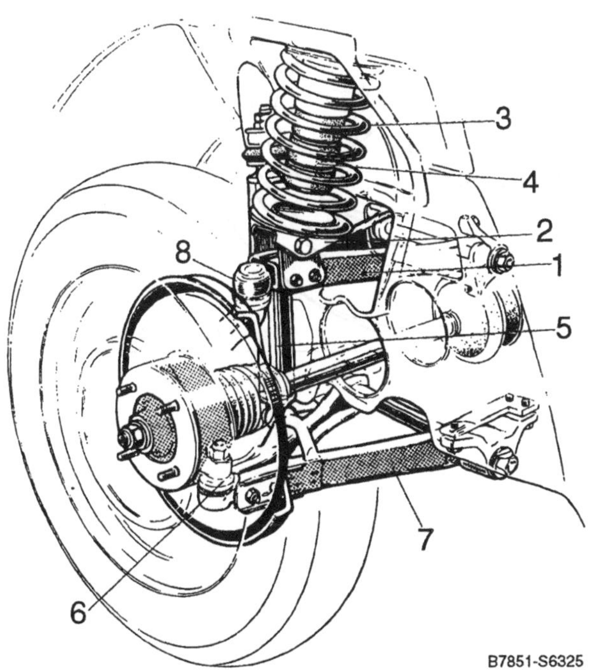

1. Upper control arm
2. Spring seat
3. Coil spring
4. Bump stop
5. Shock absorber
6. Lower ball joint
7. Lower control arm
8. Upper ball joint

Fig. 1. Front suspension.

The front wheel bearing is pressed into the steering member and the wheel bearing hub is in turn pressed into the wheel bearing. See Fig. 2.

The splined drive axle with integral constant velocity (CV) joint passes through the wheel bearing and is firmly bolted to the splined wheel bearing hub. The brake caliper is bolted to the steering swivel member and the brake disc is mounted to the wheel bearing hub. The steering rack is connected to the steering swivel member through the tie rod ends.

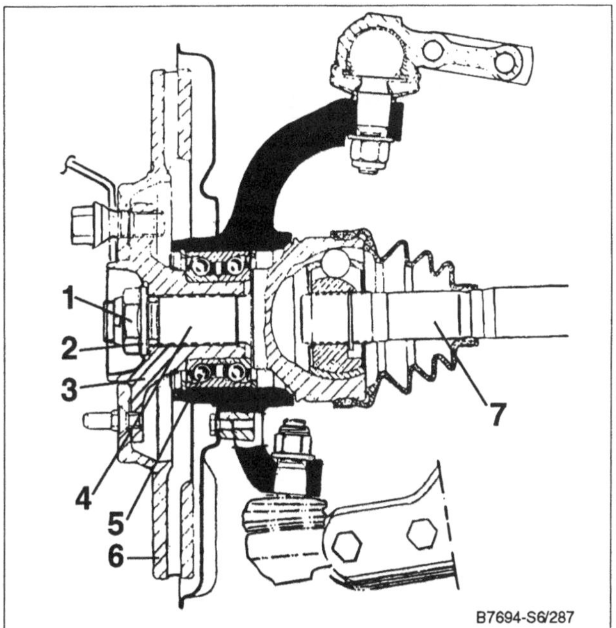

1. Locking nut (always replace)
2. Washer
3. Wheel bearing hub
4. Outer drive axle with integral CV joint
5. Wheel bearing
6. Brake disc
7. Inner drive axle

Fig. 2. Steering swivel member.

Rear Suspension

The rear axle is a lightweight tube located by five links designed to keep the rear wheels parallel and stable. See Fig. 3. The panhard rod absorbs the lateral cornering forces. The spring links and torque arms prevent the axle from rotating during braking and accelerating. On some models, a rear stabilizer bar further controls body sway or roll during cornering. Upward and downward movement of the axle is limited by the double acting shock absorbers and suspension coil springs.

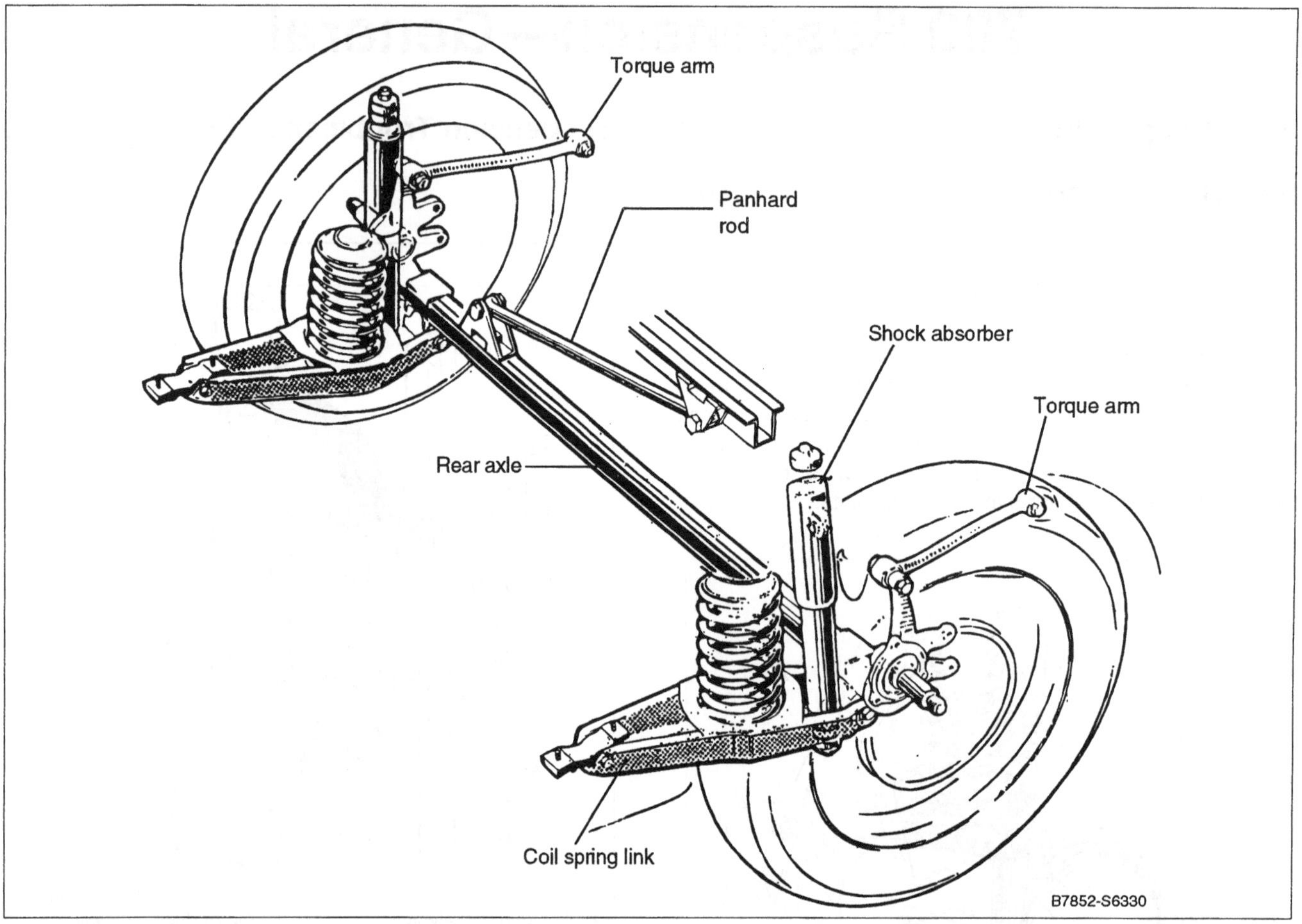

Fig. 3. Rear suspension.

SUSPENSION TROUBLESHOOTING

Stable handling and ride comfort depends on the integrity of the suspension components that gradually become worn with time and mileage. In addition to worn suspension parts, troubleshooting must consider the condition of the wheels, tires, and their alignment. General suspension and wheel alignment troubleshooting information can be found in **600 Steering and Wheel Alignment—General**.

731 Front Suspension

TABLES

GENERAL

This section covers replacement of the front shock absorbers, coil springs, and anti-roll bar only. For information on replacing the ball joints and the control arms, see **631 Ball Joints** and **632 Control Arms.**

SHOCK ABSORBERS

The function of a shock absorber, or damper, is to moderate the action of the coil spring. The shock slows the spring's bounce and allows it to return to its normal position more quickly. See Fig. 1.

Shocks require no routine maintenance and cannot be serviced. The shock absorbers are easily replaced using ordinary hand tools and should always be replaced in pairs.

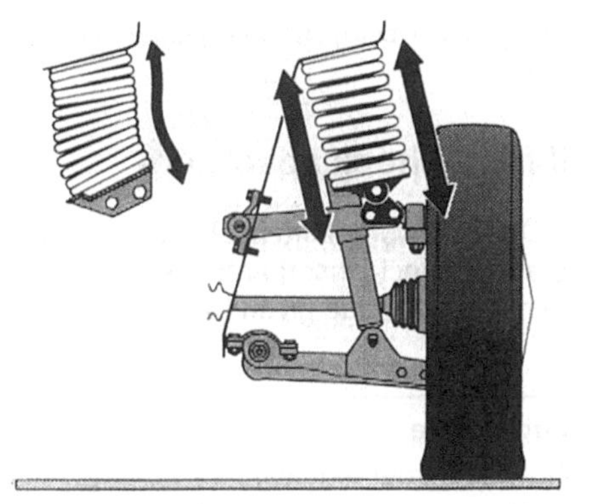

Fig. 1. Front shock absorber and pivoted suspension springs.

The best evidence of worn shock absorbers is their road handling behavior in normal driving. Worn shocks will allow skittishness over bumps and a less controlled, often wallowing, feel after bumps or in corners. Very worn shocks may reach the limits of their travel by knocking when going over bumps.

The most common—though not entirely accurate—test of shock absorber function involves vigorously bouncing each corner of the car and then releasing and observing how quickly the bouncing stops. More than one bounce usually suggests the shocks are not properly dampening and should be replaced. As a final check, look for oil leaks on the outside of the shock. If any signs of oil are present, the shocks should be replaced.

If a rattling or clunking noise is noticeable while traveling over rough road surfaces, check the lower shock absorber bushings. These bushing are available separately from an authorized Saab dealer. In addition, replacement shock absorbers come with new upper and lower bushing sets. See Fig. 2.

WARNING
Gas shock absorbers are under high pressure (400 to 600 psi) and can cause severe injury if handled improperly. Do not heat, weld, cut or deform the shock absorber in any way. Gas shock absorbers should be disposed of properly. See an authorized Saab dealer for information on gas shock disposal.

NOTE
The Saab 900 shock absorbers are of a unique design incorporating a special stop. Always replace them with Saab original equipment shock absorbers or damage to the suspension may result.

To remove front shock absorbers

1. Working in the engine compartment, remove the shock absorber upper mounting nuts. Loosen the top locknut first while holding the lower nut stationary. See Fig. 3.

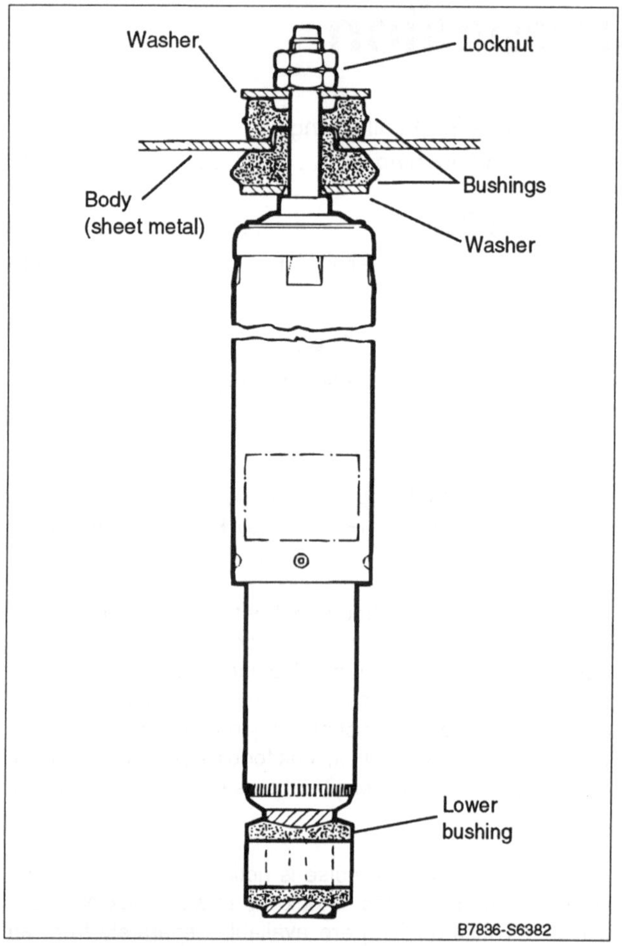

Fig. 2. Front shock absorber.

NOTE

- If the shock shaft turns while trying to loosen the nuts, cut the plastic protective tube off the shock from below and then use a locking pliers to hold the shaft stationary.
- When removing the driver's side shock, the mounting nuts can be more easily accessed if the coolant expansion tank is moved out of the way. Do not remove the hoses or cap. Remove only the fasteners holding the tank to the body.

---

2. Loosen the wheel lug bolts, then raise the front of the car and support it securely on jack stands designed for the purpose. Remove the front wheel.

WARNING

Always check that the car is stable on the jack stands before starting any work.

---

3. Remove the shock absorber lower mounting nut and washer. Remove the shock absorber. See Fig. 4. If necessary, pry the shock off its control arm mount.

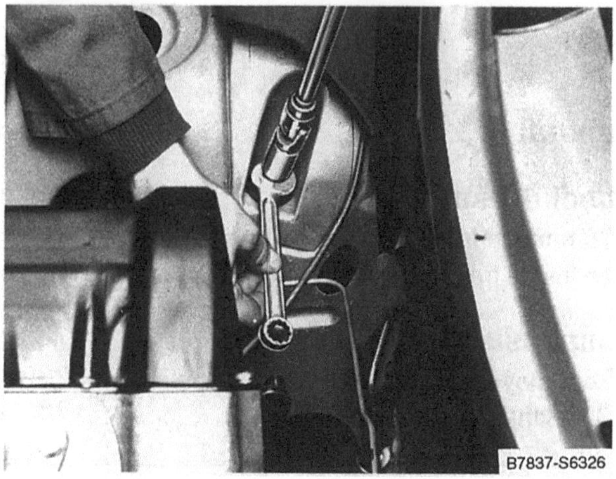

Fig. 3. Shock absorber upper mounting nut being removed. Open-end wrench is used to prevent lower nut from spinning when loosening top nut.

Fig. 4. Shock absorber being removed from lower control arm mount.

To install front shock absorbers

1. Lubricate the lower mount bushings with petroleum jelly. Install the shock absorber to the lower mount. Tighten the nut to the value given below.

Tightening torque

- front shock absorber to lower control arm . . . . 90 to 100 Nm (66 to 74 ft-lb)

---

2. Install the washer and bushing (lower) onto the top of the shock. See Fig. 5.

3. Using a jack, raise the control arm slightly while guiding the shock into its opening in the body. See Fig. 6. Working in the engine compartment, install the bushing (upper), the washer and nuts. Tighten the nuts.

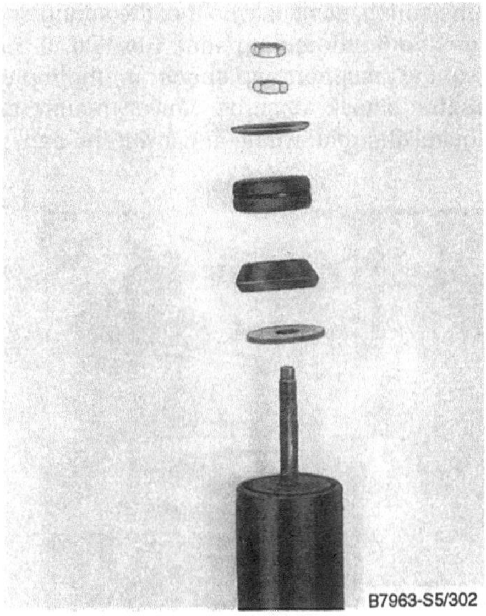

Fig. 5. Shock absorber mounting bushings, washers and nuts. Note the orientation of the upper dished washer (concave faces up) and the taper on the lower bushing.

NOTE
Tighten the shock upper mounting nuts just until the upper bushing is expanded to the outer edges of the washer.

Fig. 6. Raise lower control arm slightly using jack.

4. Lower the jack from under the control arm. Install and tighten the wheel bolts or nuts. Lower the car. Install the coolant reservoir if necessary.

Tightening torques
- wheel lug bolts ..... 105 to 125 Nm (78 to 92 ft-lb)
- wheel lug nuts ...... 90 to 110 Nm (66 to 81 ft-lb)

SUSPENSION COIL SPRINGS

Fig. 7 shows the suspension coil spring. The coil springs do not require any service or maintenance and are not normally replaced unless damaged or fatigued. Coil springs should always be replaced in sets. Removing the coil springs requires special tools. Read the procedures through before starting the job.

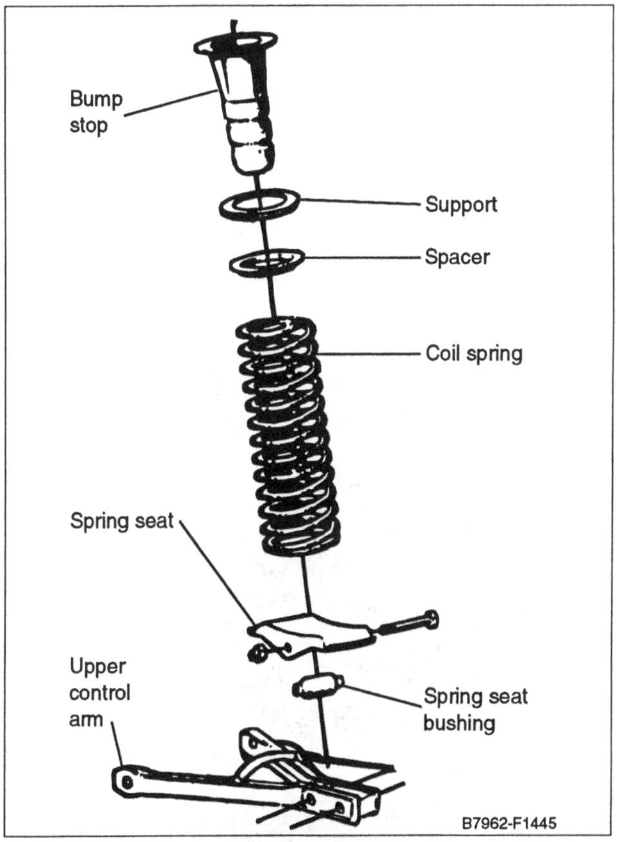

Fig. 7. Exploded view of front suspension coil spring and related components.

WARNING
Do not attempt to service the coil springs without a spring compressor designed specifically for this purpose. Attempting this job without the proper tools can result in serious injury.

To remove coil springs

1. Loosen the wheel lug bolts, then raise the front of the car and support it securely on jack stands designed for the purpose. Remove the front wheel.

WARNING
Always check that the car is stable on the jack stands before starting any work.

---

2. Remove the shock absorber lower mounting nut and washer and remove the shock from its control arm mount. If necessary, pry the shock from the mount.

3. Install the coil spring compressor to the spring. Slowly compress the spring and remove it from car. See Fig. 8.

WARNING
Work carefully as the spring is compressed. Check to make sure that the spring compressor is always securely seated on the spring.

NOTE
Mark the coil spring as an aid to installation. Used coil springs should always be installed in their original position and orientation.

Fig. 8. Compressed coil spring being removed from car. Spring compressor shown is Saab tool no. 88 18 791 and tool no. 88 18 809.

Installation of the spring is the reverse of removal. Inspect all rubber parts for wear or deterioration and replace any faulty parts found. Check the spring free length as described below. Inspect the protective coating on the spring. The springs are treated with an anti-corrosion agent, which should be touched up with paint or an equivalent sealer.

When installing the spring, position it so that the spring end is fitted against the stop on the spring seat. See Fig. 9. Be sure to install the spring's support and spacer on the top of the spring. Install the shock absorber lower mount as described above. Install the road wheel and lower the car.

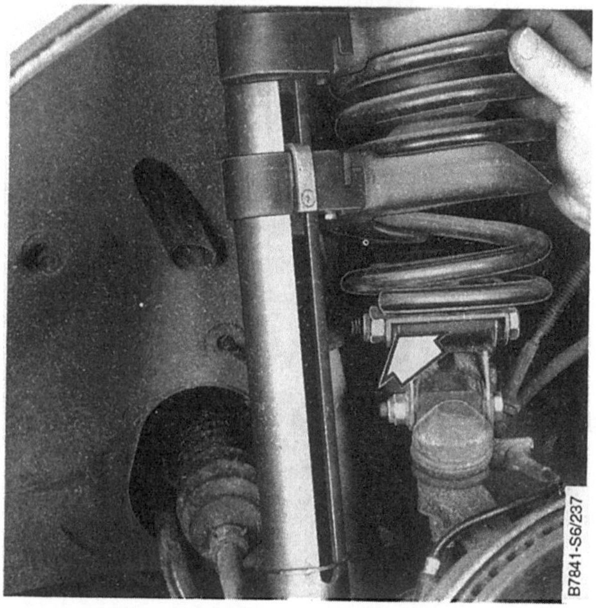

Fig. 9. Spring end (arrow) correctly positioned against spring stop on spring seat.

Tightening torques
- front shock absorber to lower control arm .... 90 to 100 Nm (66 to 74 ft-lb)
- wheel lug bolts ...... 105 to 125 Nm (78 to 92 ft-lb)
- wheel lug nuts .... 90 to 110 Nm (66 to 81 ft-lb)

Checking Coil Springs

A function of the coil spring is to control the ride height of the car. Therefore, a damaged or fatigued coil spring will usually result in a noticeable difference in ride height. **Table a** lists spring specifications. The spring must be removed from the car to measure its free length.

Various springs are installed depending on model and model year. The springs are color-coded with a paint dot. In addition, the left and right springs on some models may be different. In all cases, the new spring should match the old spring being replaced.

NOTE
For chassis ride height specifications see **601 Wheel Alignment.**

Table a. Front Coil Spring Specifications

Spring color codes	Spring free length
Standard chassis	
green, light green, black, or white	373 mm (14.7 in.)
yellow or red	380 mm (15.0 in.)
pink or brown	388 mm (15.3 in.)
blue or light blue	372 mm (14.6 in.)
Sport chassis	
silver or bronze	301 mm (11.8 in.)

To remove front coil spring seat and bushing

The front coil spring seat and bushing can be replaced after the coil spring has been removed.

1. Remove the coil spring as described above. Remove the bolt for the spring seat and remove the spring seat.

2. Press out the spring seat bushing using a suitable press (Saab tool no. 89 96 274 or equivalent). See Fig. 10.

Fig. 10. Spring seat bushing being pressed out.

3. Lubricate the new bushing with petroleum jelly and install it using a press (Saab part no. 89 96 274 or equivalent). See Fig. 11.

4. Install the spring seat with the spring stop facing out. See Fig. 12. Loosely tighten the spring seat bolt using a new self-locking nut.

Fig. 11. Spring seat bushing being pressed in.

WARNING
Do not reuse self-locking nuts. These parts are designed to be used only once and must be replaced whenever loosened or removed.

Fig. 12. Spring stop (arrow) on spring seat faces out.

Install the coil spring as described above. Make sure the end of the coil spring is against the stop on the spring seat as shown earlier in Fig. 9. When the weight of the car is on the wheels, tighten the spring seat mounting bolt.

Tightening torque
- coil spring seat to control arm .... 55 to 110 Nm (40 to 81 ft-lb)

ANTI-ROLL BAR

The front anti-roll bar is rigidly mounted to the body and to the lower control arms to control body sway or roll during cornering. The anti-roll bar bushings should be periodically inspected for deterioration, wear, or oil contamination. Special tools are needed to remove the anti-roll bar. Read the procedure through before starting the job.

To remove front anti-roll bar

1. With the car on the ground, insert a special spacer tool (Saab part no. 83 93 209) between the upper control arm and the body on each side of the car. See Fig. 13.

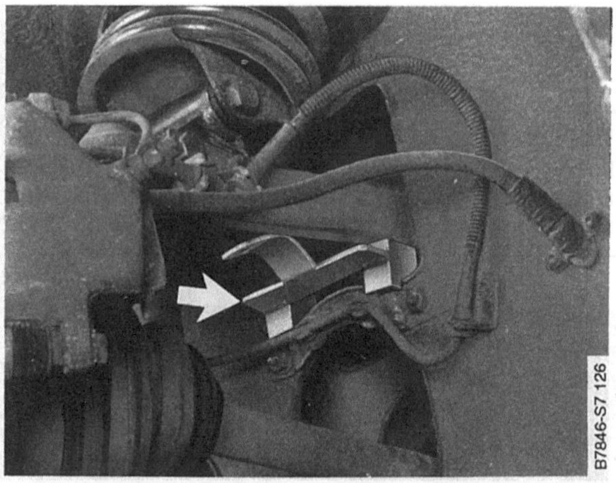

Fig. 13. Spacer tool (arrow) inserted between upper control arm and body. Insert one tool on each side of the car (2 tools needed). The spacer tool relieves suspension loading on the anti-roll bar.

2. Loosen the wheel lug bolts, then raise the front of the car and support it securely on jack stands designed for the purpose. Remove the front wheels.

WARNING
Always check that the car is stable on the jack stands before continuing any work.

---

3. Working at the lower control arms, remove the anti-roll bar mounting bolts. See Fig. 14. Remove the rubber end pieces.

4. Working from below the car, remove the two U-clamps and split rubber bushings holding the roll bar to the floor pan. See Fig. 15.

NOTE
Save the shims when the U-clamps are removed.

Fig. 14. Anti-roll bar to control arm mounting bolts (arrows).

Fig. 15. Anti-roll bar U-clamp mounting bolt being removed. Arrow shows additional nut to be removed.

5. Unbolt the steering rack and move it forward slightly. See Fig. 16.

CAUTION
Do not reuse self-locking fasteners. These parts are designed to be used only once and must be replaced whenever loosened or removed.

---

6. Turn the bar until the ends are facing up and slide the bar from the body. See Fig. 17.

Installation is the reverse of removal. Lubricate all rubber bushings with Molycote 1-4382 or equivalent. Loosely install the roll bar. Insert the U-clamp split rubber bushings with the split toward the front of the car. See Fig. 18. Loosely install the U-clamps, shims and nuts. Center the roll bar in the outer bushings and tighten the nuts. Tighten the U-clamp nuts and finally install and tighten the steering rack mounting bolts.

Fig. 16. Steering rack mounting bolts (arrows).

Fig. 17. Anti-roll bar being removed.

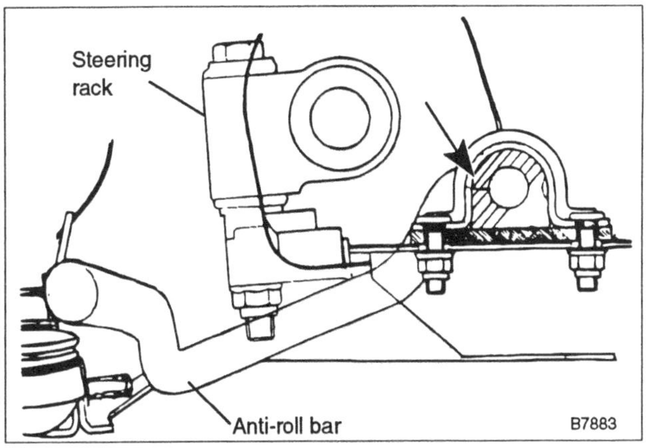

Fig. 18. Correctly installed anti-roll bar bushing with split (arrow) facing forward.

Tightening torques

- anti-roll bar bracket to lower control arm. . . . . 40 to 55 Nm (30 to 40 ft-lb)
- power steering rack to body . . . . 60 to 80 Nm (44 to 59 ft-lb)
- all other fasteners
  - M5 bolt . . . . 5 Nm (44 in-lb)
  - M6 bolt . . . . 10 Nm (89 in-lb)
  - M8 bolt . . . . 20 Nm (15 ft-lb)
  - M10 bolt . . . . 40 Nm (30 ft-lb)

732 Rear Suspension

TABLES

SHOCK ABSORBERS

The function of a shock absorber, or damper, is to moderate the action of the coil spring. See Fig. 1. The shock slows the spring's bounce and allows it to return to its normal position more quickly. For a quick check to assess the general condition of the rear shocks, see **731 Front Suspension.**

WARNING
Gas shock absorbers are under high pressure (400 to 600 psi) and can cause severe injury if handled improperly. Do not heat, weld, cut or deform the shock absorber in any way. Gas shock absorbers should be disposed of properly. See an authorized Saab dealer for information on gas shock disposal.

NOTE
The Saab 900 shock absorbers are of a unique design incorporating a special stop. Always replace them with Saab original equipment shock absorbers or damage to the suspension may result.

To remove rear shock absorbers

Fig. 2 shows the shock absorber and it related mounting hardware. The rubber bushings should always be replaced as part of shock replacement.

1. Loosen the rear wheel lug bolts or nuts, then raise the rear of the car and support it securely on jack stands designed for the purpose. Remove the rear wheels.

WARNING
Always check that the car is stable on the jack stands before starting any work.

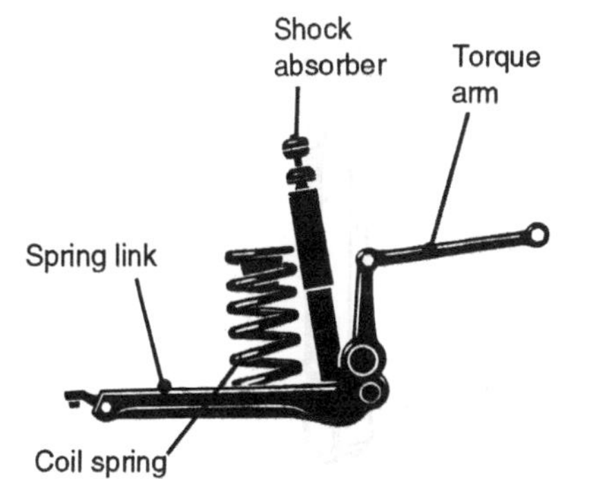

Fig. 1. Rear shock absorber and related suspension components.

2. Using a jack, raise the spring link slightly to relieve the suspension load on the shock. Remove the shock's lower mounting nut and bolt. See Fig. 3.
3. Slowly lower the jack slightly, but do not remove it.
4. Working in the wheel well, remove the upper shock absorber mounting nuts, washers, and rubber bushings. See Fig. 4. Remove the shock absorber from the body.

To install rear shock absorbers

1. Install the washer and lower bushing to the top of the shock. Use Fig. 2 as a guide to bushing and washer orientation.Guide the shock into the opening in the body. Install the upper bushing, washer and lock nuts. Tighten the nuts just until the upper bushing is expanded to the outer edges of the washer.

732

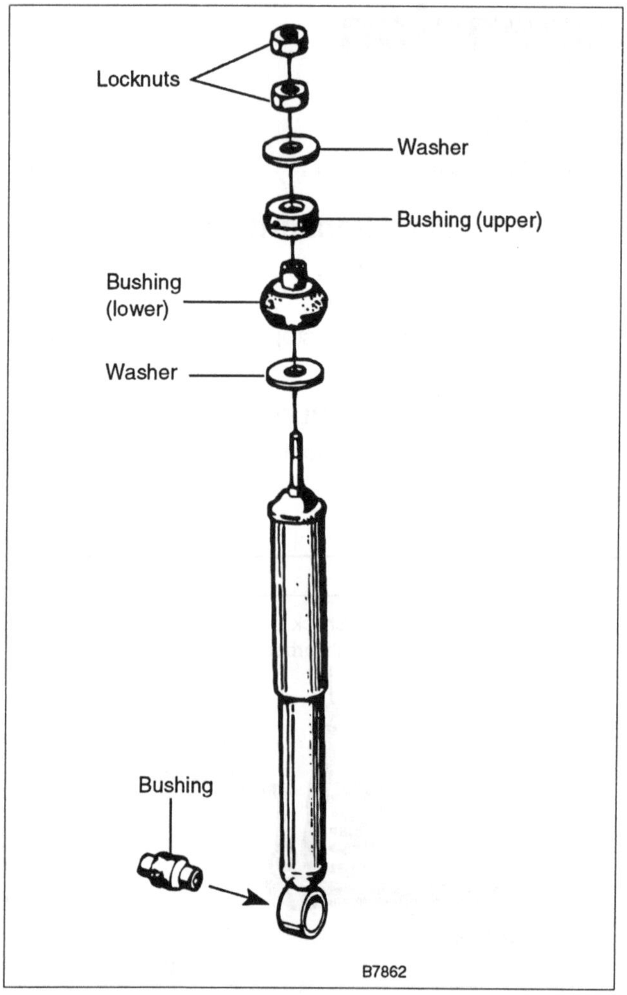

Fig. 2. Rear shock and mounting hardware.

Fig. 3. Shock absorber lower mounting bolt being removed.

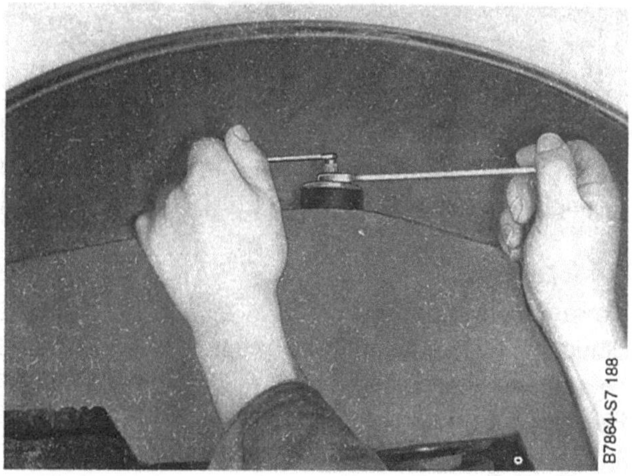

Fig. 4. Shock absorber upper mounting nuts being removed.

2. Raise the spring link slightly until the lower mounting hole in the link is aligned with the shock mounting hole. Install and tighten the bolt using a new self-locking nut. Remove the jack.

WARNING
Do not reuse self-locking nuts. These parts are designed to be used only once and should be replaced once loosened or removed.

---

3. Install the wheel and lower the car.

Tightening torques
- wheel lug bolts . . . . . . 105 to 125 Nm (78 to 92 ft-lb)
- wheel lug nuts . . . . 90 to 110 Nm (66 to 81 ft-lb)

COIL SPRINGS

Fig. 5 shows the suspension coil spring. The springs do not require any service or maintenance and are not normally replaced unless damaged or fatigued. Coil springs should always be replaced in sets.

To remove coil spring

1. Loosen the rear wheel lug bolts or nuts, then raise the rear of the car and support it securely on jack stands designed for the purpose. Remove the rear wheels.

WARNING
Always check that the car is stable on the jack stands before starting any work.

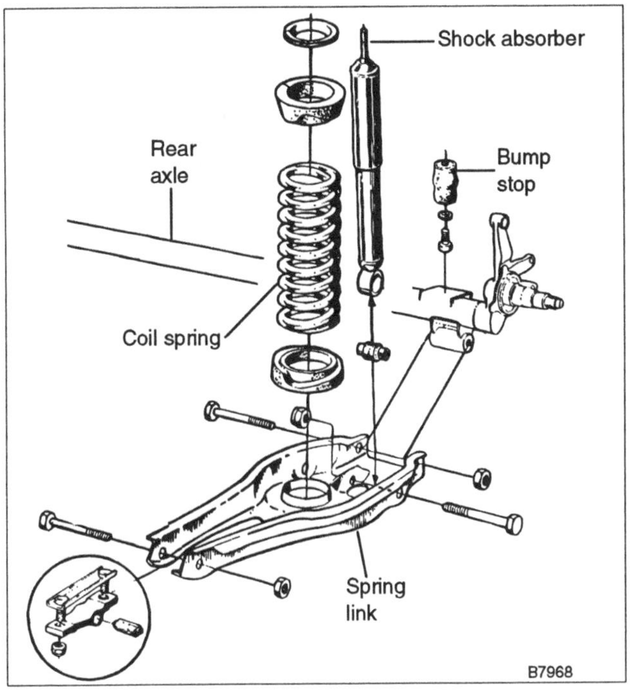

Fig. 5. Exploded view of coil spring and related suspension components.

2. On 1988 and later models, remove the handbrake cable hold down clamp from the spring link. See Fig. 6

Fig. 6. Handbrake cable hold down being removed from spring link.

3. On cars with ABS, remove the wheel speed sensor from the rear axle housing. See Fig. 7. Release the sensor harness from the clips on the spring link.

4. Using a jack, slightly raise the spring link. Then remove the shock absorber lower mounting bolt. See Fig. 3 shown earlier.

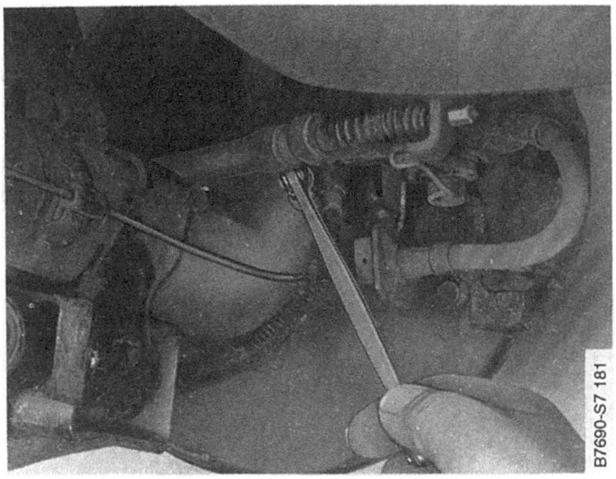

Fig. 7. ABS wheel speed sensor mounting bolt being removed from rear axle housing.

NOTE
The jack should be positioned under the rear of the spring link.

---

5. Remove the bolt and nut from the spring link rear mounting. See Fig. 8.

WARNING
Do not remove the front spring link mounting bolts while the spring is installed. Serious personal injury could result.

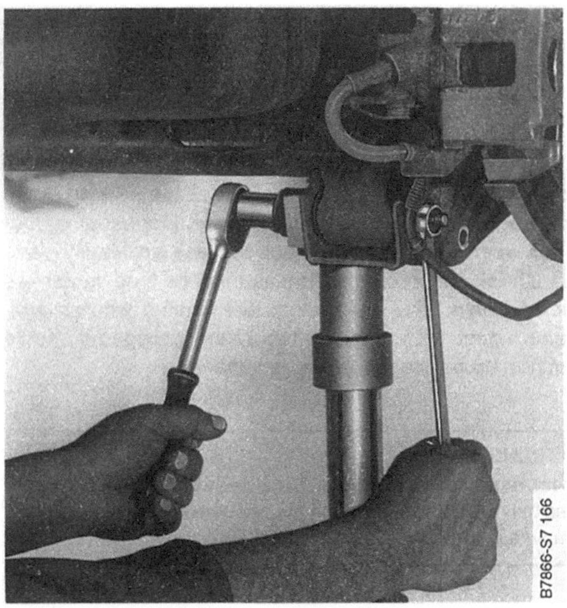

Fig. 8. Spring link rear mounting bolt being removed.

6. On cars equipped with a rear an anti-roll bar, jack up the spring link on the opposite side of the car and remove the mounting bolt from the rear of the other spring link.

7. Loosen, but do not remove, the anti-roll bar front mounting bolts and swivel the anti-roll bar down and out of the way. See Fig. 9.

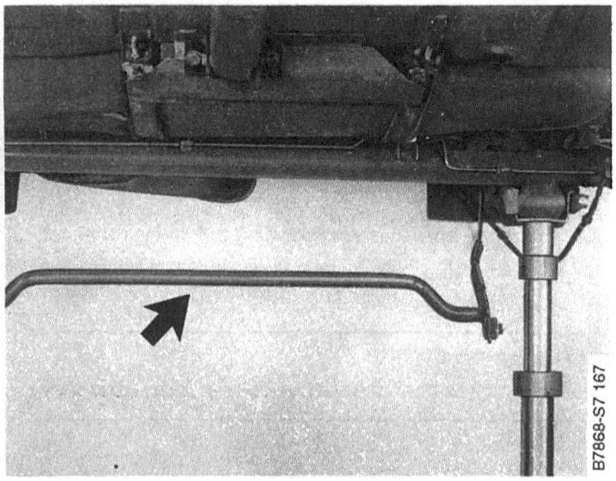

Fig. 9. Anti-roll bar (arrow) shown lowered from spring link.

8. Support the rear axle from below using a set of jack stands. Slowly lower the spring link by means of the jack. Remove the coil spring. If necessary, repeat the procedure for the other spring. See Fig. 10.

CAUTION
The rear axle must be supported to avoid damaging the flexible brake lines.

Installation is the reverse of removal. Be sure to replace all self-locking nuts. Where applicable, install the parking brake bracket and the ABS wheel speed sensor. Inspect all rubber parts for wear or deterioration and replace any faulty parts found. Check the spring by measuring its free length as described below. The coil springs are treated with an anticorrosion agent. If the coating has been damaged it should be touched up before the spring is installed.

WARNING
Do not reuse self-locking nuts. These parts are designed to be used only once and should be replaced whenever loosened or removed.

NOTE
If the bolt mounting hole in the spring link does not align with the bolt mounting hole in the rear axle, use a large water pump pliers and a screwdriver to align the holes. See Fig. 11.

Fig. 10. Coil spring being removed from car after lowering spring link.

Fig. 11. Rear axle being aligned with spring link using large water-pump pliers and screwdriver.

Tightening torques
- wheel lug bolts . . . . . . 105 to 125 Nm (78 to 92 ft-lb)
- wheel lug nuts . . . . 90 to 110 Nm (66 to 81 ft-lb)
- all other fasteners
  - M5 bolt . . . . 5 Nm (44 in-lb)
  - M6 bolt . . . . 10 Nm (89 in-lb)
  - M8 bolt . . . . 20 Nm (15 ft-lb)
  - M10 bolt . . . . 40 Nm (30 ft-lb)

Checking Suspension Coil Springs

A function of the coil spring is to control the ride height of the car. Therefore, a damaged or fatigued coil spring will usually result in a noticeable difference in ride height. **Table a** lists spring specifications. The spring must be removed from the car to measure their free length.

NOTE
For chassis ride height specifications see **601 Wheel Alignment.**

Various springs are installed depending on model and model year. The springs are color-coded with a paint dot. In addition, the left and right springs on some models may be different. In all cases, the new spring should match the old spring being replaced.

Table a. Rear Coil Spring Specifications

Spring color codes	Spring free length
Standard chassis	
black or white (left spring)	311 mm (12.24 in.)
green or light green (right spring)	308 mm (12.12 in.)
Sport chassis	
silver or bronze	293 mm (11.5 in.)

To remove rear coil spring link

1. Remove the coil spring as described above.

2. On cars equipped with an anti-roll bar, remove the spring link front mounting nuts on both sides of the car. Then remove the anti-roll bar from the spring links. See Fig. 12.

NOTE
If the spring link front mounting bolt was installed with the bolt head on the anti-roll bar side of the link, the bolt will have to be removed to remove the anti-roll bar.

---

3. Remove the two bolts that hold the spring link front mount to the body. See Fig. 13. Remove the spring link from the car.

Installation is the reverse of removal. Inspect the front and rear spring link rubber bushings and replace any parts that are worn or deteriorated. Be sure to install the spring link front mounting through-bolt so that the bolt head is towards the outside of the car.

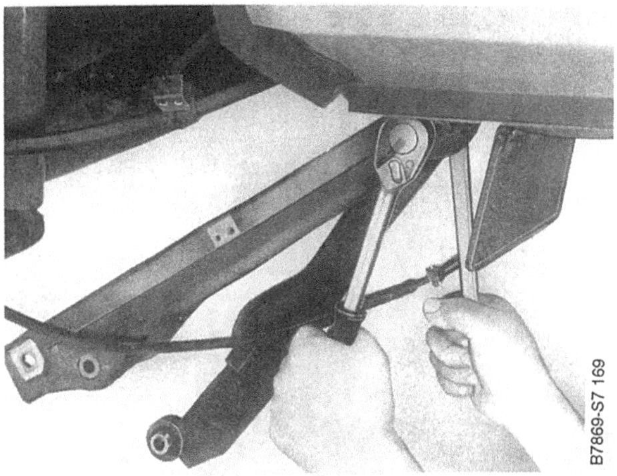

Fig. 12. Spring link front mounting nut being removed.

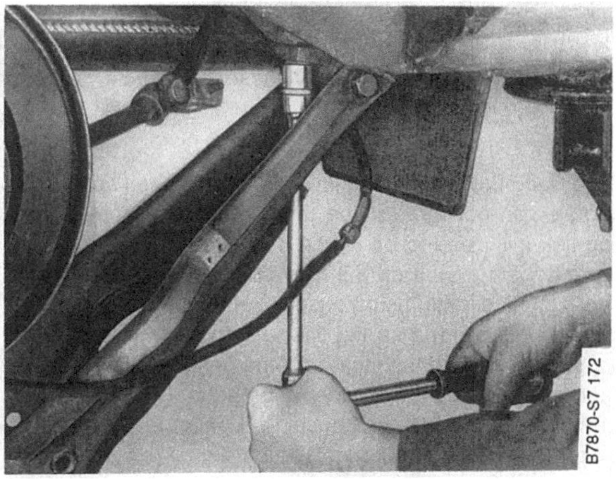

Fig. 13. Spring link front mount bolt being removed.

Tightening torques
- spring link front mount to body . . . . 70 to 90 Nm (52 to 66 ft-lb)
- all other fasteners
  - M5 bolt . . . . 5 Nm (44 in-lb)
  - M6 bolt . . . . 10 Nm (89 in-lb)
  - M8 bolt . . . . 20 Nm (15 ft-lb)
  - M10 bolt . . . . 40 Nm (30 ft-lb)

Replacing Spring Link Bushings

Replacing the spring link bushings requires special press tools.

To replace the spring link front bushing, the spring will need to be removed from the car as described above. With the link removed, Use a press and an appropriate press sleeve to remove and install the bushing. See Fig. 14. Lubricate the bushing with petroleum jelly prior to installation.

Fig. 14. Front spring link bushing being pressed out of front mount.

To replace the rear spring link bushing, the coil spring and spring link do not need to be removed, but the rear of the spring link will need to be lowered to access the bushing in the axle housing. Support the rear axle, then raise the rear of the spring link slightly using a jack. Remove the shock lower mounting bolt and the spring link rear mounting bolt. Lower the spring link (using the jack) until the bushing can be easily worked on. Then using the special press tools shown in Fig. 15, remove and install the bushing. Lubricate the bushing with petroleum jelly prior to installation.

Fig. 15. Spring link rear mount bushing (in rear axle) being pressed out using special press tools (Saab tool no. 89 96 274 shown). Same tool is used to install bushing.

ANTI-ROLL BAR

Some models are equipped with an anti-roll bar to reduce body roll while cornering. The anti-roll bar is rigidly mounted to both spring links.

To remove rear anti-roll bar

1. Loosen the rear wheel lug bolts or nuts, then raise the rear of the car and support it securely on jack stands designed for the purpose. Remove the rear wheels.

WARNING
Always check that the car is stable on the jack stands before continuing any work.

---

2. Raise the rear of the spring links slightly using a jack. Then remove the spring link rear mounting bolts on both sides of the car. See Fig. 16.

NOTE
After the anti-roll bar has been removed, reinstall the nuts for the rear mounting bolts.

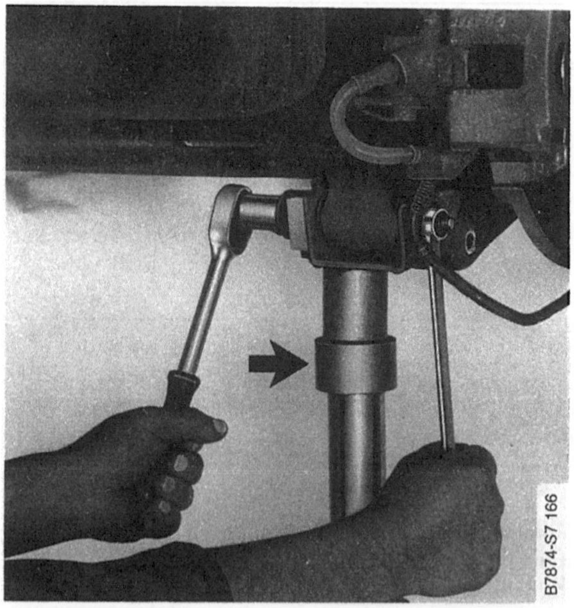

Fig. 16. Spring link rear mounting bolt being removed. Jack (arrow) shown under spring link.

3. Remove the nuts from the spring link front mounting bolts. Do not remove the bolts.

CAUTION

If the spring link front mounting bolt was installed with the bolt head on the anti-roll bar side of the link, the spring link will have to be removed as described above. Do not remove the front spring link mounting bolt without first removing the suspension coil spring.

---

4. Remove the anti-roll bar by prying it off the front mounting bolt. See Fig. 17.

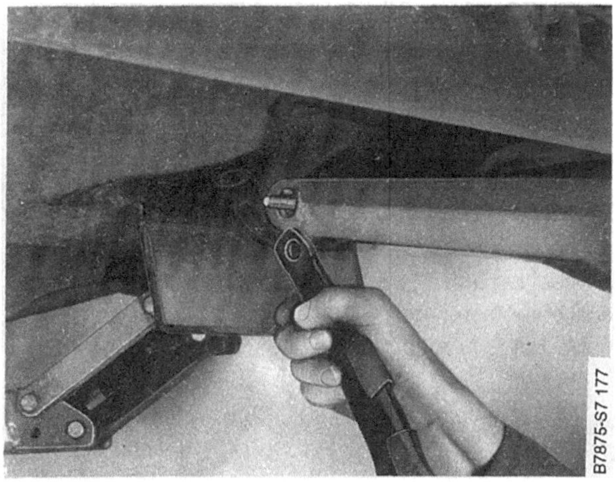

Fig. 17. Anti-roll bar being removed front mounting bolt.

Installation is the reverse of removal. Leave the spring link front mounting nuts loosely installed until the rear bolts have been installed and tightened.

REAR AXLE

Read the entire procedure before beginning any work. Extra jacks are required to support the axle during removal.

CAUTION

Do not place a jack under the rear axle to jack up the car. The axle may be damaged.

To remove rear axle

1. Loosen the rear wheel lug bolts or nuts, then raise the rear of the car and support it securely on jack stands designed for the purpose. Remove the rear wheels.

WARNING

Always check that the car is stable on the jack stands before starting any work.

---

2. Unbolt the panhard rod from the rear axle. See Fig. 18. Tie the rod up out of the way.

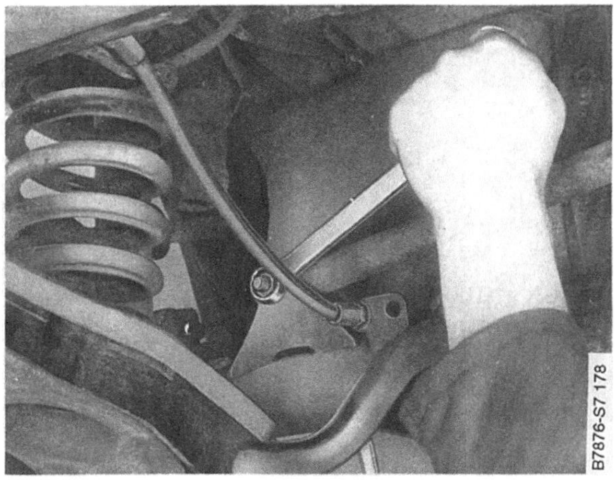

Fig. 18. Panhard rod mounting bolt being removed from rear axle.

3. Pinch off the brake line with a special hose clamp plier. As an alternative, drain the brake system as described in **520 Brake Fluid and Brake Bleeding.** Then separate the flexible brake hoses from the brake line unions on the rear axle. See Fig. 19.

WARNING

The flexible brake hoses must only be pinched off using special hose clamping pliers. Do not use any other type of clamping device or locking pliers, as the hose may become damaged and internally weakened.

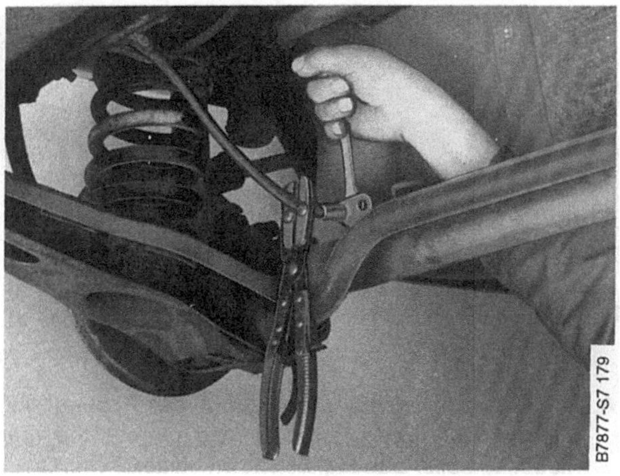

Fig. 19. Brake hoses being disconnected from brake line fitting on rear axle. Special brake hose clamping pliers shown on hose.

4. On cars with ABS, remove the wheel speed sensors from each side of the rear axle housing. See Fig. 20. Release the sensor harness from the clips on the spring link.

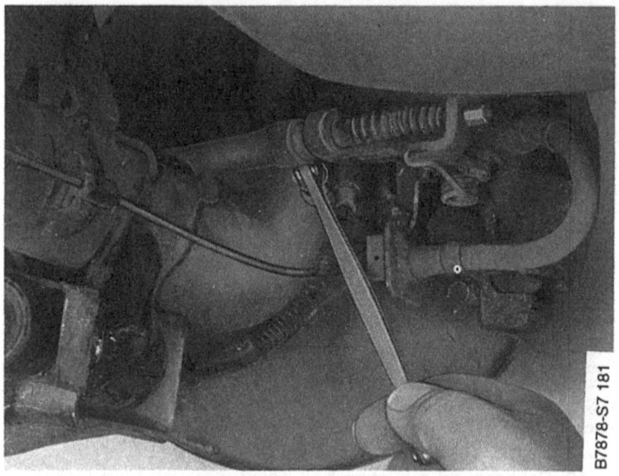

Fig. 20. ABS wheel speed sensor mounting bolt being removed from rear axle housing.

5. On 1988 and later models, disconnect the handbrake cables from the brake caliper levers. See Fig. 21. Remove the handbrake cable hold-down bracket from the axle (see Fig. 6 shown earlier).

Fig. 21. Handbrake cable (**A**) disconnected from brake lever (**B**).

6. Place a jack beneath the rear of each spring link. Slightly raise the spring links and remove the rear mounting bolts. See Fig. 16 shown earlier.

7. Place a jack under the center of the axle. Then unbolt the torque arms from both sides of the axle.

8. Lower the spring links slightly. Raise the axle up just enough to remove it from the spring link mountings and remove the axle from the car.

Installation is the reverse of removal. Replace all self-locking fasteners. Bleed the brakes as described in **520 Brake Fluid and Brake Bleeding.** Check all braking functions including the handbrake.

CAUTION
When installing the panhard rod to the axle, be certain that the bolt head is toward the fuel tank. Damage to the fuel tank may result if the bolt is installed incorrectly. Tighten the wheels to the torque values given below.

Tightening torques
- wheel lug bolts ...... 105 to 125 Nm (78 to 92 ft-lb)
- wheel lug nuts .... 90 to 110 Nm (66 to 81 ft-lb)
- all other fasteners
  - M5 bolt ....5 Nm (44 in-lb)
  - M6 bolt ....10 Nm (89 in-lb)
  - M8 bolt .... 20 Nm (15 ft-lb)
  - M10 bolt .... 40 Nm (30 ft-lb)

TORQUE ARM AND PANHARD ROD

The torque arm and panhard rod are shown in Fig. 22. The rubber bushings in the torque arms and the panhard rod can wear out, often causing a groaning or squeaking sound. The bushings are pressed into place and require special press tools to remove and install them.

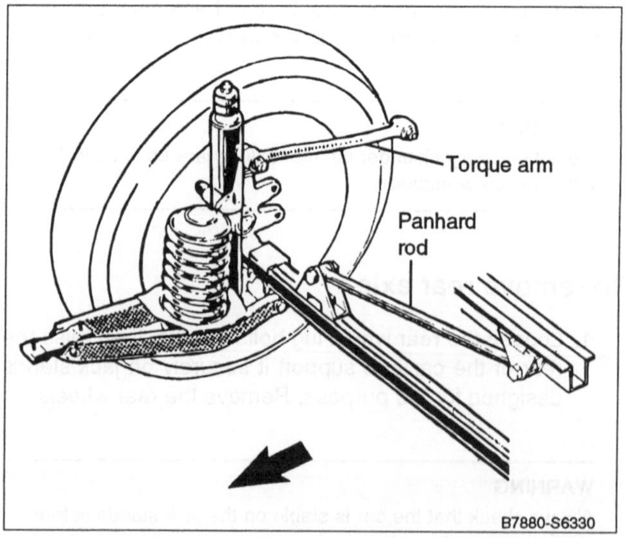

Fig. 22. View of passenger side rear suspension. Arrow points toward front of car.

To remove torque arm

1. Loosen the rear wheel lug bolts or nuts, then raise the rear of the car and support it securely on jack stands designed for the purpose. Remove the rear wheels.

WARNING
Always check that the car is stable on the jack stands before starting any work.

---

2. Remove the front and rear torque arm mounting bolts. See Fig. 23. Remove the torque arm.

Fig. 23. Torque arm front mounting bolt being removed.

3. Press out the torque arm front bushing using a 24 mm sleeve or equivalent. See Fig. 24. Press out the rear bushing using a slightly smaller sleeve.

4. Lubricate the new bushings with petroleum jelly and press them into the torque arm.

5. Install the torque arm in its mountings and loosely hand tighten the nuts.

6. Put the wheels back on the car, tighten the wheel lug bolts or nuts. Lower the car to the ground and tighten the torque arm mounting bolts.

NOTE
The torque arm mounting bolts are torqued after the suspension is loaded by the car's weight to prevent bushing distortion.

Fig. 24. Torque arm bushing being pressed out.

Tightening torques
- wheel lug bolts . . . . . . 105 to 125 Nm (78 to 92 ft-lb)
- wheel lug nuts . . . . 90 to 110 Nm (66 to 81 ft-lb)
- torque arm mounting bolts
  lubricated bushings. . . . 21 to 35 Nm (16 to 26 ft-lb)
  dry bushings . . . . 40 to 70 Nm (30 to 52 ft-lb)

To remove panhard rod

1. Raise the car.

NOTE
The rear wheels do not have to be removed.

---

2. Remove the bolts for the panhard rod mountings. Note the orientation of the panhard rod.

3. Press out the rubber bushings using a suitable size sleeve. Lubricate the new bushings with petroleum jelly and press them into the rod.

4. Install the panhard rod in its mountings and loosely hand tighten the nuts.

CAUTION
Install the panhard rod-to-axle mounting bolt so that the bolt head is toward the fuel tank. Damage to the fuel tank may result if the bolt is installed incorrectly.

---

5. Lower the car to the ground. Tighten the panhard rod mounting bolts to the value given below.

NOTE
The panhard rod mounting bolts are torqued after the suspension is loaded by the car's weight to prevent bushing distortion.

Tightening torque
- panhard rod mounting bolts . . . . 40 to 70 Nm (30 to 52 ft-lb)

To replace rubber bump stop

1. Loosen the rear wheel lug bolts or nuts, then raise the rear of the car and support it securely on jack stands designed for the purpose. Remove the rear wheels.

WARNING
Always check that the car is stable on the jack stands before continuing any work.

2. Remove the screw in the middle of the bump stop using an internal 6mm hex-head socket. See Fig. 25. Remove the bump stop and spacer.

Fig. 25. Suspension bump stop being removed.

Installation is the reverse of removal. Torque the wheel lug bolts or nuts.

Tightening torques
- wheel lug bolts . . . . . . 105 to 125 Nm (78 to 92 ft-lb)
- wheel lug nuts . . . . 90 to 110 Nm (66 to 81 ft-lb)

774 Wheel Bearings and Drive Axles

DRIVE AXLES AND FRONT WHEEL BEARINGS

The drive axle assembly is shown in Fig. 1. Each end of the axle is fitted with a swivel joint. The outer Constant Velocity (CV) joint is splined to the wheel bearing hub to drive the front wheel and the inner universal joint fits into the driver on the side of the transmission. Both joints are lubricated with a special grease and are protected from road dirt and other hazards by rubber boots. The only maintenance required of the drive axles and its joints is to periodically inspect the protective boots for damage. For information on troubleshooting a front end problem see **600 Steering and Wheel Alignment—General**.

The front wheel bearing is pressed into the steering swivel member and retained by large circlips. The wheel hub is in turn pressed into the bearing and retained by a large locking nut on the end of the outer CV joint. See Fig. 2.

CAUTION
Do not install the road wheel and lower the car to the ground while the drive axle is removed, as the wheel hub is not supported. The weight of the car can damage the wheel bearing.

To remove and install steering swivel member

1. Disconnect the battery negative (–) cable. Block the rear wheels and put the car in neutral.

2. Remove the center wheel cover from the wheel. Using a punch, tap the staked part of the drive axle nut up and away from the groove in the axle.

CAUTION
Make sure the staked part of the nut does not interfere with the threads on the CV joint. If the threads are damaged, the CV joint will have to be replaced.

---

3. With the car on the ground, loosen, but do not remove the center drive axle nut.

WARNING
Loosen the center drive axle nut only while the car is on the ground. The leverage required to do this could topple the car from a lift or jackstand.

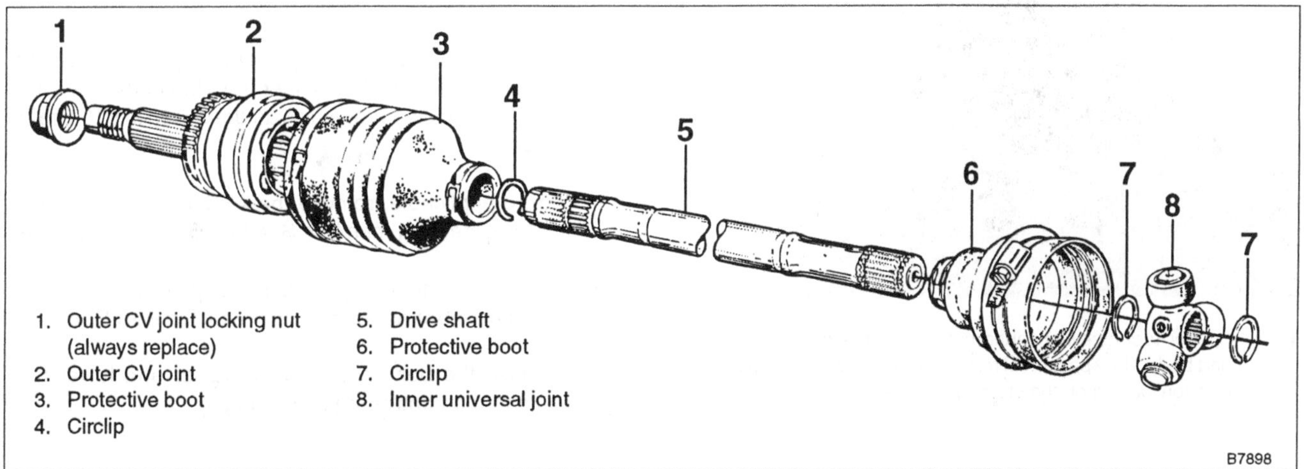

Fig. 1. Exploded view of front drive axle.

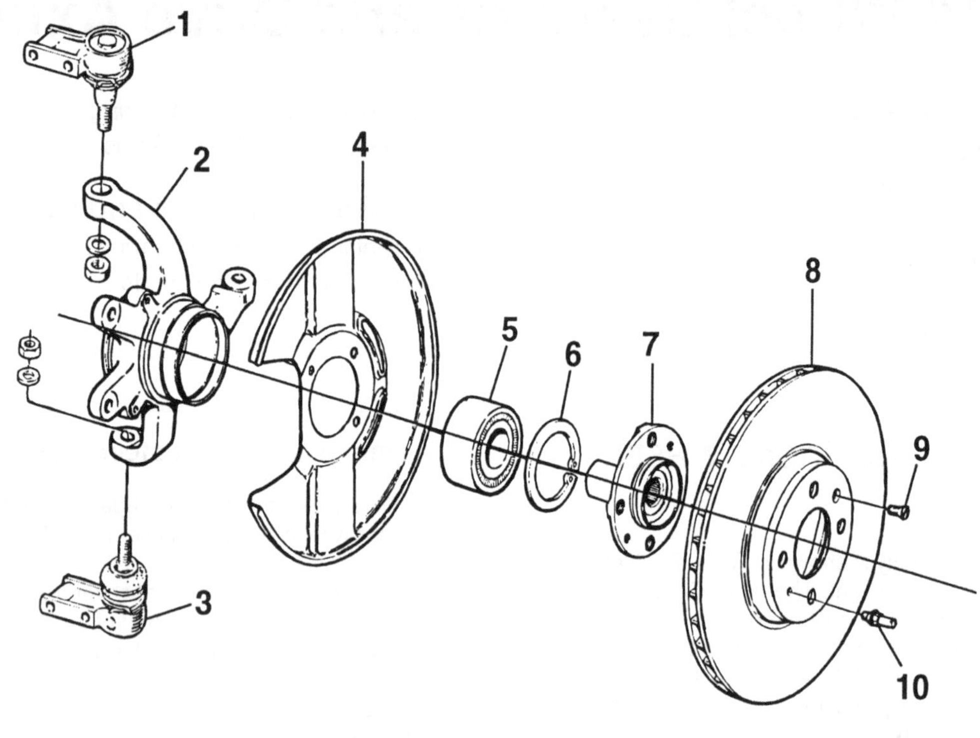

1. Upper ball joint
2. Steering swivel member
3. Lower ball joint
4. Brake dust shield
5. Wheel bearing
6. Circlip (inner circlip not shown)
7. Wheel hub
8. Brake disc
9. Brake disc mounting screw
10. Road wheel locating stud (1988 and later models)

Fig. 2. Exploded view of front wheel bearing and steering swivel member.

4. Insert the special spacer tool (Saab part no. 83 93 209) between the upper control arm and the body. See Fig. 3.

Fig. 3. Spacer tool (arrow) inserted between upper control arm and body. The spacer tool relieves upper control arm suspension loading on the steering member.

5. Raise the car and support it with jackstands. Remove the front wheel, the center drive axle nut and washer (where applicable).

WARNING

Always check that the car is stable on the jackstands before starting any work.

---

6. Remove the brake caliper as described in **523 Brake Calipers.** Suspend the caliper from the body using stiff wire. Remove the brake disc as described in **516 Brake Discs.**

7. Where applicable, remove the ABS wheel speed sensor from steering member. See **500 Brakes—General.**

8. Remove the steering rack tie rod end from the steering swivel member as described in **643 Tie Rod Ends.**

9. Remove the bolts securing the upper and lower ball joints to the control arms. See Fig. 4.

NOTE
If necessary, drive the bolts out of the control arms using a punch.

Fig. 4. Lower ball joint mounting bolt being removed. Upper ball joint mounting bolts shown at arrows.

NOTE
Some later cars are equipped with an anti-roll bar that mounts to the lower ball joint bolts.

---

10. Using stiff wire, support the drive axle by wiring it to the upper control arm. Then pull the steering swivel member from the drive axle and control arms. See Fig. 5.

CAUTION
Do not let the drive axle fall when removing the steering swivel member. The drive axle boots may be damaged.

NOTE
If necessary, use a soft-faced hammer to tap the drive axle out of the steering swivel member. See Fig. 6.

---

11. Apply grease (Molycote G or equivalent) to the splines of the drive axle. Then install the steering member to the drive axle. See Fig. 7.

12. Install the upper and lower ball joints in the control arms. Reinstall the bolts using new self-locking nuts.

Fig. 5. Steering swivel member being removed from drive axle.

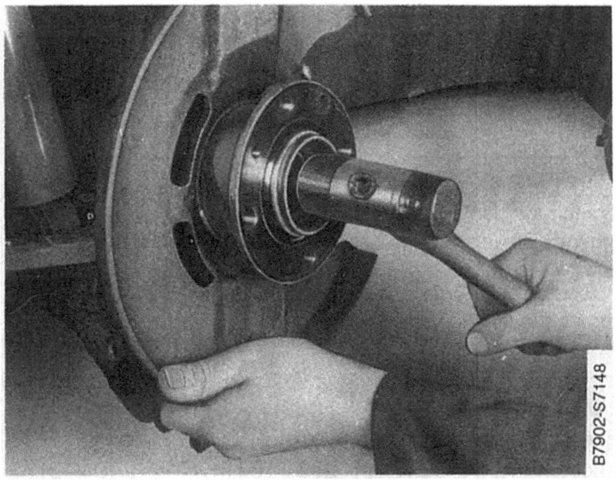

Fig. 6. Drive axle being tapped out of steering swivel member.

WARNING
Do not reuse self-locking nuts. These parts are designed to be used only once and must be replaced whenever loosened or removed.

Tightening torque
- ball joint to control arm (bolts) . . . . . 40 to 54 Nm (30 to 40 ft-lb)

---

13. Install the tie rod end, the brake caliper, and the brake disc. Where applicable, install the ABS wheel speed sensor.

14. Install the washer (where applicable) and a new center drive axle nut. Reinstall the wheel and lower the car to the ground. Tighten the drive axle nut and then stake it to the groove in the CV joint. See Fig. 8.

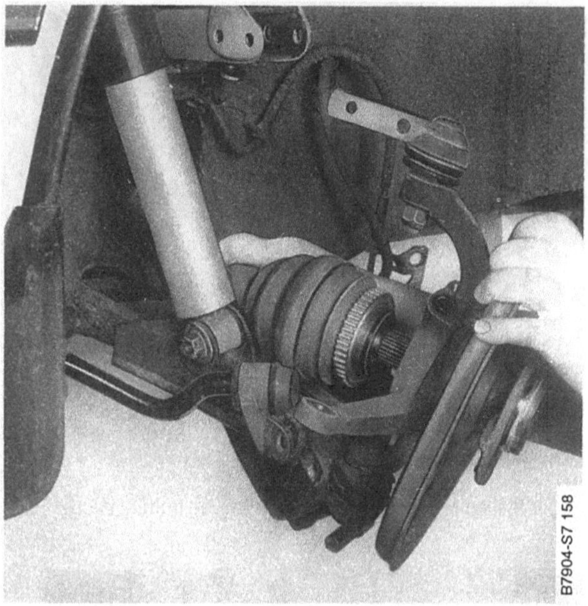

Fig. 7. Steering swivel member being installed on drive axle.

WARNING

Tighten the center drive axle nut only while the car is on the ground. The leverage required to do this could topple the car from a lift or jackstand.

Tightening torques

- wheel hub to drive axle (locking nut) . . . . .290-310 Nm (214- 229 ft-lb)
- wheel lug nuts . . . . 90-110 Nm (66-88 ft-lb)
- wheel lug bolts. . . . . .105-125 Nm (77- 92 ft-lb)

Fig. 8. Center drive axle nut being staked. Always use a new nut.

15. Remove the spacer tool. Pump the brake pedal several times to seat the brake pads.

To remove drive axle

CAUTION

Do not install the road wheel and lower the car to the ground while the drive axle is removed, as the wheel hub is not supported. The weight of the car can damage the wheel bearing.

NOTE

A new drive axle assembly was introduced on 1987 models. This drive axle is 6 mm longer than the earlier version and is not interchangeable between models. Always check with an authorized Saab dealer for the latest parts and service information.

---

1. Remove the steering swivel member as described above.

2. Loosen the boot clamp from the inner universal joint. See Fig. 9. Pull the drive axle out of the inner driver. Wrap or cover the driver.

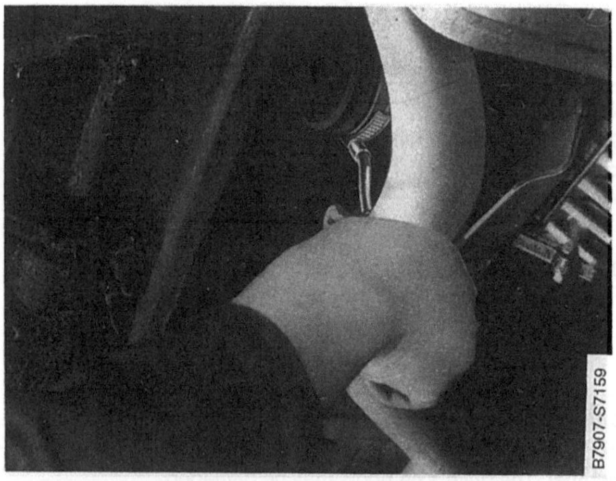

Fig. 9. Inner drive axle boot clamp being loosened.

CAUTION

Handle the drive axle and axle joints with care. Wrap or cover the joints to protect them from dirt and to prevent the bearing rollers and needle bearings from falling out of the joints.

Installation is the reverse of removal. Replace any damaged drive axle boots. Place the drive axle in the transmission driver and install the steering member as described earlier. Install the drive axle washer (where applicable) and a new center drive axle nut. Reinstall the wheel and lower the car to the ground. Tighten the drive axle nut and then stake it to the groove in the CV joint. See Fig. 8. Install and tighten the inner boot clamp.

WARNING
Do not reuse locking nuts. These parts are designed to be used only once and must be replaced whenever loosened or removed..

Tightening torques
- wheel hub to drive axle (locking nut) . . . . 290-310 Nm (214- 229 ft-lb)
- wheel lug nuts . . . . 90-110 Nm (66-88 ft-lb)
- wheel lug bolts . . . . 105-125 Nm (77- 92 ft-lb

To replace outer CV joint and joint boot

When replacing the CV joint, be sure to have on hand the special CV joint grease, a new boot, new clamps, and a new drive axle nut.

CAUTION
The outer CV joint should be lubricated with a special CV joint grease specifically designed for the purpose. Do not use any other type of lubricant. See an authorized Saab dealer for the correct grease.

1. Remove the drive axle from the car as described earlier.
2. Remove the old clamps from the boot and then cut the boot off the joint.
3. Expand the drive axle circlip shown in Fig. 10 and pull the CV joint off the shaft.
4. Thoroughly clean and degrease the joint and the drive axle. Allow the parts to dry.

CAUTION
When the grease is removed from the CV joint, the balls can fall out if the joint is positioned the right way.

5. Install the new boot clamps and the boot to the shaft. Slide the boot down the shaft.

NOTE
The old boot clamps should not be reused.

6. Repack the CV joint with the special grease. Work the grease into the joint.

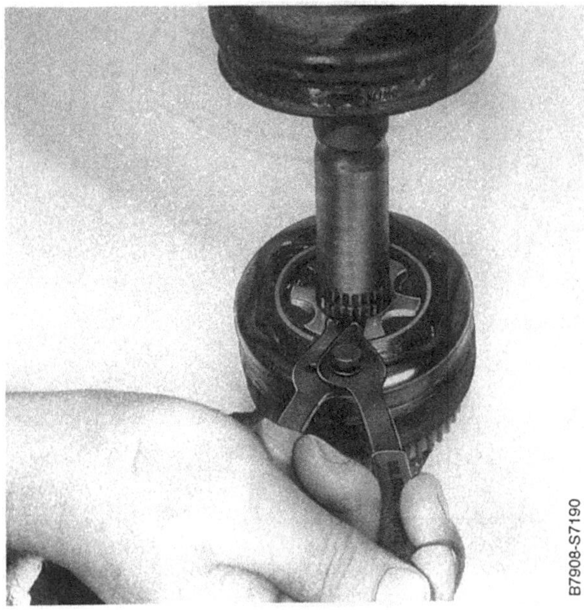

Fig. 10. Circlip pliers being used to expand CV joint circlip. Insert pliers in circlip opening. As an alternative, use a flat blade screwdriver to spread the circlip open.

Outer CV joint lubricant
- Esso nebula EP2 grease or equivalent . . . . 80 grams (approx. 3 oz.)

7. While expanding the circlip, slide the greased CV joint onto the shaft until it locks into place. Check that the CV joint is locked on the shaft by trying to pull it off.
8. Slide the boot and clamps into position. Release any trapped air in the boot by peeling up a corner of the boot and giving it a light squeeze.
9. Crimp the clamps using the appropriate crimping pliers (Knipex 1099 or equivalent). See Fig. 11.
10. Install the drive axle using a new locking nut as as described earlier. Be sure to stake the nut to the axle as shown in Fig. 8.

To replace inner universal joint and joint boot

Fig. 12 shows the inner universal joint and the transmission driver. When replacing the universal joint, be sure to have on hand the special grease, a new boot, and new clamps.

CAUTION
The universal joint should be lubricated with a special grease specifically designed for the inner joint. Do not use any other type of lubricant. See an authorized Saab dealer for the correct grease. The grease used to lubricate the inner joint is different than the grease used in the outer joint.

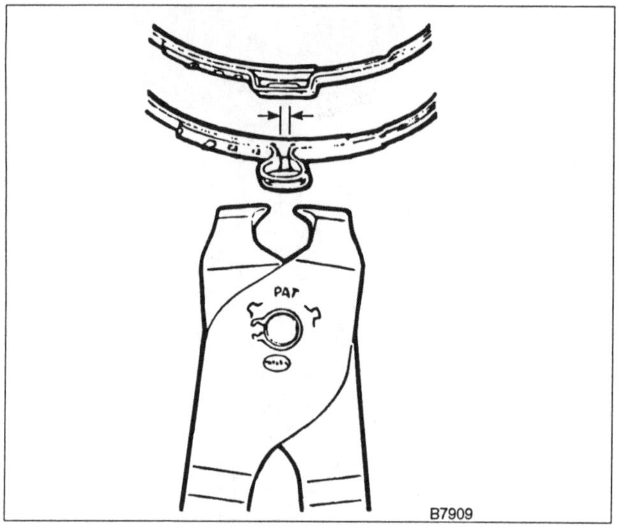

Fig. 11. CV boot clamp and crimping pliers.

NOTE
A new drive axle assembly was introduced on 1987 models. This drive axle is 6 mm longer than the earlier version and is not interchangeable between models. Always check with an authorized Saab dealer for the latest parts and service information..

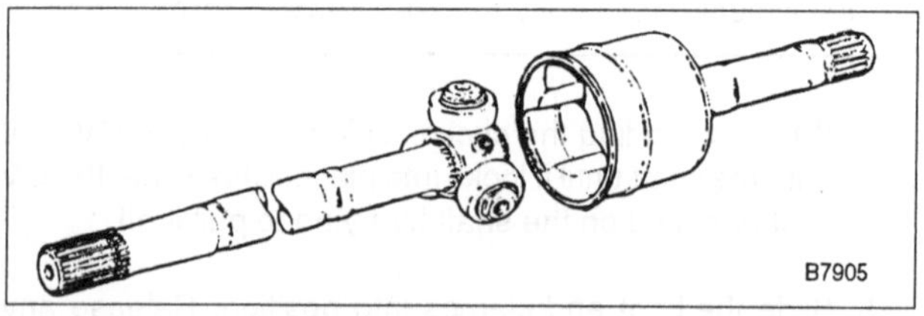

Fig. 12. Inner universal joint and transmission driver.

1. Remove the drive axle as described earlier. Mount the drive axle in a vise with the inner joint facing up.

2. Remove the circlip from the end of the drive axle. See Fig. 13. Pull the joint off the shaft while holding the roller bearings to the joint.

NOTE
Use care when removing the joint from the shaft. The rollers and small needle bearings can fall off the joint.

3. Remove the boot clamp and slide the boot off the shaft.

4. Thoroughly clean and degrease the joint, the drive axle, and the driver on the side of the transmission. Allow the parts to dry. If necessary, wipe the parts dry with a clean cloth. Where applicable, inspect the inner circlip and replace if damaged.

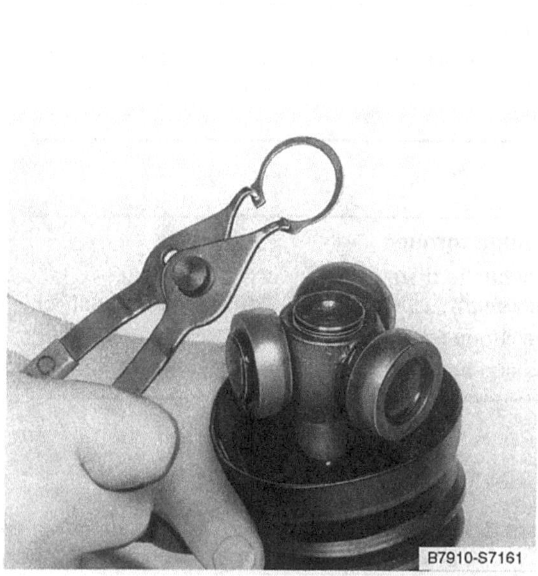

Fig. 13. Circlip being removed from inner drive axle joint.

5. Install the new boot and the new clamp on the shaft. Crimp the clamp using the appropriate crimping pliers (Knipex 1099 or equivalent). See Fig. 11 shown above.

6. Install the joint on the drive axle and install the circlip.

7. Pack the universal joint and the transmission driver with the appropriate amount of the specified grease.

Inner Universal Joint Grease
- Mobil GS57C or equivalent
- universal joint . . . . 60 g (approx. 2 oz.)
- transmission driver . . . . 115 g (approx. 4 oz.)

8. Install the drive axle as described earlier.

CAUTION
Do not let any dirt fall into the inner universal joint during installation of the drive axle. Temporarily wrap or cover the inner joint and boot during installation.

To replace front wheel bearing

Special tools, including a service press, are required to replace the front wheel bearings.

CAUTION
Do not install the road wheel and lower the car to the ground while the drive axle is removed, as the wheel hub is not supported. The weight of the car can damage the wheel bearing.

NOTE

Some cars (up through mid-1990 models) may experience a clunking or snapping noise at the wheel when backing up at low speeds with the steering at full lock (such as when parking). This noise is most likely caused by the wheel bearing sliding back and forth in the steering member. This problem can be corrected by installing a new bearing and the appropriate shims. See an authorized Saab dealer for more information on this repair.

---

1. Remove the steering swivel member from the car as described earlier in this section.

2. Using the special wheel bearing press fixture tools (Saab tool nos. 89 96 449 and 89 96 456), press the hub out of the steering member. See Fig. 14.

NOTE

Pressing the hub out of the bearing destroys the bearings. In most cases, the bearing inner race will remain on the hub.

Fig. 15. Bearing inner race being pulled off hub.

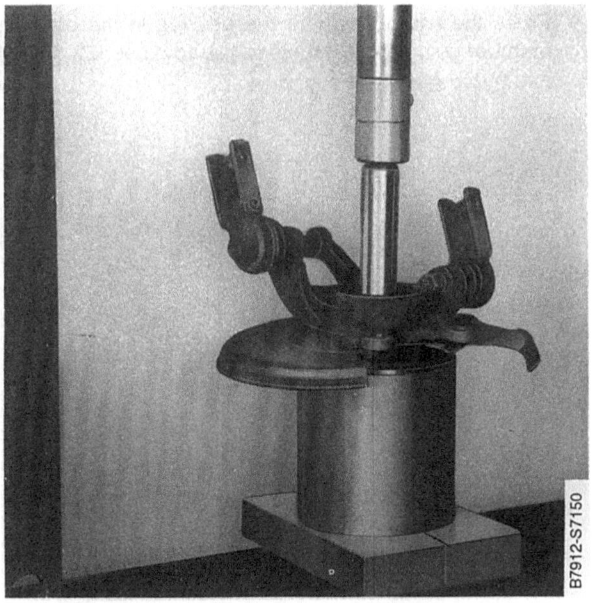

Fig. 14. Wheel hub being pressed out of wheel bearing.

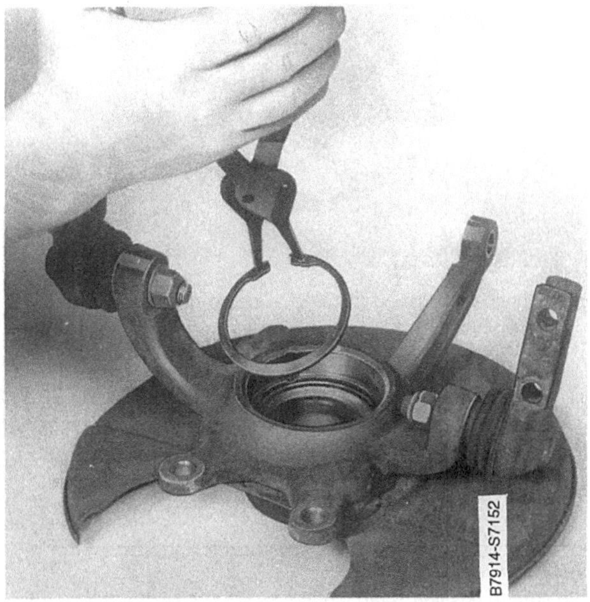

Fig. 16. Circlip being removed from rear of wheel bearing. Circlip on other side of wheel bearing must also be removed.

3. If necessary, pull the inner bearing race off the hub using a two-jaw puller. See Fig. 15.

4. Remove the circlips from the steering swivel member. See Fig. 16.

5. Using the special wheel bearing press fixture tools (Saab tool nos. 89 96 449, 89 96 456, and 83 90 114), press the wheel bearing out of the steering swivel member. See Fig. 17.

6. Clean the steering member housing. Install the inner circlip in the steering member. See Fig. 18. Lubricate the inside of the steering member with Molycote G grease or an equivalent.

NOTE

Inspect the circlip carefully. If it is deformed or will not expand into the groove, replace it with a new one.

Fig. 17. Wheel bearing being pressed out of steering member.

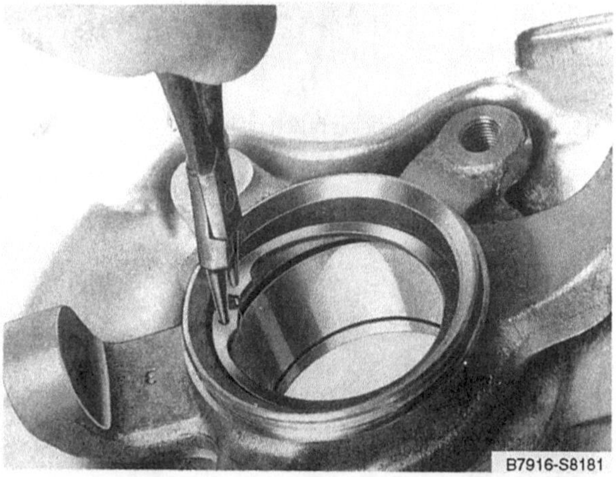

Fig. 18. Inner circlip being installed to circlip groove in steering member.

7. Using the special press tools (Saab part nos. 83 90 114 and 89 96 464), press the new bearing into the steering member until it is firmly seated on the circlip. See Fig. 19.

NOTE

Be sure to position Saab tool no. 89 96 464 so that the larger diameter collar is facing up.

---

8. Install the outer circlip.

NOTE

• Inspect the circlip carefully. If it is deformed or will not expand into the groove, replace it with a new one.

• Before pressing the hub into the steering member, inspect the wheel studs (where applicable) and replace any that are worn. Wheel studs are pressed out of the hub after the hub is removed from the steering member.

Fig. 19. Front wheel bearing being pressed into steering swivel member. Numbers shown are Saab tool numbers.

9. Press the wheel hub into the bearing in the steering member using special tools (Saab tool nos. 83 90 114 and 89 96 464). See Fig. 20.

Fig. 20. Wheel Hub (arrow) being pressed into steering member.

Install the steering swivel member as described above.

WARNING

Do not reuse self-locking nuts. These parts are designed to be used only once and must be replaced whenever loosened or removed.

Replacing Front Wheel Studs

Saab recommends that the wheel hub be pressed out of the steering member to replace the wheel studs. It may by possible in some cases to tap the stud out of the hub after removing the brake disc. See **516 Brake Discs** for information on removing the brake disc. A special press tool (Saab tool no. 89 95 920) can be used to press the studs out of the hub.

Rear Wheel Bearings and Wheel Hubs

The rear wheel bearing and hub are replaced as an assembly. Neither the bearing or the hub are available as separate replacement parts. The bearing is permanently sealed and does not require any maintenance. There are two types of rear hubs and bearings used on cars covered by this manual. See Fig. 21.

The rear wheel bearing is integral with the wheel hub. The wheel bearing hub assembly is fitted to the stub axle on the rear axle and retained by a large locking nut.

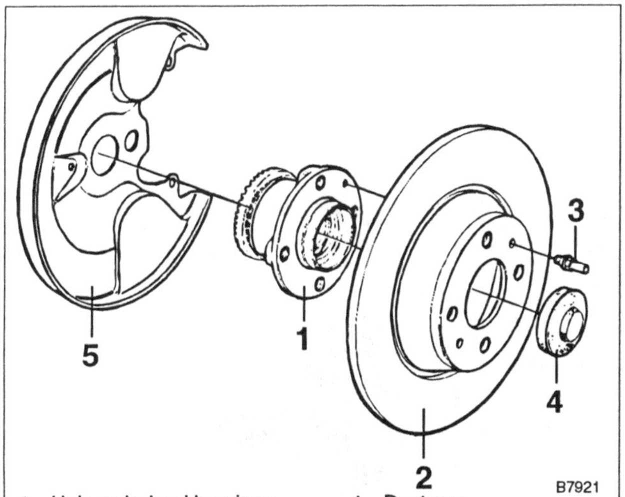

1. Hub and wheel bearings
2. Brake disc
3. Road wheel locating stud
4. Dust cap
5. Brake dust shield

Fig. 21. Exploded view of rear hub and brake disc (center axle nut not shown).

To replace rear wheel bearing

1. Working at the rear wheel, remove the hub dust cap.

2. Using a punch, pry up the staked material on the center axle nut. With the car on the ground, loosen, but do not remove the center axle nut.

WARNING

Loosen and tighten the center axle nut only while the car is on the ground. The leverage required to do this could topple the car from a lift or jackstand.

CAUTION

Make sure the staked part of the center axle nut does not interfere with the spindle threads. If the threads are stripped, the entire axle has to be replaced.

---

3. Raise the car and support it with jackstands. Remove the rear wheel, the center axle nut and washer (where applicable).

WARNING

Make sure the car is well supported with jackstands. Never use just a jack to support a car.

---

4. Remove the brake caliper as described in **523 Brake Calipers**. Suspend the caliper with stiff wire.

5. Remove the brake disc as described in **516 Brake Discs.** Where applicable, remove the ABS wheel speed sensor and handbrake cable.

6. Remove the hub.

NOTE

On 1985-1987 models press off the hub using special tool (Saab tool no. 89 96 084). See Fig. 22.

Fig. 22. Rear hub being pressed off axle on 1985 to 1987 models.

Installation is the reverse of removal. Clean the stub axle and remove any burrs or pits with an emery cloth. Lubricate the stub axle with thin oil.

NOTE
- The hub and bearings are replaced as an assembly.
- Install the hub and bearings using thumb pressure against the bearing race to prevent the race from sliding. See Fig. 23.

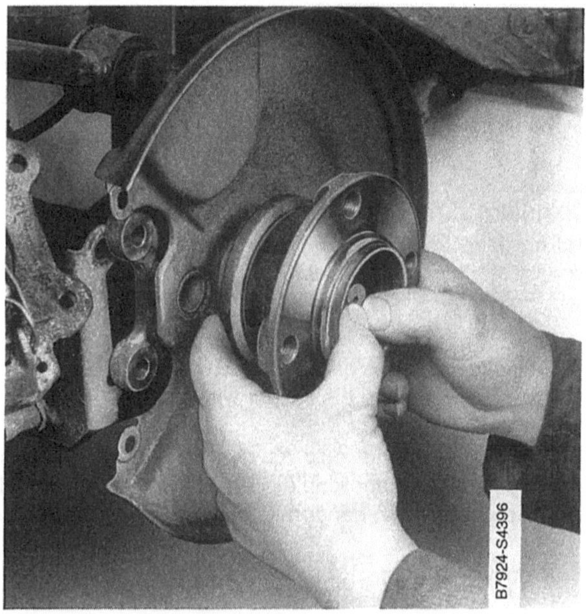

Fig. 23. Rear hub and bearing being installed.

Replace the center axle nut with a new one and torque it to the value given below with the car on the ground. Stake the nut as shown earlier.

NOTE
Early models have a replaceable seal and bearings. On later models, the bearings and seal are integral with the hub and are not available separately. See your Saab dealer for exact parts applications.

Tightening torques
- center axle nut to rear axle . . . . 300±10 Nm (220±9 ft-lb)
- wheel lug nuts . . . . 90 to 110 Nm (66 to 81 ft-lb)
- wheel lug bolts . . . . . . 105 to 125 Nm (77 to 92 ft-lb)

To replace rear wheel studs

NOTE
It may be possible in some cases to tap the wheel studs (where applicable) out of the hub using a brass or plastic hammer. If it is not possible to tap out the studs, a special press is available from Saab. The pressing procedure is given below. The new studs should be fully seated in the hub by putting the wheel on and torquing the wheel lug nuts to the correct value given elsewhere in this section.

---

1. Raise the car and support it with jackstands. Remove the rear wheel.

WARNING
Make sure the car is well supported with jackstands. Never use just a jack to support a car.

---

2. Remove the brake caliper as described in **523 Brake Calipers**. Suspend the caliper with stiff wire.
3. Remove the brake disc as described in **516 Brake Discs.**
4. Press out the wheel stud using the special tool (Saab tool no.89 95 920) and a sleeve located behind the hub. See Fig. 24.

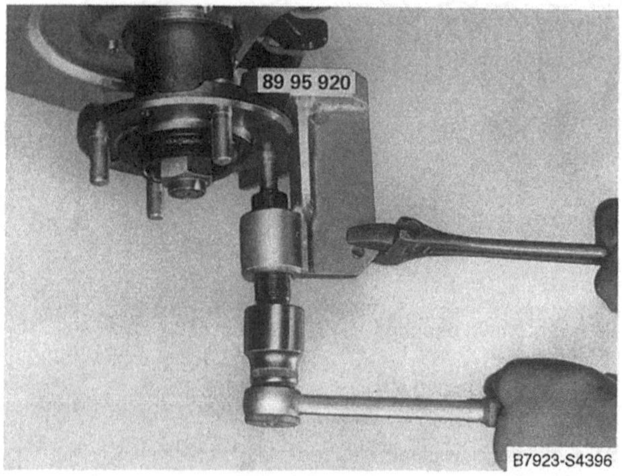

Fig. 24. Rear wheel stud being pressed out of hub. Number shown is Saab tool number.

5. Press in the new stud using the same tool. Position the sleeve on the outside of the hub.
6. Install the wheel and tighten the lug nuts.

800 Body–General

Body

Fig. 1 shows the overall body dimensions for the three 900 body styles. The hood, the grille, the doors, and the trunk lid are all removable. These bolt-on components are easily replaced even if you have little or no knowledge of auto body repair. The front fenders, however, are welded to the main body structure. Because very few screws and bolts are used in assembling the body, fewer rattles are likely to develop. The ride is quieted further by the application of sound-dampening material to the floor plates and the body panels.

The Saab 900 body is of welded, unitized construction, meaning it does not have a separate frame. This design forms a very rigid passenger compartment, with large crumple zones in the front and rear for energy absorption in the event of a collision. See Fig. 2.

During manufacture, the various body panels, subassemblies, and a number of smaller pressed-steel panels and plates are joined by electric welding. Although all body panels are available as replacement parts, many of these replacement panels must be butt-welded to the body after the damaged panels have been cut away.

For corrosion protection, all steel is treated with a multi-layer finish. The body seams are then sealed using a PVC compound. The body is undercoated and all interior cavities are flooded with a rust preventative sealant. The front fenders are open to allow trapped moisture to evaporate. See Fig. 3.

Seats, Seat Belts, and Interior

The front seats are mounted to the floor. Rear seats are bolted to the body and are easily removed for access to the rear seat belt mountings. The seat belts are typical three-point belts that cross the hips and the shoulder. Some later models use a motorized passive restraint seat belt system. Most interior trim is easily removed using ordinary hand tools.

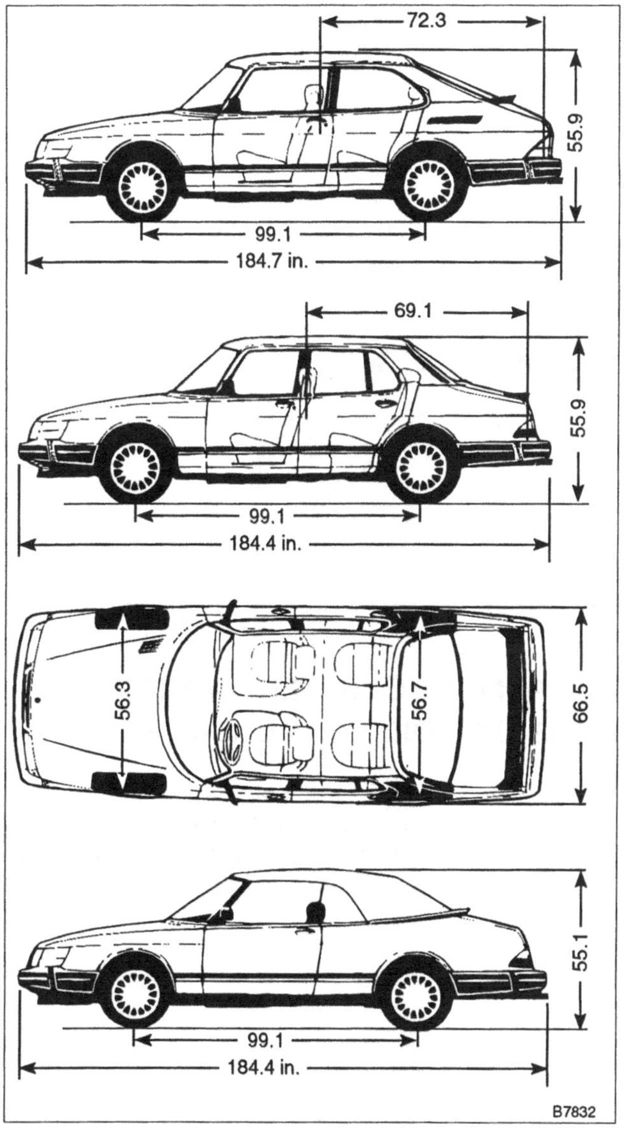

Fig. 1. Body dimensions. Dimensions on some sport models may vary slightly

800

Fig. 2. Unitized Saab 900 body shell.

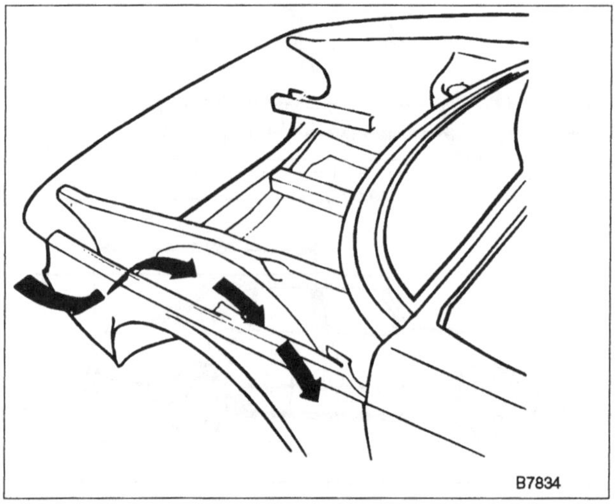

Fig. 3. Front fenders are ventilated (arrows) for corrosion protection.

INSTRUMENT CLUSTER, CENTER CONSOLE, AND INSTRUMENT PANEL

The padded instrument panel houses the instrument cluster and the ventilation and heating system. It is fastened to the body and can be removed using ordinary hand tools. The instrument cluster is removable as a unit without removing the instrument panel. All electrical repairs to the instrument cluster are covered in **3 ELECTRICAL SYSTEM**.

Some of the cars covered by this manual may be equipped with a Supplemental Restraint System (SRS) that automatically deploys an airbag. See Fig. 4.

WARNING

The SRS airbag unit uses an explosive device to electrically ignite a powerful gas. Use extreme care when working on the body, the steering wheel or steering column on cars equipped with SRS. Working involving the SRS, the steering wheel or the steering column should only be performed by an authorized Saab dealer.

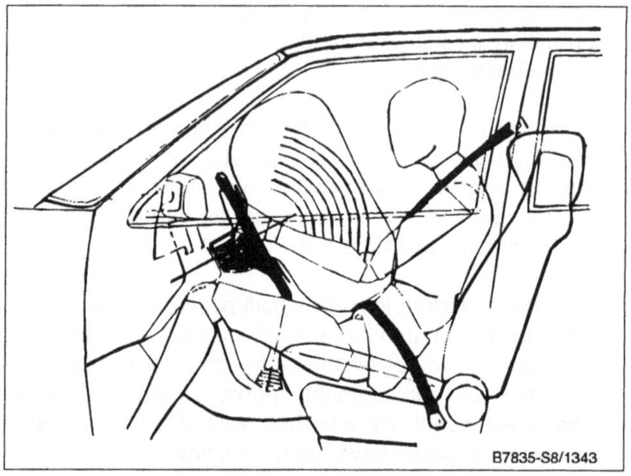

Fig. 4. Schematic of SRS (airbag) system.

812 Convertible Top

GENERAL

The convertible top consists of a top cover, a frame assembly, and an inner headliner. See Fig. 1. The top has a heated rear window element. A mercury switch prevents the heated rear window from coming on if the top is in the down position. See **371 Wiring Diagrams, Fuses and Relays** for the heated rear window circuit.

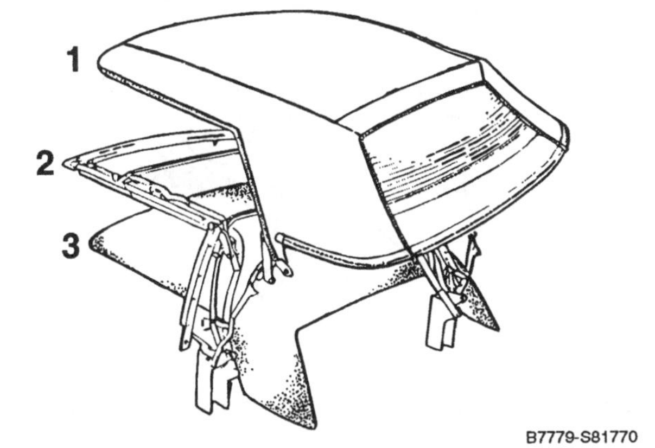

1. Top cover
2. Frame assembly
3. Headliner

Fig. 1. Convertible top.

The top is operated by an electrically-controlled hydraulic system. The system consists of an electric motor/pump assembly, two actuating pistons, a bypass valve, and connecting hoses. See Fig. 2. The motor/pump assembly is located under the rear seat and is thermally-protected (the motor shuts off after approximately 20 seconds of operation). A bypass valve enables the top to be operated manually in the event of an electrical failure. This valve is behind a trim flap at the back of the luggage compartment.

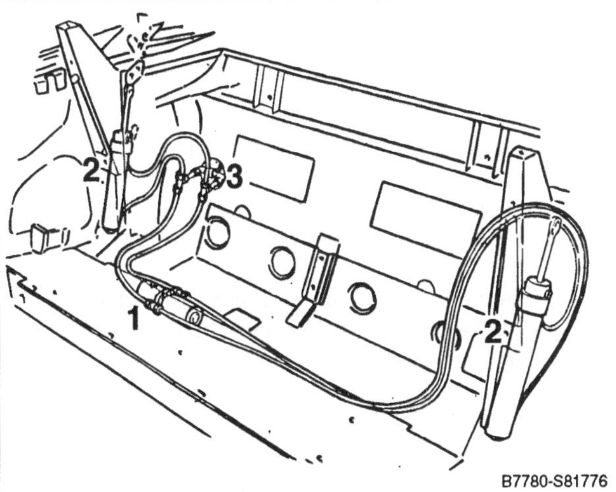

1. Electric motor (pump)
2. Actuating pistons
3. Bypass valve

Fig. 2. Hydraulic system beneath rear seat.

TOP ASSEMBLY

The top does not require any maintenance other than normal cleaning. Operate the top periodically to check that it functions correctly.

Replacement of the top requires special techniques and equipment and is beyond the scope of this manual. Information on removing and adjusting the side windows can be found in **830 Doors, Windows, and Lids**.

Adjustments

This section covers adjustments that can be made to the top. Keep in mind that doing one adjustment can affect the alignment of the top in other areas. There are five adjustments that can be made.

Water and wind leaks are usually the result of a defective seal between the weatherstripping and the top. Water leaks into the passenger compartment are most easily found using two people. One person can aim a garden hose at the car while the other person checks for the leak inside the car. Wind noise is not as easy to find. One technique is to tape over a suspected area and then test drive the car to check if the noise still exists.

The weatherstrips and seals are installed at the factory using adhesive and screws. Generally, seals are glued into place using 3M® Super Weatherstrip Adhesive. The seals are non-adjustable once installed. All fasteners should be checked and tightened before beginning any adjustment.

NOTE
A service campaign to adjust the convertible top was initiated by Saab for 1986-1987 models and some 1988 models. See an authorized Saab dealer for more information on this campaign.

To check and adjust park position

The park position allows the top to be latched easily while at the same time pulling the top taught.

1. Raise the top until it stops. The leading edge of the top should be above the windshield by the amount shown in Fig. 3.

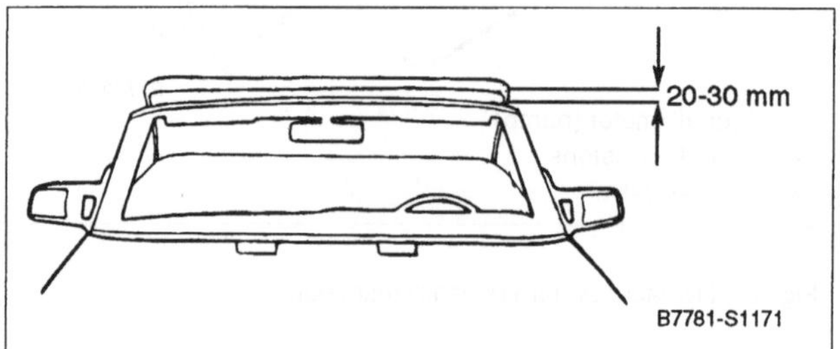

Fig. 3. Correct park position of top.

2. If the gap is incorrect, adjust the cam links on each side of the top until the gap is as specified. See Fig. 4.

NOTE
The adjustment is made using an internal hex-head key. Be sure to tighten the locknut after adjustment.

To check and adjust latched position

When the top is latched closed, there should be no side-to-side or up-down play. The guide pins, which control side-to-side movement, do not usually require adjustment. The latching hooks, which control up and down play, can be easily adjusted. See Fig. 5.

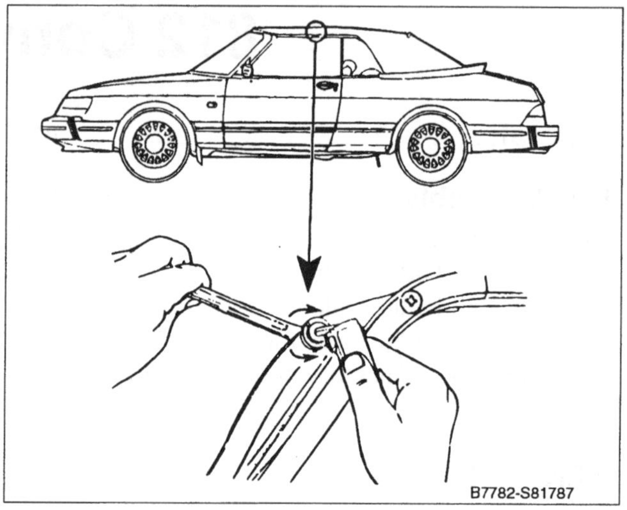

Fig. 4. Park position adjusting cam being adjusted. Cams are located above windows.

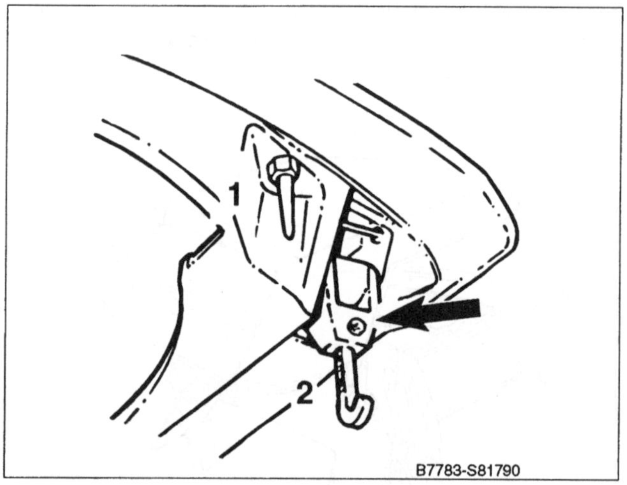

Fig. 5. Guide pin **1** and latching hook **2**. Arrow shows location of Torx adjusting screw.

1. With the top latched in position, push down on each corner of the top. If any play is found, the latching hooks should be adjusted.

2. Adjust the height of the latching hook by loosening the Torx screw and rotating the hook.

NOTE
One turn of the hook changes the height by 1 mm (0.04 in.).

To check and adjust side rail position

The side rail position controls the seal between the front windows and the top.

1. Check that the side rail is parallel with the windows. Check that no air gap exists. If any faults are found, the side rail should be adjusted.

2. Adjust the side rail by loosening the hinge screw between the front and middle rail sections. Adjust the rails sections until they are in parallel with the window and then tighten the screw. See Fig. 6.

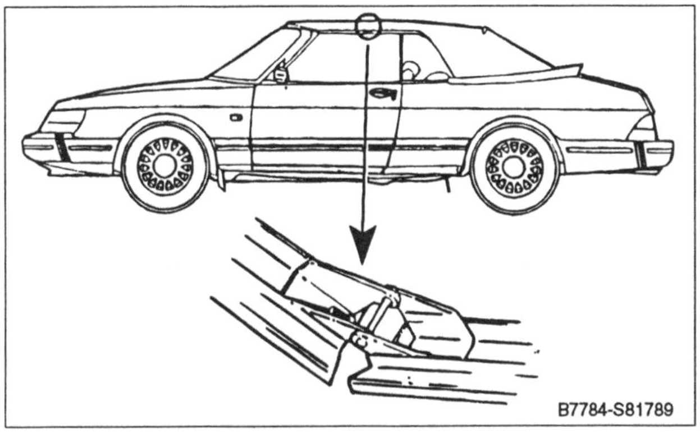

Fig. 6. Front section side rail hinge.

To check and adjust fully open position

Checking and adjusting the fully open and stowed position of the top requires a special gauge that can be made from wood or cardboard.

1. To check the open position of the top, open the top to its fully stowed position. Using the special gauge, check the top position as shown in Fig. 7.

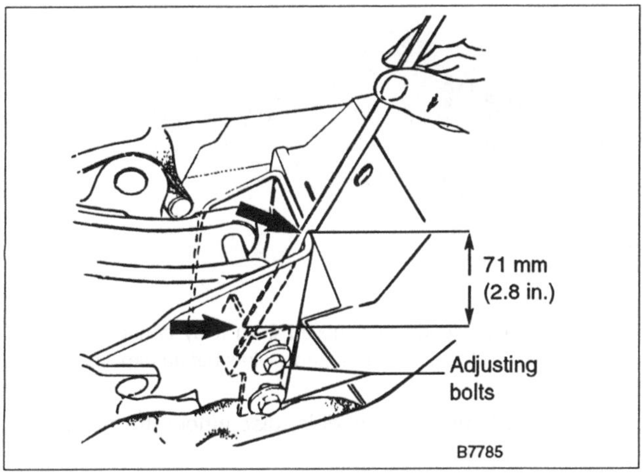

Fig. 7. Stop bracket position being checked (arrows) using special gauge. If position is incorrect loosen adjusting bolts.

2. If the position is incorrect, loosen, but do not remove, the adjusting bolts and adjust the stop bracket. Tighten the bolts.

HYDRAULIC SYSTEM

If the top does not operate, begin by checking the motor fuses. There are two fuses used in the circuit. See **371 Wiring Diagrams, Fuses and Relays** for locations. Check the fluid level in the pump reservoir. See Fig. 8. If the level is low, check for any visible leaks. Use only the special Saab hydraulic fluid when adding fluid to the system.

CAUTION
Use only the approved hydraulic fluid when adding or replacing fluid. Engine oil or ATF should not be used in the system.

Pump hydraulic fluid

- Aeroshell Fluid 4/Shell Code 60 421 (Saab part no. 30 18 694)

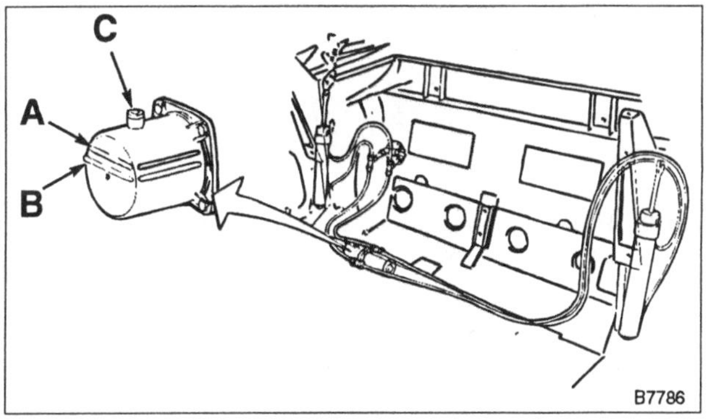

Fig. 8. Hydraulic pump fluid level. Fluid level should be at **A** with top down and at **B** with top up. Fill plug is shown at **C.**

To replace an individual component, the entire hydraulic system should be removed from the car as described below. This allows for proper bleeding of the system and also prevents any spilled hydraulic fluid from contacting and damaging painted surfaces.

WARNING
Do not allow the top to be operated when your hands are near the top mechanisms. The pistons have sufficient force to cause injury.

CAUTION
Leaking hydraulic fluid can damage paint and upholstery. Correct any leaks as soon as possible.

To remove hydraulic system

1. Remove the rear seat as described in **851 Interior Trim and Upholstery.**

2. Remove the interior trim panels below the rear side windows as described in **830 Doors, Windows, and Lids.**

3. Working in the luggage compartment, open the bypass valve to release the pressure in the system. See Fig. 9.

Fig. 9. Hydraulic system bypass valve in luggage compartment. Lift up carpet flap to access.

4. Release the handbrake to allow for slack in the handbrake cables near the pump.

5. Working in the passenger compartment, release the hydraulic hoses from the body clips.

6. Remove the split pin and clevis pin from the pistons on both sides of the car. Note the position of the washers and spacers. See Fig. 10.

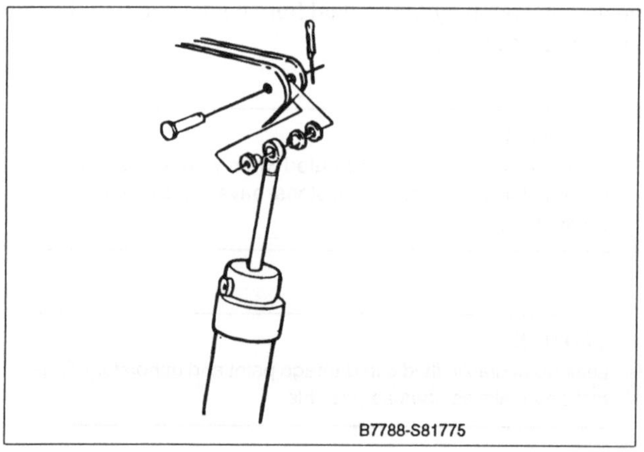

Fig. 10. Split pin, clevis pin and washers for piston.

7. Close the bypass valve. Switch on the ignition and fully retract the pistons by placing the top switch in the down position. Switch off the ignition.

8. Remove the bolts holding the piston to the side body panel. See Fig. 11.

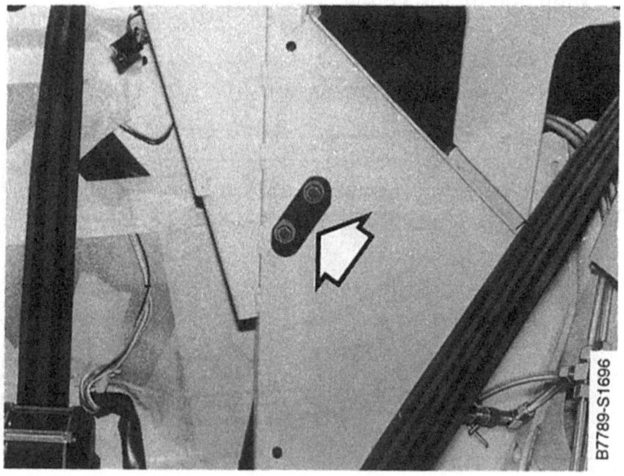

Fig. 11. Piston mounting bolts (arrow).

9. On 1987 and 1988 models, remove the pump from its mountings and disconnect the harness connectors from the relays on the bulkhead. Remove and label the pump wires from the relay connectors. See Fig. 12.

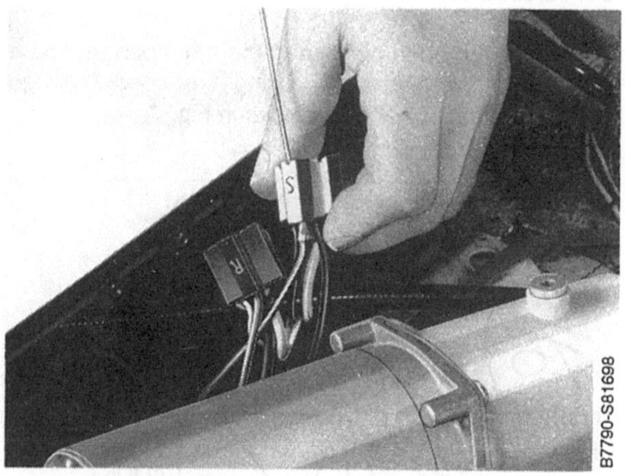

Fig. 12. Pump wiring being removed from relay connector on 1987 and 1988 moldels. Label wires before removal.

10. On 1989 and later models, disconnect the pump harness connector.

11. Remove the moisture barrier from each body panel cavity.

12. Remove the pistons by lifting them up while feeding the bottom of the piston through the cutout in the body

13. Label each piston and bracket before removing the entire hydraulic system from the car. Carefully slide the right-hand piston out from under the fuel lines and the parking brake cable during removal. See Fig. 13

NOTE

If the piston brackets are to be removed, label their positions before removal. The left and right brackets are different.

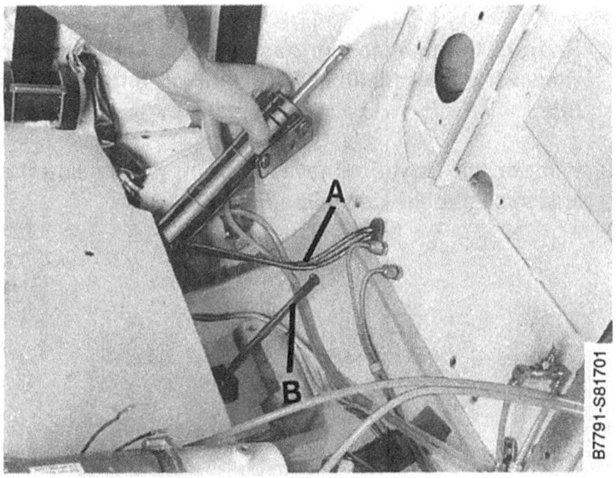

Fig. 13. Right-hand piston must pass under fuel lines **A** and handbrake cable **B** to remove.

If necessary, the individual components can be replaced once the system is removed. If any component is replaced or the system has been opened, the system must be bled as described below before installing the system.

Tightening torque

- Hydraulic hose fittings . . . . 5.1 to 6.2 Nm (45 to 54 in-lb)

To install hydraulic system

1. Make sure the pistons are fully retracted. If necessary run the pump as described below.

2. Slide the right-hand piston and hoses under the fuel lines and handbrake cable.

3. Install the pump under the handbrake cable. Plug in the pump electrical connectors. If applicable, reconnect the harness connectors to the relays. Fit the pump to its mount.

4. Install the pistons in the body cavities. Make sure the hose connections on the pistons are toward the rear of the car.

5. Reinstall the moisture barrier in each body cavity.

6. Mount the bypass valve and hoses in their clips.

7. Operate the pump to fully extend the pistons. With the top unlatched, install the washers, spacers, clevis pin and split pin in the piston. See Fig. 10.

The remaining of installation is the reverse of removal.

To bleed hydraulic system

The hydraulic system is bled with the complete system removed from the car. This will prevent any damage to painted surfaces.

1. With all lines connected, place the pump on a workbench. Place the pistons in a tray on the floor. The pistons should be at a lower level than the pump.

2. Fill the reservoir with hydraulic fluid up to the top level mark. See Fig. 14. Leave the fill plug off.

Pump hydraulic fluid

- Aeroshell Fluid 4/Shell Code 60 421 (Saab part no. 30 18 694)

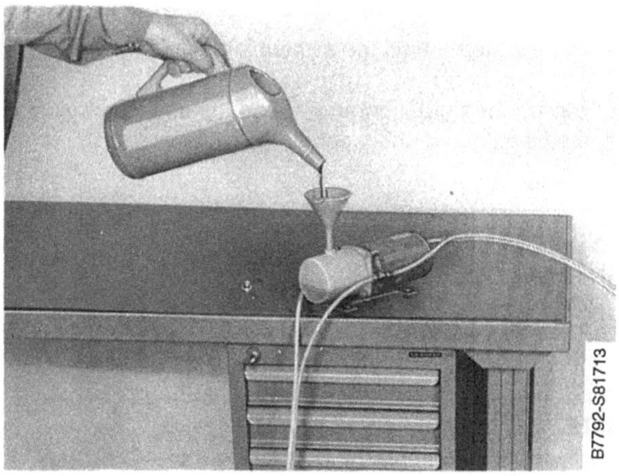

Fig. 14. Hydraulic pump being filled with hydraulic oil.

3. Loosen one hose fitting at one of the pistons. Allow fluid to run out until no air bubbles are present. Tighten the hose and repeat the procedure at the remaining three fittings. See Fig. 15.

4. Connect a battery to the pump. Run the system to fully retract or fully extend the pistons.

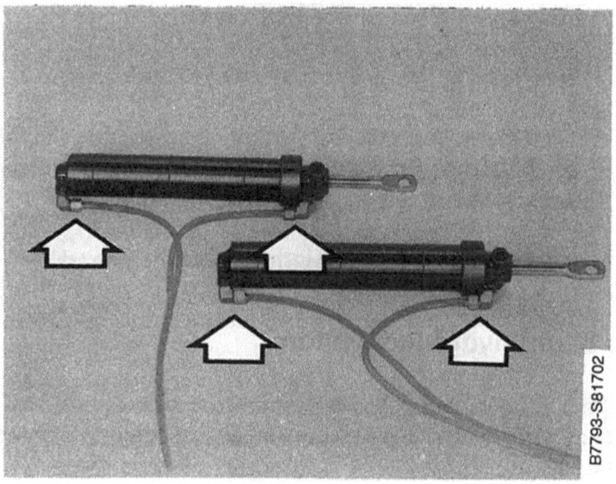

Fig. 15. Hydraulic hose fittings (arrows) at pistons used to bleed lines.

NOTE

- The electrical connection should have fuse protection. Use the same rating fuse as used in the circuit.
- Replenish the hydraulic fluid as the level drops during bleeding.

---

5. Repeat step 3 above. Run the pump and pistons in the opposite direction.

6. Repeat step 5 until the system is free of air.

7. Top up the fluid, tighten all hose connections, reinstall the fill plug and check for leaks.

Tightening torque

- Hydraulic hose fittings. . . . . 5.1 to 6.2 Nm (45 to 54 in-lb)

NOTE

The hydraulic system can be tested using a pressure gauge and fitting. The pressure gauge can be installed in place of the rear T-fitting. Be sure to bleed the system after pressure testing the pump.

Hydraulic pump

- minimum working pressure . . . . 25 bar (363 psi)

SPOILER ASSEMBLY

The spoiler assembly is held to the body using adhesive tape and nuts. There are two side pieces and one trunk lid piece.

To remove spoiler

1. Remove the rear seat as described in **851 Interior Trim and Upholstery.**

2. Remove the interior trim panels below the rear side windows as described in **830 Doors, Windows, and Lids.**

3. With the top down, remove the front nut holding the side spoiler to the body. See Fig. 16.

Fig. 16. Spoiler mounting nut (arrow) in body cavity.

4. Open the top approximately half-way and remove the center nut. See Fig. 17.

NOTE

Cars from VIN L701110 and later do not have a center nut.

WARNING

Do not allow anyone to operate the top when your hands are near the mechanism. The pistons have sufficient force to cause injury.

---

5. Working in the luggage compartment, remove the rear nut through the cutout in the carpeting. See Fig. 18.

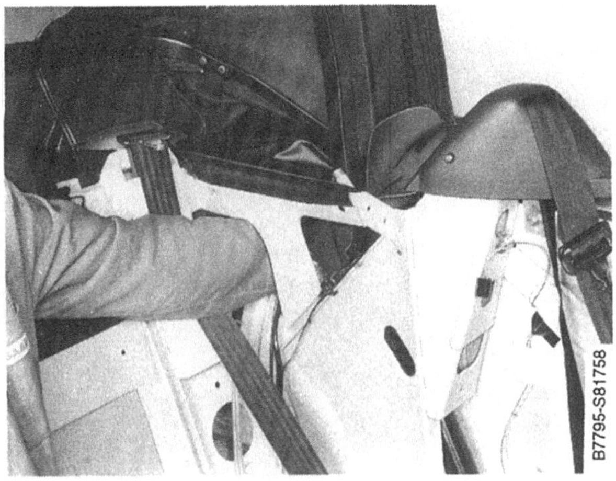

Fig. 17. Side spoiler center nut being removed through body side cavity.

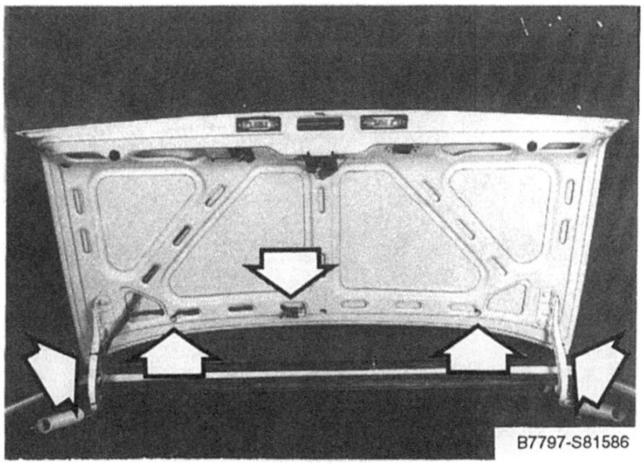

Fig. 19. Arrows show rear mount spoiler nuts and bolts to be removed from trunk lid.

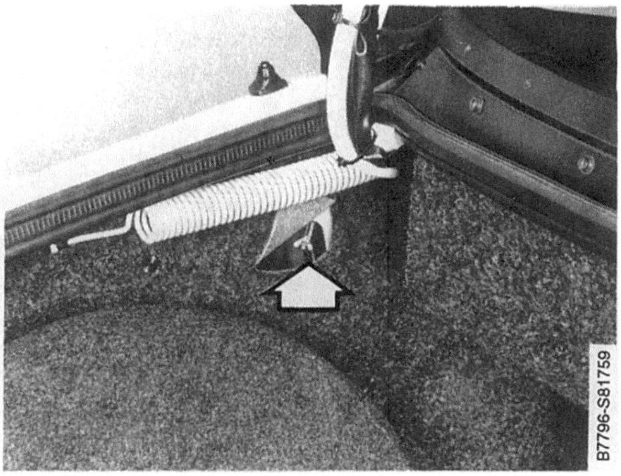

Fig. 18. Side spoiler rear nut (arrow) behind carpet cutout.

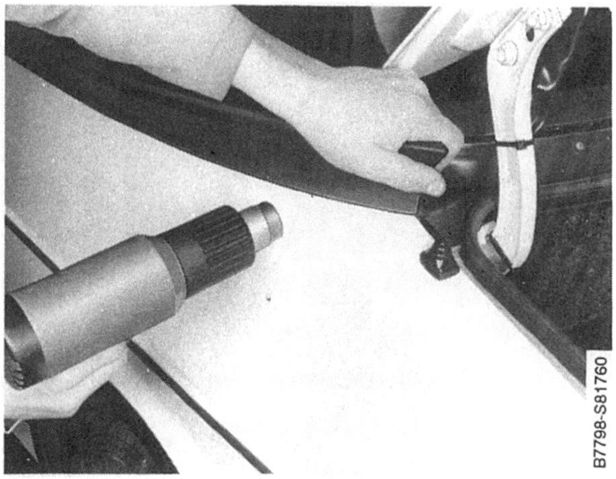

Fig. 20. Heat being applied to spoiler to loosen adhesive.

6. Unplug the hi-mount brake light connector. Remove the rubber plugs, retaining bolts and nuts for the rear mount spoiler. See Fig. 19.
7. Heat the spoiler with a heat gun to release the adhesive tape. See Fig. 20. Remove the spoiler.

CAUTION

- Use care when applying heat. Too much heat will damage the paint and spoiler.
- Pull the spoiler off gently to prevent damage to the paint.

NOTE

The side spoiler has a cable at the rear mounting hole. This cable should be saved if the spoiler is being replaced.

---

8. Remove all old adhesive from the paint and spoiler.

Installation is the reverse of removal. New double-sided adhesive tape should be installed to the spoiler. If installing a new spoiler, install the cable for the side spoiler rear mount and a rubber washer on each stud. Apply a water sealant such as Sikaflex 255 around the hole for the hi-mount brake light wire harness.

820 Hood and Fenders

GENERAL

Fig. 1 shows an exploded view of the hood and related components. The hood is bolted to hinges at the radiator crossmember. The rear of the hood has locating pins and guides. The hood release mechanism is also mounted to the radiator crossmember. When the hood is closed, the locking pin engages the release mechanism. A safety catch prevents the hood from opening fully after the release mechanism has been disengaged.

WARNING

- Welding and grinding coated steel can produce a poisonous gas containing zinc-oxide. This type of work should always be carried out by experienced body technicians having the proper respiratory equipment and work area.

- 1990 and later models are equipped with an airbag or SRS system. The airbag is an explosive device. Do not do electric (arc) welding to cars equipped with SRS (airbag) without first disarming and removing the airbag unit. You could accidentally ignite the airbag, causing serious personal injury. See an authorized Saab dealer to disarm the system.

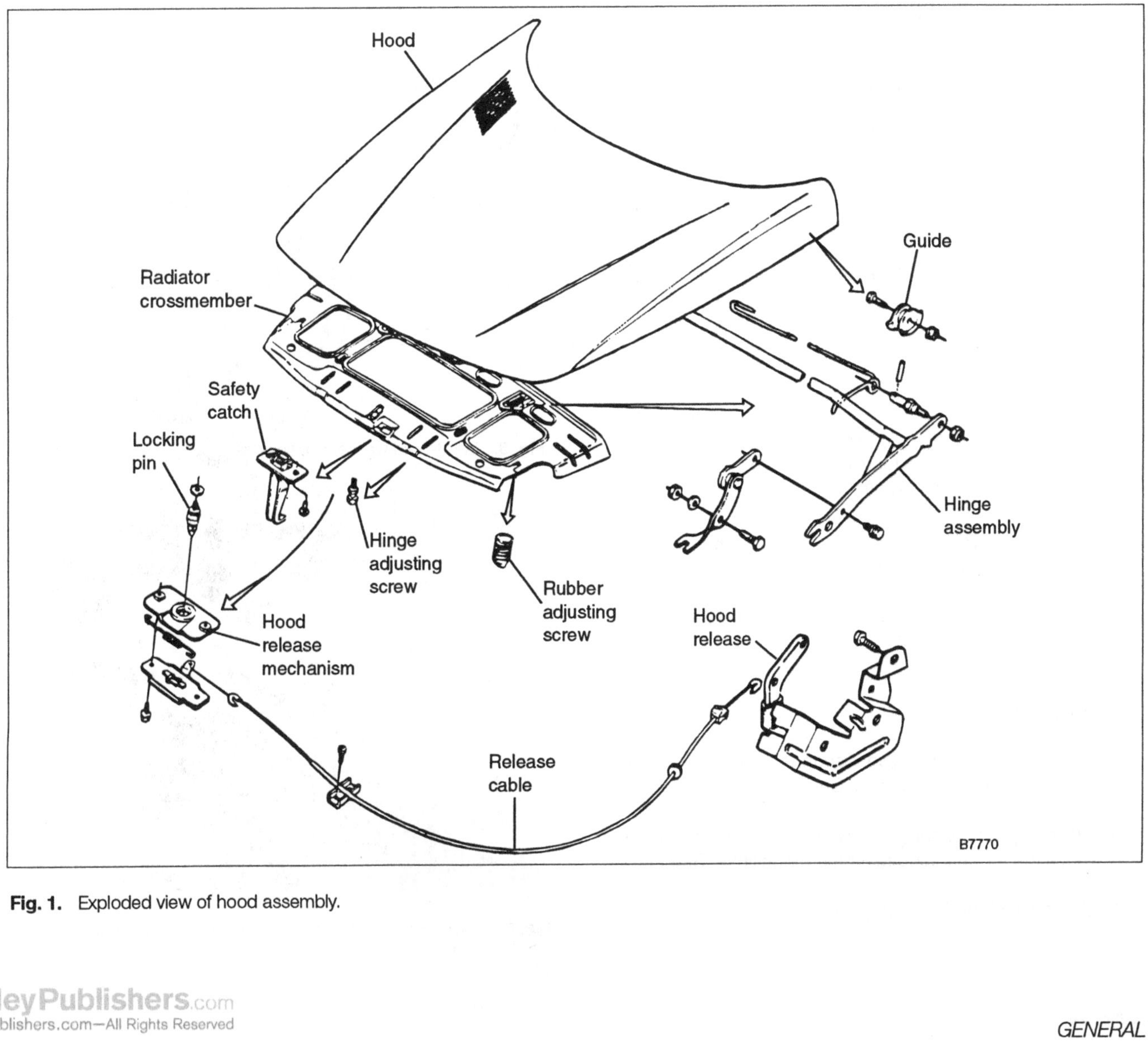

Fig. 1. Exploded view of hood assembly.

The front fenders are welded to the body to form part of the unit-body construction. See Fig. 2. Special tools and welding equipment are needed to replace a fender. Therefore, fender replacement is not covered in this manual.

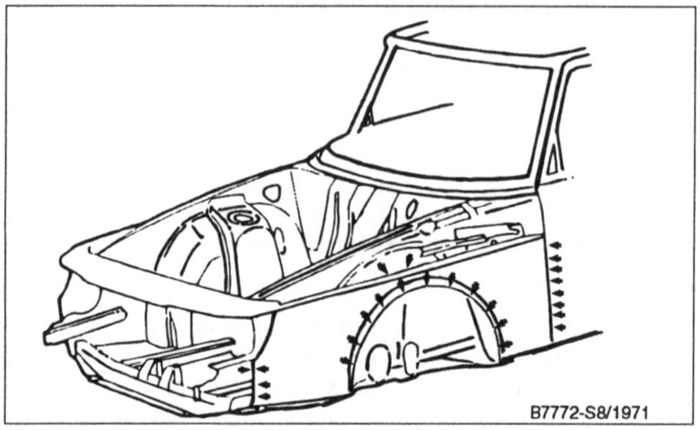

Fig. 2. Front fender welds (arrows).

To open hood if release cable is broken

If the hood cannot be opened using the hood release, a wire tool can be made to release the hood locking pin through the front grille. The tool dimensions are shown in Fig. 3.

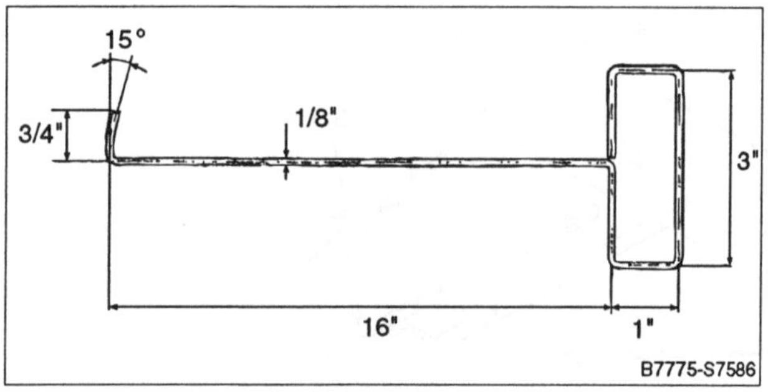

Fig. 3. Hood release tool can be made using 1/8 in. diameter wire. All dimensions are approximate.

1. Insert the tool through the grille and hook the rear of the lock mechanism. See Fig. 4.
2. Pull the latch toward the driver's side of the car until the hood releases.

To remove hood

1. Open the hood and then disconnect the hose for the windshield washer nozzles. See Fig. 5.
2. While supporting the hood with the aid of a helper, remove the mounting bolts for the hinge links. See Fig. 6.
3. Carefully lift the hood off the car.

Fig. 4. Hood lock being released using special tool. Grille shown removed for clarity.

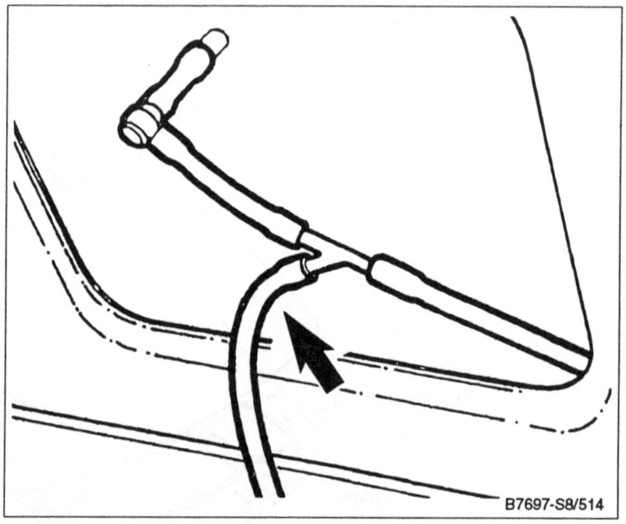

Fig. 5. Windshield washer nozzle hose (arrow).

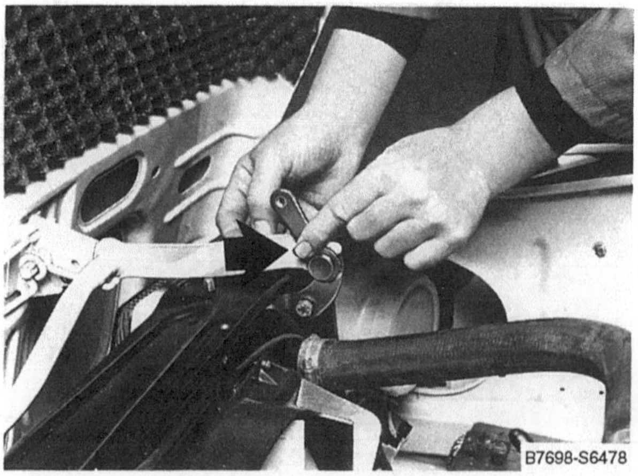

Fig. 6. Hood hinge link mounting bolt being removed (arrow).

Installation is the reverse of removal. If installing a new hood, adjust its fit as described below.

Hood Alignment and Adjustment

The hood can be adjusted up-down, left-right, and fore-aft. The front up-down adjustment may require special shims on 1985 through 1989 models.

To adjust the up-down position at the front of the hood, loosen the locking pin. See Fig. 7. On 1985 through 1989 models, add or subtract shims under the pin until the adjustment is correct. On 1990 and later models, the shims are not required owing to the return spring. When the lock pin adjustment is correct, turn the two rubber adjusting screws (shown in Fig. 1) so that there is no play when pressing down on the top of the hood.

NOTE
A maximum shim thickness of 5 mm (3/16 in.) can be added between the pin and the hood.

Fig. 7. Locking pin on 1985 through 1989 models. Adjustment shims are between pin and hood (arrow).

To adjust the up-down position and the left-right position at the rear of the hood, the guides and guide pin plates can be repositioned. The guide is mounted to the hood (see Fig. 1) and the guide pin plate is mounted to the body with the internal-hex head screws. Loosen the guide plate mounting screws and then adjust the position of the guides. When the guide position is correct, tighten the guide pin plate screws.

If the front of the hood is not centered from left to right, the release mechanism mounting bolts can be loosened to reposition the mechanism. See Fig. 8.

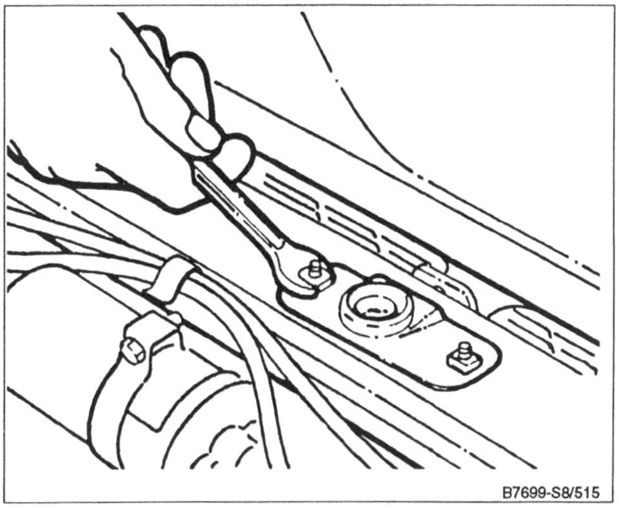

Fig. 8. Hood release mechanism mounting bolt being loosened.

Adjusting the fore/aft position of the hood can be done by changing the position of the hinge adjusting bolts at the radiator member. See Fig. 9. Making this adjustment requires removal of the front grille.

Fig. 9. Hood fore/aft adjusting screw at front radiator crossmember.

830 Doors, Windows, and Lids

Doors and Windows

This section covers the removal and installation of the doors, the interior door panels, the window regulators, and the window glass.

Doors

The doors can be easily removed once the interior trim panels have been removed and the electrical connectors disconnected. Models with Torx head hinge bolts require a special Torx socket. After installing, adjust the door so that it correctly aligns with the body allowing it to open and close properly.

To remove and install doors

1. Disconnect the negative (–) battery cable.

2. Remove the interior door panel and moisture barrier as described later under **To remove interior door panel.** Where applicable, disconnect the connectors for the power windows, power mirrors, and central locking. Pull the wiring harness from the door.

3. Have a helper hold the door, then remove the six hinge bolts for the upper and lower hinges and remove the door. See Fig. 1.

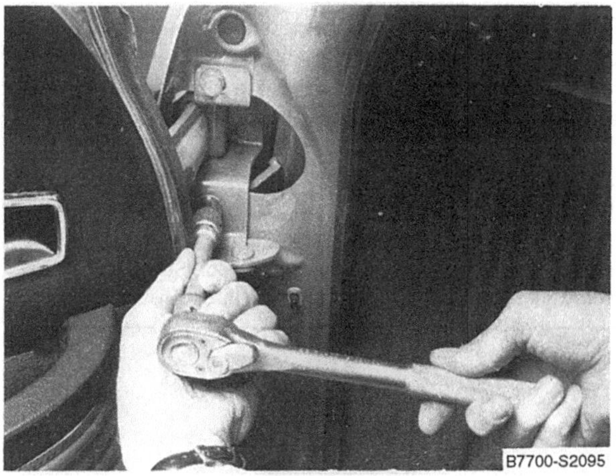

Fig. 1. Front door hinge-to-body bolts being removed.

NOTE
- Before loosening the bolts, scribe a mark showing the position of the hinge on the body so the door can be realigned in its original position.
- If a helper is not available, support the door with a floor jack and protective padding.
- On rear doors, the trim covering the B-pillar must be removed to reach the nuts. See Fig. 2.

830

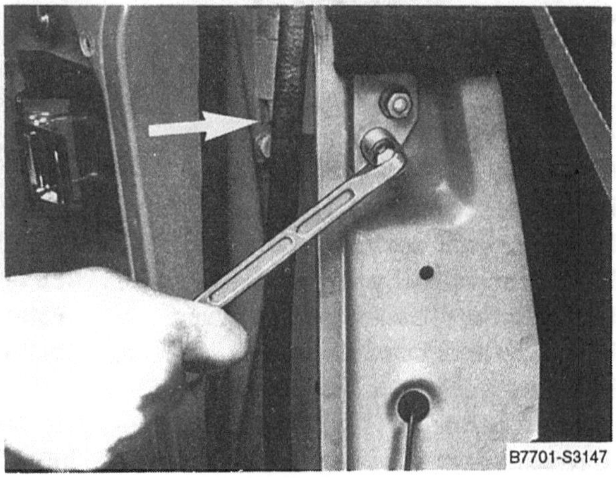

Fig. 2. Rear door hinge-to-body nuts being removed. Trim (arrow) has been removed from the B-pillar to reveal nuts.

Installation is the reverse of removal. After installing the door, adjust it as described below. When installing a new replacement door, the door parts from the old door will have to be transferred to the new door.

To adjust doors

Correct adjustment or alignment of the door allows for proper opening and closing. Adjustment also prevents wind noise. If the hinge bolts inside the door or the striker plate have not been disturbed, there is no need to adjust them.

NOTE
Some models have a locating screw in the hinge plate which must be removed before adjusting the door. See Fig. 3. This screw can be discarded after adjusting the door.

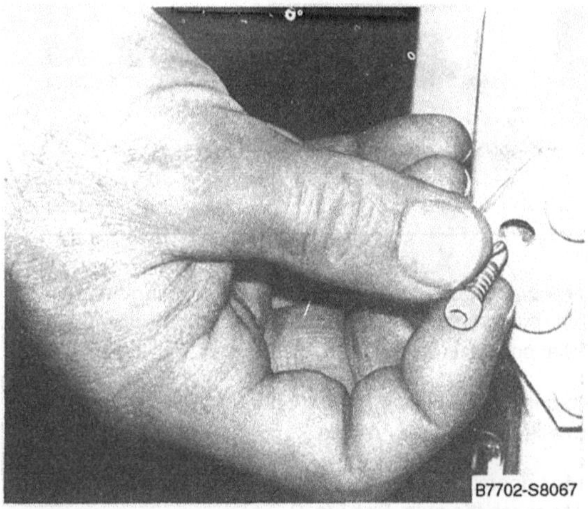

Fig. 3. Locating screw being removed before adjusting door.

1. Lubricate the hinges and latch mechanism with engine oil.

2. To adjust the flush fit of the front doors to the body, loosen the hinge-to-body bolts as shown in Fig. 1 above. The door can then be moved on the slotted holes in the hinge. Tighten the bolts, close the door and check the adjustment. Readjust as necessary.

3. To adjust the up-down or fore-aft position of the front doors, loosen the hinge-to-door bolts. See Fig. 4. This adjustment can only be done after the interior trim panel has been removed.

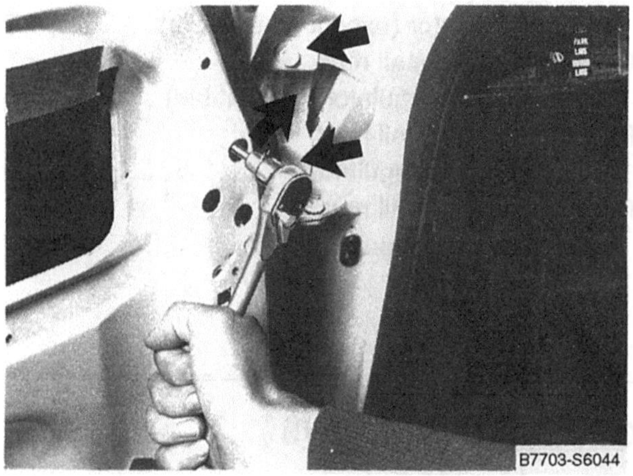

Fig. 4. Hinge-to-door bolt being loosened to adjust up-down or for-aft position of front door. The bolts for adjusting the flush fit of the door can also be seen (arrows).

4. Rear door flush fit and up-down position can be adjusted after loosening the hinge-to-door bolts. See Fig. 5.

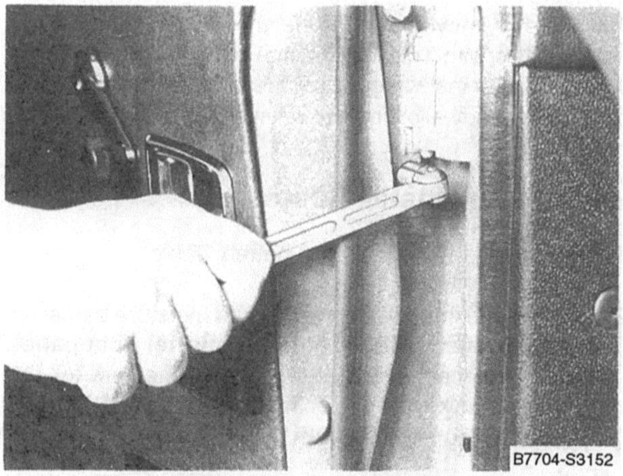

Fig. 5. Rear hinge-to-door bolt being loosened to adjust flush fit and up-down position of door.

5. Adjust the striker plate so the door is not forced up or down when closing. Also check that the door opens and closes smoothly. Loosen the screws as shown in Fig. 6 to adjust the striker plate.

Fig. 6. Striker plate being adjusted. An impact driver can be used to loosen screws.

NOTE
Additional adjustment of the rear doors can be done by loosening the hinge-to-body nuts. See Fig. 2 shown above.

To remove interior door panel

Although the procedures given below apply to the front doors, they can also be used as a guide for removing the interior trim for the rear doors.

1. Unscrew the locking knob. Where applicable, carefully pry off the disc from the window crank and remove the crank screw and crank. See Fig. 7.

NOTE
Before removing the window crank, make a note of its position with the window closed.

---

2. On sedan models, pry out the trim cover for the inner door handle. Remove the screws that hold the armrest to the door. Then carefully pry off the panel by working around the perimeter of the door. The panel is held in place by retaining clips. Lift the panel up slightly to remove it.

3. On hatchback and convertible models, remove the door pull retaining screw and remove the door pull. Slide the trim insert panel up and off its mounting clips. Remove the screws holding the panel to the door. See Fig. 8.

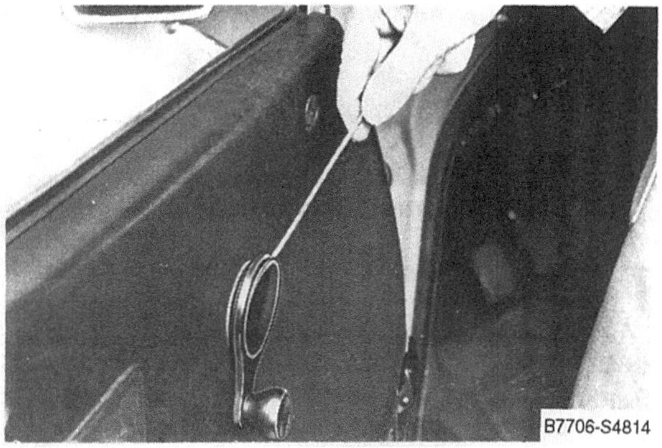

Fig. 7. Window crank trim cover being pried off.

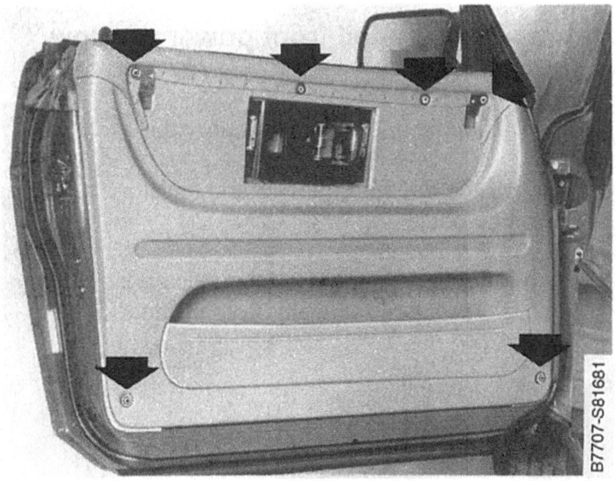

Fig. 8. Door panel retaining screws (arrows) on hatchback and convertible models. Door pull and upper trim insert panel removed.

4. Carefully remove the moisture barrier from the door.

Installation is the reverse of removal. Always test the window, mirror, and locking functions before reinstalling the panel. Replace any damaged retaining clips. If the moisture barrier is handled carefully and kept clean, it can be reused. Apply thread locking compound to the window crank handle screw.

NOTE
Always replace the moisture barrier if it becomes damaged.

Windows

This section covers removal and installation of the door windows, window regulators, and power window motors. The door windows are controlled by window regulators that move the glass in fixed channels. To replace the power window motor, the regulator assembly must first be removed from the door.

WARNING
Always disconnect the electric window motor when working on it. The motor is powerful enough to cause injury if it should come on by accident.

NOTE
Replacement of the windshield, the rear windows, and fixed side windows is not covered here. These windows are held in place with weather-tight seals. Special equipment and techniques are required to replace these windows.

To remove and install front power window regulator

1. Disconnect the negative (–) battery cable.

2. Remove the interior door trim and moisture barrier as described earlier.

3. Pry off the inner window weatherstrip spring clips and remove the inner weatherstrip from the door. See Fig. 9.

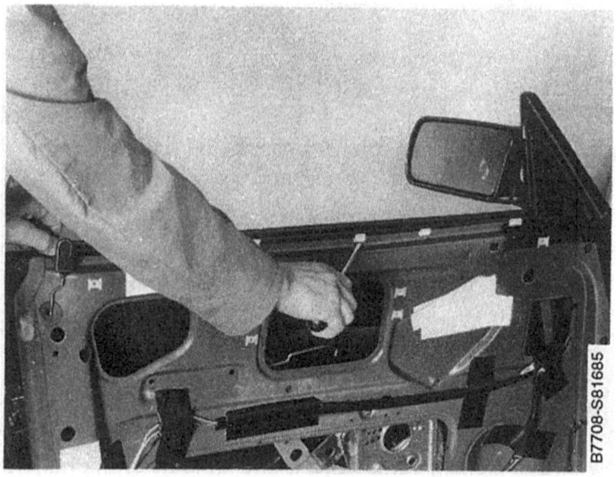

Fig. 9. Spring clip for inner weatherstrip being removed.

4. Locate the window motor electrical connector inside the door and disconnect it. See Fig. 10.

5. While supporting the window glass, remove the screws holding the glass to the regulator. See Fig. 11.

6. Remove the window glass by rotating it to the vertical position and sliding it up out of the door.

7. Remove the four bolts holding the regulator assembly to the door and remove the assembly through the opening in the door. See Fig. 12.

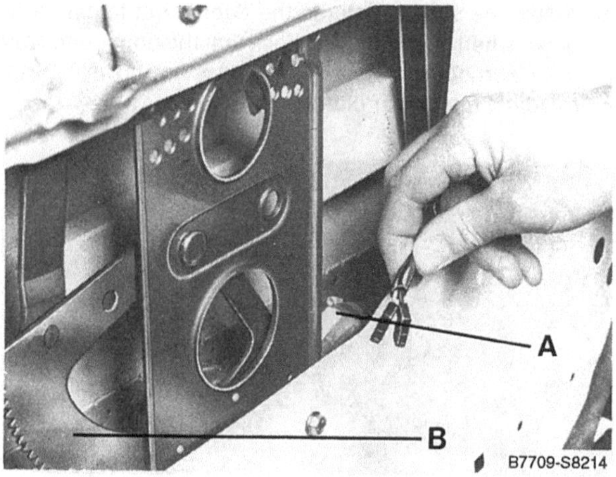

Fig. 10. Connector for power window motor disconnected. Also shown is motor **(A)** and gear sector **(B)**.

Fig. 11. Regulator-to-window glass support mounting screws (arrows). Support glass after removing screws.

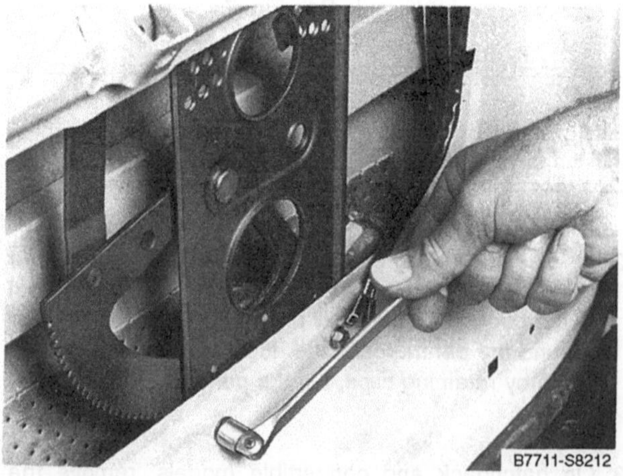

Fig. 12. Regulator-to-door mounting bolt being removed. Two bolts (not shown) at top must be removed also.

8. Remove the three bolts holding the motor to the regulator. See Fig. 13.

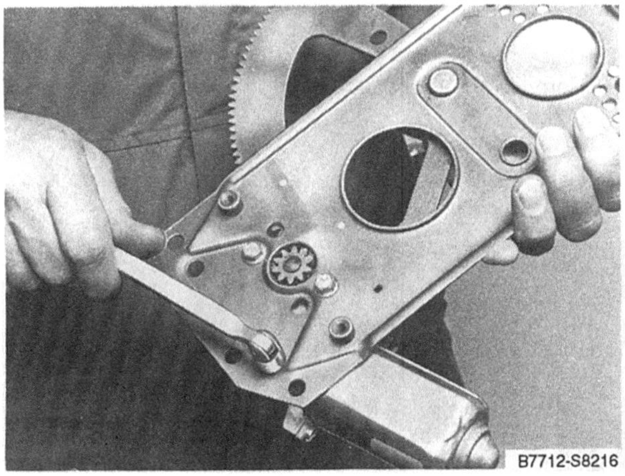

Fig. 13. Power motor mounting bolt being removed.

NOTE

The plastic moisture barrier on the old motor must be reinstalled on the new motor to keep out moisture.

Installation is the reverse of removal. Test the new motor before installing it in the door. A thin film of white grease can be applied to moving parts.

NOTE

The power motor can be tested by disconnecting the connectors at the motor and testing for voltage when the window switch is pushed in either direction (ignition key on). If power is present at the connector, the motor is faulty. If voltage is not present, the problem lies in the switch or other part of the circuit. Troubleshoot the power window circuit using the appropriate diagram in **371 Wiring Diagrams, Fuses and Relays**.

To remove and install rear power window regulator (except convertible)

1. Disconnect the negative (–) battery cable. Remove the window switch from the interior trim.

2. Remove the interior door trim and moisture barrier as described earlier. Disconnect the electrical connector for the power window located inside the door.

3. While supporting the window glass, remove the screws that hold the glass to the regulator arm. See Fig. 11 shown above. Then push the window up until it is fully closed.

CAUTION

Once the window-to-regulator screws are removed, the window should be supported. With the window fully closed, tape the glass to the door frame to prevent it from falling.

NOTE

Removal of the rear door window requires that the fixed side pane be removed so that the rear window channel can be moved out of the way. The window channel is held in place with two screws that can be accessed after removing the rubber concealing plugs in the door.

---

4. Remove the five screws that hold the regulator assembly to the door. See Fig. 14. Remove the assembly through the opening in the door.

NOTE

Slide the regulator arm forward to help in removing the assembly from the door.

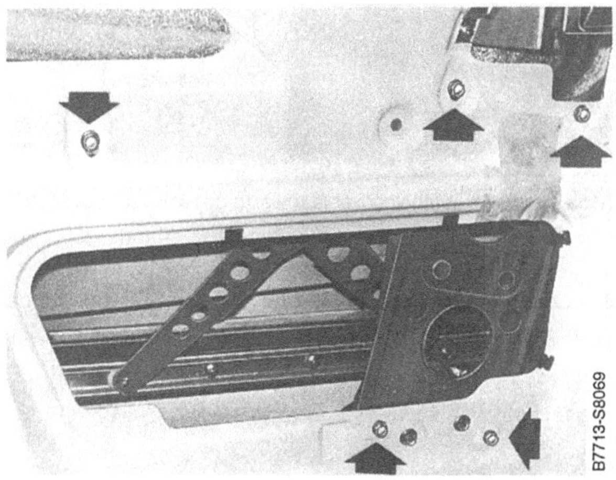

Fig. 14. Rear door regulator mounting screws (arrows).

5. Remove the screws that hold the motor to the regulator.

NOTE

Always reinstall the old protective cover to the window motor.

Installation is the reverse of removal. Test the new motor before installing it in the door. A thin film of white grease can be applied to moving parts.

To remove and install rear power window regulator (convertible)

The regulator assembly on convertibles is located in the body cavity. The regulator is removed as a complete assembly together with the window glass.

1. Disconnect the negative (–) battery cable.

2. Remove the rear seat cushion, backrest, and interior trim panel as described in **851 Interior Trim and Upholstery**.

3. Remove the seatbelt cover screws and slide the cover down. See Fig. 15.

Fig. 15. Screws for seatbelt cover (arrows) to be removed. Slide cover down seatbelt.

4. Loosen the seatbelt guide bolt and pivot the belt guide in. Remove the two window bracket bolts. See Fig. 16.

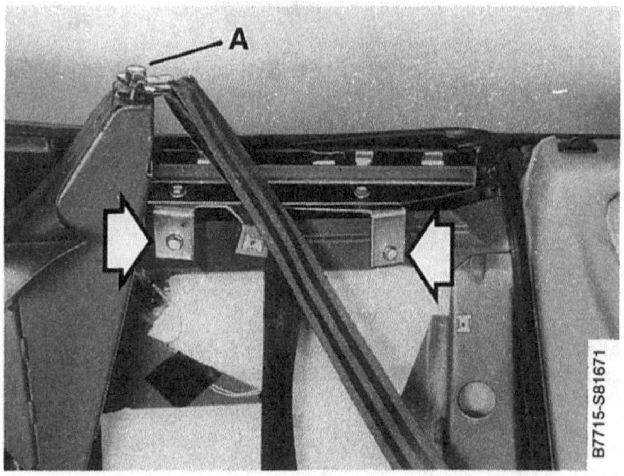

Fig. 16. Loosen seatbelt guide bolt **A**. Remove two window bracket bolts (arrows).

5. Carefully pry off the outer window seal.

NOTE

The front of the seal is glued to the body.

---

6. Raise the window up until it contacts the window stop. Remove the four nuts holding the regulator assembly to the body.

7. Using a hex key, screw the regulator studs in. See Fig. 17.

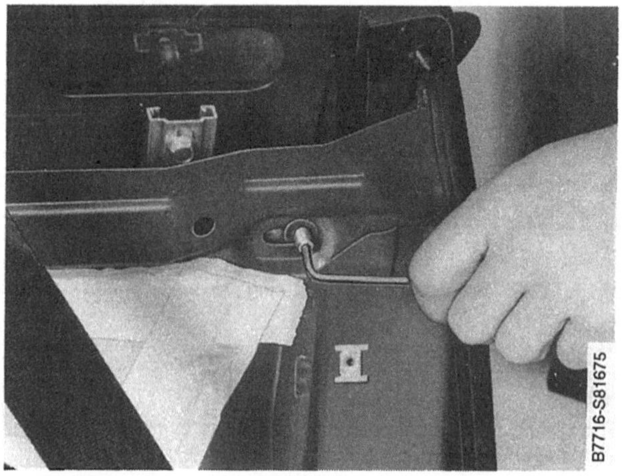

Fig. 17. Regulator mounting stud being turned in to gain clearance with body panel.

8. Pull the window glass and regulator assembly out of the body.

NOTE

The glass can be removed from the regulator by removing the three regulator-to-glass mounting bolts.

Installation is the reverse of removal. Tighten the seatbelt guide bolt to the value given below. Reinstall the outer window seal using contact adhesive around the seal and the molding.

Tightening torque

- seatbelt guide bolt to body (convertible models) . . . 24 to 40 Nm (18 to 30 lb-ft)

To remove and install front manual window regulator

1. Remove the interior door trim and moisture barrier as described above.

2. While supporting the window glass, remove the screws that hold the glass to the regulator arm. See Fig. 11 above.

NOTE
The window glass can be removed from the door after removing the inner weatherstrip from the door. See Fig. 9 above. With the weatherstrip removed, rotate the glass to a vertical position and slide it up and out of the door.

3. Remove the four bolts holding the regulator to the door and remove the regulator. See Fig. 18.

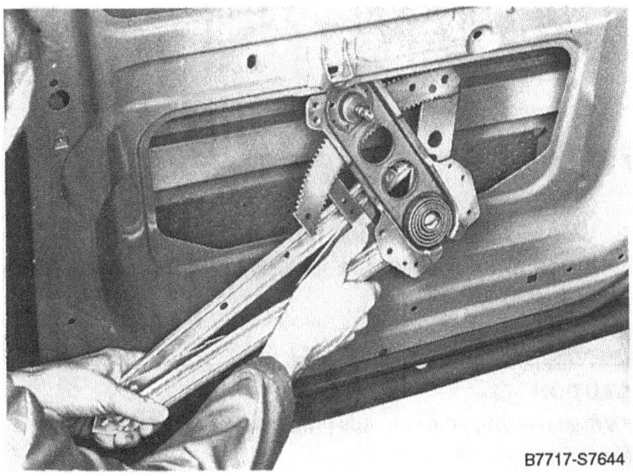

Fig. 18. Regulator being removed from front door.

Installation is the reverse of removal. A thin film of white grease can be applied to moving parts. Test the window regulator before installing the interior door trim.

To remove and install rear manual window regulator

1. Remove the door handle, interior door trim and moisture barrier as described above.

2. Support the window glass while removing the screws that hold the glass to the regulator arm. Slide the window up until it is in its fully closed position.

3. Remove the bolts holding the regulator to the door and remove the regulator. See Fig. 19.

Installation is the reverse of removal. A thin film of white grease can be applied to moving parts. Test the window regulator before installing the interior door trim.

NOTE
Removal of the rear door window requires that the fixed side pane be removed so that the rear window channel can be moved out of the way. The window channel is held in place with two screws that can be accessed after removing the rubber concealing plugs in the door.

Fig. 19. Rear door manual regulator being removed.

LOCKS, HANDLES, AND LATCH MECHANISMS

This section covers replacement of the lock cylinders, the door handles, and the latch mechanisms. Also covered here is replacement of the central locking motors.

For information on repairing the ignition lock cylinder see **432 Manual Transmission Controls**. For information on wiring for the central locking system see **371 Wiring Diagrams, Fuses and Relays.**

NOTE
- Always test all of the latch and locking functions including the child safety feature after completing any repairs.
- When working on the locking system always keep a spare key outside the car in case the system locks accidentally.

To replace door handle

1. Remove the interior door trim and moisture barrier as described earlier. Working inside the door, remove the front door handle mounting screw.

2. Working at the outside door edge, loosen the rear screw for the door handle bracket about four turns. See Fig. 20.

NOTE
On 4-door models, the rear screw for the rear door is hidden behind the weather strip.

3. Rotate the handle to the position shown in Fig. 21 and remove the handle from the door.

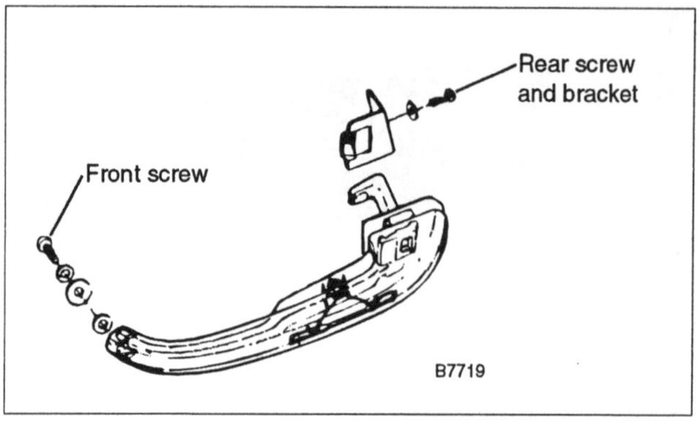

Fig. 20. Door handle assembly.

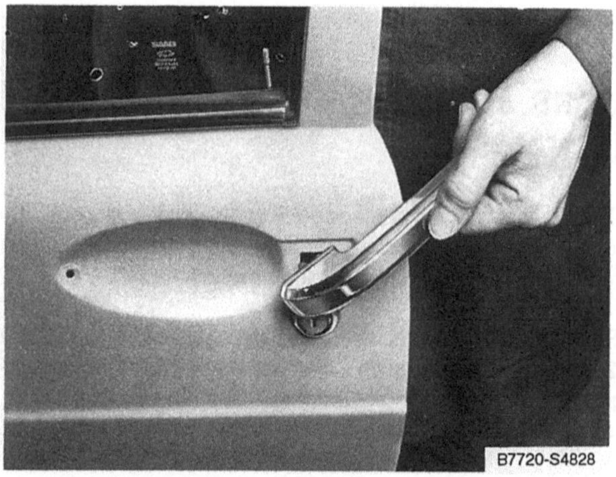

Fig. 21. Rotate door handle as shown to remove.

CAUTION
Do not force the handle when removing it from the door. The latch mechanism inside the door could be damaged.

Installation is the reverse of removal.

To replace door lock mechanism

Fig. 22 shows the front and rear door lock mechanisms.

1. Close the window. Remove the locking knob, interior door trim and moisture barrier as described above.

2. Manually rotate the latch mechanism on the door jamb to the latched (closed) position and remove the three retaining screws on the door edge.

NOTE
If the screws are frozen they can be removed with an impact driver.

Fig. 22. Front (left) and rear (right) door lock mechanisms.

3. Working inside the door, remove the link rods from the inside door handle, the lock cylinder, and the central lock motor.

CAUTION
- When removing the link rods from plastic parts do not stress the plastic as it will break.
- Do not bend or distort the link rods. The latch mechanism will not work properly when the bent links are reinstalled.

NOTE
The central lock motor can be removed to gain more working space inside the door. Remove the locking motor as described below.

---

4. Loosen, but do not remove, the screws for the outer door handle as described above.

NOTE
Loosening the outer door handle screws will allow some clearance for the latch mechanism to be removed. The outer door handle can be completely removed if necessary.

---

5. Carefully work the latch mechanism free and remove it from the door.

Installation is the reverse of removal. Apply a thin film of chassis grease (Saab special chassis grease —ESSO Nebula EP2 or equivalent) to the latch mechanism. Test all the latch functions before installing the interior trim panel.

To replace trunk lid lock mechanism

1. Remove the central lock motor as described later under the appropriate heading.

CAUTION

When removing the link rods from plastic parts do not stress the plastic as it will break.

---

2. Remove the trunk lid lock cylinder as described below.

3. Remove the two bolts for the latch mechanism, then carefully remove the latch from its mount.

Installation is the reverse of removal.

To replace hatchback lock mechanism

1. Remove the interior trim panel from the lid.

2. Working from inside the lid, disconnect the link rod from the control arm on the lock cylinder. Remove the two screws that hold the latch mechanism to the lid and remove the latch mechanism. See Fig. 23.

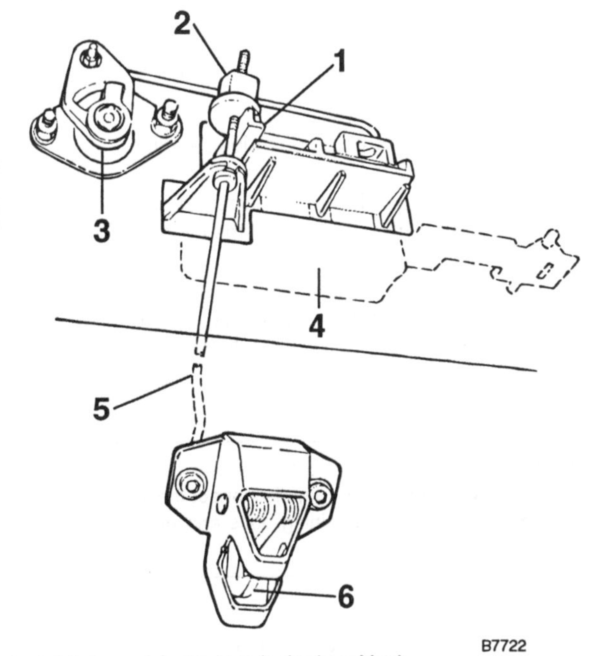

1. Slide lever (shown in unlocked position)
2. Adjusting nut
3. Lock cylinder
4. Latch handle assembly
5. Link rod
6. Latch mechanism

Fig. 23. Latch mechanism used on hatchback models.

3. Working inside the lid remove the five nuts holding the latch handle assembly to the lid. Disconnect the license plate light connectors. Remove the handle assembly.

Installation is the reverse of removal. Clean, then lubricate all sliding surfaces with a thin film of grease. The adjusting nut can be adjusted so the handle has a little slack before it releases the hatchback lid.

To replace door lock cylinder

Replacing the door lock cylinder requires removing the interior trim panel from the door. Fig. 24 shows an exploded view of the lock including the U-shaped spring clip that holds the lock to the door.

NOTE

Lock cylinders matching your key code can be specially ordered from Saab. The key code is also stamped on the lock cylinder barrel.

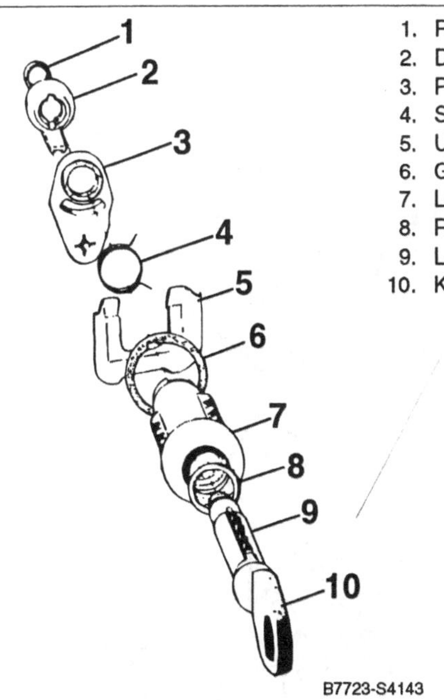

1. Ring
2. Driver
3. Plastic arm
4. Spring
5. U-shaped clip
6. Gasket
7. Lock barrel
8. Ring
9. Lock
10. Key

Fig. 24. Door lock cylinder.

1. Put the window in the closed (up) position. Remove the interior door panel trim as described earlier.

2. Unhook the link rod from the lock cylinder plastic arm.

3. Carefully pull down the U-shaped clip using pliers. Note the cutouts in the lock cylinder body for the clip.

NOTE
It is always a good idea to replace the U-shaped clip with a new one when replacing the lock. The old clip may become distorted and will not hold the lock cylinder properly.

Installation is the reverse of removal. Lubricate the lock cylinder with lock lubricant after installation.

To replace trunk lid or hatchback lid lock cylinder

Fig. 25 shows the exploded view of the trunk lid lock cylinder. The hatchback lid lock cylinder is shown in Fig 26.

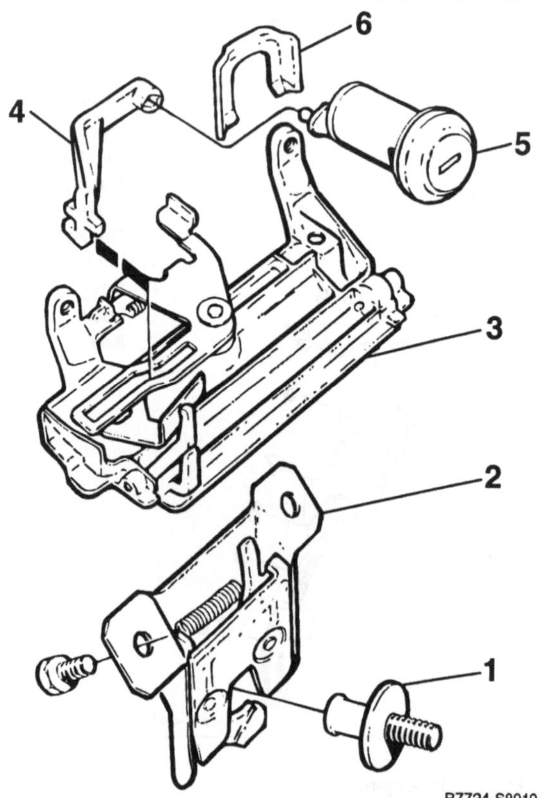

1. Striker
2. Latch assembly
3. Latch handle
4. Ball driver
5. Lock cylinder
6. U-shaped clip

Fig. 25. Exploded view of lock cylinder and latch handle on sedan trunk lid.

NOTE
Lock cylinders matching your key code can be specially ordered from Saab. The key code is also stamped on the lock cylinder barrel.

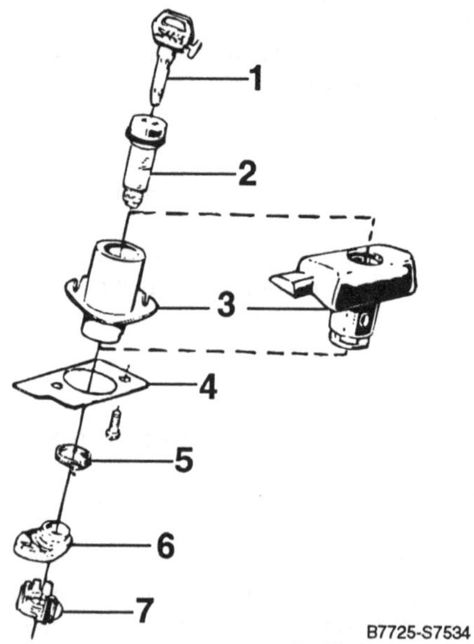

1. Key
2. Lock cylinder
3. Cylinder housing
4. Gasket
5. Spring
6. Plastic arm
7. Driver

Fig. 26. Exploded view of lock cylinder for hatchback lid.

1. Remove interior trim as described earlier.

2. On sedans, pull off the U-shaped clip and disconnect the ball driver from the back of the cylinder. Pull the lock center from the trunk lid.

3. On hatchbacks, remove the link rod from the plastic arm. Remove the two lock cylinder mounting screws and then pull the lock cylinder assembly from the lid.

NOTE
On models with central locking, disconnect the locking motor then remove the motor retaining bolts and move the motor slightly.

Installation is the reverse of removal.

To remove central locking motors

For more information on electrical tests of the central locking system, see **371 Wiring Diagrams, Fuses and Relays**.

1. Where applicable, make sure the door windows are closed. Remove the interior trim panel as described earlier.

2. Locate the door/lid lock motor and unhook the link rod from the motor.

3. Disconnect the electrical connector for the lock motor. Remove the two screws holding the motor to the door or lid.

NOTE

Front door and hatchback lock motors are mounted on brackets.

Installation is the reverse of removal. Test the central lock motor before reinstalling the interior trim panels.

CAUTION

The lock motors should not be tested using a battery or power supply. Damage to the motor may result. Test the motors only through the central locking control unit.

Hatchback and Trunk Lid

The hatchback or trunk lid can easily be removed. After installation, the fit of the lid should be carefully adjusted.

WARNING

The hatchback lid is very heavy and should not be removed unless two people, or more, are available to help.

To remove and install hatchback lid

1. Disconnect the negative (–) battery cable. Remove the ground strap near the hinge. Disconnect the electrical connectors to the heating element, central locking motor, and any lighting.

2. Mark the position of the hinges using a scribe. Loosen, but do not remove, the hinge screws.

3. With a helper supporting the rear of the lid, remove the gas assist struts on each side.

WARNING

• The hatchback lid has no support once the gas assist struts are removed. The lid could fall causing injury.

• Do not attempt to disassemble the gas assist struts. They are under very high pressure and could explode.

---

4. Remove the hinge screws that were loosened earlier, then carefully lift the lid from the body.

NOTE

After removing the lid, place small blocks of wood under the lid hinges to help position the hinge during installation. See Fig. 27.

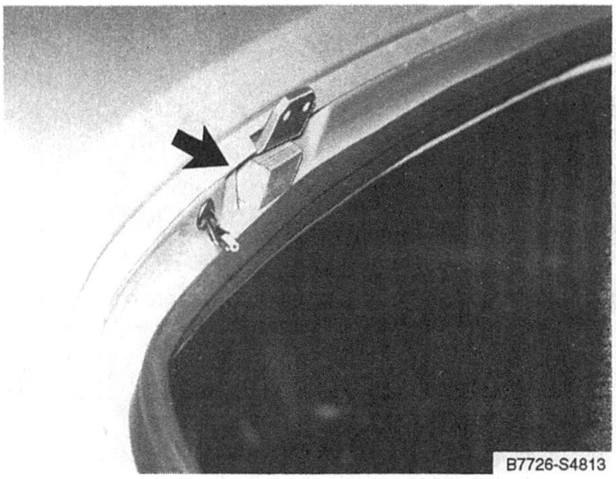

Fig. 27. Small blocks of wood (arrow) used to hold lid hinges in up position for lid installation. Be sure to remove the blocks before shutting the lid.

Installation is the reverse of removal. The lid can be adjusted for proper alignment by loosening the screws at the hinges or striker plate. Always test the heated rear window and central locking functions.

CAUTION

Do not close the hatchback lid if the gas assist struts are not attached to the lid. The struts and the body could be damaged.

To remove and install trunk lid

1. Disconnect the negative (–) battery cable. Disconnect the electrical connectors next to the left hinge and cut any wire ties holding the harness to the lid.

2. Mark the position of the hinges on the lid. Remove the two nuts holding the trunk lid to the hinge mounting. See Fig. 28.

3. Installation is the reverse of removal. Adjustment to the lid is made by loosening the nuts shown in Fig. 28.

After completing the adjustment, secure the wiring harness to the lid and test all electrical functions.

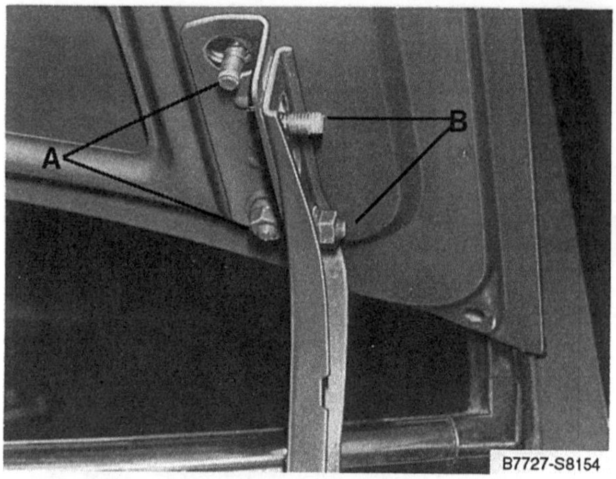

Fig. 28. Trunk lid hinge mounting nuts **(A)** to be removed. The mounting nuts **(B)** at the hinge holes do not need to be removed.

NOTE

The opening effort of the assist springs can be adjusted by repositioning the spring end into one of the three openings in the body. See Fig. 29.

Fig. 29. Trunk lid assist being adjusted by positioning spring in second cutout. The tool shown is used to stretch the

SUNROOF

Two types of sunroofs, electric and manual, are installed in cars covered by this manual. This section covers removal, installation, and adjustment of the sunroof. For troubleshooting information of the power sunroof see **371 Wiring Diagrams, Fuses and Relays**.

The electric sunroof is driven by a motor and cable assembly. The motor is located in the luggage compartment, near the spare tire. The drive cables run along each C-pillar up to the sunroof.

NOTE

To replace the drive cables, the interior headliner and the entire sunroof assembly must first be removed. On sedans, the headliner can only be removed after the rear window has been removed. Because of the nature of this procedure, it is not covered in this manual.

Water drainage hoses are located at the corners of the sunroof assembly. The front hoses run down through the corner A-pillars. The rear hoses run down through the rear pillars. These hoses must not be clogged or water may leak into the passenger compartment.

To remove sunroof

1. Slide the sunroof back and remove the four screws at the front edge of the roof panel. See Fig. 30.

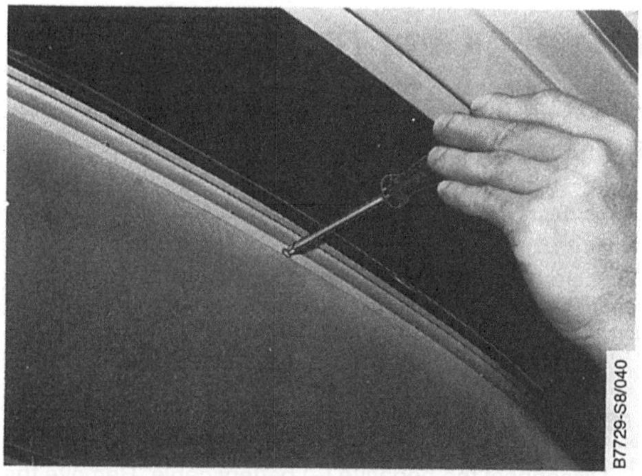

Fig. 30. Screws being removed at sunroof front edge.

2. Close the sunroof panel until a gap of 3/4 in (19 mm) remains. Then remove the top panel by raising it up at the front and pulling it forward and out.

3. Working at the front corners, remove the guide clip nuts and the clips. See Fig. 31.

4. Working at the rear corners, pry back the locking tabs for the rear guide nuts. Remove the guide nuts and pull the rear guides off of the sunroof panel frame. See Fig. 32. Where applicable, set aside the springs, mountings, bushings and sleeves.

5. Push the panel forward as far as it will go. Loosen, but do not remove the screws at the rear guide clips. See Fig. 33. Slide the clips outward as far as possible.

6. Remove the panel frame by raising the front edge while pulling it forward.

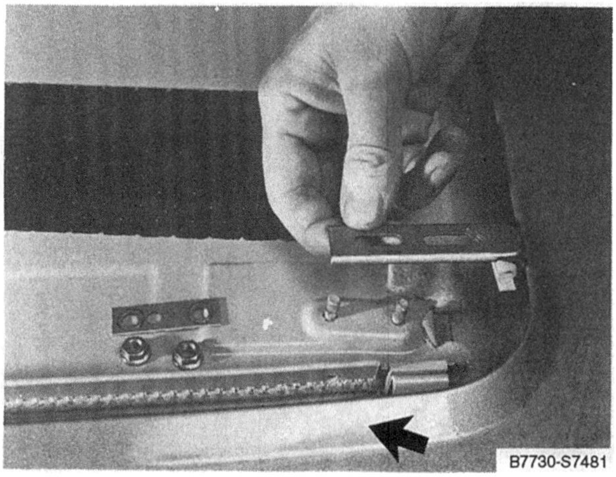

Fig. 31. Front guide clip being removed. Windbreak shown at arrow.

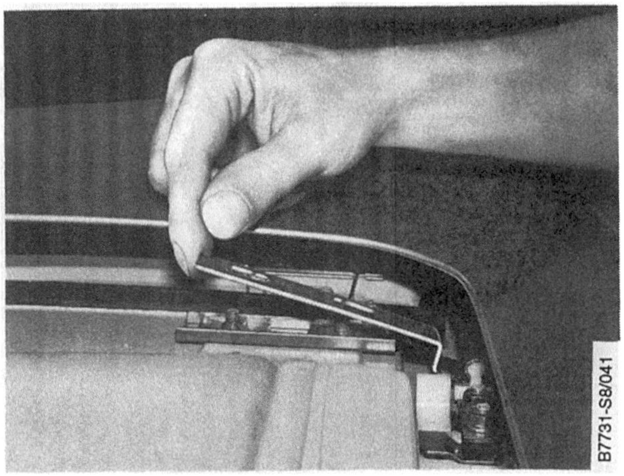

Fig. 32. Rear guide being removed. Model with power sunroof shown.

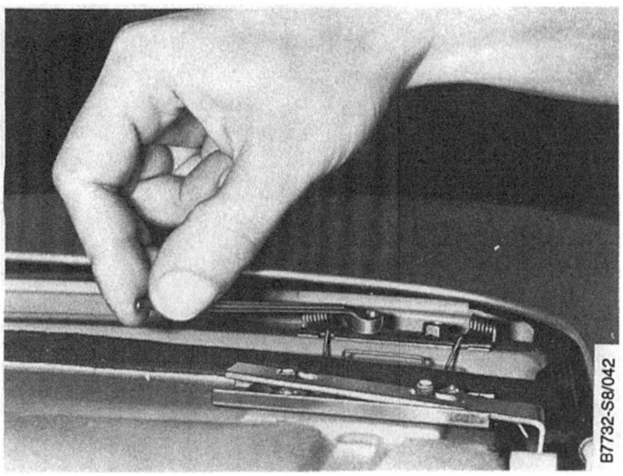

Fig. 33. Rear guide clip being loosened. Early model shown uses nuts instead of screws.

To install sunroof

Before installing the sunroof assembly, lubricate all sliding surfaces with a coat of petroleum jelly (Vaseline).

1. Install and position the panel frame against the front stops, center the frame in the sunroof opening and tighten the front guide clip nuts to obtain the clearance shown below. See Fig. 34.

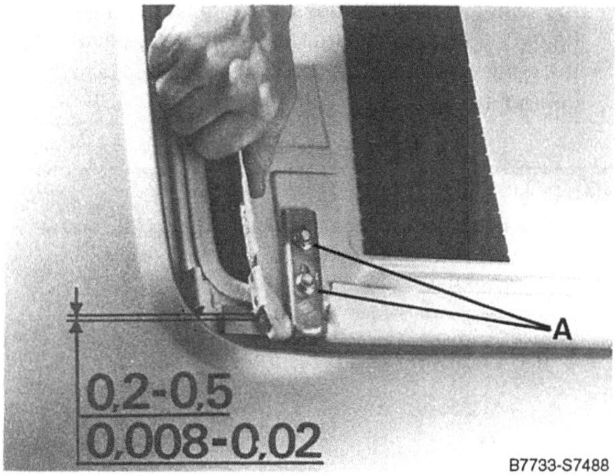

Fig. 34. Sunroof panel clearance being adjusted. Tighten nuts (**A**) when clearance is correct.

2. With the panel against the front stops, install the rear guide clips and loosely install the clip screws if they were removed earlier.
3. Have a helper gently press up on the rear edge of the panel from inside the car and then slide the guide clips into position.

NOTE
Pressing up on the rear of the panel frame will bend it slightly allowing the rear guide clips to be pushed back to their original positions.

---

4. Tighten the rear guide clip and the rear guide fasteners.
5. Position the sunroof frame as far forward as possible.
6. Install the sunroof panel into the sunroof panel frame, inserting the back first. Make sure the tab on the panel engages the spring clips on the frame. Push down and back on the panel until it locks into place.
7. Carefully slide the sunroof back while checking that the sunroof clears the panel. Install the four screws at the front edge.

CAUTION
If the sunroof top contacts the panel the paint may be scratched. Move the panel slowly to check that it clears the body sufficiently.

8. Close the sunroof top and check for a uniform gap around the opening.

9. Check that the rear of the sunroof top is flush with the roof when closed. If necessary, adjust the sunroof top height by first removing it as described above. Adjust the guide shoe screw as shown in Fig. 35. Reinstall the sunroof top and recheck the height.

CAUTION
Be careful not to scratch the paint when adjusting the screw.

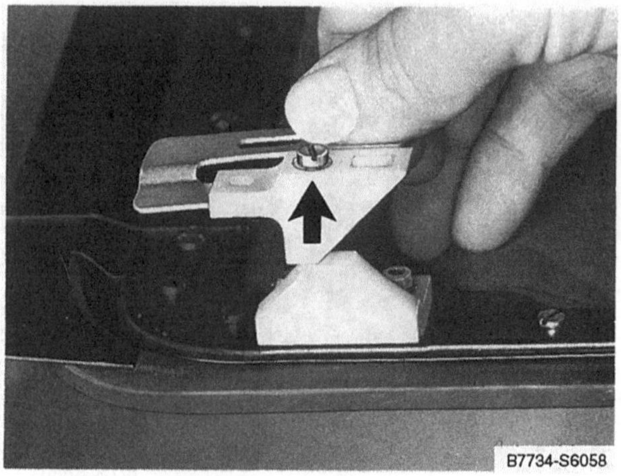

Fig. 35. Guide shoe adjusting screw (arrow) to adjust sunroof top flush with roof.

851 Interior Trim and Upholstery

SEATS AND SEATBELTS

The front seats are bolted to the floor pan via rails. The rails provide fore-aft seat adjustment. All models have heating elements in the front seats. Some models have a power seat adjustment. For information on testing the heating element see **371 Wiring Diagrams, Fuses and Relays.**

The seat belts are a combination lap-shoulder belt except for the center belt for the rear seat. The belts should be inspected periodically for webbing defects such as cuts or pulled threads. They should also be tested for proper retracting.

WARNING

- For maximum protection from injury, seat belts should be replaced as a set (including all hardware) if they are subjected to occupant loading in a collision.
- Seat belts should not be modified or repaired. The seat belt anchorage points should not be changed or modified.
- Do not install seat belts or seat belt hardware purchased from an auto-recycler or salvage yard. These parts may have been subjected to occupant loading from a collision. Only Saab original replacement equipment mounting bolts and hardware should be used. Check with an authorized Saab dealer for the latest parts information.
- Seat belt assemblies must remain together as a matched set. Improper latching of the seat belt may occur if the components have been mixed from other sets or other cars.
- Do not bleach or dye seat belt webbing. Webbing that is severely faded or redyed will not meet the strength requirements and must be replaced.

To remove front seats

1. Slide the seat to the rear. Remove the two front socket head screws holding the seat to the adjusting mechanism.
2. Disconnect the electrical connectors for the seat heater and power function (where applicable).
3. To remove the passenger's seat, slide the seat forward. Remove the nuts holding the rails to the height adjusting mechanism. Remove the seat.
4. To remove the driver's seat, release the front catch by sliding the seat back to the limit of its travel. Tilt the seat back and then lift it upward and forward, removing it from the car. See Fig. 1.

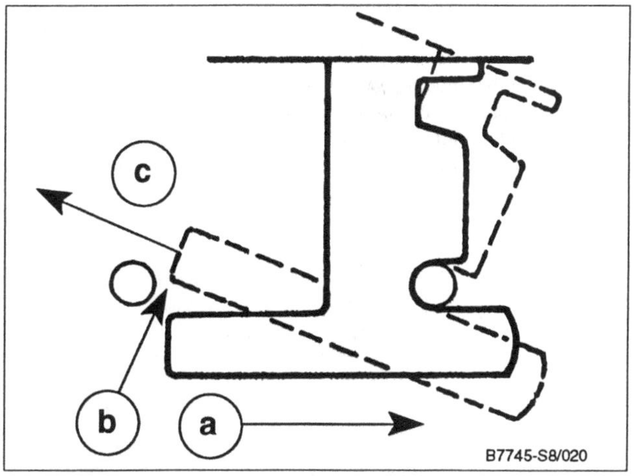

Fig. 1. To remove driver's seat, slide seat to rear (**a**), tilt seat back (**b**), and lift seat out of car (**c**).

The seat heating elements and thermostat are located under the seat covering. The bottom cover is held on with upholstery clips. The backrest cover can be removed after removing the headrest and unzipping the cover at the bottom.

To remove rear seat

1. To remove the rear seat bottom, remove the bolts along the front edge of the seat bottom or remove the seat hinge bolts. See Fig. 2. Pull the seat out of the car.

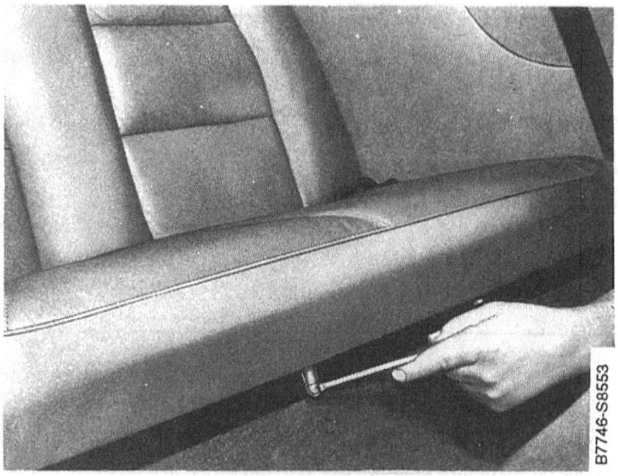

Fig. 2. Rear seat bottom cushion being removed on convertible model.

2. To remove the rear seat backrest, pivot the lower seat cushion up and into the cargo position. Release the rear seat back cushion and pivot it forward slightly. Then unbolt the hinges from the body. See Fig. 3.

NOTE
On convertible models, two anchor bolts located at the top of the backrest must also be removed. These bolts can be found behind each headrest, beneath the small cutouts in the upholstery. Two bolts at the bottom must be removed as well.

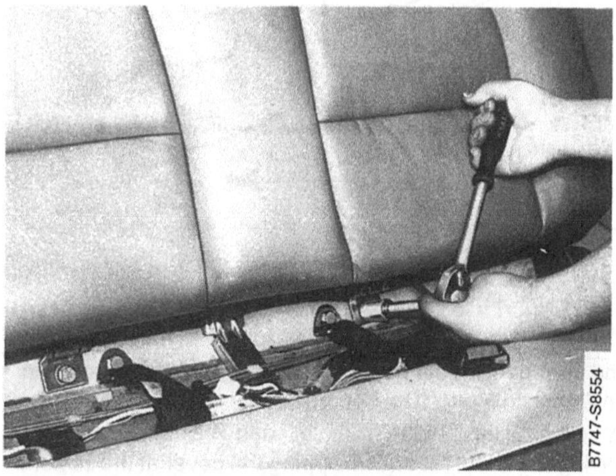

Fig. 3. Rear seat backrest being removed.

Removing Seat Belts

Removal of some of the seat belt assemblies may require removal of interior trim panels as described later in this section. When reinstalling the seat belt, always check that the seat belt retracts correctly and that the belt guides do not bind the belt. Always replace any damaged or worn components.

Tightening torque
- seat belt guide bolt to body (convertible models). . . . 24 to 40 Nm (18 to 30 lb-ft)
- all other seat belt mounting bolts . . . . 45±10 Nm (33±7 ft-lb)

NOTE
Servicing the passive (motorized) seat belts requires special equipment and techniques and is therefore beyond the scope of this manual. See an authorized Saab dealer for repairs to the passive seat belt system. For passive seat belt wiring diagrams, see **371 Wiring Diagrams, Fuses and Relays.**

NOTE
For information on removing the interior trim on convertible models, see **830 Doors, Windows, and Lids**.

CARPETS

The interior carpet is a nylon fiber on fabric. It is secured using press studs and tape fasteners.

To remove carpets

1. Remove both front seats as described earlier.
2. Remove the center console as described in **432 Manual Transmission Controls.** Remove the storage console (where applicable) as described in **853 Dashboard and Consoles.**
3. Remove the metal door sill plates. Carefully pull up the carpet and remove it from the car.

Installation is the reverse of removal. Check and correct any water leakage into the floor area before reinstalling the carpet.

INTERIOR BODY TRIM

To remove headliner

On hatchback models the headliner can be removed through the hatch opening. On two-door sedans the headliner can be very carefully removed through the right door opening. On four-door sedans, the rear window must be removed to remove the headliner.

1. Disconnect the battery negative (–) cable. On hatchbacks, remove the rear parcel shelf and put the rear seat down.
2. Remove the mirror, sunvisors, assist handles, coat hooks, and dome light. On models equipped with a sunroof, remove the edging around the sunroof opening.

CAUTION
Extra care should be taken not to stress the molded headliner particularly around the mirror and sun visors, as it will crack.

3. Remove the trim around the B-pillar seat belt. On hatchback models, remove the brackets for the rear parcel shelf. See Fig. 4. Pry off the plastic covers for the hatchback hinges. See Fig. 5.

Fig. 4. Rear parcel shelf brackets being removed.

4. Remove the trim pieces from the rear corner pillars. See Fig. 6.

NOTE
The rear corner trim pieces are held on with clips. Carefully pry the piece away from the body.

5. Pull the headliner back slightly to free it from the windshield pillar trim and remove the headliner.

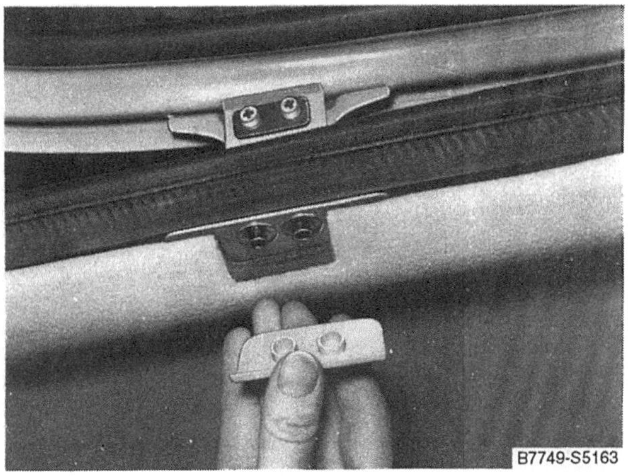

Fig. 5. Plastic trim cover for hatchback hinge.

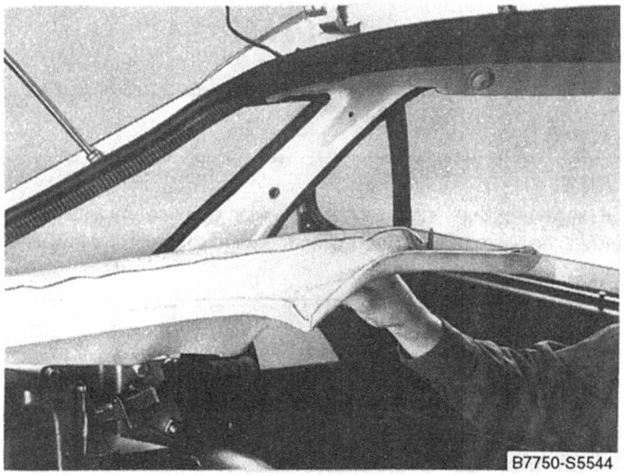

Fig. 6. Rear trim piece being removed.

Installation is the reverse of removal.

To remove lower instrument pad

1. Remove the center storage console as described in **853 Dashboard and Consoles.**
2. Remove the ashtray and the ashtray holder screws. Remove the bolt in the ashtray opening.
3. Working in the engine compartment, remove the instrument pad nuts, one near the left door hinge and the other near the right door hinge. See Fig. 7. Remove the lower instrument pad.

NOTE
Do not mistake the small nut that holds the lower instrument pad with the larger two bolts that hold the entire instrument panel assembly. The two larger bolts should not be removed.

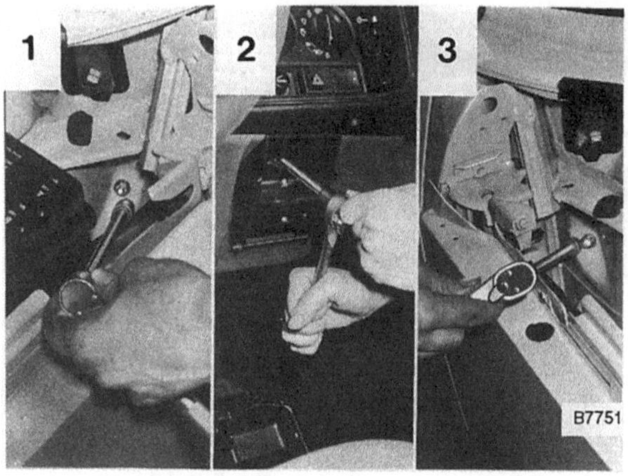

Fig. 7. Mounting points for lower instrument pad to body (**1** and **3**). Center dash pad mounting bolt shown at **(2).**

To remove interior side trim

1. Remove the back seat cushions as described earlier.

NOTE

On sedan models, the lower seat cushion does not require removal, as the cushion can be tilted forward to reach the lower trim screws.

---

2. Remove the screws holding the trim to the body. See Fig. 8. On some models, a special plastic fastener is used in place of the screws. To remove these fasteners turn them 90° counter-clockwise to unfasten.

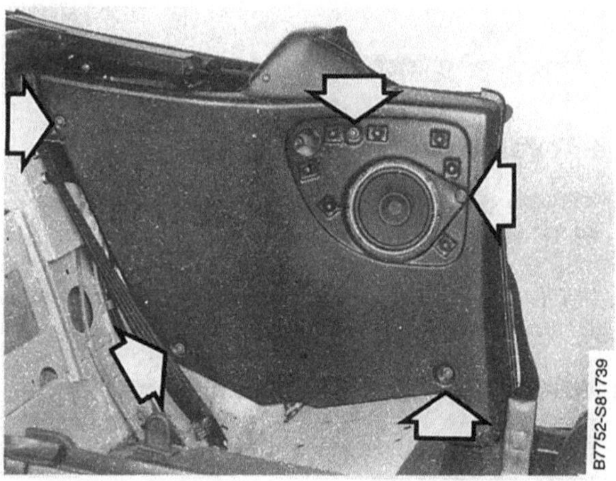

Fig. 8. Remove screws (arrows). Convertible model shown.

NOTE

On convertibles, the speaker and reading light connectors must be disconnected.

To remove luggage compartment side trim (hatchback models)

1. Remove the rear floor panel and floor plate.

2. Remove the side carpet sections.

NOTE

Some models use a molded plastic trim piece in place of the carpet section. This piece is held in with special plastic fasteners which are removed by turning them 90°.

ANTENNA

To test electric antenna

1. Remove the luggage compartment side trim as described above.

2. Check that the ground wire to the chassis is tight.

3. Test for battery voltage at the red wire (motor supply).

4. Test for battery voltage at the green wire (up/down signal) when the radio is first turned on.

If voltage is as specified, the antenna is faulty. If voltage is not as specified, check the wiring and relay using the appropriate wiring diagram. listed in **371 Wiring Diagrams, Fuses and Relays.**

To replace electric antenna mast

1. Disconnect the battery negative (–) cable. Remove the nut holding the antenna to the body.

2. Working inside the car, pull back the luggage compartment side carpeting. Remove the screws holding the antenna assembly to the body. Disconnect the wires.

3. Remove the screws holding the antenna tube to antenna housing. See Fig. 9. Remove the tube after making a note of its position.

4. Connect the motor leads to a battery. Black wire to ground (–). Red and green wires to positive (+). Wait until the motor has stopped with the antenna in the extended position.

NOTE

The motor will stop after several seconds. The gear rack/mast assembly can now be pulled out of the motor.

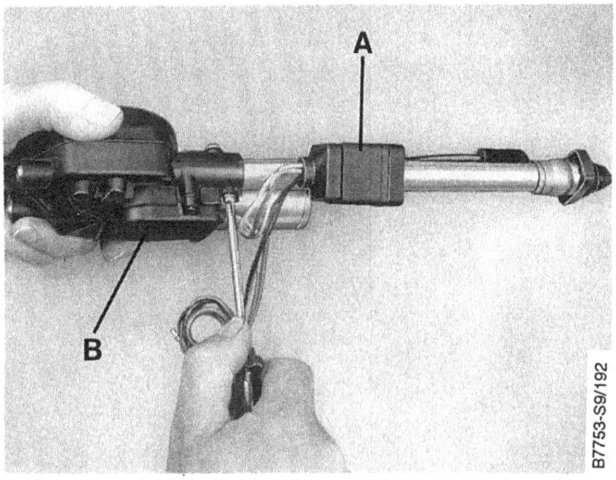

Fig. 9. Antenna tube screws being removed. Also shown is the relay **(A)** and the park position switch cover **(B)**.

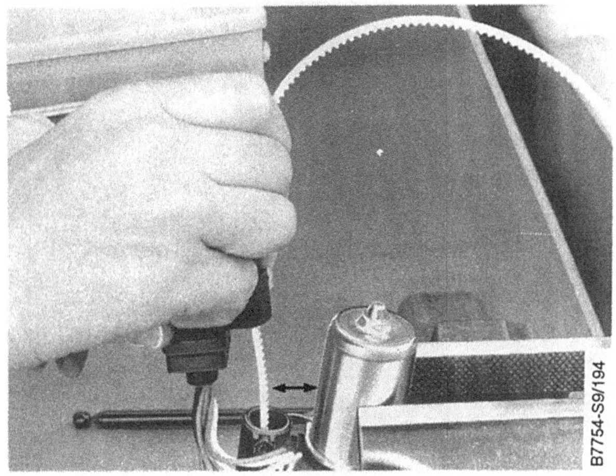

Fig. 10. Antenna gear rack correctly installed with teeth facing motor (arrow).

5. Insert the new mast/gear rack into the motor with the teeth facing the motor. See Fig. 10.

NOTE
The new gear rack should be greased with a multipurpose grease.

---

6. Connect the electrical leads as described above. Gently engage the rack teeth into the motor.

NOTE
To help engage the rack, momentarily touch the motor red lead to the battery several times.

---

7. When the gear rack is all the way in, disconnect the leads.

8. Replace the antenna tube in its correct position. Reinstall the antenna in the car.

853 Dashboard and Consoles

DASHBOARD

The one-piece padded dashboard panel is removed after the instrument panel, radio and steering wheel have been removed. Removal of the steering wheel is given in **641 Steering Column.** Removal of the instrument panel that holds the dashboard switches is covered in **364 Electric Controls and Switches.**

WARNING
1990 and later models are equipped with an airbag or SRS system. The airbag is an explosive device. Do not attempt to remove the steering wheel or service the wiring in the steering column assembly. You could accidentally ignite the airbag, causing serious personal injury. See an authorized Saab dealer for any repairs to the SRS system or to disarm the system prior to removing the steering wheel.

NOTE
Removal of the center console between the front seats is covered in **432 Manual Transmission Controls.**

To remove padded dashboard

1. Remove the battery negative (–) cable.

2. Remove the instrument panel as described in **364 Electric Controls and Switches.**

3. Remove the left and right front speaker grilles.

NOTE
Use an offset screwdriver to remove the speaker grille mounting screws.

4. Working through the speaker grille openings, remove the left and right upper dashboard mounting screws. See Fig. 1

5. Working below the glove box, remove the two screws shown in Fig. 2.

6. Working under the left-hand side of the dash, remove the corner mounting screw shown in Fig. 3.

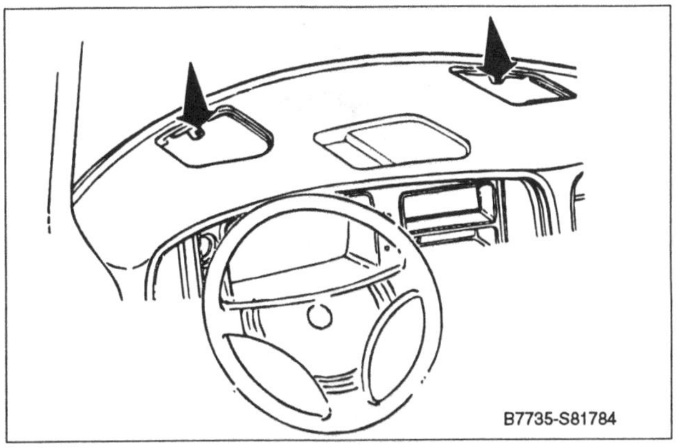

Fig. 1. Dashboard upper mounting screws (arrows).

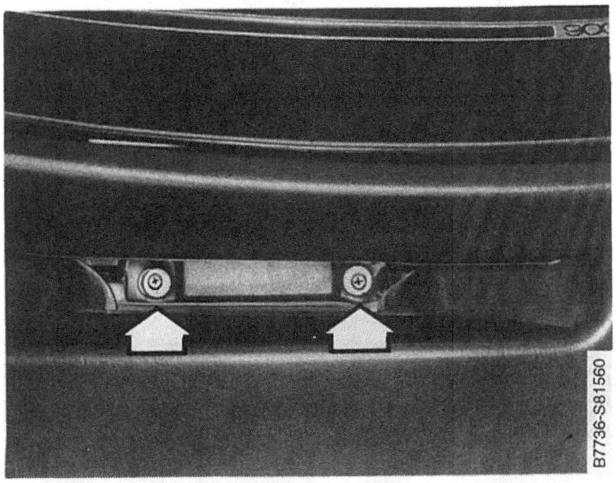

Fig. 2. Lower dashboard mounting screws below glovebox (arrows).

7. Remove the dashboard as shown in Fig. 4.

Installation is the reverse of removal. Check that the defroster hoses are attached to the speaker grilles. Check all instrument panel functions.

To remove storage console

1. Remove the rubber boot in front of the shift lever by squeezing it in pulling it out. See Fig. 5.

2. Remove the ashtray. Remove the screws holding the ashtray bracket to the console. See Fig. 6.

853

Fig. 3. Mounting screw on under left-hand side of dash (arrow).

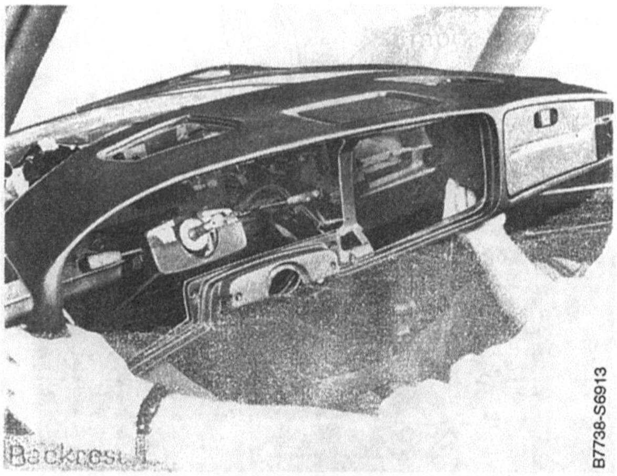

Fig. 4. Padded dashboard being removed.

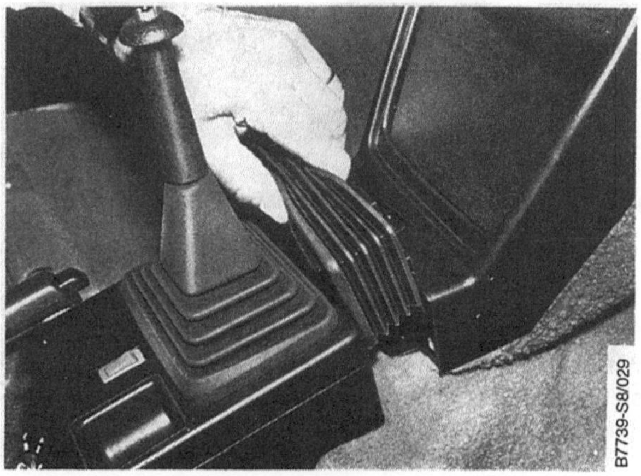

Fig. 5. Rubber boot being removed.

3. Working through the ashtray opening, remove the bolt holding the console to the dashboard.

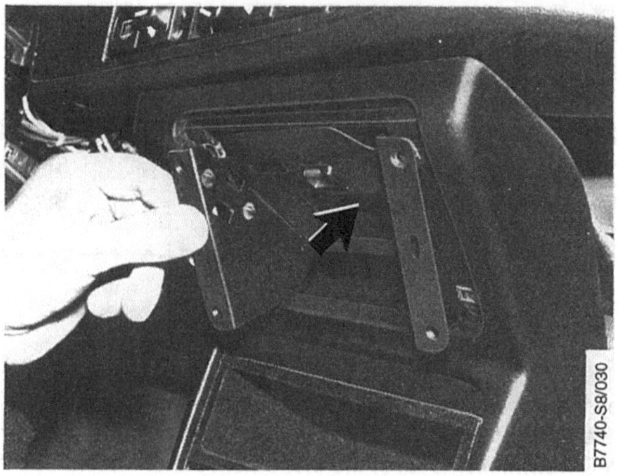

Fig. 6. Ashtray bracket being removed. Console bolt (not shown) is removed through opening (arrow).

4. Disconnect the ashtray illumination wires and remove the bulb holder.

5. Where applicable, remove the gauges or the radio equipment in the center of the console. Disconnect and label the wiring.

6. Remove the six screws holding the console to the car body and dash. Remove the console.

INSTRUMENT CLUSTER

The instrument cluster can be easily removed once the instrument panel is removed as described in **364 Electric Controls and Switches.** To replace bulbs in the cluster, remove the left speaker grille and reach into the opening. See Fig. 7. The bulbs can be removed and tested by checking for continuity at the socket terminals.

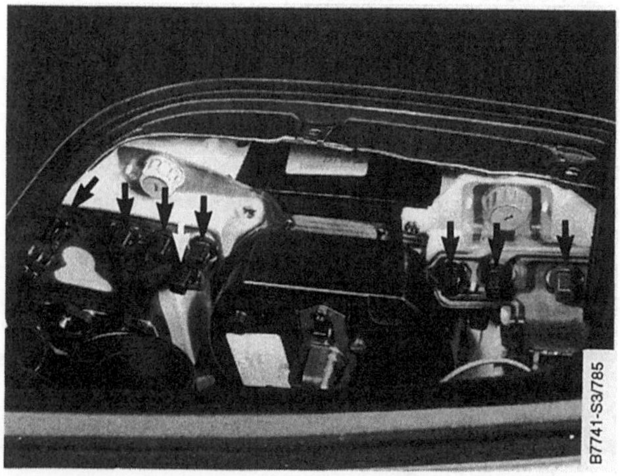

Fig. 7. Instrument cluster bulbs (arrows) as viewed through left speaker opening

Instrument cluster bulb rating

- instrument illum ination . . . . 3 watt
- charge warning lamp . . . . 2 watt
- other warning/indicator lamps . . . . 1.2 watt

To remove instrument cluster

1. Remove the battery negative (–) cable. Remove the steering wheel as described in **641 Steering Column**.

WARNING
1990 and later models are equipped with an airbag or SRS system. The airbag is an explosive device. Do not attempt to remove the steering wheel or service the wiring in the steering column assembly. You could accidentally ignite the airbag, causing serious personal injury. See an authorized Saab dealer for any repairs to the SRS system or to disarm the system prior to removing the steering wheel.

---

2. Remove the instrument panel as described in **364 Electric Controls and Switches.**
3. Remove the left speaker grille.
4. Working through the speaker grille, disconnect the electrical connectors at the back of the cluster. Disconnect the speedometer cable by bending the spring out of the way and pulling the cable from the instrument. See Fig. 8.

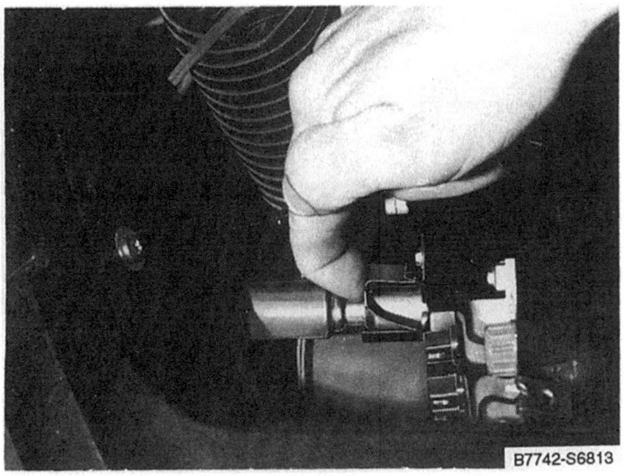

Fig. 8. Speedometer cable retaining clip being pried up to removed cable from speedometer.

NOTE
If the speedometer cable is being replaced, free the cable from the grommet in the bulkhead then pull the cable into the engine compartment. Remove the other end of the cable from the transmission.

---

5. Remove the screws holding the cluster to the dash. See Fig. 9. Remove the cluster.

Fig. 9. Instrument cluster retaining screw being removed.

NOTE
- Replacement instruments are available from an authorized Saab dealer. Fig. 10 shows the individual instruments removed from the cluster housing.
- Laws in some areas require that you fill out an odometer verification sticker when replacing a speedometer.

Fig. 10. Instruments.

854 Heating and Air Conditioning

TABLES

General

This section of the manual covers repairs and troubleshooting for the heating and air conditioning (A/C) systems. This includes repair information for the heater core, the heater valve, the ventilation blower, the dash-mounted controls, and the A/C components.

CAUTION
The A/C system is a sealed system and should only be serviced by an authorized Saab dealer or other qualified repair shop. Special tools and equipment are necessary to service the A/C system.

NOTE
- Repairs to the cooling system, including heater hose replacement, are covered in **261 Radiator and Cooling System**.
- Additional electrical troubleshooting is covered in **371 Wiring Diagrams, Fuses and Relays.**

Fig. 1 shows the heating and air conditioning system's air distribution system. Fresh air enters the car through the vent in the hood and passes through the A/C evaporator and through the heater core. Air exits through the rear slots near the luggage area. Air flow through the system is directed through vacuum-actuated dampers.

Temperature, air flow, and ventilation fan speeds are controlled by separate instrument panel knobs. The air distribution knob also contains electric contacts for the ventilation blower.

Heating is controlled by regulating coolant flow through the heater core via the heater valve. The temperature regulating knob is directly connected to the heater valve, using a control rod.

All of the cars covered by this manual are equipped with factory-installed A/C. The major components of the A/C system are shown in Fig. 2.

The A/C system is protected by a pressure switch, a fuse for the compressor, an anti-frost switch and an A/C coolant temperature switch. See **261 Radiator and Cooling System** for information on the A/C coolant temperature switch. These components switch off the compressor to prevent damage to the A/C system.

WARNING
- Adding refrigerant or servicing the A/C system without the proper tools, equipment, or knowledge may cause severe personal injury as well as damage to the A/C components.
- Wear eye protection when inspecting the system. R-12 at normal atmospheric pressures can evaporate and freeze anything it contacts.
- At normal operating temperature the heating system is pressurized with hot engine coolant. Allow the system to cool as long as possible before opening—a minimum of an hour—then release the coolant expansion cap very slowly to allow safe release of pressure.
- Releasing cooling system pressure lowers the coolant's boiling point, and the coolant may boil suddenly. Use heavy gloves and wear eye and face protection to guard against scalding.

854

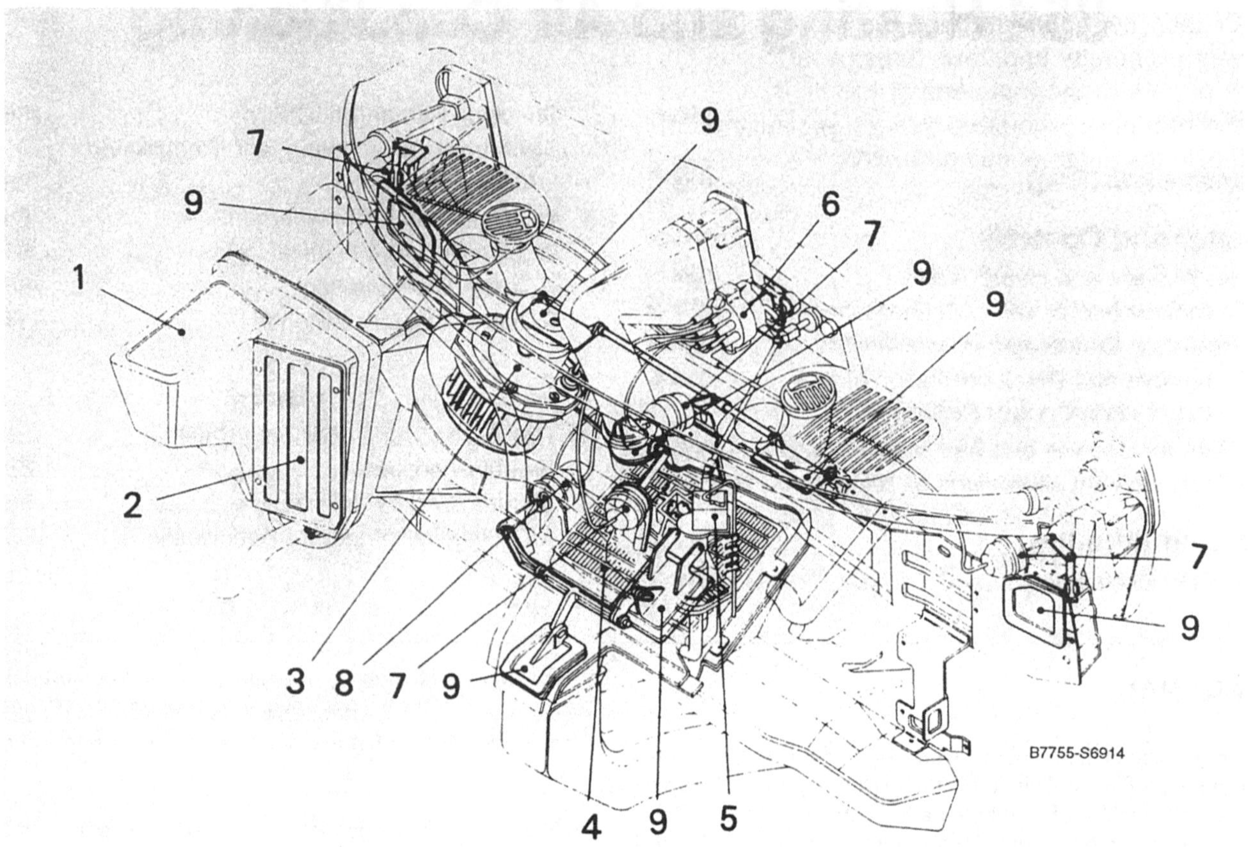

1. Air intake
2. Evaporator
3. Blower motor
4. Heater core
5. Heater valve
6. Air distribution switch
7. Vacuum servo (single stage)
8. Vacuum servo (dual stage)
9. Vacuum controlled flaps

Fig. 1. Cut-away view of air distribution system.

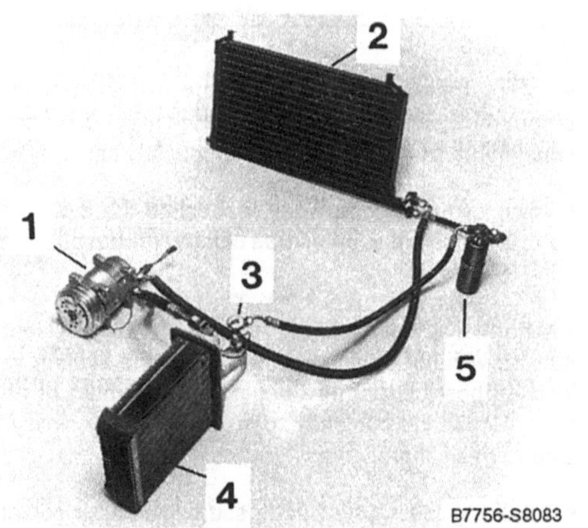

1. Compressor
2. Condenser
3. Expansion valve
4. Evaporator
5. Receiver/drier

Fig. 2. A/C components.

CAUTION

Avoid adding cold water to the coolant while the engine is hot or overheated. If it is absolutely necessary to add coolant to a hot system, do so only with the engine running and coolant pump turning.

TROUBLESHOOTING

Table a shows common heating and air conditioning problems. Each system has its own causes and corrective actions.

Insufficient heater output may be caused by a faulty heater valve, an air flap not opening, or a cooling system fault. Coolant leaking visibly into the passenger compartment is a sign of a faulty heater core. A sweet, anti-freeze odor in the car's interior, or a constantly fogged windshield may also indicate a faulty core. Inspect the carpet and the area near the footwell vents for any moisture or coolant. Water appearing on the carpets or near the footwells could be a clogged A/C evaporator condensation drain hose.

Table a. Heating and Air Conditioning System Troubleshooting

Symptom	Probable cause	Corrective action
1. Heater output inadequate (temperature gauge reading normal)	**a.** Heater hose restricted **b.** Heater core or heater valve clogged or faulty **c.** Heater controls broken	**a.** Replace hose. See **261 Radiator and Cooling System** **b.** Clean or replace heater valve. Have heater core cleaned **c.** Inspect and replace any faulty controls
2. Heat always present at side vents, temperature regulation erratic—cannot be regulated to cooler settings	**a.** Heater valve not fully closed due to disconnected or damaged control rod **b.** Internal seals in heater valve faulty	**a.** Check control rod operation **b.** Replace heater valve
3. A/C cooling output inadequate	**a.** System charge (refrigerant) low **b.** Anti-frost switch faulty **c.** Expansion valve faulty	**a.** Check system charge at sight glass. Have system recharged if necessary. Check for leaks **b.** Test switch and replace if necessary **c.** Test system pressure and replace expansion valve if necessary

HEATER AND CONTROLS

Heater Core and Heater Valve

Coolant flow through the heater core is controlled by the heater valve. The heater valve is connected directly to the temperature knob. See Fig. 3. There are no adjustments that can be made to the valve or control rod.

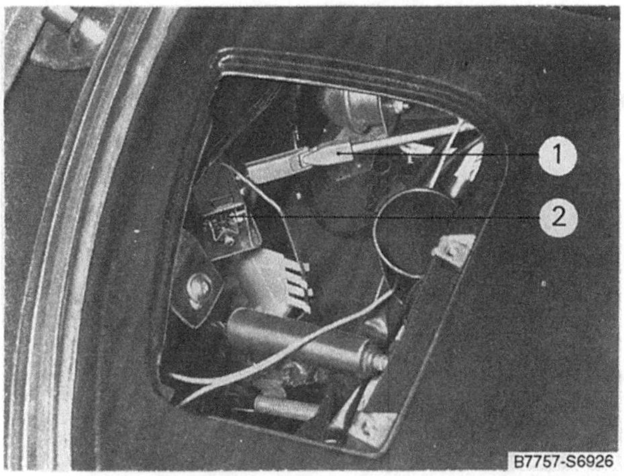

Fig. 3. Heater valve (**2**) and control rod (**1**) as viewed through left-front speaker grille opening.

To replace heater valve and heater core

1. Disconnect the negative (–) cable from the battery.
2. Drain the coolant as described in **261 Radiator and Cooling System.**
3. Working in the engine compartment, label and remove the heater hoses at the heater valve fittings. Plug the ends of the fittings on the valve.
4. Remove the lower steering column cover, the storage console (where applicable), and lower dashboard pad as described in **851 Interior Trim and Upholstery.**

WARNING

1990 and later models are equipped with an airbag or SRS system. The airbag is an explosive device. Do not attempt to remove the steering wheel or service the wiring in the steering column assembly. You could accidentally ignite the airbag, causing serious personal injury. See an authorized Saab dealer for any repairs to the SRS system or to disarm the system prior to removing the steering wheel.

---

5. Remove the left speaker grille, then slide the heater control rod forward to disengage it from the knob. Slide rod backwards to remove it from the heater valve. See Fig. 3 above.
6. If removing just the heater valve use a long 4-mm hex-head driver to remove the screws securing the heater valve to the heater core. See Fig. 4.

NOTE

A ball type (swivel) hex-head driver will help make removal and installation of the heater valve screws easier.

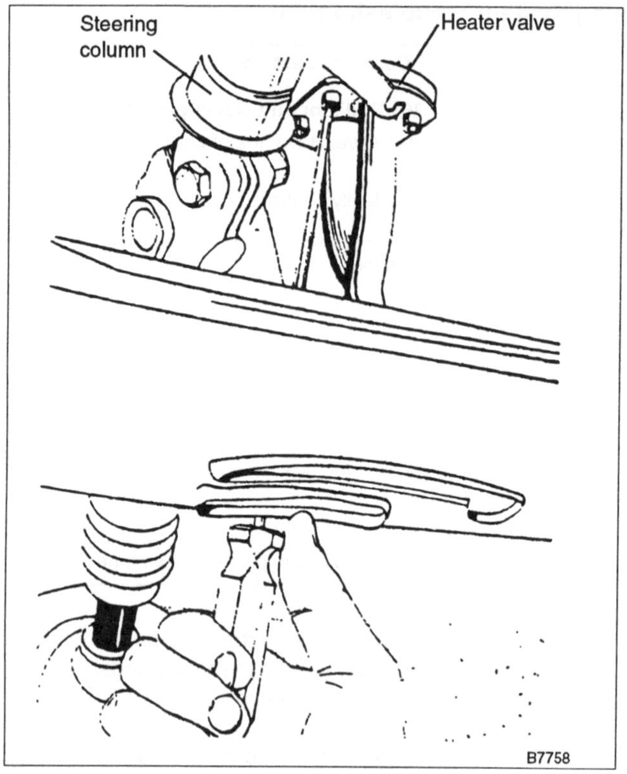

Fig. 4. Heater valve being removed from heater core.

WARNING

- **Replacement heater valves may interfere with throttle pedal operation and require a new throttle pedal arm. See an authorized Saab dealer for correct parts information.**
- **On cars with cruise control, be sure to refit the actuator chain guard to the heater valve to prevent interference with the chain.**

7. Remove the ventilation air diffuser in the footwell. See Fig. 5.

8. Remove the screws securing the lower section of the heater box and remove the lower section. See Fig. 6.

9. Unhook the brake pedal return spring from the pedal and depress the pedal slightly. Lower the heater core and heater valve assembly out of the car.

10. Where applicable, remove the heater valve capillary tube from the heater core. Separate the heater valve from the heater core. See Fig. 7.

CAUTION

Handle the capillary tube carefully. Do not crease or bend the tube.

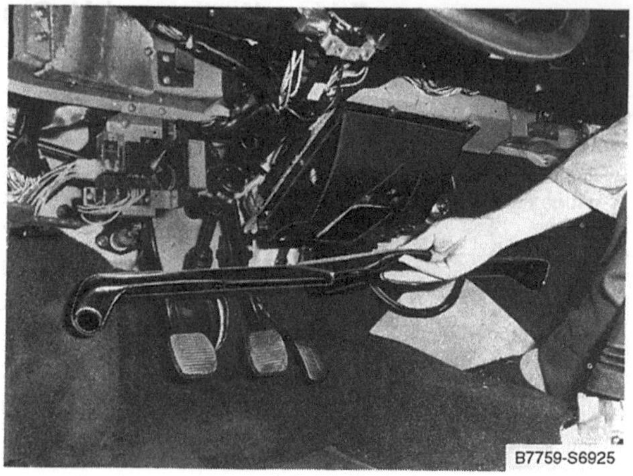

Fig. 5. Ventilation air diffuser being removed.

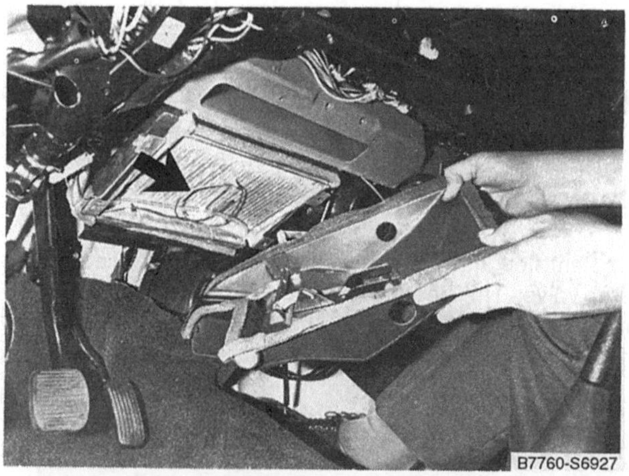

Fig. 6. Lower heater box section being removed. Also visible is the heater valve capillary tube (arrow).

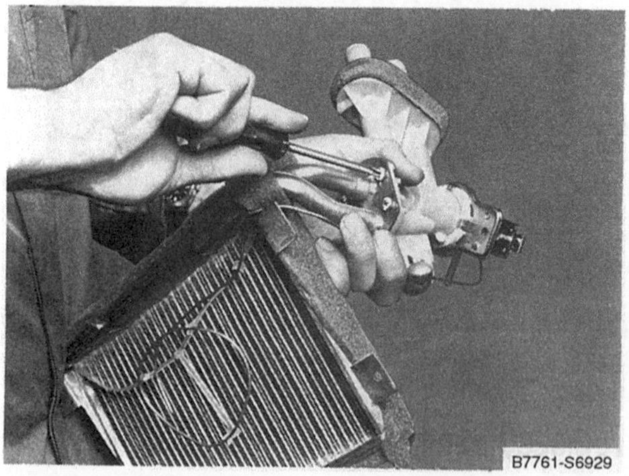

Fig. 7. Heater valve being removed from heater core.

Installation is the reverse of removal. Always replace the heater core-to-heater valve O-rings. The heater valve should be in the off position (flat surface up) before connecting the control rod. Reconnect the heater hoses to the valve. Check the operation of all foot pedals before installing the lower dash pad. Refill and bleed the cooling system as described in **261 Radiator and Cooling System** and check the operation of the heater functions.

NOTE
Reinstall the heater hoses to the heater valve so that the top hose leads to the water pump and the lower hose comes from the intake manifold.

Ventilation Blower and Blower Switch

The blower assembly is mounted in the passenger compartment footwell and runs except when the air distribution switch is in the 0 (six o'clock) position. A blower resistor pack, located below the left speaker grille, controls the speed of the blower motor. There are three resistors for three speeds of the motor. If the resistor pack overheats, an internal switch closes and blows the fuse on the main fuse/relay panel.

To quickly test the blower motor, remove the right speaker grille and probe the connector with a test lamp. If there is no voltage reaching the motor with the ignition on and the blower switch on high, check the fuse on the main fuse/relay panel and check the wiring. See **371 Wiring Diagrams, Fuses and Relays.** If voltage is present, the motor is probably faulty.

NOTE
- A blower motor that runs only when the air distribution switch is in position 1 (7 o'clock position) usually indicates a fault with the blower motor resistors or blower switch.
- To replace the blower switch, remove the switch panel as described in **364 Electric Controls and Switches**, then remove the clip holding the switch and the wire connector.

To remove and install ventilation blower

1. Disconnect the negative (–) cable from the battery.
2. Remove the dashboard instrument panel as described in **364 Electric Controls and Switches.**
3. Remove the padded dashboard as described in **853 Dashboard and Consoles.**
4. Disconnect the blower wiring from the blower.
5. Remove right defroster housing mounting screws.
6. Remove the three blower mounting screws and remove the blower. See Fig. 8.

Fig. 8. Blower motor being removed.

Installation is the reverse of removal.

NOTE
A sound insulation kit is available from Saab to reduce blower motor noise. See an authorized Saab dealer for the correct application and part number.

Air Distribution Control Switch, Vacuum Servos and Air Flaps

Vacuum operated flaps direct air flow to the various air outlets. The panel-mounted air distribution switch routes vacuum to the correct vacuum servo to open and close the air flaps. See Fig. 9. The vacuum for the system is supplied via a vacuum storage tank. The switch also functions to close microswitches for the ventilation blower motor and the recirculation valve.

NOTE
To replace the air distribution switch, the instrument panel must be removed. See **364 Electric Controls and Switches.**

There are six vacuum servos that open and close air flaps. The recirculation switch on the instrument panel controls an electric solenoid that routes vacuum to the recirculation door servo. See Fig. 10.

To test vacuum servos and air flaps

Many vacuum related faults are the result of pinched or disconnected hoses. Always inspect the hoses carefully when experiencing problems with the servos.

1. With the engine running, check for vacuum at the vacuum tank on the passenger side wheel liner.

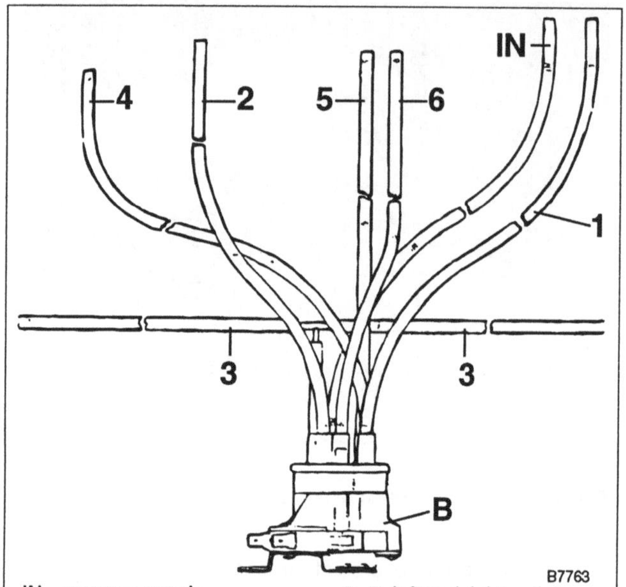

IN – vacuum supply
B – switch
1– to recirculation damper
2– to center outlet
3– to left and right outlets
4– to defroster
5– lefthand footwell outlet
6– righthand footwell outlet

Fig. 9. Air distribution switch showing vacuum hose routing.

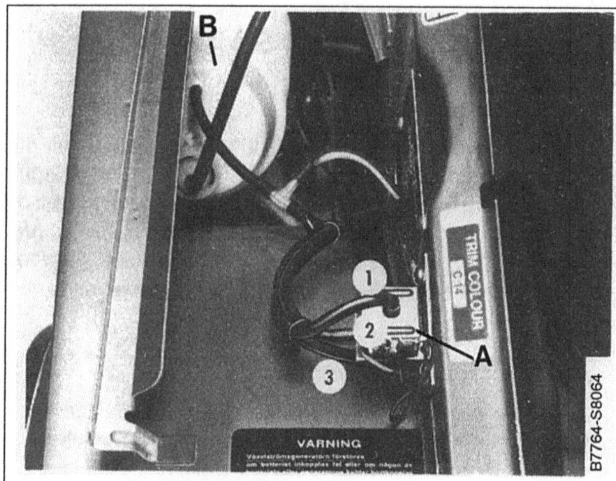

1. Yellow to heater box vacuum box.
2. Blue to recirculation servo
3. White to T connection vacuum tank

Fig. 10. Air recirculation solenoid (**A**) on wheel housing next to vacuum tank (**B**) on 1985 through 1987 models. Recirculation solenoid on 1988 and later models is in passenger compartment, behind right footwell kickpanel.

2. Check the operation of the system servos and the air distribution switch using the information found in **Table b.**

NOTE

The servos should be visible when making the tests outlined in the table below. Remove the two speaker grilles to gain access to the outer vent and defroster servos. The lower instrument panel should be removed as described in **853 Dashboard and Consoles** to check the lower servos and flaps. The switch panel must be removed to test the air distribution switch.

Table b. Vacuum Servo Fault Tracing

	blower speeds 1-3 flaps open—defroster flaps partial—none flaps closed—floor, center, outer A/C compressor—selectable
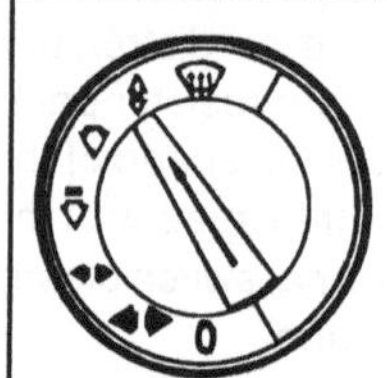	blower speeds 1-3 flaps open—floor flaps partial—defroster flaps closed—center, outer A/C compressor—selectable
	blower speeds 1-3 flaps open—floor flaps partial—none flaps closed—defroster, center, outer A/C compressor—selectable
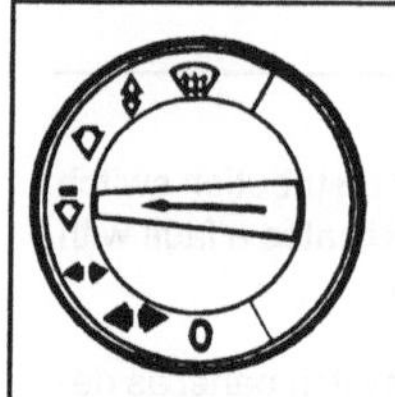	blower speeds 1-3 flaps open—center flaps partial—floor flaps closed—defroster, outer A/C compressor—selectable
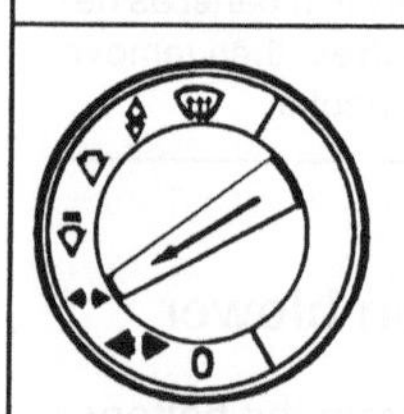	blower speeds 1-3 flaps open—outer, center flaps partial—floor flaps closed—defroster A/C compressor—selectable
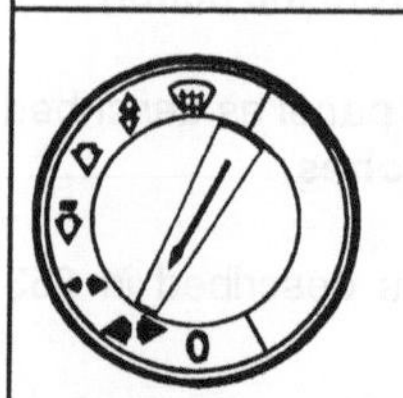	blower speed—high flaps open—floor, outer, center flaps partial—none flaps closed—defroster A/C compressor—max. or off
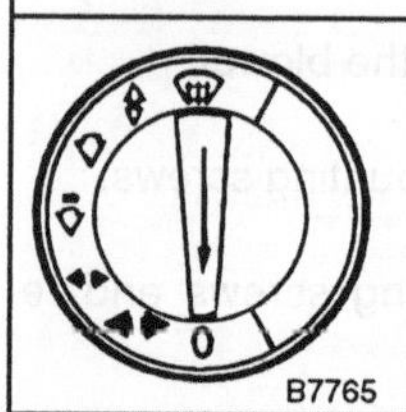 B7765	blower speed—off flaps open—none flaps partial—none flaps closed—all A/C compressor—off

Vacuum Servo Location

- outer vent — below speaker grilles
- defroster vents — behind center of dash
- center vent — above heater box
- footwell — passenger side of heater box
- fresh air intake — in engine compartment at A/C evaporator housing

3. If any faults are found, use a hand vacuum pump to check the operation of the servos. If the servo operates correctly, check the hose from the switch. If no faults are found, the switch is faulty and should be replaced.

NOTE
The vacuum tank, the switch and the lines can also be checked for leaks using the hand pump.

AIR CONDITIONING

It is recommended that all service to the A/C system be left to an authorized Saab dealer or other qualified repair shop. For information on the A/C selector switch as well as other electrical components, see **371 Wiring Diagrams, Fuses and Relays.**

Dismounting of the A/C compressor and the condenser, without disconnecting the hoses, is covered as part of the engine removal procedure in **201 Engine Removal and Installation.** If any of the hoses or components are disconnected and the system is opened, special equipment will be needed to remove moisture from the system before it is closed back up.

WARNING
Air conditioning (A/C) service and repair requires special equipment and knowledge. Incorrect procedures may not only damage the system, but may also be hazardous. Pressures in excess of 300 psi are created in the system when it is operating. The refrigerant R-12 is poisonous. In its vapor form it can accumulate in areas with poor ventilation and cause suffocation. Also, in vapor or liquid form R-12 can immediately freeze anything it contacts, including eyes and skin.

System Description

Fig. 11 is a schematic view of a typical Saab A/C system. In an A/C system, the heat from the passenger compartment is absorbed by the refrigerant (R12) in the evaporator, causing ther refrigerant to boil (evaporate). This heat is then released into the atmosphere when the R-12 is cooled and condensed into a liquid at the condenser. Moisture is removed at the evaporator in the same way that water drops form on a cold glass. The moisture drips onto the water tray in the evaporator housing and is routed away via a drain hose. This is the reason a water puddle is often be seen under the car when the A/C is operating.

The compressor forces the R-12 through the system and at the same time pressurizes it, raising the R-12's boiling point to make it more easily condensed. The compressor is engaged by an electro-magnetic clutch that is actuated when the A/C is turned on.

The condenser, which looks like a small radiator, is located in front of the engine radiator. The receiver/drier, located behind the battery, removes small amounts of moisture and dirt from the system. The evaporator and expansion valve are located in the engine compartment on the passenger side at the firewall.

WARNING
Wear eye protection when inspecting the system. R-12 at normal atmospheric pressures can evaporate and freeze anything it contacts.

NOTE
- Run the air conditioning system for a few minutes, every few weeks, to keep the seals lubricated.
- Check that the condensation drain hose from the evaporator housing is clear. This can be accomplished by blowing it out with a low pressure compressed air nozzle.

Checking Refrigerant Charge

Inspect the refrigerant charge by starting the engine and turning the air conditioner on to MAX A/C. With the compressor running (clutch cycled on) view the sight glass on the top of the receiver/drier (located near the battery). There should be few or no bubbles visible in the glass. See Fig. 12.

Discharging, Evacuating and Recharging the A/C System

Discharging the system is necessary any time a defective component is replaced. To prevent component damage, a totally discharged system must be evacuated (sometimes called pulling a vacuum) using special equipment before recharging. This removes any moisture from the lines and components, which can damage the system. For system discharging, evacuating and recharging, see an authorized Saab dealer or other shop qualified to do A/C repair work.

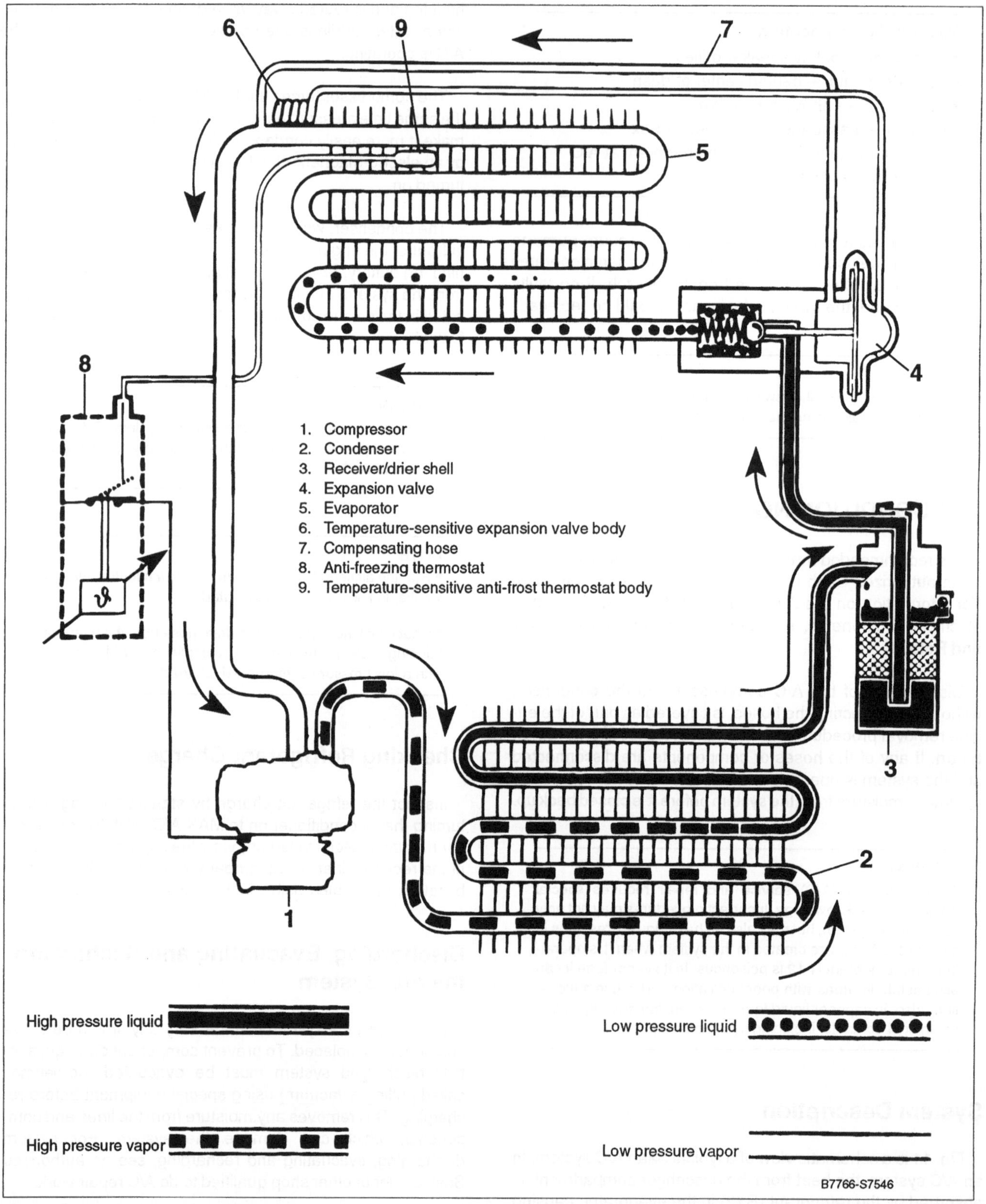

Fig. 11. Schematic view of A/C system.

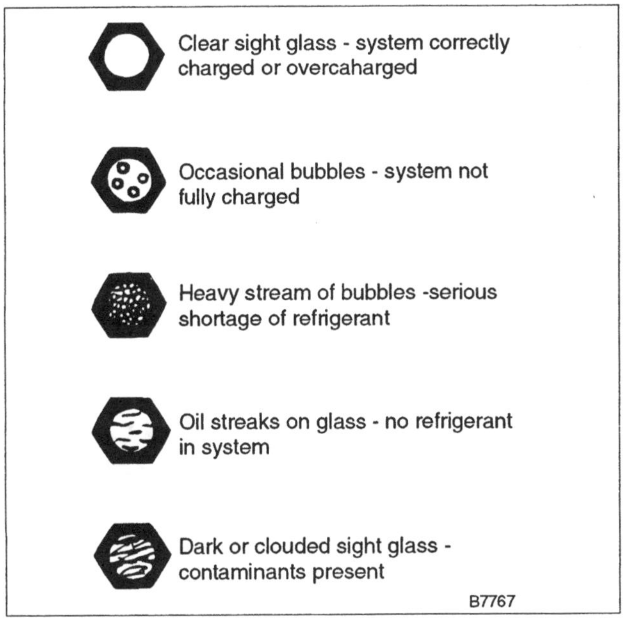

Fig. 12. Air conditioning sight glass indicators.

WARNING
Discharging the system or adding refrigerant without the proper tools, equipment, or knowledge may cause severe personal injury as well as damage to the A/C components.

A/C Component Replacement

The information given below is intended to be used by those experienced in A/C service. When replacing components, always use new O-rings and gaskets. Lubricate the fittings and O-rings with refrigerant oil during installation. **Table c** lists the *appropriate amount* of refrigerant oil that should be added to the system when replacing individual components. Fig. 13 shows tightening torques for the various hoses and components.

WARNING
The system must be discharged before replacing any A/C components or hoses. Always evacuate the system before recharging it.

CAUTION
The oil filler plug on the A/C compressor is not an oil level plug. Do not check the compressor oil level through this plug.

NOTE
If the A/C system has been opened up to the atmosphere for more than five minutes the receiver/drier must be replaced.

Table c. A/C System Refrigerant Specifications

Refrigerant capacity (R12)	1.0 kg. (2.2 lb)
Refrigerant oil capacity (compressor replacement)	1.75 dl (4.9 oz.)
Refrigerant oil capacity per component (due to sudden leakage)	
Evaporator	0.5 dl (1.4 oz.)
Receiver/drier	0.2 dl (0.57 oz.)
Condenser	0.2 dl (0.57 oz.)
Hose	0.2 dl (0.57 oz.)
Oil fill plug torque	8–12 Nm (6–9 ft-lb)

Air Conditioning Electrical Tests

Covered under this heading are tests for the A/C compressor clutch, pressure and temperature safety switches, and the A/C selector switch.

NOTE
Additional wiring information can be found in **371 Wiring Diagrams, Fuses and Relays**.

To test A/C compressor

1. Turn the ignition on and turn the temperature regulating knob to its full cold position. Turn the A/C switch on and listen for the compressor clutch to click on. If no click is heard, the compressor clutch may not be receiving voltage or the clutch itself may be faulty.

2. Check for voltage between the single-wire connector leading out of the compressor and ground. If voltage is present, the compressor clutch is faulty and should be replaced.

3. If voltage is not present, check for a faulty compressor fuse in the main fuse/relay panel. Check the A/C coolant temperature switch, the low pressure switch, and the anti-frost switch as described below. Check for damaged wiring or a faulty (open) suppression diode using an ohmmeter.

NOTE
The A/C diode is in the wiring harness at the rear of the camshaft cover. See **371 Wiring Diagrams, Fuses and Relays** for A/C wiring diagrams.

---

4. If no faults are found up to this point, test the anti-frost valve as described below.

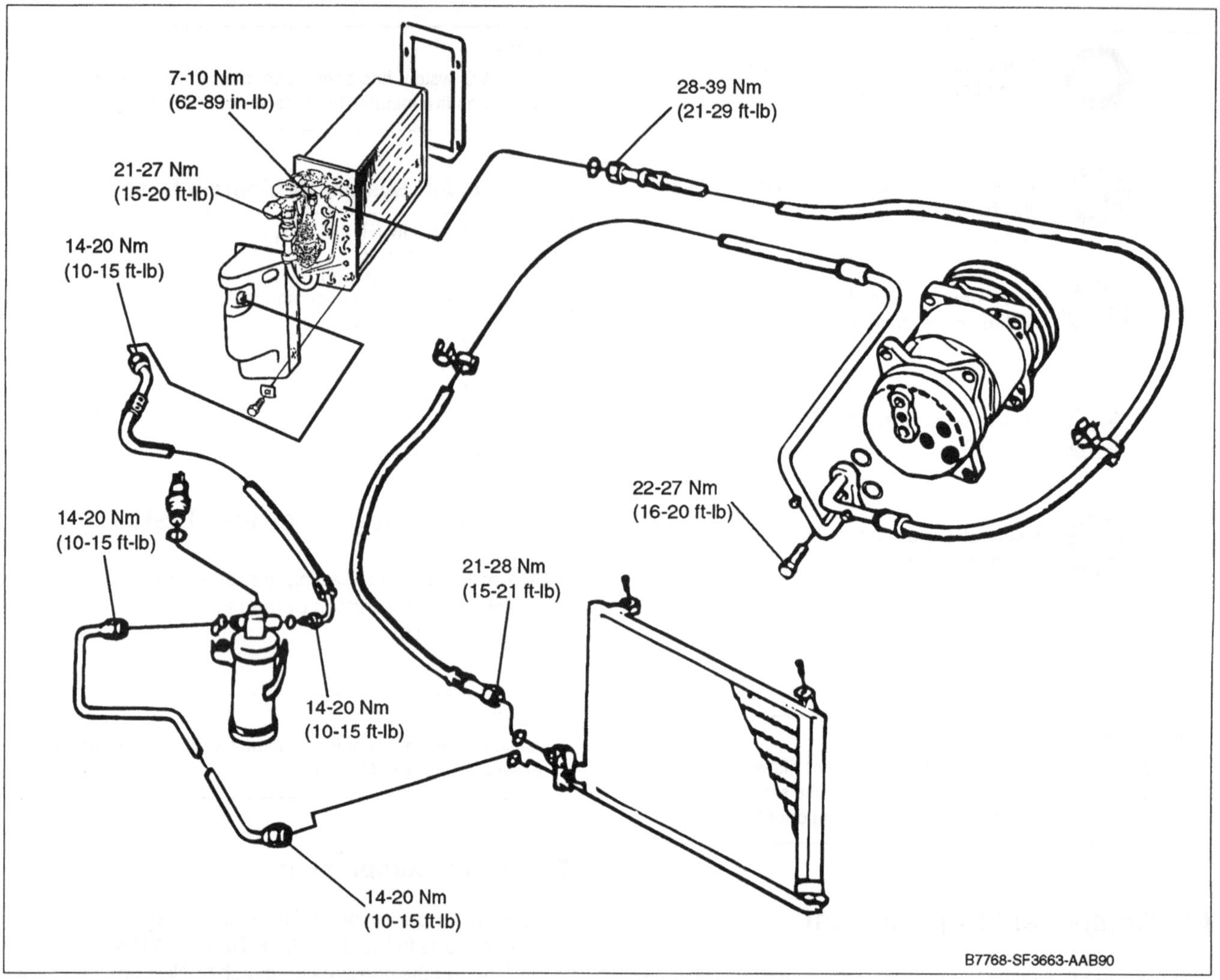

Fig. 13. A/C hose and component tightening torques.

To test A/C safety switches

A low pressure switch in the receiver/drier interrupts power to the compressor clutch if the pressure is too low. An anti-frost switch automatically disengages the compressor clutch when the temperature at the evaporator drops below a certain level. An A/C coolant temperature switch on the hose to the thermostat interrupts the compressor if the coolant temperature is excessive.

NOTE

The radiator cooling fan should run when the A/C is switched on (key on). When the compressor clutch cycles (engine running), the additional cooling fan should also run.

A/C Safety Switches

- Low pressure switch
  opening pressure . . . . 2.9 bar (42 psi)
- Coolant temperature switch
  opening temperature. . . . . 115°C (239°F)
- Anti-frost switch
  switch closed. . . . . . . 4.0 to 6.2 °C (39.2 to 43.1 °F)
  switch open . . . . .0.4 to 2.6 °C (33 to 36.6 °F)

1. Disconnect the connector from the low pressure switch and connect and ohmmeter across the switch terminals. With the engine running and the A/C on, there should continuity. If not, either the refrigerant charge is low or the switch is faulty.

WARNING
- Do not loosen or remove the low pressure switch. Refrigerant will be expelled. The system must be discharged before this switch is removed.
- Operate the A/C only long enough to make the tests.

2. Check the anti-frost switch by disconnecting the switch connector and connecting and ohmmeter across the switch terminals. There should continuity with the switch at room temperature. If not the switch is faulty and should be replaced. See Fig. 14.

NOTE
- When installing a new anti-frost switch, be sure the capillary tube is in its original position.
- Check that the capillary tube does not interfere with the recirculation door.

3. Check the A/C coolant temperature switch by disconnecting the switch connector and connecting and ohmmeter across the switch terminals. There should be continuity across the terminals at all temperatures up to 239°F (115°C). If the switch is open, it is faulty and should be replaced.

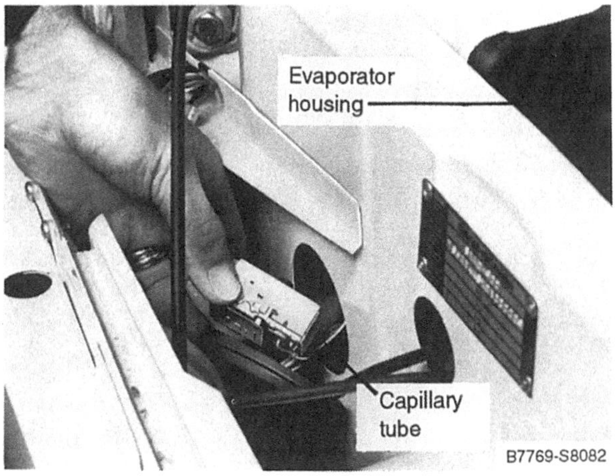

Fig. 14. Anti-frost switch being removed from evaporator housing.

NOTE
To replace the A/C coolant temperature switch, drain the coolant as described in **261 Radiator and Cooling System**.

860 Bumpers

Removing and Installing Bumpers

Two types of bumpers are used on cars covered by this manual, depending on model year. See Fig. 1 and Fig. 2.

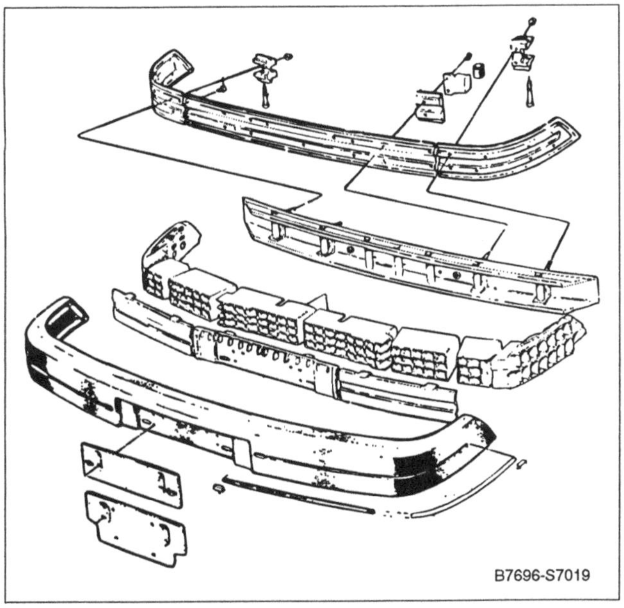

Fig. 1. Exploded view of front bumper used on 1985 and 1986 models. Rear bumper is similar.

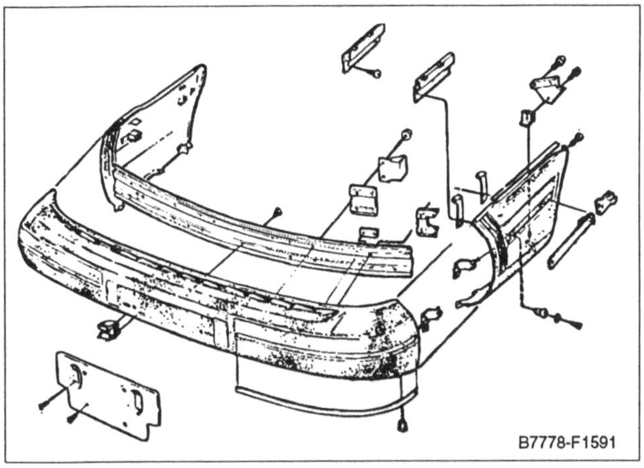

Fig. 2. Exploded view of front bumper used on 1987 and later models. Rear bumper is similar.

To remove bumpers

1. To remove the front bumper, remove the two internal-hex head screws holding the bumper to the bumper brackets. See Fig. 3. Remove the bumper.

NOTE
On cars with headlight wipers, remove the wipers before removing the front bumper as described in **363 Wipers and Washers.**

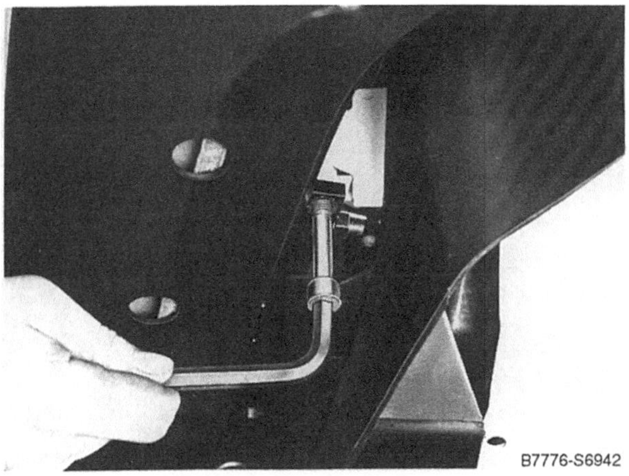

Fig. 3. Front bumper internal-hex head screw as viewed from below. Early style bumper shown.

2. To remove the rear bumper, remove the four nuts holding the bumper to the bumper mounting brackets. See Fig. 4.

860

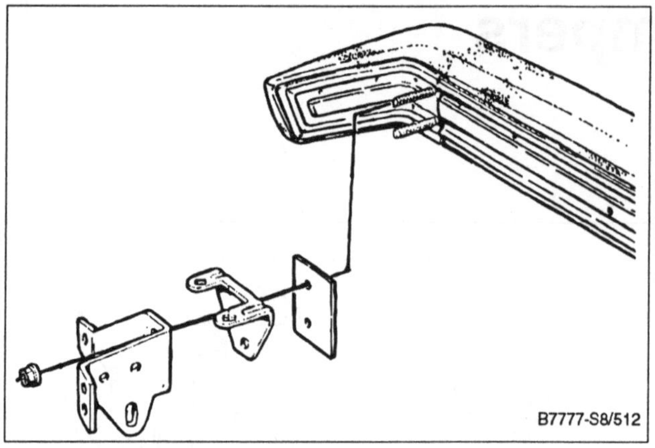

Fig. 4. Rear bumper mounting bracket (right-hand side). Early style bumper shown.

Installation is the reverse of removal. Individual bumper replacement parts are available from an authorized Saab dealer.

INDEX

Subjects are indexed by repair group number, followed by the page number(s) within the repair group where the subject can be found. For example **210**-4 refers to repair group **210 Engine Disassembly and Assembly**, page 4.

> ***WARNING —***
>
> *• Automotive service and repair is serious business. You must be alert, use common sense, and exercise good judgement to prevent personal injury and complete the work safely.*
>
> *• Before beginning any work on your vehicle, thoroughly read the safety notice, disclaimer, and all the Cautions and Warnings listed near the front of this manual.*
>
> *• Always read the complete procedure before you begin the work. Pay special attention to any Cautions and Warnings that accompany that procedure, or other information on a specific topic.*

INDEX

Subjects are indexed by repair group number, followed by the page number(s) within the repair group where the subject can be found. For example **210**-4 refers to repair group **210 Engine Disassembly and Assembly**, page 4.

> ***WARNING —***
> *• Automotive service and repair is serious business. You must be alert, use common sense, and exercise good judgement to prevent personal injury and complete the work safely.*
> *• Before beginning any work on your vehicle, thoroughly read the safety notice, disclaimer, and all the Cautions and Warnings listed near the front of this manual.*
> *• Always read the complete procedure before you begin the work. Pay special attention to any Cautions and Warnings that accompany that procedure, or other information on a specific topic.*

INDEX

Subjects are indexed by repair group number, followed by the page number(s) within the repair group where the subject can be found. For example **210**-4 refers to repair group **210 Engine Disassembly and Assembly**, page 4.

> ***WARNING —***
> *• Automotive service and repair is serious business. You must be alert, use common sense, and exercise good judgement to prevent personal injury and complete the work safely.*
> *• Before beginning any work on your vehicle, thoroughly read the safety notice, disclaimer, and all the Cautions and Warnings listed near the front of this manual.*
> *• Always read the complete procedure before you begin the work. Pay special attention to any Cautions and Warnings that accompany that procedure, or other information on a specific topic.*

6 INDEX